Communications
in Computer and Information Science   **662**

Commenced Publication in 2007
Founding and Former Series Editors:
Alfredo Cuzzocrea, Dominik Ślęzak, and Xiaokang Yang

More information about this series at http://www.springer.com/series/7899

Tieniu Tan · Xuelong Li
Xilin Chen · Jie Zhou
Jian Yang · Hong Cheng (Eds.)

Pattern Recognition

7th Chinese Conference, CCPR 2016
Chengdu, China, November 5–7, 2016
Proceedings, Part I

 Springer

Editors
Tieniu Tan
Chinese Academy of Sciences
Institute of Automation
Beijing
China

Xuelong Li
Xi'an Institute of Optics and Precision
  Mechanics
Chinese Academy of Sciences
Xi'an
China

Xilin Chen
Chinese Academy of Sciences
Institute of Computing Technology
Beijing
China

Jie Zhou
Tsinghua University
Beijing
China

Jian Yang
Nanjing University of Science
  and Technology
Nanjing
China

Hong Cheng
University of Electronic Science
  and Technology
Chengdu, Sichuan
China

ISSN 1865-0929   ISSN 1865-0937   (electronic)
Communications in Computer and Information Science
ISBN 978-981-10-3001-7   ISBN 978-981-10-3002-4   (eBook)
DOI 10.1007/978-981-10-3002-4

Library of Congress Control Number: 2016950420

Printed on acid-free paper

This Springer imprint is published by Springer Nature
The registered company is Springer Nature Singapore Pte Ltd.
The registered company address is: 152 Beach Road, #22-06/08 Gateway East, Singapore 189721, Singapore

Preface

Welcome to the proceedings of the 7th Chinese Conference on Pattern Recognition (CCPR 2016), which was held in Chengdu! Over the past decade, CCPR had been hosted by Beijing (2007, 2008, and 2012), Nanjing (2009), Chongqing (2010), and Changsha (2014) with great success. After the emergence of the Asian Conference on Pattern Recognition (ACPR) in 2011, CCPR and ACPR started being held on alternating years since 2012, making CCPR a biennial conference ever since.

Today, pattern recognition is applied in an increasing number of research domains such as autonomic understanding of vision, speech, language, and text. The recent innovative developments in big data, robotics, and multimodal interface have forced the area of pattern recognition to face both new opportunities and unprecedented challenges. Furthermore, like other flourishing new techniques, including deep learning and brain-like computation, we believe that pattern recognition will certainly exhibit greater-than-ever advances in the future. With the aim of promoting the research and technical innovation in relevant fields domestically and internationally, the fundamental objective of CCPR is defined as providing a premier forum for researchers and practitioners from academia, industry, and government to share their ideas, research results, and experiences. The selected papers included in the proceedings not only address challenging issues in various aspects of pattern recognition but also synthesize contributions from related disciplines that illuminate the state of the art.

This year, CCPR received 199 submissions, all of which are written in English. After a thorough reviewing process, 121 papers were selected for presentation as full papers, resulting in an acceptance rate of 60.53 %. An additional nine papers concerning emotion recognition were included. We are grateful to Prof. Mu-ming Poo from the Chinese Academy of Sciences, Prof. Mark S. Nixon from the University of Southampton, and Prof. Matthew Turk from the University of California for giving keynote speeches at CCPR 2016.

The high-quality program would not have been possible without the authors who chose CCPR 2016 as a venue for their publications. We are also very grateful to the members of Program Committee and Organizing Committee, who made a tremendous effort in soliciting and selecting research papers with a balance of high quality, new ideas, and new applications. We appreciate Springer for publishing these proceedings; and we are particularly thankful to Celine (Lanlan) Chang, Leonie Kunz, and Jane Li from Springer for their effort and patience in collecting and editing these proceedings.

We sincerely hope that you enjoy reading and benefit from the proceedings of CCPR 2016.

November 2016

Tieniu Tan
Xuelong Li
Xilin Chen
Jie Zhou
Jian Yang
Hong Cheng

Organization

Steering Committee

Tieniu Tan	Institute of Automation, Chinese Academy of Sciences, China
Chenglin Liu	Institute of Automation, Chinese Academy of Sciences, China
Jian Yang	Nanjing University of Science and Technology, China
Hongbin Zha	Peking University, China
Nanning Zheng	Xi'an Jiaotong University, China
Jie Zhou	Tsinghua University, China

General Chairs

Tieniu Tan	Institute of Automation, Chinese Academy of Sciences, China
Xuelong Li	Xi'an Institute of Optics and Precision Mechanics, Chinese Academy of Sciences, China

Program Chairs

Xilin Chen	Institute of Computing Technology, Chinese Academy of Science, China
Jie Zhou	Tsinghua University, China
Jian Yang	Nanjing University of Science and Technology, China
Hong Cheng	University of Electronic Science and Technology of China, China

Local Arrangements Chairs

Lu Yang	University of Electronic Science and Technology of China, China
Kai Liu	Sichuan University, China
Lei Ma	Southwest Jiaotong University, China

Workshop Chairs

Ce Zhu	University of Electronic Science and Technology of China, China
Lei Zhang	Sichuan University, China

Publicity Chairs

Hua Huang	Beijing Institute of Technology, China
Xiao Wu	Southwest Jiaotong University, China

Program Committee

Haizhou Ai	Tsinghua University, China
Xiaochun Cao	Institute of Information Engineering, Chinese Academy of Sciences, China
Hong Chang	Institute of Computing Technology, Chinese Academy of Sciences, China
Xiaotang Chen	Institute of Automation, Chinese Academy of Science, China
Yingke Chen	Sichuan University, China
Badong Chen	Xi'an Jiaotong University, China
Cunjian Chen	West Virginia University, USA
Xuewen Chen	Wayne State University, USA
Fei Chen	Italian Institute of Technology, Italy
Long Cheng	Institute of Automation, Chinese Academy of Sciences, China
Jian Cheng	Institute of Automation, Chinese Academy of Sciences, China
Rongxin Cui	Northwestern Polytechnical University, China
Daoqing Dai	Sun Yat-Sen University, China
Cheng Deng	Xidian University, China
Weihong Deng	Beijing University of Posts and Telecommunications, China
Yongsheng Dong	Chinese Academy of Sciences, China
Jing Dong	National Laboratory of Pattern Recognition, China
Junyu Dong	Ocean University of China, China
Leyuan Fang	Hunan University, China
Jun Fang	University of Electronic Science and Technology of China, China
Yachuang Feng	Xi'an Institute of Optics and Precision Mechanics, Chinese Academy of Sciences, China
Jufu Feng	Peking University, China
Jianjiang Feng	Tsinghua University, China
Shenghua Gao	ShanghaiTech University, China
Xinbo Gao	Xidian University, China
Yue Gao	Tsinghua University, China
Quanxue Gao	Xidian University, China
Yongxin Ge	Chongqing University, China
Xin Geng	Southeast University, China
Zhenhua Guo	Tsinghua University, China
Junwei Han	Northwestern Polytechnical University, China
Hongsheng He	The University of Tennessee, USA
Chenping Hou	National University of Defense Technology, China
Jiangping Hu	University of Electronic Science and Technology of China, China
Dewen Hu	National University of Defense Technology, China
Kaizhu Huang	Xi'an Jiaotong-Liverpool University, China
Qinghua Huang	South China University of Technology, China
Kaiqi Huang	Institute of Automation of the Chinese Academy of Sciences, China

Rongrong Ji	Xiamen University, China
Wei Jia	Chinese Academy of Sciences, China
Jia Jia	Tsinghua University, China
Dongmei Jiang	Northwestern Polytechnical University, China
Lianwen Jin	South China University of Technology, China
Qin Jin	Renmin University of China, China
Xin Jin	Tsinghua University, China
Wenxiong Kang	South China University of Technology, China
Jianhuang Lai	Sun Yat-sen University, China
Xirong Li	Renmin University of China, China
Ya Li	Institute of Automation, Chinese Academy of Sciences, China
Ming Li	SYSU-CMU Joint Institute of Engineering, Sun Yat-sen University, China
Xi Li	Zhejiang University, China
Wujun Li	Nanjing University, China
Chun-Guang Li	Beijing University of Posts and Telecommunications, China
Yanan Li	Imperial College London, UK
Jia Li	Beihang University, China
Yufeng Li	Nanjing University, China
Zhouchen Lin	Peking University, China
Liang Lin	Sun Yat-Sen University, China
Huaping Liu	Tsinghua University, China
Xinwang Liu	National University of Defense Technology, China
Qingshan Liu	Nanjing University of Information Science and Technology, China
Heng Liu	Anhui University of Technology, China
Kang Liu	Xi'an Institute of Optics and Precision Mechanics, Chinese Academy of Sciences, China
Min Liu	Hunan University, China
Chenglin Liu	Institute of Automation, Chinese Academy of Sciences, China
Dong Liu	University of Science and Technology of China, China
Wenju Liu	Institute of Automation, China Academy of Sciences, China
Jiwen Lu	Tsinghua University, China
Guangming Lu	Harbin Institute of Technology, China
Yue Lu	East China Normal University, China
Huchuan Lu	Dalian University of Technology, China
Xiaoqiang Lu	Xi'an Institute of Optics and Precision Mechanics, Chinese Academy of Sciences, China
Bin Luo	Anhui University, China
Siwei Lyu	State University of New York, University at Albany, USA
Zhanyu Ma	Beijing University of Posts and Telecommunications, China
Yajie Miao	Carnegie Mellon University, USA
Zhenjiang Miao	Beijing Jiaotong University, China
Jing Na	University of Bristol, UK
Feiping Nie	University of Texas, USA

Yanwei Pang	Tianjin University, China
Yu Qiao	Shenzhen Institutes of Advanced Technology, Chinese Academy of Sciences, China
Hongliang Ren	National University of Singapore, Singapore
Nong Sang	Huazhong University of Science and Technology, China
Björn Schuller	Imperial College London, UK
Shiguang Shan	Institute of Computing Technology, Chinese Academy of Sciences, China
Feng Shao	Ninbo University, China
Linlin Shen	Shenzhen University, China
Li Su	University of Chinese Academy of Sciences, China
Ning Sun	Nankai University, China
Zhenan Sun	Institute of Automation, Chinese Academy of Sciences, China
Jun Sun	Fujitsu R&D Center Company Ltd., China
Jian Sun	Xi'an Jiaotong University, China
Fuchun Sun	Tsinghua University, China
Xiaoyang Tan	Nanjing University of Aeronautics and Astronautics, China
Sheng Tang	Institute of Computing Technology, Chinese Academy of Sciences, China
Huajin Tang	Sichuan University, China
Masayuki Tanimoto	Nagoya Industrial Science Research Institute, Japan
Jianhua Tao	Institute of Automation, Chinese Academy of Sciences, China
Wenbing Tao	Institute for Pattern Recognition and Artificial Intelligence, Huazhong University of Science and Technology, China
Qi Wang	Northwestern Polytechnical University, China
Hongxing Wang	Nanyang Technological University, Singapore
Qiao Wang	Southeast University, China
Meng Wang	Hefei University of Technology, China
Yunhong Wang	Beihang University, China
Jinjun Wang	Xi'an Jiaotong University, China
Zengfu Wang	University of Science and Technology of China, China
Liwei Wang	Peking University, China
Hanzi Wang	Xiamen University, China
Ruiping Wang	Institute of Computing Technology, Chinese Academy of Sciences, China
Liang Wang	National Laboratory of Pattern Recognition, Institute of Automation of the Chinese Academy of Sciences, China
Xiangqian Wu	Harbin Institute of Technology, China
Yihong Wu	Institute of Automation, Chinese Academy of Sciences, China
Jianxin Wu	Nanjing University, China
Ying Wu	Northwestern University, China
Shiming Xiang	National Laboratory of Pattern Recognition, Institute of Automation, Chinese Academy of Sciences, China
Mingxing Xu	Tsinghua University, China
Zenglin Xu	University of Electronic Science and Technology, China

Yong Xu	Harbin Institute of Technology, China
Long Xu	Chinese Academy of Sciences, China
Qianqian Xu	Institute of Information Engineering of Chinese Academy of Sciences, China
Hui Xue	Southeast University, China
Haibin Yan	Beijing University of Posts and Telecommunications, China
Jinfeng Yang	Tianjin Key Lab for Advanced Signal Processing, Civil Aviation University of China, China
Chenguang Yang	South China University of Technology, China
Meng Yang	Shenzhen University, China
Wankou Yang	Southeast University, China
Gongping Yang	Shandong University, China
Yujiu Yang	Tsinhua University, China
Jucheng Yang	Tianjin University of Science and Technology, China
Wenming Yang	Tsinghua University, China
Mao Ye	University of Electronic Science and Technology of China, China
Ming Yin	Guangdong University of Technology, China
Xucheng Yin	University of Science and Technology Beijing, China
Zhou Yong	China University of Mining and Technology, China
Shiqi Yu	Shenzhen Institute of Advanced Technology, Chinese Academy of Sciences, China
Yuan Yuan	Xi'an Institute of Optics and Precision Mechanics, Chinese Academy of Sciences, China
Dechuan Zhan	Nanjing University, China
Shishuai Zhang	Guangxi Normal University, China
Zhaoxiang Zhang	Institute of Automation, Chinese Academy of Sciences, China
Daoqiang Zhang	Nanjing University of Aeronautics and Astronautics, China
Tianzhu Zhang	Institute of Automation, Chinese Academy of Sciences, China
Changshui Zhang	Tsinghua University, China
Minling Zhang	Southeast University, China
Hongzhi Zhang	Harbin Institute of Technology, China
Lijun Zhang	Nanjing University, China
Lin Zhang	Tongji University, China
Weishi Zheng	Sun Yat-sen University, China
Wenming Zheng	Southeast University, China
Ping Zhong	National University of Defense Technology, China
Guoqiang Zhong	Ocean University of China, China
Xiuzhuang Zhou	Capital Normal University, China
Jie Zhou	Tsinghua University, China
Jun Zhu	Tsinghua University, China
Liansheng Zhuang	University of Science and Technology of China, China
Yuexian Zou	Peking University, China
Wangmeng Zuo	Harbin Institute of Technology, China

Contents – Part I

Robotics

Computer Vision

Basic Theory of Pattern Recognition

Contents – Part II

Speech and Language

Emotion Recognition

Robotics

Constrained Spectral Clustering on Face Annotation System

Jiajie Han$^{(\boxtimes)}$, Jiani Hu, and Weihong Deng

School of Information and Communication Engineering,
Beijing University of Posts and Telecommunications,
Xi Tu Cheng Road, 10, Beijing 100876, China
{dxs,jnhu,whdeng}@bupt.edu.cn

Abstract. Face clustering is a common feature in face annotation system like intelligent photo albums and photo management systems. But unsupervised clustering algorithms perform poorly and researchers turn to work with constrained clustering algorithms that take the user interactions as constraints. Mostly, the constraints are pairwise constraints in the form of Must-Link or Cannot-Link, which can be easily integrated in spectral clustering algorithm. In this paper, we propose a design of face annotation system that can generate more informative constraints and better use constraints with constrained spectral clustering. And we examine the system in a lab situation dataset and a real-live dataset, of which results demonstrate the effectiveness of our method.

Keywords: Face clustering · Spectral clustering · Constrained clustering · User interactions · Pairwise constraints

1 Introduction

In nowadays face annotation systems, such as intelligent photo albums and photo management systems, photo clustering or face clustering based on identity is a common feature. However, even though the technology and clustering theories have rapidly developed in past decades, there is still no proper method for face clustering in the real world because it is challenging to decide the parameters of clustering algorithms and the clustering result is far from satisfactory, telling the fact that the involvement of users is necessary [1]. Thus, researchers and companies turn to design systems [2] that assist users to annotate faces or figures in the photos, in order to construct better clusters with user interactions.

Obviously, most of the user interactions can transform into pairwise constraints between faces. When the user tagged more than one face with the same identity, these faces should be in the same cluster, called Must-Link (ML) constraints. Meanwhile, faces tagged different identities should to different clusters, called Cannot-Link (CL) constraints. Sometimes, users should answer the question from the program that "Whether this two faces are the same individual? ", and "Yes" for ML, "No" for CL. Therefore, the face clustering problem switches to a constrained clustering problem.

© Springer Nature Singapore Pte Ltd. 2016
T. Tan et al. (Eds.): CCPR 2016, Part I, CCIS 662, pp. 3–12, 2016.
DOI: 10.1007/978-981-10-3002-4_1

Constrained clustering has been studied in past years. Wagstaff et al. [3] are the first to consider constrained clustering by encoding available domain knowledge in the form of pairwise ML and CL constraints, proposing the COP-Kmeans. Following them, several other variants of the original K-means algorithm have been developed [4,5]. Semi-supervised or constraint extensions have been developed for other types of clustering algorithms, including density-based methods [6,7] and spectral clustering algorithms [8–13].

In this paper, we mainly work on spectral clustering for its convenience to integrate pairwise constraints. And we propose a method to generate informative pairwise constraints with a designed face annotation system.

The remainder of the paper is organized as following. Section 2 presents the current state of art on constrained spectral clustering. Section 3 introduces the designed face annotation system and describes our method in detail. Section 4 describes the experiments and our analysis on the results while Sect. 5 presents conclusions.

2   Constrained Spectral Clustering

The common steps of spectral clustering algorithm can be described as following [14]:

1. Given n points that we want to cluster into k clusters, form the affinity matrix $A \in \mathbb{R}^{n \times n}$.
2. Define D to be the diagonal matrix with $D_{ii} = \sum_j A_{ij}$.
3. Normalize: $N = D^{-1/2} A D^{-1/2}$.
4. Find $x_1, x_2, ..., x_k$, the k largest eigenvectors of N and form the matrix $X = [x_1 x_2 ... x_k] \in \mathbb{R}^{n \times k}$ by stacking the eigenvectors in columns.
5. Form the matrix Y from X by normalizing each of X's rows to be unit length (i.e. $Y_{ij} = X_{ij}/(\sum_j X_{ij}^2)^{-1/2}$).
6. Treating each row of Y as a point in $\mathbb{R}^k$, cluster them into k clusters via Kmeans or any other algorithm.
7. Assign the original point i to cluster c if and only if row i of the matrix Y was assigned to cluster c.

In the face topic of computer vision, affinity can be treated as similarity. Therefore, we initialize A by computing the cosine similarity between faces (i.e. $A_{ij} = cosine\_similarity(i, j)$) and set $A_{ii} = 0$.

There are many ways to make spectral clustering to be constrained [15]. In [9], inspired by Spectral Learning (SL) [8], authors add the constraints into the affinity matrix by setting MLs valued 1 and CLs valued 0. Differently, "Flexible Constrained Spectral Clustering" (CSP) [10] introduces the constraints at the eigenspace computation step. "Spectral Clustering with Linear Constraints" (SCLC) [11] does not use Kmeans but allows binary clustering. "Constrained Clustering via Spectral Regularization" (CCSR) [12] adds constraints at an intermediate stage that modifies the eigenspace. Among them, modifying the affinity matrix is the simplest, which is our choice.

It is widely believed that the more constraints, the better performance, encouraging adding more constraints. However, [16] shows that the inappropriate constraints may degrade the performance and introduces metrics to identify the relevance of a constraint. Thus, we would discuss the constraints generated with our method in Sect. 4.

3   Constraints Generating

The generation of constraints depends on the design of face annotation system. Figure 1 shows the framework of our proposed system.

After the face detection and feature exaction, all faces are represented by feature vectors, which are the input of our system. Firstly, an unsupervised clustering algorithm such as kmeans or spectral clustering would run on the input data. The system would organize the clustering results and display the largest cluster to the user. Later, the users are encouraged to mark an error on this displayed cluster. If there is no error in the largest cluster, the second largest cluster would be displayed and so forth. If none of the clusters has error, the clustering task would be completed.

For our purpose, we hope the users can mark the wrong faces that should not appear in this cluster. The amount of marked wrong faces does not matter but the first one should be marked, reducing the user workload in some way. And we sort the cluster elements with the distance to the cluster center in ascending order so that the closest wrong face would be clear enough for users to mark.

With the marked face, the generation of constraints would begin. Figure 2 illustrates the strategy. As we sort the cluster elements, it is clear that the faces that before the first wrong face in this cluster are right ones. So ML constraints would be built between each of them. Moreover, as we notice the wrong one, CL constraints can be made between the wrong one with the decided right ones. In all, we can obtain no less than one constraints with one marked constraint.

The next step is our constrained spectral clustering. To simply our approach, we adopt the constraints to the affinity matrix directly [8]. For each pair of faces

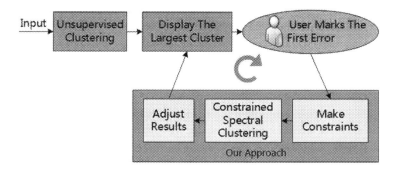

Fig. 1. The framework of our face annotation system

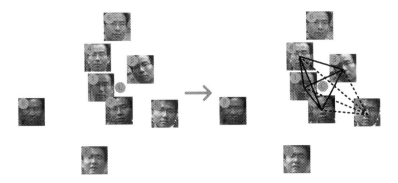

Fig. 2. Faces in the cluster are sorted by the distance to the cluster center (yellow point) in ascending order and the index is shown on the point. The same color represents the same individual and the blue points clearly do not belong to this orange cluster. The 5th point is the first error so that there are ML constraints (solid black line) among points 1,2,3 and 4, CL constraints (dotted black line) between 5 to 1,2,3 and 4 respectively.

(i, j) that is ML, assign the values $A_{ij} = A_{ji} = 1$. Likewise, for each pair of faces (i, j) that is CL, assign the values $A_{ij} = A_{ji} = 0$. Besides, our constrained spectral clustering initialize centers for Kmeans step with centers of the previous clustering result.

We adjust clustering results after the constrained spectral clustering to ensure the quality of user annotation. As we only display the largest cluster, we force the faces in the largest cluster to satisfy the CL constraints, which can avoid users marking the same wrong face when the clustering results happen to be same as the previous clustering and lead to at least a new-coming CL constraint. The faces that violate CL constraints in the largest cluster would be moved to other clusters that satisfy the current CL constraints and close to them. If there are no proper clusters, the face would be sent to its farthest cluster and set as that cluster's center. It is reasonable because there must be at least two clusters need to be merged and such faces must belong to some new clusters in this situation.

After the adjustment, the new largest cluster would be display and the system would work in a loop. The core idea is iteratively adding and using constraints from user interactions by modifying the affinity matrix for spectral clustering.

4   Experiments

To valid our approach on the proposed face annotation system described in Sect. 3, a series of experiments are performed and reported in this section. Firstly, the face features we used are described. Then, the datasets and evaluation methods are referred. Experimental results and discussion come last.

4.1 Features

As the development of deep learning methods in recent years, we construct a DCNN for our face feature extraction after face detection to our raw photos. Inspired by [17], the designed network contains eleven layers, including nine convolutional layers and two fully-connected layers. Every convolutional layer is followed by ReLU. And the last layer is a n-way softmax layer which is set in training phase. We train the network with a large self-collected dataset and use face identification signals as supervision. The hidden layer output which contains 3000 neural units before the softmax layer is taken as the extracted features of input faces. Therefore, every face can be presented with a vector of 3000 dimensions and the similarity of each pair can be computed with cosine similarity.

4.2 Datasets

Usually, face annotation system works on small datasets rather than big datasets because the system is mostly a personal tool. And there is no need to test on big datasets that are million level.

Our first dataset is the Olivetti Face Database [18], which consists of 400 face images from 40 individuals. All images are cropped to 64×64 with merely face and there are 10 face images for each individual.

A self-collected dataset is considered as well. The photos are taken in a trip so that the dataset is much closer to real world than Olivetti and closer to the daily demand of personal photo management. Different from the Olivetti, the Trip dataset contains photos with multi poses and various lighting, while the Olivetti is all frontal faces in the same situation. Besides, the photos have complex background and the face is just part of the image and there are more than one face in some photos.

For the Olivetti, we extract face features after alignment with SDM [19] because all the faces have been perfectly cropped, which can be regarded as output faces from face detection. But for Trip dataset, we firstly run a face detection with Yu's work[1] and there are 181 faces detected as a result. Later, the detected faces are cropped and aligned before feature extraction. Table 1 shows the statistical data of the datasets. It is believed that the Trip dataset is harder than Olivetti for its unbalanced distribution and the complex face status, which is shown in our experiments.

Table 1. Statistical data of the datasets.

Dataset	Faces	Individuals	Max	Min	Avg
Olivetti	400	40	10	10	10
Trip	181	12	33	6	≈ 15

---

[1] https://github.com/ShiqiYu/libfacedetection.

4.3   Evaluation Methods

To evaluate the clustering results, the accuracy (AC) [20] and the Rand Index (RI) [21] are used in our experiments.

The AC is defined by counting the number of correctly assigned faces and dividing by m as follows:

$$AC = \frac{1}{m} \sum_{i=1}^{m} \delta(s_i, map(r_i)) \qquad (1)$$

where m is the total number of faces, s_i and r_i denote the ground truth label and predicted label from the clustering algorithms, $\delta(x, y)$ equals 1 if $x = y$ and equals 0 otherwise, and $map(r_i)$ is the mapping function that maps each cluster label r_i to the equivalent label from the ground truth label. The best mapping can be found by using the Kuhn-Nunkres algorithm.

The RI is a measure of agreement between two partitions, i.e., P_1 and P_2, of the same dataset D. Each partition is viewed as a collection of $n * (n - 1)/2$ pairwise decisions, where n is the size of D. For each pair of points d_i and d_j in D, P_x either assigns them to the same cluster or to different clusters. Let a be the number of decisions where d_i is in the same cluster as d_j in P_1 and in P_2. Let b be the number of decisions where the two instance are placed in different clusters in both partitions. Total agreement can then be calculated using:

$$RI = \frac{a + b}{n * (n - 1)/2} \qquad (2)$$

4.4   Experimental Results

We test the annotation times vs. AC to examine our approach in boosting the clustering performance. Figure 3 shows the curves which are the average of 10 separated experiments. The horizontal axis is the annotation time, counting from

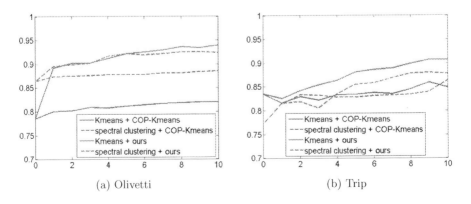

(a) Olivetti   (b) Trip

Fig. 3. Annotation times vs. AC

0 to 10. The zero time of annotation is the first clustering period in our design which is unsupervised clustering because we get no constraints at the beginning. We compare two different unsupervised clustering algorithms: Kmeans and spectral clustering, assuming the number of individuals is known before clustering. Although there is a gap between the performances at the beginning both on the two datasets, all of the results reach a better level as the annotation going on, proving the effectiveness of our approach.

Experiment results with COP-Kmeans are treated as the baseline. Specially, to avoid the violations of constraints that make COP-Kmeans fail easily, we are only concerned about the CL constraints. And similar to our spectral clustering, the faces that violate all CL constraints would be put into their farthest cluster. Using COP-Kmeans as the constrained clustering algorithm instead of our proposed spectral clustering still shows slight improvements thanks to our designed system but ours is quite better.

We regard the constraints as a partition and observe with RI, shown in Table 2 in percentage. The RI of clustering results is shown in the parentheses. For the perfect partition or all constraints of the set, the RI is 100 percent. We can conclude from the results that we just use a small proportion of the perfect partition and achieve high performance.

Table 2. RI of constraints (clustering results)

T	Kmeans +ours (Olivetti)	Spectral clustering +ours (Olivetti)	Kmeans +ours (trip)	Spectral clustering +ours (trip)
0	0(98.87)	0(99.32)	0(97.08)	0(95.66)
1	0.01(99.42)	0.01(99.43)	6.02(96.89)	5.33(96.21)
2	0.02(99.47)	0.04(99.47)	7.44(97.24)	6.50(96.47)
3	0.05(99.44)	0.07(99.46)	8.29(97.30)	6.97(96.34)
4	0.06(99.48)	0.09(99.51)	8.60(97.49)	7.15(96.98)
5	0.10(99.54)	**0.13(99.55)**	8.91(97.72)	7.94(97.22)
6	0.14(99.55)	0.16(99.52)	9.30(97.88)	8.22(97.22)
7	0.18(99.56)	0.19(99.54)	9.40(97.84)	8.37(97.49)
8	0.21(99.57)	0.21(99.55)	9.46(98.00)	8.66(97.60)
9	0.24(99.57)	0.23(99.53)	9.72(98.05)	**8.90(97.63)**
10	**0.26(99.58)**	0.25(99.49)	**9.85(98.09)**	8.94(97.61)

4.5   Discussion

A main factor that decides the performance is the amount of constraints. Consequently, we count the average amount of constraints and the result is shown in Fig. 4. Different from other researches that work on the quantity of constraints,

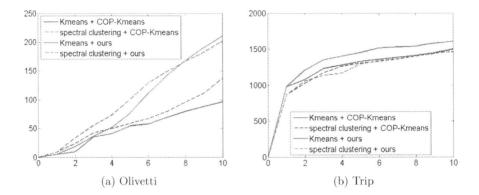

Fig. 4. Annotation times vs. constraints amount

we concern the quantity changes at every iteration. In Olivetti, the constraints quantity of our approach increases faster than COP-Kmeans, while it is less obvious in Trip. Although getting close amount of constraints, our approach outperforms the COP-Kmeans in Trip. Had been found in some research, the amount of constraints does not always lead to better result, which is proved in our experiment too. Though the amount keep increasing, the AC would drop sometimes. In addition, the increase amount of constraints is decreasing as annotation time increasing which is clear in Trip because we aim at adding at least 1 CL constraints while most ML constraints may have been made in previous iterations.

Another factor is the informative of constraints, which is a metric introduced in [16]. To prove our approach is adding more informative constraints, we compared with using random constraints. We randomly add the same amount of constraints as our approach in every iteration and the result is shown in Fig. 5. We take the average result of 10 experiments in every random process. The effect

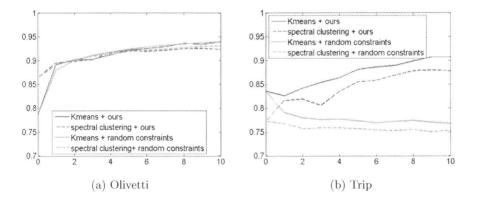

Fig. 5. Times vs. AC

is not clear in Olivetti for its easiness but highly proved in Trip. It can be concluded from our experiment results that in such a face annotation system, more informative constraints can be found easily.

In a word, our approach wins by getting more informative and more constraints and making better use of constraints.

5   Conclusion

To fulfill the demand of face clustering, we propose a face annotation system that can make more informative constraints, and better use constraints in spectral clustering. The experiment results in two datasets that one is from lab situation and the other is from real live prove the efficiency. To step forward, there are several other constrained clustering algorithms that can involve in our design, which may be our future work. And as the continuous development of clustering algorithm, there would be more constrained algorithms in the future.

Acknowledgments. This work was partially sponsored by supported by the NSFC (National Natural Science Foundation of China) under Grant No. 61375031, No. 61573068, No. 61471048, and No.61273217, the Fundamental Research Funds for the Central Universities under Grant No. 2014ZD03-01, This work was also supported by Beijing Nova Program, CCF-Tencent Open Research Fund, and the Program for New Century Excellent Talents in University.

References

1. Cohn, D., Caruana, R., McCallum, A.: Semi-supervised clustering with user feedback. Constrained Clustering: Adv. Algorithms Theor. Appl. 4(1), 17–32 (2003)
2. Cui, J., Wen, F., Xiao, R., Tian, Y., Tang, X.: EasyAlbum: an interactive photo annotation system based on face clustering and re-ranking. In: Proceedings of the SIGCHI Conference on Human Factors in Computing Systems, 367–376. ACM (2007)
3. Wagstaff, K., Cardie, C., Rogers, S., Schrodl, S.: Constrained k-means clustering with background knowledge. In: ICML, vol. 1, pp. 577–584 (2001)
4. Basu, S., Banerjee, A., Mooney, R.: Semi-supervised clustering by seeding. In: Proceedings of 19th International Conference on Machine Learning (2002)
5. Bilenko, M., Basu, S., Mooney, R.J.: Integrating constraints and metric learning in semi-supervised clustering. In: Proceedings of the twenty-first international conference on Machine learning, p. 11. ACM (2004)
6. Lelis, L., Sander, J.: Semi-supervised density-based clustering. In: Ninth IEEE International Conference Data Mining, ICDM 2009, pp. 842–847. IEEE (2009)
7. Ruiz, C., Spiliopoulou, M., Menasalvas, E.: C-DBSCAN: density-based clustering with constraints. In: An, A., Stefanowski, J., Ramanna, S., Butz, C.J., Pedrycz, W., Wang, G. (eds.) RSFDGrC 2007. LNCS (LNAI), vol. 4482, pp. 216–223. Springer, Heidelberg (2007)
8. Kamvar, K., Sepandar, S., Klein, K., Dan, D., Manning, M., Christopher, C.: Spectral learning. In: International Joint Conference of Artificial Intelligence. Stanford InfoLab (2003)

9. Xiong, C., Johnson, D., Corso, J.J.: Active clustering with model-based uncertainty reduction. arXiv preprint arXiv:1402.1783 (2014)
10. Wang, X., Davidson, I.: Flexible constrained spectral clustering. In: Proceedings of the 16th ACM SIGKDD International Conference on Knowledge Discovery and Data Mining, 563–572. ACM (2010)
11. Xu, L., Li, W., Schuurmans, D.: Fast normalized cut with linear constraints. In: IEEE Conference on Computer Vision and Pattern Recognition, pp. 2866–2873. IEEE (2009)
12. Li, Z., Liu, J., Tang, X.: Constrained clustering via spectral regularization. In: IEEE Conference Computer Vision and Pattern Recognition, pp. 421–428. IEEE (2009)
13. Rangapuram, S.S., Hein, M.: Constrained 1-spectral clustering. arXiv preprint arXiv:1505.06485 (2015)
14. Ng, A.Y., Jordan, M.I., Weiss, Y.: On spectral clustering: analysis and an algorithm. Adv. Neural Inf. Process. Syst. **2**, 849–856 (2002)
15. Voiron, N., Benoit, A., Filip, A., Lambert, P., Ionescu, B.: Semi-supervised spectral clustering with automatic propagation of pairwise constraints. In: 2015 13th International Workshop Content-Based Multimedia Indexing (CBMI), pp. 1–6. IEEE (2015)
16. Davidson, I., Wagstaff, K.L., Basu, S.: Measuring constraint-set utility for partitional clustering algorithms. In: Fürnkranz, J., Scheffer, T., Spiliopoulou, M. (eds.) PKDD 2006. LNCS, vol. 4213, pp. 115–126. Springer, Heidelberg (2006)
17. Simonyan, K., Zisserman, A.: Very deep convolutional networks for large-scale image recognition. arXiv preprint arXiv:1409.1556 (2014)
18. Samaria, F.S., Harter, A.C.: Parameterisation of a stochastic model for human face identification. In: Proceedings of the Second IEEE Workshop on Applications of Computer Vision. IEEE, pp. 138–142 (1994)
19. Xiong, X., Torre, F.: Supervised descent method and its applications to face alignment. In: Proceedings of the IEEE Conference on Computer Vision and Pattern Recognition, pp. 532–539 (2013)
20. Xu, W., Liu, X., Gong, Y.: Document clustering based on non-negative matrix factorization. In: Proceedings of the 26th Annual International ACM SIGIR Conference on Research and Development in Information Retrieval, pp. 267–273. ACM (2003)
21. Rand, W.M.: Objective criteria for the evaluation of clustering methods. J. Am. Stat. Assoc. **66**(336), 846–850 (1971)

Axial-Decoupled Indoor Positioning Based on Location Fingerprints

Wei Yanhua[⊠], Zhou Yan, Wang Dongli, and Wang Xianbing

Institute of Control Engineering, Xiangtan University, Xiangtan 411105, China
358494653@qq.com

Abstract. Indoor positioning using location fingerprints, which are received signal strength (RSS) from wireless access points (APs), has become a hot research topic during the last a few years. Traditional pattern classification based fingerprinting localization methods suffer high computational burden and require a large number of classifiers to determine the object location. To handle this problem, axial-decoupled indoor positioning based on location-fingerprints is proposed in this paper. The purpose is to reduce the decision complexity while keeping localization accuracy through computing the position on X- and Y-axis independently. First, the framework of axial-decoupled indoor positioning using location fingerprints is given. Then, the training and decision process of the proposed axial-decoupled indoor positioning is described in detail. Finally, pattern classifiers including the least squares support vector machine (LS-SVM), support vector machine (SVM) and traditional k-nearest neighbors (K-NN) are adopted and embedded in the proposed framework. Experimental results illustrate the effectiveness of the proposed axial-decoupled positioning method.

Keywords: Location fingerprint · Axial-decoupled · Indoor positioning · Pattern classification

1 Introduction

With the popularity of wireless networks, the rapid growth of intelligent mobile phones and increasing maturity of pervasive computing technology, location based services (LBS) have attracted more and more attention and shown great popularity in many applications, such as indoor positioning, tracking, navigation, and location-based security [1–3]. The common positioning system such as the Global Positioning System (GPS) doesn't perform very well in urban settings especially inside buildings with a limited line-of-sight (LOS) from satellites. Because of the complexity of the indoor environment, it is usually difficult to provide a satisfactory level of accuracy in most applications. Therefore, one of the major challenges is to design real-time and accurate indoor positioning systems that can be easily deployed on commercially available

This work was supported by the National Science Foundation (NNSF) of China (under Grant 61100140 and 61104210) and the Construct Program of the Key Discipline in Hunan Province.

T. Tan et al. (Eds.): CCPR 2016, Part I, CCIS 662, pp. 13–26, 2016.
DOI: 10.1007/978-981-10-3002-4_2

mobile devices without any hardware installation or modification. Indoor positioning systems could be used to give access to an interactive map of a building. For instance, they could locate a person through an airport to the boarding gate, help a person find her room or facilitate the way of finding items of a shopping list in a supermarket.

There have been a variety of studies on indoor localization. As far as the indoor localization method is concerned, it can be categorized by the measurable quantities obtained from the transmitted signals. Received signal strength (RSS)-based localization methods have been extensively studied as an inexpensive solution for indoor positioning in recent years [4–6]. Compared with other methods based on algorithms (e.g., time-of-arrival (TOA) or angel-of-arrival (AOA) methods of UWB signals), RSS can be easily obtained by a Wi-Fi integrated mobile device, without any hardware installation or modification [7]. Various RSS-based indoor positioning and tracking algorithms have been proposed using the location information of access points (APs), which may not be available or hard to obtain in practice [8]. The major challenge for accurate RSS-based location comes from the variations of RSS due to the dynamic and unpredictable nature of radio channel by the structures within the building, such as shadowing, multipath, the orientation of wireless device, etc. [9]. Then, another approach is to pre-built radio map, termed as fingerprinting, to localize a mobile device [10], instead of using a propagation model to describe the relationship between RSS and position [11, 12]. Therefore, an implementation of an indoor positioning system based on fingerprinting signals of wireless local area networks (WLANs) has been proposed to estimate a location for indoor areas.

The location fingerprinting technique connects location-dependent characteristics (e.g. RSS) with grids in the region of interest (ROI) through measuring signals from available APs without knowing their location in advance, and uses these characteristics to infer the location [13–15]. At present, the location fingerprint positioning has attracted great attention of many researchers. The localization problem under this framework can be modeled as a pattern classification problem since at each time instant the user is located at a specific point in space [16, 17]. Commonly, the service area is pre-partitioned into a set of regions (e.g., a grid of cells); each serves as a class and a multi-class classification tool is used to assign a given fingerprint into one of these classes. A class of popular localization algorithms is based on pattern classification techniques including k-nearest neighbors (K-NNs), neural networks (NNs) and support vector machines (SVMs) etc. To name a few, Zhu et al. [18] introduce grid concept, changed position matching into multi-class classification problem and obtain the object location by SVM. Feng et al. [5] propose an accurate RSS-based indoor positioning system using the theory of compressive sensing, which is a method to recover sparse signals from a small number of noisy measurements by solving an l_1-minimization problem. In [19], Shin et al. propose a fingerprint positioning multi-classifier model on WLAN, making use of many results based on Bayesian combination rule [20] and majority vote [21] to obtain the fingerprint position. In [22], Xiang et al. propose a scalable semi-supervised learning (3SL) technique for building accurate fingerprinting from a small portion of labeled samples. Dortz et al. [23] propose a new method that compares online and offline signal strength probability distributions in order to find the nearest offline locations.

Traditional pattern classification based fingerprinting localization methods suffer high computational burden and require a large number of classifiers to determine the object location [24–26]. To handle this problem, axial-decoupled classification model is proposed in this paper for indoor positioning using RSS fingerprinting. The purpose is to reduce the decision complexity while keeping localization accuracy through computing the position on X- and Y axis independently, which can reduce the number of classifiers. Experimental results illustrate the effectiveness of the proposed axial-decoupled positioning method.

The rest of the paper is organized as follows. In Sect. 2, the problem of fingerprint based indoor localization framework is formulated. In Sect. 3, the proposed axial-decoupled method for localization is given, which followed by experimental results on both small-size and large-size dataset are included in Sect. 4. Section 5 concludes the paper with future researching directions given.

2 Problem Formulation

Considering a typical indoor positioning scenario, where a user carries a mobile device equipped with a WLAN adapter, using only RSS measurements from available APs. The location of these APs is unknown. The main task of the positioning system is to estimate the user's current location and illustrate it on a map (floor plan) on the device.

The location of the mobile is estimated by comparing the current RSS reading with a restored database called fingerprints, which is a table of measured RSS form a similar device over a grid of points. Several methods can be adopted to compare the RSS reading with the fingerprints. In this paper, axial-decoupled indoor positioning based on location-fingerprints is proposed. As depicted in Fig. 1, the proposed positioning system of axial-decoupled consists of two stages: offline phase (also known as training phase) and online phase (also known as positioning phase). In the offline phase, according to the site-surveying, the RSS from multiple APs at different grid points are collected and stored in a fingerprint database. The vector of mean RSS values at point on the grid is called the location fingerprint of that point. The fingerprint sample is then compared with fingerprints stored in the radio map for determining the location of the mobile devices on the grid.

Offline phase includes the following steps:

I. Partition the ROI into a grid of cells, each cell receives RSS samples from wireless APs in order to build RSS feature vectors, and the sample is divided into X- and Y-axis training samples;

II. Using a normalized X- and Y-axis training sample to train classifier independently and obtained X-axis and Y-axis classifier.
   Online phase includes the following steps:

III. A user carries a mobile device equipped with a WLAN adapter and enters the ROI, collecting RSS sample from wireless APs at the current position. Then, using the collected RSS sample as offline trained X- and Y-axis classifier input. Therefore, the X- and Y-axis decision results are obtained independently;

IV. Combining the X- and Y-axis decision results to locate mobile device.

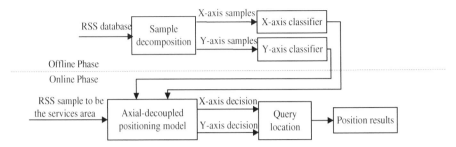

Fig. 1. Axial-decoupled for indoor positioning classification model based on location fingerprints

3   Axial-Decoupled Positioning

3.1   Offline Phase

During the offline phase, the samples of RSS readings are collected from known locations, referring to the reference points (RPs), by pointing the mobile device to different orientations. When considering an indoor positioning system covered with a WLAN in a single floor inside a building, we assume that there are N APs in the area and they are all visible throughout the area under consideration. A ROI is defined over the two dimensional floor plan. Assuming the ROI is partitioned as a $l_x \times l_y$ grid according to the X- and Y-axis, we have $l = l_x \cdot l_y$ grids in the area.

In each grid, the mobile users collect the RSS fingerprints from different APs and represented as a vector $\{\bar{f}_i = (rss_1^i, rss_2^i, \ldots rss_n^i, \ldots, rss_N^i)\}_{i=1}^w$, consisting of w finger-prints sample at location with known coordinates, where rss_n^i is a RSS value corresponding to the i-th sample of the n-th AP's RSS value and N is the total number of available wireless APs. The sample of RSS feature vectors is denoted as $f_{(m_i, n_i)} = (\bar{f}_i, m_i, n_i)$, where m_i and n_i are respectively corresponded to the i-th sample of the grid in the X- and Y-axis of the class number, and $m_i = 1, 2, \ldots, l_x, n_i = 1, 2, \ldots, l_y$. The sample fingerprint collection of the X- and Y-axis, i.e. $f_{x_i} = (\bar{f}_i, m_i), f_{y_i} = (\bar{f}_i, n_i)$. Next, using the above two kinds of samples to train the multi-classifier respectively, and the axial-decoupled indoor positioning classification model is obtained. This lays a foundation for the online positioning phase.

3.2   Online Phase

During the online phase, a user carries a mobile device equipped with a Wi-Fi adapter and enters the ROI, collecting RSS sample from wireless APs at the current position. The online RSS reading is compared with fingerprints stored in the database to determine the current localization by X- and Y-axis classifier independently. Following the positioning result of the X- and Y-axis is obtained jointly by the classifying decision.

For example, at the current position (x_k, y_k), the collected RSS fingerprint is $\bar{f}_k = (rss_1, rss_2, \ldots rss_n, \ldots rss_N)$. The RSS fingerprints as the input of the trained offline on X- and Y-axis classifier, and to become the decision result on X- and Y-axis respectively. Next combine the results on both axes to locate the mobile device. The procedure steps are as follows:

(i) The predicted class $\left(m_x^k, n_y^k\right)$ of the test sample $\bar{f}_k$ is obtained by X- and Y-axis classifiers, which are trained in the offline phase.

(ii) The grid is determined by predicted class and the grid centroid is the predicted coordinate $\hat{P}_k$, $\hat{P}_k = (\hat{x}_k, \hat{y}_k)$.

(iii) Adopting the 2-norm to calculate the deviation between prediction coordinate and the actual coordinate. The location accuracy A is denoted as:

$$A = \left\| \hat{P}_k - P_k \right\| \tag{1}$$

where P_k means the actual coordinates of the k-th test sample, $P_k = (x_k, y_k)$; $\|\cdot\|$ means a vector of 2-norm, $\left\| \hat{P}_k - P_k \right\| = \sqrt{(\hat{x}_k - x_k)^2 + (\hat{y}_k - y_k)^2}$.

3.3   Procedure of the Proposed Axial-Decoupled Method

The procedure of the proposed axial-decoupled positioning method can be presented as follows:

I. Firstly, partitioning ROI into a grid of cells, then collecting RSS data in each cell and decomposing into X- and Y-axis training samples. Lastly, determining the X- and Y-position according to decomposed samples (in terms of Step 2–5), respectively.

II. Preprocessing: Normalizing the training samples to the [−1, 1].

III. Training phase: (i) classifier parameters: when using LS-SVM, SVM, first select the parameters (c, g) by grid search method; (ii) train classifiers by One-Against-All (OAA) [27] or One-Against-One (OAO) [28] approach independently.

Axial-decoupled positioning technique classifies the fingerprint samples of X- and Y-axis with OAO or OAA independently to locate the target. One of the differences of OAO and OAA approaches lies in the required classifiers [27, 28]: as for k class problem, $k(k − 1)/2$ binary classifiers are needed for OAO approach; however, OAA approach requires k binary classifiers.

IV. Online phase: normalizing the test samples and obtaining the X- and Y-axis categories according to classifier results in the Step 3.

V. Calculating the positioning output according to IV.

3.4 Performance Analysis

The common pattern classifier based fingerprint approach is to partition the ROI into a number of regions, each representing a class. Typically, grid partitioning is used, a unit cell of this partition is chosen as one class. If the 2D positioning area is portioned as $l_x \times l_y$ grids, each being a cell of size $1/l_x \times 1/l_y$. Thus, there are $l_x \cdot l_y$ classes. Using a multiclass classifier, a fingerprint can be classified into one of these classes and the center of the corresponding cell is the location estimate. It refers to grid method. However, rather than estimating both X- and Y-coordinate simultaneously, the proposed axial-decoupled method estimates the coordinates independently. For the X-axis, the area is partitioned into l_x column stripes, each serving as a class. Therefore there are l_x classes for the X-axis. Using a multiclass classifier, we can classify the X-coordinate into one of these stripes. Similarly, in a separate classification procedure, l_y classes are obtained by partitioning the area into row stripes and used to estimate the Y-coordinate. It can be seen from the above analysis, this method uses only $l_x + l_y$ classes, which is much fewer than $l_x \cdot l_y$ (commonly for multi-label classifier based indoor positioning, $l_x, l_y \gg 1$). The corresponding training time is reduced because the number of classifier decreased.

4 Experiment and Discussion

4.1 Test Dataset and Model Parameter Selection

To evaluate the performance of our axial-decoupled positioning method, we conduct experiments using two benchmarks:

(a) Small-size dataset: This dataset is from an indoor experiment used in [25] (University of Trento), containing a collection of 257 RSS fingerprints at 257 sample locations in a WLAN with 6 APs (Fig. 2). The sample locations are regular-grid points of the floor. Each fingerprint is measured at a sample location by a person carrying a personal digital assistant (PDA), as a receiver receiving signals from the APs. The PDA always points at north. A random 90 % of this collection (232 samples) is used for training and the rest of the samples (25 samples) are for testing.

(b) Large-size dataset: This dataset is from a real-world large-scale RSS dataset (From Xiangtan University Building of College of Information Engineering). The ROI is a room with an area of 14 m $\times$ 6 m. In this room area, 84 partitioned grids of size 1 m $\times$ 1 m are used. A fully regular grid could not be followed due to the presence of various obstacles such as tables and other furniture. RSS fingerprints are collected from 84 grids in a WLAN with twelve APs. It should be pointed out that only 4 of APs' positions are given previously. Measurements on some grids are in NLOS condition due to the obstacles such as wall, desk and other devices while others are in LOS condition. For each grid, we collect 40 RSS measurements from all APs by a person carrying a PDA, and the PDA points at four different heading orientations (east/west/south/north). This results totally 3360

Fig. 2. The small-size dataset's map: 30 m × 25 m (after [25]); the blue diamond represents the position of the six access points

samples, from which we randomly use 90 % dataset as the training set (3024 samples) and the rest of the samples (336 samples) for testing.

The experimental operation environment is Windows XP operating system, CPU G645, 3.47G RAM, MATLAB R2009a. In order to compare the advantages of the axial-decoupled indoor positioning method to traditional indoor positioning method, pattern classifiers like LS-SVM, SVM, and K-NN are applied to location fingerprint positioning framework. When using LS-SVM and SVM, we need to choose kernel function of the condition to satisfy Mercer [29]. There are a variety of kernel functions such as polynomial functions, radial basis functions (RBF) and sigmoid kernel. We take RBF as kernel function in the all experiments. Before the application of LS-SVM and SVM classification, the regularization and kernel parameter (c, g) should be determined. Regularization parameter c affects the generalization ability of the classifier by controlling the misclassification rate. Parameter c, which is too high, will cause the fact that the accuracy of training set classification is too high while the accuracy of test set classification is too low. Parameter g determines the complexity of sample feature subspace distribution. Therefore, the parameter c and g jointly influence classifier generalization ability and the final location accuracy. In this section, the apartments are determined through grid searching from $[2^{-10}, 2^{10}]$. For K-NN, K (e.g., 1, 3 or 5) represents the number of nearest-neighbor fingerprints used to estimate the unknown location, while distance-type can be "E" and "M" that presents Euclidean distance and Manhattan distance, respectively.

For ease of presentation, classifiers under axial-decoupled framework are denoted as AD-LS-SVM, AD-SVM, AD-1NN(E), AD-1NN(M), AD-3NN(E), AD-3NN(M), AD-5NN(E), AD-5NN(M), respectively.

4.2   Results and Discussion

A. Small-size Dataset

In the small-size dataset experiment, the grid size parameter is $l_x \times l_y \in \{7 \times 7, 8 \times 8, \ldots, 15 \times 15\}$. To compare the proposed method with traditional method, we use LS-SVM, SVM and K-NN classifier for the positioning of axial-decoupled or traditional grid method. Figure 3 is the experimental result (various cases of LS-SVM and SVM adopts OAO combination approach) of LS-SVM and SVM classifier under the condition of decoupled and non-decoupled in the different grid size.

From Fig. 3, we can see that the following conclusions:

(i) Decoupled vs. Non-decoupled: For LS-SVM and SVM, location accuracy and computation time of the decoupled are obviously better than the non-decoupled. In terms of location accuracy, the grid size shows slightly influence on decoupled positioning method. With the successive increase of grid density, location

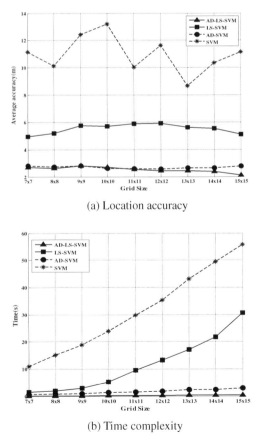

(a) Location accuracy

(b) Time complexity

Fig. 3. Location accuracy and time complexity of different methods

accuracy changes in positioning method based on decoupled classifier relatively tends to be slow, while the change of the non-decoupled classifier is much more severe. As we can see from Fig. 3(a), When the grid size is 15×15, AD-LS-SVM will get a relatively good location accuracy (2.1071 m), it is a half of LS-SVM (5.1051 m). When it comes to computational costs, the time needed for decoupled classifier is far less than non-decoupled classifier. As the grid density increased, time needed for two kinds of classification methods is also increased accordingly. But the growing speed in non-decoupled is much faster than the decoupled positioning technique proposed in this paper. It is observed from Fig. 3(b), the time of LS-SVM (30.62 s) needed is 80 times longer than that of AD-LS-SVM (0.38 s) in the grid size of 15×15. Therefore, the location fingerprint positioning method of axial-decoupled has higher location accuracy and a lower computational cost than non-decoupled.

(ii) LS-SVM vs. SVM: In non-decoupled, LS-SVM and SVM differ largely both in location accuracy and computation time. When concerning location accuracy, LS-SVM is higher than SVM, the location error of which is almost twice than that of the former. As the increase of grid density, LS-SVM location accuracy is relatively stable, while the SVM is volatile. In terms of computation time, LS-SVM under the same grid is much lower than SVM, the cost of which is almost twice to three times than that of the former. It is observed that the time of SVM method increase faster when the grid becomes denser, the time of SVM is even longer than that of LS-SVM. Because the number of classes in each method increases with the grid becomes denser. Therefore, LS-SVM has both a better positioning effect and a lower time complexity than SVM in non-decoupled.

(iii) AD-LS-SVM vs. AD-SVM: For the evaluation, the AD-LS-SVM is comparable to AD-SVM both in location accuracy and computation time. When the grid becomes larger, the location accuracy of AD-SVM is slightly higher than AD-LS-SVM. In the case $(l_x, l_y) = (10, 10)$, where the location accuracy remains the same. With the grid density increased, not only the location accuracy of AD-LS-SVM is slightly better than AD-SVM but also the computation time needed for AD-LS-SVM in any grid density is less than AD-SVM.

In order to further analyze the decoupled positioning technique. Figure 4 shows the average location accuracy and time complexity varying with the grid density. Here we adopt OAO and OAA approach for AD-LS-SVM and AD-SVM classifier.

From Fig. 4, we can see that the following conclusions:

(i) Location accuracy of OAO and OAA: For a specific classifier (e.g., LS-SVM or SVM), positioning accuracy of the classifier based on OAO approach is slightly higher than that of OAA approach. With the increase of grid density, the positioning accuracy changes of OAO and OAA is similar, in other words, the location accuracy is worse with the grid density increased. But when the grid density reaches a certain value (e.g., 11×11 for AD-SVM-OAA, 12×12 for AD-SVM-OAO), the location accuracy becomes rather poor (except AD-LS-SVM-OAO).

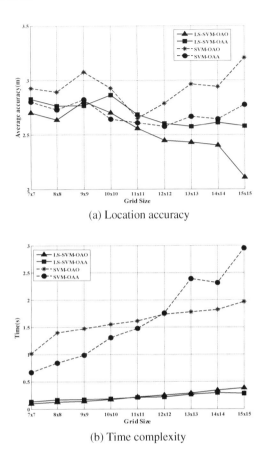

(a) Location accuracy

(b) Time complexity

Fig. 4. AD-LS-SVM, AD-SVM: OAO vs. OAA

(ii) Computational complexity of OAO and OAA: For LS-SVM, we can see that the time of OAO approach increases faster when the grid becomes denser, especially when the grid increases to a certain extent (e.g., 11×11 or 12×12) or more, the time of OAO is even longer than that of OAA. Because classification function increases with classes, and lower its speed in the decision-making process. Similarly, the same conclusions can be obtained as for SVM.

From the above analysis, we can see that the positioning precision of OAO approach is superior to OAA for a specific classifier. The computation time has relationship with the size and class of training samples. When the grid density is small, the computation time is lower than that of OAA. When the grid density is increased to a certain extent, costs of the former will surpass the latter, and both computational costs are increasing. Figure 5 shows the average location accuracy of K-NN classifier in decoupled and non-decoupled conditions when the grid parameter is 7×7.

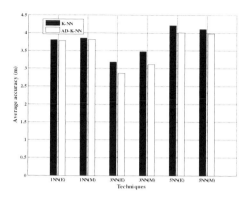

Fig. 5. K-NN's location accuracy varying with different K under decoupled and non-decoupled conditions

From Fig. 5, it is clearly see that the location accuracy of decoupled K-NNs is superior to non-decoupled ones, but both of them are worse than AD-LS-SVM and AD-SVM. Furthermore, the number of neighbors also affects the performance of technique and the value of K must be chosen carefully. When $K = 3$, the location accuracy is commonly superior to $K = 1$ and $K = 5$, whether using Euclidean distance or Manhattan distance. In addition, we can see that the precision is higher in using Euclidean distance when $K = 3$. It is because that taking only one neighbor into consideration may obtain a good accuracy but poor robustness. On the other hand, considering too many neighbors, even if it limits the risk of *wrong* neighbor, widens the potential area for the estimate location and thus leads to a lower accuracy.

B. Large-size Dataset

To show how our algorithm is also suitable for large-size dataset, the following experiment is made. In the large-size dataset experiment, the grid size parameter is $l_x \times l_y = 14 \times 6$. When comparing the proposed method with traditional method, we also use LS-SVM, SVM and K-NN classifier for the positioning of axial-decoupled and traditional grid method. Figure 6 is the experimental result (various cases of LS-SVM and SVM adopts OAO combination approach) of LS-SVM and SVM classifier under the condition of decoupled and non-decoupled.

Figure 6 shows the average localization accuracy and time complexity varying in decoupled and non-decoupled conditions. As noticed, the location accuracy and computation time of proposed decoupled method are obviously better than traditional grid method. Comparing four methods in the Fig. 6, we observe that AD-LS-SVM is the one with the best location accuracy (1.7217 m) and the lowest computational burden (0.4649 s), while SVM is the contrary. As we can see from Fig. 6, it's obvious that LS-SVM has a better location accuracy and the lowest time compared with the SVM. This experiment strongly indicates that, for the large-size dataset, our proposed method is substantially better than traditional grid methods.

Figure 7 shows the average location accuracy of K-NN classifier in decoupled and non-decoupled when the grid parameter is 14×6. From Fig. 7, we can clearly see that the

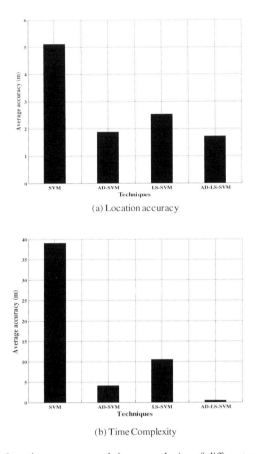

(a) Location accuracy

(b) Time Complexity

Fig. 6. Location accuracy and time complexity of different methods

location accuracy of decoupled K-NNs is superior to non-decoupled ones. Both can obtain similar accuracy with decoupled LS-SVM and decoupled SVM. It is observed that the Euclidean distance and Manhattan distance provide similar results. In addition, the number of neighbors also affects the performance of technique and the value of K must be chosen carefully. Figure 7 shows that for K-NNs a smaller K corresponding to better location accuracy. The best location accuracy (1.6288 m) is achieved by 1NN(E).

C. Remarks

Our experiment with the small and large datasets has shown that: (1) axial-decoupled method is much better than traditional grid method, both in location accuracy and computation complexity; (2) AD-LS-SVM and AD-SVM are comparable with each other, the former is slightly more accurate and faster; (3) Among the traditional grid methods, LS-SVM is obvious better than SVM and with the best location accuracy and the lowest computational burden; (4) LS-SVM and SVM are substantially less accurate than K-NN in the traditional grid methods; (5) The location accuracy of OAO approach is superior to OAA for a specific classifier.

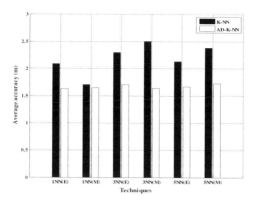

Fig. 7. K-NN's location accuracy varying with different K under decoupled and non-decoupled conditions

5   Conclusion and Future Work

Axial-decoupled indoor positioning based on location fingerprints is proposed. The experiment results of traditional methods such as LS-SVM, SVM and K-NN show that axial-decoupled ones are much better than traditional grid ones, both in location accuracy and computation complexity. In the near future, we would focus on studying the feature selection of RSS and the dynamic target tracking that based on the axial-decoupled indoor positioning.

Acknowledgements. This work is supported by the National Natural Science Foundation of China (No. 61100140 and 61104210).

References

1. Sertthin, C., Fujii, T., Ohtsuki, T.: Multi-band received signal strength fingerprinting based indoor location system. IEICE Trans. Commun. **93**(8), 1993–2003 (2010)
2. Akyildiz, I.F., Su, W., Sankarasubramaniam, Y., Cayirci, E.: A survey on sensor networks. IEEE Commun. Mag. **40**(8), 102–114 (2002)
3. Popov, L.: iNav: a hybrid approach to Wi-Fi localization and tracking of mobile devices. In: Samuel, M. (ed.) Computer Science and Engineering, pp. 59–60. MIT, Massachusetts (2008)
4. Chiang, S.Y., Kan, Y.C., Lin, H.C.: Precise RSSI models for practical indoor WSN localization. Inf. Int. Interdisc. J. **16**(12), 8869–8885 (2013)
5. Feng, C., Au, W.S.A., Valaee, S.: Received-signal-strength-based indoor positioning using compressive sensing. IEEE Trans. Mob. Comput. **11**(12), 1983–1993 (2012)
6. Genming, D., Zhenhui, T.A.N., Jinsong, W.U.: Efficient indoor fingerprinting localization technique using regional propagation model. IEICE Trans. Commun. **97**(8), 1728–1741 (2014)
7. Hwang, J., Yoe, H.: Design and implementation of the Livestock activity monitoring system using RSSI of ZigBee and ratiometric. Inf. Int. Interdisc. J. **17**(3), 1047–1052 (2014)

8. Paul, A.S., Wan, E.A.: Wi-Fi based indoor localization and tracking using sigma-point Kalman filtering methods. In: Proceedings of the IEEE/ION Symposium on Position, Location & Navigation, pp. 646–659, May 2008
9. Goldsmith, A.: Wireless Communications. Cambridge University Press, New York (2005)
10. Kaemarungsi, K., Krishnamurthy, P.: Modeling of indoor positioning systems based on location fingerprinting. In: Proceedings of the 23th IEEE Computer and Communications Societies, vol. 2, pp. 1012–1022, March 2004
11. Bahl, P., Padmanabhan, V.N.: RADAR: an in-building RF-based user location and tracking system. In: Proceedings of the 19th IEEE Annual Joint Conference on Computer and Communications Societies, vol. 2, pp. 775–784, March 2000
12. Yang, Q., Pan, S.J., Zheng, V.W.: Estimating location using wi-fi. IEEE Intell. Syst. **23**(1), 8–13 (2008)
13. Ahriz, I., Oussar, Y., Denby, B.: Full-band GSM fingerprints for indoor localization using a machine learning approach. Int. J. Navig. Obs. **2010**(2010), 7 (2010)
14. Kjærgaard, M.B.: A taxonomy for radio location fingerprinting. In: Hightower, J., Schiele, B., Strang, T. (eds.) LoCA 2007. LNCS, vol. 4718, pp. 139–156. Springer, Heidelberg (2007)
15. Genming, D., Zhenhuim, T.A.N., Jinsong, W.U.: Indoor Fingerprinting Localization and Tracking System Using Particle Swarm Optimization and Kalman Filter. IEICE Trans. Communications **98**(3), 502–514 (2015)
16. Mautz, R.: Overview of current indoor positioning systems. Geodezija ir Kartografija **35**(1), 18–22 (2009)
17. Figuera, C., Rojo-Álvarez, J.L., Wilby, M.: Advanced support vector machines for 802.11 indoor location. Sig. Process. **92**(9), 2126–2136 (2012)
18. Zhu, Y.J., Deng, Z.L.: Multi-classification algorithm for indoor positioning based on support vector machine. Comput. Sci. **39**(4), 32–35 (2012)
19. Shin, J., Han, D.: Multi-classifier for WLAN fingerprint-based positioning system. In: Proceedings of World Congress on Engineering 2010, London, UK, vol. 1, June 2010
20. Xu, L., Krzyzak, A., Suen, C.Y.: Methods of combining multiple classifiers and their applications to handwriting recognition. IEEE Trans. Syst. Man Cybern. **22**(3), 418–435 (1992)
21. Kuncheva, L.I., Bezdek, J.C., Duin, R.P.W.: Decision templates for multiple classifier fusion: an experimental comparison. Pattern Recogn. **34**(2), 299–314 (2001)
22. Xiang, L., Wang, D., Wei, Y.H.: Location-fingerprint based indoor localization via scalable semi-supervised learning. Information-An Int. Interdisc. J. **18**(2), 641–652 (2015)
23. Dortz, N.L., Gain, F., Zetterberg, P.: Wi-Fi fingerprint indoor positioning system using probability distribution comparison. In: Proceedings IEEE Acoustics, Speech and Signal Processing, pp. 2301–2304 (2012)
24. Duda, R.O. (ed.): Pattern Classification. Wiley, Hoboken (2012)
25. Battiti, R., Brunato, M.: Statistical learning theory for location fingerprinting in wireless LANs. Comput. Netw. **47**(6), 825–845 (2005)
26. Tagashira, S., Kanekiyo, Y., Arakawa, Y., et al.: Collaborative filtering for position estimation error correction in WLAN positioning systems. IEICE Trans. Commun. **94**(3), 649–657 (2011)
27. Anand, R., Mehrotra, K., Mohan, C.K.: Efficient classification for multiclass problems using modular neural networks. IEEE Trans. Neural Netw. **6**(1), 117–124 (1995)
28. Hastie, T., Tibshirani, R.: Classification by pairwise coupling. Ann. Stat. **26**(2), 451–471 (1998)
29. Sun, Z., Sun, Y.: Fuzzy support vector machine for regression. IEEE Trans. Fuzzy Syst. **4** (2), 3336–3341 (2003)

AdaUK-Means: An Ensemble Boosting Clustering Algorithm on Uncertain Objects

Lei Xu[1], Qinghua Hu[2], Xisheng Zhang[1], Yanshuo Chen[3],
and Changrui Liao[4(✉)]

[1] School of Electronic and Communication Engineering,
Shenzhen Polytechnic, Shenzhen, China
[2] School of Computer Science and Technology, Tianjin University, Tianjin, China
[3] Department of Economics, University of Wisconsin Madison, Madison, USA
[4] Key Laboratory of Optoelectronic Devices and Systems of Ministry of Education
and Guangdong Province, College of Optoelectronic Engineering,
Shenzhen University, Shenzhen, China
cliao@szu.edu.cn

Abstract. This paper considers the problem of clustering uncertain objects whose locations are uncertain and described by probability density functions (pdf). Though K-means has been extended to UK-means for handling uncertain data, most existing works only focus on improving the efficiency of UK-means. However, the clustering quality of UK-means is rarely considered in existing works. The weights of objects are assumed same in existing works. However, the weights of objects which are far from their cluster representatives should not be the same as the weights of objects which are close to their cluster representatives. Thus, we propose an AdaUK-means to group the uncertain objects by considering the weights of objects in this article. In AdaUK-means, the weights of objects will be adjusted based on the correlation between objects by using Adaboost. If the object pairs are must-link but grouped into different clusters, the weights of the objects will be increased. In our ensemble model, AdaUK-means is run several times, then the objects are assigned by a voting process. Finally, we demonstrate that AdaUK-means performs better than UK-means on both synthetic and real data sets by extensive experiments.

Keywords: Uncertain data · UK-means · Weighted objects · Ensemble clustering · Adaboost

1 Introduction

Due to the limitation of instruments, the collected data (for example, in the case of sensor data) cannot provide correct or precise information. Thus, data uncertainty attracts more attentions. Existential uncertainty, value uncertainty and relationship uncertainty are three types of data uncertainty. Value uncertainty

T. Tan et al. (Eds.): CCPR 2016, Part I, CCIS 662, pp. 27–41, 2016.
DOI: 10.1007/978-981-10-3002-4_3

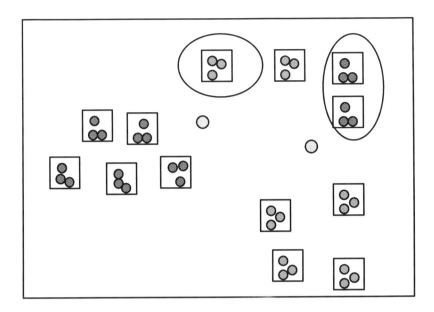

Fig. 1. The limitation of UK-means. (Color figure online)

on a dimension is caused by instrument limitation or delay. Existential uncertainty [1–4] is caused by not sure the tuple exists in a database. Relationship uncertainty [5] learns the order on a dimension. In our clustering problem, value uncertainty [6–10] is studied.

Several clustering methods have been extends to uncertain data, i.e. Uncertain K-means (UK-means) [6]. Existing clustering methods only focus on improving the efficiency of clustering uncertain data [6–8]. The weights of objects are assumed same in existing methods. However, the weights of objects which are far from their cluster representatives should not be the same as the weights of objects which are close to their cluster representatives.

For example, as shown in Fig. 1 the uncertain objects in the same color should be grouped to the same cluster. The yellow points present the location of cluster representatives of UK-means. But the objects in the circle would likely to be grouped into other cluster in UK-means. The uncertain object which is grouped uncorrectly should be close to its corresponding cluster representative. Thus, the weights of objects which are must-link but grouped into different clusters should be increased. Then the cluster representative will shift to the objects which are grouped uncorrectly. AdaUK-means is proposed to group the uncertain objects by considering the weights of objects in this article. AdaUK-means is based on the ensemble model, and the decision is made by a voting process. In AdaUK-means, the weights of objects will be adjusted based on the correlation between objects by using Adaboost [11]. If the object pairs are must-link but grouped into different clusters, the weights of the objects will be increased. In our ensemble model, AdaUK-means is run several times, then the objects are assigned by a voting process. Our contributions of this work include:

1. Different from UK-means, the weights of objects are considered in our method. The weights of objects will be adjusted during clustering process. The weights of objects which are must-link but grouped to different clusters will be increased in next iteration in AdaUK-means.
2. The ensemble model is used to assign uncertain objects in AdaUK-means. The objects are assigned by a voting process in our method, which is different from existing works assigning the objects directly.
3. The cluster representative is updated by the weighted objects, which is different from UK-means. In UK-means, the cluster representative is the average of the mean vectors of objects assigned to the cluster. The cluster representative considers the weights of objects in AdaUK-means.

The remainder of this paper is organized as follows. Related work of clustering uncertain data is briefly reviewed in Sect. 2. The preliminary knowledge of clustering uncertain objects is introduced in Sect. 3. The ensemble method of UK-means (AdaUK-means) is introduced in Sect. 4. Section 5 shows the experimental evaluation on the performance of AdaUK-means. Finally, our work is concluded in Sect. 6.

2   Related Work

Uncertain K-means (UK-means) is proposed to handle uncertain objects based on K-means. UK-means is computationally expensive because of a large amount of expected distance calculations. Some pruning technics [7,8] are proposed to improve the efficiency of UK-means by pruning farther candidate clusters. The clustering quality is rarely considered in UK-means. UK-means is based on K-means, the weights of objects equals to each other. UK-means is limited in handling non-spherical data. In our work, the weights of must-link objects are adjusted depends on the clustering results of last iteration.

Some density-based clustering methods have also been extended to handle uncertain objects, i.e. FDBSCAN [9] and FOPTICS [10]. FDBSCAN is based on DBSCAN [12], and FOPTICS is based on OPTICS [13]. Fuzzy distance function is used in FDBSCAN and FOPTICS. The assumption of independent pairwise distances between objects may cause wrong results. For the purpose of making reliable results, a small number of cluster representatives are returned with quality guarantees in [14]. Then the user selects the results from the clusterings. The work of [14] just focuses on generating cluster representatives without improving the clustering quality. The weights of objects are the same in previous work. Different from previous work, the weights of objects is considered in our ensemble clustering model.

Besides the studies of clustering uncertain objects, the problem of classification uncertain data is also considered in [15–20]. Other mining technics on uncertain data are also studied, i.e. outlier detection [21–24], frequent pattern mining [25–30] and domain orders learning [31–36].

3   Preliminary Knowledge

As we have known that the learning model based on similarity metrics can perform better when classifying multimedia data [37–40], expected Euclidean distance is used in UK-means [6] and our work. Before introducing the preliminary knowledge, the notations used in this paper are listed in Table 1.

Table 1. The notations used in this paper.

Parameter	Description
ED	Expected distance
$UD(o_i)$	The uncertain domain of o_i
o_i	Uncertain object o_i
x	Possible location of o_i
f_i	The probability density function of o_i
$F_i(s_{i,t})$	The probability of sample $s_{i,t}$ of o_i
$\overline{o}_i$	The mean vector of o_i
p_{c_j}	The cluster representative of j-th cluster
$\overline{p}_{c_j}$	The mean vector of p_{c_j}
V_u	The feature on the u-th dimension
$dom(V_u)$	The value range of the u-th dimension

In our work, the uncertainty is defined in a multi-dimensional space. In the defined space, the uncertain region of an object is represented by a probability distribution function (PDF) or a probability density function. PDF (probability distribution function) describes the distribution of the probabilities of possible locations. In our work, the uncertain objects o_i falls within a finite region $UD(o_i)$. The probability density of the possible location within $UD(o_i)$ is represented by a probability density function (pdf), f_i. There is $\int_{UD(o_i)} f_i(x)\, dx = 1$ [41].

The uncertain domain $UD(o_i)$ is divided into T grid cells. $s_{i,t}$ is the location (vector) of the t-th sample of o_i. The expected Euclidean distance (EED) between object o_i (represented by f_i) and the cluster representative p_{c_j} is calculated as Eq. (1).

$$ED(o_i, p_{c_j}) = \sum_{t=1}^{T} F_i(s_{i,t}) ED(s_{i,t}, \overline{p}_{c_j}) \tag{1}$$

$$= \sum_{t=1}^{T} \int_{UD(o_i)} f_i(s_{i,t}) ED(s_{i,t}, \overline{p}_{c_j}) dx.$$

The mean vector of o_i ($\overline{o_i}$) is the weighted mean of all T samples calculated by Eq. (2). In UK-means, the mean vector of cluster representative p_{c_j} ($\overline{p}_{c_j}$) is the average of the mean vectors of objects assigned to the cluster Eq. (3). In Eq. (3), $|c_j|$ represents the number of objects assigned to cluster c_j, and $C(o_i) = c_j$ describes the cluster that o_i assigned to.

$$\overline{o_i} = \sum_{t=1}^{T} s_{i,t} \times F_i(s_{i,t}). \tag{2}$$

$$\overline{p}_{c_j} = \frac{1}{|c_j|} \sum_{o_i \in \{o_i | C(o_i) = c_j\}} \overline{o_i}. \tag{3}$$

4   Ensemble Clustering Model for Uncertain Data

UK-means is a kind of EM (expected maximum) algorithm, but it is likely to attain an local optimal value. In this section, we introduce our method AdaUK-means, which is an ensemble clustering method based on Adaboost. Given a small number of objects (no more than 10) that should be in the same cluster, and the weights of objects in the same cluster are grouped in different clusters will be increased. The objects will be assigned to the clusters by a voting process. We first define the problem of ensemble clustering of uncertain objects, then the ensemble clustering method (AdaUK-means) is introduced in detail.

4.1   Problem Definition

In our model, there are N objects, and m numerical (real-valued) feature attributes $V_1, ..., V_m$. The range value of the u-th dimension $V_u (1 \leq u \leq m)$ is denoted as $dom(V_u)$. Each o_i is associated with a probability density function (pdf $f_i(x)$), where x is a possible location of o_i, and $UD(o_i)$ is the uncertain domain of o_i. Each tuple x is associated with a feature vector $x = (\tilde{x}_1, \tilde{x}_2, ..., \tilde{x}_m)$, where $\tilde{x}_u \in dom(V_u)(1 \leq u \leq m)$. Given N uncertain objects, the goal of ensemble clustering is to minimize sum of the distances between uncertain objects and cluster representatives Eq. (4). The distance is related to the weight of each UK-means. During clustering process, UK-means is executed u times, and $u \times K$ cluster representatives are obtained.

$$\min_{\overline{p}_{1.c_1}, ..., \overline{p}_{u.c_K}} \sum_{i=1}^{N} \frac{1}{\alpha_u} \int_{UD(o_i)} f_i(x) D(x, \overline{p}_{l,c_j}) dx. \tag{4}$$

Where D is Euclidean distance between objects, and α_u is the weight of the uth UK-means. The problem of ensemble clustering of uncertain objects is finding $u \times K$ cluster representatives by minimizing Eq. (4).

4.2   Ensemble Method

UK-means is considered as a kind of EM (Expectation Maximization) algorithm. UK-means is likely to attain an local optimal value. For the purpose of improving the performance of UK-means, AdaUK-means is proposed for handling uncertain objects based on Adaboost. Different from [14], our method considers uncertain objects.

AdaBoost is short for Adaptive boosting [11], which is proposed for improving the performance of learning methods. Adaboost is an adaptive method that subsequent classifiers are built focusing on objects misclassified by previous classifiers. AdaBoost calls a weak classifier repeatedly in a series of classifiers. During each iteration, the weights of objects is updated to indicate the importance of examples in a data set. The weights of incorrectly classified examples are increased (or alternatively, the weights of correctly classified examples are decreased), so that the new classifier focuses more on those incorrectly classified examples.

4.3   AdaUK-Means

AdaUK-means executes UK-means u times on the weighted uncertain objects. The objects will be assigned to the cluster by the minimum distance between the uncertain objects and cluster representatives during each UK-means. $u \times K$ cluster representatives will be returned in Algorithm 1. There are three steps in AdaUK-means as shown in Fig. 2. First, the weights of uncertain objects are calculated by Adaboost. Second, the cluster representative is updated by the weighted objects. Step 1 and 2 will be repeated U times to get an ensemble model. Third, all the objects are assigned based on ensemble UK-means.

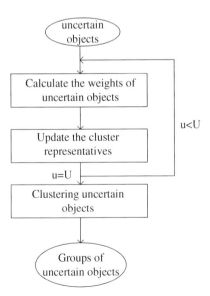

Fig. 2. The flowchart of AdaUK-means.

Weights of Uncertain Objects. The weights of uncertain objects are computed in Line 9 Algorithm 1. In Algorithm 1, $h_u(o_i) \neq C(o_i)$ means that o_i is assigned uncorrectly by h_u ($C(o_i)$ is the cluster that object o_i should be assigned, $h_u(o_i)$ is the cluster o_i assigned to by h_u). The subsequent cluster representatives are obtained in favor of those objects which are misclassified in previous iteration in Adaboost [11]. For example, if o_l and o_l' are in the same cluster and are assigned to the same cluster ($h_u(o_l) = h_u(o_l')$), the value of $M(o_l)$ is -1. Otherwise, the value of $M(o_l)$ is 1. If the objects are clustered correctly, the weights of them will be decreased. n is the number of objects which are labeled by the correlation with other $n - 1$ objects. n is a very small number in our method, i.e. no more than 10.

Update of Cluster Representative. The weights of objects are updated by weighted objects (Algorithm 1 Line 28). The weights of objects which are must-link but assigned to different clusters will be increased. Algorithm 2 describes the calculation of cluster representatives. The cluster representatives are updated by weighted uncertain objects Eq. (5). The cluster representative is the weighted average of the mean vectors of objects which are assigned to the cluster. The update of cluster representative considers the correlation between uncertain objects, which is different from UK-means.

$$
\overline{p}_{c_j} = \frac{1}{\sum\limits_{o_i \in c_j} w_u(o_i \in c_j)} \sum_{i=1}^{|c_j|} (w_u(o_i \in \{o_i | C(o_i) = c_j\})
$$
$$
\times \overline{o}_i \in \{o_i | C(o_i) = c_j\}); \tag{5}
$$

Clustering Uncertain Objects. In Algorithm 1, o_i is assigned by the minimizing the objective function of Eq. (4). Algorithm 3 shows the clustering process of AdaUK-means. In each UK-means, the uncertain objects are assigned to the cluster by Eq. (6) first, then the objects are assigned by a voting process Eq. (7).

$$
argmin_{c_l}(ED(o_i, \overline{p}_{u,c_1}), ED(o_i, \overline{p}_{u,c_2}), ..., ED(o_i, \overline{p}_{u,c_K})). \tag{6}
$$

$$
argmax_{c_q}(w_q). \tag{7}
$$

5   Experimental Evaluation

In this section, the AdaUK-means is evaluated by comparing with UK-means. All codes were written in Matlab and were run on a Windows machine with an Intel 2.66 GHz Pentium(R) Dual-Core processor and 4 GB of main memory. The clustering quality of our method is evaluated on both synthetic data sets and real data sets.

The clustering quality of a method is clustering quality, which is measured by precision. $\mathbb{G}$ is denoted as the ground truth of clustering results, and $\mathbb{R}$ is the clustering results obtained by a clustering algorithm. If there are two objects

---

Algorithm 1. AdaUK-means

Input: N uncertain objects $\{(o_j), j = 1, 2, ..., N\}$, the rounds of u, the number of clusters K.

$n/2$ object pairs $\{(o_m, o_z)\}$

Output: clusters of objects (i.e. A_k).

1: **for** $i = 1; i \leq N; i + +$ **do**
2:     compute $\overline{o_i}$ of labeled objects by Eq.(2);
3: **end for**
4: **for** $i = 1; i \leq n; i + +$ **do**
5:     initialize $w_1(i) = \frac{1}{n}$;
6: **end for**
7: **for** $u = 1; u \leq U; u + +$ **do**
8:     **for** $i = 1; i \leq n; i + +$ **do**
9:         compute all normalized weight $w_u(i) = w_u(i)/\sum w_u(i)$;
10:     **end for**
11:     $h_t = learn(d_u)$;
12:     **for** $i = 1; i \leq n; i + +$ **do**
13:         **for** $j = 1; j \leq K; j + +$ **do**
14:             **for** $t = 1; t \leq T; t + +$ **do**
15:                 compute expected Euclidean distance between o_i and $\overline{p}_{c_j}$ by
$ED(o_i, p_{c_j}) = w_i \times \sum_{t=1}^{T} F_i(s_{i,t}) ED(s_{i,t}, \overline{p}_{c_j})$;
16:             **end for**
17:         **end for**
18:         assign o_i to the cluster by Eq.(6);
19:     **end for**
20:     $\epsilon_u = \sum_{i=1}^{n} w_u(i)[h_u(o_i) \neq C(o_i)]$;
21:     **if** $\epsilon_u \geq \frac{1}{2}$ **then**
22:         stop;
23:     **else**
24:         $\alpha_u = ln\frac{1-\epsilon_u}{\epsilon_u}$;
25:     **end if**
26:     update all $w_{u+1}(i) = w_u(i)exp(-\alpha_u M(o_i))$;
27: **end for**
28: **for** $i = 1; i \leq N; i + +$ **do**
29:     $c_j = clustering(o_i)$;
30: **end for**

---

appearing in the same cluster in a clustering, they are a pair. True positive (TP) represents the set of common pairs of objects in both $\mathbb{G}$ and $\mathbb{R}$; False positive (FP) is the set of pairs of objects in $\mathbb{R}$ but not in $\mathbb{G}$; False negative (FN) is the set of pairs of objects in $\mathbb{G}$ but not in $\mathbb{R}$. The precision of a clustering algorithm $\mathbb{R}$ are calculated as follows:

$$precision_{\mathbb{R}} = |TP|/(|TP| + |FP|),\qquad(8)$$

where $|.|$ is the number of objects in the set. 5-fold cross-validation for each data set are generated to evaluate the effectiveness of the algorithms.

Algorithm 2. Learn h_u [11]

Input: a set of n labeled objects $\{(o_i, c_j), i = 1, 2, ..., n\}$,
class labels c_j ($j \in \{1, 2, ..., K\}$),
the weight of n objects $\{d_t(o_i), i = 1, 2, ..., n\}$.
Output: weak learner h_u.

1: **for** $j = 0$; $j < K$; $j + +$ **do**
2:      calculate $\overline{p}_{c_j}$ by Eq.(5);
3: **end for**
4: $h_u = (\overline{p}_{u,c_1}, \overline{p}_{u,c_2}, ..., \overline{p}_{u,c_K})$.

Algorithm 3. Clustering $c(o_i)$

Input: unlabeled uncertain object o_i,
the mean vector of cluster representatives $\left\{ \overline{p}_{c_j}, j = 1, 2, ..., K \right\}$,
the weights of learners $\{\alpha_u, u = 1, ...U\}$, constant T, K and U.
Output: cluster representatives (i.e. c_j).

1: **for** $j = 1$; $j \leq K$; $j + +$ **do**
2:      initialize $w_j = \frac{1}{K}$;
3: **end for**
4: **for** $u = 1$; $u \leq U$; $u + +$ **do**
5:      **for** $j = 1$; $j \leq K$; $j + +$ **do**
6:          **for** $t = 1$; $t \leq T$; $t + +$ **do**
7:              compute expected Euclidean distance between o_i and $\overline{p}_{c_j}$ by
$EED(o_i, p_{c_j}) = \sum_{t=1}^{T} F_i(s_{i,t})ED(s_{i,t}, \overline{p}_{c_j})$;
8:          **end for**
9:      **end for**
10:     assign o_i to the cluster c_l by Eq.(6);
11:     $w_j = w_j + \alpha_u$;
12: **end for**
13: assign o_i to the cluster c_q by Eq.(7);
14: **for** $j = 1$; $j \leq K$; $j + +$ **do**
15:     calculate $\overline{p}_{c_j}$ by Eq.(3);
16: **end for**

5.1   Synthetic Data Sets

For the purpose of demonstrating the performance of AdaUK-means, 110 data sets with Gaussian distribution are generated. The N uncertain objects in a data set were equally grouped into K clusters. For each cluster, the centers of $\frac{N}{K}$ uncertain objects were generated from a Gaussian distribution. The baseline values of parameters used for the experiments on Gaussian data sets are summarized in Table 2. In each data set, T samples are used to represented an uncertain objects in a D-dimensional space. The uncertain domain of an object on each dimension is $un \times un$.

Table 2. Baseline values of parameters for experiments using Gaussian data sets.

Parameter	Description	Baseline value
N	Number of uncertain objects	100
T	Number of samples per object	100
K	Number of clusters	2
D	Number of dimensions	2
un	Uncertainty of each object	0.05
σ	The standard deviation of each cluster	4

The precision of AdaUK-means compared with UK-means is shown in Tables 3, 4, 5 and 6. In Table 3, u represents the number of UK-means. Tables 3, 4, 5 and 6 show that our ensemble method (AdaUK-means) performs better than that of UK-means on average. The precision of AdaUK-means with different us does not change significantly.

In the experiments, the performance of UK-means decreases dramatically when K is increasing. AdaUK-means improves the precision of UK-means by 11.9 % on average with varying K. AdaUK-means improves the clustering quality of UK-means by 4.5 %–5 % on average. The experimental results demonstrate that AdaUK-means can improve the precision of UK-means when clustering uncertain objects.

Table 3. Precision on synthetic data sets with varying object number N.

Specified parameter	AdaUK-means $(u = 2)$	AdaUK-means $(u = 3)$	AdaUK-means $(u = 4)$	UK-means
$N = 100$	**0.67**	0.66	0.66	**0.67**
$N = 200$	0.63	**0.65**	**0.65**	**0.65**
$N = 300$	0.64	**0.66**	**0.66**	**0.66**
$N = 400$	**0.67**	0.65	0.65	0.63
$N = 500$	**0.65**	0.64	0.64	0.64
Average	**0.652**	**0.652**	**0.652**	0.65

Table 4. Precision on synthetic data sets with varying cluster number K.

Specified parameter	AdaUK-means $(u = 2)$	AdaUK-means $(u = 3)$	AdaUK-means $(u = 4)$	UK-means
$K = 2$	**0.67**	0.66	0.66	**0.67**
$K = 3$	0.5	**0.51**	0.5	0.46
$K = 4$	**0.54**	**0.54**	0.53	0.43
$K = 5$	**0.46**	**0.46**	**0.46**	0.38
Average	**0.543**	**0.543**	0.538	0.485

Table 5. Precision on synthetic data sets with varying dimension number D.

Specified parameter	AdaUK-means $(u = 2)$	AdaUK-means $(u = 3)$	AdaUK-means $(u = 4)$	UK-means
$D = 2$	**0.67**	0.66	0.66	**0.67**
$D = 3$	**0.68**	**0.68**	0.67	0.62
$D = 4$	**0.66**	0.64	0.65	0.6
$D = 5$	**0.65**	0.63	**0.65**	0.6
Average	**0.665**	0.653	0.658	0.623

Table 6. Precision on synthetic data sets with varying sample number T.

Specified parameter	AdaUK-means $(u = 2)$	AdaUK-means $(u = 3)$	AdaUK-means $(u = 4)$	UK-means
$T = 50$	**0.66**	0.65	**0.66**	0.62
$T = 100$	**0.66**	0.65	**0.66**	0.62
$T = 200$	**0.67**	0.66	0.65	0.66
$T = 300$	0.66	0.66	**0.67**	0.64
$T = 400$	0.66	**0.67**	**0.67**	0.62
Average	**0.662**	0.658	**0.662**	0.632

5.2   Real Data Sets

The experiments are also done on real data sets. The parameters of the chosen data sets used for the experiments are summarized in Table 7. The attributes of all the data sets are numerical obtained from measurements. We follow the common practice in the research work of this area [6–8, 18–20] to generate the uncertainty of real data sets. For object o_i on the u-th dimension (i.e. the attribute A_u), the point value $v_{i,u}$ reported in a data set is used as the mean of a pdf $f_{i,u}$, defined over an interval $[a_{i,u}, b_{i,u}]$. The range of values for A_u (over the whole data set) is noted and the width of $[a_{i,u}, b_{i,u}]$ is set to $un \times |A_u|$, where $|A_u|$ denotes the width of the range for A_u and un is a parameter to control the uncertainty of data set. Gaussian distribution is used to generate pdf $f_{i,u}$, which

Table 7. Selected data set from the UCI machine learning repository.

Data set	No. of tuples	No. of attributes	No. of classes
Iris	150	4	3
Wine	178	13	3
Hear	270	13	2
Breast cancer	569	30	2
Ionosphere	311	32	2

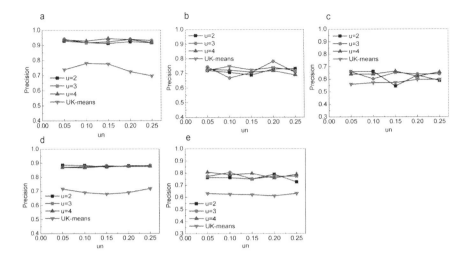

Fig. 3. (a) Comparison in precision with UK-means on Iris (b) comparison in precision with UK-means on Wine (c) comparison in precision with UK-means on Hear (d) comparison in precision with UK-means on Breast Cancer (e) comparison in precision with UK-means on Ionosphere.

the standard deviation is set to be $\frac{1}{4} \times (b_{i,u} - a_{i,u})$ (the same as that in [19]). T samples are used to generate pdf over the interval. The point value is transformed into uncertain samples on Gaussian distribution by using the controlled parameter un and T samples. To compare with UK-means, T is set to be 100 and un is from 5 % to 25 %. The point value data become uncertain when the error model (un and pdf) is applied for them.

Figure 3 shows the precision of our ensemble method compared with UK-means on the selected UCI data sets. The precision may be affected by varying un. However, the precision of our method is higher than that of UK-means on the five selected data sets. The experimental results show that the error model (un) can affect the clustering results of algorithm slightly.

Our ensemble method has been done on both synthetic and real data sets. The experimental results on those data sets demonstrate that the ensemble method can improve the clustering performance of UK-means on both synthetic and real data sets.

6   Conclusions

In this paper, the problem of clustering uncertain objects whose locations are described by probability density functions (pdf) is studied. For the purpose of improving the performance of clustering quality, an ensemble clustering method is proposed to assign the objects by minimizing the distance between objects and cluster representatives. For the purpose of overcoming the limitation of

UK-means, Adaboost is combined to enhance the learning strength of the proposed model. The experimental evaluation demonstrates that our ensemble semi-supervised UK-means outperforms UK-means on both synthetic and real data sets. We will consider clustering uncertain heterogeneous data in further work.

Acknowledgments. The work described in this paper was supported by the grants from National Natural Science Foundation of China (Nos. 61308027, 61575128), Guangdong Provincial Department of Science and Technology (2015A030313541), Science and Technology Innovation Commission of Shenzhen (KQCX20140512172532195).

References

1. Dalvi, N.N., Suciu, D.: Efficient query evaluation on probabilistic databases. VLDB J. **16**(4), 523–544 (2007)
2. Barbará, D., Garcia-Molina, H., Porter, D.: The management of probabilistic data. IEEE Trans. Knowl. Data Eng. **4**(5), 487–502 (1992)
3. Cheng, R., Kalashnikov, D.V., Prabhakar, S.: Querying imprecise data in moving object environments. IEEE Trans. Knowl. Data Eng. **16**(9), 1112–1127 (2004)
4. Ruspini, E.H.: A new approach to clustering. Inf. Control **15**(1), 22–32 (1969)
5. Jiang, B., Pei, J., Lin, X., Cheung, D.W., Han, J.: Mining preferences from superior, inferior examples. In: Proceedings of the 14th ACM SIGKDD International Conference on Knowledge Discovery and Data Mining, pp. 390–398 (2008)
6. Chau, M., Cheng, R., Kao, B., Ng, J.: Uncertain data mining: an example in clustering location data. In: Ng, W.-K., Kitsuregawa, M., Li, J., Chang, K. (eds.) PAKDD 2006. LNCS (LNAI), vol. 3918, pp. 199–204. Springer, Heidelberg (2006)
7. Ngai, W.K., Kao, B., Chui, C.K., Cheng, R., Chau, M., Yip, K.Y.: Efficient clustering of uncertain data. In: Proceedings of the 6th IEEE International Conference on Data Mining, pp. 436–445 (2006)
8. Kao, B., Lee, S.D., Cheung, D.W., Ho, W.-S., Chan, K.F.: Clustering uncertain data using voronoi diagrams. In: Proceedings of the 8th IEEE International Conference on Data Mining, pp. 333–342 (2008)
9. Kriegel, H.-P., Pfeifle, M.: Density-based clustering of uncertain data. In: Proceedings of the 11th ACM SIGKDD International Conference on Knowledge Discovery and Data Mining, pp. 672–677 (2005)
10. Kriegel, H.-P., Pfeifle, M.: Hierarchical density-based clustering of uncertain data. In: Proceedings of the 5th IEEE International Conference on Data Mining, pp. 689–692 (2005)
11. Freund, Y., Schapire, R.E.: A decision-theoretic generalization of on-line learning and an application to boosting. In: Vitányi, P. (ed.) Computational Learning Theory. LNCS, vol. 904, pp. 23–37. Springer, Heidelberg (1995)
12. Ester, M., Kriegel, H.-P., Sander, J., Xu, X.: A density-based algorithm for discovering clusters in large spatial databases with noise. In: Proceedings of the 2nd International Conference on Knowledge Discovery and Data Mining, pp. 226–231 (1996)
13. Ankerst, M., Breunig, M.M., Kriegel, H.-P., Sander, J.: Optics: ordering points to identify the clustering structure. In: Proceedings ACM SIGMOD International Conference on Management of Data, pp. 49–60 (1999)

14. Züfle, A., Emrich, T., Schmid, K.A., Mamoulis, N., Zimek, A., Renz, M.: Representative clustering of uncertain data. In: The 20th ACM SIGKDD International Conference on Knowledge Discovery and Data Mining, KDD 2014, New York, NY, USA, 24–27 August 2014, pp. 243–252 (2014)

15. Bi, J., Zhang, T.: Support vector classification with input data uncertainty. In: Advances in Neural Information Processing Systems 17 [Neural Information Processing Systems] (2004)

16. Qin, B., Xia, Y., Prabhakar, S., Tu, Y.-C.: A rule-based classification algorithm for uncertain data. In: Proceedings of the 25th International Conference on Data Engineering, pp. 1633–1640 (2009)

17. Qin, B., Xia, Y., Prabhakar, S.: Rule induction for uncertain data. Knowl. Inf. Syst. 29(1), 103–130 (2011)

18. Tsang, S., Kao, B., Yip, K.Y., Ho, W.-S., Lee, S.D.: Decision trees for uncertain data. In: Proceedings of the 25th International Conference on Data Engineering, pp. 441–444 (2009)

19. Tsang, S., Kao, B., Yip, K.Y., Ho, W.-S., Lee, S.D.: Decision trees for uncertain data. IEEE Trans. Knowl. Data Eng. 23(1), 64–78 (2011)

20. Ren, J., Lee, S.D., Chen, X., Kao, B., Cheng, R., Cheung, D.W.-L.: Naive bayes classification of uncertain data. In: The 9th IEEE International Conference on Data Mining, pp. 944–949 (2009)

21. Jiang, B., Pei, J.: Outlier detection on uncertain data: objects, instances, and inferences. In: Proceedings of the 27th International Conference on Data Engineering, pp. 422–433 (2011)

22. Aggarwal, C.C., Yu, P.S.: Outlier detection with uncertain data. In: Proceedings of the SIAM International Conference on Data Mining, pp. 483–493 (2008)

23. Wang, B., Xiao, G., Yu, H., Yang, X.: Distance-based outlier detection on uncertain data. In: 12th IEEE International Conference on Computer and Information Technology, pp. 293–298 (2009)

24. Matsumoto, T., Hung, E.: Accelerating outlier detection with uncertain data using graphics processors. In: Tan, P.-N., Chawla, S., Ho, C.K., Bailey, J. (eds.) PAKDD 2012, Part II. LNCS, vol. 7302, pp. 169–180. Springer, Heidelberg (2012)

25. Aggarwal, C.C., Li, Y., Wang, J., Wang, J.: Frequent pattern mining with uncertain data. In: Proceedings of the 15th ACM SIGKDD International Conference on Knowledge Discovery and Data Mining, pp. 29–38 (2009)

26. Chui, C.-K., Kao, B.: A decremental approach for mining frequent itemsets from uncertain data. In: Washio, T., Suzuki, E., Ting, K.M., Inokuchi, A. (eds.) PAKDD 2008. LNCS (LNAI), vol. 5012, pp. 64–75. Springer, Heidelberg (2008)

27. Chui, C.-K., Kao, B., Hung, E.: Mining frequent itemsets from uncertain data. In: Zhou, Z.-H., Li, H., Yang, Q. (eds.) PAKDD 2007. LNCS (LNAI), vol. 4426, pp. 47–58. Springer, Heidelberg (2007)

28. Leung, C.K.-S., Brajczuk, D.A.: Mining uncertain data for constrained frequent sets. In: International Database Engineering and Applications Symposium, pp. 109–120 (2009)

29. Zhang, Q., Li, F., Yi, K.: Finding frequent items in probabilistic data. In: Proceedings of the ACM SIGMOD International Conference on Management of Data, pp. 819–832 (2008)

30. Bernecker, T., Kriegel, H.-P., Renz, M., Verhein, F., Züfle, A.: Probabilistic frequent itemset mining in uncertain databases. In: Proceedings of the 15th ACM SIGKDD International Conference on Knowledge Discovery and Data Mining, pp. 119–128 (2009)

31. Lawrence, R.D., Almasi, G.S., Kotlyar, V., Viveros, M.S., Duri, S.: Personalization of supermarket product recommendations. Data Min. Knowl. Disc. **5**(1/2), 11–32 (2001)

32. Govindarajan, K., Jayaraman, B., Mantha, S.: Preference queries in deductive databases. New Gener. Comput. **19**(1), 57–86 (2000)

33. Chomicki, J.: Querying with intrinsic preferences. In: Jensen, C.S., Šaltenis, S., Jeffery, K.G., Pokorny, J., Bertino, E., Böhn, K., Jarke, M. (eds.) EDBT 2002. LNCS, vol. 2287, pp. 34–51. Springer, Heidelberg (2002). doi:10.1007/3-540-45876-X_5

34. Chomicki, J.: Database querying under changing preferences. Ann. Math. Artif. Intell. **50**(1–2), 79–109 (2007)

35. Kießling, W., Köstler, G.: Preference SQL - design, implementation, experiences. In: Proceedings of 28th International Conference on Very Large Data Bases, pp. 990–1001 (2002)

36. Lacroix, M., Lavency, P.: Preferences: putting more knowledge into queries. In: Proceedings of 13th International Conference on Very Large Data Bases, pp. 217–225 (1987)

37. Wang, D., Kim, Y.-S., Park, S.C., Lee, C.S., Han, Y.K.: Learning based neural similarity metrics for multimedia data mining. Soft. Comput. **11**(4), 335–340 (2007)

38. Wang, D., Ma, X.: Learning pseudo metric for multimedia data classification and retrieval. In: Negoita, M.G., Howlett, R.J., Jain, L.C. (eds.) KES 2004. LNCS (LNAI), vol. 3213, pp. 1051–1057. Springer, Heidelberg (2004)

39. Cong, Y., Wang, S., Fan, B., Yang, Y., Yu, H.: UDSFS: unsupervised deep sparse feature selection. Neurocomputing

40. Cong, Y., Yuan, J., Luo, J.: Towards scalable summarization of consumer videos via sparse dictionary selection. IEEE Trans. Multimedia **14**(1), 66–75 (2012)

41. Hung, E., Xiao, L., Hung, R.Y.S.: An efficient representation model of distance distribution between uncertain objects. Comput. Intell. **28**(3), 373–397 (2012)

A Vehicle Trajectory Analysis Approach Based on the Rigid Constraints of Object in 3-D Space

Wen Jiang, Zhang Zhaoyang[✉], Song Huansheng, and Pang Fenglan

School of Information Engineering, Chang'an University, Xi'an 710064, China
{2014224015, zhaoyang_zh, 2014124037}@chd.edu.cn,
1192750414@qq.com

Abstract. A reliable and effective trajectories similarity metric is one of key factors for vehicle trajectories clustering problem. A trajectory clustering algorithm based on the rigid constraints of vehicles in 3-D space is proposed in this paper, which conducts vehicle trajectories clustering effectively and precisely by using a new 3-D trajectories similarity metric. Based on two key procedures, camera calibration and a reconstruction of 2-D trajectories in 3-D space, a valuable principle that the heights of the trajectories have a linear relationship between them is found through using the kinematic properties of vehicle rigid body in moving. A more valuable information need to be pay attention is that the height of two trajectories that with displacement difference satisfies a plane surface character in 3-D space when conducts a height enumeration. The experimental results show that the trajectories are very stable and reliable for clustering and event detection when reconstructing their relative position in 3-D world coordinate system.

Keywords: Trajectory clustering · Rigid constraints · Camera calibration · 3-D reconstruction

1 Introduction

Trajectory clustering is an important topic in video analysis, the purpose of which is to assign a common cluster label to individual trajectories, and it is widely used in activity surveillance, traffic flow estimation and emergency response [1, 2].

A trajectory is typically defined as a data sequence, which is consisting of several concatenated state vectors from tracking, and that means it is an indexed sequence of positions and velocities in a given time window [3]. Trajectory data can provide effective spatial-temporal information of objects for event analysis [4]. Clustering is one of the most effective approaches for trajectory data analysis [5]. It has been studied for several decades, and many methods, such as k-means and Density-Based Spatial Clustering of Applications with Noise (DBSCAN), are proposed for trajectory data analysis. For trajectory based event analysis, a key and important step is defining a similarity measurement for two trajectories with different lengths, which has a greater impact on the result. The reason is that length difference induced by the kinematic properties of moving objects makes assessing similarity between two or more trajectories exceedingly difficult [4]. To address the problem of similarity measurement

© Springer Nature Singapore Pte Ltd. 2016
T. Tan et al. (Eds.): CCPR 2016, Part I, CCIS 662, pp. 42–52, 2016.
DOI: 10.1007/978-981-10-3002-4_4

between trajectories, Fu et al. [6] proposed a method to resample trajectories with different lengths to ones with equal lengths. Then the similarity between two trajectories can be computed by the distances between corresponding points on two trajectories. Stefan Atev et al. [7] proposed a trajectory-similarity measure based on the Hausdorff distance, and a modification strategy is given to improve its robustness through accounting for the fact that trajectories are ordered collections of points. Abraham et al. [8] deal the length difference by applying a spatial-temporal similarity measure with given Points of Interest (POI) and Time of Interest (TOI), in which the spatial similarity is treated as a combination of structural and sequence similarities and evaluated by using the techniques of dynamic programming. Thus the similarity set is formed and is used to broadcast trigger-based messages to vehicles in a neighborhood area for future route and information-sharing activities. Hao et al. [9] proposed a length scale directive Hausdorff (LSD-Hausdorff) trajectory similarity measure. Wang et al. [4] proposed a novel descriptor named trajectory kinematics descriptor to represent trajectories based on the kinematic properties from the point-of-view of Frenet-Serret frames. Madhuri Debnath et al. [10] proposed a new framework to cluster sub-trajectories based on a combination of their spatial and non-spatial features. This algorithm combines techniques from grid based approaches, spatial geometry and string processing. Yang Fan et al. [11] proposed a novel 3D visualization tool, in which the Agglomerative Information Bottleneck (AIB) based clustering scheme is illustrated comprehensively, to help users understand the clustering approach vividly and clearly.

The advantage of the methods discussed above is that the similarity between two trajectories with different length can be calculated directly and easily. This is due to some preprocessing steps, such as shortening trajectories with longer length or stretching trajectories with shorter lengths to the ones with same length, formulating trajectories by parameter models to perform the similarity measurement. In other words, the original trajectories are modified and only the position information of the trajectories is used for similarity calculation in these methods. However, the kinematic property of trajectories, which is a typical feature for moving objects, is still neglected [4].

To further improve the robustness of trajectory clustering and to obtain an optimal cluster number, how to define a proper and effective distance measurement for trajectories is still a key and challenging work for trajectory clustering schemes.

The emphasis in this paper is on obtaining a complete trajectories information, which includes points and kinematic property, through a test method of height enumeration. Two properties, the heights of the trajectories satisfy linear properties and two trajectories with displacement difference satisfies a plane surface character in 3-D space when conduct a height enumeration, are found. Based on these properties, the 3-D information of trajectories in 2-D image can be reconstructed completely, and a stable and reliable data for trajectories clustering and events detecting is established. The rest of this paper is organized as follows: Sect. 2 briefly introduces the camera calibration. Section 3 introduces the 3-D reconstruction of 2-D trajectories. Section 4 shows the procedures of trajectories clustering. In the last section, the experimental results and conclusions are presented.

2 Camera Calibration

Camera calibration can help to deal with perspective distortion of object appearance on 2-D image plane which induces a very difficult problem for most feature based 2-D image processing methods. A calibrated camera makes it possible to recover discriminant metrics robust to scene or view angle changes, and it is greatly helpful for some applications like classification and tracking among multi-cameras. Additionally, with camera calibrated, we can use of prior information of 3-D models to estimate real 3-D poses of objects in videos and make object detection or tracking more robust to noise and occlusions [12].

Assume that $P_W(x_W, y_W, z_W)$ is a 3-D coordinate point on target in the 3-D world coordinates system, $P_C(x_C, y_C, z_C)$ and $p_I(u_I, v_I)$ are its corresponding points in the camera coordinates system and the imaging plane respectively. Then the relationship between the world coordinate system and the camera coordinate system can be formulated as

$$\begin{bmatrix} x_C \\ y_C \\ z_C \\ 1 \end{bmatrix} = \begin{bmatrix} R & T \\ 0^T & 1 \end{bmatrix} \begin{bmatrix} x_W \\ y_W \\ z_W \\ 1 \end{bmatrix} \tag{1}$$

where R is a rotation matrix with size 3×3, and can be marked as $R \triangleq \begin{bmatrix} r_1 & r_2 & r_3 \\ r_4 & r_5 & r_6 \\ r_7 & r_8 & r_9 \end{bmatrix}$.

Based on Eq. (1), a transformation from world coordinate to image plane can be formulated as

$$z_C \begin{bmatrix} u_I \\ v_I \\ 1 \end{bmatrix} = K \begin{bmatrix} R & T \\ 0^T & 1 \end{bmatrix} \begin{bmatrix} x_W \\ y_W \\ z_W \\ 1 \end{bmatrix} \tag{2}$$

where K is the camera internal parameters with a matrix form $K = \begin{bmatrix} \alpha & 0 & u_0 \\ 0 & \beta & v_0 \\ 0 & 0 & 1 \end{bmatrix}$,

T is the translation parameters and can be marked as a 3×1 matrix with a form $T = \begin{bmatrix} t_1 & t_2 & t_3 \end{bmatrix}^T$, R and T compose the external parameter matrix of a camera.

The Eqs. (1) and (2) represent the model of a given camera, and the internal parameters and external parameters can be calculated accurately using the vanishing points based recovery method [12]. Then the camera calibration is done and the relationship between world coordinate and image plane is established exactly based on (1) and (2). Figure 1 shows a calibration results for a given scene, the original point of world coordinate is set as the perpendicular foot point of the camera, the direction perpendicular to road plane (blue), the vanishing direction along the road direction

Fig. 1. The result of camera calibration (Color figure online)

(green) and the vanishing direction orthogonal to the road direction (red) are set as z-direction, x-direction and y-direction of the world coordinate respectively.

3   3-D Reconstruction of the 2-D Trajectories

Trajectories, providing spatial-temporal information, are the best feature to describe the kinematic properties of foreground objects. A trajectory is typically defined as a data sequence, which is consisting of several concatenated state vectors from tracking, and that means it is a indexed sequence of positions and velocities in a given time window. The drawback of trajectories is each node belongs to it is described by a 2-D kinematic model. Generally, a trajectory can be represented as a point data set as follows:

$$Tra_{2D}(m, t, t - k) = \{p_I(m, t), p_I(m, t - 1), \cdots, p_I(m, t - (k - 1))\} \qquad (3)$$

where m is the number of a given trajectory in trajectory data set, k is the number of nodes on the given trajectory, $p_I(m, t)$ and $p_I(m, t - k)$ is the image coordinate of trajectory node in the current frame and the previous k frame, respectively.

In order to obtain complete 3-D information of a trajectory, the height information is necessary. However, the real height information is hard to get directly from 2-D trajectory in image plane. Fortunately, for a rigid moving object, the points on its surface have a same property, the nature of synchronous motion. Thus, we can get trajectories information under 3-D description, as follows:

$$Tra_{3D}(m, t, t - k) = \{P_W(m, t), P_W(m, t - 1), \cdots, P_W(m, t - (k - 1))\} \qquad (4)$$

and two trajectories belong to a same rigid moving object body, with sequence number m and n, meet two constraints as their specific properties, as follows:

$$DisDif(m, n, t - i) = Dis(Tra_{3D}(m, t, t - i)) - Dis(Tra_{3D}(n, t, t - i)) = 0 \qquad (5)$$

and

$$Dis(Tra_{3D}(m, t, t - i)) = |P_W(m, t)P_W(m, t - i)|$$
$$= \sqrt{(x_W(m, t) - x_W(m, t - i))^2 + (y_W(m, t) - y_W(m, t - i))^2 + (z_W(m, t) - z_W(m, t - i))^2}$$
$$(6)$$

where $P_W(m,t)$ and $P_W(m,t-i)$ is the 3-D world coordinate point of trajectory node in the current frame and the previous i frame, $|P_W(m,t)P_W(m,t-i)|$ is their distance in 3-D space, and $DisDif(m,n,t,i)$ is the displacement difference between two trajectories, $Dis(Tra_{3D}(m,t,t-i))$ is the Euclidean distance between two points on the current frame and the previous i frames for the trajectory with sequence number m in 3-D world coordinate system, x_W, y_W and z_W are the 3-D coordinate value of a node.

Based on the Eq. (2), if the world coordinates of a point on target, $P_W(x_W, y_W, z_W)$, is given, we can get its image coordinates $p_I(u_I, v_I)$ directly, and the relationship between them can be expressed as a transformation as follows:

$$p_I = f \bullet P_W, \tag{7}$$

where f can be calculated by Eqs. (1) and (2). Furthermore, if the height of a point on the image plane is given, we can get its 3-D world coordinates using the relationship as follows:

$$P_W = F^{-1} \bullet (p_I \oplus h) \tag{8}$$

Where F^{-1} is the inverse transformation implied in (2). That means, we can calculate the corresponding 3-D coordinates of a point in 2-D image plane when give its height in 3-D space. However, the real height value of a point in 2-D image is unknown. Fortunately, the height of an object is a bounded value. Thus, for two trajectories belong to the same rigid moving object, we can conduct a test using the method of height numeration to estimate the real value of this height. In this test, we can calculate a displacement difference between two trajectories under two supposed trajectory heights. When enumerating all possible combination of two heights, a series of displacement difference can be obtained using the following formulation:

$$\begin{cases} Trajectory\ m : h_m = 0, 1, 2, \cdots, 199cm \\ Trajectory\ n : h_n = 0, 1, 2, \cdots, 199cm \\ DisDiff(m,n,t,i)|_{h_m,h_n} = Dis(Tra_{3D}(m,t,t-i))|_{h_m} - Dis(Tra_{3D}(m,t,t-i))|_{h_n} \end{cases} \tag{9}$$

where the height enumeration value is choose between 0 cm and 199 cm due to the fact that the height of vehicle is smaller than 2 meters in most of monitoring scenes. Figure 2(a) shows the trajectory set in 2-D image and Fig. 2(b) shows the displacement difference between two trajectories (the red and green trajectory in Fig. 2(a)) that belong to a same rigid object under height enumeration. The white plane is a reference plane, which is given by the property of Eq. (5), and represents the displacement difference, $DisDif(m,n,t-i)$, is zero. The gray oblique plane is a calculated data plane using (9), which represents the displacement difference between two trajectories under height enumeration. In Fig. 2(b), a point on the gray plane means a value of displacement difference between the red and green trajectories in Fig. 1, which is calculated by giving the two trajectories a height values respectively. The red axis and green axis represent the height of red trajectory and green trajectory in Fig. 2(a), and the blue axis represents the displacement difference value. Figure 2 shows that if the

(a)   (b)

Fig. 2. Trajectories and its displacement difference under height enumeration. (a) trajectories in 2-D image; (b) displacement difference under height enumeration for two trajectories(red one and green one in (a)) (Color figure online)

trajectories belong to a same rigid object, the heights of which satisfy a linear relationship

$$A \cdot h_m + B \cdot h_n + C = 0 \tag{10}$$

which is the intersecting line of the white plane and the gray plane, and the displacement difference between two trajectories satisfy

$$DisDif(m, n, t, t - i) = A \cdot h_m + B \cdot h_n + C \tag{11}$$

and show in the gray plane.

To illustrate the correctness of Eqs. (10) and (11) as a quantitative way, we set two supposed trajectories in the traffic sense that Fig. 2(a) shows. Based on calibrated camera data, for the two supposed trajectory, their coordinates of beginning point in 3-D space are $(-4.2, 12.8, 1.0)$ and $(0.4, 13.2, 2.0)$ respectively. The speed of the two trajectories is 50 km/H as the vehicle going forward, and the video frame rate is 25 frames per second (fps). The two supposed trajectories are shown in Fig. 3(a) and their displacement difference under height enumeration is shown in Fig. 3(b), in which their height satisfy the linear relationship $-0.317Hg + 0.37 h - 0.423 = 0$, where Hg and Hr is the height of first trajectory (the green one) and the second trajectory (the red one), respectively. It is worth noting that, for trajectory difference metric, Eq. (5), in 3-D space, the coordinates in x-axis and y-axis has no influence for the result. This means that trajectory Eq. (10) only depends on the relative height difference of two given trajectories when the speed is unchanged.

Actually, the real height for any trajectory in 2-D image is unknown. Hence the heights of trajectories cannot be calculated based on Eq. (10). Fortunately, the similarity measurement of two trajectories can be solved effectively as long as the relative height of them is given. That means if a reference trajectory can be found, the height of other trajectories can be calculated. In this paper, the heights of all the trajectories extracted from the 2-D image are set to zero, the corresponding trajectories in 3-D

(a)   (b)

Fig. 3. Two supposed ideal trajectories and its displacement difference under height enumeration. (a) Two supposed ideal trajectories in 2-D image; (b) displacement difference under height enumeration for two supposed ideal trajectories (the red on and the green one) (Color figure online)

space can constructed and each trajectory has a calculated moving distance. The trajectory that with a minimum moving distance has a relative small height value in 3-D space, thus the height of this trajectory can be set to zero and the heights of the other trajectories will be calculated exactly under the zero-value height trajectory assumption. Figure 4 shows the trajectories and their relative height calculated under zero-value assumption of the reference trajectory. In Fig. 4(a), the red trajectory is the one has the smallest height. Figure 4(b) shows the relative height of the other trajectories by setting the height of red trajectory in Fig. 4(a) to zero.

The Fig. 4 shows a valuable property that a speed difference of two trajectories belongs to two different vehicles induced a height difference under the rigid constrains. This property is useful for trajectory clustering because the vehicles are rigid moving objects on the road. And above all, the Euclidean distances between the corresponding points belong to two trajectories which from a same vehicle is a fixed value in 3-D space and the variance of those distances is zero. While for two trajectories from two different vehicles, the variance is not zero when a speed difference exists between two vehicles. Based on those properties and the 3-D coordinates of trajectory points, the trajectory similarity metric and clustering can be improved.

(a)   (b)

Fig. 4. The result of trajectories relative height calculation. (a) Trajectories set with the red one denoted as the trajectory has the smallest height value; (b) relative heights of trajectories calculation under the height of red one in (a) is reset to 0. (Color figure online)

4 Trajectory Clustering

Based on the trajectory information in 3-D space and choosing the variance of the Euclidean distances between the corresponding points belong to two trajectories as a trajectory similarity metric, a clustering result can be obtained by the DBSCAN method [13]. Actually, the vehicle similar to a cuboid, and the size of the most vehicles in city with in the size 5.0m × 1.8m × 1.6m (long × width × height), thus we conduct a scale transformation to the trajectory points in 3-D space for a better performance of DBSCAN by considering the vehicle as a cube-like structure. The Table 1 shows the scale parameter we used, and the clustering result for Fig. 2 based on DBSCAN is shown in Fig. 5.

Table 1. Scale parameter for cuboid-like vehicles

Size of vehicle (cuboid-like structure)	Long	Width	Height
	5.0 m	1.8 m	1.6 m
Scale parameter	1.0	2.778	3.125
Size after transformed (cube-like structure)	5.0	5.0	5.0

Fig. 5. Trajectory clustering result based on DBSCAN for Fig. 2

Based on the Eq. (8), the 3-D information of all trajectories in 2-D image can be reconstructed in a 3-D data space and a trajectory clustering can be done. Figure 6(a) shows a 2-D trajectory set and its clustering result, Fig. 6(b) shows their corresponding 3-D data set with normalized height information, in which the yellow trajectory is the one with the smallest height value (Fig. 6(a)) and the heights of the other trajectories is calculated by set yellow one's height to 0 (Fig. 6(b)).

As that is shown in Figs. 4 and 6, if the speeds of different rigid moving objects are not consistent with each other, the heights of the trajectories will present an obvious gap, and the trajectories can be clustering based on the gap difference. If the speeds of different rigid moving objects are consistent with each other, the trajectory clustering is also easy to take through using the lane-constrained, vehicle size and location. Additionally, if the trajectories come from a same rigid moving object, the height difference between them is stability, while if the trajectories belong to different rigid moving objects, the height difference between them has a large variation when the speed

(a)   (b)

Fig. 6. Trajectories in 2-D image and their 3-D reconstruction. (a) The trajectories in 2-D image, the one with the smallest height is denoted as yellow; (b) the 3-D reconstruction of trajectories in (a). (Color figure online)

difference exists. For different realistic application environment, if feature descriptor for tracking is rich, DBSCAN (Density Based Spatial Clustering of Application with Noise) clustering method can achieve good results. If trajectories are sparse, it is best to make full use of lane-constrained, vehicle size and location information for clustering.

5   Experimental Results and Conclusions

Two video data were conducted to examine the effectiveness of the trajectory analysis method proposed in this paper. One is the video data of monitor scene for a beltway in city of Xi'an, and another is a monitor scene of Sanlu Outer Ring Road in city of Shanghai. Based on the proposed trajectory analysis method, the 3-D information of trajectories set can be obtained and described in a 3-D data space. Then combining with DBSCAN clustering algorithm, the vehicle segmentation can be realized effectively. In this processing scheme, the cuboid-like vehicle structure feature is using in DBSCAN clustering method. In this way, the vehicle segmentation result is a robust and accurate one. The rigid constrain used in the proposed method are chosen to estimate whether two trajectories belong to a same rigid moving vehicle or not. The estimation method based on Eq. (9) is an accurate numerical calculation problem, thus the results is accurate and reliable and can be used to trajectory based vehicle counting application. The vehicle counting accuracy is chosen as a quantitative index to evaluate the performance of the proposed method in realistic application environment (Fig. 7 shows two monitor scenes we used). The video data are tested on a Windows XP platform with a Pentium 4 3.2-GHz central processing unit (CPU) and 2-GB random access memory (RAM). The proposed algorithm is implemented with Visual C++ on a raw video format. Table 2 shows the video information and vehicle counting results based on the proposed method.

The result in Table 2 shows that in the situation with a larger scene and a lower resolution, the vehicle counting accuracy cannot up to 100 % even the 3-D information of trajectories are provided. It does not means the rigid constrains and 3-D information is low value information. Actually, the main factor induces the accuracy less

(a)   (b)

Fig. 7. Two test monitor scene. (a) A monitor scene for a beltway in city of Xi'an; (b) a monitor scene for Sanlu Outer ring road in city of Shanghai.

Table 2. Vehicle counting result for video data

Monitor scene	A beltway in Xi'an	Sanlu road in Shanghai
Image size	1280*720	720*288
Descriptor for tracking	ORB tracking	Corner feature tracking
Total frame	755 frames	9320 frames
Total number of vehicles	69 vehicles	503 vehicles
Number of counted vehicles	67	497
Relative accuracy	97.1 %	98.8 %
Number of right counted	67	459
Absolute accuracy	97.1 %	91.2 %

than100 % is the provided trajectories are not good enough in some cases, such as incomplete trajectories induces by serious barrier, a lower distance resolution of height between trajectories which is caused by large scene, the segmentation error for large container trucks. Additionally, the proposed method is helpless for segmentation of two vehicles that moving synchronously with very small traffic spacing.

Fortunately for us, the value of the proposed method is that two important properties are found (based on Fig. 2) by using the rigid constrains for moving objects, and it provides a method for obtaining complete 3-D information of the trajectories. We test the effectiveness of the important properties by applying them for vehicle counting in two realistic monitor situations. The future work will concentrate on two aspects: one is attempting to use a richer image feature descriptor to vehicle tracing, the other one is reconstructing the surface of vehicles in 3-D space based on the 3-D coordinates of the feature and given a 3-D space model for moving objects.

Acknowledgement. This work was supported by the National Natural Science Foundation of China under Grants 61572083, Key Program of Natural Science of Shannxi under Grants 2015JZ018 and the Fundamental Research Founds for Central University under Grants 310824151034.

References

1. Morris, B.T., Trivedi, M.M.: A survey of vision-based trajectory learning and analysis for surveillance. IEEE Trans. Circ. Syst. Video Technol. **18**(8), 1114–1127 (2008)
2. Hu, W.M., Tian, T., Wang, L., Maybank, S.: A survey on visual surveillance of object motion and behaviors. IEEE Trans. Syst. Man Cybern. Part C: Appl. Rev. **34**(3), 334–337 (2004)
3. Sivaraman, S., Trivedi, M.M.: Looking at vehicles on the road: a survey of vision-based vehicle detection, tracking, and behavior analysis. IEEE Trans. Intell. Transp. Syst. **14**(4), 1773–1796 (2013)
4. Wang, W.C., Chung, P.C., Cheng, H.W., Huang, C.R.: Trajectory kinematics descriptor for trajectory clustering in surveillance video. In: IEEE International Symposium on Circuits Systems, pp. 1198–1201 (2015)
5. Mitsch, S., Müller, A., Retschitzegger, W., Salfinger, A., Schwinger, W.: A survey on clustering techniques for situation awareness. In: Ishikawa, Y., Li, J., Wang, W., Zhang, R., Zhang, W. (eds.) Web Technologies and Applications. LNCS, vol. 7808, pp. 815–826. Springer, Heidelberg (2013)
6. Fu, Z., Hu, W., Tan, T.: Similarity based vehicle trajectory clustering and anomaly detection. In: Proceedings of the IEEE International Conference on Image Processing, September 11–14, vol. 2, pp. II-602-605 (2005)
7. Atev, S., Miller, G., Papanikolopoulos, N.P.: Clustering of vehicle trajectories. IEEE Trans. Intell. Transp. Syst. **11**(3), 647–657 (2010)
8. Abraham, S., Lal, P.S.: Spatio-temporal similarity of network-constrained moving object trajectories using sequence alignment of travel locations. Transp. Res. Part C **23**, 109–123 (2012)
9. Hao, J.Y., Gao, L., Zhao, X.: Trajectory clustering based on length scale directive Hausdorff. In: Proceedings of the 16th International IEEE Annual Conference on Intelligent Transportation Systems (ITSC 2013), The Hague, The Netherlands, 6–9 October 2013
10. Debnath, M., Tripathi, P.K., Elmasri, R.: A novel approach to trajectory analysis using string matching and clustering. In: Proceedings of IEEE 13th International Conference on Data Mining Workshops, The Dallas, TX, December 7–10, pp. 986-993 (2013)
11. Fan, Y., Xu, Q., Guo, Y.J., Liang, S.: Visualization on agglomerative information bottleneck based trajectory clustering. In: Proceedings of 2015 IEEE 19th International Conference on Information Visualisation, Barcelona, July 22–24, pp. 557-560 (2015)
12. Zhang, Z.X., Li, M., Huang, K., Tan, T.: Practical camera auto-calibration based on object appearance and motion for traffic scene visual surveillance. In: proceedings of IEEE Conference on Computer Vision and Pattern Recognition, the Anchorage, AK, June 23–28, pp. 1-8 (2008)
13. Uncu, O., Gruver, W.A., Kotak, D.B., Sabaz, D.: GRIDBSCAN: GRId density-based spatial clustering of applications with noise. In: Proceedings of 2006 IEEE International Conference on Systems, Man, and Cybernetics, Taipei, Taiwan, October 8–11, vol. 4, pp. 2976-2981 (2006)

Robust Features of Finger Regions Based Hand Gesture Recognition Using Kinect Sensor

Fengyan Wang[1] and Zengfu Wang[1,2,3(✉)]

[1] Department of Automation, University of Science and Technology of China,
Hefei 230026, China
fyw@mail.ustc.edu.cn, zfwang@ustc.edu.cn
[2] Institute of Intelligent Machines,
Chinese Academy of Sciences, Hefei 230031, China
[3] National Engineering Laboratory for Speech and Language Information Processing,
University of Science and Technology of China, Hefei 230026, China

Abstract. Thanks to the emergence of commercial depth cameras, e.g., Kinect, hand gesture recognition has attracted great attention in recent years. In this context, we present a novel Kinect based hand gesture recognition system which focuses on the features of finger regions. A hand cropping approach is proposed to extract the useful finger regions from a noisy hand image including palm, wrist and arm obtained by Kinect. Furthermore, an original dissimilarity metric, called Balanced Finger Earth Movers Distance (BFEMD), is used to classify hand gestures along with the hierarchical recognition strategy. Finally, the 12 popular gestures recognition experiments have been done to illustrate the effectiveness of the proposed gesture recognition system, and the experimental results show that the proposed system can achieve high recognition accuracy at a high speed.

Keywords: Balanced Finger Earth Mover's Distance · Hierarchical recognition · Human computer interaction · Hand gesture recognition · Kinect

1 Introduction

Due to hand gestures are meaningful part of spoken language, vision based hand gesture recognition is playing an important role for human computer interaction (HCI), and its potential applications have been involved in every aspect of the society and economic life, such as sign language recognition, computer games and virtual reality [1].

Because of the nature of optical cameras, traditional hand gesture recognition approaches are usually affected by lighting conditions or cluttered environment, and unable to separate hand robustly from image, which significantly affects the recognition performance, so a number of attempts have been made to locate and track hand region from the image sequence in a bid to achieve high recognition

T. Tan et al. (Eds.): CCPR 2016, Part I, CCIS 662, pp. 53–64, 2016.
DOI: 10.1007/978-981-10-3002-4_5

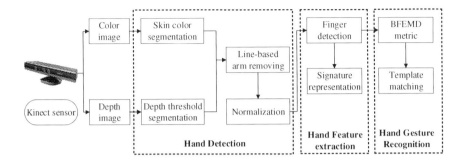

Fig. 1. Overview of the proposed finger-only based hand gesture recognition system

accuracy [2,3]. For instance, some of them employ data gloves to avoid the influence of uncontrolled environment, but the prohibitive price and inconvenient usage hinder it from becoming a popular way for gesture recognition [4]. Another class of methods applies color markers to replace electronic gloves but it requires complex calibration [5]. There are also some other approaches, like skin color model [6] and hand shape model [7]. These methods are easy to implement, but they are usually not robust to dynamic background or illumination variation.

In the last few years, the emergence of depth cameras with commodity price, e.g., Kinect sensor, opens up new frontiers in vision based hand gesture recognition [8,9]. Depth information captured by Kinect provides much help to hand region detection. But it is still a great challenge for hand gesture recognition using Kinect, since it is difficult to accurately detect and segment a human hand which occupies a very small portion of the image only by means of depth information.

In this paper, we study hand gesture recognition and present a complete hand gesture recognition system with Kinect. After obtaining hand shapes with Kinect, we propose a novel distance metric named Balanced Finger Earth Mover's Distance (BFEMD), along with the hierarchical recognition strategy, to distinguish hand gestures. The introduced method naturally combines the information from color image and depth map. By jointly using these two kinds of information, the proposed gesture recognition system is very robust to environment change. Furthermore, it is very efficient and can be implemented in real time. The architecture of the proposed hand gesture recognition system is schematized in Fig. 1. Three main stages are incorporated: (1) hand detection, (2) hand feature extraction and (3) hand gesture recognition.

2 Finger-Only Based Hand Gesture Recognition System

2.1 Hand Detection

As we know, hand detection is of great importance to gesture recognition, it is the foundation of the whole system and much related to the recognition accuracy.

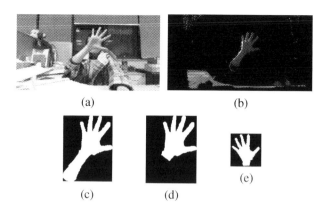

(a)   (b)

(c)   (d)   (e)

Fig. 2. Hand detection. (a) The input image. (b) The foreground region segmented by depth thresholding. (c) The skin color detection result. (d) The result of removing the arm area. (e) The final normalized pure hand region. (Color figure online)

Generally, hand detection consists of three parts: hand localization, segmentation and normalization. In this work, we utilize Kinect sensor as the image acquisition unit, which can capture color image with size of 1920×1080 pixels and depth map with size of 512×480 pixels respectively.

Since users make gestures with various clothes, it is inevitable to obtain the redundant arm area more or less when using skin color model to detect the hand regions. To address this problem, some methods require experimenters to wear long sleeves [10], and others demand users to wear a black belt on the gesturing hand's wrist to segment the hand shape more accurately [11]. However, aforementioned methods adopt wearable devices to simplify this issue. In this paper, to handle the noisy hand shapes with a certain length of the arm obtained from Kinect, we propose a markless hand extraction method which focuses on removing arm exactly to get the hand region, which can be seen from Fig. 2. Firstly, by setting a suitable depth threshold, the foreground and background are divided, thus we can obtain a foreground region, as shown in Fig. 2(b). Although the depth map is usable to segment hand from cluttered background, it is not very accurate in details because of the low resolution and inaccuracy of the depth information, so skin color model and filtering operation are introduced to obtain skin color region. Nevertheless, Fig. 2(c) illustrates that the acquired skin color region usually not only contains hand, but also the redundant arm.

To circumvent this problem, we propose a line-based arm removal algorithm to search the best hand cropping position. The detailed steps are listed as follows.

(1) Remove fingers roughly by means of morphological opening operation, only remain the palm and arm parts.
(2) Use Hough Transform to detect all possible line segments, select the correct lines on either side of the arm, then the direction of the arm is represented by the average slope of these lines.

(3) Obtain the slope of cutting line which is perpendicular to the arm, fit an arm-cutting line.
(4) Set an appropriate stride, the cutting line is moved step by step along the direction of the arm, and record the width of arm recursively to determine the best cutting location, i.e., the position of wrist. Therefore, the objective that segmenting the hand region accurately is achieved (see Fig. 2(d)).

In order to facilitate the subsequent processing, some appropriate normalization techniques, including scale, translation and rotation normalizations are employed. The scale and translation normalizations are implemented by resize the hand shape into 100×100 pixel resolution. As to rotation normalization, we propose two schemes to determine the rotation angle according to two different cases. First, when users make gestures without long sleeves, the average slope of the lines on either side of the arm can be regarded as the rotation angle. Second, it is unable to detect the lines along the arm when user wears long sleeves, so we adapt an alternative approach which is used in [10] to obtain the rotation angle, Fig. 3 presents some hand detection and normalization results.

2.2   Hand Feature Extraction

Generally speaking, given an accurate normalized hand shape, the most distinct features of hand gestures are finger shapes, on the contrary, the palm does not play an critical role for gesture recognition [12]. Based on this principle, we propose a new finger detection method to obtain the finger parts accurately, Fig. 4 explains the flowchart of the proposed finger detection algorithm.

(1) Get the fingertips: divide the hand shape into fingers and palm by morphological opening, the center of palm is calculated as the momentum of palm and the positions of outstretched fingertips are obtained by adapting finger detection approach in [10].
(2) Locate the fingerwebs: both fingertips and fingerwebs are on the hand contour, suppose that there are n fingertips, there must be $n-1$ fingerwebs, so we separate the hand contour into $n-1$ sub-contours, i.e., the contours between two adjacent fingertips, then the curvature analysis method [13] is employed to find fingerwebs.

Fig. 3. Some results of hand segmentation and normalization

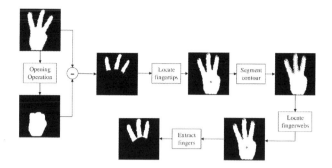

Fig. 4. The flowchart of the proposed finger detection algorithm

(3) Fit a segment line: according to the relative locations of fingertips and fingerwebs, fit a segment line to separate fingers from the hand shape robustly. The finger cropping results can be seen in Fig. 5.

Fig. 5. The accurate fingers detected by the proposed finger detection algorithm

After obtaining the accurate finger shapes, we discuss how to extract the most typical features of each gesture. Instead of representing the finger shapes by time-series curve [11], we propose to use three simple features: the angle and the area of a finger, as well as the number of pixels on the hand contour between a fingertip with the reference point, to simplify the hand shape but retain as much information as possible. Suppose that there are m $(m \leq 5)$ visible fingers Q_i, the corresponding fingertips denoted as F_i (i = 1, 2, $\cdots$, m). The angle θ_i, the area S_i, the number of pixels on the hand contour N_i between the fingertip F_i with the reference point C_2 can be calculated as follows:

$$\begin{cases} \theta_i = < C_0 C_1, C_0 F_i > /2\pi \\ S_i = pn(Q_i)/ \max\{pn(Q_1), pn(Q_2), ...pn(Q_m)\} \\ N_i = pn_c(F_i - C_2)/PN_c \end{cases} \quad (1)$$

The reference points are illustrated in Fig. 6. Note that the angle θ_i is calculated between each fingertip F_i and the reference point C_1 relative to palm center C_0, and $pn(Q_i)$ means the number of pixels within the finger Q_i. $pn_c(F_i - C_2)$

Fig. 6. The reference points that are used in hand feature extraction

denotes the number of pixels on the hand contour between the fingertip F_i to the reference point C_2, and PN_c means the total number of pixels on the hand contour. These three features are normalized, and then combined into different signatures in the next section which will be of great importance to our novel Balanced Finger Earth Movers Distance metric.

2.3 Hand Gesture Recognition

In this section, for better understanding the proposed hand gesture recognition approach, we first briefly introduce classical Earth Mover's Distance (EMD) [14], which is widely used in many problems such as image retrieval [15] and pattern recognition [16]. Then we introduce the proposed Balanced Finger Earth Mover's Distance metric.

Earth Mover's Distance Algorithm. The original EMD method is a measure of the distance between two probability distribution over a region, it is named after a conventional transportation problem, i.e., the process of moving piles of earth spread around one set of locations into another set of holes in the same space. Given a set of suppliers I, a set of consumers J, and the cost c_{ij} to ship a unit of supply from $i \in I$ to $j \in J$, the aim is to find an optimal set F of flow f_{ij}, i.e., the amount of supply shipped from i-th supplier to j-th consumer. To minimize the overall cost,

$$WORK(I, J, F) = \sum_{i=1}^{m} \sum_{j=1}^{n} c_{ij} f_{ij} \tag{2}$$

subject to the following constraints:

$$\begin{cases} f_{ij} \geqslant 0, & 1 \leq i \leq m, 1 \leq j \leq n \\ \sum_{j=1}^{n} f_{ij} = s_i, & 1 \leq i \leq m \\ \sum_{i=1}^{m} f_{ij} = t_j, & 1 \leq j \leq n \\ \sum_{i=1}^{m} \sum_{j=1}^{n} f_{ij} = \min(\sum_{i=1}^{m} s_i, \sum_{j=1}^{n} t_j) \end{cases} \tag{3}$$

where s_i is the supply of i-th supplier and t_j is the capacity of j-th consumer.

Balanced Finger Earth Mover's Distance. Our work is inspired by FEMD which is proposed in paper [11], however there are some major limitations of FEMD algorithm. First, though FEMD approach can address the problem of partial matches by means of introducing penalty on empty holes, it may bring about mismatching because the fingers in those gestures are not correctly segmented due to the distortion. What is more, different hand sizes also change the area of each finger in FEMD, and it is difficult and challenging to obtain the accurate finger extraction in real-time. Therefore, we put forward a unique Balanced Finger Earth Mover's Distance (BFEMD) to address above-mentioned problems.

Formally, given two input hand gestures with the same number m of outstretched fingers ($m \leq 5$), let $A = \{(c1_{A1}, c2_{A1}, c3_{A1}), \cdots, (c1_{Am}, c2_{Am}, c3_{Am})\}$ be the first hand signature, where $c1_{Ai}$ is the first feature of i-th finger, similarly, $c2_{Ai}$ is the second feature and $c3_{Ai}$ is the third feature; $B = \{(c1_{B1}, c2_{B1}, c3_{B1}), \cdots, (c1_{Bm}, c2_{Bm}, c3_{Bm})\}$ is the second signature. $D = [d_{ij}]$ is the ground distance matrix of signature A and B, where d_{ij} is the ground distance from i-th finger of A to j-th finger of B. In this paper, the finger feature $c1$ is intuitively the amount of supply or capacity, and $c2$, $c3$ are used to determine the ground distance d_{ij} using $L1$ distance, that is,

$$d_{ij} = |c2_{Ai} - c2_{Bj}| + |c3_{Ai} - c3_{Bj}|. \tag{4}$$

Then a new type of EMD which adds three parameters $c1$, $c2$ and $c3$ can be expressed:

$$EMD(A, B, c1, c2, c3) = \arg\min \sum_{i=1}^{m} \sum_{j=1}^{m} d_{ij} f_{ij}$$

$$= \arg\min \sum_{i=1}^{m} \sum_{j=1}^{m} (|c2_{Ai} - c2_{Bj}| + |c3_{Ai} - c3_{Bj}|) * f_{ij}$$

subject to :

$$\begin{cases} f_{ij} \geqslant 0, & 1 \leq i \leq m, 1 \leq j \leq m \\ \sum_{j=1}^{m} f_{ij} = c1_{Ai}, & 1 \leq i \leq m \\ \sum_{i=1}^{m} f_{ij} = c1_{Bj}, & 1 \leq j \leq m \\ \sum_{i=1}^{m} \sum_{j=1}^{m} f_{ij} = \min(\sum_{i=1}^{m} c1_{Ai}, \sum_{j=1}^{m} c1_{Bj}) \end{cases} \tag{5}$$

According to the previous description in Sect. 2.2, we extract three distinct features (θ_i, S_i, N_i) of each finger. By selecting different finger feature as the amount of supply or capacity $c1$ and the other two features as the $c2$, $c3$ that are used to determine the ground distance d_{ij}, so we can ge three different types of EMD, which can be formalized as follows:

$$\begin{cases} EMD(A, B, c1, c2, c3) = EMD(A, B, S, N, \theta), & \text{if } c1 = S, c2 = N \text{ and } c3 = \theta \\ EMD(A, B, c1, c2, c3) = EMD(A, B, \theta, S, N), & \text{if } c1 = \theta, c2 = S \text{ and } c3 = N \\ EMD(A, B, c1, c2, c3) = EMD(A, B, N, \theta, S), & \text{if } c1 = N, c2 = \theta \text{ and } c3 = S \end{cases} \tag{6}$$

To balance the three values of EMD, we propose a novel metric named Balanced Finger Earth Mover's Distance (BFEMD)

$$
\begin{aligned}
BFEMD(A, B) = &\, \alpha EMD(A, B, S, N, \theta) \\
&+ \beta EMD(A, B, \theta, S, N) \\
&+ \gamma EMD(A, B, N, \theta, S)
\end{aligned}
\tag{7}
$$

where $\alpha, \beta, \gamma \in \{0, 1\}$. Parameters α, β, γ modulates the importance of three terms, we will investigate the effects of α, β, γ in Sect. 3.2.

Hierarchical Recognition and Template Matching. Hierarchical recognition means the whole standard hand gesture templates are divided into several subsets in advance according to the number of outstretched fingers. For an input hand image, the number of its visible fingers is first recognized using the finger detection algorithm, then the input gesture is further identified with the corresponding gesture template subsets by means of the proposed BFEMD metric. We adapt the template matching approach to hand gesture recognition, that is, the input image is recognized as the class with which it has the minimum dissimilarity distance:

$$
c = \arg\min BFEMD(H, T_c)
\tag{8}
$$

where H is the input hand gesture, T_c is the standard template of class c, and $BFEMD(H, T_c)$ denotes the proposed Balanced Finger Earth Mover's Distance between the input hand gesture and each template.

3   Experiments

3.1   Database

In order to demonstrate the performance of both the line-based arm removal algorithm and BFEMD metric hierarchical recognition algorithm, a hand gesture database that contains gestures without long sleeves and artificial marks is needed. Therefore, we collect a new color-depth hand gesture database using Microsoft Kinect 2.0. It contains 12 gesture signs with 10 different poses from 10 people, thus our database has 12 gestures $\times$ 10 subjects/gesture $\times$ 10 poses/gesture $= 1200$ cases for testing totally. Figure 7 shows the defined 12 hand gestures to be recognized in the experiments, which are labeled from 0 to 11. Our database is a challenging real-life database, which is collected from uncontrolled environment. For each gesture, the subject posed with variations in hand orientation, scale, articulation, etc.

3.2   Performance Evaluation

The proposed hand recognition method is implemented with C++ and all experiments are done on an Intel Core i3-3220 3.30 GHz CPU with 4 GB of RAM. Next we evaluate the performance of the proposed system from mean accuracy, time efficiency and comparison with FEMD.

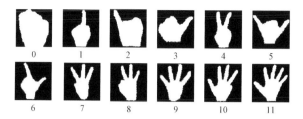

Fig. 7. Gesture signs in our experiments captured by Microsoft Kinect

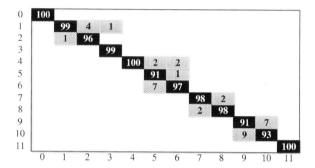

Fig. 8. The confusion matrix of hand gesture recognition using BFEMD with default parameters ($\alpha = \beta = \gamma = 1$, unit: %)

Mean Accuracy and Time Efficiency: We fix $\alpha = \beta = \gamma = 1$ as default parameters, Fig. 8 shows the confusion matrix of the proposed hand gesture recognition system on our database. It can be found that BFEMD based method achieves an mean accuracy rate of 96.8 %. As it shows, the two most confused gesture categories are gesture 5 and 9, the relative recognition rates are a little bit lower than other postures, and the reason may be that the gesture 5 and 6, the gesture 9 and 10 have similar contours and finger features after normalization and discarding the palm. Besides, due to the few number of extracted finger features, BFEMD operates efficiently (around 0.087 s) for the recognition process with our C++ implementation. As a result, the proposed BFEMD similarity metric with hierarchical recognition strategy is not only robust but also has the potential to be used in real-time applications.

Comparison with FEMD: To further illustrate the superiority of our system, we compare it with FEMD [11] on our database. The hand shapes are segmented and preprocessed using the same method described in Sect. 2.1. It should be noticed that two different finger decomposition methods are adopted for FEMD in [11], and the reported average accuracies are very close. In particular, we apply the thresholding decomposition based method in our database.

Figure 9 presents the confusion matrix of FEMD, the obtained recognition rates in our experiments by using FEMD are lower than the original method in [11], a major reason is that employing different databases. To be specific,

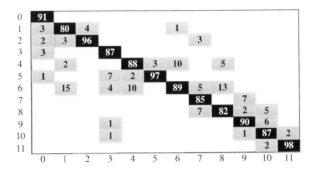

Fig. 9. The confusion matrix of hand gesture recognition using FEMD (unit: %)

our database contains several gestures of the same outstretched fingers which means the segmented fingers have very similar global structures, so only adopting one simple feature that used to determine the ground distance d_{ij} in FEMD methods can not classify these gestures correctly. Moreover, the palm size and the finger length has great impact on the recognition accuracy using thresholding decomposition.

3.3   Parameter Sensitivity

To demonstrate the effectiveness of the proposed system, we evaluate three important parameters (α, β, γ) in BFEMD formulation, the results are shown in Table 1. It can conclude that for different number of visible fingers, the optimal value of α, β, γ are different. For instance, $\alpha = 1, \beta = 0, \gamma = 0$ is the optimal values for one outstretched finger and $\alpha = 1, \beta = 0, \gamma = 1$ is best for two outstretched fingers.

Table 1. Parameter sensitivity on α, β, γ in BFEMD formulation

Rate \ α, β, γ Fingers	1,0,0	0,1,0	0,0,1	1,1,0	0,1,1	1,0,1	1,1,1
1	**0.99**	0.63	0.95	0.92	0.79	0.92	0.98
2	0.96	0.88	0.96	0.95	0.93	**0.97**	0.96
3	0.93	0.97	0.97	0.95	0.97	0.96	**0.98**
4	**0.99**	0.85	0.89	0.92	0.85	0.95	0.91

By using the corresponding optimal α, β, γ for the specific number of visible fingers, the average recognition accuracy rate of 12 hand gestures can arise from

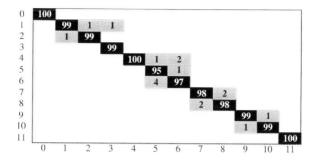

Fig. 10. The confusion matrix of hand gesture recognition using BFEMD with optimal parameters (α, β, γ)

96.8 % to 98.6 % which can be seen from Fig. 10. It suggests that for different values of coefficients, the gesture recognition rate varies a lot and we can select optimal parameters according to the number of outstretched fingers to enhance the mean recognition accuracy.

4   Conclusions

In this paper, we present an robust static hang gesture recognition system based on Earth Mover's Distance metric. A markless hand cropping approach has first been presented to handle noisy hand shapes with a certain length of the arm that obtained by Kinect. Then we introduced a novel technique for finger extraction. Finally, an efficient finger-only based hand gesture recognition system has been proposed, which is based on a novel Balanced Finger Earth Mover's Distance. The partial matching issue is addressed by introducing the hierarchical recognition strategy. 12 popular gestures recognition experiments including comparison with FEMD has been done to illustrate the robustness and effectiveness of our hand gesture recognition system based on BFEMD. High recognition rate and high computational speed for hand gesture recognition are achieved. We plan to further investigate robust depth and texture features for BFEMD and extend it to dynamic hand gesture recognition.

Acknowledgements. This work was supported by the National Natural Science Foundation of China (No. 61472393).

References

1. Wachs, J.P., Kolsch, M., Stern, H., Edan, Y.: Vision-based hand gesture applications. Commun. ACM **54**, 60–71 (2011)
2. Rautaray, S.S., Agrawal, A.: Vision based hand gesture recognition for human computer interaction: a survey. Artif. Intell. Rev. **43**, 1–54 (2015)
3. Murthy, G.R.S., Jadon, R.S.: A review of vision based hand gesture recognition. Int. J. Inf. Technol. Knowl. Manag. **2**, 405–410 (2009)

4. Weissmann, J., Salomon, R.: Gesture recognition for virtual reality applications using data gloves and neural network. In: International Joint Conference on Neural Networks, vol. 3, pp. 2043–2046 (1999)
5. Kry, P.G., Pai, D.K.: Interaction capture and synthesis. ACM Trans. Graph. **25**, 872–880 (2006)
6. Teng, X., Wu, B., Yu, W., et al.: A hand gesture recognition based on local linear embedding. J. Vis. Lang. Comput. **16**, 442–454 (2005)
7. Suarez, J., Murphy, R.R.: Hand gesture recognition with depth images: a review. In: The 21st IEEE International Symposium on Robot and Human Interactive Communication, pp. 411–417 (2012)
8. Wu, Y., Lin, J., Huang, T.S.: Analyzing and capturing articulated hand motion in image sequences. IEEE Trans. Pattern Anal. Mach. Intell. **27**, 1910–1922 (2005)
9. Shotton, J., Sharp, T., Kipman, A., Fitzgibbon, A., Finocchino, M., Blake, A., Cook, M., Moore, R.: Real-time human pose recognition in parts from single depth images. Commun. ACM **56**, 116–124 (2013)
10. Liu, S.P., Liu, Y., Yu, J., Wang, Z.F.: A static hand gesture recognition algorithm based on Krawtchouk moments. In: 6th Chinese Conference on Pattern Recognition, pp. 321–330 (2014)
11. Ren, Z., Yuan, J., Meng, J., Zhang, Z.: Robust part-based hand gesture recognition using kinect sensor. IEEE Trans. Multimed. **15**, 1110–1120 (2013)
12. Cheng, H., Dai, Z., Liu, Z., Yang, Z.: An image-to-class dynamic time warping for both 3D hand gesture recognition and trajectory hand gesture recognition. Pattern Recogn. **55**, 137–147 (2016)
13. Trigo, T.R., Pellegrino, S.R.M.: An analysis of features for hand-gesture classification. In: Proceedings of the 17th International Conference on Systems, Signals and Image Processing, pp. 412–415 (2010)
14. Rubner, Y., Tomasi, C., Guibas, L.J.: The earth movers distance as a metric for image retrieval. Int. J. Comput. Vis. **40**, 99–121 (2000)
15. Datta, R., Joshi, D., Li, J., Wang, J.Z.: Image retrieval: ideas, influences, ans trends of the new age. ACM Comput. Surv. **40**, 1–60 (2008)
16. Hao, H., Wang, Q.L., Li, P.H., Zhang, L.: Evaluation of ground distances and features in EMD-based GMM matching for texture classification. Pattern Recogn. **57**, 152–163 (2016)

Circular Object Detection in Polar Coordinates for 2D LIDAR Data

Xianen Zhou$^{(\boxtimes)}$, Yaonan Wang, Qing Zhu, and Zhiqiang Miao

College of Electrical and Information Engineering,
Hunan University, Changsha 410082, China
{zhouxianen,yaonan,zhuqing,miaozhiqiang}@hnu.edu.cn

Abstract. This paper presents a new circular object detection method based on geometric property and polynomial fitting in polar coordinates instead of implementing it in Cartesian coordinates for 2-Dimension (2D) lidar data. There are three procedures of the algorithm. Firstly, a simple and fast segmentation method is proposed. Then, according to the circle property, five robust and effective features in natural lidar coordinates for each segment are defined. Finally, these features are normalized and fed into Support Vector Machine (SVM) to detect the target circular object. Three videos containing 1330 frames data are manually labeled and used to test the performance of the proposed algorithm. The best accuracy is 99.79 % and the execution time is lower than 16.93 ms. Experimental results demonstrate that circular object can be detected efficiently and accurately by the proposed method.

Keywords: 2D LIDAR · Circle detection · Feature extraction · Polar coordinates

1 Introduction

Circle detection is an important and fundament operation in pattern recognition and image processing [1]. The developed circle detection methods are generally based on Hough transform [2,3], Least squares [4,5] and randomized strategy [6,7] for camera images. The problem is studied by proposing solutions that rely on a modeling of scene in Cartesian coordinates. However, the images captured by camera are easy to be disturbed by view point, angle of camera, image pixels, the natural environment and so on. So, many special applications such as mobile robot navigation and Simultaneous Localization And Mapping (SLAM) use 2D laser radar sensors to get information. 2D lidar, which low cost, small size and low power consumption, is sensitive to small jumping changes [8]. And it can obtain the target object direction and distance conveniently. Using 2D lidar to detect circular objects is a popular problem. So far, there are a great many of line, curve and circle detection approaches for 2D lidar data. All these methods depend on models in Cartesian coordinates.

© Springer Nature Singapore Pte Ltd. 2016
T. Tan et al. (Eds.): CCPR 2016, Part I, CCIS 662, pp. 65–78, 2016.
DOI: 10.1007/978-981-10-3002-4_6

The traditional object detection methods for 2D lidar usually consist of four main parts: (1) Laser data acquisition and pre-processing, (2) data segmentation, (3) feature extraction and (4) object detection by classification methods. Zhang et al. [9] used the Gaussian-Newton optimization method to fit circle and line parameters. Feng et al. [10] described a geometrical feature detection frame that curvature values are used to detect curve and line segments. When the segment of curvature function has a value larger than the specified value, the segment is detected as a curve segment. Otherwise, it is considered as a line segment. Zhao and Chen [11] used the similar principle to detect circle and line. In [12], authors denoted three different circle detection algorithms for 2D lidar data including the least square sense presented in [13]. For all these three methods, the circle parameters could be expressed by definition formula. Many other researchers proposed various methods to detect and track people by range finder [14–16]. And there are other applications by extracting the features of circle. Felipe et al. [17] presented principled techniques to obtain semantic feature detection statistics in set based SLAM for laser range data. They focus on the example of the extraction of circular cross-sectioned features, such as trees, pillars and lampposts, in outdoor environments. A circle is fitted to each segment by minimizing the squared error of the fit. All these methods implemented in Cartesian coordinates. However, the output data of 2D lidar is usually denoted by polar coordinates. If all data processing could do in polar coordinates. It could reduce execution time and improve the robustness. Alen the first proposed a high-speed feature extraction in polar coordinates for laser rangefinders (LRF) [18]. He noted that line detection could be implemented in its natural coordinates effectively. To reduce the complexity, he applied Log-Hough transform to detect lines. Because of advances processed in polar coordinates, some other feature extraction based on this algorithm appeared [19,20]. However, there no circle detection algorithm implemented in polar coordinates for lidar data.

In this paper, we proposed a new circle detection method based on geometric property and polynomial fitting in polar coordinates for 2D lidar data. Assume circular object radius is known and there is no occlusion. We define 5 invariant and robust features in natural lidar coordinates including fitness, symmetry, estimate radius error, circularity and straightness denoted by F_{fit}, F_{sym}, F_{ere}, F_{cir} and F_{str} respectively. Then, use SVM to detect the target circular object. Three videos[1] containing 1330 frames data labeled by us are used to test the performance of the proposed algorithm. In a few words, the main contributions of this paper are presented as follows: (1) we firstly proposed 5 features in polar coordinates and proved that these features are effective and robust to detect circular objects by experiments. (2) Redefined a new simple and fast segmentation method. (3) Labeled 3 videos of scan rangefinder that can be used to test performance of algorithms.

The remainder of this paper is structured as follows. Section 2 briefly describes the principles of the proposed method and explains the 5 features

---

[1] Videos and codes download: https://yunpan.cn/cSngC6y6rjZ8I, download-code: 6242.

definition. Then, collect, label 3 videos, do some tests and analyze these experimental results. Finally, the conclusions of this paper are discussed in Sect. 4.

2   Proposed Method

Figure 1 is the flowchart of the proposed algorithm. It mainly consists of 3 parts, i.e. data segmentation, feature extraction, circular object recognition. The numbers in the small circles denote the procedures of the proposed method. For instance, the numbers, both of 1 and 2, respectively denote the segmentation and feature extraction operation. The meanings of all these numbers are noted at the top of Fig. 1. Contents including Fig. 1(a)–(d) and the corresponding procedures in the big dashed frame are implemented only in training stage. Segmentation is the first step after the acquisition stage. We improved the well known segment approach proposed by Dietmayer et al. [21] (DIET) and called it Improved Dietmayer method (IDIET). Then, pick out those valid segments whose number of pixels larger than a preset threshold. Considering the lidar data property of circular object in polar coordinates, resorting to polynomial fitting, we define and extract 5 features from valid segments. Finally, these features are fed into SVM to detect specific circular object.

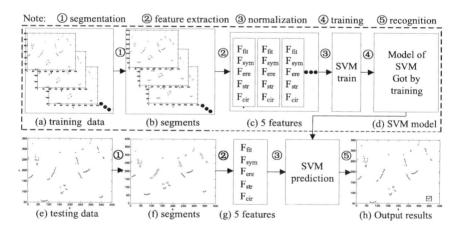

Fig. 1. Overview of the proposed circular object detection method in sensor coordinates.

2.1   Data Acquisition and Segmentation

As Fig. 2 shown, the direction of the black line between two circles is the forward direction of LRF. LRF data provided by 2D laser sensors is typically in polar coordinates and can be denoted as

$$\{(\rho(i), \theta(i))|i = 1, ..., N\} \tag{1}$$

where $\rho(i)$ is the measured distance of an obstacle to the sensor rotating axis at direction $\theta(i)$. The scan points are got by LRF with a given angular resolution $\theta(i) = \theta(i) - \theta(i-1)$. Each LRF data frame contains N scan points. Usually, to reduce the influences of systematic error and motion error, correcting the systematic error and motion error is a necessary procedure. We ignore these errors because the movement speed of the LRF and detected objects is very slow. Sometimes, both of them are motionless. So the systematic error and motion error are very low and almost have no influence to the circular objects detection.

Our 2D LRF is a RPLidar shown as Fig. 3 doing planar range scans with the minimal and maximal measurement is corresponding 150 (mm) and 6000 mm. Detail specification of RPLidar is shown in Table 1.

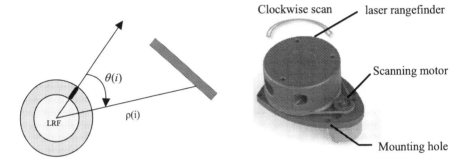

Fig. 2. LRF data in polar coordinates.   **Fig. 3.** RPLidar.

Table 1. Specification of RPLidar

Parameters	Values
Scanning distance	150–6000 mm
Scanning accuracy (150–1500 mm)	0.5 mm
Scanning accuracy (1501–6000 mm)	1 % of the actual distance
Scanning angle	360 (deg)
Angle resolution	1 deg
Time of scan	100 (ms)

One frame data of RPLidar which has not been processed is shown as Fig. 4 (a). It can be transformed to Cartesian coordinates according to $x = \rho \cos(\theta)$ and $y = \rho \sin(\theta)$ shown as Fig. 4(b).

After getting a frame of LRF data, segmentation operation is performed. Segmentation is the process of transforming the raw LRF scan points into primal groups of segments. And one segment is a set of LRF scan points in the plane close to each other and probably belonging to one single object. There are many

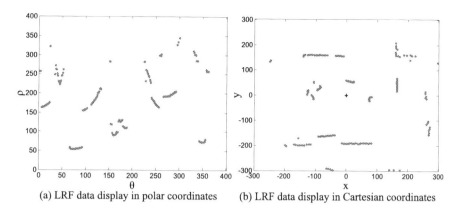

(a) LRF data display in polar coordinates   (b) LRF data display in Cartesian coordinates

Fig. 4. LRF data in polar and Cartesian coordinates. The black + in figure (b) denotes the location of RPLidar.

segmentation methods usually classified into 2 kinds: Point-Distance-Based Segmentation Methods (PDBS) and Kalman Filter Based Segmentation Methods (KFBS) [5]. We improved a PDBS method based DIET to segment because it is very simple and easy to implement in polar coordinates. The basic form of DIET is shown as follows [21]:

   if $D(\rho(i), \rho(i+1)) > D_{th}$, {LRF scan points belong to the same segment}

   else { LRF scan points belong to different segment },

where D_{th} is the threshold condition and given by

$$D_{th} = C_0 + C_1 \min(\rho(i), \rho(i+1)), \{i = 1, ..., N-1\} \qquad (2)$$

where C_0 is a constant parameter used for decreasing noise and C_1 is computed as

$$C_1 = \sqrt{2(1 - \cos(\Delta\theta))} \qquad (3)$$

and $D(\rho(i), \rho(i+1))$ written as Eq. 4 is the Euclidean distance between two consecutive scanned points.

$$D(\rho(i), \rho(i+1)) = \sqrt{\rho(i)^2 + \rho(i+1)^2 - 2\rho(i)\rho(i+1)\cos(\Delta\theta)}, \{i = 1, ..., N-1\} \qquad (4)$$

where $\Delta\theta$ is the difference angle degree between two consecutive scanned points. i.e. the angle resolution.

We improved DIET method and called the redefined method IDIET. There are two parts refinements.

(1) When the first and last of LRF scan points belong to the same segment, these points are divided into two different segments if we use DIET to segment. It may lead to the circular object can not be detected. To overcome the problem, a determination condition appended is whether the first and last scan points belong to the same segment. If the both points satisfy the decision condition, we copy the first segment to the end of last segment and merge them together.

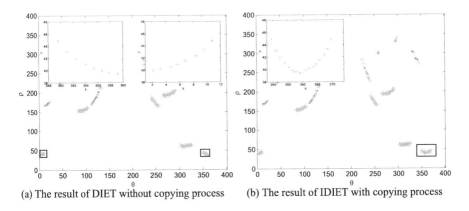

(a) The result of DIET without copying process   (b) The result of IDIET with copying process

Fig. 5. Segments obtained with copying process and without it.

Meanwhile, the first segment is marked to make it ignored in the next step since it has been copied and will be handled with the last segment as a whole. The result of IDIET without improvement and the result of IDIET modified by the skill are shown as Fig. 5. Two adjacent segments are displayed by diverse shape. These scan points in the black rectangle located at the left-bottom of Fig. 5(a) are magnified and shown as the small Figure located at the right-up of Fig. 5(a). And these scan points in the black rectangle located at the right-bottom of Fig. 5(a) are magnified and shown as the small Figure located at the left-up of Fig. 5(a). The first and last segments are merged and shown as Fig. 5(b). It's obvious that the first and last segments are from the same object, but they would be separated if the DIET is used to segment. Obviously, the refinement method can solve the problem.

(2) It is impossible that $\Delta\theta$ is exactly equal to angle resolution in reality. And if $\Delta\theta$ larger than a specified threshold, the corresponding two LRF scan points belong to different segments. So one angle threshold condition C_2 is presented as follows.

if $D(\rho(i), \rho(i+1)) > D_{th}$ or $\Delta\theta > C_2$, {LRF scan points belong to the same segment}

else { LRF scan points belong to different segment}

For one single object, the corresponding segment which is a set of scan points close to each other. In theory, the angle difference of each adjacent of two points of the segment obtained by one single object is exactly equal to angle resolution, i.e. $\Delta\alpha$. However, there are errors exist, actually. Generally, it is about equal to $\Delta\alpha$. And the corresponding difference of two adjacent scan points come from two different objects is much larger than $\Delta\alpha$. C_2 is just used to distinguish the adjacent scan points which are came from one same single object or two different objects. To reserve a certain margin, C_2 is generally set equals to 3-5 times the $\Delta\alpha$.

The difference of DIET and IDIET is shown as Fig. 6. Two adjacent segments are displayed by diverse shape. The scan points in the black rectangle region

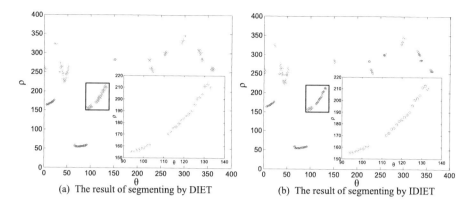

(a) The result of segmenting by DIET   (b) The result of segmenting by IDIET

Fig. 6. The results segmented by DIET and IDIET.

which are magnified and shown as in the right-bottom of Fig. 6(a) belong to the same segment when DIET is applied. When IDIET is used, these points are separated two different segments shown as Fig. 6(b). Obviously, we are more willing to get the result just like the result of IDIET.

2.2 Feature Extraction and Circular Object Recognition

There are two assumptions for the proposed method. (1) Assumption 1: Circular object radius is known. In other words, we want to detect the specific circular object and with clear radius. (2) Assumption 2: There is no occlusion. The output of LRF is sparse information. There is only small amount information about the target object. If there have any occlusion, it is too difficult to detect the target object without depending on other information. Under the condition of these two assumptions, we define 5 features in polar coordinates for circular object containing fitness, symmetry, estimate radius error, circularity and straightness denoted as F_{fit}, F_{sym}, F_{ere}, F_{cir} and F_{str} respectively.

To distinguish circle and straight line, consider line and circle and the corresponding observations shown as Fig. 7.

Firstly, consider the scan points of circle. Shown as Fig. 7 (a), all observations lie on the circle arc $\overarc{ABC}$. The actual radius of the circle and the distance between LRF and circle are denoted by Rr and $\rho_c(i)$ respectively. Consider the triangle $\triangle OO_1P_i$, where $O_1B = O_1P = R_r$, $OO_1 = O_1B + BO$. In theory, according to the law of cosines, $\rho_c(i)$ can be calculated by Eq. 5.

$$\rho_c(i) = \sqrt{(\rho_{\min} + R_r)^2 + R_r^2 - 2(\rho_{\min} + R_r)R_r \cos(\theta'(i))}, \{0 \leq \theta'(i) \leq \frac{\pi}{2} - \alpha\}$$
(5)

where the range of $\theta(i)'$ is determined by Rr and θ_{min}. i.e. $\alpha = \arcsin \frac{R_r}{R_r + \rho_{\min}}$ and $\alpha \in (0, 90)$.

For the straight line, similarly, shown as Fig. 7 (b), consider the right triangle $\triangle OBC$, the corresponding distance between LRF and line, $\rho_l(i)$, can be computed by

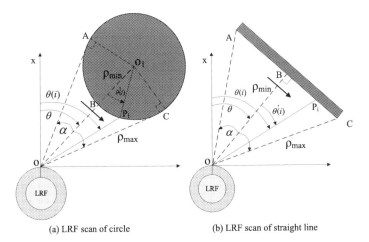

(a) LRF scan of circle   (b) LRF scan of straight line

Fig. 7. LRF scan of circle and straight line. The black circle and rectangle respectively denote circular and straight objects. LRF scan along the clockwise direction and start from the x-axis. A,B,C and P_i are observations. A and C are corresponding to the maximal distance between LRF and target object, the corresponding output of LRF are $(\rho_{max}, \theta + \alpha)$ and $(\rho_{max}, \theta - \alpha)$ respectively. B is corresponding to the minimal distance between LRF and target object, LRF output is (ρ_{min}, θ). P_i is the $i-th$ observation and the corresponding LRF output is $(\rho(i), \theta(i))$. To conveniently denote some geometric features, $\theta(i)'$ is defined and shown as in Figure (a) and (b).

$$\rho_l(i) = \frac{\rho_{min}}{\cos(\theta'(i))}, \{0 \le \theta'(i) \le \alpha\} \tag{6}$$

where the range of $\theta(i)'$ can be obtained by $\alpha = \arccos \frac{\rho_{min}}{\rho_{max}}$ and $\alpha \in (0, 90)$. Certainly, if the actual length of the straight line L_r is known. α can be got by $\alpha = \arctan \frac{L_r}{2\rho_{min}}$.

An example is taken to show the distinction of the function $\rho_c(i)$ and $\rho_l(i)$. To describe the symmetrical characteristic, coordinate origin and x-axis are redefined. For circle, O_1 and O_1B are taken as coordinate origin and x-axis respectively. And the clockwise direction is taken as the positive direction. Set $\rho_{min} = 100$, $R_r = 10$, because they satisfy Eq. (5) and $\theta(i)' \in (\alpha - \pi/2, \pi/2 - \alpha)$, the figure of this function is can be got and shown as Fig. 8(a). For straight line, O and OB are taken as coordinate origin and x-axis respectively. Set $\rho_{min} = 100$, $L_r = 26$, according to Eq. 6 and $\theta(i)' \in [-\alpha, \alpha]$, the corresponding figure could be drawn as Fig. 8(b).

Obviously, these two functions can be fitted by quadratic model described as

$$\rho_q = p_0 + p_1\theta_q + p_2\theta_q{}^2 \tag{7}$$

If the number of LRF scan points, i.e. $N(N > 3)$, are received and the corresponding polar coordinates are $(\rho(i), \theta(i))$ and $i \in [1, N]$. According to Eq. 7,

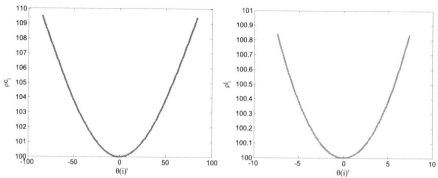

(a) The figure of distance between LRF and circle   (b) The figure of distance between LRF and line

Fig. 8. The figures of $\rho_c(i)$ and $\rho_l(i)$.

Instead of θ_q and ρ_q using all $\theta(i)$, $\rho_q(i)$. Least squares method followed to compute the parameters p_0, p_1 and p_2. $min\{\rho_q(i)|i \in [1,N]\}$, $max\{\rho_q(i)|i \in [1,N]\}$ are denoted as $min(\rho_q)$ and $max(\rho_q)$ respectively. Meanwhile, $min\{\theta_q(i)|i \in [1,N]\}$, $max\{\theta_q(i)|i \in [1,N]\}$ are denoted as $min(\theta_q)$ and $max(\theta_q)$ respectively. Then fitness could be defined as

$$F_{fit} = \frac{1}{max\,(\theta_q) - min\,(\theta_q)} \sum_{i=1}^{N} |\rho_q(i) - \rho(i)| \tag{8}$$

This feature could be used to extract circles and lines. If target object is a circle or line, the corresponding F_{fit} is a small number. Otherwise, F_{fit} is a large number.

According to the difference of theoretical calculation value and observed value. Three features used to distinguish circles from lines are defined. Two features among them denote the possibility that the corresponding object is more likely to be circle or straight line. So they are named circularity (F_{cir}), straightness (F_{str}). Another feature applied to estimate the radius size is called estimate radius (F_{er}).

For circle, shown as Fig. 7(a), consider the right triangle $\triangle OCO_1$. Two cases are:

(1) α denotes the measured value of $\angle OO_1C$ and it can be calculated by the observation degrees of the both scan points B and C in LRF nature coordinates. α_{ct} denotes the theoretical calculation value of $\angle O_1OC$ and it is computed by

$$\alpha_{ct} = \arcsin \frac{R_r}{\rho_{min} + R_r} \tag{9}$$

The difference of α_{ct} and α is defined circularity formulized as

$$F_{cir} = |\alpha_{ct} - \alpha| \tag{10}$$

(2) R_{est} denotes the theoretical calculation value of the circle radius. Consider the right $\triangle OO_1C$ in Fig. 7(a), and it can be obtained as

$$R_{est} = \frac{\rho_{\min} \sin \alpha}{1 - \sin \alpha} \tag{11}$$

And the estimate radius error (F_{ese}) is defined as

$$F_{ese} = |R_r - R_{est}| \tag{12}$$

In theory, for a circle, F_{cir} and F_{ese} are equal to zero. In practice, there are measurement errors, so F_{cir} and F_{ese} are not exactly equal to zero, but are still close to zero. If the detected object is not a circle, F_{cir} and F_{ese} are large values.

For straight line, shown as Fig. 7(b), consider the right triangle $\triangle OBC$ and define the theoretical calculation value of $\angle BOC$ denoted α_{lt} computed by

$$\alpha_{lm} = \arccos \frac{\rho_{\min}}{\rho_{\max}} \tag{13}$$

Then, F_{str} can be defined as

$$F_{str} = |\alpha_{lm} - \alpha| \tag{14}$$

As similar as the situation of circle, if the detected object is a straight line. F_{str} is a small value and it is still close to zero too. Otherwise, F_{str} is a large value. So circle and straight line can be recognized in polar coordinates resorting to F_{cir} and F_{str}.

The last feature is defined symmetry denoted by F_{sym} because that all LRF scan points locate on the circle arc $\overgroup{ABC}$ and are asymmetrical about the straight line OO_1 shown as Fig. 7(a). Generally, symmetry level is measured by the Hausdorff distance [1]. However, in this paper, F_{sym} is simply computed by

$$F_{sym} = \left| \frac{\max(\theta_q) + \min(\theta_q)}{2} - \theta_{fit} \right| \tag{15}$$

where θ_{fit} is the axis of symmetry of Eq. 7 and $\theta_{fit} = -\frac{p_1}{2p_2}$.

To verify the validity of these 5 features, we use RPLidar to collect data, segment and pick out the valid segments, then extract features. The 5 features of 20 segments of circle and other shapes are shown as Table 2. The 2–th column denotes the serial number (SN) of these 20 features. Obviously, these features could be used to distinguish circle and other shapes effectively.

After calculating all features of each valid segment, group them as a feature vector and normalization. And the normalization method abides the following two rules: (1) set F_{ere}/R_r, $F_{cir}/100$ and $F_{str}/100$, other features remain the same. (2) Followed the first step, set the feature to zero if its value is lower than zero. Set it to one if its value is larger than one. Then, SVM is applied to identify whether the detected object is a circular object or a straight line or other shapes. Here we choose the standard C-SVM with a radial-base function as the kernel function. Before the recognition, use the manual labeled training data to generate the model of SVM.

Table 2. Five features of circle and other shapes

Parameters	SN	F_{fit}	F_{sym}	F_{ere}	F_{cir}	F_{str}
Circle	1	0.19	0.02	1.02	0.79	17.02
	2	0.19	0.05	0.62	0.48	18.25
	3	0.20	0.01	1.24	0.97	16.89
	4	0.30	0.08	0.58	0.45	19.21
	5	0.30	0.06	0.76	0.59	19.11
	6	0.29	0.05	0.91	0.71	18.84
	7	0.25	0.06	0.67	0.52	18.64
	8	0.19	0.03	1.04	0.81	17.52
	9	0.36	0.01	1.49	1.04	17.52
	10	0.26	0.02	0.84	0.68	18.89
Other shapes	11	2.94	0.04	19.37	4.14	18.21
	12	0.43	0.77	58.58	14.44	0.09
	13	0.79	0.48	149.18	26.09	8.90
	14	2.90	0.23	9.42	4.24	7.12
	15	0.64	0.42	3.20	1.33	11.86
	16	1.88	0.12	540.67	46.75	0.20
	17	0.35	0.20	288.12	38.35	3.41
	18	1.02	0.35	13.72	2.36	5.89
	19	0.24	0.32	18.25	4.54	0.95
	20	0.19	0.05	10.02	8.76	6.09

3   Experiments and Analysis

We captured 3 videos containing 1330 frames data by RPLidar, labeled circular objects and used the labeled data to evaluate the performance of the proposed algorithm. The proposed method is implemented with Matlab2010R and the experiments are conducted on a ThinkPad E430 with a Intel(R) Core i3-3110M 2.4 GHz CPU and 4 GB RAM. And we implement SVM using the libsvm. The performances of the proposed method are evaluated by 3 parameters i.e. execution time (T_{exe}), accuracy (P_{acc}) and recall (P_{rec}). Texe denotes the required time starting from the segmentation and to output the detection result and it is measured by milliseconds. A plastic cast whose radius is 110 m and the corresponding size in polar space Rr = 11 and a rectangle paper box are the detected objects. The number of frames in a video is denoted by N_{fra}. Assume we have got the frames containing no-occlusion target circular (N_{NO}), part-occlusion (N_{PO}), complete-occlusion (N_{CO}). And for each frame, only one circular object with in polar space $R_r = 11$ exist. For the no-occlusion frames, the proposed method detects Ncos circular object correctly in one frame containing the circle and other shapes. There are only N_{cir} frames that just containing one target

Table 3. The consist of 3 videos

Video	N_{fra}	N_{NO}	N_{CO}	N_{PO}
Video1	375	375	0	0
Video2	481	468	0	13
Video3	474	426	18	30

circle just detected correctly. And for complete-occlusion frames, N_{com} frames are detected correctly i.e. null target circle is found in the frame. Then P_{acc} and P_{rec} are defined as $P_{acc} = \frac{N_{cir}+N_{com}}{N_{NO}+N_{CO}}$ and $\mathrm{P}_{rec} = \frac{N_{cos}}{N_{NO}}$. There are 4 steps for these experiments as follows.

Firstly, resorting to RPLidar, we collect testing data consist of 3 videos contained 1330 frames in total and each frame is saved in txt file format. The first, second and third testing data denoted by Video1, Video2 and Video3, are respectively consisted of 375, 481 and 474 frames. Detailed information about these videos is shown as Table 3.

For Video1, RPLidar placed on a movement robot and the target plastic cask remains stationary. For others, RPLidar keeps static, the plastic cask or the rectangle paper box keep moving. The plastic cask and the rectangle paper box are respectively considered as circle and line for the LRF. Secondly, all testing data is manually labeled. Use the IDIET method to segment each frame and pick out those segments with more than 3 scan points as valid segments. Plotting all these valid segments, manually find the circular object and store the corresponding order of the segment in the excel file format. For the same video, all the orders of circular object save in the same excel file. There are 3 videos, so 3 excel files are got to store all labeled data. Then, one of the labeled video are chosen as the training data and fed into SVM. Because we use the standard C-SVM with a radial-base function as the kernel function. There are two parameters i.e. the penalty parameter of the error term and kernel parameter respectively denoted by c and g. To determine the optimal parameters for training, we employ a parameter selection tool, i.e. grid.py, provided by libsvm. It uses cross validation technique to estimate the accuracy of each parameter combination in the specified range and helps us to decide the best parameters. The model of SVM is got and saved. Finally, the model and other 2 videos are taken as the input parameters of SVM. Repeat the third and last step, until all the 3 videos have been feed into SVM for training model.

Table 4 shows the results of the proposed circular object detection method on 3 testing videos. It can be found that the method achieves an accuracy rate of 92.53 % by using different videos to train SVM. The accuracy is not very high but the best recall reaches up to 100 %. So we can combine the proposed method with other features or use other better classification methods to further optimize the results in the future. The average time processed one frame is about 16 ms in the experiments. In other words, the implemented algorithm performs calculations at above 50 HZ which is currently much higher than the LRF data sampling rate. In one word, the proposed circular object detection method implemented in polar

Table 4. The results of the proposed method

Training data	Testing data	$P_{acc}(\%)$	$P_{rec}(\%)$	N_{cir}	N_{com}	N_{cos}	$T_{exe}(ms)$
Video1 (c: 512, g: 8)	Video1	98.40	100.00	360	0	375	16.93
	Video2	97.86	99.36	458	0	465	14.54
	Video3	98.20	98.36	418	18	419	14.56
Video2 (c:512, g: 8)	Video1	92.53	94.93	347	0	356	15.93
	Video2	99.79	99.79	467	0	467	14.64
	Video3	96.85	97.42	412	18	415	14.38
Video3 (c: 2, g: 8)	Video1	92.00	97.60	345	0	366	16.08
	Video2	99.36	99.79	465	0	467	14.62
	Video3	99.32	99.53	423	18	424	14.26

coordinates is very simple, fast and effective. And it can be used in various fields, such as movement robot navigation, SLAM and other application too.

4   Conclusions

This paper presented a new circle detection method processing the measurements in polar coordinates and given a novel perspective to implement circular object detection without using any temporal information. Five features are defined in polar coordinates for each segments obtained by the proposed IDIET method. And then these features are fed into SVM to detect circular objects. The proposed algorithm evaluated on actual sequences from a 2D scanning laser. The average execution time is about 16 ms. The accuracy of the proposed method is higher than 92.53 %. The results demonstrate that the proposed method is effective and robust to detect circle in polar coordinates. However, when the circular object has partial occlusion, the accuracy would decrease dramatically. In the future, we will consider choose better classification method or combine it with tracking algorithm such as Kalman Filter to improve accuracy and overcome the partial occlusion problem.

Acknowledgments. This work was supported in part by the National Natural Science Foundation of China under Grant (No. 61175075, No. 61573134), the National Science and technology support program (No. 2015BAF11B01).

References

1. Huang, Y.H., Chung, K.L., Yang, W.N., Chiu, S.H.: Efficient symmetry-based screening strategy to speed up randomized circle-detection. Pattern Recogn. Lett. **33**(16), 2071–2076 (2012)
2. Hough, C.: Method and means for recognizing complex patterns. US Patent 3,069,654 (1962)
3. Yuen, H., Princen, J., Illingworth, J., Kittler, J.: Comparative study of Hough transform methods for circle finding. Image Vis. Comput. **8**(1), 71–77 (1990)
4. Chernov, N., Lesort, C.: Least squares fitting of circles. J. Math. Imaging Vis. **23**(3), 239–252 (2005)

5. Chaudhuri, D.: A simple least squares method for fitting of ellipses and circles depends on border points of a two-tone image and their 3-D extensions. Pattern Recogn. Lett. **31**(9), 818–829 (2010). LNCS: Authors Instructions 9

6. Chen, T.C., Chung, K.L.: An efficient randomized algorithm for detecting circles. Comput. Vis. Image Underst. **83**(2), 172–191 (2001)

7. Zhang, H., Wiklund, K., Andersson, M.: A fast and robust circle detection method using isosceles triangles sampling. Pattern Recogn. **54**, 218–228 (2016)

8. Wang, X., Cai, Y., Shi, T.: Road edge detection based on improved RANSAC and 2D LIDAR data. In: 2015 International Conference on Control, Automation and Information Sciences (ICCAIS), pp. 191–196 (2015)

9. Zhang, S., Adams, M., Tang, F., Xie, L.: Geometrical feature extraction using 2D range scanner. In: Proceedings of 4th International Conference on Control and Automation, ICCA 2003, pp. 901–905 (2003)

10. Feng, X., He, Y., Huang, W., Yuan, J.: Natural landmarks extraction method from range image for mobile robot. In: 2nd International Congress on Image and Signal Processing, CISP 2009, pp. 1–5 (2009)

11. Zhao, Y., Chen, X.: Prediction-based geometric feature extraction for 2D laser scanner. Rob. Auton. Syst. **59**(6), 402–409 (2011)

12. Premebida, C., Nunes, U.: Segmentation and geometric primitives extraction from 2D laser range data for mobile robot applications. Robotica **2005**, 17–25 (2005)

13. Gander, W., Golub, G.H., Strebel, R.: Least-squares fitting of circles and ellipses. BIT Numer. Math. **34**(4), 558–578 (1994)

14. Lee, J.H., Tsubouchi, T., Yamamoto, K., Egawa, S.: People tracking using a robot in motion with laser range finder. In: 2006 IEEE/RSJ International Conference on Intelligent Robots and Systems, pp. 2936–2942 (2006)

15. Arras, K.O., Mozos, M., Burgard, W.: Using boosted features for the detection of people in 2D range data. In: 2007 IEEE International Conference on Robotics and Automation, pp. 3402–3407 (2007)

16. Weinrich, C., Wengefeld, T., Volkhardt, M., Scheidig, A., Gross, H.-M.: Generic distance-invariant features for detecting people with walking aid in 2D laser range data. In: Menegatti, E., Michael, N., Berns, K., Yamaguchi, H. (eds.) Intelligent Autonomous Systems. AISC, vol. 13, pp. 735–747. Springer, Heidelberg (2016)

17. Inostroza, F., Leung, K.Y., Adams, M.: Semantic feature detection statistics in set based simultaneous localization and mapping. In: 2014 17th International Conference on Information Fusion (FUSION), pp. 1–8 (2014)

18. Alempijevic, A., Dissanayake, G.: High-speed feature extraction in sensor coordinates for laser rangefinders. In: Proceedings of the 2004 Australasian Conference on Robotics and Automation, pp. 1–6 (2004)

19. Lherbier, N., Fortin Noyer, J., Lherbier, R., Fortin, B.: Automatic feature extraction in laser rangefinder data using geometric invariance. In: IEEE Forty Fourth Asilomar Conference on Signals, Systems and Computers, pp. 199–203 (2010)

20. Fortin, B., Lherbier, R., Noyer, J.C.: Feature extraction in scanning laser range data using invariant parameters: application to vehicle detection. IEEE Trans. Veh. Technol. **61**(9), 3838–3850 (2012)

21. Dietmayer, K.C., Sparbert, J., Streller, D.: Model based object classification and object tracking in traffic scenes from range images. In: Proceedings of IV IEEE Intelligent Vehicles Symposium (2001)

Intensity Estimation of the Real-World Facial Expression

Yan Gao[(✉)], Shan Li[(✉)], and Weihong Deng[(✉)]

Beijing University of Posts and Telecommunications, Beijing 102209, China
gaoyanne@gmail.com, Queenie3@live.com,
whdeng@bupt.edu.cn

Abstract. Affect computing or Automatic affect sensing has aroused extensive interests of researchers in the area of machine learning and pattern recognition. Most previous research focused on face detection and emotion recognition while our research explores facial intensity estimation, which cares more about the dynamic changes on a face. CK+ database and Real-world Affective Face Database (RAF-DB) are used to test and implement the algorithms in this paper. To settle the problem of intensity estimation, classification and ranking algorithms are used for training and testing intensity levels. Meanwhile, the performance of five different feature representations is evaluated using the accuracy results obtained from classification approach. By using the optimum feature representation as the input to the next designed training model, ranking results can be attained. Techniques of Learning to Rank in the area of information retrieval are utilized to combat the situation of intensity ranking. RankSVM and RankBoost are used as frameworks to estimate the ranking scores based on sequences of images. The experimental results of scoring are evaluated by the indexes used in information retrieval. Algorithms used in the research are well organized and compared to generate an optimal model for the ranking task.

Keywords: Intensity level · Learning to rank · RankSVM

1 Introduction

Researchers have developed very effective and successful algorithms and systems to solve the problem of affect computing and also built widely spread databases (most available for non-commercial use), such as CK, MMI, etc. In our research, we use the CK+ database as a representation of lab-based and posed databases. Images of this kind of databases are collected under controlled lab conditions. Human pose deliberately in front of the camera under the instructions and the illumination, head poses, affect elicitation method and diversity of subjects are all controllable. As a result, the results that generated using these databases may be limited. On the contrary, images exist in the real world vary in age, gender, ethnicity, illumination condition, image resolution and even present complex face gesture and ambiguous boundary of emotions. RAF-DB (Real-world Affective Face Database), collected and established by Shuo Zhang et al. of PRIS lab of BUPT, is also used in our research. This database contains roughly

© Springer Nature Singapore Pte Ltd. 2016
T. Tan et al. (Eds.): CCPR 2016, Part I, CCIS 662, pp. 79–92, 2016.
DOI: 10.1007/978-981-10-3002-4_7

30,000 real-world facial images posted on the social network Flickr and we have added thousands of facial images downloaded from Instagram to the database.

Facial expression intensity estimation is an extension of emotion recognition. Intensity is more about the dynamics of the facial movement and subtle changes of the face muscle. Estimating the intensity of facial expressions and getting scoring results are the ultimate objective of this research. Instead of considering all of the categories of emotions, we just focus on the happy/joy emotion to conduct our research. With the application of CK+ and RAF-DB, we design learning models to obtain expected results and evaluate these applied methods. We conduct several experiments to explore characteristic of used datasets, different feature representations and designed algorithms. To combat the situation of receiving intensity score, we use both classification methods and learning to rank (L2R) methods.

The remainder of this report is structured as follows. In Sect. 2, we give preliminaries. An overview of previous facial recognition model (related to intensity) is summarized to illustrate previous works in the area of facial intensity estimation. A description and detailed statistics of the two databases used in the research are also presented. In Sect. 3, four types of feature extraction are presented and a brief introduction of dimensionality reduction is also presented. In Sect. 4, a framework of the Facial Intensity Estimation System is demonstrated. In Sect. 5, experimental results of three-level model and L2R model are presented and evaluated.

2 Preliminaries

2.1 Previous Works

In previous works, methods to settle the problem of intensity estimation include classification-based method, regression-based method and boosting method. Most frequently used classification method is Support Vector Machine (SVM). Former researchers like Bartlett et al. calculated the distance from the testing objects to the SVM separated hyperplane to estimate the level of intensity. The longer the distance is, the higher the intensity will be. Specifically, Mahoor et al. [3] settled the problem using 6-class classification with binary classifiers. He evaluated the AU intensity from several representative action units of infant facial images. However, the class overlap among classes cannot be avoided, which makes these approaches to be vulnerable. Recently in 2013, a database called DISFA [5] was established by Mavadati et al. The authors evaluated thirteen action units for intensity classification. It is a through and huge (16.5 GB) database based on video analysing.

In the study of Delannoy et al. [1], an image-based classification approach was developed for the three levels (low, medium, and high) expression intensity estimation. However, they have not well employed the ordinal relationship among labels. Therefore, Chang et al. [4] enhanced the performance using their own-improved RED-SVM approach, which combined ordinal regression. Yang et al. used the method of Rank-Boost to do intensity estimation and met good results. So it came to me that the method of learning to rank should be another approach to settle the problem of classification.

The approach of L2R can improve the shortage of multi-class classification approach, which overlooks the ordinal consistency of the predicted intensities.

Table 1 lists recent affect analysis systems related to intensity estimation by categorizing them in terms of reference, model and their performance. In our research, we apply SVM, Adaboost and learning to rank (L2R) methods to do the task.

Table 1. Overview of automatic affect recognition systems (related to intensity)

Reference	Model	Performance	Registration	Representation
Delannoy et al. [1] '08	SVM	CK	AAM	Facial points
Yang et al. [2] '09	RankBoost	CK	2p	Dynamic Haar
Mahoor et al. [3] '09	SVM	CK	AAM	LPI
Chang et al. [4] '13	RED-SVM	CK+	AAM	Gabor
Mavadati et al. [5] '13	SVM	DISFA	AAM	LBP, HOG, Gabor
FERA baseline [6] '15	SVM, SVR	BP4D, SEMAINE	AAM	LGBP, Geometric

2.2   Expression Databases

The RAF-DB was built by the efforts of the PRIS lab of BUPT for facial expression analysis in 2014–2016. Images were collected from the social network - Flicker and Instagram. The image sequences of 7 kinds of emotions in the database were first imbalanced. The amount of neutral and happiness were the most and the amount of fear and disgust were the least. We then equilibrated the amount of emotions in the database to solve the imbalance condition by adding emotion categories in minority. The addition to the database equilibrated the unbalance of the number of seven emotion sequences (six basic emotions and neutral) to form a near-balanced database.

Some examples of RAF-DB (aligned and of happy category) are depicted in Fig. 1. It can be seen that the collected images are of great diversity in subjects' age, gender, head poses, illumination conditions and image resolutions. This diversity increases the difficulty of training but on the other hand reflects the condition of real-world environment. RAF-DB is more suitable and feasible for real-world facial expression analysis.

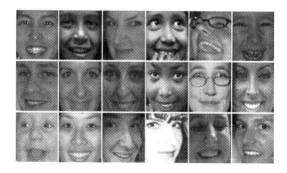

Fig. 1. Example images from RAF-DB of happy category

It is assumed that each category of emotions is independent and the internal variation of each emotion is to be considered. Figure 2 depicts the variation and the amount of facial images (aligned) of each category of emotions in RAF-DB. Figure 3 depicts the emotion proportion in CK+ database and RAF database. The bars in light orange represent the category of joy. It can be seen that the joy images hold 22 % of the whole image pools in CK+ database. There are total 327 sequences that have emotion labels in CK+ database while each sequence consists roughly 20–50 images. The amount of joy category in RAF-DB occupies 18 % of the whole image group, which is in the number of 4607. These images of joy category are used for training and testing in the research.

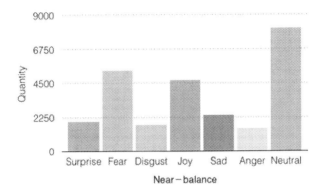

Fig. 2. Distribution of emotions of RAF-DB

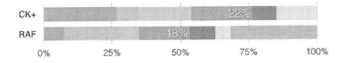

Fig. 3. Comparison of emotion proportion between databases

The amount of images of RAF-DB is largely more than CK+ database. It is known that the more of training subjects, the higher of accuracy is. Different from CK+ whose subjects have consecutive several images of same persons, the subjects in RAF database are independent from each other. Considering the regularity of CK+ and diversity of RAF-DB, it may require more training subjects for RAF-DB to get the same accurate results as CK+.

3 Feature Extraction

3.1 Feature Visualization

In our research, four kinds of feature representations are employed as the inputs to the next step of facial intensity estimation task. Features including LBP, HOG and Gabor are extracted as low-level information and used in the training task of building the model while the facial points are used in the face alignment task. They perform differently when applying to distinct classification or ranking algorithms, which will be proved in the experiments. Figure 4 shows these four kinds of feature representation as an illustration of their extraction process and representation principle.

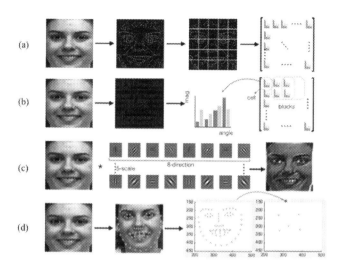

Fig. 4. Feature Representations (a) LBP; (b) HOG; (c) Gabor; (d) Facial points

3.2 Dimensionality Reduction

Dimensionality reduction technique is often used after feature extraction process. It can be expressed as the problem of getting optimal representation of the data vectors in a lower-dimensional subspace.

Principal component analysis (PCA) is a highly recommended algorithm, which is often applied to reduce the complexity of dimensionality when analyzing facial images into a lower dimension space. The idea of doing reduction is the technique of using an orthogonal projection base, which is a process of de-correlation.

Discrete Cosine Transform (DCT) is another useful technique in the area of image compression. The cosine transform, like the Fourier transform, uses a sinusoidal basis functions. An obvious difference between two kinds of transforms is that the base functions of cosine transform are not complex.

In comparison with PCA, DCT is computationally less burdensome. The idea of using DCT is a zigzag matrix transformation. First the facial image is shifted into the DCT space. Next an inverse transformation is conducted to realize the purpose of dimensionality reduction. The transformation is about deserting coefficients of heist frequencies respectively. Different from PCA, which decomposes the covariance matrix using eigenvalue, DCT approach is not data-dependent. This is the reason that the computation magnitude of DCT is less than PCA.

4   Facial Intensity Estimation Framework

Combining the feasibility and performance of previous models, we work out a whole process of Facial Intensity Estimation (FIE) system (show in Fig. 5). Considering the volume of data and processing speed, the 3D data and video data are ignored; instead we only employ image data as the analyzing objects. Moreover, the emotion detection process is negligible in our research because it is assumed that the emotion category is pre-known before estimating the intensity. The purpose of focusing on specific emotion intensity estimation is to assure the continuity and accuracy of using feature representations. When aligning facial images, facial points features are used to localize the position of face. Specifically, five points (center of eyes, nose and corner of lips) are used for localization. Three types of feature representations (LBP, HOG, Gabor) are applied as the input of training models. Dimension reduction techniques are also used to reduce the extensive dimensionality of Gabor in order to save time and speed up. After getting the feature vectors, the algorithms designed for intensity estimation are implemented. In general, four algorithms are used to form corresponding training models, shown in Fig. 5. SVM and Adaboost are used as three-level intensity classification while RankSVM and RankBoost as L2R methods.

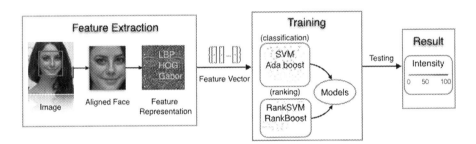

Fig. 5. Flowchart of FIE system

4.1   Three-Level Intensity Classification

Most previous method used in Table 1 is SVM, which is a widely used and classical algorithm for classification. In our research, this algorithm is also applied for intensity classification. First, we extend the application of SVM from expression emotion classification to intensity classification results. In the training phase, we employed the category-label data as inputs to multi-classifiers. The estimated intensity result is directly expressed as a value according to the distance from the testing objects to the classification hyperplanes. In SVM, the distances to the boundary of SVM are used to evaluate the level of intensities. In the research, we divide the intensity estimation results into three levels (light, medium and strong). The problem is regarded as a multi-class classification issue and one-against-all SVM is applied to training classifiers. The three intensity levels are considered independent to each other. However, the result may be inaccurate with some errors because it is not necessary just assign the distance to the SVM boundary as the expression intensity. Meanwhile, the algorithm of Adaboost is also applied for classification.

The classification is conducted using CK+ database, whose advantage is that facial images of each subject are contingent, thus the intensity labels are easy to be annotated. Intensity levels are defined as light (near-neutral), medium (onset) and strong (apex). We take 9 images from each happiness sequences within 327 sequences to construct the training set and testing set, whose labels are annotated manually. When representing the intensity labels as ranking orders, only the relative order between labels is considered. That is, it is not necessary to concern what the exact value of the intensity difference is, but pay attention to which expression is more intense.

We conduct the experiment using both SVM and Adaboost for classification. The performances of feature representations are also evaluated. The SVM method uses multi-class classifier and in one-versus-one approach. The Adaboost method uses 256 boosting rounds.

4.2   Intensity Estimation of L2R Method

The classification method is classical and basic. It can be noticed from Table 1 that RankBoost was also used as a method of intensity estimation. We continue this line to apply the learn-to-rank approach (usually used in information retrieval) into expression intensity estimation. As far as we know, no one has extended the application of RankSVM into the subject of facial intensity estimation. As a result, we implement L2R method into the research to attain intensity scores that can be expressed from 0 to 100. These score results are obviously an extension of three-level model in terms of range and accuracy. Moreover, the estimation indexes to estimate the performance of information retrieval can also be applied to estimate the performance of intensity estimation results.

This L2R model is conducted both in CK+ and RAF-DB. Since we are considering the relative intensity, images of one specific sequence in CK+ are selected to construct a training subset. So the quantity of training is limited (43 images). But the arrangement from neutral to apex is quite accurate and well ordered. On the contrary, the training subset of RAF-DB is huge in quantity, but images are collected from different subjects, which are totally independent from each other. This increases the difficulty of annotation and the possibility of errors.

5   Experimental Results

5.1   Three-Level Intensity Results

We test the SVM and Adaboost algorithms for classification and conduct five types of feature representations: LBP, HOG, Gabor, gabPCA (Gabor + PCA) and gabDCT (Gabor + DCT) in each algorithm. Intensity is divided into three levels and represented as labels from 1 to 3 (light, medium and strong) for classification. Table 2 presents the dimension of each feature representation and their corresponding training time when doing multi-class SVM classification. The dimensionality of Gabor representation is

Table 2. Dimension and training time using various features

	LBP	HOG	Gabor	GabPCA	GabDCT
Dimension	2891	4356	10000	300	300
Time	0.410471	0.359895	0.226641	0.007636	0.009406

the highest and can be compressed to 300 dimensions (self-defined) after dimension reduction. As a result, the dimensions of GabPCA and GabDCT are less than others in a large extent. It is obvious that this evident gap in dimensions leads to a sharp decline in training time (roughly 40 times), which is real useful when estimating large data volume and can avoid the problem of out of memory.

Figure 6 illustrates the accuracy of SVM and Adaboost when using six feature representations respectively. It can be seen that the general performance of SVM is superior to Adaboost. On the aspect of feature representation analysis, LBP and gabPCA are fair to each other and perform the best both using SVM and Adaboost. Considering the factor of training time, gabPCA is more appropriate to conduct the experiment. HOG and Gabor representation perform just opposite using SVM and Adaboost. Considering the dimension reduction technique, gabPCA performs pretty much better than gabDCT. Therefore, although PCA and DCT can both carry the point of quick compression, the effect of PCA is the optimum.

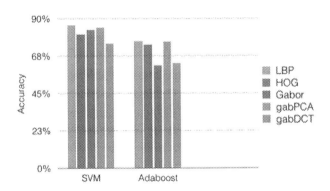

Fig. 6. Accuracy of SVM and Adaboost classification

5.2 L2R Results

A. Ranking on CK+

When training the ranking algorithms on CK+ database, 43 facial images corresponding to intensity labels from 1 to 43 of the same subject are trained, which extends the intensity scale to a large extent comparing with three-level model. So the method is more sensitive to the response of little. Figure 7 compares the testing performance of using SVM and RankSVM with respect to the true intensity of 9 facial images in a sequence to show the improvement. It can be seen that the intensity curve of

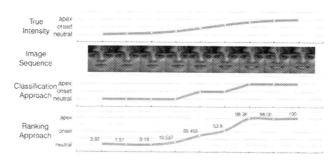

Fig. 7. Performance of SVM and RankSVM relative to true intensity

classification approach is completely discrete and the curve of ranking approach is nearly continuous. Since RankSVM is a pairwise method, the intensity labels are converted to a sparse matrix that demonstrates the relative relation with [1, −1] between each pair of the training subject. By using RankSVM, we can get the strength grade of each dimension of feature vectors. For example, strength grade of each dimension of total 2891 dimensions can be obtained using LBP representation. Then multiplying the grade vector with corresponding feature vector matrix generates the intensity score of each testing subjects. In order to present the scores, we normalize the intensity scores and stretch them to the scale from 0 to 100. Experimental results using RankSVM can be shown on Fig. 7. It can be seen that although the rising trend of using SVM and RankSVM are the same with true intensity, but the intensity curve of ranking approach is continuous and the intensity estimation scores are more precise being able to distinguish the minute happiness variation between adjacent images.

B. Ranking on RAF-DB

We also train the ranking algorithms on RAF-DB. Facial images of RAF database are annotated according to the judgment of 42 persons from the PRIS lab and the intensity probabilities of seven emotions are calculated from dividing the number of reliable annotation by the total reliable annotation. We filter out all the facial images with most votes of joy and use this subset to conduct the experiment. After the collection of experimental subjects and model training, we test the models on 99 testing subset images. We select some examples from the results using RankSVM and demonstrate them in Fig. 9. It can be seen that the ranking method works well when testing on RAF-DB in which facial images are various in age, gender, ethnicity and illumination conditions existing in the real world. The predicted scores vary from neutral to apex and they are approximately consistence with the true intensity that we can observe in naked eyes. Note that actually the intensity score is evaluated in relative order. Facial images with higher score are happier than images with lower score. The scores are not independent because the model being trained from relative intensity labels tests them. However, the scores can be fairly used as a judgment of intensity estimation results.

When conducting RankBoost method, we use 50 boosting rounds to build the model – enough but not costing. Figure 8 shows the distribution of intensity scores of

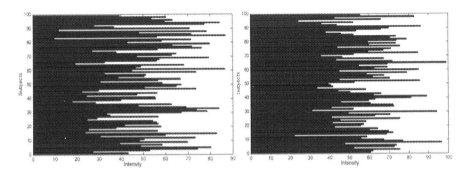

Fig. 8. Intensity distribution of testing subjects (RankSVM on the left, RankBoost on the right)

99 testing subjects using RankSVM and RankBoost respectively. Scoring results demonstrated in Fig. 9 are selected from this distribution of 99 testing objectives. When doing comparisons horizontally, it can be seen that disparity exists considering the intensity levels of the same testing subset. But when paying attention on specific intensity score (especially the peak values and bottom values), the relative magnitudes are consistent comparing with adjacent intensity scores.

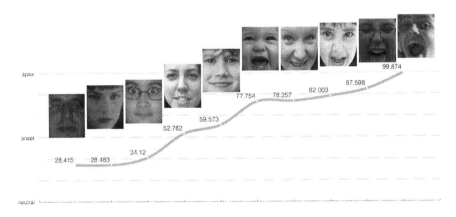

Fig. 9. Performance of RankSVM on RAF-DB

C. Evaluation

In ranking problems, the models are evaluated based on the relevance. In analogy with document retrieval, the facial images are treated as documents and labels of intensity are treated as different relevance. Higher value of intensity label refers to higher relevance. Mean Average Precision (MAP) and Normalized Discount Cumulative Gain (NDCG) are both evaluation indexes for ranking that are usually used in the field of information retrieval. Here I use them for intensity estimation.

The measurement of $P@n$ is about evaluating the top n documents of the ranking list given a pre-defined query. In the problem of intensity estimation, top results are considered as correctly ranked intensity labels. So $P@n$ is defined as:

$$P@n = \frac{\# \, correctly \, ranked \, intensity \, labels \, in \, n \, results}{n} \quad (1)$$

Mean Average Precision (MAP) reflects the degree of precision in single query. The more relevant the document is, the level of ranking is higher. MAP here refers to AP and is defined as:

$$AP = \frac{\sum_{n-1}^{N}(P@n * rel(n))}{\# total \, images} \quad (2)$$

The notation of $rel(n)$ is a binary function:

$$rel(n) = \begin{cases} 1 \; if \; n^{th} \, img \, is \, ranked \, correctly \\ 0 \; otherwise \end{cases} \quad (3)$$

Note that MAP suits for the binary judgment: relevant or irrelevant. Normalized Discount Cumulative Gain (NDCG) can handle multiple levels of relevance. The level of relevance is represented as r.

$$N(n) = Z_n \sum_{j=1}^{n} \frac{2^{r(j)} - 1}{\log(1 + j)} \quad (4)$$

The notation of r(j) is the rating of the jth image in the list, and Z_n stands for normalization constant. Steps for computing NDCG can be described as: compute the gain of each image (on the base of 2), discount the gain by the ranking position, cumulate the discounted gain of the list and normalize the discounted cumulative gain.

The performance using different ranking methods on CK+ database and RAF database can be evaluated by the evaluation indexes – MAP and NDCG. Figure 10 shows the evaluation results. These judgment criteria emphasize the relative ranking

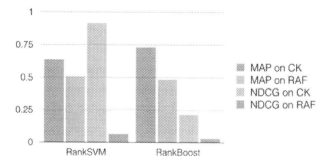

Fig. 10. Evaluations of RankSVM and RankBoost

result: $x_i - x_j$ instead of the decision accuracy. Before evaluation, sort the intensity levels from the least to the greatest. Then the ith subject is said to be accurate if its score is larger than the $(i-1)^{th}$ subject. When analyzing MAP who reflects the ranking accuracy, testing on CK+ is more effective than testing on RAF-DB. The reasons are: number of images of testing subset using RAF-DB is larger; testing sequence of facial images of CK+ come from the same person with various joy levels, rather than independent with each other in RAF-DB; manually annotated labels may be inaccurate because of the different sensing standard between people when using RAF-DB; variation patterns exist to represent the extent of joy considering the polymorphism of facial images of RAF-DB, for example, it is hard to judge the relevant happy extent between a slight open mouth and up-pulled lip corners; errors are more easily to happen using RAF-DB because of the varying illumination condition, image resolution and the possible existence of obscured properties like glasses and fingers. The evaluation of NDCG even enlarges the difference between CK+ and RAF comparing to MAP. NDCG defines several relevance levels instead binary level. NDCG is calculated using dividing the minimum DCG (real ranking index) by the DCG of predicted ranking intensity index. Moreover, the relevance levels used on RAF are quite more than that on CK+ (three levels), which also increases the gap between cumulative gain of real ranking results and predicted ranking results.

In general, the optimum solution of the facial intensity estimation problem can be summarized after doing these experiments. For both CK+ database and RAF-DB, feature extraction techniques of LBP and GabPCA perform better than other representations included in the research. Using Learning to Rank method to settle the ranking problem of facial images achieved expected effects. Comparing the performance of RankSVM and RankBoost, results obtained by the algorithm of RankSVM are superior taking consideration of the evaluation indexes.

6   Conclusion

A thorough designed Facial Intensity Estimation system with regard to real-world facial images is implemented and presented. The steps of the system include: collecting and crawling images to equilibrate RAF-DB; doing surveys about state-of-the-art databases used in the area of face detection and recognition; summarizing previous models related to intensity estimation; selecting and designing the algorithms used for implement; training models, testing results and evaluate results of experiments.

First, the classic algorithm of SVM is used to do intensity classification according to the annotated intensity labels. Employing different feature representations to the algorithm gives different classification accuracies. Among five features that we used for training and testing, the performances of LBP and GabPCA are almost the same, rank the highest. As a result, these two types of feature representations are used for further training and testing tasks. To meet the basic ranking regularity, we implemented three-level model, using which facial images can be classified with discrete labels from light, medium to strong. In order to pursue more accurate and continuous ranking results, L2R methods are implemented.

Then, the problem of intensity estimation is settled using learning to rank methods including RankSVM and RankBoost. Images of the category of happiness in CK+ and RAF-DB are mainly investigated. Because of the difference of image quantity and characteristics between two databases, the performance of two databases varies. Considering L2R algorithms, the objective of using these methods is extending the discrete intensity level of classification to a large scale of ranking scores. Although errors exist, experiments tested on CK+ and RAF-DB all prove the feasibility of ranking method. The curves of the ranking results selected from testing objects of two databases are consistent with true intensity. Moreover, when comparing the performance of two pairwise ranking algorithms, the experimental results show that the most suitable algorithm to do this task is RankSVM. Apart from previous evaluation indexes like ROC used in affect recognition, indexes (MAP, NDCG) used in the area of information retrieval are accepted to conduct the differences and performances of RankSVM and RankBoost. Actually, it is proved that the optimal designed approach for implementing FIE system is RankSVM algorithm with feature representations of LBP and GabPCA.

Challenges still exist in the research area of facial intensity estimation. It is known that the effectiveness of feature representation determines largely the accuracy of results. A challenge in the future is to expend the research subjects to non-frontal images, which requires analysis of head poses. Many researchers study head pose problem as an in-depth and whole subject of research. Moreover, another aspect that can be studied as future work is the analysis of video sequences, which is a great challenge because extensive video frames require a large amount of data to be evaluated. However, many researchers have put their eyes on this kind of data source adding the temporal information.

Acknowledgement. This work was partially sponsored by supported by the NSFC (National Natural Science Foundation of China) under Grant No. 61375031, No. 61573068, No. 61471048, and No. 61273217, the Fundamental Research Funds for the Central Universities under Grant No. 2014ZD03-01, This work was also supported by Beijing Nova Program, CCF-Tencent Open Research Fund, and the Program for New Century Excellent Talents in University.

References

1. Delannoy, J., McDonald, J.: Automatic estimation of the dynamics of facial expression using a three-level model of intensity. In: IEEE International Conference on Automatic Face & Gesture Recognition (2008)
2. Yang, P., Liu, Q., Metaxas, D.N.: IEEE rankboost with l-1 regularization for facial expression recognition and intensity estimation. In: International Conference of Computer Vision (ICCV) (2009)
3. Mahoor, M., Cadavid, S., Messinger, D., Cohn, J.: A framework for automated measurement of the intensity of non-posed facial action units. In: IEEE CVPR Workshop on Human Communicative Behaviour Analysis (2009)

4. Chang, K.Y., Chen, C.S., Hung, Y.P.: Intensity rank estimation of facial expressions based on a single image. In: IEEE International Conference on Systems, Man, and Cybernetics, pp. 3157–3162 (2013)
5. Mavadati, S., Mahoor, M., Bartlett, K., Trinh, P., Cohn, J.: DISFA: a spontaneous facial action intensity database. IEEE Trans. Affect. Comput. **4**(2), 151–160 (2013)
6. Valstar, M.F., Almaev, T., et al.: FERA 2015 - second facial expression recognition and analysis challenge. In: 2015 IEEE International Conference on Automatic Face & Gesture Recognition and Workshops (FG 2015). IEEE (2015)

An Emotional Text-Driven 3D Visual Pronunciation System for Mandarin Chinese

Lingyun Yu[1], Changwei Luo[1], and Jun Yu[1,2(⊠)]

[1] Department of Automation, University of Science and Technology of China,
Hefei, China
{yuly,luocw}@mail.ustc.edu.cn, harryjun@ustc.edu.cn
[2] State Key Laboratory for Novel Software Technology,
Nanjing University, Nanjing, People's Republic of China

Abstract. This paper proposes an emotional text-driven 3D visual pronunciation system for Mandarin Chinese. Firstly, based on an articulatory speech corpus collected by Electro-Magnetic Articulography (EMA), the articulatory features are trained by Hidden Markov model (HMM), and the fully context-dependent modeling is taken into account by making full use of the rich linguistic features. Secondly, considering the fact that the emotion is more remarkably adjusted in the articulatory domain owing to the independency in the manipulation of articulators, the differences between articulatory movements in different emotions are investigated. Thirdly, the emotional speech is generated by adjusting the speech parameters, such as fundamental frequency (F0), duration and intensity, based on Praat. Then when playing the generated emotional speech, the corresponding articulatory movements are synthesized by the HMM prediction rules simultaneously which is used to drive the head mesh model along with emotional speech. The experiments demonstrate the system can synthesize accurate emotional speech synchronized animation of articulators at phoneme level.

Keywords: Articulatory movement · Hidden Markov model · Fully context-dependent modeling · Emotional speech

1 Introduction

According to the mechanism of human speech, the movements of articulators generate the acoustic signal. Therefore, articulatory features can provide an effective and important description of speech as an alternative to acoustic representation [1,2]. The articulatory movements trajectories generated by text can be regarded as an important supplement parameters to improve and modify the characteristics of synthesized voice [3,4], which have a significant role in real life. In audio-visual system, articulatory features can be used to synthesize realistic facial animation to learn the correct pronunciation for learners [5,6], which improve the application in human computer interaction [7,8].

© Springer Nature Singapore Pte Ltd. 2016
T. Tan et al. (Eds.): CCPR 2016, Part I, CCIS 662, pp. 93–104, 2016.
DOI: 10.1007/978-981-10-3002-4_8

However, only synthesizing text-driven 3D visual pronunciation system is far from enough to satisfy the demand, it is more important to create a more realistic and emotional system. Owing to the independency in the manipulation of each individual articulator, it provides more degrees of freedom and less overlaps among them in the articulatory space [9]. So that emotions are more effectively adjusted in the articulatory domain. However, the analyses of articulatory movements in different emotions are more difficult than acoustics. Because the expression in different emotions is complex, and the recording of articulatory movements by EMA in emotional speech is relatively difficult. Nevertheless, how the movement of articulators when people speak emotional speech is still significant to build a 3D visual pronunciation system.

1.1   Related Works

In order to build an emotional 3D visual pronunciation system, the features of articulatory movements in different emotions should be investigated at first. Erickson et al. [11] find that the emotion 'anger' may involve more jaw opening, the emotion 'suspicion' involves a raising and fronting of the tongue, and 'admiration' involves a lowering and backing of the tongue. Li et al. [12] analyze totally 9 vowels in Mandarin Chinese in different emotions in details. 'Happy' has a high rising boundary tone, 'Angry' has a high falling boundary tone, and 'sad' and 'Neutral' have a low boundary tone. Besides the tongue body rises higher for 'Angry' and 'Sad' than 'Happy' and 'Neutral', the tongue shape changes more greatly for 'Sad' than other emotions. Lip protrudes for 'Sad' and 'Angry' and spreads for 'Happy'. Lee et al. [9] research the connection with velocity of articulatory movements. Angry speech shows the largest movement range and velocity of tongue tip vertical movement with jaw opening and tongue tip forwarding, while it shows smaller tongue tip velocity for sadness. Lee et al. [13] investigate that the tangential velocity and the curvature of articulatory movements trajectories follow the one-third power law. Erickson et al. [10] find 'Sad' speech has lower lip and jaw, and more tongue tip protrusion than 'Happy'.

In addition to the articulatory movements, the acoustic features in different emotions are also researched. The term is relatively mature and Murray and Arnott [14] summarize the relationship between emotions and speech parameters (Table 1 in [14]) detailedly. It is abbreviated for triviality.

1.2   Paper Contribution

In theory, the articulatory movements can be predicted by HMM, when the emotional dataset is available. In practice, the articulatory movements in different emotions is difficult to be collected by EMA. How to synthesize an emotional 3D visual pronunciation system by the articulatory movements and the corresponding emotional speech will be exploited when only neutral speech is available in the paper.

The paper mainly includes as follows: (1) The accuracy of articulatory movements predicted by HMM from Mandarin Chinese text is increased when making full use of fine-grained set of linguistic context features. (2) The emotional speech is generated by adjusting the acoustic parameters when only neutral speech is available. (3) The variation of acoustic and articulatory features, especially the articulatory movements, are investigated in different emotions, and then they are used to adjust the articulatory movements which is predicted by HMM to generate the trajectories of articulators in different emotions. (4) An accurate and realistic emotional speech synchronized animation of articulators is synthesized by the emotional speech and the trajectories of articulators.

2   Method

The overall process of an emotional text-driven 3D visual pronunciation system as follows:

Firstly, the fully context-dependent HMM is trained with the articulatory features and rich linguistic features. Secondly, the emotional speech is generated by modifying the acoustic parameters such as F0, duration and intensity. Because the duration of emotional speech are different owing to the differences between the speech rate in different emotions. In order to make the duration of emotional speech be equal to the length of the articulatory frames, the text is segmented, combining with the emotional speech, to obtain the label. The phoneme sequences are obtained from text, while the duration of each phoneme is obtained from the emotional speech. And the articulatory movements can be predicted by the HMM which has already been trained before. Thirdly, the differences between articulatory movements in different emotions are investigated and used to modify the articulatory movements to obtain the trajectories of articulators in different emotions. In the end, the trajectories of articulators, with the emotional speech, are used to drive the head mesh model to synthesize the emotional speech synchronized animation.

Each part will be described in details as follows.

2.1   A HMM-Based Text-to-Articulatory Movement Prediction

Figure 1 shows the framework of HMM-based text-to-articulatory prediction method. During training, the articulatory features with D_x dimensions are extracted from the text. Then a set of context-dependent HMMs λ is estimated to maximize the likelihood function $P(X|\lambda)$, $X = \left[x_1^T, x_2^T, x_3^T, \ldots, x_N^T\right]^T$ is the observed feature sequence $X \in R^{(3ND_x \times ND_x)}$. Here we define each observation probability as a single Gaussian distribution. N denotes the length of the sequence, $(.)^T$ represents the matrix transpose. The observation feature vector of the t frame is $x_t \in R^{3D_x}$. $x_{S_t} \in R^{D_x}$ is the static articulatory parameters, Δx_{S_t} and $\Delta^2 x_{S_t}$ are respectively the velocity and acceleration components as [3,17]:

$$x_t = \left[x_{S_t}^T, \Delta x_{S_t}^T, \Delta^2 x_{S_t}^T\right]^T \tag{1}$$

Where

$$\Delta x_{S_t} = 0.5\Delta x_{S_{t+1}} - 0.5\Delta x_{S_{t-1}} \quad \forall t \in [2, N-1] \tag{2}$$

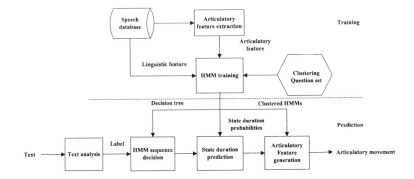

Fig. 1. The flowchart of HMM-based text-to-articulatory prediction method.

$$\Delta x_{S_1} = \Delta x_{S_2}, \Delta x_{S_N} = \Delta x_{S_{N-1}} \tag{3}$$

$$\Delta^2 x_{(S_t)} = x_{(S_{(t+1)})} - 2x_{(S_t)} + x_{(S_{(t-1)})}, \forall t \in [2, N-1] \tag{4}$$

$$\Delta^2 x_{S_1} = \Delta^2 x_{S_2}, \quad \Delta^2 x_{S_N} = \Delta^2 x_{S_{N-1}} \tag{5}$$

The articulatory movements can get by the maximum likelihood parameter generation (MLPG) algorithm [15], such as

$$X_S^* = arg \max_{X_s} P(X|\lambda) \quad = arg \max_{X_s} P(W_X X_S|\lambda) \tag{6}$$

$$= arg \max_{X_s} \sum_{\forall q} P(W_X X_S, q|\lambda) \tag{7}$$

$$[X_S^*, q] \approx arg \max_{X_s, q} P(W_X X_S, q|\lambda) = arg \max_{X_s, q} P(W_X X_S, \lambda|q) P(q|\lambda) \tag{8}$$

X can be regarded as the liner transform of static articulatory feature sequence $X_S = \left[X_{S_1}^T, X_{S_2}^T, X_{S_3}^T, \ldots, X_{S_N}^T\right]$, that means $X = W_X X_S$, where $W_X \in R^{3ND_x \times ND_x}$ is determined by the velocity and acceleration calculation function in Eqs. (2)–(5) [3], $q = \{q_1, q_2, q_3, \ldots, q_N\}$ is the state sequence of articulatory feature, According to the Eqs. (7)–(8), that means the optimal state sequence is chosen to realize the approximation via a two-step optimization. The first step is to obtain the optimal state sequence:

$$q^* = arg \max_q P(q|\lambda) \tag{9}$$

The state sequence is determined by the state duration probabilities [16]. The next step is that the optimal state sequence is fixed to obtain the optimal articulatory feature sequence:

$$X_S^* = arg \max_{X_S} P(W_X X_S|\lambda, q^*) \tag{10}$$

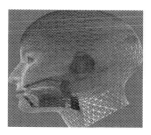

Fig. 2. The head mesh model.

2.2 3D Articulatory Animation

A 3D head mesh model of articulators (Fig. 2) including mandible, lip, tongue, palate and pharynx, skin, eyes and skeleton is used to realize highly realistic articulatory animation [1]. Next the lip animation and tongue animation are explained briefly for conciseness.

According to the principle of facial motion dynamics, the anatomical model for lip animation includes skeleton, skin and muscle. The skeleton includes skull and jaw. The skin is approximated by an elastic mesh [18] and is connected with muscle by two classes of spring to simulate the elastic of skin and forbid the skin to be split. In the end, the muscles are modeled by the Water model [19] which is used to divide the facial muscle into kinds of muscles to represent the tension and shrinkage severally [20]. In the situation, the movement trajectories of LI, LL, UL are used to drive the lip model to generate realistic animation.

The tongue model can be divided into muscle and soft tissues. The muscle is endowed with constitutive model which includes the active and passive properties of muscle fibers [21]. The constitutive model can be derived by a strain energy approach [1]. Similarly, the articulatory movements of TI, TB, TR drive the tongue model to get the animated simulation.

3 Experiments

3.1 Database

In the experiment, six sensors are located at the Tongue Rear (TR), tongue Blade (TB), Tongue Tip (TT), Lower Incisor (LI), Upper Lip (UL) and Lower Lip (LL) to record the trajectories of articulatory movements by EMA depicted in Fig. 3. Each sensor records the data in 3 dimensions: x-axis (front-to-back), y-axis (left-to-right), z-axis (bottom-to-up). Since the change in the y-axis is small, only the data in x-axis and z-axis, making a total of 12 EMA features at each frame, is used in the experiment respectively. The dataset consists of 380 sentences and the corresponding acoustic waveforms recorded by a Chinese woman. The audio data is record at 16 kHz, while the articulatory feature data is sampled at 200 Hz. Besides, another dataset, consisting of 695 emotional sentences, is used to investigate the movements of articulators in different emotions.

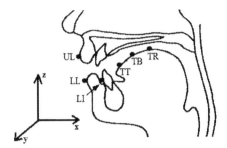

Fig. 3. Sensor adhering position.

3.2   System Construction

In order to construct the context-dependent HMM system, a model structure of 5-state, left-to-right with no skip is adopted to train HMM. The training and prediction procedures are implemented by HTS toolkits [22].

The emotional speech is generated via adjusting the acoustic features such as F0, duration and intensity by Praat [23]. Using the different duration of the emotional speech, the corresponding articulatory movements of speech are predicted simultaneously by HMM which has been trained before. Then the articulatory movements are modified based on the characteristics of articulators in different emotions. Connecting the emotional speech and articulatory movements, an accurate emotional speech synchronized animation of articulators at phoneme level is synthesized based on a head mesh model.

3.3   Articulatory Movement Predicted from Text

Fully context-dependent HMM is trained by articulatory features and rich linguistic features. For instance, the chinese sentence 'ming ling zheng zai xiu jia de quan ti zhi zhan yuan li ji fan hui suo shu bu dui, bing hao zhao suo you ai guo zhe can jia bao wei zu guo', the root mean square error (RMSE) of articulatory movements predicted from text is shown in the Table 1. Compared with the RMSE of the triphone and monophone models, that of the fully context-dependent model decreases obviously. The reason is that the fully context-dependent model takes the rich linguistic and prosodic features into account besides the phone identities used in the quinphone models.

Using the same sentence as Table 1, the trajectories of the natural and predicted articulatory movements of the TT_z are shown in Fig. 4, which shows that the trajectory predicted from fully context-dependent model is more close to that of the natural. Therefore, the performance of the fully context-dependent model with rich linguistic and prosodic features outperforms the triphone model, and our result precedes the result in [1].

Table 1. The RMSE of different articulators respectively.

	Monophone	Triphone	Fully context-dependent
The average RMES of all the articulators	2.3631	2.2241	1.8509
TR_x	2.2111	2.0299	1.8444
TR_z	2.4316	2.1546	1.5204
TB_x	2.3226	2.2692	1.9821
TB_z	2.3209	2.1205	1.9804
TT_x	3.0845	2.9221	2.4575
TT_z	2.6938	2.5579	1.8752
LI_x	1.9713	1.8651	1.5512
LI_z	1.9647	1.8815	1.4322
UL_x	1.2123	1.1639	1.0377
UL_z	2.5093	2.3917	2.4423
LL_x	2.4617	2.3169	1.7898
LL_z	2.6624	2.5315	1.8135

3.4   The Emotional 3D Visual Pronunciation System

The Position of Articulators in Different Emotions. To synthesize the emotional 3D visual pronunciation system, we firstly need to analyze and summarize the variation of the articulatory movements by the emotional dataset to observe whether it fulfills the similar rules introduced in Sect. 1.1. For example, we analyze the word 'nv hai' (girl), which is extracted from sentences in different emotions. The differences between articulators in different emotions like 'Neutral', 'Happy', 'Sad' and 'Angry' are compared in Fig. 5.

From the Fig. 5, the articulatory movements vary obviously when emotionally charged. The tongue body of emotional speech is higher and more forward than that in neutral, especially the 'Sad'. The tongue rear is low and back for 'Sad'. The lower lip protrudes when the emotion is 'sad', while spreads when the emotion is 'happy'. The upper lip just protrudes slightly for 'Sad'. In a word, the variation of articulatory movements for 'Sad' is more obvious than those of other emotions. The conclusion that we obtains via the analysis of data is mainly similar to that of the Sect. 1.1.

In order to built a complete emotional text-driven 3D visual pronunciation system, the emotional speech plays an important role in the system. The emotional speech is closely linked with the F0, duration and intensity [14]. We need to modify these parameters by Praat at first according to [14].

After the emotional speech generated, the articulatory movements are predicted via the HMM and adjusted based on the variation of articulatory movements in different emotions. The head mesh model has been built in [1]. By combining emotional speech with articulatory movements, an accurate emotional speech synchronized animation of articulators at phoneme level is synthesized based on a head mesh model. The results are shown in Fig. 6. The Fig. 6 shows

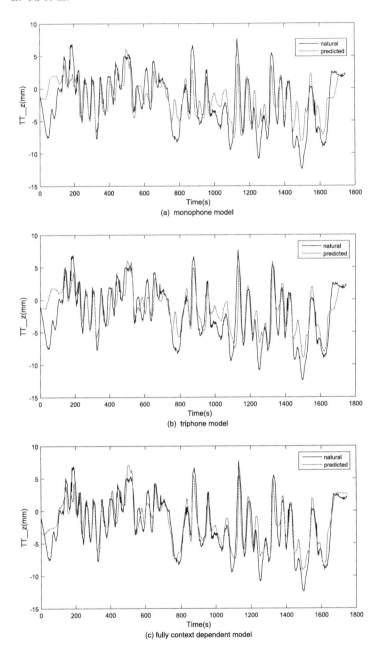

Fig. 4. The comparison with the natural and predicted articulatory movements of the TT_z by (a) monophone model (b) triphone model and (c) fully context-dependent model.

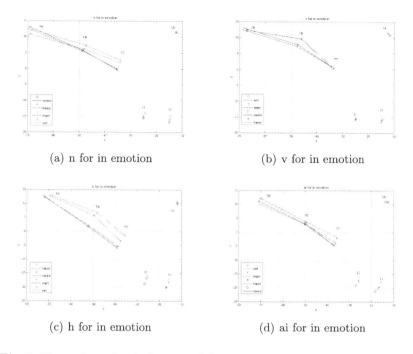

(a) n for in emotion

(b) v for in emotion

(c) h for in emotion

(d) ai for in emotion

Fig. 5. The positon of articulators in different emotion for 'n', 'v', 'h', and 'ai'.

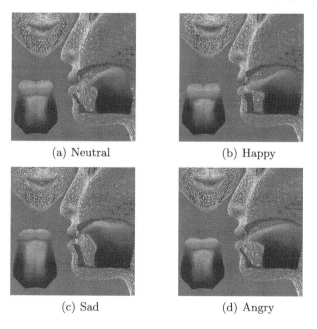

(a) Neutral

(b) Happy

(c) Sad

(d) Angry

Fig. 6. The articulatory movements for the word 'chu'.

the articulatory movements for the word 'chu' in different emotions. Compared with the emotion 'Neutral', the lip spreads and the tongue body is forwards in the emotion 'Happy'. The emotions 'Sad' and 'Angry' involve a raising and forward of the tongue tip, especially for the emotion 'Sad'.

Perceptual Evaluation of the Emotional 3D Visual Pronunciation System. In accordance with the analysis of emotional dataset in Sect. 3.4 and the summary from [9,11,12,14], human vocal emotional effects and the characters of articulatory movements relative to 'Neutral' are summarized in Table 2. The accuracy of the emotional speech synchronized animation of articulators is evaluated based on Table 2.

The videos of 'Neutral', 'Sad', 'Angry' and 'Happy' are recorded respectively. Eight volunteers are invited to evaluate the system from the accuracy of articulatory movements, the naturalness of speech and the synchronization between speech and articulators by the videos. The highest score is five and the lowest score is zero. Figure 7 lists the average scores of the system.

From Fig. 7, the performance of the 'Sad' and 'Happy' outperform that of the 'Angry' in the accuracy of articulatory movements and the naturalness of speech. Perhaps one of the reasons is that the speed rate of 'Angry' is too fast to determinate the articulatory movements. Moreover, it should be stated that

Table 2. The summary of vocal emotional effects and articulatory movements relative to the 'Neutral'.

	Acoustic			Articulatory movements		
	F0	Speech rate	Intensity	Tongue	Jaw	Lip
Sad	Slightly lower	Slightly slower	Lower	TT/TB:forward/higher TR:back/lower	Closing	Protruding
Happy	Much higher	Normal/faster	Higher	TT:higher TR:forward	Opening	Spreading
Angry	Much higher	Much faster	Higher	TT/TB:forward/higher	Opening	Protruding

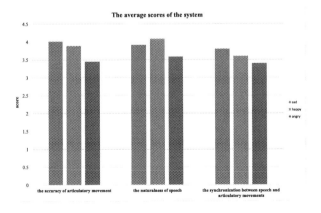

Fig. 7. The average scores of the system.

the videos exist inconsistence between the audio and video. The phenomenon may be caused by the computer configuration or video software which making the radio delays for about one second relative to video.

4   Conclusion

The paper mainly generates an emotional 3D visual pronunciation system for Mandarin Chinese. Firstly, the fully context-dependent label, considering the rich linguistic features, is used to train the HMM. The RMSE of articulatory trajectory predicted from the fully context-dependent model is lower than that of the triphone or monophone model. Secondly, the emotional speech is generated by modifying the acoustic parameters by Praat and then the corresponding articulatory movements are predicted by HMM. Thirdly, the differences between articulatory movements for different emotions are investigated and used to modify the articulatory movements to obtain the trajectories of articulators in different emotions. The variation for tongue and jaw movements is especially significant for 'Sad' speech. In the end, connecting the emotional speech with the corresponding articulatory movements, an accurate emotional speech synchronized animation of articulators at phoneme level is synthesized based on a head mesh model. In future, facial expression will be added to make the system more realistic.

Acknowledgement. This work is supported by the National Natural Science Foundation of China (No. 61572450 and No. 61303150), the Open Project Program of the State Key Lab of CAD & CG, Zhejiang University (No. A1501), the Fundamental Research Funds for the Central Universities (WK2350000002), the Open Funding Project of State Key Laboratory of Virtual Reality Technology and Systems, Beihang University (No. BUAA-VR-16KF-12).

References

1. Yu, J., Li, A.: 3D visual pronunciation of Mandarine Chinese for language learning. In: 2014 IEEE International Conference on Image Processing (ICIP), pp. 2036–2040. IEEE (2014)
2. Ling, Z.-H., Richmond, K., Yamagishi, J.: An analysis of HMM-based prediction of articulatory movements. Speech Commun. **52**(10), 834–846 (2010)
3. Ling, Z.H., Richmond, K., Yamagishi, J., et al.: Integrating articulatory features into HMM-based parametric speech synthesis. IEEE Trans. Audio Speech Lang. Process. **17**(6), 1171–1185 (2009)
4. Toda, T., Black, A.W., Tokuda, K.: Statistical mapping between articulatory movements and acoustic spectrum using a Gaussian mixture model. Speech Commun. **50**(3), 215–227 (2008)
5. Ben-Youssef, A., Shimodaira, H., Braude, D.A.: Speech driven talking head from estimated articulatory features. In: The International Conference on Acoustics, Speech and Signal Processing, pp. 4573–4577 (2014)

6. Zhu, P., Xie, L., Chen, Y.: Articulatory movement prediction using deep bidirectional long short-term memory based recurrent neural networks and word/phone embeddings. In: Sixteenth Annual Conference of the International Speech Communication Association (2015)

7. Yu, J., Wang, Z.F.: A video, text and speech driven realistic 3D virtual head for human-machine interface. IEEE Trans. Cybern. **45**(5), 977–988 (2015)

8. Jun, Y., Wang, Z.F.: 3D facial motion tracking by combining online appearance model and cylinder head model in particle filtering. Sci. Chin. - Inf. Sci. **57**(2), 274–280 (2014)

9. Lee, S., Yildirim, S., Kazemzadeh, A., et al.: An articulatory study of emotional speech production. In: INTERSPEECH, pp. 497–500 (2005)

10. Erickson, D., Zhu, C., Kawahara, S., et al.: Articulation, acoustics and perception of Mandarin Chinese emotional speech

11. Erickson, D., Abramson, A., Maekawa, K., et al.: Articulatory characteristics of emotional utterances in spoken English. In: INTERSPEECH, pp. 365–368 (2000)

12. Li, A., Fang, Q., Hu, F., et al.: Acoustic and articulatory analysis on Mandarin Chinese vowels in emotional speech. In: 2010 7th International Symposium on Chinese Spoken Language Processing (ISCSLP), pp. 38–43. IEEE (2010)

13. Lee, S., Kato, T., Narayanan, S.S.: Relation between geometry and kinematics of articulatory trajectory associated with emotional speech production. In: Ninth Annual Conference of the International Speech Communication Association (2008)

14. Murray, I.R., Arnott, J.L.: Toward the simulation of emotion in synthetic speech: a review of the literature on human vocal emotion. J. Acoust. Soc. Am. **93**(2), 1097–1108 (1993)

15. Odell, J.J.: The use of context in large vocabulary speech recognition. Am. J. Math. **75**(2), 241–259 (1996)

16. Yoshimura, T.: Duration modeling for HMM-based speech synthesis. In: ICSLP, vol. 90, no. 3, pp. 692–693 (1998)

17. Tokuda, K., Yoshimura, T., Masuko, T., et al.: Speech parameter generation algorithms for HMM-based speech synthesis. In: IEEE International Conference on ICASSP, pp. 1315–1318 (2000)

18. Lee, Y, Terzopoulos, D, Waters, K.: Realistic modeling for facial animation. In: Proceedings of the 22nd Annual Conference on Computer Graphics and Interactive Techniques, pp. 55–62. ACM (1995)

19. Marcos, S, Bermejo, J.G.G., Zalama, E.: A realistic facial animation suitable for human-robot interfacing. In: 2008 IEEE/RSJ International Conference on Intelligent Robots and Systems, IROS 2008, pp. 3810–3815. IEEE (2008)

20. Ekman, P., Friesen, W.V.: Manual for the Facial Action Coding System. Psychologists Press, Palo Altom (1978)

21. Tang, C.Y., Zhang, G., Tsui, C.P.: A 3D skeletal muscle model coupled with active contraction of muscle fibres and hyperelastic behaviour. J. Biomech. **42**(7), 865–872 (2009)

22. Zen, H., Nose, T., Yamagishi, J., et al.: The HMM-based speech synthesis system (HTS) version 2.0. Ieice Technical report Natural Language Understanding and Models of Communication, vol. 107, no. 406, pp. 301–306 (2002)

23. Praat speech processing softward. http://www.fon.hum.uva.nl/praat/

Interactive Banknotes Recognition for the Visual Impaired With Wearable Assistive Devices

Dian Huang, Hong Cheng$^{(\boxtimes)}$, and Lu Yang

School of Automation Engineering, Center for Robotics,
University of Electronic Science and Technology of China, Chengdu, China
hcheng@uestc.edu.cn

Abstract. In this paper, we develop a new system, named WVIAS (Wearable Vision Impaired Assistive System), using camera-based computer vision technology to recognize banknote in natural scene aim to help visually impaired people. WVIAS is made up of two mainly parts. In the front, there is a micro camera, set on the glass or mounted on the helmet, to acquire video sequence. In the back, a high performance portable computer is planted to run processing algorithm. To make the system robust to variety conditions including occlusion, rotation, scaling, cluttered background, illumination change, viewpoint variation, and worn or wrinkled banknotes during recognition, we propose a method that using finger pointing as HCI to point out potential targeting district which we call region of interest (ROI), thereafter, we can sharply reduce the processing time by using ROI to replace original image combining with effective ORB feature. The HCI-based framework is effective in collecting more class-specific information and robust in dealing with partial occlusion and viewpoint changes. To authenticate the robustness and generalizability of the proposed approach, we have collected a large dataset of banknotes from natural scene. The proposed algorithm improved the mean average precision from 20.3 % to 61.6 %. The experiments result has shown the effectiveness of our proposal both on the natural scene static dataset and the dynamic video sequence.

Keywords: Banknote recognition · ORB · Hand detection · Visual impaired · HCI

1 Introduction

World Health Organization (WHO) approximates that there were 285 million visually impaired[1] people around the world in 2014, about 2.6 % of the total population. According to these statistics, 246 million had low vision and 39 million were blind [20]. Approximately 90 % of these people live in developing countries

---

[1] This is a general term for people with vision based disabilities, people who are blind have no vision while those with low vision have limited sight.

© Springer Nature Singapore Pte Ltd. 2016
T. Tan et al. (Eds.): CCPR 2016, Part I, CCIS 662, pp. 105–118, 2016.
DOI: 10.1007/978-981-10-3002-4_9

and 82 % of blind people are aged 50 and above [22]. Visually impaired persons adapt to life by using various assistive methods such as the white cane, sensory substitution and electronic devices [3, 21]. The white cane is a common mobility tool used by the visually impaired [1]. Sensory substitution is accomplished by using one sense to compensate for the lack of another. Research has shown that individuals which are blind from an early age have enhanced hearing compared to those with late blindness and to sighted individuals due to early initiation of sensory substitution [7, 21].

Visually impaired people face a number of challenges when interacting with the environments because according to the perception research more than 80 % of information is encoded visually in daily life. One specific difficulty is that a blind person would encounter to know the value of the banknotes he or she is holding in their daily life. Traditionally, there are two mainly ways for the blind to recognize the value of paper bills, one is judging according to the different size and the ratio of length and width [19] with some assistive devices. The other is using the braille on the corner of the banknote. But, these ways are unreliable when the banknote is folded or klunky.

Banknotes in different countries have texture, color, size, and appearance differences, so feature extraction and identification approaches that is used in one country usually is not usable in other countries or does not work properly over there. Moreover, lots of previous work in concept of banknotes recognition is restricted to specific standard conditions. For example many of the developed approaches do not support rotation or noisy backgrounds, or in some other approaches the whole banknote must be visible in taken picture. By considering visually impaired limitations to take pictures with special physical characteristics, these approaches are not user friendly enough.

In this study we consider many complexity and variety for taken pictures and as a result the developed system can support rotation, scaling, complex or noisy background, camera angel changes, collision and even the variation of illumination. Since the photo takers in proposed methods are visually impaired, these complexities should be considered to result in an efficient system. For explication we define these concept below and show some of them show in Fig. 1.

The extraction of distinctive, sufficient, stable features is significance for accuracy and robustness of banknote recognition algorithm. The development of interest points detectors and descriptors in computer vision such as SIFT, SURF, especially, ORB enable the request of feature extraction by using local image features. The sophisticated designs of local image features make them robust to geometric and photometric variations such as rotation, scaling, viewpoint variation, and illumination change. These properties make them well adapted for detecting objects appearing with different scales and orientations in images. Recent advances in theories, sensors, and embedded computing hold the promise to enable computer vision technique to address their needs. Although a number of studies on camera-based banknotes recognition have been published in literatures [5, 9, 11, 14], some of them are restricted to specific and standard environment. For example, the whole bill must be visible without occlusion, wrinkles,

Non-uniform illumination

Self occlusion

Different pointview

Scale variety

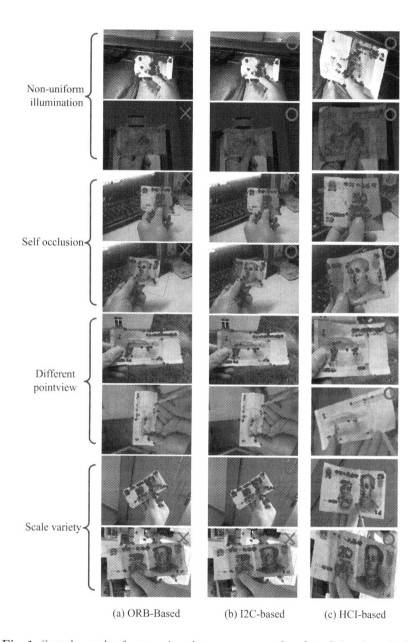

(a) ORB-Based   (b) I2C-based   (c) HCI-based

Fig. 1. Sample result of comparison between proposed and traditional methods.

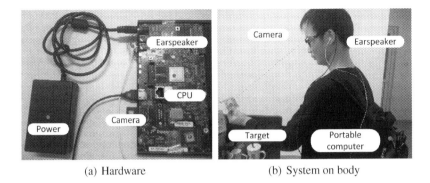

(a) Hardware   (b) System on body

Fig. 2. Implementation of the WVIAS.

etc. An automatic banknote recognition system for blind people should be able to recognize banknotes in a wide variety of real world environments, such as occlusions, cluttered background, varied illumination, different viewpoints, and worn or wrinkled banknotes, as well as to provide the feedback of aiming guidance. In this paper, we propose a novel HCI-based banknote recognition system by using ORB feature which encoded with color, texture, spatial information, to achieve high recognition accuracy and to handle various challenging conditions in real world environments. Figure 2 (the system diagram) shows our system diagram. First, the region of interest is extracted by using the fingertip as a HCI to reduce the search window size. The features of each query image are and extracted by ORB and color. These features are then matched with the pre-computed ORB features of reference regions of the ground truth image in each banknote category. The numbers of matched features are compared with automatic thresholds of each reference region to determine the banknote category. Furthermore, the spatial relationship of matched features is employed to avoid false recognition with negative images and provide aiming guidance to blind users for bill image capture. Then the system outputs the recognition result.

2   Related Work

The existing banknotes recognition approaches in literature, mainly use image processing and neural network techniques. Moreover, there is some developed devices which use physical characteristics of cashes like size or color. The most considerable point about banknotes recognition is that many developed devices or methods in one country are not usable in other countries. In this section we will introduce some of these devices and methods.

Money Talker [10] is a device that recognizes Australian banknotes electronically, using the reflection and transmission properties of light. This device uses the largely different colors and patterns on each Australian banknote. Different color lights are transmitted through the inserted note and the corresponding sensors detect distinct ranges of values depending on the color of the note. Cash

Test [16] is another device determines the value of banknote by a mechanical means relying on the different lengths of each notes. The downfall of the Cash Test is that it does not allow for the shrinking of notes with time nor the creases or rips that are common in our notes. Moreover, while Cash Test is cheap and extremely portable by the reports of users, it is inaccurate and difficult to use. Kurzweil reader [13], iCare [12] (uses a wearable camera for imaging and a CPU chip for computation) and some other devices have been developed too, each has their advantages and disadvantages, but the common property of them is that the user most carry a device everywhere. Some of them are bulky and expensive too.

In [9], Hassanpour and Farahabadi proposed a paper note recognition method which uses Hidden Markov Model (HMM). This method models banknotes using texture characteristics including size, color and a texture-based feature from banknote and compares extracted vector with the instants in a data base of paper cashes. But the main purpose of this method is to distinguish national banknotes from different countries. In [6], Garcia-Lamont et al. introduced a recognition method for Mexican banknotes. This method uses color and texture features. For classification, the LVQ network and G statistic are employed. The color feature is modelled under the RGB color space and the texture feature in grey scale, is modelled with Local Binary Patterns method. This method originally developed for automatic banknote detection is stable conditions. So do not support illumination variation. Also this method cannot overcome the existence of background in the image and is not usable as visually impaired assistance. In [14], Lui proposed a background subtraction and perspective correction algorithm using Ada-boost framework and trained it with 32 pairs of pixels to identify the banknotes values. This system uses video recording and works on snapshots. This method does not support rotation and noisy background and strongly relays on the white and straight edges of banknotes.

Several techniques have been developed to identify banknotes. Lee and Jeon [23] utilized a distinctive point extraction that used a coordinate data extraction method from specific parts of a Euro banknote representing the same color. In order to recognize banknotes, they used two key properties of banknotes: direction including front, rotated front, back, and rotated back and different value (5 10, 20, 50, 100, 200 and 500). A neural network was trained for banknote classification. The results showed a high recognition rate about 95 % and a low training period. Forsini et al. [5] presented a neural network based bill recognition and verification method. For the first time, they extracted features from the image and then input them to a neural network for training and testing. Kosaka and Omatu proposed the learning vector quantization (LVQ) method to recognize 8 kinds of Italian Liras [11,15]. Most banknote recognition methods employed neural network techniques for classification [4,5,8,10–12,15,17,23]. In the above methods, the whole banknote must be visible.

As we can see, while all these methods have their advantages, we cannot use them in a user friendly and accurate manner for banknotes recognition. So we are going to develop a user friendly method which is portable with no difficulties, have enough accuracy, and considering impaired people limitation for taking pictures we will induct our method begin with the introduction of system overview in the next section.

3   System Overview

We shall describe in this section the main hardware and software components of the WVIASs current implementation. Our system is robust and effective with the following advantages. Its both easy using and robust to variety environment to give users a natural using experience. In our work, a method has been proposed that is a part of a wearable aiding system for the visually impaired. A standard video camera has been used in this system as the image collecting device. This method is based on identification of the geometrical patterns that characterize the different denominations of banknotes and does not require training or building the data base. This method is bound by the requirements of a wearable system. This system require the flowing components: a wearable frame similar to conventional eyeglasses that a camera are mounted on it, a portable computer which provide a user interface and can process the image acquired by the camera, a set of headphone to announce the monetary value to the user.

3.1   Hardware

The increase in computation capability of mobile devices gives motivation to develop applications that can assist visually impaired persons. The proposed wearable vision assistive system is designed to take advantage of the portability of mobile devices and provide a simple user interface that makes use of the system easy for the visually impaired. Key design requirements for a portable system including small device size, low weight and easy to use [17]. To achieve this goal, a mini USB camera, earspeaker and light weigh portable computer are used to form a compact, lightweight and efficient system (Fig. 2). We take the i.MX6 Quad core processor of Freescale Semiconductor's as our portable computer, which is multimedia-focused processor offering high performance processing optimized for lowest power consumption.

4   Robust Interactive Banknotes Recognition

4.1   ORB Based Matching

In the baseline of banknote recognition framework ORB [18] feature match is used. ORB is becoming one of the most popular feature detector and descriptor in computer vision field. It is able to generate scale-invariant and rotation-invariant interest points with descriptors. Evaluations show its superior performance in terms of repeatability, distinctiveness, and robustness. The calculation and matching of ORB is also very fast, which is desirable in the real-time applications. ORB is selected as the interest point detector and descriptor based on the following reasons: (1) Banknote image could be taken under the conditions of rotation and scaling change. Interest points with descriptors generated by ORB are invariant to rotation and scaling changes. (2) Computational cost of ORB is small, which enable fast interest point localization and matching. In this part we provide a brief summary of the construction process of ORB. ORB is the combination of oFAST and rBRIEF. As we know FAST keypoints detector is widely used because of their computational properties which really fits for the embedded applications. However, FAST do not have an orientation component. ORB add an efficiently computed orientation during detecting keypoints stage which called oFAST. The orientation is computed using a simple but effective measure of corner orientation, the intensity centroid. The intensity centroid assumes that a corners intensity is offset from its center, and this vector may be used to impute an orientation.

In order to validate the effectiveness of ORB in banknote recognition, we evaluate its performance on interest points matching of banknotes. We collect a testing dataset of nearly 2400 banknote images containing about 200 images for each class of banknotes (¥1, ¥5, ¥10, ¥20, ¥50, and ¥100). The banknotes dataset covers a wide variety of conditions, such as occlusion, rotation, changes of scaling, illumination and viewpoints. Then 12 images of six classes of bills with front and back side are taken as reference images, which are used to match with images in our banknotes dataset by ORB.

4.2   Image to Class Based Approach

During we are analyzing the miss recognition samples we found that the baseline algorithm can effectively detect the indoor scene banknotes, but as for the outdoor scenarios detection result is relatively poor. We found that mainly caused by the reference banknotes are taken under the stable controlled indoor environment, while the a lot of test samples are acquired in variety condition. At the same time, we found some defiled banknotes cant be recognized correctly, due to the commodity circulation the appearance of the banknotes becoming coarser, from the point view of image processing, lots of corner, edge, texture information, which are critical to the effectiveness of recognition algorithm, have lost.

Thinking further about it, all the reference templates are acquired in specific controlled situations, while the using environment is unrestricted natural scene, which lead to poor generalization of the algorithm facing with the inconsistent illumination. In order to reduce the influence of the variety illumination, we applied the histogram equalization during the image preprocessing. In addition, based on the baseline of we introduce I2C (image to class) [2] distance metric algorithm to handle the problem. The I2C distance is a kind of similarity measurement can be used to handle multiple templates problem. The mainly idea of I2C-based algorithm is to use multiple reference templates to simulate of different illumination, collecting several groups of old and new, different illumination conditions added to the reference set in the collection, and during the recognition process, we inducted KNN vote to get the best match result according the I2C distance between classes. In this way it can effectively solve the problem of in the similar illumination.

4.3   HCI Based Approach

To achieve computational efficiency and memory savings, a robust finger detection algorithm is used, to reduce the search area to a smaller search window centered around predictions of the fingertip position. Its enables a user to specify, segment the region of interest, recognize targeting objects, such as banknotes, by simply pointing at and encircling them with the user's fingertip. Our system accepts input from a color camera, and segments the input video stream based on color. The color segmented image is then fed into a skin/non-skin discrimination algorithm to detect skin tones and extract the user's hand. Once the hand is extracted, shape and curvature analysis is used to determine the coordinate position of the fingertip. Finally, the region of interest is calculate according to the position of fingertip.

Robust Fingertip Detection. We address the task of pixel-level hand detection in the context of ego-centric cameras. Extracting hand regions in ego-centric videos is a critical step for understanding handobject manipulation and analyzing hand-eye coordination. However, in contrast to traditional applications of hand detection, such as gesture interfaces or sign-language recognition, ego-centric videos present new challenges such as rapid changes in illuminations, significant camera motion and complex hand-object manipulations. To quantify the challenges and performance in this new domain, we present a fully labeled indoor/outdoor ego-centric hand detection benchmark dataset containing over 200 million labeled pixels, which contains hand images taken under various illumination conditions (Fig. 3).

We are interested in understanding how local appearance and global illumination should be modeled to effectively detect hand regions over a diverse set of imaging conditions. To this end we examine the use of global appearance features as a means of representing changes in global illumination. We explain our local features and global appearance-based mixture model below. To account for different illumination conditions induced by different environments (e.g. indoor,

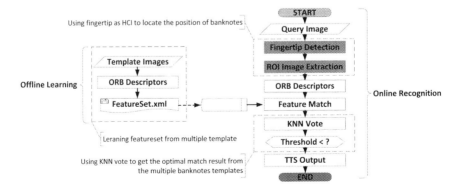

Fig. 3. Flow chart of the proposed HCI-based banknotes recognition algorithm

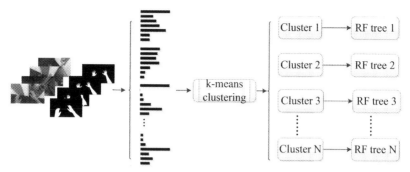

(a) Labeled hand pixels (b) HSV Histogram (c) K-means clustering (d) Train regression trees

Fig. 4. Flow chart of the proposed hand detection algorithm

outdoor, stairway, direct sunlight) we train a collection of regressors indexed by a global color histogram. The posterior distribution of a pixel x given a local appearance feature l and a global appearance feature g, is computed by marginalizing over different scenes c_i,

$$p(x|l,g) = \sum_{i=1}^{k} p(x|l,c_i)p(c_i|g) \tag{1}$$

where $p(x|l,g)$ is the output of a discriminative global appearance-specific regressor, c_i is the ith scene in the k scenes, and $p(c_i|g)$ is a conditional distribution of a scene given a global appearance feature g. Different global appearance models are learned using kmeans clustering on the HSV histogram of each training image and a separate random tree regressor is learned for each cluster (Fig. 4). By using a histogram over all three channels of the HSV colorspace, each scene cluster encodes both the appearance of the scene and the illumination of the scene. Intuitively, we are modeling the fact that hands viewed under similar global appearance will share a similar distribution in the feature space. At test time, the conditional $p(x|l,c_i)$ is approximated using an uniform distribution over the n nearest models(in our test we use n = 5) learned at training.

The result we get from above is a probability map of hand region (Fig. 5b) and after threshold the binary map shows in Fig. 5c. We extract a contour (Fig. 5d) consist by a sequences of points, and our task is to find which point in that sequences is fingertip. The concept here is that, the point that represents the fingertips is the peaks along the contour perimeters. We use both distance (Fig. 5e) and angle constrains (Fig. 5f) to find the fingertip at each point P_i in the contour, we calculate the cosine of the angle between 2 vectors $\overrightarrow{P_{i-r}P_i}$ and $\overrightarrow{P_iP_{i+r}}$, where P_i is the ith point, P_{i-r} is the $(i-r)th$ point, and P_{i+r} is $(i+r)th$ in the hand contour. We use $r = 50$ in $640 * 480$ resolution. The idea here is that contour point P_i with the cosine close to 1 will represent potential peaks of valleys along the perimeters. We dispel valleys by convert the $\overrightarrow{P_{i-r}P_i}$ and $\overrightarrow{P_iP_{i+r}}$ vector in to 3D vectors lying on the $xy-plane$, and calculate the cross product.

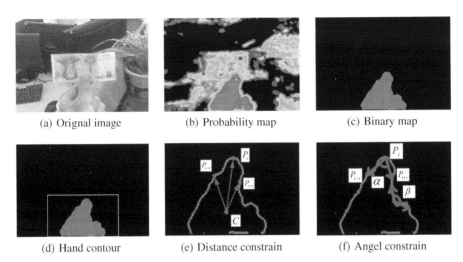

(a) Orignal image   (b) Probability map   (c) Binary map

(d) Hand contour   (e) Distance constrain   (f) Angel constrain

Fig. 5. Result of illumination invariant hand detection and fingertip extraction.

If the z component of the cross product is positive then it the valley. We reserve those potential peaks as fingertips if they meet both of two conditions below: (1) $\cos < \overrightarrow{P_{i-r}P_i}, \overrightarrow{P_iP_{i+r}} >$ is above 0.5. (2) $\cos < \overrightarrow{P_{i-r}P_i}, \overrightarrow{P_iP_{i+r}} >$ is the maximum in its neighborhood of the contour. After that we may get several alternative options, we just select the point farthest from center of palm contour.

ROI Extraction. On the condition that the fingertip has been detected, we can decide the potential area of the target object according to the fingertip and the image boundaries, which we called region of interest (ROI) or reduced search window. We draw the ROI based on the hypothesis that users can roughly point out the target through their finger. The following paragraph is a brief introduction of how we design the ROI.

$$\begin{cases} x_0 = x - 1/2 * w \\ y_0 = y - h \\ w_0 = \alpha * \min(x, w - x) \\ h_0 = \beta * \max(y, h - y) \end{cases} \tag{2}$$

Lets denote width of the frame is w, the height is h, the fingertip $P(x, y)$. We know rectangle can be defined, in 2D space, by the struct which include a beginning point with its height h_0 and width w_0. The parameters α and β, which be used to control the range of ROI, are defined by some segment algorithm or simply using the empirical value. Here, we pursuit for the computational effectiveness so that we just take empirical in our system. The original ROI is defined by Eq. (2), which is an horizontal rectangle.

$$\begin{bmatrix} x_1' \ y_1' \\ x_2' \ y_2' \\ x_3' \ y_3' \\ x_4' \ y_4' \end{bmatrix} = \begin{bmatrix} x_0 & y_0 \\ x_0 & y_0 + h_0 \\ x_0 + w_0 & y_0 \\ x_0 + w_0 \ y_0 + h_0 \end{bmatrix} * \begin{bmatrix} \cos\theta \ -\sin\theta \\ \sin\theta \ \cos\theta \end{bmatrix} \tag{3}$$

To make our ROI more precisely, we conduct Givens Rotation to adjust it. The details show in Eq. (3), where the θ it angle between horizon and direction of fingertip (Fig. 6a). The final ROI is the result of Givens Rotation (Fig. 6b). Using the ROI image (Fig. 6c) as a input, then we get HCI-based recognition algorithm.

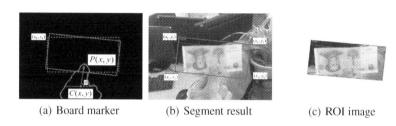

(a) Board marker   (b) Segment result   (c) ROI image

Fig. 6. Result of HCI-based ROI image extraction

5 Experimental and Analysis

5.1 Experimental Setup

The dataset collected from a wide variety of environments includes banknotes taken under the conditions of occlusion, cluttered background, rotation, and changes of illumination, scaling, and viewpoints. Figure 1 demonstrates several sample images from each condition. The dataset presented in our experiment is more challenging than that from other banknote recognition papers. For example, the dataset in [23] was collected by using a scanner to scan the bills which were taken under restricted or standard conditions. Thus, our dataset generalizing the conditions of taking banknote images is more challenging and more approximates to the real world application environment (Table 1).

To test the performance of the proposed banknote recognition algorithm, a dataset of including about 2400 images, taken from a wide variety of conditions to approximate the environment of real world application, which containing 200 images for each class of bill (¥1, ¥5, ¥10, ¥20, ¥50, and ¥100) is created for evaluation.

5.2 Experimental Result

The confusion matrix of three methods show in Fig. 7. We address seven classes in the test, including 6 kinds of banknotes and natural scene image without any

Table 1. Natural scene banknote dataset description

	Description
Environment	Natural scene
Capture device	Logitech C150
Image size	640 * 480
Num. of class	6
Samples of each class	About 400
Application	Banknote recognition or Fingertip/hand detection

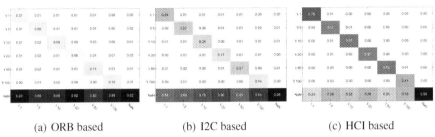

(a) ORB based   (b) I2C based   (c) HCI based

Fig. 7. Confusion matrix comparison of three methods.

banknotes. To make our statics more reasonable, the false positives are categorized to natural scene. As it show in the Fig. 7, our HCI-based banknote recognition algorithm significantly improve the accuracy compared with the traditional ORB-based method.

The experimental results have shown the effectiveness of ORB and our HCI-based framework for banknote recognition in Fig. 8(a). The scale-invariant and rotation-invariant interest point detector and descriptor provide by ORB is robust to handle the image rotation, scaling change and illumination change.

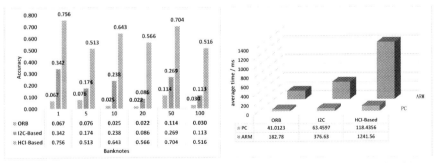

(a) Accuracy of proposed algorithm   (b) Time consume of proposed algorithm

Fig. 8. Results of banknote recognition

Meanwhile, our HCI-based framework generates more class-specific information which is more distinctive to distinguish a query image from cluttered background and effective to deal with partial occlusions. What more, further experiments show that scaling change also can be settled. This is because ORB descriptor is invariant to scaling and our HCI based framework guarantee enough points in banknotes. In real application, it is obviously that a blind person takes the banknotes in a wide range with scaling change.

The computational cost for recognition include extracting ORB features from testing images and matching them with features of all the reference templates, and displaying the classification output. As it shown in Fig. 8b the average speed of the baseline and proposed algorithm on a testing image at the resolution of 640×480 pixels on different platform PC (Pentium(R) Dual-core E6500 @2.93 GHz 4 GB RAM) and the embed (i.MX6Quad application Processor @1 GHz, 1 GB RAM). As for our algorithm is evaluated in a more challenging dataset, our algorithm achieves 61.6 % accuracy and little lower than the existing banknote recognition algorithms. For instance, the average recognition rate of the algorithm [17] based on SIFT is 76 %. Some neural network based banknote recognition systems achieved recognition rate about 90 % [11,15]. But all of them are test on extremely restricted environment.

6 Future Work

Our future work will focus on optimization of the whole system and enlarging the testing dataset by incorporating more banknotes of different countries. The computation speed will be reduced to less than 1 s for each testing image on ARM platform. In addition, we will study the significant human interface issues including auditory output and system configuration. Whats more, well improve the accuracy of algorithm to make it more efficient and fast to provide the visual impaired users a better using experience. Some other assistive applications also will take into consideration, such as daily using object recognition and navigation etc., to integrate on the WVIAS.

Acknowledgement. The authors would like to thank all the reviewers for their insightful comments. This work was supported by the National Natural Science Foundation of China (Grant Nos. 61305033, 61273256 and 6157021026), Fundamental Research Funds for the Central Universities (ZYGX2014Z009).

References

1. Betts, K.: National perspective: Q&A with national federation of the blind & association of higher education and disability. Online Learn. J. **17**(3) (2013)
2. Cheng, H., Dai, Z., Liu, Z., Zhao, Y.: An image-to-class dynamic time warping approach for both 3D static and trajectory hand gesture recognition. Pattern Recogn. **55**, 137–147 (2016)
3. Cimarolli, V.R., Boerner, K.: Social support and well-being in adults who are visually impaired. J. Vis. Impair. Blind. **99**(9), 521 (2005)

4. Domínguez, A.R., Lara-Alvarez, C., Bayro, E.: Automated banknote identification method for the visually impaired. In: Bayro-Corrochano, E., Hancock, E. (eds.) CIARP 2014. LNCS, vol. 8827, pp. 572–579. Springer, Heidelberg (2014)

5. Frosini, A., Gori, M., Priami, P.: A neural network-based model for paper currency recognition and verification. Neural Netw. **7**, 1482–1490 (1996)

6. García-Lamont, F., Cervantes, J., López, A.: Recognition of mexican banknotes via their color and texture features. Expert Syst. Appl. **39**, 9651–9660 (2012)

7. Gougoux, F., Belin, P., Voss, P., Lepore, F., Lassonde, M., Zatorre, R.J.: Voice perception in blind persons: a functional magnetic resonance imaging study. Neuropsychologia **47**(13), 2967–2974 (2009)

8. Grijalva, F., Rodriguez, J., Larco, J., Orozco, L.: Smartphone recognition of the US banknotes' denomination, for visually impaired people. In: Andean Council International Conference, pp. 1–6. IEEE (2010)

9. Hasanuzzaman, F.M., Yang, X., Tian, Y.: Robust and effective component-based banknote recognition for the blind. In: International Conference on Systems, Man, and Cybernetics, vol. 42, pp. 1021–1030 (2012)

10. Hinwood, A., Preston, P., Suaning, G., Lovell, N.: Bank note recognition for the vision impaired. Aust. Phys. Eng. Sci. Med. **29**, 229–233 (2006)

11. Kosaka, T., Omatu, S., Fujinaka, T.: Bill classification by using the LVQ method. In: International Conference on Systems, Man, and Cybernetics. IEEE (2001)

12. Krishna, S., Little, G., Black, J., Panchanathan, S.: A wearable face recognition system for individuals with visual impairments. In: International ACM SIGACCESS Conference on Computers and Accessibility. ACM (2005)

13. Kurzweil, R.C., Albrecht, P., Gashel, J., Gibson, L.: Portable reading device with mode processing, US Patent 8,711,188, April 2014

14. Liu, X.: A camera phone based currency reader for the visually impaired. In: International Conference on Computers and Accessibility, pp. 305–306. ACM (2008)

15. Omatu, S., Fujinaka, T., Kosaka, T., Yanagimoto, H., Yoshioka, M.: Italian Lira classification by LVQ. Neural Netw. **4**, 2947–2951 (2001). IEEE

16. RBoA: How the RBA assists people with a vision impairment to differentiate notes. http://banknotes.rba.gov.au/resources/for-people-with-vision-impairment/. Accessed 02 May 2016

17. Reiff, T., Sincak, P.: Multi-agent sophisticated system for intelligent technologies. In: International Conference on Computational Cybernetics, pp. 37–40. IEEE (2008)

18. Rublee, E., Rabaud, V., Konolige, K., Bradski, G.: ORB: an efficient alternative to sift or surf. In: International Conference on Computer Vision. IEEE (2011)

19. Springer, K., Subramanian, P., Turton, T.: Australian banknotes: assisting people with vision impairment. RBA Bull. 01–12 (2015)

20. Varma, R., Bressler, N.M., Doan, Q.V., Danese, M., Dolan, C.M., Lee, A., Turpcu, A.: Visual impairment and blindness avoided with ranibizumab in Hispanic and non-Hispanic whites with diabetic macular edema in the United States. Ophthalmology **122**(5), 982–989 (2015)

21. Wan, C.Y., Wood, A.G., Reutens, D.C., Wilson, S.J.: Early but not late-blindness leads to enhanced auditory perception. Neuropsychologia **48**(1), 344–348 (2010)

22. WHO: 10 facts about blindness and visual impairment. http://www.who.int/features/factfiles/blindness/en/. Accessed 02 May 2016

23. Youn, S., Choi, E., Baek, Y., Lee, C.: Efficient multi-currency classification of CIS banknotes. Neurocomputing **156**, 22–32 (2015)

Spontaneous Smile Recognition for Interest Detection

Zhenzhen Luo, Leyuan Liu, Jingying Chen[✉], Yuanyuan Liu, and Zhiming Su

National Engineering Research Center for E-Learning,
Central China Normal University, Wuhan 430079, China
AndreaLoves@163.com, {lyliu,chenjy}@mail.ccnu.edu.cn,
jane19840701@hotmail.com, happyszm@foxmail.com

Abstract. "Interest"is a critical bridge between cognitive and effective issues in learning. Student's interest has great impact on learning performance. Hence, it's necessary to detect student's interest and make them more engaged in the learning process for productive learning. Student's interest can be detected based on the facial expression recognition, e.g., smile recognition. However, various head poses, different illumination, occlusion and low image resolution make smile recognition difficult. In this paper, a conditional random forest based approach is proposed to recognize spontaneous smile in natural environment. First, image patches are extracted within the eye and mouth regions instead of the whole face to improve the robustness and efficiency. Then, the conditional random forests based approach is presented to learn the relations between image patches and the smile/non-smile features conditional to head poses. Furthermore, a K-means based voting method is introduced to improve the discrimination capability of the approach. Experiments have been carried out with different spontaneous facial expression databases. The encouraging results suggest a strong potential for interest detection in natural environment.

Keywords: Smile recognition · Conditional random forest · Interest detection · Head poses

1 Introduction

"Interest"is a critical bridge between cognitive and affective issues in learning. Student's interest has great impact on learning performance. Hence, it's necessary to detect student's interest and make them more engaged in the learning process for productive learning. Student's interest can be detected based on the facial expression recognition. Among various facial expressions, smile is very informative. *Senechal et al.* [1] and *Chen et al.* [2] used the smile as important visual feature to detect one's affective state. In class, smile reflects the student's emotion associated with learning interesting, hence, smile recognition is crucial to analyze the student's learning interest.

© Springer Nature Singapore Pte Ltd. 2016
T. Tan et al. (Eds.): CCPR 2016, Part I, CCIS 662, pp. 119–130, 2016.
DOI: 10.1007/978-981-10-3002-4_10

In recent years, lots of works have been done on smile recognition [3–8]. *Whitehill et al.* [3] used Gabor and Box filters to extract features, and then employed the GentleBoost and Support Vector Machine (SVM) to detect smile. *Shan et al.* [4] used the intensity difference of two randomly selected facial patches as features, and trained a smile classifier by AdaBoost. These approaches achieve good results on frontal face images. *Li et al.* [6] distinguished smile from neutral facial expressions using a neural architecture that combines fixed and adaptive non-linear 2-D filters. The proposed approach had been tested on JAFFE database with deliberately posed expressions on frontal face. *An et al.* [7] used a holistic flow-based face registration, and extracted Local Binary Pattern (LBP), Local Phase Quantization (LPQ), and Histogram of Oriented Gradients features are for Extreme Learning Machine (ELM) classifier training. 88.5 % accuracy has been achieved on GENKI-4K database. *Abd et al.* [8] proposed an expression classifier consisting of a set of Random Forests (RF) paired with SVM labelers for spontaneous smile recognition. Their work is based on a 3D facial expression dataset (BU-3DFE) with high-resolution and clean background images. *Dapogny et al.* [9] presented Pairwise Conditional Random Forests (PCRF) to recognize dynamic facial expressions, they achieved good results on posed expressions. Actually, in classroom, student' smile is often happened spontaneously with various head poses, different illuminations, occlusions and low image resolution which makes the smile recognition difficult. In this paper, a conditional random forest based approach is proposed to recognize spontaneous smile in a natural classroom.

Random forest (RF) is an ensemble classifier that consists of many decision trees. It has been proven to be effective to solve computer vision problems [10–13]. Random forest learns the probability over the parameter space from the entire training set, while conditional random forest (C-RF) learns several conditional probabilities over parameter space instead. Conditional random forest is a probability estimation model under constraint conditions of random forest. It has been widely used on classification and regression in computer vision works. *Sun et al.* [14] proposed conditional regression forests for human pose estimation. *Dantone et al.* [11] used the conditional regression forests for real-time facial feature detection. Their works achieved better performance than RF.

To detect students' learning interest in a natural classroom, spontaneous smile under various head poses is recognized with the following efforts:

(1) A conditional random forests based approach is presented to learn the relations between image patches and the smile/non-smile features conditional to head poses.

(2) Image patches are extracted within the discriminative regions for expression recognition (e.g., eye and mouth regions) instead of the whole face to build trees in the forest to improve the robustness and efficiency.

(3) A K-Means clustering based voting method is introduced to improve the discrimination capability of the approach.

Fig. 1. An exemplar result of spontaneous smile recognition in a natural classroom for interest detection.

Experiments have been carried out with spontaneous facial expression datasets. An example result of spontaneous smile recognition in a natural classroom is shown in Fig. 1, encouraging results suggest a strong potential for interest detection in natural environment.

The rest of the paper is described as follows. The details of the proposed smile recognition approach are presented in Sect. 2. Section 3 gives the experimental results and discussions. The conclusions are made in Sect. 4.

2   Smile Recognition Using Conditional Random Forests

The proposed smile recognition approach is composed of two stages as illustrated in Fig. 2. In the first stage, head poses are estimated using Dirichlet-tree distribution enhanced Random Forests approach (D-RF) [15]. The head poses are divided into 5 classes in the horizontal direction, i.e.,$\{-90°, -45°, 0°, +45°, +90°\}$. Mouth and eye regions are extracted based on geometric features of human face. In the second stage, smile is classified using the conditional random forests based approach. First, multiple texture features, i.e., Gabor, LBP and image intensity are extracted from patches within the mouth and eye regions to build trees in random forest conditional to head poses. As the eye and mouth regions have different distinctive expression information, two layers of conditional random forests trained respectively using mouth and eye regions are combined to strengthen the classification performance. Then the K-means based voting method is introduced to improve the discrimination capability of the approach. Finally, the face can be classified as smile or non-smile.

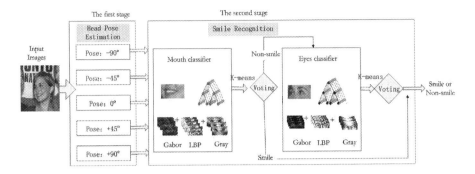

Fig. 2. The flowchart of the proposed approach.

2.1 Head Pose Estimation

Face is first detected using a boosting algorithm and a set of Haar-like features [16]. Then, the Dirichlet-tree distribution enhanced Random Forests approach (D-RF) [15] is used to estimate head poses in the horizontal direction. In the forest each child layer is related to his parent. Hence, the D-RF only computes the probability of a tree in the child layer instead of all trees' probabilities in the forest. Therefore, D-RF can provide high accuracy and efficiency [15].

In this work, head poses are classified into 5 categories, i.e., $\{-90^0, -45^0, 0^0, +45^0, +90^0\}$. Gabor and histogram distribution features are used. In leaves of random forest, there are 5 probabilistic models of head poses and the distributions of the models are represented by an adaptive Gaussian mixture model (GMM) [11]:

$$P(a|l(\mathbf{P})) = N(a; \overline{a_l}, \Sigma_l) \tag{1}$$

Where $\overline{a_l}$ and Σ_l are the mean and covariance matrix of head pose a, while $P(a|l(\mathbf{P}))$ indicates the probability that the patch $\mathbf{P}$ in the leaf $l(\mathbf{P})$ of random forest. The estimated result of head pose is used as the prior probability in conditional random forest for smile recognition.

2.2 Smile Recognition

While a random forest aims to learn the probability $p(B_i/P)$ given an image patch P. A conditional random forest aims to learn the several conditional probabilities $p(B_i/a_j, P)$, where $B_i, (i = 0, 1)$ represent the expression, a_j, $(1 <= j <= 5)$ represent the head poses which are estimated as described in the previous section. To recognize smile under different head poses, conditional random forest is built, the head poses is first estimated, then the trees conditional to the head poses are learned.

In order to model the conditional probability, the training set is split into five subsets corresponding to five head poses. then, $p(B_i/a_j, P)$ can be obtained by considering the probabilities over all trees from subsets.

Training. Each tree T in the forest is built from a set of patches $P_i = \{I_i, B_i\}$ randomly sampled from mouth/eye sub-regions of the training images. $I_i = (I_i{}^1, I_i{}^2, I_i{}^3)$ represents multiple channels of appearance features including the image intensity, Gabor and LBP features. $B_i = (0, 1)$ represents smile and non-smile face.

We define a patch comparison feature as our binary tests φ, similar to [10–12]:

$$\varphi = |R_1|^{-1} \sum_{k \epsilon R_1} I^\alpha(k) - |R_2|^{-1} \sum_{k \epsilon R_2} I^\alpha(k) \tag{2}$$

Where R_1 and R_2 are two rectangular patches randomly extracted from the image patch, and $I^\alpha(k)$ is the feature channel $\alpha \in \{1, 2, 3\}$.

A splitting candidate that best splits the feature set P into two subsets P_L and P_R is selected, which maximizes the evaluation function Information Gain (IG),

$$P_L = \{P|\varphi < \tau\}, P_R = \{P|\varphi \geq \tau\} \tag{3}$$

$$IG(\varphi) = \arg\max(H(P|a_j) - \sum_{s \in L, R} \frac{P_s}{P}(H(P_s|a_j))) \tag{4}$$

Where τ is threshold, $H(P|a_j)$ is the defined class uncertainty measure.

$$H(P|a_j) = -\sum_{i=1}^{2} \frac{\sum_{\mathbf{P_n} \epsilon \mathbf{p}} p(B_i/a_j, P_n)}{|P|} log(\frac{\sum_{\mathbf{P_n} \epsilon \mathbf{p}} p(B_i/a_j, P_n)}{|P|}) \tag{5}$$

Where $p(B_i/a_j, P_n)$ indicates the probability that the patch P_n belongs to the smile expression B_i under the head poses estimation probability model a_j. $|P|$ represents the number of patches in the leaf.

Create leaf l when IG is below a predefined threshold or when a maximum depth is reached. Each leaf stores head pose and expression probabilities. Mouth classifier and eye classifier are trained individually in the similar way.

Testing. During testing, each tree is a weaker classifier. The testing patches are evaluated by all the weaker classifiers. At each node of a tree, the patches are evaluated by the stored binary test and passed either to the right or left child until a leaf node is reached. At each leaf l, the classification probability is stored. By passing all the patches down all the trees in the conditional random forest, the probability can be obtained by averaging over all trees [10]:

$$p(B_i/a_j, P) = \frac{1}{T} p((B = B_i/a_j, l(P)) \tag{6}$$

Where T represents the number of trees are used. As the output of the forest is the probability for the presence of the smile expression, in this study, a K-means [18] clustering based method is introduced to vote the probabilities stored in leaves of the forest.

K-Means Clustering for Voting. The classification probability $p(B = 1|P_n)$ of the patch P_n classified to the smile expression can be obtained as described above. Finally, the test image can be classified based on the classification probabilities of all the patches

$$p_i = \frac{1}{N}p(B = 1|P_n) > \tau_p \tag{7}$$

While τ_p is defined as classifier's decision boundary (classification threshold), N represents the patch numbers. The decision boundary τ_p of classifier (7) is obtained from training set. Some approaches obtain the decision boundary τ_p by minimizing the error rate of training set, i.e., stump [17]. However, this method does not consider the data distribution in the decision space, the error rate minimum of the training set does not guarantee the error rate minimum of the test set, and also may cause over fitting. Gaussian model based voting method is used in [14,15], which consider the data distribution in the decision space, but it requires the data to satisfy Gaussian distribution. Therefore, this paper proposes a boundary decision method based on K-means clustering.

The probabilities $\{p_1, p_2, p_3 \cdots p_n\}$ learned from the C-RF are used as the input data for K-Means clustering to obtain the decision boundary:

(1) Initialize clustering center: $c_0 = \min\{p_1, p_2, p_3 \cdots\}$, $c_1 = \max\{p_1, p_2, p_3 \cdots\}$.
(2) Calculate the Euclidean distance between each p_i and clusters' centers c_0 and c_1, and then assign the data to the nearest cluster. Two classes of data are represented by $C_0 = \{p_1^0, p_2^0, p_3^0 \cdots\}$, $C_1 = \{p_1^1, p_2^1, p_3^1 \cdots\}$.
(3) Calculate the mean of the classes as new clustering centers.
(4) Repeat (2), (3) until the clustering centers remain the same.
(5) Output the clustering results C_0, C_1.

Decision boundary is obtained from training sets, which is determined by finding the furthest point from two cluster centers.

$$\tau_p(a_j) = \frac{1}{2}(\max\{p_1^0, p_2^0, p_3^0 \cdots \} + \min\{p_1^1, p_2^1, p_3^1 \cdots \}) \tag{8}$$

Finally, the mouth and eye classifiers are combined to determine the face is smile/non-smile. First, the image is classified using the mouth classifier, if it is non-smile, then the image is classified with the eye classifier as well. Otherwise, the image is classified as smile face, it doesn't need to use the eye classifier, the computational time can be saved in this way.

3   Experiments

Databases. The proposed approach has been evaluated with two publically available databases (i.e. LFW[19] and GENKI-4K) and our lab database (CCNU-Classroom) collected from an overhead camera in a classroom. The LFW database consists of 5749 individual facial images. The images have been collected in real scenario and vary in poses, resolutions, races, expressions, lighting conditions, occlusion, and make-up. There are totally 13,233 faces in LFW database.

GENKI-4K is a public database collected in unconstrained scenarios. This database has 4,000 face images with 1,838 "non-smile"and 2,162 "smile"faces. The head poses range (approximately from $-20°$ to $20°$).

CCNU-Classroom contains smile and non-smile face images with different head poses (from -90^0 to $+90^0$ in the horizontal direction) of 3500 subjects under different lighting conditions in the classroom.

Training. To train trees in forest, we have chosen some parameters based on empirical observations, e.g., the maximum depth of each tree is 15, the number of patches extracted from each image is 350, 150 patches from the mouth sub-region and 100 patches from the eyes sub-region; the size of the normalized face is 125*125, the size of the extracted patch is 0.25*face size. Each tree is grown based on a randomly selected subset of the LFW database.

Testing. Testing parameters are similar to the training ones. In order to evaluate the proposed approach, the recognition accuracy is defined as follows:

$$Accuracy = \frac{correct_{Num}}{Total_{Num}} \times 100\% \qquad (9)$$

Where the $correct_{Num}$ represents the number of correct estimated images in the testing set. The $Total_{Num}$ represents the number of all the testing images.

Results on Different Databases. The example results of the proposed approach tested on the LFW and CCNU-Classroom databases are shown in Figs. 3 and 4.

The smile recognition accuracy using the proposed method is shown in Table 1. The head pose estimation accuracy is about 83.52 % on LFW database. From the result, one can see that the smile recognition accuracy is not highly dependent on head pose estimation accuracy.

Fig. 3. The example results on LFW database.

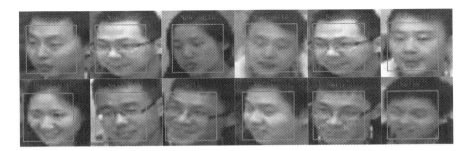

Fig. 4. The example results on CCNU-Classroom.

Table 1. The recognition accuracy achieved using the proposed approach on LFW database.

Poses	-90^0	-45^0	0^0	$+45^0$	$+90^0$
Accuracy (%)	86.84	90.63	92.86	90.67	89.04

In order to test the proposed method on images captured in a natural classroom to understand students' learning interest, we collect the CCNU-Classroom database. The recognition results on CCNU-Classroom database are shown in the Table 2.

Table 2. The recognition accuracy achieved using the proposed approach on CCNU-classroom.

Poses	-90^0	-45^0	0^0	$+45^0$	$+90^0$
Accuracy (%)	73.21	84.31	88.86	86.70	80.28

We compare the proposed approach with the RF [10], the comparison results on both LFW and CCNU-Classroom database are given in Table 3. On LFW database, the proposed approach and RF achieve 90.73 % and 81.74 % average (Avr) accuracy, respectively. On CCNU-Classroom database, the proposed approach and RF achieve 82.67 % and 71.58 % average accuracy. The results show the proposed approach has a better performance than the RF.

Results Using Different Regions. The smile recognition results using different regions (i.e., whole face, mouth and eye sub-regions individually, and combined mouth and eye sub-regions) are given in Table 4. From the results, one can see that the combined mouth and eye sub-regions provide better performance than other regions.

Table 3. The comparison results on LFW and CCNU-Classroom databases using the RF and proposed approach (%)

Poses	LFW						CCNU-classroom					
	-90^0	-45^0	0^0	$+45^0$	$+90^0$	Avr	-90^0	-45^0	0^0	$+45^0$	$+90^0$	Avr
Proposed	86.84	90.63	92.86	90.67	89.04	90.01	73.21	84.31	88.86	86.70	80.28	82.67
RF	82.00	73.27	78.00	81.00	87.98	81.74	51.79	82.00	78.89	78.57	66.67	71.58

Table 4. The recognition accuracy achieved using different regions on LFW database (%).

Poses	Face	Mouth	Eyes	Mouth + Eyes
0^0	78.00	91.08	67.74	95.09
$+45^0$	75.50	88.50	64.50	90.05
$+90^0$	72.08	86.86	62.08	86.86

Results Using Different Voting Methods. The comparison results using the K-means clustering based method and GMM [11,14] on LFW and CCNU-classroom databases are given in Table 5. From the results, one can see that the K-means clustering based method provides better performance than GMM.

Table 5. The recognition accuracy using different voting methods on LFW and CCNU-Classroom databases (%).

Poses	LFW		CCNU-classroom	
	K-means	GMM	K-means	GMM
0^0	95.09	90.78	88.89	87.78
$+45^0$	90.50	88.50	86.96	85.04
$+90^0$	86.86	85.23	79.66	77.94
Avr	90.81	88.17	85.17	83.59

Comparisons with State of the Art. We compare the proposed method with state of the art approaches [4,7] on the same databases, GENKI-4K. The comparison results are given in Table 6. In [4], pixel intensity difference (PID) is extracted as feature, while AdaBoost is used as classifier. In [7], Local Binary Pattern (LBP) is extracted as feature, while respectively Linear Discriminant Analysis (LDA) and SVM are used as classifiers. And Histogram of Oriented Gradients (HOG) is extracted as feature, while ELM is used as classifier. The results show that the proposed method provides higher accuracy than those methods do in [4,7].

Some example results on GENKI-4K database using the proposed method are given in Fig. 5.

Table 6. Comparisons to the method in [4,7] on the GENKI-4K database

Method	Feature	Classifier	Accuracy (%)
An [7]	LBP	LDA	76.60
An [7]	LBP	SVM	84.20
An [7]	HOG	ELM	88.50
Shan [4]	PID	AdaBoost	89.70
Proposed	LBP, gray, gabor	C-RF	91.14

Fig. 5. The example results on GENKI-4K database.

4   Conclusions

In this paper, a conditional random forest based approach is proposed to recognize spontaneous smile in natural environment. First, image patches are extracted within the eye and mouth regions instead of the whole face to improve the robustness and efficiency. Then, the conditional random forests based approach is presented to learn the relations between image patches and the smile/non-smile features conditional to head poses. Finally, a K-means clustering based voting method is introduced to improve the discrimination capability of the approach.

Experiments have been carried out with spontaneous expression databases, e.g., LFW, GENKI-4K and CCNU-Classroom databases. The proposed approach achieve 90.73 %, 91.14 % and 82.67 % average accuracy on these datasets. The encouraging results suggest a strong potential for students' interest detection in natural environment.

Acknowledgment. This work was supported by Research funds from the Humanities and Social Sciences Foundation of the Ministry of Education (No. 14YJAZH005), National Social Sciences Foundation (No. 16BSH107), Research funds from Ministry of Education and China Mobile (MCM20130601), Research Funds of CCNU from the Colleges' Basic Research and Operation of MOE (No. CCNU16A02020), National Key Technology Research and Development Program (NO. 2014BAH22F01), Research Funds of CCNU from the Colleges Basic Research and Operation of MOE (No.

CCNU14A05019), Research Funds of CCNU from the Colleges Basic Research and Operation of MOE (No. CCNU14A05020).

References

1. Senechal, T., Turcot, J., Kaliouby, R.: Smile or smirk? Automatic detection of spontaneous asymmetric smiles to understand viewer experience. In: 10th IEEE International Conference and Workshops on Automatic Face and Gesture Recognition (FG), vol. 10(1), pp. 323–330 (2013)
2. Chen, J., Chen, D., Li, X.: Toward improving social communication skills upon multimodal sensory information. IEEE Trans. Ind. Inform., 264–269 (2013)
3. Whitehill, J., Littlewort, G., Fasel, I.: Toward practical smile detection. IEEE Trans. Pattern Anal. Mach. Intell. (PAMI) **31**(11), 2106–2111 (2009)
4. Shan, C.: Smile detection by boosting pixel differences. IEEE Trans. Image Process. **21**(1), 431–436 (2012)
5. Shimada, K., Matsukawa, T., Noguchi, Y., Kurita, T.: Appearance-based smile intensity estimation by cascaded support vector machines. In: Computer Vision Asian Conference on Computer Vision (ACCV), pp. 277–286 (2010)
6. Li, P., Phung, S., Bouzerdom, A., Tivive, F.: Automatic recognition of smiling and neutral facial expressions. In: International Conference on Digital Image Computing: Techniques and Applications (DICTA), pp. 582–586 (2010)
7. An, L., Yang, S., Bir, B.: Efficient smile detection by extreme learning machine. Neurocomputing **149**, 354–363 (2015)
8. Abd, M.K., Levine, M.D.: Fully automated recognition of spontaneous facial expressions in videos using random forest classifiers. IEEE Trans. Affect. Comput. **5**(2), 141–154 (2014)
9. Dapogny, A., Bailly, K., Dubuisson, S.: Pairwise conditional random forests for facial expression recognition. In: IEEE International Conference on Computer Vision (ICCV), pp. 3783–3791 (2015)
10. Breiman, L.: Random forests. Mach. Learn. **45**(1), 5–32 (2001)
11. Dantone, M., Gall, J., Fanelli, G., Gool, L.: Real-time facial feature detection using conditional regression forests. In: International conference on Computer Vision and Pattern Recognition (CVPR), pp. 3394–3401 (2012)
12. Fanelli, G., Gall, J., Gool, L.: Real time head pose estimation with random regression forests. In: IEEE Conference on Computer Vision and Pattern Recognition (CVPR), pp. 617–624 (2011)
13. Huang, C., Ding, X., Fang, C.: Head pose estimation based on random forests for multiclass classification. In: International Conference on Pattern Recognition (ICPR), pp. 934–937 (2010)
14. Sun, M., Kohli, P., Shotton, J.: Conditional regression forest for human pose estimation. In: International conference on Computer Vision and Pattern Recognition(CVPR), pp. 3394–3401 (2012)
15. Liu, Y., Chen, J., Su, Z., Luo, Z., Luo, N., Liu, L., Zhang, K.: Robust head pose estimation using Dirichlet-tree distribution enhanced random forests. Neurocomputing **173**(2), 42–53 (2016)
16. Viola, P., Jones, M.: Robust real-time object detection enhanced random forests. Int. J. Comput. Vis. **4**(2), 34–47 (2001)
17. Wayne, I., Langley, P.: Induction of one-level decision trees. In: Proceedings of the Ninth International Conference on Machine Learning (ML), pp. 233–240 (1992)

18. Na, S., Liu, X.: Research on K-means clustering algorithm. In: Third International Symposium on Intelligent InformationTechnology and Security Information, pp. 63–67 (2011)
19. Huang, G.B., Ramesh, M., Berg, T., Learned-Miller, E.: Labeled faces in the wild: a database for studying face recognition in unconstrained environments. Technical report, University of Massachusetts, Amherst (2007)

Road Extraction Based on Direction Consistency Segmentation

Lei Ding[1(✉)], Qimiao Yang[2], Jun Lu[1], Junfeng Xu[1], and Jintao Yu[1]

[1] Department of Photogrammetry and Remote Sensing,
Zhengzhou Institute of Surveying and Mapping, Zhengzhou 450001, China
dingleil4@outlook.com
[2] 61206 Troops, Dalian 116000, China

Abstract. A common strategy for road extraction from remote sensing images is classification based on spectral information. However, due to a common phenomenon that different objects can be with similar spectral characteristics, classification results usually contain many interference regions which do not correspond to any road entity. To solve this problem, a road extraction method based on direction consistency segmentation is proposed in this paper. In binary road classification images, considering that road regions in these images usually have consistent local directions, pixels with similar main directions are merged into objects. After acquiring these objects, geometric measurements such as *LFI* (Linear Feature Index) and region area are calculated and a segment-linking algorithm is used to recognize and extract road objects among them. Various test images are used to verify the effectiveness of this method and contrast experiments are performed between the proposed binary image processing method and two existing methods. Experimental results show that this method has advantages in both accuracy, computational efficiency and stability, which can be used to extract road regions in remote sensing images at different resolutions.

Keywords: Road extraction · Direction consistency segmentation · Geometric feature · Segment-linking

1 Introduction

Road, as an important kind of ground objects, its automatic recognition, location and extraction have always been a hot topic in the field of photogrammetry and remote sensing. A common strategy for road extraction from remote sensing images is first using spectral and texture features for classification, then analyzing the geometrical features of the classified result to remove those regions with features not corresponded to roads.

Existing methods for classification are rich. For example, while processing color images or multi-spectral images, supervised classification method such as SVM (Support vector machine) and Artificial Neural Network can be used to process. In terms of grayscale images, automatic or semi-automatic methods like threshold method, Ostu segmentation and K-Means clustering method can be used. Textual feature based methods such as MRF (Markov Random Fields) works as well. However,

© Springer Nature Singapore Pte Ltd. 2016
T. Tan et al. (Eds.): CCPR 2016, Part I, CCIS 662, pp. 131–144, 2016.
DOI: 10.1007/978-981-10-3002-4_11

due to a common phenomenon that different objects can be with similar spectral characteristics, classification results usually contain many interference regions. These regions are linked with road regions and are difficult to remove. Additionally, due to the influence of vehicles and marker lines on the roads and shadows of trees and buildings along the roads, road segments in the classified images are commonly interrupted and discontinue. In consequence, it is necessary to further process the image and use geometric features to reduce or eliminate these non-road interference regions.

Among the existing road extraction papers, much work has been done to solve this problem. The RPS (Road Part Segmentation) algorithm proposed by Bennamoun et al. [1] and improved by Sukhendu Das et al. [2] works on classified road binary results, the fundamental principle of which is segmenting the interference regions linked with roads by using curvature information and removing these regions with a given area threshold. Path-opening was proposed as a road extraction algorithm by Valero et al., which can detect narrow and linear structural elements, thus eliminating non-road patches which are not linear. Zhang and Couloigner [3] simplified the ATS (Angular Texture Signature) method to analyze the shape features of objects on binary images and proposed shape measurements including mean, compactness, eccentricity and direction, which can be applied on ATS polygons to recognize road regions. Zhu et al. [4, 5] built a parallel pairs model to analyze and detect roads in SAR images. Song and Civco [6] introduced smoothness and density as similarity criterions to segment the road group image into geometrically homogeneous objects. Jia et al. [7] designed adjacency, direction similarity, overlapping and length as criterions in her study on segment-linking. Lei et al. [8] segmented the image based on grayscale consistency and used shape feature such as length-width ratio and density, but this method was limited while processing images with much interference or when the gray values of road regions variate. In conclusion, shape features such as main direction and length-width ratio can reflect the similarity between the measured items and the shape of road, while the bottleneck of road extraction lies in separating the potential road segments from the complex classified image with many broken road parts and non-road patches. Existing methods simply use connective segmentation (for classified images) or gray value based segmentation (for original images), thus incapable of processing images with great interference.

In this case, this paper presents a road extraction method based on direction consistency segmentation. Existing classification methods are first used to acquire binary road interest regions, after which a segmentation method for binary images based on direction consistency is proposed. This segmentation method calculates a main direction for each pixel and merge pixels with similar main directions into items, thus separating road segments from their surroundings. Then, calculates *LFI* and a certain other geometric features of each separated item to decide if it is road segment or not. Mathematical morphology and the segment-linking method are used to achieve the final road extraction result.

2   Road Extraction Based on Direction Consistency Segmentation

As mentioned above, a 'classification-extraction' strategy is adopted in this proposed road extraction method. Existing classification algorithms are first used to acquire binary images with road interest regions, after which a direction consistency segmentation method is proposed to separate these regions into various separated items. Geometric features are analyzed to identify road segments and the segment-linking and morphology method are used to achieve the final result. The framework for the proposed method is shown in Fig. 1.

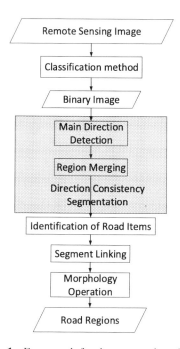

Fig. 1. Framework for the proposed method

Image classification methods using spectral information are rich, thus not further studied in this paper. Existing classification methods are directly used, the specific kind of which is alternative according to the image to be processed. This paper is mainly focused on further processing of the achieved classification result, which is using geometric features to separate and identify road regions. Therefore, geometric features of roads in classified images are analyzed first to decide what method to be used.

2.1   Geometric Features of Roads in Classified Images and the Corresponding Identification Methods

Classified images are simplified reflection of the complex road situations. In a classified image where the spectral information is erased, the extraction of roads mainly relies on geometric features. The geometric features of road regions and non-road regions are clearly different, while the former can be concluded as follow [9–11]:

(1) Since areas like parking lots, courtyards and buildings have similar spectral features with roads, they can hardly be distinguished by classification methods. Therefore, roads in classified images are often surrounded by these interference areas;
(2) The curvature of road is limited. Severe twists are rare in a road segment;
(3) The shape of roads is linear or stripped;
(4) The width of roads in the same image can be different, while the maximum of it is usually under a certain threshold;
(5) The width of a road keeps approximately the same, while some parts of it can be broken or narrow influenced by occlusions;
(6) Edges of regions can be uneven due to the drawbacks of classification method, where irregular convex or concave often appears;

These geometric features should be thoroughly considered while designing processing methods. Specifically, feature (1) determines that segmentation should be done first to separate potential road parts from their surroundings. Feature (2) makes it possible to segment regions by calculating the main direction of each pixel, since the general direction of a road segment keeps the same, with no fierce variations. Feature (3) makes it possible to use length-ratio or *LFI* to judge if a certain item belongs to a road segment or not. Feature (4) makes width a useful feature measurement to exclude non-road patches with large parameters. Feature (5) can be used to detect broken road parts or non-road patches connected to road regions. This feature is often used in some methods based on parallel lines detection. Feature (6) decides that a commonly used feature measurement such as shape index [12] (or length-area ratio) is not much valid since non-road regions often have large perimeters due to their irregular edges.

Table 1 shows the discussed geometric features of roads and their corresponding solutions for identification.

Table 1. Road geometric features and their corresponding identification solutions

Road geometric features	Identification solutions
Non-road regions are connected to road regions	Separate non-road regions and road regions
General direction keeps the same	Segmentation based on main direction
Shapes are often linear or stripped	Length-width ratio or *LFI*
The maximum width of roads in a certain image is limited	Set width threshold
Width of a road keeps approximately the same	Broken road or non-road patches can be identified; Parallel lines detection based methods
Irregular edges	Measurements like shape index have limitations

Therefore, this paper proposes a road segmentation and identification method based on direction consistency. The segmentation is done by merging pixels with similar main directions into regions, while the identification of road items is accomplished by using several feature measurements.

2.2 Main Direction Detection for Binary Images

As mentioned above, compared with interference regions such as buildings and parking lots, one of the important geometric features of roads is that their general direction in the same segment keeps unchanged, namely the local part of a road shows clear directionality. Considering this, the proposed segmentation method merges the pixels with similar directions into regions so that segmentation can be accomplished.

As the first stage of the proposed segmentation method, the calculation of main direction for each pixel is enlightened by the ATS (Angular Texture Signature) method. Traditional ATS is designed for grayscale images [13, 14]. The fundamental principle of it is using a rectangular template to rotate on the pixel at a certain angle step, in which way a set of templates with different angles can be achieved. Then the texture features within each template areas are calculated. Since this process should be done for each pixel and the texture features should be analyzed by calculating the variance in each rectangular model, it costs much time. Provided that the value of the target class in a binary image is 1 and the value of the background class is 0, a simple way of applying ATS on binary images is to replace the variance of gray values by the mean value of pixels within each model [15]. Another advantage of it in decreasing the computation time is that only non-zero pixels are needed for this calculation.

To further increase the accuracy for main direction detection and meanwhile decreasing the computation costs, the proposed method improves the binary image based ATS method. The model at each angle is replaced from a rectangular area to a pixel-width 'probe', meanwhile the number of angles for detection is increased. For a certain pixel $P(x, y)$, provided that its number of angles for detection is n and the maximum length of each probe is l, the angle of each detected direction can be described as $\alpha_i = i \cdot \pi/n \ (0 \leq i < n)$, thus the pixel coordinates set corresponded to this probe can be described as $\{A_1(\cos \alpha_1, \sin \alpha_1), A_2(2 \cdot \cos \alpha_1, 2 \cdot \sin \alpha_1), \ldots, A_l(l \cdot \cos \alpha_1, l \cdot \sin \alpha_1)\}$. Sum up the number of non-zero pixels within each probe to get a value T. The main direction of the pixel P is the direction corresponded to the probe with the maximum T. As shown in Fig. 2(a), 36 probes are generated to detect the main direction of pixel P, which are numbered from 1 to 36. Since each pair of directions at the difference of π correspond to the same practical direction of a road segment, these detected directions should be recorded from 1 to 18. Figure 2(b) shows the number of non-zero pixels at each direction. It can be seen that the maximum number of non-zero pixels are at the directions numbered as 1 and 19, which are practically the same direction, thus the main direction of P should be recorded as 1. Apply this method on each non-zero pixel in the image to get its main direction. While processing low-resolution images, the number of probes should be set to 72 or higher to make the detection more accurate.

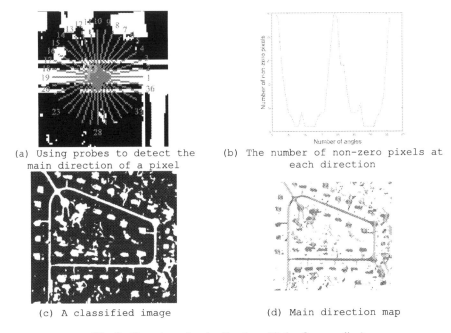

(a) Using probes to detect the main direction of a pixel

(b) The number of non-zero pixels at each direction

(c) A classified image

(d) Main direction map

Fig. 2. Detection of main directions (Color figure online)

Figure 2(c) is a classified image of a suburb street. It can be seen that this image contains many non-road patches which are actually buildings and shadows of trees. Main directions of pixels in this image are detected and rendered with colors as shown in Fig. 2(d), in which the similarity of colors implies the similarity of main directions.

2.3 Merging Pixels with Similar Directions Based on Region Growing

After detecting the main direction of each pixel, further steps are needed to merge pixels with similar directions into regions. As a commonly used method in the area of image processing, region growing has the advantages of high computation speed and flexible start point selection and terminal condition setting, thus it is adopted to merge pixels with similar directions.

Region growing is a commonly used semi-auto segmentation method proposed by Zucker [16], the fundamental principle of which is using a certain similarity criterions to merge pixels adjacent to a given start points so that a homogeneous region can be acquired. After choosing one or several seed points as the start points for growing, pixels in the neighborhood of the given seed point are merged into the growing region once they meet certain similarity criterions. Then, these pixels are taken as the seed points in a new round of growing. This process is repeated until no more pixels satisfying the criterions are merged or until the growing region satisfies a certain terminal condition [17]. In this paper, the similarity criterion for merging pixels is set as the similarity of main directions of pixels, namely the direction consistency. Any pixel that has not been merged yet can be chosen as a start point. In this way, all non-zero pixels can be merged and various homogenous regions are acquired.

Two details of road directions are taken into account while designing criterions for measuring the direction consistency: (1) The direction of a road segment stays generally the same, thus the difference between the direction of any pixel in this segment and the main direction of this segment should not be huge; (2) The main directions of pixels in a road witnesses a progressive change at the curves.

Therefore, the criterion for measuring direction consistency is set as follow: (1) The difference between the main direction of the detected pixel and the main direction of the growing region (namely the main direction for most pixels within this growing region, calculated and updated in each round of growing) should be no more than a given threshold T_1; (2) The difference of the main direction of the detected pixel and the main direction of the seed point in this round should be no more than a given threshold T_2. The specific value of T_1 and T_2 should be set according to the number of detected directions. Usually T_2 is set corresponded to the angle of $5°$ and make $T_1 = 3 \cdot T_2$.

Apply this method on the detected main direction map as shown in Fig. 2(d). The segmentation result is shown in Fig. 3(a), in which each color represents a homogenous region.

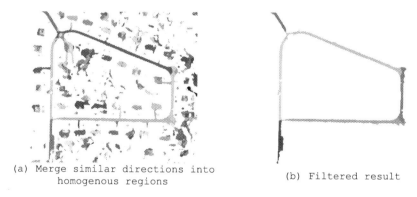

(a) Merge similar directions into homogenous regions

(b) Filtered result

Fig. 3. Direction consistency segmentation

2.4   Identification of Road Items and the Further Processing

After the segmentation process, potential road segments and non-road patches are separated. Therefore, it is now possible to identify these road items and remove non-road items based on their geometric features.

Wang et al. concluded 11 geometric features which are commonly used in road extraction, including area, length-width ratio, shape index, geometric density, main direction, asymmetry, compactness, rectangularity, etc. After a thorough analysis on the validity of each feature, it was concluded that the length-width ratio and the main direction are very effective in identifying road items, while area can be used to exclude large non-road patches. Based on this study, the following geometric features are adopted to decide if a separated item belongs to a road segment or not:

(1) Area A: the number of pixels that the measured item contains;
(2) Width W_{MER}: the width of minimum enclosing rectangle of the measured item;
(3) *LFI* (Linear Feature Index) of the measured item [18], which is an improved version of length-width ratio. Traditional length-width ratio measurement is defined as the ratio of the length of minimum enclosing rectangle (L_{MER}) to the width of it (W_{MER}), as is shown in formula (1).

$$R = \frac{L_{MER}}{W_{MER}} \tag{1}$$

Miao et al. improved it and proposed a *LFI* measurement to describe the extent that the outline of the measured item being close to a rectangle. The basic principle of it is to stretch the interest region that contains the measured item to build a new rectangle, after which calculating the length-width ratio of this new rectangle.

$$L \cdot W = n_p$$
$$\text{s.t. } L^2 = L_{MER}^2 + W_{MER}^2 \tag{2}$$

As is shown in formula (2), L and W is the length and width of the new triangle and n_p is the area of the measured item. The definition of *LFI* is shown in formula (3).

$$LFI = \frac{L}{W} = \frac{L}{n_p/L} = \frac{L^2}{n_p} \tag{3}$$

Set a threshold for each geometric measurement. Only those items that meet all threshold conditions can be considered as road segments or road parts. All non-road items should be erased.

Apply this method to identify the road items in Fig. 3(a), the corresponding filtered result is shown in Fig. 3(b).

Due to the limitation of classification methods and the influence of occlusions, it is sometimes inevitable that the roads which are practically consistent shown as interrupted or broken in its classified binary images. Additionally, the proposed method is based on the consistency of road direction, thus has some limitations at the curves for the direction changes greatly in these parts, where interruptions of the segmented result may happen. To offset these bad influences, broken road parts should be connected. For roads without severe interruption, it is possible to calculate the geometric directions at each end of the broken part to repair the original road information. Nikzad et al. [19] had proposed a method for detecting and separating worms in medical images, which was used by Sukhendu et al. [2] to address the issue of broken road connection in road extraction. The main idea of this method is to analyze the thinned outlines of the classified road result and connect road ends when their direction difference is no more than a given threshold.

This method is adopted in this paper to connect broken roads if their interruptions are not severe. Finally, the morphological opening and small regions removing operation is conducted to get a comparatively smooth road extraction result with regular edges.

3   Experiments and Analysis

To prove the effectiveness of the proposed method, several groups of experiments are conducted on images at different resolutions and with different road distributions. The part of processing for binary images in the proposed method is compared with two existing methods which have shown effectiveness in road extraction, namely the RPS (Road Part Segmentation) method [2] and the path-opening method [1]. In this way, the characteristics of each method can be analyzed.

3.1   Experiment 1: Road Extraction from High-Resolution Images

As shown in Fig. 4, four groups of contrast experiments are conducted on four high-resolution images. The original images are shown in Fig. 4(a). The first and forth group are done on 0.3 m resolution google earth images. The second group is done on 0.15 m resolution orthographic images. The third group use grayed aerial images [20]. The reference roads which are manually drawn are shown in Fig. 4(b). Bayesian classification method is conducted on color images, while the Otsu method is conducted on the grayscale image. The classification results are shown in Fig. 4(c). The path-opening method, RPS method and the corresponding part of the proposed method are separately conducted on the Fig. 4(c), the corresponding results of which are shown in Fig. 4(d)–(f).

The cost of computation time for each method is shown in Table 2.

3.2   Experiment 2: Road Extraction from Low Resolution Images

As shown in Fig. 5, three groups of contrast experiments are conducted on low resolution images. Original images are shown in Fig. 5(a). In the first group, a ZY-3 image is used for experiment. The second and third group use low-resolution aerial images for experiments. Reference roads are shown in Fig. 5(b). The classification results are shown in Fig. 5(c). The three methods are separately conducted on Fig. 5(c), the corresponding results of which are shown in Fig. 5(d)–(f).

The cost of computation time for each method is shown in Table 3.

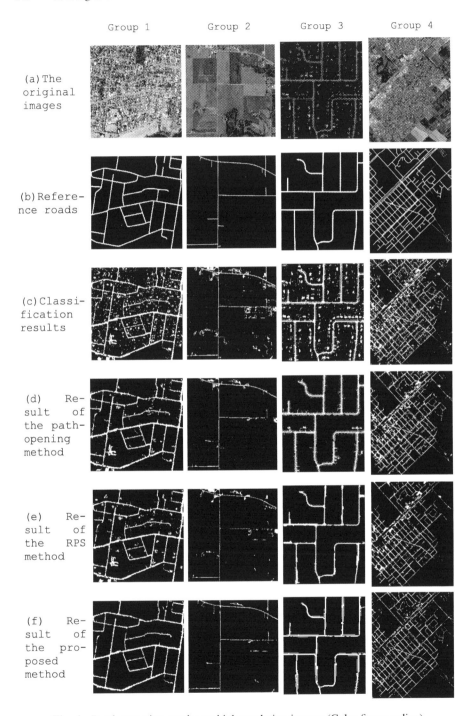

Fig. 4. Road extraction results on high-resolution images (Color figure online)

Table 2. Computation time

	Group 1	Group 2	Group 3	Group 4
Picture size/pixels	3372 × 3225	893 × 893	586 × 672	2000 × 2467
The path-opening method/s	2377.92	197.2	192.1	1037.8
The RPS method/s	4569.8	45.5	26.3	3429.0
The proposed method (the part of processing on classified results)/s	188.6	2.4	3.3	36.8

3.3 Analysis on Experimental Results

The results of the proposed method and two comparative methods can be analyzed to evaluate their effectiveness. In terms of the two comparative methods, it can be seen that:

(1) The path-opening method is effective in removing isolated small regions, though some road parts are removed at the same time. However, this method is unable to remove non-road patches connected to roads. Therefore, it is more suitable for processing low-resolution images in which roads are linear and comparatively disperse.

(2) The RPS method can remove non-road patches connected to roads and can to some extent repair the incomplete road segments to achieve a smoothed result, but its processing is unstable. Since this method simply uses area of each region as the geometric feature to remove non-road patches, some road segments are also erased, causing some road information to be missed in the final result. Additionally, some non-road regions are even expanded as a result of the repairing of incomplete regions.

Compared with these two methods, the proposed method can achieve comparatively good extraction results on either high or low resolution images with different distributions of roads. It can remove both non-road patches connected to roads and isolated small regions, thus achieving a good result even on some complex images (such as group 4 in experiment 1 where the classified result contains many interrupted or broken road segments). An unsolved problem of the proposed method is that linear non-road regions can hardly be filtered since their shapes are similar with road segments. In addition, there is room for improvement in terms of recovering road information from the broken parts of roads, although the proposed method shows advantages compared with the two comparative methods.

In terms of computation time, the cost of these three tested methods are all closely related with pixels to be processed, which is determined by the image size and the density and width of roads. In general, the proposed method shows clear advantage in computation time compared with the path-opening method and the RPS method, though it also costs considerable time while processing high-resolution images with wide roads. The specific comparison between these three methods are shown in Table 4.

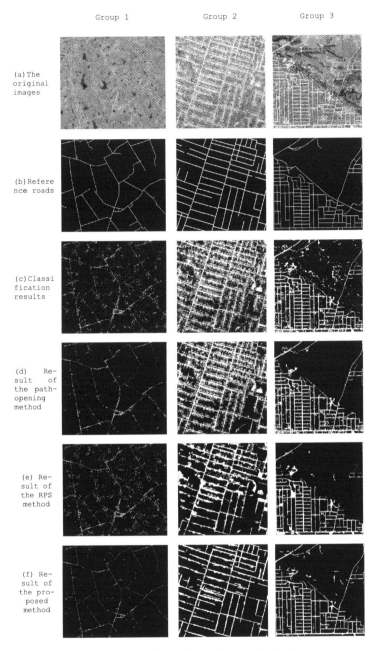

Fig. 5. Road extraction results on low resolution images

Table 3. Computation time

	Group 1	Group 2	Group 3
Picture size/pixels	2533×2179	539×539	600×612
The path-opening method/s	2861.22	169.1	111.3
The RPS method/s	542	344.8	49.7
The proposed method (the part of processing on classified results)/s	23.3	4.59	2.7

Table 4. Comparison between three road extraction methods

Methods	Performance of removing non-road patches connected to roads	Performance of removing isolated areas	Performance of computation speed	Application scope
Path-opening	Poor	Comparatively good, unable to remove regions with large area	Poor	Low-resolution images with disperse roads
RPS	Fair, with errors	Unstable, road segment are often removed	Poor	High-resolution images with edge-smooth roads
The proposed method	Good	Fair	Fair	Both high-resolution and low-resolution remote sensing images

4 Conclusions

A direction consistency segmentation method is proposed in this paper to separate linear items from classified images and a set of road extraction method is designed based on it. According to the experimental result, the proposed method shows a wider application scope compared with two representative existing methods. The proposed road extraction method based on direction consistency segmentation shows good performance on either high-resolution or low-resolution images, capable of processing images with complex road situations, with great advantages in computation speed. However, it still worth further study to address the problem of repairing road information from interrupted or broken road classification result.

References

1. Bennamoun, M., Mamic, G.J.: Fundamentals and Case Studies. Springer Science & Business Media, Heidelberg (2012)

2. Das, S., Mirnalinee, T.T., Varghese, K.: Use of salient features for the design of a multistage framework to extract roads from high-resolution multispectral satellite images. IEEE Trans. Geosci. Remote Sens. **49**(10), 3906–3931 (2011)

3. Zhang, Q., Couloigner, I.: Benefit of the angular texture signature for the separation of parking lots and roads on high resolution multi-spectral imagery. Pattern Recogn. Lett. **27**(9), 937–946 (2006)

4. Zhu, C.S., Zhou, W., Guan, J.: Main roads extraction from SAR imagery based on parallel pairs detection. J. Image Graph. **10**, 1908–1917 (2011)

5. Cheng, J.H., Gao, G., Ku, X.S., et al.: Review of road network extraction from SAR images. J. Image Graph. **01**, 11–23 (2013)

6. Song, M., Civco, D.: Road extraction using SVM and image segmentation. Photogram. Eng. Remote Sens. **70**(12), 1365–1371 (2004)

7. Jia, C.L., Zhao, L.J., Wu, Q.C., et al.: Automatic road extraction from SAR imagery based on genetic algorithm. J. Image Graph. **13**(6), 1134–1142 (2008)

8. Lei, X.Q., Wang, W.X., Lai, J.: A method of road extraction from high-resolution remote sensing images based on shape features. Acta Geodaetica et Cartographica Sinica **05**, 457–465 (2009)

9. Ding, L., Yao, H., Guo, H.T., et al.: Using neighborhood centroid voting to extract road centerlines from classified images. J. Image Graph. **20**(11), 1534–2526 (2015). doi:10. 11834/jig.20151112

10. Zhang, R., Zhang, J.X., Li, H.T.: Semi-automatic extraction of ribbon roads from high resolution remotely sensed imagery based on anguar texture signature and profile match. J. Remote Sens. **02**, 224–232 (2008)

11. Lin, X.G., Zhang, J.X., Li, H.T., et al.: Semi-automatic extraction of ribbon road from high resolution remotely sensed imagery by a t-shaped template match-ing. Geomatics Inf. Sci. Wuhan Univ. **03**, 293–296 (2009)

12. Huang, X., Zhang, L.: Road centreline extraction from high-resolution imagery based on multiscale structural features and support vector machines. Int. J. Remote Sens. **30**(8), 1977–1987 (2009)

13. Haverkamp, D.: Extracting straight road structure in urban environments using IKONOS satellite imagery. Opt. Eng. **41**(9), 2107–2110 (2002)

14. Gibson, L.: Finding road networks in Ikonos satellite imagery. In: Proceedings of the ASPRS Annual Conference, ASPRS, 2003, Anchorage, Alaska, pp. 05–09 (2003)

15. Zhang, R., Zhang, J.X., Li, H.T.: Semi-automatic extraction of ribbon roads from high resolution remotely sensed imagery based on anguar texture signature and profile match. J. Remote Sens. **02**, 224–232 (2008)

16. Zucker, S.W.: Region growing: childhood and adolescence. Comput. Graph. Image Process. **5**, 382–399 (1976)

17. Peng, F.P., Bao, S.S., Zeng, B.Q.: Segmentation of liver based on adaptive region growing. Comput. Eng. Appl. **46**(33), 198–200 (2010)

18. Miao, Z., Shi, W., Zhang, H., et al.: Road centerline extraction from high-resolution imagery based on shape features and multivariate adaptive regression splines. IEEE Geosci. Remote Sens. Lett. **10**(3), 583–587 (2013)

19. Rizvandi, N.B., Pizurica, A., Philips, W., et al.: Edge linking based method to detect and separate individual c. elegans worms in culture. In: Digital Image Computing: Techniques and Applications (DICTA), pp. 65–70. IEEE (2008)

20. Computer Vision Lab, Data (2013). http://cvlab.epfl.ch/data/delin

Fingertip in the Eye: An Attention-Based Method for Real-Time Hand Tracking and Fingertip Detection in Egocentric Videos

Xiaorui Liu, Yichao Huang, Xin Zhang[(⊠)], and Lianwen Jin

School of Electronic and Information Engineering,
South China University of Technology, Guangzhou, China
{eexinzhang,eelwjin}@scut.edu.cn

Abstract. The hand and fingertip tracking is the crucial part in the egocentric vision interaction, and it remains a challenging problem due to various factors like dynamic environment and hand deformation. We propose a convolutional neural network (CNN) based method for the real-time and accurate hand tracking and fingertip detection in RGB sequences captured by an egocentric mobile camera. Firstly, we build a large scale dataset, Ego-Finger, containing plenty of scenarios and human labeled ground truth. Secondly, we propose a two stage CNN pipeline, i.e., the human vision inspired Attention-based Hand Tracker (AHT) and the hand physical constrained Multi-Points Fingertip Detector (MFD). Comparing with state-of-the-art methods, the proposed method achieves very promising results in the real-time fashion.

Keywords: Attention-based hand tracking · Multiple points fingertip detection · Large scale ego-finger dataset

1 Introduction

With the development of smart wearable cameras such as Microsoft HoloLens and Google Glass, the egocentric vision and its applications have drawn lots of attention of people from various fields. We are particularly interested in its vision-based interaction with human. As discussed in [4], the hand and fingertip-based gesture interaction could be the most appropriate approach because it acts like a virtual mouse to which people get used.

The accurate robust and real-time hand tracking and fingertip detection in RGB egocentric videos remain great challenges due to many factors like background complexity, illumination variation, hand shape deformation, motion blur etc. Recent developments are using depth sensor [17,18], but we only use color sequences for both indoor and outdoor situation. Hand segmentation algorithms [2,12,13] are sensitive to background disturbance and motion blur. Several tracking-based methods [8,10] can be employed for hand tracking but still face difficulties on the fingertip tracking since it's too small. In [14], the skin detector, DPM-based context and shape detector are combined to detect hands,

© Springer Nature Singapore Pte Ltd. 2016
T. Tan et al. (Eds.): CCPR 2016, Part I, CCIS 662, pp. 145–154, 2016.
DOI: 10.1007/978-981-10-3002-4_12

but their method is time-consuming due to the sliding window strategy. Convolutional neural network (CNN) is applied in [1] to detect hands but this framework is too slow due to redundant proposals. In [3], the detector returns whether hands appear without their locations. A two stages CNN-based hand and fingertip detection framework is proposed in [9] but it is trained and tested on a small and limited data set.

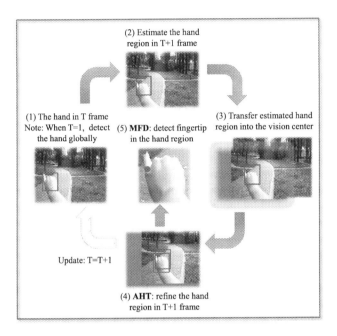

Fig. 1. Hand tracking and fingertip detection pipeline

2   Dataset: Ego-Finger

Convolutional neural network, which is adopt as the basic model in this paper, has been empirically verified to be a powerful feature extractor for objects in images and videos [11,15]. To achieve satisfied feature representation learning, we establish a new large scale dataset named Ego-Finger[1], which has been released for academic research.

The Ego-Finger dataset contains 93,729 RGB frames (640×480 pixels) from 24 egocentric videos capturing the hand poses pointing at different directions by index finger. These videos are taken and labeled by individual participants in 24 various scenarios. Besides, the labels include the bounding box of the hand, locations for index fingertip and index finger joint. Correspondingly, it involves

---

[1] Ego-Finger dataset is available at:
http://www.hcii-lab.net/data/SCUTEgoFinger/index.htm.

Fig. 2. Examples of the dataset frames captured in 24 scenarios

challenges from several aspects as follows: background complexity, including scenario diversity, illumination variation, skin-like object clutters and fast motion; hand appearance diversity, including skin colour variation, shape deformation, rotation and even motion blur. Figure 2 shows some samples in 24 scenarios of the dataset.

3   Hand Tracking and Fingertip Detection

Directly detecting or tracking the tiny fingertip in RGB videos is intractable. Huang [9] empirically confirms that two-stage pipeline is effective for fingertip detection because the hand region helps filter out most part of complicated background disturbance. Therefore, this paper follows the two-stage pipeline with significant improvements. In the first stage, we propose the human vision mechanism inspired Attention-based Hand Tracker (AHT) to return hand region. The second stage is the Multi-Points Fingertip Detector (MFD) that imposes the finger shape constraint.

3.1   Attention-Based Hand Tracker: AHT

Baseline Hand Detector: BHD. The baseline hand detector is used for the first frame hand location and further experimental comparisons. Similar to the hand detector in [9], we employ an 8-layer CNN to extract features, followed by a 3-layer fully connected layer as regressor to predict labels of the hand bounding box, which is denoted by a vector (x, y, w, h). Notation (x, y) is the box center while w and h denote width and height. All labels are normalized to 0 to 1. The network, originated from AlexNet [11], is shown in Fig. 3.

The objective function is the Euclidean distance between (x_i, y_i, w_i, h_i) and $(\bar{x}_i, \bar{y}_i, \bar{w}_i, \bar{h}_i)$, which denote the predicted and ground truth labels of training sample X_i respectively. The objective function is optimized by SGD algorithm.

After training, the hand features such as skin color, hand edge, their hierarchical combinations and even semantic information, can be extracted by the

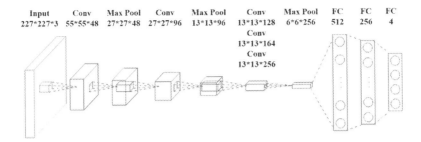

Fig. 3. Network for hand detection

convolutional and max pooling layers. These feature maps represent the information of hand size and location. The following fully connected layers, considered as regressor, will map those features to parameters of the hand bounding box. In this paper, we refer this framework as the baseline model.

Attention-Based Hand Tracker: AHT. Based on our experiments, we discover that hand detection performance varies while the hand locates in different areas in the frame, and we believe it is essentially due to the non-uniform distribution of the training data that causes central bias. To further explore this discovery, on one hand, we apply the baseline model and find that the detected hand bounding box in Surrounding Zone, as Fig. 4(a) indicates, is prone to drift towards the Attention Focus Zone, as Fig. 4(b) shows.

On the other hand, we analyze the hand location distribution of the training data, and reveal that it fits the Gaussian distribution as Fig. 4(a) shows. The discovered distribution is essentially consistent with human eye visual mechanism that the hand locates in the vision center more often. According to human vision research, people unconsciously focuses on the central part of vision, which academically named "center bias" [5,19] or "center prior" [6,7]. While gazing at the target object, human eyes will relocate target by transferring it to the vision center.

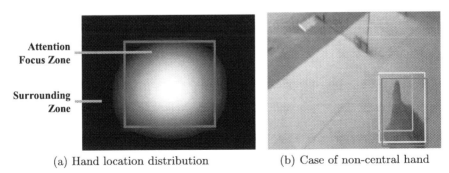

(a) Hand location distribution   (b) Case of non-central hand

Fig. 4. (a) Hand location distribution: the intensity indicates frequency. (b) Case of non-central hand: green, yellow and red box indicate the hand region returned by BHD, AHT and ground truth label respectively. (Color figure online)

Above analyses inspire us that we could remove the center bias influence to achieve equally accurate hand detection in both Attention Focus Zone and Surrounding Zone by transferring the hand region to the vision center. It is the key idea of the attention-based hand tracking method.

It is worth noticing that while transferring the vision center, we keep the size of video frames input into CNN while filling those part exceeding the original image with mean values to avoid feature response. To adapt to this change, we fine tune the CNN parameters with synthesized training samples which are roughly centralized on hand and filled with mean values for exceeding part. We refer this fine-tuned CNN as the AHT model.

Due to temporal consistence, the hand position and size change slightly in adjacent video frames. Therefore, we can roughly estimate the target position in the current frame by the tracking result from the previous frame. We translate the estimated hand region into the vision center to generate a new image, and apply AHT model to refine the hand bounding box in current frame. The whole hand tracking process is shown in Fig. 1 and can be formalized as Algorithm 1.

Algorithm 1. Attention-based Hand Tracking Framework

$T = 1$, detect the hand globally by BHD and get initial hand region B^1
loop
    1. Estimate the hand region $\hat{B}^{T+1} = B^T$ in $T + 1$ frame according to tracking result in T frame
    2. Translate the roughly estimated hand region into the "vision center"
    3. Refine the hand bounding box B^{T+1} in $T + 1$ frame by fine-tuned CNN
    4. Update $T = T + 1$
end loop

By this attention-based tracking method, we can detect the hand accurately in arbitrary position of the video frames regardless of detection bias as Fig. 4(b) shows. Experiments show that the AHT significantly promotes overlap rate by 11.1 % compared to the BHD and outperforms two cutting-edge vision tracking systems.

3.2 Multi-points Fingertip Detector: MFD

Obtaining the hand region by AHT helps to filter out most background disturbance. To detect the fingertip, it's intuitive to apply hand segmentation algorithms [12,13] to parse the hand. However, in the egocentric vision captured by a mobile camera, motion blur caused by fast motion, rapid illumination change are intractable for most segmentation algorithms. In this specific situation, the global contextual information of the hand is of great concern.

According to [16], researchers usually describe a hand with multiple points models, indicating the fact that multiple key points of hand form spatial constraints. In our case, the index fingertip location is biologically adjacent to index

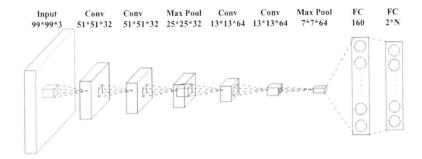

Fig. 5. Network for fingertip detection

finger joint. Shuffling, swaying, twisting or flipping of hand would never change the structural constraint among these hand points.

Convolutional neural network, as an powerful appearance model, can extract not only low level features of the fingertip but also global contextual information of the whole hand. Therefore, we design an network which takes the hand region as input and output the coordinates of key points as Fig. 5 shows. N (1 or 2) denotes the number of key points to predict and the objective function is the Euclidean distance between predictions of all points and the ground truth labels.

4   Experiments and Discussions

4.1   Data Preparation

We randomly choose 21 videos as training set containing 80,755 frames and the rest 3 videos are test set containing 12,974 frames. For AHT, we randomly flip the images horizontally. For MFD, we make random affine transformation in scale, translation and rotation for hand region. Data augmentation enlarges coverage of data, which is beneficial to reduce overfitting.

4.2   Hand Tracking Performance

We conduct a comparison with two cutting-edge vision tracking systems named Kernelized Correlation Filters (KCF) [8] and "HOG+LR" [20]. As to compare the One Pass Evaluation (OPE) which is a common evaluating protocol in vision tracking tasks, we select 3 video clips with continuous frames in the test set because KCF and HOG+LG methods are not designed for long sequence tracking.

Figure 6 shows the hand tracking results for different methods in 3 video clips: (1) Baseline Hand Detector (BHD); (2) Attention-based Hand Tracker (AHT); (3) Kernelized Correlation Filters (KCF); (4) HOG+Logistic Regression (HOG+LR). In video clip 1, all methods work well and our proposal ranks number one. KCF fails in video clip 2 and HOG+LR fails in video clip 3, but our proposal still tracks hand very well.

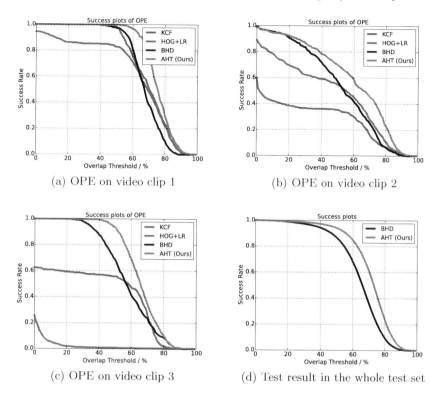

Fig. 6. Performance comparisons with one pass evaluation (OPE) criteria on 3 short video clips (a–c) and test result on the whole test set (d).

By using the whole long test sets, we can only compare performances of BHD and AHT since others fail shortly. The BHD and AHT achieve 0.658 and 0.731 average overlaps with ground truth hand region respectively as Fig. 6(d) shows. Relatively, the AHT significantly improves the hand region overlap for 11.1 % on average. Considering 0.5 overlap as threshold, our proposal AHT achieves success rate of 94.3 % comparing to 87.0 % of BHD [9].

4.3 Fingertip Detection Performance

To explore the interrelation of hand tracking and fingertip detection, we compare 3 strategies for hand tracking: (a) Baseline Hand Detector (BHD); (b) Attention-based Hand Tracker (AHT); (c) Ground Truth Hand Bounding Box (GT). And for the fingertip detection module, Single-Point Fingertip Detector (SFD) is compared with Multi-Points Fingertip Detector (MFD). In MFD, we choose two key points including index fingertip and finger joint as Fig. 2 shows, because they are biologically adjacent and form inherent structural constraint. Fingertip detection error is showed in Fig. 7.

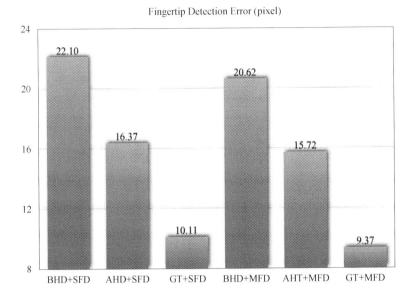

Fig. 7. Comparison of different fingertip detection methods

From above comparisons, it can be concluded that inner mismatching of cascaded structure would cause extra error for fingertip detection. In other words, once the hand region is not cropped appropriately, the fingertip detector will accordingly predict poorly, which indicates the importance of accurate hand tracking. It confirms the effectiveness of AHT and MFD. The proposed AHT+MFD method enormously improves the fingertip detection accuracy for 28.9 % compared to BHD+SFD method.

Applying Caffe framework on single GPU GTX 980, our system consumes about 2 ms and 1 ms for AHT and MFD respectively. Combining extra time on image processing, the overall pipeline costs about 6 ms and can run in real-time.

5   Conclusion

This paper presents a new CNN-based attention-based hand tracking and fingertip detection framework. The Attention-based Hand Tracker is inspired by visual attention prior, and the Multi-Points Fingertip Detector models the physical constraint of hand. Based on the large egocentric fingertip dataset we built, the proposed AHT+MFD pipeline produces very promising results comparing with state-of-the-art methods. The hand detector achieves 0.731 average overlap rate on hand tracking, and integrated pipeline achieves 15.72 pixels error on average out of 640 × 480 video sequences. Future work will focus on multiple hands tracking and fingertip detection.

Acknowledgement. This research is supported in part by MSRA University Collaboration Fund (No.: FY16-RES-THEME-075), Science and Technology Planning Project of Guangdong Province (Grant No.: 2015B010130003, 2015B010101004, 2016A010101014), Fundamental Research Funds for the Central Universities (Grant No.: 2015ZZ027).

References

1. Bambach, S., Lee, S., Crandall, D.J., Yu, C.: Lending a hand: detecting hands and recognizing activities in complex egocentric interactions. In: Proceedings of the IEEE International Conference on Computer Vision, pp. 1949–1957 (2015)
2. Baraldi, L., Paci, F., Serra, G., Benini, L., Cucchiara, R.: Gesture recognition in ego-centric videos using dense trajectories and hand segmentation. In: IEEE Conference on Computer Vision and Pattern Recognition Workshops (CVPRW), pp. 702–707 (2014)
3. Betancourt, A., Morerio, P., Marcenaro, L., Rauterberg, M., Regazzoni, C.: Filtering SVM frame-by-frame binary classification in a detection framework. In: IEEE International Conference on Image Processing (ICIP), pp. 2552–2556 (2015)
4. Betancourt, A., Morerio, P., Regazzoni, C.S., Rauterberg, M.: The evolution of first person vision methods: a survey. IEEE Trans. Circ. Syst. Video Technol. **25**(5), 744–760 (2015)
5. Bindemann, M.: Scene and screen center bias early eye movements in scene viewing. Vis. Res. **50**(23), 2577–2587 (2010)
6. Cheng, M., Mitra, N.J., Huang, X., Torr, P.H., Hu, S.: Global contrast based salient region detection. IEEE Trans. Pattern Anal. Mach. Intell. **37**(3), 569–582 (2015)
7. Goferman, S., Zelnik-Manor, L., Tal, A.: Context-aware saliency detection. IEEE Trans. Pattern Anal. Mach. Intell. **34**(10), 1915–1926 (2012)
8. Henriques, J.F., Caseiro, R., Martins, P., Batista, J.: High-speed tracking with kernelized correlation filters. IEEE Trans. Pattern Anal. Mach. Intell. **37**(3), 583–596 (2015)
9. Huang, Y., Liu, X., Zhang, X., Jin, L.: Deepfinger: a cascade convolutional neuron network approach to finger key point detection in egocentric vision with mobile camera. In: The IEEE Conference on System, Man and Cybernetics (SMC), pp. 2944–2949 (2015)
10. Kalal, Z., Mikolajczyk, K., Matas, J.: Tracking-learning-detection. IEEE Trans. Pattern Anal. Mach. Intell. **34**(7), 1409–1422 (2012)
11. Krizhevsky, A., Sutskever, I., Hinton, G.E.: Imagenet classification with deep convolutional neural networks. In: Advances in Neural Information Processing Systems, pp. 1097–1105 (2012)
12. Li, C., Kitani, K.M.: Model recommendation with virtual probes for egocentric hand detection. In: IEEE International Conference on Computer Vision (ICCV), pp. 2624–2631 (2013)
13. Li, C., Kitani, K.M.: Pixel-level hand detection in ego-centric videos. In: IEEE Conference on Computer Vision and Pattern Recognition (CVPR), pp. 3570–3577 (2013)
14. Mittal, A., Zisserman, A., Torr, P.H.: Hand detection using multiple proposals. In: BMVC, pp. 1–11. Citeseer (2011)
15. Russakovsky, O., Deng, J., Su, H., Krause, J., Satheesh, S., Ma, S., Huang, Z., Karpathy, A., Khosla, A., Bernstein, M., et al.: Imagenet large scale visual recognition challenge. Int. J. Comput. Vis. **115**, 1–42 (2014)

16. Sun, X., Wei, Y., Liang, S., Tang, X., Sun, J.: Cascaded hand pose regression. In: The IEEE Conference on Computer Vision and Pattern Recognition (CVPR) (2015)
17. Supancic, J.S., Rogez, G., Yang, Y., Shotton, J., Ramanan, D.: Depth-based hand pose estimation: data, methods, and challenges. In: The IEEE International Conference on Computer Vision (ICCV) (2015)
18. Tompson, J., Stein, M., Lecun, Y., Perlin, K.: Real-time continuous pose recovery of human hands using convolutional networks. ACM Trans. Graph. (TOG) **33**(5), 169 (2014)
19. Tseng, P.H., Carmi, R., Cameron, I.G., Munoz, D.P., Itti, L.: Quantifying center bias of observers in free viewing of dynamic natural scenes. J. Vis. **9**(7), 4 (2009)
20. Wang, N., Shi, J., Yeung, D.Y., Jia, J.: Understanding and diagnosing visual tracking systems. In: The IEEE International Conference on Computer Vision (ICCV) (2015)

Multiple-Classifiers Based Hand Gesture Recognition

Simin Li, Zihan Ni, and Nong Sang[✉]

Key Laboratory of Image Processing and Intelligent Control
(Huazhong University of Science and Technology),
Ministry of Education, School of Automation,
Huazhong University of Science and Technology, Wuhan 430074, China
nsang@hust.edu.cn

Abstract. Gesture recognition technology is important in the field of human-computer interaction (HCI), the gesture recognition technology which is based on visual is sensitive to the impact of the environment. We proposed a multiple-classifiers based gesture recognition algorithm that recognizes ten kinds of gesture. The algorithm gets the size and the direction of the gesture by hand tracking algorithm which has the ability of segmentation and can give us the rough outline of the tracked hand. Based on this information, we rotate the image and get the upright image of the hand gesture. Then we extract the HOG feature of the upright hand image, and use multiple classifiers to classify the gesture. The algorithm has a better recognition rate whether the background color is similar to the skin or complex.

Keywords: Human computer interaction · Gesture recognition · Contour features · Multiple classifiers

1 Introduction

1.1 Motivation

As a way of interpersonal communication, gesture has the property of simple and direct. So gesture and gesture recognition terms are heavily encountered in human-computer interaction [1]. Hand gesture recognition is an important subtask in many computer vision applications. Its natural ability to represent the ideas makes it a good choice of interface to the computer system.

In our work, we proposed a multiple-classifiers based gesture recognition algorithm that recognizes ten kinds of gesture. The algorithm acts well even the angle and the scale change.

1.2 Related Work

To enable vision based hand gesture recognition, numerous approaches have been proposed, which can be classified into geometric features based approaches and

© Springer Nature Singapore Pte Ltd. 2016
T. Tan et al. (Eds.): CCPR 2016, Part I, CCIS 662, pp. 155–163, 2016.
DOI: 10.1007/978-981-10-3002-4_13

statistical learning based approaches. Panwar [1] accomplished the task of recognition depending on the shape parameters of the hand gesture. Ahuja [2] proposed a scheme using a database-driven hand gesture recognition based upon skin color model approach and thresholding approach along with an effective template matching using PCA. Dardas [3] recognized hand gestures via bag-of-features and multiclass support vector machine (SVM). Joshi [4] used HOG features to identify the static hand gestures and learned these features by SVM. Guo [5] tried to get the samples in six directions through rotation, which is to adapt to the direction of the gesture in practical application.

The approach given in [1] can hardly recognize the shape parameters of the hand gesture when the distance between the hand and the camera is not close enough. The skin color model approach in [2] is not reliable. The SIFT features in [3] are not good enough to perform the gesture. The HOG features in [4] work well with SVM, but it doesn't allow for multi-angle of the gesture. The approach given in [5] has samples in six directions to adapt to the direction of the gesture, but it's complicated to compute and the intra-class distance between the positive samples is too large, which reduce the accuracy of the classifier.

In this paper, we get the contours of the hand by tracking and segmentation, then estimate the direction of the hand and rotate the hand patch to get an upright gesture which allows for multi-angles of the gesture, and recognize the gesture based on the HOG features of the gesture and SVM.

2   Multiple-Classifiers Based Hand Gesture Recognition

Our gesture recognition system contains three parts, hand tracking and segmentation, hand rotation and recognition. The system can recognize ten kinds of gesture as shown in Fig. 1.

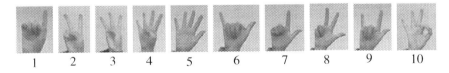

1   2   3   4   5   6   7   8   9   10

Fig. 1. The gestures we recognized.

2.1   Hand Tracking and Segmentation

In this part, we get the position and the contour of the hand. Our hand tracking algorithm is based on the segmentation [6]. That is to say, we can get the tracking result and the segmentation result of the hand after tracking.

In our tracking algorithm, we use a spatial model representing the hand, the spatial model is an ellipse M_I, as shown in Fig. 2.

M_I is represented by an elliptic region as mentioned above. M_I is the inscribed ellipse of the hand detection window or the track result of the previous frame.

Fig. 2. The spatial model of the previous frame.

We use $d(x)$ presenting the model where $d(x)$ is the Mahalanobis distance between the pixel x and center of the hand, $d(x)$ is computed as

$$d(\mathbf{x}) = \sqrt{(\mathbf{x} - \mathbf{x_m})^T \mathbf{C}^{-1} (\mathbf{x} - \mathbf{x_m})} \tag{1}$$

where $\mathbf{x}$ is the spatial coordinate of the pixel, $\mathbf{x_m}$ is the center of M_I, and $\mathbf{C}$ is the covariance matrix of M_I, $d(\mathbf{x}) = d_o$ is the initial model of hand.

Then we get the color model of the object and the background through GMM. The GMM for the object is established based on the pixels where

$$0 < d(\mathbf{x}) < 0.5d_o \tag{2}$$

Similarly, the GMM for the background is established based on the pixels where

$$1.25d_o < d(\mathbf{x}) < 1.5d_o \tag{3}$$

Then we use the min-cut theory [7] and construct the graph to divide the points where

$$0.5d_o < d(\mathbf{x}) < 1.25d_o \tag{4}$$

into object points and non-object points, and all the object points constitute the object area T. In the end, we compute the spatial mean and covariance matrix of the pixels in the object region to construct the correct object spatial model M_I as shown in Fig. 3.

We compute the upright minimum bounding rectangle enclosing the object region and then a concentric rectangle region expanded by 60 pixels in width and height is defined as the search range.

Fig. 3. The segmentation result of the frame.

The GMM for the hand is established based on the pixels in hand region and the GMM for the background is established based on the pixels in the background region within the search range.

The weight image, indicating the probability of each pixel belonging to the object, is computed using the log ratio as

$$w(\mathbf{x}) = log(\frac{p(\mathbf{y}|G_O)}{p(\mathbf{y}|G_B)} + 1) \tag{5}$$

where y is the RGB value at the pixel x in the new frame. p(y|F) is the Gaussian mixture distribution as

$$p(\mathbf{y}|F) = \sum_{i=1}^{k} \frac{\pi_i}{\sqrt{|\mathbf{\Sigma_i}|}} \exp\{-\frac{1}{2}(\mathbf{y} - \mathbf{u_i})^T \mathbf{\Sigma_i}^{-1}(\mathbf{y} - \mathbf{u_i})\} \tag{6}$$

where k is the number of the components (Fig. 4).

Fig. 4. The weight image of the frame.

We run the EM-like iterations [6] on the weight image to estimate the center and shape of the object in the new frame which determine the object spatial model M_I, as

$$\mathbf{m^k} = \frac{\sum_i \mathbf{x_i} w(\mathbf{x_i}) g(\mathbf{x_i}|\mathbf{m^{k-1}}, \mathbf{C^{k-1}})}{\sum_i w(\mathbf{x_i}) g(\mathbf{x_i}|\mathbf{m^{k-1}}, \mathbf{C^{k-1}})} \tag{7}$$

$$C^k = \beta \frac{\sum_i (\mathbf{x_i} - \mathbf{m}^{k-1})^T (\mathbf{x_i} - \mathbf{m}^{k-1}) w(\mathbf{x_i}) g(\mathbf{x_i}|\mathbf{m^{k-1}}, \mathbf{C^{k-1}})}{\sum_i w(\mathbf{x_i}) g(\mathbf{x_i}|\mathbf{m^{k-1}}, \mathbf{C^{k-1}})} \tag{8}$$

2.2   Hand Patch Rotation

The rotation of the hand is to rotate the image block which contains the hand to get an upright hand as the input of the classifiers (Fig. 5).

In this part, we first use an ellipse to fit the hand contour which we get in the hand tracking module. Here we use the least squares estimation method [8] to get the ellipse.

Because not all of the gesture points are in the ellipse, we compute the minimum bounding rectangle which can enclose the ellipse and then a concentric rectangle region

Fig. 5. The spatial model of the frame.

expanded by certain rate in width and height is defined as the hand patch. The circumscribed rectangle is an angled rectangle, and inside the rectangle is the complete gesture need to be recognized.

To improve the recognition rate of the system, we rotate the hand patch to get an upright hand. We suggest (x_E, y_E) is the center point of the ellipse, θ_E is the angle between the long axis of the ellipse and the vertical direction, L_E is the length of the major axis, and S_E is the length of the short axis. Among them θ_E is the degree to rotate. After rotating, we get the upright hand patch (Fig. 6).

Fig. 6. The segmentation result, the concentric rectangle and the upright hand patch.

After rotation, the hand patch needs to be cut. We remain the width of the patch, and cut the patch until the ratio of the length and the width is 1.5. By doing this, we ensure the size of the hand patch is as same as that of the samples which are used to train the classifiers. Until now, we acquire the final hand patch to be recognized.

2.3   Gesture Recognition

The recognition of the gesture in this part is to recognize the type of the gesture in the hand patch which we get in the hand extraction module.

We extract the HOG feature of the hand patch as the input of classifiers. The classifiers are trained by SVM. There are 10 gestures to be recognized in this paper, so we need to train 10 classifiers. But there are some similarities between the different gestures, so a same gesture may be classified into more than one classes, in which case, the final recognition result can't be confirmed. So we can't directly take the class, which corresponding to the largest non-negative score, to be the final recognition result.

So the score of each class needs to be normalized according to the formula as

$$d_i = \frac{y_i(\mathbf{w}_i\mathbf{x} + b)}{||\mathbf{w}_i||} \tag{9}$$

Here, d_i is the distance between the feature vector and the hyper-plane of the classifier i. The class which gets the biggest is the recognition result of the hand patch.

3   Experiments

3.1   Training

To train the classifiers, we collect the positive and the negative samples, the ratio of the length and the width of these samples is 1.5. Each classifier's positive samples' number is 4000. Its negative samples have two parts, one is samples of background which contains no gesture, and the other contains gestures which are not corresponding to the classifiers. The total number of the negative samples is 12000.

Before training, we resize each sample to 48*72 to ensure that the system can process the image when the distance between the hand and the camera is within 3 m.

We extract the HOG features of the positive and negative samples. The cell size is 6*8, the block size is 12*16, the stride is 6*8, and the dimensions of the HOG are 2016.

3.2   Experiment Result

Figure 7 shows the recognition result of our system in different background. It shows that our system performs well whether the background is simple, complex or even very similar to the color of the skin.

Fig. 7. The recognition result in different background. The patch in the upper right corner is the hand patch that we get in the hand extraction module.

To verify the robustness of the algorithm, we collect three videos in different background as shown in Fig. 8.

(a)simple background   (b)complex background   (c)similar background

Fig. 8.   Three different background.

The experimenter in the video will make ten kinds of gestures, and they change the angle of their hands.

We compare the recognition rates with the algorithm proposed in [4, 5]. Fig. 9 shows the hand patch that will be recognized by our algorithm and algorithms in [4, 5]. Figure 9(b) is the patch to be recognized by our algorithm, and Fig. 9(c) is the patch to be recognized by algorithm in literature [4, 5]. The hand has an obvious angle for the gesture in image Fig. 9(a), the proposed algorithm rotates the hand patch to an upright one, but the [4, 5] recognize the hand patch without rotation.

(a)   (b)   (c)

Fig. 9.   The hand patch.

Table 1 shows the recognition result of the three algorithms in different background, the algorithm proposed in this paper has the highest recognition rate. Because

Table 1.   Our approach in comparison to two relevant algorithms in [4, 5] using the number of successfully recognized frames.

Sequence	Frames	[4]	[5]	Our
a	1153	654	928	1119
b	970	586	723	805
c	1248	984	1009	1078

the algorithm in this paper rotates the patch, it can adapt to changes of the angle. Moreover, it fits the contour of the hand so that reduce the impact of the background and gets a higher recognition rate.

The algorithm in [4], its positive gesture samples are upright, and it doesn't correct the direction when recognizing, so when the angle of the gestures changes, the recognition rate of the method will reduce.

The algorithm in [5], the upright samples are rotated to get the samples in six directions. Using these samples to train the classifiers can make the classifier have certain ability to recognize the gestures with angle. But the samples for one classifier contain gesture with different angles, so the intra-class distance between the positive samples is larger than that of the algorithm proposed in this paper. So the accuracy of the classifier trained in [5] is lower than the accuracy of the classifier in this paper.

On the one hand, the training samples of the classifier proposed in this paper are all upright, so their intra-class distance is lower than that of the samples which contain multi-angles gestures. On the other hand, the paper rotates the hand patch to ensure the gesture is upright before the recognition, so the recognition algorithm based on multi-classifiers proposed in this paper has higher recognition rate.

4   Conclusion

The gesture recognition algorithm base on multi-classifiers proposed in this paper performs well in simple background, complex background, or even in background which has the similar color with skin. Through rotation and sizing, we can get an upright gesture with the same size of the samples used to train the classifiers, which successfully improve the performance of the gesture recognition.

In the further work, we plan to figure out more complex situations, such as the situation when the gesture is across the face, to improve the availability of the system. Moreover, notice that our system has only 10 kinds of gesture. Hence, it's meaningful to expand the type of the gestures.

References

1. Panwar, M.: Hand gesture recognition based on shape parameters. In: 2012 International Conference on Computing, Communication and Applications (ICCCA), pp. 1–6. IEEE (2012)
2. Ahuja, M.K., Singh, A.: Static vision based Hand Gesture recognition using principal component analysis. In: IEEE International Conference on Moocs, Innovation and Technology in Education. IEEE (2015)
3. Dardas, N.H., Georganas, N.D.: Real-time hand gesture detection and recognition using bag-of-features and support vector machine techniques. IEEE Trans. Instrum. Meas. **60**(11), 3592–3607 (2011)
4. Joshi, G., Vig, R.: Histograms of orientation gradient investigation for static hand gestures. In: 2015 International Conference on Computing, Communication & Automation (ICCCA), pp. 1100–1103. IEEE (2015)

5. Guo, W., Wang, X.: Hand gesture detection technology based on HOG and SVM. Electron. Sci. Technol. **27**(8), 15–18 (2014)
6. Wang, H., Sang, N., Yan, Y.: Real-time tracking combined with object segmentation. In: International Conference on Pattern Recognition, pp. 4098–4103. IEEE Computer Society (2014)
7. Rother, C., Kolmogorov, V., Blake, A.: Grabcut: interactive foreground extraction using iterated graph cuts. ACM Trans. Graph. (TOG) **23**(3), 309–314 (2004). ACM
8. Fitzgibbon, A., Pilu, M., Fisher, R.B.: Direct least square fitting of ellipses. IEEE Trans. Pattern Anal. Mach. Intell. **21**(5), 476–480 (1999)

Recognition of Social Touch Gestures Using 3D Convolutional Neural Networks

Nan Zhou and Jun Du[(⊠)]

University of Science and Technology of China, Hefei, Anhui,
People's Republic of China
francis7999@outlook.com, jundu@ustc.edu.cn

Abstract. This paper investigates on the deep learning approaches for the social touch gesture recognition. Several types of neural network architectures are studied with a comprehensive experiment design. First, recurrent neural network using long short-term memory (LSTM) is adopted for modeling the gesture sequence. However, for both handcrafted features using geometric moment and feature extraction using convolutional neural network (CNN), LSTM cannot achieve satisfactory performances. Therefore, we propose to use the 3D CNN to model a fixed length of touch gesture sequence. Experimental results show that the 3D CNN approach can achieve a recognition accuracy of 76.1 % on the human-animal affective robot touch (HAART) database in the recognition of social touch gestures challenge 2015, which significantly outperforms the best submitted system of the challenge with a recognition accuracy of 70.9 %.

Keywords: Deep learning · Social touch gesture · 3D CNN

1 Introduction

In recent years there has been an increasing interest on human-robot interaction studies that use touch modality. In social human-robot interaction, the correct interpretation of touch gestures provides additional information about affective contents in touch, and can be used together with audio-visual cues to improve affect recognition performance [1]. Some well-known robots, such as AIBO (1999), Paro (2001), Nao (2006) and Reeti (2011) are equipped with touch sensors. Some researchers have investigated skin-like sensing, i.e. lots of sensors spreading all over the robot body [2–4].

In the international conference on multimodal interaction (ICMI) last year, the Recognition of Social Touch Gestures Challenge 2015 was launched. In this challenge, organizers provided participants with pressure sensor grid datasets of various touch gestures, namely Corpus of Social Touch (CoST) database and Human-Animal Affective Robot Touch (HAART) database [5]. Many conventional classification approaches were adopted, including support vector machine (SVM) [6], logistic regression [7], random forest [6, 8, 9], and multiboost [8].

Recently, the deep learning techniques are successfully used in many research areas. Convolutional neural networks (CNNs) [10] are one type of feedforward neural networks, which make promising results especially for the computer vision area.

© Springer Nature Singapore Pte Ltd. 2016
T. Tan et al. (Eds.): CCPR 2016, Part I, CCIS 662, pp. 164–173, 2016.
DOI: 10.1007/978-981-10-3002-4_14

CNN receives raw images as the inputs, uses trainable kernels to extract features and pooling layers to down-sample feature maps, and makes output feature maps highly invariant to specific input transformation. Researchers have found that with appropriate parameters and regularization terms, CNNs can outperform methods with manually extracted features [11–13]. Recurrent neural networks are another type of deep neural networks with directed circles between units which make it "deep in time". RNNs can model the temporal actions by changing their outputs through time. But this simple RNN is suffering from problems like gradient vanishing and explosion, easy to lose the track of long term connections [14]. However, by using RNN with the long short-term memory (LSTM) structure, the gradient vanishing problem can be alleviated [15].

In this study, we investigate on the deep learning approaches for the social touch gesture recognition which are rarely mentioned for 2015 challenge. Several types of neural network architectures are studied with a comprehensive experiment design. First, recurrent neural network using long short-term memory (LSTM) is adopted for modeling the gesture sequence. However, for both handcrafted features using geometric moment and feature extraction using CNN, LSTM cannot achieve satisfactory performances. Therefore, we propose to use the 3D CNN to model a fixed length of touch gesture sequence. Experimental results show that the 3D CNN approach can achieve a recognition accuracy of 76.1 % on the human-animal affective robot touch (HAART) database in the recognition of social touch gestures challenge 2015, which significantly outperforms the best submitted system of the challenge with a recognition accuracy of 70.9 %.

The remainder of the paper is organized as follows. In Sect. 2, LSTM with geometric moment features (denoted as GM-LSTM) is introduced. In Sect. 3, LSTM with CNN-based feature extraction (denoted as LRCN) is presented. In Sect. 4, 3D CNN is elaborated. In Sects. 5 and 6, we report experimental results and analysis. Finally we conclude the paper in Sect. 7.

2 GM-LSTM

Geometric moments represent geometric features of an image and are invariant to rotation, transition and scaling, which are also called invariant moments [16]. In image processing, geometric moments can be used as important features to represent objects. The zeroth-order moment, $\mu_{0,0}$ and the first-order moments, $\mu_{1,0}$ and $\mu_{0,1}$ are given by

$$\mu_{0,0} = \sum_{x=0}^{w} \sum_{y=0}^{h} I(x, y) \tag{1}$$

$$\mu_{1,0} = \frac{1}{\mu_{0,0}} \sum_{x=0}^{w} \sum_{y=0}^{h} x I(x, y) \tag{2}$$

$$\mu_{0,1} = \frac{1}{\mu_{0,0}} \sum_{x=0}^{w} \sum_{y=0}^{h} y I(x, y) \tag{3}$$

where w and h represent the width and height of an image, respectively. $I(x, y)$ is the intensity of pressure at (x, y). Higher-order moments are calculated from the following equation

$$\mu_{i,j} = \frac{1}{\mu_{0,0}} \sum_{x=0}^{w} \sum_{y=0}^{h} (x - \mu_{1,0})^i (y - \mu_{0,1})^j I(x, y) \qquad (4)$$

In our implementation, the zeroth, first, second and third order moments for each frame are calculated, resulting in a 10-dimension feature vector.

In GM-LSTM architecture, the input layer receives a sequence (432 frames for each sample) of GM feature vectors while the output layer has 7 units, each corresponding to a type of gesture labels. The activation function of output layer is the softmax function, and the loss function is cross-entropy error function. The hidden layers are represented by the LSTM layers as shown in Fig. 1.

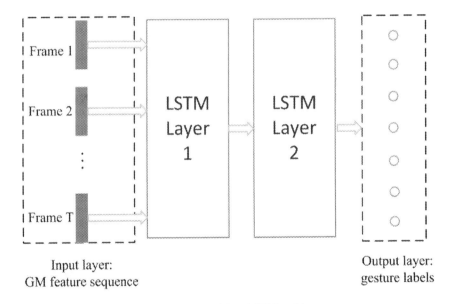

Fig. 1. A diagram of GM-LSTM architecture.

3   LRCN

Donahue et al. proposed long-term recurrent convolutional networks (LRCNs) to deal with problems like activity detection, image captioning and video captioning in 2014. It performed well for many datasets, becoming a type of leading deep learning methods [17]. LRCNs are one kind of deep neural networks that can make end-to-end classification of videos. It uses CNN to extract hierarchical features for each frame, and LSTM to model the feature sequence. Then the parameters of the whole network are updated using backpropagation through time (BPTT) algorithm.

As a combination of CNN and LSTM, LRCN has its own architecture including the input layer, several convolutional layers and pooling layers for CNN, several LSTM layers, and the output layer, as shown in Fig. 2. For our task, the input layer receives

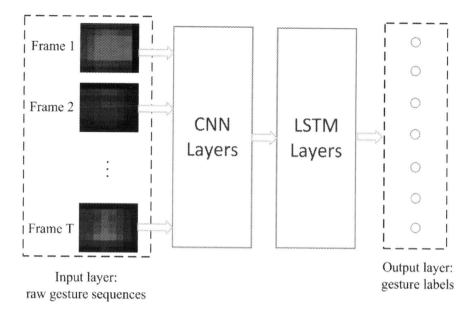

Input layer:
raw gesture sequences

Output layer:
gesture labels

Fig. 2. A diagram of LRCN architecture.

raw input of gesture samples, namely a 8×8 image for each frame with the pixel value ranging from 0 to 1023. The output layer has 7 units, each corresponding to a type of gesture labels. The activation function of output layer is the softmax function, and the loss function is cross entropy error function.

4   3D CNNs

Ji et al. proposed 3D convolutional neural networks to tackle the problem like human action detection in videos and achieved excellent performance, indicating the superiority of 3D CNNs compared with other approaches [18].

In 2D CNNs, convolutions are applied on the 2D feature maps to compute features from the spatial dimensions only. When applied to video analysis problems, it is desirable to capture the motion information encoded in multiple contiguous frames. To this end, the proposed 3D CNNs can simultaneously compute features from both spatial and temporal dimensions. By the use of 3D CNNs, contiguous frames in a gesture sequence are first stacked up, reshaped into a 3D cube. Then 3D convolutions with 3D kernels are applied on the cube. That's why feature maps of 3D CNN are relevant to both spatial and temporal information, and can capture motion information. The value at position (x, y, z) of the j-th feature map in the i-th layer is given by

$$v_{ij}^{xyz} = \tanh\left(b_{ij} + \sum_m \sum_{p=0}^{P_i-1} \sum_{q=0}^{Q_i-1} \sum_{r=0}^{R_i-1} w_{ijm}^{pqr} v_{(i-1)m}^{(x+p)(y+q)(z+r)}\right) \quad (5)$$

where $\tanh(\cdot)$ is the activation function, b_{ij} is the value at position (i, j) of the bias vector, w_{ijm}^{pqr} is the value at (p, q, r) of the m-th convolution kernel, P_i and Q_i are the height and width of the kernel, and R_i is the temporal length of the kernel.

The overall architecture of 3D CNNs is shown in Fig. 3. Apart from the 3D convolutional layers and max pooling layers, another main difference from the GM-LSTM and LRCN is the fully connected layers are adopted before the output layer.

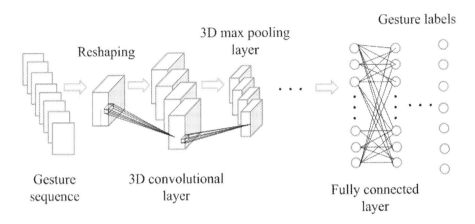

Fig. 3. A diagram of 3D CNN architecture.

5 Experiments and Results

Our experiments are conducted on the HAART dataset of the recognition of social touch gestures challenge 2015. The main purpose of HAART dataset design is to find methods of recognizing human emotion by gestures. The sampling rate of HAART dataset is 54 Hz and the time duration is 8 s. The number of participants is 10. The number of gestures types is 7. The size of sensor grid is 10×10, which was trimmed to 8×8 to match that of CoST dataset in the challenge [5]. Other details can refer to [5].

5.1 GM-LSTM and LRCN Experiments

In the GM-LSTM experiments, two LSTM layers with 50 units for each layer are good configurations to balance the model capability and the over-fitting problem. After 500 epochs of training, the test set accuracy is 65.3 %.

In the LRCN experiments, we can achieve a fair performance when using three convolutional layers and two 128-unit LSTM layers. The first convolutional layer has 4 feature maps, and its kernel size is 3×3. The second convolutional layer has 8 feature maps, its kernel size is 3×3, and the second convolutional layer is followed by an average pooling layer. The third convolutional layer has 16 feature maps, and its kernel size is 2×2. The test set accuracy at the 500-th epoch is 60.6 % (Table 1).

Table 1. The accuracy of test set at the 500-th epoch in GM-LSTM and LRCN experiments.

Classifier	Accuracy (%)
GM-LSTM	65.3
LRCN	60.6

5.2   3D CNN Experiments

The input layer of 3D CNN receives a raw gesture sequence with 432 frames. For each frame, the size of the image is 8×8 with the pixel value ranging from 0 to 1023. The output layer has 7 units, each corresponding to a type of gesture label. The activation function of output layer is the softmax function, and the loss function is cross entropy error function. The number of training epochs is 600.

5.2.1   Experiments on the Number of Feature Maps
The number of feature maps in every convolutional layer determines the dimension of features extracted and is important for recognition performance. In this set of experiments, the number of convolutional layers was fixed to 4, and we found that when the number configuration of feature maps for 4 convolutional layers was set as 16-32-64-128, 3D CNNs achieved the best performance, compared with 8-16-32-64 and 32-64-128-256 setting, as shown in Table 2. The kernel sizes in every convolutional layer were all set to be $3 \times 3 \times 3$.

Table 2. The accuracy of test set in feature map configuration experiments

Feature map configuration	Accuracy (%)
8-16-32-64	72.9
16-32-64-128	75.1
32-64-128-256	68.1

Table 3. The accuracy of test set for different numbers of fully connected layers.

The number of fully connected layers	Accuracy (%)
1	66.9
2	71.3
3	72.5
4	70.9

5.2.2   Experiments on the Number of Fully Connected Layers
The number of fully connected layers determines the complexity and generalization capability of 3D CNNs. The optimal number can vary substantially in different tasks. In this set of experiments, the number of units in every fully connected layer was 256 (Table 3).

Table 4. The accuracy of test set in the dropout value experiments

Dropout value	Accuracy (%)
0(no dropout)	68.9
0.2	64.5
0.5	71.3

5.2.3 Experiments on the Dropout

When training deep neural networks, if the dataset is not large enough, we should use dropout as a trick to prevent over-fitting. Dropout was proposed by Hinton et al. in 2012. It randomly screens some weights of the units in hidden layers and improves neural networks by preventing co-adaptation of feature detectors [19]. Our experiments achieved a relatively good performance when the dropout value was 0.5, compared with no dropout case and the dropout value of 0.2. In these experiments, the number configuration of feature maps for 4 convolutional layers was set as 8-16-32-64, and the number of units in every fully connected layer was 256 and the number of fully connected layers was 3 (Table 4).

5.3 Overall Comparison

The final configuration of 3D CNN is as follows. The number of convolutional layers is 4. The number configuration of feature maps for 4 convolutional layers is 16-32-64-128. The kernel sizes in every convolutional layer are all set to be $3 \times 3 \times 3$. The pooling sizes in four convolutional layers are $5 \times 1 \times 1$, $3 \times 2 \times 3$, $3 \times 2 \times 2$, $3 \times 2 \times 2$, respectively. The number of fully connected layers is 3. The number of units in the fully connected layers is 1024. The dropout value is 0.5 and learning rate is 0.001. The initialization method is he_normal [20], and the batch size is 20.

With this configuration, we achieved the best test set accuracy of 76.1 %, which is significantly better than the first ranked result (70.9 % in [6]) in the challenge. And 3D CNN becomes the state-of-art method for social touch gestures recognition task. Table 5 shows an overall performance comparison with other approaches.

Table 5. An overall performance comparison with other approaches.

Approach	Classifier	Accuracy (%)
Proposed	3D CNN	**76.1**
Proposed	GM-LSTM	65.3
Proposed	LRCN	60.6
[6]	Random forest	70.9
[6]	SVM	68.5
[7]	Logistic regression	67.7
[8]	Random forest	66.5
[8]	Multiboost	64.5
[9]	Random forest	61.0

6 Discussion

GM-LSTMs and LRCNs did not perform well on the touch gesture recognition task. It might be partially due to that there were not enough training data for building LSTM layers, which usually needed longer sequence data compared with the fullly connected layers. And LRCNs even performed worse than GM-LSTMs, indicating that the use of CNN for feature extraction is not always helpful if the sequence model (LSTM) is not well enough.

3D CNNs did perform well on the touch gesture recognition task, which implied that 3D CNNs were really good at extracting motion information from contiguous frames, even on a small data set compared with other approaches.

As we can see in Table 6, two types of gestures—constant and no touch—have better accuracies with few false recognitions. The rub gesture is similar to scratch and they are easily incorrectly classified to each other. The stroke is similar to rub. The tickle has the lowest accuracy and is often incorrectly classified to scratch. All those problems are apparently related with the way the gesture samples were collected and the interconnections between gestures. And we can make more studies in the future.

Table 6. The confusion matrix for HARRT dataset

Counts	Constant	No touch	Pat	Rub	Scratch	Stroke	Tickle
Constant	34	1	0	0	0	0	0
No touch	0	36	0	0	0	0	0
Pat	0	0	29	2	2	1	2
Rub	0	0	0	27	2	5	2
Scratch	0	0	1	6	26	1	2
Stroke	0	0	1	5	1	27	2
Tickle	0	0	0	2	20	2	12

7 Conclusion

This paper analyzed different neural network structures including 2D CNNs, 3D CNNs and LSTMs. GM-LSTMs, LRCNs and 3D CNNs are compared on the social touch gestures recognition task. The conclusion is that the 3D CNN approach achieves the best result, beyond the first ranked result in the challenge—70.9 %, and 3D CNN is the state-of-art method for touch gestures classification task. In comparing 3D CNN experiments with GM-LSTM and LRCN experiments, we observe that LSTMs require larger dataset and 3D CNNs are more robust to the data size. Also we analyzed the similarities of gestures based on confusion matrix.

In the future, on one hand, we will further verify the effectiveness on large gesture databases. On the other hand, we will apply 3D CNNs to other tasks involving temporal information.

References

1. Balli Altuglu, T., Altun, K.: Recognizing touch gestures for social human-robot interaction. In: Proceedings of the 2015 ACM International Conference on Multimodal Interaction (ICMI), pp. 407–413 (2015)
2. Cooney, M.D., Nishio, S., Ishiguro, H.: Recognizing affection for a touch-based interaction with a humanoid robot. In: IEEE/RSJ International Conference on Intelligent Robots and Systems (IROS), pp. 1420–1427 (2012)
3. Billard, A., Bonfiglio, A., Cannata, G., Cosseddu, P., Dahl, T., Dautenhahn, K., Mastrogiovanni, F., Metta, G., Natale, L., Robins, B., et al.: The ROBOSKIN project: challenges and results. In: Padois, V., Bidaud, P., Khatib, O. (eds.) Romansy 19–Robot Design, Dynamics and Control. CISM International Centre for Mechanical Sciences, pp. 351–358. Springer, Vienna (2013)
4. Knight, H., Toscano, R., Stiehl, W.D., Chang, A., Wang, Y., Breazeal, C.: Real-time social touch gesture recognition for sensate robots. In: IEEE/RSJ International Conference on Intelligent Robots and Systems (IROS), pp. 3715–3720 (2009)
5. Jung, M.M., Cang, X.L., Poel, M., MacLean, K.E.: Touch challenge '15: recognizing social touch gestures. In: Proceedings of the 2015 ACM on International Conference on Multimodal Interaction (ICMI), pp. 387–390 (2015)
6. Ta, V.-C., Johal, W., Portaz, M., Castelli, E., Vaufreydaz, D.: The Grenoble system for the social touch challenge at ICMI 2015. In: Proceedings of the 2015 ACM International Conference on Multimodal Interaction (ICMI), pp. 391–398 (2015)
7. Hughes, D., Farrow, N., Profita, H., Correll, N.: Detecting and identifying tactile gestures using deep autoencoders, geometric moments and gesture level features. In: Proceedings of the 2015 ACM International Conference on Multimodal Interaction (ICMI), pp. 415–422 (2015)
8. Falinie, Y. Gaus, A., Olugbade, T., Jan, A., Qin, R., Liu, J., Zhang, F., Meng, H., Bianchi-Berthouze, N.: Social touch gesture recognition using random forest and boosting on distinct feature sets. In: Proceedings of the 2015 ACM International Conference on Multimodal Interaction (ICMI), pp. 399–406 (2015)
9. Altuglu, T.B., Altun, K.: Recognizing touch gestures for social human-robot interaction. In: Proceedings of the 2015 ACM International Conference on Multimodal Interaction (ICMI), pp. 407–413 (2015)
10. LeCun, Y., Bottou, L., Bengio, Y., Haffner, P.: Gradient-based learning applied to document recognition. Proc. IEEE **86**(11), 2278–2324 (1998)
11. Yu, K., Xu, W., Gong, Y.: Deep learning with kernel regularization for visual recognition. In: NIPS, pp. 1889–1896 (2008)
12. Ahmed, A., Yu, K., Xu, W., Gong, Y., Xing, E.: Training hierarchical feed-forward visual recognition models using transfer learning from pseudo-tasks. In: Forsyth, D., Torr, P., Zisserman, A. (eds.) ECCV 2008. LNCS, vol. 5305, pp. 69–82. Springer, Heidelberg (2008). doi:10.1007/978-3-540-88690-7_6
13. Mobahi, H., Collobert, R., Weston, J.: Deep learning from temporal coherence in video. In: ICML, pp. 737–744 (2009)
14. Bengio, Y., Simard, P., Frasconi, P.: Learning long-term dependencies with gradient descent is difficult. IEEE Trans. Neural Netw. **5**(2), 157–166 (1994)
15. Hochreiter, S., Schmidhuber, J.: Long short-term memory. Neural Comput. **9**(8), 1735–1780 (1997)
16. Teh, C.-H., Chin, R.T.: On image analysis by the methods of moments. IEEE Trans. Pattern Anal. Mach. Intell. **10**(4), 496–513 (1988)

17. Donahue, J., Hendricks, L.A., Guadarrama, S., Rohrbach, M., Venugopalan, S., Saenko, K., Darrell, T.: Long-term recurrent convolutional networks for visual recognition and description (2014). CoRR, abs/1411.4389
18. Ji, S., Xu, W., Yang, M., Yu, K.: 3d convolutional neural networks for human action recognition. IEEE Trans. Pattern Anal. Mach. Intell. **35**(1), 221–231 (2013)
19. Hinton, G.E., Srivastava, N., Krizhevsky, A., Sutskever, I., Salakhutdinov, R.: Improving neural networks by preventing co-adaptation of feature detectors (2012). CoRR, abs/1207.0580
20. He, K., et al.: Delving deep into rectifiers: surpassing human-level performance on imagenet classification. arXiv preprint arXiv:1502.01852 (2015)

Object Property Identification Using Uncertain Robot Manipulator

Kunxia Huang[1], Chenguang Yang[1,2($\boxtimes$)], and Hong Cheng[3]

[1] Key Laboratory of Autonomous Systems and Networked Control,
College of Automation Science and Engineering,
South China University of Technology, Guangzhou 510640, China
cyang@ieee.org
[2] Zienkiewicz Center for Computational Engineering,
Swansea University, Swansea SA1 8EN, UK
[3] Center for Robotics, University of Electronic Science and Technology of China,
Chengdu 611731, Sichuan, China

Abstract. This paper presents a learning control algorithm to identify object properties by an uncertain robot manipulator. On one hand, for the robot system with unknown (or immeasurable) parameters, the manipulator dynamics properties are uncertain so we use adaptive parameter learning law to estimate the practically unknown dynamics. On the other hand, a reference model is specified to be followed. In order to identify the geometry and elasticity of the interacting object, the reference point and feedforward force in reference model is adapted in each trial. Because the updating of the reference model utilizes the estimated parameters, the learning law of parameter estimation is thus designed to guarantee the convergence of parameter estimation in finite time (FT). Simulation studies demonstrate the effectiveness of our proposed method.

Keywords: Unknown parameter estimation · Object property identification · Reference model · Adaptive control

1 Introduction

With the rapid development of advanced technologies of robotics, the application of robots is more and more extensive and most of times robot is supposed to interact with an object, such that the robot needs to understand the object's properties, such as texture, elasticity, etc. Visual sensing has been widely used to recognize object's properties [1]. However, there are limitations to the visual

This work was partially supported by National Nature Science Foundation (NSFC) under Grant 61473120, Guangdong Provincial Natural Science Foundation 2014A030313266 and International Science and Technology Collaboration Grant 2015A050502017, Science and Technology Planning Project of Guangzhou 201607010006 and the Fundamental Research Funds for the Central Universities under Grant 2015ZM065.

© Springer Nature Singapore Pte Ltd. 2016
T. Tan et al. (Eds.): CCPR 2016, Part I, CCIS 662, pp. 174–188, 2016.
DOI: 10.1007/978-981-10-3002-4_15

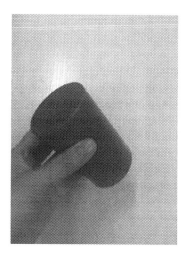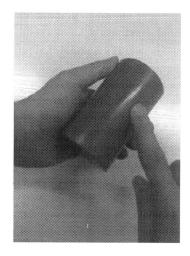

Fig. 1. Illustrations of human's haptic exploration of an object

sensor, such as the need for appropriate light and no covered. Sometimes the robot has to work in the dark environment without light, or has to pick an object covered by other things. Robots equipped with only visual sensors cannot accomplish tasks under the conditions mentioned above, thus we need to develop other sensing technology. Our human have tactile sensation assisting the vision, which inspires us that we can develop the robot's haptic identification.

Tactile sensing is similar to the visual sensing in the sense that both of them emulate human's sensing. Human beings can estimate the shape and texture of different objects using the tactile sensation of the hand (Fig. 1). In contrast, the main method for robot manipulator to detect the object properties is to use specialized tactile sensor. However, it means that we should figure a corresponding tactile sensor for a robot before having a sense of touch, which is strict for most of the existing robots. To find a simple way to estimate the object property, the haptic algorithm appears and it generally does not demand much on the hardware specification.

It has been shown in [2] that human adapt the force and impedance when they interact with external environments skillfully. The human central nervous system (CNS) selectively adapt the endpoint impedance of the limbs to the unstable environment [3]. As studied in [4], the interaction force is tuned to counteract a compliant object, and the reference trajectory is controlled around an external object. According to the mechanism in [4], an algorithm adapting the reference trajectory of robot arms interacting with a novel environment has been proposed in [5], in which the feedforward force was not adapted. The paper [6] presents a human like learning algorithm adapting the reference point and force to estimate the interacting object's geometry and elasticity. It studies the case when the model of robot manipulator is known and fixed. But the property of robot is complex and its physical parameters are uncertain or time-varying in many practical situations. There are many uncertainties in the robot system, especially in the robot dynamics.

Adaptive control has little demand on the knowledge of the inertial parameters and variations in payload which makes it necessary to control the unknown or time-varying systems like robot system. The active joints are controlled using adaptive control approach with optimized performance in [7]. Adaptive model reference control and NN based trajectory planner have been designed on WIP systems for dynamic balance and motion tracking of desired trajectories in [8]. In addition, adaptive control [9,10] has been widely studied to estimate the unknown parameters of controlled system and track the desired trajectory. The paper [11] presents several novel adaptation laws for parameter estimation in non-linear systems, especially robotic systems with unknown parameters. And the convergence of parameter estimation error can be guaranteed in finite time. The adaptive laws use the filter operations on the system dynamics to avoid the robot joint accelerations measurements, which are sensitive to measurement noises in practical.

In brief, to identify the object's property using uncertain robot manipulator, we need to estimate robotic unknown parameters firstly, and then detect the external interacting object's properties. Refering to the adaptation of interaction force and reference trajectory to estimate the object's geometry and elasticity [6], we consider the model in [6] as a reference model whose desired point and feedforward force should be updated iteratively.

In this paper, a learning control algorithm for robot to identify object geometry and elasticity (Fig. 2) is proposed. The dynamics properties of robot manipulator to interact with object are uncertain (or immeasurable). Thus firstly, we utilize adaptive parameter estimation to estimate the practical unknown parameters. Secondly, we set a reference model whose reference point and feedforward force are adapted in each trial to identify the properties of the object. Besides, the convergence of parameter estimation error should be guaranteed in finite time (FT). Simulation resultes demonstrate the performance of the proposed algorithm.

Fig. 2. A robot manipulator in contact with an unknown object on the chair [Screenshot of V-REP]

2 Robot Kinematics, Dynamics and Controller

2.1 Kinematics and Dynamics

The forward kinematics of the robot is described as

$$x(t) = \phi(q) \tag{1}$$

where $x(t), q \in \mathbb{R}^n$ and n are the positions of robot end effector in Cartesian space, joint coordinates and the degree of the robotic freedom, respectively. Differentiating the forward kinematics relation $x = \phi(q)$ with respect to time

$$\dot{x} = J(q)\dot{q}, \ddot{x} = J(q)\ddot{q} + \dot{J}(q)\dot{q} \tag{2}$$

where $J(q) = \dfrac{\partial \phi}{\partial q}$ is the Jacobian matrix denoting the relation of velocity between the Cartesian space and joint coordinate.

The manipulator dynamics is given by

$$M(q)\ddot{q} + C(q, \dot{q})\dot{q} + G(q) = \tau - J^T(q)f_I \tag{3}$$

where $q, \dot{q}, \ddot{q} \in \mathbb{R}^n$, n are the robot joint position, velocity, acceleration, and the degree of the freedom, respectively; $M(q)$ is the (symmetric, positive definite) inertia matrix; $C(q, \dot{q})$ is the Coriolis/centrifugal matrix; $G(q)$ represents the gravitational torque; τ is the vector of control torque applied on the robot joint; f_I is the interaction force applied by external environment.

Due to the object identification task is usually defined in the Cartesian space, we need to transfer the dynamics (3) to the operational space. To do so, by considering the kinematics (1) and (2) and dynamics (3), the manipulator dynamics in the Cartesian space can be obtained as follows.

$$M_x(q)\ddot{x} + C_x(q, \dot{q})\dot{x} + G_x(q) = \tau_x - f_I \tag{4}$$

where

$$
\begin{aligned}
M_x(q) &= J^{-T}(q)M(q)J^{-1}(q), \\
C_x(q, \dot{q}) &= J^{-T}(q)(C(q, \dot{q}) - M(q)J^{-1}(q)\dot{J}(q))J^{-1}(q), \\
G_x(q) &= J^{-T}G(q),
\end{aligned}
$$

$\tau_x = J^{-T}(q)\tau$ which is the wrench applied on the robot end effector in the Cartesian space. And in this paper, τ_x is the control input to be designed follows. The matrix $M_x(q)$ is also symmetric and positive definite. According to [12], we have the following properties.

Property 1. The matrix $2C_x(q, \dot{q}) - \dot{M}_x(q)$ is skew-symmetric.

Property 2. The left side of robotic dynamics (4) can be represented in a linearly parameterized form as

$$M_x(q)\ddot{x} + C_x(q, \dot{q})\dot{x} + G_x(q) = Y(x, \dot{x}, \ddot{x})\theta \tag{5}$$

where $\theta \in \mathbb{R}^{n_\theta}$ is a constant parameter vector of the manipulator; n_θ is a positive integer representing the number of these parameters; $Y(x, \dot{x}, \ddot{x}) \in \mathbb{R}^{n \times n_\theta}$ is the regression matrix which is independent of the real robotic system and the current joint positions.

2.2   Reference Model

Let us consider a desired reference model for the robot manipulator to follow

$$M_m \ddot{e} + C_m \dot{e} + K_m e = f_I - f_d \qquad (6)$$

where $e = x_d - x$ with x_d the designed point in the operational space and f_d a feedforward force to maintain a contact force between robot end effector and the object; M_m, C_m, K_m represent the desired mass, damping and stiffness matrices, respectively. The selection of M_m, C_m, K_m depends on the specific task. Taking the griding task as an example, a large stiffness is required in the vertical direction to working surface, while a small stiffness is requested in the horizontal direction along the surface.

In the following, we are going to design a controller in the presence of unknown parameters θ, such that the closed-loop controller robot system could match with the reference model (6). In addition, by the means of adapting the reference point x_d and feedforward force f_d, the robot end effector can interact with an unknown object and explore its geometry and elasticity as illustrated in Fig. 2. Let us consider the following matching error as in [13].

$$w = -M_m \ddot{e} - C_m \dot{e} - K_m e + f_I - f_d \qquad (7)$$

Therefore, the learning control design is to develop an parameter learning law so that the system response satisfies the behavior of the specified target model (6) in finite time, equivalently,

$$\lim_{t \to t_a} w(t) \to 0, \qquad (8)$$

where t_a represents a finite time specified by the designer. After the succeed in parameter estimation in the first trial, that is the matching error $w = 0$ after the first trial, then from the second trial we can adapt the reference point x_d and feedforward force f_d in reference model (6) using the estimated parameters obtained in the first trial which guarantees the convergence of matching error.

3   Robotic Learning Control

In this section, we give details of the learning control design. For the convenience of the following analysis, we define an augmented matching error as below:

$$\bar{w} = K_f w = -\ddot{e} - K_d \dot{e} - K_p e + K_f (f_I - f_d) \qquad (9)$$

where $K_d = M_m^{-1} C_m, K_p = M_m^{-1} K_m, K_f = M_m^{-1}$.

Denote L and L^{-1} as Laplace transformation operator and inverse Laplace transformation operator, respectively. Let us introduce a filtered matching error z as in [7].

$$z = L^{-1}\{(1 - \frac{\Gamma}{s+\Gamma})L\{\dot{e}\} + \frac{1}{s+\Gamma}L\{\epsilon\}\} \tag{10}$$

where ϵ is the practical implementation of $\bar{w}$, since the acceleration measurement is normally unavailable in practical situation:

$$\epsilon = -K_d\dot{e} - K_p e + K_f(f_I - f_d) \tag{11}$$

For convenience of following computation, we rearrange z:

$$z = -\dot{e} + e_h + \epsilon_l \tag{12}$$

where $e_h = L^{-1}\{\frac{\Gamma s}{s+\Gamma}L\{e\}\}$, $\epsilon_l = L^{-1}\{\frac{1}{s+\Gamma}L\{\epsilon\}\}$, represent high-pass and low-pass filtered respectively. Substituting (9), (10), (11) and (12), the augmented matching error $\bar{w}$ can be written as

$$\bar{w} = \dot{z} + \Gamma z \tag{13}$$

which implies that $\dot{z} = 0$ and $z = 0$ will lead to $w = 0$.

As to [14], we set the control input torque as below

$$\tau_x = \tau_{ct} + \tau_{fb} + \tau_\delta + \hat{f}_I \tag{14}$$

where $\tau_{ct}, \tau_{fb}, \tau_\delta$ are the computed torque vector denoting the adaptive control item, feedback torque vector, and compensation torque vector used to compensate the force measurement error, respectively and $\hat{f}_I$ is the measurement of external force f_I. In addition, the force measurement noise $\tilde{f}_I = \hat{f}_I - f_I \neq 0$ is bounded by a known bound, that is $||\tilde{f}_I|| \leq \delta$.

In particular, the computed torque vector τ_{ct} is given by

$$\tau_{ct} = \hat{M}_x\ddot{x}_r + \hat{C}_x\dot{x}_r + \hat{G}_x = Y(\ddot{x}_r, \dot{x}_r, \dot{x}, x)\hat{\theta} \tag{15}$$

where $\hat{\theta}$ is the estimate of θ and

$$\dot{x}_r = \dot{x}_d - e_h - \epsilon_l$$
$$\ddot{x}_r = \ddot{x}_d - \dot{e}_h - \dot{\epsilon}_l \tag{16}$$

The feedback torque vector and the compensation torque are given by

$$\tau_{fb} = -Kz \tag{17}$$
$$\tau_\delta = -K_\delta\text{sgn}(z) \tag{18}$$

where K is a symmetric positive definite matrix, and $K_\delta > \delta$.

Integrating the control input (14) into the manipulator dynamics in Cartesian space (4), the closed-loop dynamics is obtain as below

$$M_x(q)\dot{z} + C_x(q,\dot{q})z + Kz = Y(\ddot{x}_r, \dot{x}_r, \dot{x}, x)\tilde{\theta} - (K_\delta\text{sgn}(z) - \tilde{f}_I) \tag{19}$$

where $\tilde{\theta} = \hat{\theta} - \theta$.

Because for each trial we need to update the reference point x_d and feedforward force f_d, it indicates that the matching error w should converge to 0 in finite time. Thus we design the adaptive law for $\hat{\theta}$ according to [11].

Define alternative vectors as $F_x(q, \dot{q}) = M_x(q)\dot{x}$ and $H_x(q, \dot{q}) = -\dot{M}_x(q)\dot{x} + C_x(q, \dot{q})x + G_x(q)$, then based on Property 2, we obtain

$$\begin{cases} F_x(q, \dot{q}) = M_x(q)\dot{x} = Y_1(q, \dot{q})\theta \\ H_x(q, \dot{q}) = -\dot{M}_x(q)\dot{x} + C_x(q, \dot{q})x + G_x(q) = Y_2(q, \dot{q})\theta \end{cases} \tag{20}$$

where $Y_1(q, \dot{q}), Y_2(q, \dot{q}) \in \mathbb{R}^{n \times N}$ are new regressor matrices without the joint acceleration $\ddot{q}$ which is sensitive to measurement noise. Consequently the system (4) can be described as

$$\dot{F}_x(q, \dot{q}) + H_x(q, \dot{q}) = \tau_x - f_I \tag{21}$$

where $\dot{F}_x(q, \dot{q}) = \frac{d}{dt}[M_x(q)\dot{x}] = \dot{Y}_1(q, \dot{q})\theta$.

Denote $Y_{1f}(q, \dot{q}) \in \mathbb{R}^{n \times N}, Y_{2f}(q, \dot{q}) \in \mathbb{R}^{n \times N}$ and $\tau_f \in \mathbb{R}^n$ as the filtered $Y_1(q, \dot{q}), Y_2(q, \dot{q})$ and τ_x, respectively:

$$\begin{cases} l\dot{Y}_{1f}(q, \dot{q}) + Y_{1f}(q, \dot{q}) = Y_1(q, \dot{q}), Y_{1f}(q, \dot{q})|_{t=0} = 0 \\ l\dot{Y}_{2f}(q, \dot{q}) + Y_{2f}(q, \dot{q}) = Y_2(q, \dot{q}), Y_{2f}(q, \dot{q})|_{t=0} = 0 \\ l\dot{\tau}_f + \tau_f = \tau_x, \tau_f|_{t=0} = 0 \\ l\dot{f}_f + f_f = f_I, f_f|_{t=0} = 0 \end{cases} \tag{22}$$

Then, (21) can be represented as follows

$$\dot{F}_x(q, \dot{q}) + H_x(q, \dot{q})\,[\dot{Y}_{1f}(q, \dot{q}) + Y_{2f}(q, \dot{q})]\theta = \tau_f - f_f \tag{23}$$

Substituting the first equation of (22) into (23), we obtain:

$$[\frac{Y_1(q, \dot{q}) - \dot{Y}_{1f}(q, \dot{q})}{l} + Y_{2f}(q, \dot{q})]\theta = Y_f(q, \dot{q})\theta = \tau_f - f_f \tag{24}$$

where $Y_f(q, \dot{q}) = \frac{Y_1(q, \dot{q}) - \dot{Y}_{1f}(q, \dot{q})}{l} + Y_{2f}(q, \dot{q}) \in \mathbb{R}^{n \times N}$ is the new regressor matrix, which is the function of joint position q and velocity $\dot{q}$ without $\ddot{q}$. And it will be used for the parameter estimation.

To accommodate the parameter estimation, we define matrix $P \in \mathbb{R}^{N \times N}$ and the vector $Q \in \mathbb{R}^N$ as

$$\begin{cases} \dot{P} = -\ell P + Y_f{}^T Y_f, P(0) = 0 \\ \dot{Q} = -\ell Q + Y_f{}^T \tau_f, Q(0) = 0 \end{cases} \tag{25}$$

where $\ell > 0$ is a design parameter.

Define an auxiliary vector $W \in \mathbb{R}^N$ computed from P, Q in (25) as

$$W = P\hat{\theta} - Q \tag{26}$$

According to [11], we obtain

$$W = P\hat{\theta} - P\theta = -P\tilde{\theta} \tag{27}$$

To guarantee that the parameter estimation error $\tilde{\theta}$ converge to 0 in finite time, we choose the second adaptive law for $\hat{\theta}$ in [11] as follows

$$\dot{\hat{\theta}} = -\Lambda \frac{P^T W}{\|W\|} \tag{28}$$

where $\Lambda > 0$ is a design constant learning gain matrix.

Select the Lyapunov function as $V = \frac{1}{2} W^T P^{-1} P^{-1} W$; then it have been established in [11] that $\lim\limits_{t \to \infty} W = 0$ holds in finite time $t_a \leq 2\sqrt{V(0)}/\mu$, where $\mu = (\lambda_{\min}(\Lambda) - \|P^{-1}\psi'\|)\sigma\sqrt{2}$ is a positive scalar, chosen larger than a prespecified constant, and $\psi = -\int_0^t e^{-\ell(t-r)} Y_f{}^T(r) f_f(r) \mathrm{d}r$ is bounded as long as f_I and $Y_f(q, \dot{q})$ are bounded.

The parameter error $\tilde{\theta}$ converges to a compact set in finite time t_a satisfying $\lim\limits_{t \to t_a} P\tilde{\theta} = \psi$, thus when $t > t_a$ the matching error $w = 0$, that is the system (4) tracking the reference model (6) accurately. Thus in the end of the first trial, we have learnt the robotic unknown parameters θ^0, and we can simply set the future estimation of parameter $\hat{\theta}^k = \theta^0(T)(k \geq 1)$, where T is the period of the first trial and $T > t_a$, which always guarantees the matching error $w = 0$. Then we can adapt the reference point x_d and feedforward force f_d in reference model from the second trial and achieve the object property identification.

4   Reference Model Adaptation

For the convenience, we suppose below that the robot manipulator detects and interacts with objects only in one direction along the x axis of the operational space. However the algorithms can be developed in the same way in the three axes x, y and z. Assume that the values of damping C_m, stiffness K_m and initial reference point x_t are well chosen in each trial so that the end effector smoothly approaches and then presses the unknown object without destroying the object. Consider the elastic property of the object, when the robot end effector approaches and presses the object, the external force can be described as

$$f_I = K_0(x - x_b) \tag{29}$$

where K_0 is the object's stiffness and x_b denotes the boundary of the object.

Considering that when $t > t_a$, the matching error $w = 0$. We combine (6) and (29) and obtain the closed-loop dynamics during contact:

$$M_m \ddot{e} + C_m \dot{e} + K_m e - K_0(x - x_b) = -f_d \tag{30}$$

Consider for each trial a finite time t_f, which is large enough so that at the instant t_f the robot end effect has reached the equilibrium point x_* interacting with the object. From the above analysis, it can be seen that t_f should be $t_f \geq t_a$. At the equilibrium point one can derive

$$K_0(x_* - x_b) = -K_m(x_* - x_d) + f_d \tag{31}$$

Denoting $v^k = x_*^k - x_d^k$, and according to [6], we use the adaptation law of reference point x_d and feedforward force f_d from one trial k to the next as follows

$$
\begin{cases}
x_d^0 = x_d^1 = x_t, \ i = 2, 3, \cdots \\
x_d^{k+1} = x_d^k + \alpha^k v^k + (1 - \alpha^k)(x_t - x_d^k), \\
\quad \text{if } v^k \leq 0 \\
x_d^{k+1} = x_d^k + (1 - \alpha^k)(x_t - x_d^k), \ \text{otherwise}
\end{cases}
\tag{32}
$$

$$
\begin{cases}
f_d^0 = f_d^1 = F_0, \ i = 2, 3, \cdots \\
f_d^{k+1} = f_d^k + K_m(v^k - v^{k-1}) + \beta^k(F_0 - f_d^k), \\
\quad \text{if } v^k \leq 0 \\
f_d^{k+1} = f_d^k + \beta^k(F_0 - f_d^k), \ \text{otherwise}
\end{cases}
\tag{33}
$$

where $0 < \alpha^k < 1$ is a compliance factor which the larger it is the reference point would change to be more compliant to the object, and $0 < \beta^k < 1$ is a relax factor tending to push f_d back to the value of F_0. They both will be defined in (39).

5   Learning the Object Properties

By defining

$$
a_1 \equiv \frac{K_m}{K_0}, \ a_2 \equiv x_b, \ s^k \equiv \frac{x_*^k}{x_*^k - x_d^k}
$$
$$
\phi_1^k \equiv \frac{f_d^k}{K_m(x_*^k - x_d^k)} - 1, \ \phi_2^k \equiv \frac{1}{x_*^k - x_d^k}
\tag{34}
$$

Integrating equations (34) into the closed-loop dynamics (30), we obtain

$$
A^T \Phi^k = s^k
\tag{35}
$$

where $A = [a_1, a_2]^T$ and $\Phi = [\phi_1, \phi_2]^T$.

As (35) is linear in the parameters, we use a weighted least square (WLS) [15] for fast convergence as follows

$$
\hat{A}^{k+1} = \hat{A}^k + L^k(s^{k+1} - (\hat{A}^k)^T \Phi^k)
$$
$$
L^k = \frac{P^k \Phi^k}{\sigma^{k-1} + (\Phi^k)^T P^k \Phi^k}
\tag{36}
$$
$$
P^{k+1} = P^k - \frac{P^k \Phi^k (\Phi^k)^T P^k}{\sigma^{k-1} + (\Phi^k)^T P^k \Phi^k}
$$
$$
i = 2, 3, \cdots
$$

where $\hat{A}^k = [\hat{a}_1^k, \hat{a}_2^k]^T$ with $\hat{a}_1^k$ and $\hat{a}_2^k$ the estimates of a_1 and a_2. Then, the estimates of K_0 and x_b at the kth trial, that are denoted as $\hat{K}_0^k$ and $\hat{x}_b^k$ can be obtained as

$$
\begin{cases}
\hat{K}_0^k \equiv \frac{K_m}{\hat{a}_1^k}, \ \hat{x}_b^k \equiv \hat{a}_2^k \ \text{if } e^k \leq 0 \\
\hat{K}_0^k \equiv \beta^k K_0^{k-1}, \ \hat{x}_b^k \equiv \alpha^k \hat{x}_b^{k-1} + (1 - \alpha^k) x_t \ \text{if } e^k > 0
\end{cases}
\tag{37}
$$

Mathematically, the initial value of $\hat{A}$, that is $\hat{A}^0$ can be chosen arbitrarily, however to be practical, we choose the initial estimate of K_0 and x_b as $\hat{K}_0^1 = \hat{K}_0^0 = \bar{K}_0$ and $\hat{x}_b = x_s$, where $\bar{K}_0$ is the largest possible stiffness of the detected object and x_s is the initial position of robot end effector. In addition, the initial value of P in (36) can be simply chosen as $P^0 = I$ with I the identity matrix. The weighting sequence σ^k is given according to [15]

$$\sigma^k = \frac{1}{\log^{1+\gamma}(1 + \sum_0^k \|\Phi^k\|^2)} \tag{38}$$

With $\hat{K}_0$ available, now the values of compliance factor α^k and the relax factor β^k in (32) and (33), can be decided as follows

$$\alpha^k = \lambda \frac{\hat{K}_0}{\bar{K}_0}, \ \beta^k = 1 - \alpha^k \tag{39}$$

where λ is a constant specified by the designer.

6 Simulation Studies

To demonstrate the effectiveness of the proposed control method, we consider an ideal model which a 2-degrees-of-freedom robot manipulator contacts with a spring, as shown in Fig. 3. The robot manipulator can move in the xy plane while the spring produces a force to the robot manipulator only along the x axis.

In the following, $m_i, l_i, I_i, l_{ci}, i = 1, 2$ represent the mass, the length, the inertia about the z-axis that comes out of the page passing through the center of mass, and the distance from the previous joint to the center of mass of link

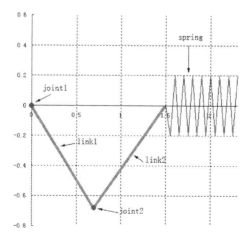

Fig. 3. A robot manipulator in contact with a spring (plane view)

i, respectively. And we set $m_1 = m_2 = 10.0\,kg, l_1 = l_2 = 1.0\,m, I_1 = I_2 = 0.83\,kgm^2, l_{c1} = I_{c2} = 0.5\,m$. Using the following abbreviation:

$$s_{12} = \sin(q_1 + q_2), c_{12} = \cos(q_1 + q_2), c_1 = \cos(q_1),$$
$$s_1 = \sin(q_1), s_2 = \sin(q_2), c_2 = \cos(q_2) \tag{40}$$

then the kinematic constraints mentioned in (2) and the dynamics (3) are given by

$$J(q) = \begin{bmatrix} -(l_1 s_1 + l_2 s_{12}) & -l_2 s_{12} \\ l_1 c_1 + l_2 c_{12} & l_2 c_{12} \end{bmatrix} \tag{41}$$

$$J^{-1}(q) = \frac{1}{l_1 l_2 s_2} \begin{bmatrix} l_2 c_{12} & l_2 s_{12} \\ -(l_1 c_1 + l_2 c_{12}) & -(l_1 s_1 + l_2 s_{12}) \end{bmatrix} \tag{42}$$

and

$$M(q) = \begin{bmatrix} M_{11} & M_{12} \\ M_{21} & M_{22} \end{bmatrix}, C(q, \dot{q}) = \begin{bmatrix} C_{11} & C_{12} \\ C_{21} & C_{22} \end{bmatrix}, G(q) = 0 \tag{43}$$

where

$$M_{11} = m_1 l_{c1}^2 + m_2 (l_1^2 + l_{c2}^2 + 2l_1 l_{c2} c_2) + I_1 + I_2,$$
$$M_{12} = M_{21} = m_2 (l_{c2}^2 + l_1 l_{c2} c_2) + I_2, M_{22} = m_2 l_{c2}^2 + I_2,$$
$$C_{11} = -m_2 l_1 l_{c2} s_2 \dot{q}_2, C_{12} = -m_2 l_1 l_{c2} s_2 (\dot{q}_1 + \dot{q}_2),$$
$$C_{21} = m_2 l_1 l_{c2} s_2 \dot{q}_1, C_{22} = 0 \tag{44}$$

The initial positions of the robot manipulator in Cartesian space are

$$x^k(0) = 1.0\,m, y^k(0) = 0.0\,m \tag{45}$$

To obtain a great tracking effect, we set a linear trajectory along x axis specified in Cartesian space as

$$x_{dt}(t) = 1 + 0.5(6t^5 - 15t^4 + 10t^3), (x_{dt} \leq x_d); \quad y_d(t) = 0 \tag{46}$$

where $t \in [0, T]$ and in the first trial for parameter estimation, $T = 30\,s > t_a$ in which the parameter estimation converges to a compact set, while in future trials, $T = 10\,s$ which is enough to make the matching error $w = 0$ and the robot manipulator has reached the equilibrium point x_*, and x_d the saturation of x_{dt} adapted iteratively.

The rest position of the spring is at x_b, and the spring exerts a force to the manipulator in the following manner:

$$f_I^k = \begin{cases} 0, & x^k < x_b; \\ K_0(x^k - x_b), & x^k \geq x_b. \end{cases}$$

where K_0 is the elasticity of the spring. The force measurement noise $\tilde{f}_I$ is set as the uniform-random-number signal with the known amplitude as 1.

The parameters in reference model (6) are set as

$$M_m = 0.5I(2), C_m = 10I(2), G_m = 40I(2) \tag{47}$$

where $I(2)$ represents a 2×2 unit matrix. With $\hat{\theta}^0(0) = 0$, the parameters in controller (14) and learning law (28) are given as

$$K = 50I(2), K_\delta = 2I(2), l = 0.01,$$
$$\Gamma = 1, \ell = 20, \Lambda = 20 \tag{48}$$

The default feedforward force is set as $F_0 = 0\,N$ and the largest possible object stiffness is set as $\bar{K}_0 = 500\,N/m$.

The estimated parameter $\hat{\theta}^0$ and matching error w^0 in the first trial are shown in Figs. 4 and 5, respectively. It can be seen that at the end of the first trial the system can obtain an estimated parameter stably and match the reference model with a small error. Then in later trial, we just let $\hat{\theta}^k = \hat{\theta}^0(30), k \geq 0$ which guarantees the convergence of matching error w to 0.

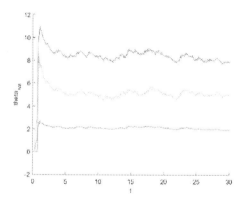

Fig. 4. The estimated parameter in the first trial $\hat{\theta}^0$

The estimation of K_0 and x_b when $K_0 = 200\,N/m, x_b = 1.3\,m; K_0 = 200\,N/m, x_b = 1.35\,m; K_0 = 230\,N/m, x_b = 1.35\,m$ are shown in Figs. 6, 7 and 8, respectively. For each situation, the robot has 20 trials for exploration. It can be seen from the Figs. 6, 7 and 8 that the stiffness and boundary of the spring can be estimated and approached after few trials. But there are some chattering phenomenons such as in the 18th trial in Fig. 6, which may be caused by the singular point due to the error in y direction. Thus to avoid this situation, our next work should eliminate the error in y direction.

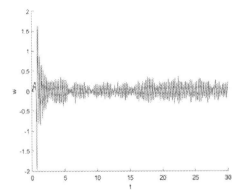

Fig. 5. The matching error in the first trial w^0

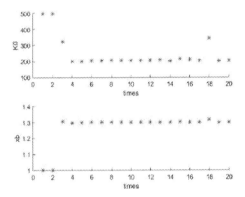

Fig. 6. The estimation of K_0, x_b when $K_0 = 200\,N/m, x_b = 1.3\,m$

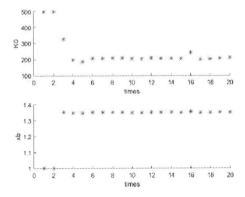

Fig. 7. The estimation of K_0, x_b when $K_0 = 200\,N/m, x_b = 1.35\,m$

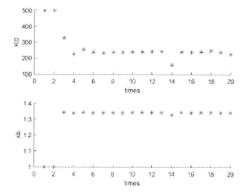

Fig. 8. The estimation of K_0, x_b when $K_0 = 230 \, N/m, x_b = 1.35 \, m$

7 Conclusion

This paper develops an adaptive parameter estimation and controller guaranteeing the convergence of parameter estimation in finite time for uncertain robotic system. In addition, the real robotic system tracks and matches a reference model whose reference point and feedforward force adapt iteratively. During the interaction control, the algorithm is able to identify the properties of the object, in particular, the object's geometric border and stiffness. The controller in this paper does not required the accurate force measurement. Simulations studies demonstrate the efficiency of this controller in practical robotic application.

References

1. Forrest, A.K.: Robot vision. Phys. Technol. **17**(17), 5–9 (1986)
2. Burdet, E., Osu, R., Franklin, D.W., et al.: The central nervous system stabilizes unstable dynamics by learning optimal impedance. Nature **414**(6862), 446–449 (2001)
3. Franklin, D.W., Liaw, G., Milner, T.E., et al.: Endpoint stiffness of the arm is directionally tuned to instability in the environment. J. Neurosci. Official, J. Soc. Neurosci. **27**(29), 7705–7716 (2007)
4. Chib, V.S., Patton, J.L., Lynch, K.M., et al.: Haptic identification of surfaces as fields of force. J. Neurophysiol. **95**(2), 1068–1077 (2006)
5. Yang C., Burdet E.: A model of reference trajectory adaptation for interaction with objects of arbitrary shape and impedance. In: 2011 IEEE/RSJ International Conference on Intelligent Robots and Systems (IROS), pp. 4121–4126. IEEE (2011)
6. Yang, C., Li, Z., Burdet, E.: Human like learning algorithm for simultaneous force control and haptic identification. In: IEEE/RSJ International Conference on Intelligent Robots and Systems, pp. 710–715. IEEE (2013)
7. Yang, C., Li, J., Li, Z., Chen, W., Cui, R.: Adaptive control of robot system of up to a half passive joints. In: Natraj, A., Cameron, S., Melhuish, C., Witkowski, M. (eds.) TAROS 2013. LNCS, vol. 8069, pp. 264–275. Springer, Heidelberg (2014)

8. Yang, C., Li, Z., Jing, L.: Trajectory planning and optimized adaptive control for a class of wheeled inverted pendulum vehicle models. IEEE Trans. Syst. Man Cybern. Part B Cybern. A Publ. IEEE Syst. Man Cybern. Soc. **43**(1), 24–36 (2012)
9. Ioannou, P., Sun, J.: Robust Adaptive Control. Prentice Hall, New Jersey (1996)
10. Sastry, S., Bodson, M.: Adaptive Control: Stability, Convergence, and Robustness. Prentice Hall, New Jersey (1989)
11. Na, J., Mahyuddin, M.N., Herrmann, G., et al.: Robust adaptive finite-time parameter estimation and control for robotic systems. Int. J. Robust Nonlinear Control **25**(16), 3045–3071 (2015)
12. Ge, S.S., Lee, T.H., Harris, C.J.: Adaptive Neural Network Control of Robotic Manipulators. World Scientific, London (1998)
13. Wang, D., Cheah, C.C.: An interative learning control scheme for impedance control of robotic manipulators. Int. J. Rob. Res. **17**(10), 1091–1104 (1998)
14. Li, Y., Yang, C., Ge, S.: Learning compliance control of robot manipulator in contact with the unknown environment. In: Proceedings of the 2010 IEEE Conference on Automation Science and Engineering (CASE), pp. 644–649. IEEE (2010)
15. Guo, L.: Self-convergence of weighted least-squares with applications to stochastic adaptive control. IEEE Trans. Autom. Control **41**(1), 79–89 (1996)

Computer Vision

Pose-Invariant Face Recognition Based on a Flexible Camera Calibration

Xiaohu Shao, Cheng Cheng$^{(\boxtimes)}$, Yanfei Liu, and Xiangdong Zhou

Chongqing Institute of Green and Intelligent Technology,
Chinese Academy of Sciences, Chongqing, China
{shaoxiaohu,chengcheng,liuyanfei,zhouxiangdong}@cigit.ac.cn

Abstract. In this paper, we present a flexible camera calibration for pose normalization to accomplish a pose-invariant face recognition. The accuracy of calibration can be easily influenced by errors of landmark detection or various shapes of different faces and expressions. By jointly using RANSAC and facial unique characters, we explore a flexible calibration method to achieve a more accurate camera calibration and pose normalization for face images. Our proposed method is able to eliminate noisy facial landmarks and retain the ones which best match the undeformable 3D face model. The experimental results show that our method improves the accuracy of pose-invariant face recognition, especially for the faces with unsatisfied landmark detection, variant shapes, and exaggerated expressions.

Keywords: Camera calibration · 3D alignment · Face recognition

1 Introduction

Face recognition plays an important role in pattern recognition and computer vision applications. In recent years, face recognition has made great progress with deep learning technique developing. Methods using deep learning and large training dataset [1–4] have almost achieved super-human accuracy on the LFW benchmark [5,6]. However, it remains a difficult problem for faces in the wild due to the variations in pose, illumination and expression. More specifically, different poses of the same face have dramatically different appearances, causing fatal problems to most of current face recognition systems.

In order to solve the aforementioned problems, many approaches have been explored, they can be categorized into feature-based methods and normalization-based methods.

The pose insensitive feature-based methods are widely used, they try to extract specific features which are invariant or insensitive to different poses. Wiskott *et al.* [7] collapse face variance of pose and expression by extracting concise face descriptions in the form of image graphs. Gross *et al.* [8] develop the theory of appearance-based face recognition from light-field, which leads directly to a pose-invariant face recognition algorithm that uses as many images of the

© Springer Nature Singapore Pte Ltd. 2016
T. Tan et al. (Eds.): CCPR 2016, Part I, CCIS 662, pp. 191–200, 2016.
DOI: 10.1007/978-981-10-3002-4_16

face as are available. Lai *et al.* [9] use wavelet transform and multiple view images to determine the reference image representation. Restricted to capacity of these representations and limited dataset, above mentioned methods are not able to get satisfied features which is insensitive to pose of faces in the wild. DCNN based face recognition have been widely reported in recent studies, because features trained by DCNN with huge size of dataset have a strong representation for variant of object, they achieve state-of-the-art performances on recognition of different poses of faces. Taigman *et al.* [1] derive a face representation form a nine-layer deep neural network. Sun *et al.* [2] propose to learn a set of high-level feature representations which called DeepID feature through deep learning for face verification. In 2014, they proposed two very deep neural network architectures to achieve a higher face identification accuracy [3]. Liu *et al.* [4] combine a multi-patch deep CNN and deep metric learning to extract low dimensional but very discriminative feature for face recognition.

Normalize-based method tries to normalize different faces to a unified frontal face to improve the accuracy of recognition. Chai *et al.* [10] use locally linear regression (LLR) to generate the virtual frontal view from a given non-frontal face image, this method is not able to always preserve the identity information. Berg [11] takes advantage of a reference set of faces to perform an identity-preserving alignment, warping the faces in a way that reduces differences due to pose and expression. Hu *et al.* [11] reconstruct a 3D face model from a single frontal face image, and synthesize faces with different PIE to characterize face subspace. Wang [12] proposes a fully automatic, effective and efficient framework for 3D face reconstruction based on a single face image in an arbitrary view. Asthana *et al.* [13] build a 3D Face Pose Normalization system which improves the recognition accuracy of face variation up to $\pm 45°$ in yaw and $\pm 30°$ in pitch angles. Zhu *et al.* [14] present a pose and expression normalization method to recover the neutral frontal faces without little artifact and information loss. Hasser *et al.* [15] use an unmodified 3D reference to approximate shape of all query faces and synthesize frontal faces. These 3D-based methods estimate the normalization transformations from correspondence between 2D and 3D facial landmarks, they are often efficient but suffers from errors and variety of landmarks which are caused by landmark detection, various shapes and exaggerated expressions.

Inspired by the above approaches, we present a flexible camera calibration for 3D alignment in order to improve pose-invariant face recognition. Different with work [14], we present a flexible camera calibration based on RANSAC [16] and facial unique characters to estimate poses of faces for pose normalization of faces. Our flexible camera calibration is insensitive to outliers of landmarks caused by landmark detection or variant of shape and expressions. The experimental results show that our method improves the accuracy of pose-invariant face recognition, especially for the faces with unsatisfied landmark detection, variant shapes, and exaggerated expressions.

Our pose-invariant face recognition includes three steps: First, we estimate the pose of a face using our proposed flexible camera calibration from

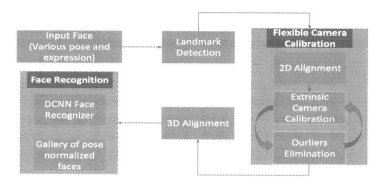

Fig. 1. Framework of the pose-invariant face recognition system

correspondence between 2D landmarks and an undeformable 3D face model. Second, we calculate the transformations of 3D alignment based on the estimated pose. Finally, we get the pose-normalized face and use them to train DCNN model for face recognition. The framework of our pose-invariant face recognition system is shown in Fig. 1.

The remainder of this paper is organized as follows: Sect. 2 introduces the details of flexible camera calibration and framework of our pose-invariant face recognition. Section 3 provides the experimental results of proposed method compared with other methods on face recognition. The conclusion and future work is provided in Sect. 4.

2 Facial Pose Normalization

Previous work of face recognition have witnessed the efficiency of the pose-normalized face and 3D face. In this section, we normalize poses of faces by proposed flexible camera calibration from correspondence between 2D landmarks and an undeformable 3D face model.

The problem of camera calibration can be described as follows: Given a mean 3D model of face $\mathbf{S} \in \Re^{3 \times n}$ with total n vertices, landmarks on the 2D face $\mathbf{s} \in \Re^{2 \times n}$, the goal is to estimate the intrinsic camera parameters $\mathbf{A} \in \Re^{3 \times 3}$, rotation matrix $\mathbf{R} \in \Re^{3 \times 3}$ and translation vector $\mathbf{t} \in \Re^{3 \times 1}$. $[\mathbf{R}, \mathbf{t}]$ is also known as extrinsic camera parameters. To find the parameters that best project the 3D face model to the 2D landmarks, we solve the nonlinear least squares optimization problem:

$$\{\mathbf{A}^*, \mathbf{R}^*, \mathbf{t}^*\} = \min_{\mathbf{A}, \mathbf{R}, \mathbf{t}} \|\mathbf{f}(\mathbf{A}, \mathbf{R}, \mathbf{t}, \mathbf{S}) - \mathbf{s}\|_F^2, \tag{1}$$

$$\mathbf{f} = \mathbf{f}_1 \circ \mathbf{f}_2, \tag{2}$$

$$\mathbf{f}_1(\mathbf{A}, \mathbf{R}, \mathbf{T}, \mathbf{S}) = \mathbf{A}(\mathbf{R}\mathbf{S} + \mathbf{T}), \tag{3}$$

$$\mathbf{f}_2(\mathbf{S}) = \begin{bmatrix} \mathbf{S}_1^\top \oslash \mathbf{S}_3^\top \\ \mathbf{S}_2^\top \oslash \mathbf{S}_3^\top \end{bmatrix} \tag{4}$$

where $\mathbf{T} = [\mathbf{t}, \mathbf{t}, ...] \in \Re^{3 \times n}$ consists of n copies of $\mathbf{t}$, $\mathbf{f}_2$ projects 3D vertices into 2D image, $\oslash$ denotes element-wise division, $\mathbf{S}_i$ is the row vector of i.

In order to get the correspondence of 3D face model and 2D landmarks, we get a mean 3D face model obtained from USF Human ID 3D face [18] and 2D landmarks by recent methods of facial landmark detection. We select 49 vertices from 70000 vertices to reconstruct a simple 3D face model. Automatic facial landmark detection on face images has been well studied [17–22], We select the method [19] for its satisfied accuracy on faces with large poses and its efficiency. Similarly with work [15], we retain 49 facial landmarks and exclude the contour landmarks, because different poses would change the matching relationship of contour landmarks and vertices of the 3D model.

2.1 Intrinsic Parameter Unit by 2D Alignment

Estimating the intrinsic parameters $\mathbf{A}$ and extrinsic parameters $[\mathbf{R}, \mathbf{t}]$ at the same time for a single image is an ill-pose problem. Work [22] estimates $\mathbf{A}$ by using many frames as its initialization. Work [15] uses a fixed $\mathbf{A}$ for aligned LFW images. The sizes and locations of faces on LFW images are almost the same, they can be seen sharing the same intrinsic matrix. But for an arbitrary image, its unsuitable to use the supposed intrinsic parameters. An approximate $\hat{\mathbf{A}}$ can be fixed when a face image I is aligned into coordinate of standard LFW dataset by similarity transformation. The source is the 2D facial landmarks, and target shape is the reference landmarks $\bar{\mathbf{s}}$, which can be calculated from the mean shape of all shapes in LFW images. The aligned landmarks $\hat{s}$ and image $\hat{I}$ is shown in Fig. 2.

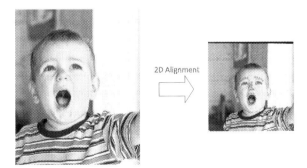

Fig. 2. An example of 2D alignment.

2.2 Flexible Extrinsic Camera Calibration

After the face is 2D aligned by similarity transformation, $[\mathbf{R}, \mathbf{t}]$ is to be estimated from the 2D facial landmarks and the 3D face model:

$$\{\mathbf{R}^*, \mathbf{t}^*\} = \min_{\mathbf{R}, \mathbf{t}} \left\| \mathbf{f}(\hat{\mathbf{A}}, \mathbf{R}, \mathbf{t}, \mathbf{S}) - \hat{s} \right\|_F^2 \qquad (5)$$

The above problem is known as 3D pose estimation, which is usually solved by iterative method based on *Levenberg-Marquardt Algorithm* (LMA) [23]. This optimization is efficient and accurate when the vertices of 3D face are able to match the 2D landmarks very well. However, as noises often exist in landmark detection and different person with expressions have various shapes of landmarks, it is impossible to match the various 2D landmarks with the undeformable 3D model accurately. These matching errors decrease the accuracy of pose estimation, so we need to eliminate these large errors of landmarks before the iteration.

RANSAC is an iterative method to estimate parameters of a mathematical model from a set of data which contains outliers [24]. However, when the number of iteration computed is limited, the solution may not be optimal. Considering efficiency and accuracy of pose normalization, we cannot afford no limited iterations.

When we use RANSAC to eliminate the outliers of facial landmarks on a large dataset, we observe that outliers often appear as landmarks of particular parts, such as eyebrow, top and bottom of mouth. It seems that the accuracy of these landmarks location is less than other landmarks, or these landmarks are not able to match the undeformable 3D model very well caused by variant of person and expressions. The probability distribution of each landmark which is labeled as an outlier in dataset by general RANSAC is shown in Fig. 3.

In order to speed up outlier elimination of landmarks, we separate all N landmarks in two pools according with their probability distribution labels as an inlier in training dataset: inliers pool $\mathbf{\Phi} = \{\phi_1, \phi_2, ..., \phi_p\}$, outliers pool $\mathbf{\Psi} = \{\psi_1, \psi_2, ..., \psi_q\}$, where ϕ_i denotes the ith landmark which is labeled an inlier with large probability, ψ_j denotes the jth landmark which is labeled as an outlier with large probability. In the process of eliminating outliers, landmarks belonged to $\mathbf{\Phi}$ are selected to calculate the pose using *LMA* optimization with less probability, landmarks belonged to $\mathbf{\Psi}$ are selected as inliers with more chance. The process of flexible extrinsic camera calibration is summarized in Algortihm 1. First, we use all of landmarks to estimate the initial $[\mathbf{R}, \mathbf{t}]$. Second, we project the 3D model into the 2D image and calculate the distance

Fig. 3. Probability distribution of each landmark which are labeled as outliers in dataset. All landmarks are drawn by red circles with different sizes. The larger size of circle represents that the current landmark is labeled as an outlier with larger probability. (Color figure online)

between each projected landmark and the corresponding real landmark. Third, landmark noises are eliminated by comparing the threshold and the normalized distance. We control the opportunity of elimination by setting the threshold θ_1 for landmarks belong to $\mathbf{\Phi}$ larger than threshold θ_2 for landmarks belong to $\mathbf{\Psi}$.

Algorithm 1. Flexible Extrinsic Camera Calibration

Input: 2D aligned facial landmarks $\hat{s}$, 3D face model $\mathbf{S}$, instrinsic camera parameter $\hat{\mathbf{A}}$, index pool of inliers and outliers $\mathbf{\Phi}$, $\mathbf{\Psi}$.

Output: rotation matrix $\mathbf{R}^*$, translation vector $\mathbf{t}^*$.

 1: **while** not converged **do**
 2:     Calculate $\mathbf{R}$ and $\mathbf{t}$ by using LMA with $\hat{s}$ and $\mathbf{S}$.
 3:     Project $\mathbf{S}$ to 2D landmarks $\mathbf{s_{proj}}$ by using $\hat{\mathbf{A}}$, $\mathbf{R}$ and $\mathbf{t}$.
 4:     Cacluate distance $\mathbf{D} = \{d_1, d_2, ..., d_N\}$ between each landmark of $\hat{s}$ and $\mathbf{s}_{proj}$.
 5:     Obtain $\mathbf{D_{sort}} = \{d_{sort,1}, d_{sort,2}, ..., d_{sort,L}\}$ by sorting elements of $\mathbf{D}$ in a descending order.
 6:     Find the index $\mathbf{E} = \{e_1, e_2, ..., e_N\}$ of the first L elements of $\mathbf{D_{sort}}$ in $\hat{s}$.
 7:     **for** $i = 0 \rightarrow L - 1$ **do**
 8:         **if** $(e_i \in \mathbf{\Phi}$ and $d_{sort,i} > \theta_1)$ **or** $(e_i \in \mathbf{\Psi}$ and $d_{sort,i} > \theta_2)$ **then.**
 9:             Eliminate the landmark of index e_i.
10:         **end if**
11:     **end for**
12: **end while**
13: Generate the final parameters $[\mathbf{R}^*, \mathbf{t}^*]$.

In our experiments, we set $\theta_1 = 0.08$, $\theta_2 = 0.05$, $L = 10$ when $N = 49$, outlier elimination quickly converges in only 1 or 2 stages.

2.3   3D Alignment and Face Recognition

After the extrinsic parameters are calculated, we caculate the normalization transformation based on the estimated poses $[\hat{\mathbf{A}}, \mathbf{R}^*, \mathbf{t}^*]$. Then, we get the 3D aligned faces (more details can be found in [15,25]) and use them to train models for face recognition. An example of our 3D alignment result can be seen in Fig. 4.

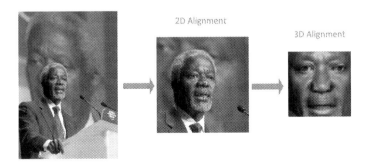

Fig. 4. An example of our 3D alignment result.

3   Experiments

In this section, we present experimental results of our proposed method on face recognition.

3.1   Database

LFW: LFW dataset consists of more than 13000 faces of 5749 celebrities. Each face has been labeled with the name of the person pictured. The number of faces varies from 1 to 530 for one person. 1680 of the people pictured have two or more distinct photos in the dataset. It is the most commonly used database for unconstrained face recognition.

CASIA-WebFace [26]**:** It contains 10575 subjects and 49414 images, which are collected from Internet by a semi-automatical way. CASIA-WebFace is prepared for training, LFW is used to evaluate our alignment compared with other alignment methods.

3.2   2D Alignment and 3D Alignment

After detecting faces [27] and landmarks [19] in an image, we use facial landmarks to normalize faces. 2D affine transformation is often used to align faces for improving face recognition. It is used to approximately scale, rotate and translate the image into a new warped image. It is also called 2D-alignment, pose normalization is often called 3D-alignment, which can be applied to compensate out-of-plane rotation. In this paper, we compare the three methods of alignment in face recognition: 2D alignment, 3D alignment of [15], and our proposed method.

3.3   Performance Analysis

We get aligned faces by applying the three alignment methods for training and test datasets, then we train three DCNN models on the training dataset. To evaluate the discriminative capability of the face representation, we compare the cosine distance of a pair of a normalized features which are transformed by PCA. The comparison of face recognition results on LFW by applying standard protocols and BLUFR protocols [26] are listed in Tables 1 and 2. The results show that our method is better than the other two normalization methods. We train models with the *BN-inception v1* network [28] on *Caffe* platform [29] from scratch for DCNN models training.

Because the limitation of GPU resources and the scale of training set, and our goal is only to show that face recognition can benefit from our 3D alignment method, we do not get the best result compared with the recent results on LFW. We believe that we can get state-of-the-art face recognition performance using our proposed method if we continued to adjust parameters, enlarge dataset and train deeper models.

Table 1. The performance of our proposed method compared with other methods on LFW under standard protocol.

Method	Accuracy ± SE
2D alignment	0.9623 ± 0.0107
3D alignment of [15]	0.9660 ± 0.01
Proposed method	**0.9673 ± 0.0081**

Table 2. The performance of our proposed method compared with other methods on LFW under standard protocol.

Method	VR@FAR $= 0.1\%$	DIR@FAR $= 1\%$, Rank $= 1$
2D alignment	68.64%	34.74%
3D alignment of [15]	73.27%	37.47%
Proposed method	**74.72%**	**38.57%**

4   Conclusion and Future Work

In this paper, we present a flexible camera calibration for 3D alignment to improve pose-invariant face recognition. Compared with previous normalization work, our method based on RANSAC and facial unique characters is insensitive to outliers of landmarks caused by landmark detection or variant of person and expressions. Experiments show that it the best performance on recognition of faces under complicated environment.

In the future, we will continue to improve our 3D alignment method to overcome the difficulty brought by various poses and expressions of faces. We will also get a further study to solve this problem by applying the deep learning method.

Acknowledgement. This work was supported by the Project from National Science Foundation for Youths of China (No. 61502444), National Natural Science Foundation of China (No. 61472386), Strategic Priority Research Program of the Chinese Academy of Sciences (No. XDA06040103). Two Titan X GPUs applied for this research were donated by the NVIDIA Corporation.

References

1. Taigman, Y., Yang, M., Ranzato, M., Wolf, L.: Deepface: closing the gap to human-level performance in face verification. In: Proceedings of the IEEE Conference on Computer Vision and Pattern Recognition, pp. 1701–1708 (2014)
2. Sun, Y., Wang, X., Tang, X.: Deep convolutional network cascade for facial point detection. In: Proceedings of the IEEE Conference on Computer Vision and Pattern Recognition, pp. 3476–3483 (2013)

3. Sun, Y., Wang, X., Tang, X.: Deep learning face representation from predicting 10,000 classes. In: Proceedings of the IEEE Conference on Computer Vision and Pattern Recognition, pp. 1891–1898 (2014)
4. Liu, J., Deng, Y., Huang, C.: Targeting ultimate accuracy: face recognition via deep embedding (2015). arXiv preprint arXiv:1506.07310
5. Huang, G.B., Ramesh, M., Berg, T., Learned-Miller, E.: Labeled faces in the wild: a database for studying face recognition in unconstrained environments. Technical report 07–49, University of Massachusetts, Amherst (2007)
6. Huang, G.B., Learned-Miller, E.: Labeled faces in the wild: updates and new reporting procedures. Technical report 14–003, Department of Computer Science, University Massachusetts Amherst, Amherst, MA, USA (2014)
7. Wiskott, L., Fellous, J.M., Kuiger, N., Von Der Malsburg, C.: Face recognition by elastic bunch graph matching. IEEE Trans. Pattern Anal. Mach. Intell. **19**(7), 775–779 (1997)
8. Gross, R., Matthews, I., Baker, S.: Appearance-based face recognition and light-fields. IEEE Trans. Pattern Anal. Mach. Intell. **26**(4), 449–465 (2004)
9. Lai, J.H., Yuen, P.C., Feng, G.C.: Face recognition using holistic fourier invariant features. Pattern Recogn. **34**(1), 95–109 (2001)
10. Chai, X., Shan, S., Chen, X., Gao, W.: Locally linear regression for pose-invariant face recognition. IEEE Trans. Image Process. **16**(7), 1716–1725 (2007)
11. Hu, Y., Jiang, D., Yan, S., Zhang, L., Zhang, H.: Automatic 3D reconstruction for face recognition. In: 2004 Proceedings of the Sixth IEEE International Conference on Automatic Face and Gesture Recognition, pp. 843–848. IEEE (2004)
12. Wang, C., Yan, S., Li, H., Zhang, H., Li, M.: Automatic, effective, and efficient 3D face reconstruction from arbitrary view image. In: Aizawa, K., Nakamura, Y., Satoh, S. (eds.) PCM 2004. LNCS, vol. 3332, pp. 553–560. Springer, Heidelberg (2004). doi:10.1007/978-3-540-30542-2_68
13. Asthana, A., Marks, T.K., Jones, M.J., Tieu, K.H., Rohith, M.: Fully automatic pose-invariant face recognition via 3D pose normalization. In: Proceedings of the IEEE Conference on International Conference on Computer Vision (ICCV), pp. 937–944. IEEE (2011)
14. Zhu, X., Lei, Z., Yan, J., Yi, D., Li, S.Z.: High-fidelity pose and expression normalization for face recognition in the wild. In: Proceedings of the IEEE Conference on Computer Vision and Pattern Recognition, pp. 787–796 (2015)
15. Hassner, T., Harel, S., Paz, E., Enbar, R.: Effective face frontalization in unconstrained images. In: Proceedings of the IEEE Conference on Computer Vision and Pattern Recognition, pp. 4295–4304 (2015)
16. Fischler, M.A., Bolles, R.C.: Random sample consensus: a paradigm for model fitting with applications to image analysis and automated cartography. Commun. ACM **24**(6), 381–395 (1981)
17. Zhu, X., Ramanan, D.: Face detection, pose estimation, and landmark localization in the wild. In: Proceedings of the IEEE Conference on Computer Vision and Pattern Recognition, pp. 2879–2886. IEEE (2012)
18. Cao, X., Wei, Y., Wen, F., Sun, J.: Face alignment by explicit shape regression. Int. J. Comput. Vis. **107**(2), 177–190 (2014)
19. Xiong, X., Torre, F.: Supervised descent method and its applications to face alignment. In: Proceedings of the IEEE Conference on Computer Vision and Pattern Recognition, pp. 532–539 (2013)
20. Ren, S., Cao, X., Wei, Y., Sun, J.: Face alignment at 3000 fps via regressing local binary features. In: Proceedings of the IEEE Conference on Computer Vision and Pattern Recognition, pp. 1685–1692 (2014)

21. Kazemi, V., Sullivan, J.: One millisecond face alignment with an ensemble of regression trees. In: Proceedings of the IEEE Conference on Computer Vision and Pattern Recognition, pp. 1867–1874 (2014)
22. Cao, C., Hou, Q., Zhou, K.: Displaced dynamic expression regression for real-time facial tracking and animation. ACM Trans. Graph. (TOG) **33**(4), 43 (2014)
23. Zhang, Z.: A flexible new technique for camera calibration. IEEE Trans. Pattern Anal. Mach. Intell. **22**(11), 1330–1334 (2000)
24. Dementhon, D.F., Davis, L.S.: Model-based object pose in 25 lines of code. Int. J. Comput. Vis. **15**(1–2), 123–141 (1995)
25. http://www.openu.ac.il/home/hassner/projects/frontalize. Accessed 12 Feb 2015
26. Liao, S., Lei, Z., Yi, D., Li, S.Z.: A benchmark study of large-scale unconstrained face recognition. In: 2014 IEEE International Joint Conference on Biometrics (IJCB), pp. 1–8. IEEE (2014)
27. Viola, P., Jones, M.J.: Robust real-time face detection. Int. J. Comput. Vis. **57**(2), 137–154 (2004)
28. Laurent, C., Pereyra, G., Brakel, P., Zhang, Y., Bengio, Y.: Batch normalized recurrent neural networks (2015). arXiv preprint arXiv:1510.01378
29. Jia, Y., Shelhamer, E., Donahue, J., Karayev, S., Long, J., Girshick, R., Guadarrama, S., Darrell, T.: Caffe: convolutional architecture for fast feature embedding (2014). arXiv preprint arXiv:1408.5093

Pedestrian Detection Aided by Deep Learning Attributes Task

Chao Qiu, Yinhui Zhang[✉], Jieqiong Wang, and Zifen He

Faculty of Mechanical and Electrical Engineering,
Kunming University of Science and Technology, Kunming 650500, China
chaoqiu23@foxmail.com, yinhui_z@163.com

Abstract. Deep Learning methods have achieved great successes in pedestrian detection owing to their ability of learning discriminative features from pixel level. However, most of the popular methods only consider using the deep structure as a single feature extractor (one attribute) which may confuse positive with hard negative samples. To address this ambiguity, this work jointly learns three different attributes, including parts, deformation and similarity attributes. This paper proposes a new deep network which jointly optimizes the three attributes and formulates them to form a binary classification task. Extensive experiments show that the proposed method outperforms competing methods on the challenging Caltech and ETH benchmarks.

Keywords: Deep learning · Similarity attributes · Pedestrian detection

1 Introduction

Pedestrian detection has attracted wide attention in automotive safety, robotics, and intelligent video surveillance. The main challenges of this task are caused by large variation and confusion between human body and background and the variation in the intra-class of pedestrians, including clothing and occlusion.

In order to detect pedestrian in an image, popular methods can generally be divided in to two types, well designed features [1–7] and features extracted by deep structures [8–12,14]. In the former type, conventional methods extracted Harr [1], HOG [2], or HOG-LBP [3] from images to train a template by applying SVM [5] or boosting [4] algorithm. Then, the learned template can be applied as a filter for detecting an entire human body. Recently, the popular deformable part models (DPM) are studied to deal with the large variation of a human pose and viewpoint. The mixed model learns the deformation attributes between parts and root and the edge information of the human.

In the latter type, the deep learning methods achieve great successes by extracting discriminative features from pixel level. Ouyang and Wang [11] proposed a joint deep convolution structure, which optimizes the deformable parts and occlusion, jointly, by introducing a deformation layer and a specific hidden layer. However, these methods only consider the features extracted from a single

© Springer Nature Singapore Pte Ltd. 2016
T. Tan et al. (Eds.): CCPR 2016, Part I, CCIS 662, pp. 201–210, 2016.
DOI: 10.1007/978-981-10-3002-4_17

image which may cause confusion or misclassify. The features whether obtained from well-designed structures or deep structures ignore the relationship or shared attributes of the same category, since these methods extract the features from images independently.

To address the problem, this work proposes to explore the relation among the images which contain humans.

This paper introduces a deep structure aided by a binary coding algorithm to learn the similarity attributed that are shared among the positive samples. And this work also constructs a parallel deep network structure for jointly learning feature extraction, parts deformation and occlusion information. Then, outputs from the two parallel deep structures are feed into the final decision layer to decide whether the object is human.

This work has the following main contributions. (1) To our knowledge, this work is the first one to learn the similarity attributes by a deep learning method for pedestrian detection. In this paper, similarity attributes are obtained by applying a deep structure and an image coding algorithm. (2) This work proposed a parallel deep structure which concurrent learns the similarity attributes, parts edge attributes and parts deformation attributes. Aided by this parallel structure, the results on the Caltech and ETH datasets outperform other competing algorithms.

The rest paper will be arranged as follows. Section 2 introduces the construction of similarity attributes. A joint deep network for learning deformation attributes is introduced in Sect. 3. Our proposed parallel deep structure is discussed in Sect. 4. Section 5 demonstrates the performance of our approach through experimental evaluation on popular datasets. Finally, Sect. 6 gives a conclusion of this work. Figure 1 shows the pipeline of our proposed approach.

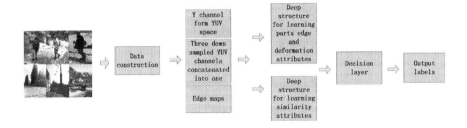

Fig. 1. The pipeline of the proposed approach

2   Similarity Attributes

In this section, we address the task of learning similarity attributes. In this work the similarity attributes are learned by a weak supervisory manner. Because, we only need to group positive samples and feed them into the algorithm for learning attributes. Therefore, we only use the positive labels for grouping samples and drop the labels when training. We will fist review some notations of binary code and introduce our idea of learning similarity attributes in Sect. 2.1. The learning process can be divided into two stages. The first stage, discussed in Sect. 2.2, is to initialize the parameters in the similarity deep network (the bottom one showed in Fig. 2). The second stage, discussed in Sect. 2.3, describes the proposed deep network for training.

2.1   A Brief Introduction of Binary Code and Our Similarity Attributes

In this subsection, we will first review some notations of image binary coding and introduce our idea of learning similarity attributes.

Learning binary codes was first introduced for representing large scale image collections [20–23]. Because, encoding high dimensional image descriptor as compact binary strings can enable large efficiency gains in storage and computation speed for image retrieval.

Hence, the goal of this task is to learn a binary code matrix $B \in \{-1, 1\}^{n \times c}$, where c denotes the length of the code when encoding an image and n is the number of the code. Assume we have a set of n data points $\{x_1, x_2, \cdots, x_n\}$, $x_i \in \mathbb{R}^d$, that forms the rows of the data matrix $\{x_1, x_2, \cdots, x_n\}$, $X \in \mathbb{R}^{n \times d}$, the binary coding function can be formulated as $h_k(x) = \text{sign}(xw_k)$, where $k = 1, 2, \cdots, c$ represents the bite of the binary code and w_k is a column vector of hyperplane coefficients. Here, $\text{sign}(v) = 1$ if $v \geq 0$ and 0 otherwise. For a matrix or a vector, $\text{sign}(\cdot)$ represents a element-wise operation of above formula. Therefore, the binary encoding process can be denoted as $B = \text{sign}(XW)$.

In order to retrieve images in database, the binary code should preserve the key information of images. And the key information (the binary code) for the same category should be same (similar), ideally, so that image retrieval can accurately be done. Inspirit by this fact, we believe that images contain same category (pedestrian) should have same (similar) key information (binary code). Hence, we proposed a new method to learn similarity attribute based on binary code to represent the similarity between images that contain the same category.

2.2   Initialize the Parameters

Before training stage, we initialize the hyper parameters of the similarity deep network.

First, we sample all the positive instances from training set to from a dataset that only contains one category (pedestrian). After we get the one category dataset, we encode each image from the dataset by applying a binary code method. And in this binary coding method we follow [13] to us a iterative quantization (ITQ) approach to reduce quantization error for learning a better binary code. Because ITQ approach enables the binary coding method to learn the codes that preserve the locality structure of the data.

Since the binary code only contain 0 and 1. The code itself can be used as labels. For example, assume we have a ten length code, 1101001011, each bit in this code can be seen as a label of an attribute, 1 for positive 0 otherwise. And as discussed in Sect. 2.1, the binary codes of same category should contain the same key information of an image, in other words, the binary codes of same category should be same (similar). Thus, the codes from the same category should have the same (similar) number of 1 (positive attribute) at same (similar) bit location and have different number of 1 at different bit location otherwise. This means they share same (similar) attributes if they come from the same category. Therefore, we can think the binary code as labels of a multi-classification task, each bit represents a class. And we apply support vector machine (SVM) for classification. After training the multi-class SVM, we obtain the weights of this multi classification task. The weights can be considered as a template that contains all attributes of a category (pedestrian). Hence, we call the features filtered by this template as similarity attributes.

2.3   Deep Network for Learning Similarity Attributes

Our proposed similarity deep network for learning the similarity attributes consist of four layers including an input layer. The first layer is an input layer which input the image data to the deep network. The following layer is a convolutional layer which convolute 64 $9 \times 9 \times 3$ kernels with input data. The third layer is an average pooling layer which pools the output of the previous layer using 4×4 boxcar filters with a 4×4 sub-sampling step. Since, the convolution and pooling preserve the spacial structure of original images, the output of these two layer (deep features) can also be considered as deep 'images'. Therefore, we construct a similarity attributes layer following by the average pooling layer. This layer apply the template we discussed in previous subsection and compute scores of attributes.

Finally the scores of attributes computed in similarity layer are feed into a decision system with the deformation and edge attributes discussed in Sect. 3 to classify the objects final label.

3   Deformation and Edge Attributes

In order to learn the part deformation and edge attributes in our work. We construct a deep convolution neural network. This deep structure has four layers including an input layer and a deformation layer similar to the definition in [11]. The input data were feed into the first convolutional layer by the input layer. Then, the following convolutional layer convolutes the data with kernels and output 64 filtered data maps. And the structure passes the maps into a pooling layer with a subsampling step. In the third layer and forth layer the structure compute the edge and part deformation attributes and form the part scores which are finally passed into the decision system for deciding the label of the object.

4   Parallel Deep Network

Our parallel deep structure consists of two convolution neural networks and a decision system. The convolutional neural network has four layers including an input layer, two convolutional layers and an average pooling layer. These two deep structure parallel learn the part deformation, edge and similarity attributes. They have the same input data and the output maps of the two deep structures are sent into the decision system for assigning the label. Figure 2 illustrates the parallel deep structure proposed in this paper.

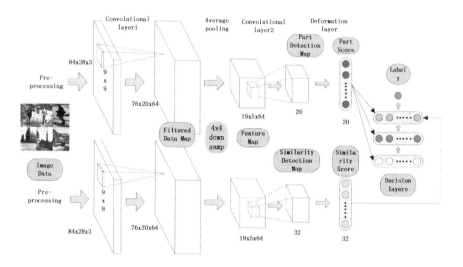

Fig. 2. The proposed parallel deep structure

4.1   Decision System

The decision system is similar to the Restricted Boltzmann Machine (RBM) [15] and Deep Belief Net (DBN) [16, 17]. Figure 3 illustrates the model of RBM, DBN and our proposed decision system.

As illustrated in Fig. 3, our decision system has three layers and layer two and three have extra hidden units which prevent the disturbance form the previous layer. The part scores are passed to the three layers of the decision system with weights similar to the way in [11]. However, the similarity scores are only passed to the top layers non-hidden units with weights. This is because the similarity attributes only contribute to the final decision of objects label, they do not attempt to decide the way of occlusion defined in [11]. And finally, after passing to the top layer the part scores and the similarity scores are jointly decide the label of object.

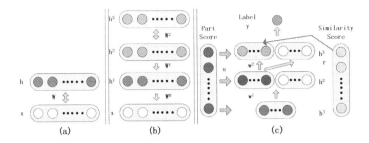

Fig. 3. (a) RBM, (b) DBN, (c) our decision system

Similar to the definition in [11], we denote the model of part and deformation attributes for back-propagation (BP) as follows:

$$hp_j^1 = \sigma(c_j^1 + u_j^1 s_j^1)$$
$$hp_j^{l+1} = \sigma(\mathbf{hp}^{l\mathrm{T}}\mathbf{w}_{*,j}^l + c_j^{l+1} + u_j^{l+1} s_j^{l+1}), l = 1, 2 \tag{1}$$

Where hp_j^l and s_j^l denote the hypothesis and score of the jth part at level l, respectively. Here, $\mathbf{hp}^l = [hp_1^l, hp_2^l, \cdots, hp_p^l]$, c_j^i is bias, $\mathbf{W}^l$ models the correlation between $\mathbf{hp}^l$ and $\mathbf{hp}^{l+1}$, $\mathbf{w}_{*,j}^l$ is the jth column of, $\sigma(\cdot)$ is the sigmoid function.

In this system, the model of similarity for BP is defined as follows:

$$hs_j = \sigma(c_j + r_j ss_j) \tag{2}$$

Where hs_j and ss_j denotes the hypothesis and score of the jth attributes, respectively. The parameter r_j is the jth similarity attributes weight in the top layer. And the model of final output hypothesis (label) is defined as:

$$y = \sigma(\tilde{\mathbf{h}}\mathbf{w}^{\mathbf{cls}} + b) \tag{3}$$

Where $\tilde{\mathbf{h}} = \phi(\mathbf{hp}^3, \mathbf{hs})$ is a matrix, $\mathbf{w}^{cls}$ is considered as the linear classifier for the units in the top layer.

In order to learn the parameters in the convolution neural networks, prediction error is back-propagated through **s** and **ss**. The gradients for **s** and **ss** are:

$$\frac{\partial L}{\partial s_i^l} = \frac{\partial L}{\partial hp_i^l}\frac{\partial hp_i^l}{\partial s_i^l} = \frac{\partial L}{\partial hp_i^l}hp_i^l(1 - hp_i^l)u_i^l \tag{4}$$

$$\frac{\partial L}{\partial ss_i} = \frac{\partial L}{\partial hs_i}\frac{\partial hs_i}{\partial ss_i} = \frac{\partial L}{\partial hs_i}hs_i(1 - hs_i)r_i \tag{5}$$

$$\frac{\partial L}{\partial \tilde{\mathbf{h}}_i} = \frac{\partial L}{\partial y}y(1 - y)w_i^{cls} \tag{6}$$

$$\frac{\partial L}{\partial hp_i^2} = w_{i,*}^2[\frac{\partial L}{\partial \mathbf{hp}^3} \odot \mathbf{hp}^3 \odot (1 - \mathbf{hp}^3)] \tag{7}$$

$$\frac{\partial L}{\partial hp_i^1} = w_{i,*}^1 [\frac{\partial L}{\partial \mathbf{hp}^2} \odot \mathbf{hp}^2 \odot (1 - \mathbf{hp}^2)] \tag{8}$$

$\odot$ operates the Hadamard product, which is a element-wise production. L denotes the loss function and we use the log loss function in this work.

5   Experiments

The proposed approach is evaluated on the Caltech dataset [18] and the ETH dataset [19]. Similar to [11] we use HOG + CSS and Linear SVM for reducing candidate detection windows at both training and testing stages. And we apply the evaluation code provided by Dollar et al. online are used for evaluation following the criteria proposed in [18] which is the log-average miss rate. In the experiments, we follow the popular methods that evaluating the performance on the reasonable subset of the evaluated datasets. And we compare our proposed algorithm with several competing methods including VJ [24], CovNet [25], HOG [26], MultiFtr [27], LatSvm-V2 [28], MultiFtr+CSS [29], FPDW [30], CrossTall [31], MultiFtr+motion [29], MultiResC [32] and JDN. The result of JDN is obtained by running the code provided by the author in [11] on our Laptop for fairly comparing with our method. Therefore the result is different to the one provided in original paper. And other results are obtained from Caltech datasets homepage.

5.1   Results of the Caltech-Test Dataset

First we train our model and JDN model on the Caltech-Training datasets. Then, we test our algorithm and the state-of-the-art JDN method on the Caltech-Test datasets. And finally, we plot the results of all the methods for comparing the performance. Figure 4 shows the results on Caltech datasets. And the result of

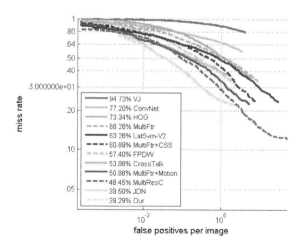

Fig. 4. The results on Caltech-Test datasets

our approach gets a similar but better performance than the state-of the-art JDN model and outperformances other competing algorithms.

5.2 Results of the ETH Dataset

For a fair comparison on the ETH dataset, we follow the training setting commonly adopted by state-of-the-art approaches; that is, using the INRIA training dataset in [1] to train our models and JDN. The results of our approach and other competing methods are illustrated in Fig. 5. The results show that our method outperforms the state-of-the-art method (JDN) nearly 2.

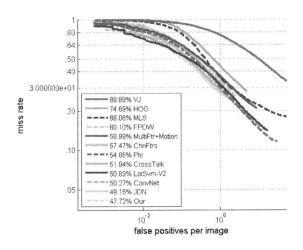

Fig. 5. The results on ETH datasets

6   Conclusion

This paper proposes a pedestrian detection method which is aided by the similarity attributes. In this paper, a parallel deep structure is introduced which helps to learn the similarity attributes. We evaluate the proposed method on two popular datasets, Caltech and ETH. And we compare our method with competing methods on both two datasets. The experiment results show that our method outperforms the state-of-the-art method and other competing methods. In the future work, we are going to optimize the parameters of the method which are the same to JDN method because we believe the optimization of our parameters will further improve the performance.

Acknowledgments. This work was supported by the National Natural Science Foundation of China under Grants Nos. 61302173, 61461022.

References

1. Viola, P., Jones, M.J., Snow, D.: Detecting pedestrians using patterns of motion and appearance. IJCV **63**(2), 153–161 (2005)
2. Dalal, N., Triggs, B.: Histograms of oriented gradients for human detection. In: CVPR (2005)
3. Wang, X., Han, T.X., Yan, S.: An HOG-LBP human detector with partial occlusion handling. In: ICCV (2009)
4. Dollar, P., Tu, Z., Perona, P., Belongie, S.: Integral channel features. In: BMVC (2009)
5. Dollar, P., Appel, R., Belongie, S., Perona, P.: Fast feature pyramids for object detection. TPAMI **36**, 1532–1545 (2014)
6. Zhang, S., Bauckhage, C., Cremers, A.: Informed haarlike features improve pedestrian detection. In: CVPR, pp. 947–954 (2013)
7. Felzenszwalb, P.F., Girshick, R.B., McAllester, D., Ramanan, D.: Object detection with discriminatively trained part-based models. TPAMI **32**(9), 1627–1645 (2010)
8. Ouyang, W., Wang, X.: A discriminative deep model for pedestrian detection with occlusion handling. In: CVPR (2012)
9. Ouyang, W., Zeng, X., Wang, X.: Modeling mutual visibility relationship in pedestrian detection. In: CVPR (2013)
10. Sermanet, P., Kavukcuoglu, K., Chintala, S., LeCun, Y.: Pedestrian detection with unsupervised multi-stage feature learning. In: CVPR (2013)
11. Ouyang, W., Wang, X.: Joint deep learning for pedestrian detection. In: ICCV (2013)
12. Luo, P., Tian, Y., Wang, X., Tang, X.: Switchable deep network for pedestrian detection. In: CVPR, pp. 899–906 (2014)
13. Gong, Y., Lazebnik, S.: Iterative quantization: a procrustean approach to learning binary codes. In: Proceedings of IEEE Conference on Computer Vision and Pattern Recognition (2011)
14. Tian, Y., Luo, P., Wang, W.: Pedestrian detection aided by deep learning semantic tasks (2014). arXiv preprint arXiv:1412.0069
15. Hinton, G.E.: Training products of experts by minimizing contrastive divergence. Neural Comput. **14**, 1771–1800 (2002)
16. Hinton, G.E., Osindero, S., Teh, Y.: A fast learning algorithm for deep belief nets. Neural Comput. **18**, 1527–1554 (2006)
17. Hinton, G.E., Salakhutdinov, R.R.: Reducing the dimensionality of data with neural networks. Science **313**(5786), 504–507 (2006)
18. Dollar, P., Wojek, C., Schiele, B., Perona, P.: Pedestrian detection: an evaluation of the state of the art. IEEE Trans. PAMI **34**(4), 743–761 (2012)
19. Ess, A., Leibe, B., Gool, L.V.: Depth and appearance for mobile scene analysis. In: ICCV (2007)
20. Torralba, A., Fergus, R., Weiss, Y.: Small codes and large image databases for recognition. In: CVPR (2008)
21. Wang, J., Kumar, S., Chang, S.-F.: Semi-supervised hashing for large-scale image retrieval. In: CVPR (2010)
22. Wang, J., Kumar, S., Chang, S.-F.: Sequential projection learning for hashing with compact codes. In: ICML (2010)
23. Weiss, Y., Torralba, A., Fergus, R.: Spectral hashing. In: NIPS (2008)
24. Viola, P., Jones, M.J., Snow, D.: Detecting pedestrians using patterns of motion and appearance. IJCV **63**(2), 153–161 (2005)

25. Sermanet, P., Kavukcuoglu, K., Chintala, S., Lecun, Y.: Pedestrian detection with unsupervised and multi-stage feature learning. In: CVPR (2013)
26. Dalal, N., Triggs, B.: Histograms of oriented gradients for human detection. In: CVPR (2005)
27. Wojek, C., Schiele, B.: A performance evaluation of single and multi-feature people detection. In: Rigoll, G. (ed.) DAGM 2008. LNCS, vol. 5096, pp. 82–91. Springer, Heidelberg (2008)
28. Felzenszwalb, P., Grishick, R.B., McAllister, D., Ramanan, D.: Object detection with discriminatively trained part based models. IEEE Trans. PAMI **32**, 1627–1645 (2010)
29. Walk, S., Majer, N., Schindler, K., Schiele, B.: New features and insights for pedestrian detection. In: CVPR (2010)
30. Dollar, P., Belongie, S., Perona, P.: The fastest pedestrian detector in the west. In BMVC, 2010
31. Dollár, P., Appel, R., Kienzle, W.: Crosstalk cascades for frame-rate pedestrian detection. In: Fitzgibbon, A., Lazebnik, S., Perona, P., Sato, Y., Schmid, C. (eds.) ECCV 2012, Part II. LNCS, vol. 7573, pp. 645–659. Springer, Heidelberg (2012)
32. Park, D., Ramanan, D., Fowlkes, C.: Multiresolution models for object detection. In: Daniilidis, K., Maragos, P., Paragios, N. (eds.) ECCV 2010, Part IV. LNCS, vol. 6314, pp. 241–254. Springer, Heidelberg (2010)

Crowd Collectiveness Measure via Path Integral Descriptor

Wei-Ya Ren[1(⊠)], Guo-Hui Li[2], and Yun-Xiang Ling[1,2]

[1] Department of Management Science and Engineering,
Officers College of Chinese Armed Police Force,
Chengdu 610213, Sichuan, China
weiyren.phd@gmail.com, Yxling@tom.com
[2] College of Information System and Management,
National University of Defense Technology, Changsha 410073, Hunan, China
gli2010a@l63.com

Abstract. Crowd collectiveness measuring has attracted a great deal of attentions in recently years. We adopt the path integral descriptor idea to measure the collectiveness of a crowd system. A new path integral descriptor is proposed by exponent generating function to avoid parameter setting. Several good properties of the proposed path integral descriptor are demonstrated in this paper. The proposed path integral descriptor of a set is regard as the collectiveness measure of a set, which can be a moving system such as human crowd, sheep herd and so on. Self-driven particle (SDP) model and the crowd motion database are used to test the ability of the proposed method in measuring collectiveness.

Keywords: Crowd collectiveness · Exponent generating function · Path integral

1 Introduction

In recent years, the collectiveness motions of crowd systems have attracted a great deal of attentions in both fundamental and application researches because of its universe, intuitionistic and macroscopic properties [2–4, 7, 16, 17]. Measure the collectiveness motions of crowd systems is wildly used in crowd surveillance [5, 6]. In [11], a novel method is proposed to quantify the structural properties of collective manifolds of crowds and compare crowd behaviors across different scenes.

The key of method in [11] is to compute the path integral descriptor [1] between individuals. The concept of path integral was first introduced in statistical mechanics and quantum mechanics [1, 8–10, 12], where it summed up the contributions of all possible paths to the evolution of a dynamical system. Path integral is one of the linkage algorithms in agglomerative clustering, which are free from the restriction on data distributions [18]. On the basis of path integral, [1] define a way to compute the path integral descriptor, which measures the stability of a set and is used to define the

This work is subsidized by National Natural Science Foundation of China under Grant 71673293.

T. Tan et al. (Eds.): CCPR 2016, Part I, CCIS 662, pp. 211–224, 2016.
DOI: 10.1007/978-981-10-3002-4_18

affinity measure between two sets. [1] use a generating function to produce the path integral descriptor, which is adopted in [11]. Inspired by [1], we proposed a new method to compute the path integral descriptor of an edge, a node and a set, respectively. Exponent generating function is used to sum the path integral. We show several good properties of the proposed path integral descriptor. Different from [1, 11], the proposed method has no parameter in computing path integral descriptor, which is more suitable to compare crowd behaviors across different scenes.

Large value of the path integral descriptor means that nodes in the set have a high coherence motion. Thus, the path integral descriptor is actually the measurement of collectiveness of a set. The proposed collectiveness descriptor can be used to quantify the structural properties of collective manifolds of crowds. We compare the proposed method with the state-of-the-art method on self-driven particle (SDP) model [15] and crowd motion database [11].

The paper is organized as follows. The proposed path integral descriptor is proposed in Sect. 2. Then, the collectiveness measure is presented in Sect. 3. Conclusion is given in Sect. 4.

2   Path Integral Descriptor

2.1   Neighborhood Graph

Given a set C of samples $X = [x_1, x_2, \ldots, x_n] \in R^{D \times N}$, which D is the dimension of data and $N = |C|$ is the number of samples. Then, we build a directed graph $G = (V, E)$, where V is the set of vertices corresponding to the samples in X, and E is the set of edges connecting vertices. The graph is associated with a weighted adjacency matrix W, where $w_{ij} \in [0, 1]$ is the weight of the edge from vertex i to vertex j. $w_{ij} = 0$ if and only if there is no edge from i to j. The famous K-NN graph is adopted as the edge connection strategy, which means each vertex has K edges pointing from itself to its K nearest neighbors.

2.2   Path Integral and Collectiveness via Exponent Generating Function

The affinity measure between two clusters is the key of an agglomerative clustering algorithm [1]. Following [1], we compute affinity measure by the path integral descriptor.

2.2.1   Path Integral

For simplicity, we consider all samples as a cluster set C, and assume its weighted adjacency matrix is W. The path integral is a generalization of path counting, where paths in a cluster represent the connectivity [13]. We adopt the path integral definition in [1]: Compute path integral by selecting the starting and ending vertices of the path, let $\gamma_l(i_0, i_l) = \{i_0 \to i_1 \to i_2 \to \cdots \to i_l\}$ denote a directed path of length l through nodes $i_0, i_1, \ldots i_l$ on W between nodes i_0 and i_l. The path integral on a specific path $\gamma_l(i_0, i_l)$ [1] is defined as

$$\tau_{\gamma_l}(i_0, i_l) = \prod_{k=0}^{l} w(i_k, i_{k+1}). \tag{1}$$

Let the set $\mathcal{P}_l(i_0, i_l)$ denotes all the possible paths of length l between nodes i_0 and i_l, then the l-path similarity $\tau_l(i_0, i_l)$ between nodes i_0 and i_l is defined as [1]

$$\tau_l(i_0, i_l) = \sum_{\gamma_l(i_0, i_l) \in \mathcal{P}_l(i_0, i_l)} \tau_{\gamma_l}(i_0, i_l). \tag{2}$$

According to the algebraic graph theory [13], $\tau_l(i_0, i_l)$ can be efficiently computed as [1]

$$\tau_l(i_0, i_l) = W^l(i_0, i_l). \tag{3}$$

Then, l-path integral τ_l between all pairs of nodes can be computed as $\tau_l = W^l$ [13].

2.2.2   Generating Function Regularization

Give node i and node j, we prefer to consider its all l-path integrals $\tau_l(i, j)$ ($l = 1, 2, \ldots, \infty$) by generating function regularization. Due to the value of $W^l(i, j)$ becomes larger when l become larger. We have to assign a meaningful value for the sum of a possibly divergent series [14] of $W^l(i, j)$ ($l = 1, 2, \ldots, \infty$) [1]. The value that consider all l-path integrals $\tau_l(i, j)$ ($l = 1, 2, \ldots, \infty$) can be regard as the integral descriptor of node i and node j [1]. Different from [1], we use the exponent generating function to define the path integral descriptor of an edge as follows

$$z(i, j) = \frac{1}{e^H} [\delta(i, j) + \sum_{l=1}^{\infty} \frac{W^l(i, j)}{l!}]. \tag{4}$$

where $\delta(i, j)$ is the Kronecker delta function defined as $\delta(i, j) = 1$ if $i = j$ and $\delta(i, j) = 0$ otherwise. $H = ||A||_{\infty}$ is a constant, where $A = [a]_{ij} = (W > 0)$, A is a binary matrix that $a_{ij} = 1$ if $w_{ij} > 0$ and $a_{ij} = 0$ otherwise.

Theorem 1. $z(i, j)$ is the (i, j) entry of matrix Z, where $Z = \frac{1}{e^H} e^W$ and $0 \leq z(i, j) \leq \frac{1}{H} + \frac{\delta(i, j)}{e^H} - \frac{1}{He^H}$.

Proof. From (4), we have

$$Z = \frac{1}{e^H} \left[I + \sum_{l=1}^{\infty} \frac{W^l}{l!} \right] = \frac{1}{e^H} e^W.$$

According to (4), it is obvious that $z(i, j) \geq 0$. Now we prove $z(i, j) \leq \frac{1}{H} + \frac{\delta(i, j)}{e^H} - \frac{1}{He^H}$. Given $A = (W > 0)$, then we have

$$z(i,j) = \frac{1}{e^H} \lim_{l \to \infty} \left[\delta(i,j) + \frac{W(i,j)}{1!} + \frac{W^2(i,j)}{2!} + \cdots + \frac{W^l(i,j)}{l!} \right]$$

$$\leq \frac{1}{e^H} \lim_{l \to \infty} \left[\delta(i,j) + \frac{A(i,j)}{1!} + \frac{A^2(i,j)}{2!} + \cdots + \frac{A^l(i,j)}{l!} \right] \triangleq \mathcal{M}(i,j).$$

Since $A = (W > 0)$ and $H = ||A||_\infty$, we have $A^2(i,j) \leq H$, $A^3(i,j) \leq H^2$, $\ldots, A^l(i,j) \leq H^{l-1}$. We have

$$z(i,j) \leq \mathcal{M}(i,j) \leq \frac{1}{e^H} \lim_{l \to \infty} \left[\delta(i,j) + \frac{1}{1!} + \frac{H}{2!} + \cdots + \frac{H^{l-1}}{l!} \right]$$

$$= \frac{1}{He^H} \lim_{l \to \infty} \left[\delta(i,j)H - 1 + 1 + \frac{H}{1!} + \frac{H^2}{2!} + \cdots + \frac{H^l}{l!} \right] = \frac{1}{He^H} \left[\delta(i,j)H - 1 + e^H \right]$$

$$= \frac{1}{H} + \frac{\delta(i,j)}{e^H} - \frac{1}{He^H}.$$

∎

2.2.3   Path Integral Descriptor

After defining the path integral descriptor of an edge, we can define the path integral descriptor of a node and a set by using method in [1]. Path integral descriptor of node i at l-path scale is defined as

$$\varphi_l(i) = \sum_{j \in C} \tau_l(i,j) = [W^l \mathbf{1}]_i. \tag{5}$$

where $\mathbf{1} \in R^{|C| \times 1}$ is a vector with all elements as one, $[\cdot]_i$ denotes i-th element of a vector.

Then the path integral descriptor of set at l-path scale is defined as the mean of all node collectiveness [1]

$$\Phi_l = \frac{1}{|C|} \sum_{i=1}^{|C|} \varphi_l(i) = \frac{1}{|C|} \mathbf{1}^T W^l \mathbf{1}. \tag{6}$$

Path integral descriptor of node [1] on all the path integral is written as

$$\varphi(i) = \frac{1}{e^H} [1 + \sum_{l=1}^{\infty} \frac{\varphi_l(i)}{l!}] = [Z\mathbf{1}]_i. \tag{7}$$

Path integral descriptor of set [1] on all the path integral is written as

$$\Phi = \frac{1}{|C|} \sum_{i=1}^{|C|} \varphi(i) = \frac{1}{|C|} \mathbf{1}^T Z\mathbf{1} = \frac{1}{|C|e^H} \mathbf{1}^T e^W \mathbf{1}. \tag{8}$$

Then, one can denote the path integral descriptor of any set by (8). In this instance, $\Phi = \Phi_{\mathcal{C}}$ for set $\mathcal{C}$. In Sect. 3, we will further explain the path integral descriptor in physical intuition.

Property 1 (Convergence). Z *always converges.*

Proof. Since $0 \leq W_{ij} \leq 1$, we have $W \leq U$ by defining a matrix $U \in R^{|\mathcal{C}| \times |\mathcal{C}|}$ with all elements as one. $Z = \frac{1}{e^H} e^W \leq \frac{1}{e^H} e^U$. Thus, Z always converges. ∎

Property 2. *If $W_{i,i} = 0$, then $\Phi(W) = \Phi(W + I)$. If $W + I$ is also block diagonal then $\Phi(W) \leq 1$. The equality only stands when $W = (W > 0)$ and all blocks in $W + I$ have same size.*

Proof. Given $A = (W > 0)$ and $H = \|A\|_\infty$. Since $W_{i,i} = 0$, then $\|A + I\|_\infty = H + 1$, we have $Z(W) = \frac{e^W}{e^H}$ and $Z(W + I) = \frac{e^{W+I}}{e^{H+1}}$. It is easy to know that $\frac{Z(W)}{Z(W+I)} = 1$, thus $Z(W) = Z(W + I)$ and $\Phi(W) = \frac{1}{|\mathcal{C}|} \mathbf{1}^T Z(W) \mathbf{1} = \frac{1}{|\mathcal{C}|} \mathbf{1}^T Z(W + I) \mathbf{1} = \Phi(W + I)$.

If $W + I$ is block diagonal, and has c blocks with size of $k_1, k_1, \ldots, k_c$, respectively. We have

$$(A + I)^l(i,j) = \begin{cases} k_i^{l-1}, & \text{if } (i,j) \text{ belongs to block } i, \ i \in \{1, 2, \ldots, c\} \\ 0, & \text{otherwise.} \end{cases}$$

Since $H + 1 = \|A + I\|_\infty \geq k_i \geq 1, i \in \{1, 2, \ldots, c\}$, $\sum_i k_i = |\mathcal{C}|$, we have

$$\Phi(W) = \Phi(W + I) = \frac{1}{|\mathcal{C}|} \mathbf{1}^T Z(W + I) \mathbf{1} = \frac{1}{|\mathcal{C}|} \mathbf{1}^T \frac{e^{W+I}}{e^{H+1}} \mathbf{1}$$

$$= \frac{1}{|\mathcal{C}| e^{H+1}} \mathbf{1}^T \left[I + (W + I) + \frac{(W+I)^2}{2!} + \cdots + \frac{(W+I)^\infty}{\infty!} \right] \mathbf{1}$$

$$\leq \frac{1}{|\mathcal{C}| e^{H+1}} \mathbf{1}^T \left[I + (A + I) + \frac{(A+I)^2}{2!} + \cdots + \frac{(A+I)^\infty}{\infty!} \right] \mathbf{1}$$

$$= \frac{1}{|\mathcal{C}| e^{H+1}} \left[|\mathcal{C}| + \sum_i k_i^0 * k_i^2 + \frac{\sum_i k_i^1 * k_i^2}{2!} + \cdots + \frac{\sum_i k_i^\infty * k_i^2}{\infty!} \right]$$

$$\leq \frac{1}{|\mathcal{C}| e^{H+1}} \left[|\mathcal{C}| + (H + 1) \sum_i k_i + \frac{(H+1)^2 \sum_i k_i}{2!} + \cdots + \frac{(H+1)^\infty \sum_i k_i}{\infty!} \right]$$

$$= \frac{|\mathcal{C}|}{|\mathcal{C}| e^{H+1}} e^{H+1} = 1.$$

The equality only stands when $W = (W > 0)$ and all blocks in $W + I$ have same size such that $k_1 = k_2 = \cdots = k_c = H + 1$. ∎

Property 3 (Bounds of Φ). $0 \leq \Phi \leq 1$.

Proof. Since we have $w_{ij} \in [0,1]$ and $W\mathbf{1} \leq A\mathbf{1} \leq H\mathbf{1}$ by giving $A = (W > 0)$ and $H = ||A||_{\infty}$. Then,

$$
\begin{aligned}
\Phi &= \frac{1}{|\mathcal{C}|e^H} \mathbf{1}^T e^W \mathbf{1} = \frac{1}{|\mathcal{C}|e^H} \mathbf{1}^T \left[I + \sum_{l=1}^{\infty} \frac{W^l}{l!} \right] \mathbf{1} = \frac{1}{|\mathcal{C}|e^H} \lim_{l \to \infty} \left[\mathbf{1}^T \mathbf{1} + \mathbf{1}^T W \mathbf{1} + \frac{\mathbf{1}^T W^2 \mathbf{1}}{2!} + \cdots + \frac{\mathbf{1}^T W^l \mathbf{1}}{l!} \right] \\
&\leq \frac{1}{|\mathcal{C}|e^H} \lim_{l \to \infty} \left[\mathbf{1}^T \mathbf{1} + H \mathbf{1}^T \mathbf{1} + \frac{H^2 \mathbf{1}^T \mathbf{1}}{2!} + \cdots + \frac{H^l \mathbf{1}^T \mathbf{1}}{l!} \right] \\
&\leq \frac{|\mathcal{C}|}{|\mathcal{C}|e^H} \lim_{l \to \infty} \left[1 + H + \frac{H^2}{2!} + \cdots + \frac{H^l}{l!} \right] = \frac{|\mathcal{C}|}{|\mathcal{C}|e^H} e^H = 1. \quad \blacksquare
\end{aligned}
$$

Property 4 (Approximate error bound of Z). $||Z - Z_{1 \sim n}|| \leq \frac{1}{e^H} \sum_{l=n+1}^{D-1} \frac{W^l}{l!} + \frac{1}{e^H} \frac{||W||^D}{D!} \frac{D+1}{D+1-||W||}$ if $n < D - 2$, and $||Z - Z_{1 \sim n}|| \leq \frac{1}{e^H} \frac{||W||^{(n+1)}}{(n+1)!} \frac{n+2}{n+2-||W||}$ otherwise. $Z_{1 \sim n}(n)$ denotes the sum of first n terms of Z, the matrix norm of matrix $B(b_{ij} \in [0,1])$ is defined as $||B|| = \sum_{ij} B_{ij}$, and $||A|| = D > 0$.

Proof. (See approximate error bound details in [19]).

If $n < D - 2$, let $R = Z - Z_{1 \sim n} - \frac{1}{e^H} \sum_{l=n+1}^{D-1} \frac{W^l}{l!} = \frac{1}{e^H} \sum_{l=D}^{\infty} \frac{W^l}{l!}$. We can find that $||B_1 B_2|| \leq ||B_1|| \cdot ||B_2||$, where B_1, B_2 are two matrices that range from $[0,1]$. Thus, $||W^l|| \leq ||W||^l$. Notice that $||W|| \leq ||A|| = D \leq |\mathcal{C}|H$, then we have

$$
\begin{aligned}
||R|| &= \frac{1}{e^H} \sum_{l=D}^{\infty} \frac{||W^l||}{l!} \leq \frac{1}{e^H} \sum_{l=D}^{\infty} \frac{||W||^l}{l!} = \frac{1}{e^H} \lim_{l \to \infty} \left[\frac{||W||^D}{D!} + \frac{||W||^{(D+1)}}{(D+1)!} + \cdots + \frac{||W||^l}{l!} \right] \\
&= \frac{1}{e^H} \frac{||W||^D}{D!} \lim_{l \to \infty} \left[1 + \frac{||W||}{D+1} + \cdots + \frac{||W||^{(l-D)}}{(D+1)\ldots(l-1)l} \right] \\
&\leq \frac{1}{e^H} \frac{||W||^D}{D!} \lim_{l \to \infty} \left[1 + \frac{||W||}{D+1} + \cdots + \frac{||W||^{(l-D)}}{(D+1)^{(l-D)}} \right] = \frac{1}{e^H} \frac{||W||^D}{D!} \frac{1}{1 - ||W||/(D+1)} \\
&= \frac{1}{e^H} \frac{||W||^D}{D!} \frac{D+1}{D+1-||W||}.
\end{aligned}
$$

If $n \geq D - 2$, let $R = Z - Z_{1 \sim n} = \frac{1}{e^H} \sum_{l=n+1}^{\infty} \frac{W^l}{l!}$. Then we can find that

$$
||R|| \leq \frac{1}{e^H} \frac{||W||^{(n+1)}}{(n+1)!} \frac{n+2}{n+2-||W||}. \quad \blacksquare
$$

Property 5 (Bounds of W^l). $0 \leq W^l(i,j) \leq H^{l-1}$, $l = 1, 2, \ldots, \infty$.

Proof. Obviously, $W^l(i,j) \geq 0$ since $0 \leq W(i,j) \leq 1$. Given $A = (W > 0)$, it is easy to know that $W^l(i,j) \leq A^l(i,j)$, $l = 1, 2, \ldots, \infty$. Denote $H = \|A\|_\infty$, we have $A^2(i,j) \leq H, A^3(i,j) \leq H^2, \ldots, A^l(i,j) \leq H^{l-1}$. Finally,

$$W^l(i,j) \leq A^l(i,j) \leq H^{l-1}, \quad l = 1, 2, \ldots, \infty.$$

∎

We rewrite Z as follows

$$Z = \frac{1}{e^H} e^W = \frac{1}{e^H}\left[1 + W + \frac{W^2}{2!} + \cdots + \frac{W^\infty}{\infty!}\right] = \alpha_0 I + \alpha_1 W + \alpha_2 W^2 + \cdots + \alpha_\infty W^\infty.$$

(9)

Thus, Z can be seen as the linear combination of $I, W, W^2, \ldots, W^\infty$. Obviously, $\alpha_l = \frac{1}{l!e^H}$ and $\alpha_0 = \alpha_1 > \alpha_2 > \cdots > \alpha_\infty$. It seems Z pay less and less attention to W^l when l become larger. However, $W, W^2, \ldots, W^\infty$ take values from different scales. If we normalized W^l by making $W^l(i,j)$ ranges from $[0,1]$, i.e., $\tilde{W}^l = W^l/H^{l-1}$. We have

$$Z = \alpha_0 I + \alpha_1 W + \alpha_2 W^2 + \cdots + \alpha_\infty W^\infty = \tilde{\alpha}_0 I + \tilde{\alpha}_1 \tilde{W} + \tilde{\alpha}_2 \tilde{W}^2 + \cdots + \tilde{\alpha}_\infty \tilde{W}^\infty. \quad (10)$$

where $\tilde{\alpha}_0 = \frac{1}{e^H}$, and $\tilde{\alpha}_l = \frac{H^{l-1}}{l!e^H}$, $l = 1, 2, \ldots, \infty$.

Theorem 2. $\frac{1}{e^H} \leq \tilde{\alpha}_l \leq \frac{H^{H-1}}{H!e^H}$ for $l = 1, 2, \ldots, H$. The left equality stands when $l = 1$ while the right equality stands when $l = H$ or $l = H - 1$. $0 \leq \tilde{\alpha}_l < \frac{H^{H-1}}{(H)!e^H}$ for $l = H + 1, H + 2, \ldots, \infty$.

Proof. When $l \leq H$, $\frac{\tilde{\alpha}_l}{\tilde{\alpha}_{l-1}} = \frac{H^{l-1}}{l!e^H}\frac{(l-1)!e^H}{H^{l-2}} = \frac{H}{l} \geq 1$. The equality only stands when $l = H$. Thus, $\tilde{\alpha}_H \geq \tilde{\alpha}_{H-1} > \cdots > \tilde{\alpha}_1 = \frac{1}{e^H}$.

When $l \geq H - 1$, $\frac{\tilde{\alpha}_l}{\tilde{\alpha}_{l+1}} = \frac{H^{l-1}}{l!e^H}\frac{(l+1)!e^H}{H^l} = \frac{l+1}{H} \geq 1$. The equality only stands when $l = H - 1$. Thus, $\tilde{\alpha}_{H-1} \geq \tilde{\alpha}_H > \cdots > \tilde{\alpha}_\infty = 0$.

Then, we have $\tilde{\alpha}_H = \tilde{\alpha}_{H-1} > \tilde{\alpha}_l$, $l = 1, 2, \ldots, \infty$ and $l \neq H, H - 1$. ∎

From Theorem 2, we know that Z be seen as the linear combination of $I, \tilde{W}, \tilde{W}^2, \ldots, \tilde{W}^\infty$. $\tilde{\alpha}_H$ and $\tilde{\alpha}_{H-1}$ are two largest values among all coefficients. If $\tilde{W}$ is a K-NN graph, then $H = \|A\|_\infty = K$ and each node points to K neghbors. It means that, Z pay more attention to $\tilde{W}^H$, $\tilde{W}^{H-1}$ and other $\tilde{W}^l$ if l is close to $H - 1$ or H. We show an example in Fig. 3. The value of $\tilde{W}^l$ will also affect the value of $\tilde{\alpha}_l \tilde{W}^l$. If $\tilde{W}^l = 0$, we have $\tilde{\alpha}_l \tilde{W}^l = 0$ even when l is close to $H - 1$ or H. Notice that $max(W^l(i,j))/H^{l-1}$ become smaller when l become larger. Thus, $\|\tilde{\alpha}_l \tilde{W}^l\|_2$ usually take the largest value when $l < H$.

3 Collectiveness

In Sect. 2.2, we introduce the path integral descriptor Φ of a set. If nodes in a set have a high coherence, Φ will have a large value. If nodes in a set are similar with each other, Φ will also be large. Thus, the path integral descriptor can be comprehended as the measurement of collectiveness of the set. In fact, reference [11] already uses this idea. However, the difference between the proposed paper and [11] is the different method to compute Φ.

The self-driven particle (SDP) model [15] is a famous model for studying collective motion and shows high similarity with various crowd systems in nature [16]. In this section, we use this model to test the proposed path integral descriptor, or named collectiveness, same as [11]. The ground-truth of collectiveness in SDP is known for evaluation. SDP model produce a system of moving particles that are driven with a constant speed [11]. SDP gradually turns into collective motion from disordered motion [11]. Each particle will update its direction of velocity to the average direction of the particles in its neighborhood at each frame [11]. The update of velocity direction θ [15] for every particles i in SDP is

$$\theta_i(t+1) = <\theta_j(t)>_{j\in\mathcal{N}(i)} + \Delta\theta. \tag{11}$$

where $<\theta_j(t)>_{j\in\mathcal{N}(i)}$ denotes the average direction of velocities of particles within the neighborhood $\mathcal{N}(i)$ of i. $\Delta\theta$ is a random angle chosen with a uniform distribution within the interval $[-\eta\pi, \eta\pi]$, where η tunes the noise level of alignment [15].

3.1 Neighborhood Graph

Given N moving particles in SDP, we can measure the similarity of particles by K-NN graph. For simplicity, we adopt the method in [11] to compute the K-NN graph for comparison. At time t, the weight value on edge between particle i and particle j are defined by [11]

$$w_t(i,j) = \begin{cases} max(\frac{v_i v_j^T}{||v_i||_2||v_j||_2}, 0), & if \ j \in N(i) \\ 0, & otherwise \end{cases}. \tag{12}$$

where v_i is the velocity vector of particle i.

3.2 Numerical Analysis

We compare the proposed method with the state-of-the-art method in [11]. As seen in Fig. 1, we show an example of measuring collective motion in SDP.

We compute the collectiveness of all frames and then compute the relevant coefficient between the measured collectiveness and the ground truth. As seen in Fig. 2, we show an example that analyze the relevant between the ground truth (GT) and the method in [11] and the proposed method, and the proposed method has a higher

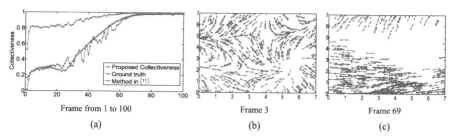

Fig. 1. An example of measuring collective motion in SDP, and $N = 400$, $K = 20$, size of ground $L = 7$, the absolute value of velocity $||v|| = 0.03$, the interaction radius $r = 1$ and $\eta = 0$. At the beginning, collectiveness Φ is low since the spatial locations and moving directions of individuals are randomly assigned. The behaviors of individuals gradually turn into collective motion from random movements. (a) Compare the collectiveness ground-truth with collectiveness measured by the proposed method and by [11]. (b) The SDP model on frame 3. (c) The SDP model on frame 69.

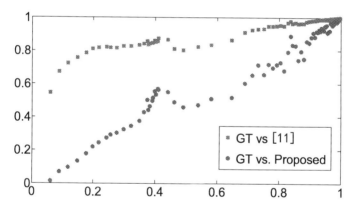

Fig. 2. An example. Relevant analysis of the measured collectiveness and the ground truth collectives (system order). The x-axis denotes the ground truth value and the y-axis denotes the collectiveness computed by algorithms. The relevant coefficient between the ground truth (GT) and the method in [11] is 0.88, while the relevant coefficient between the ground truth (GT) and the proposed method 0.94. Parameters: $N = 400$, $K = 20$, size of ground $L = 7$, the absolute value of velocity $||v|| = 0.03$, the interaction radius $r = 1$ and $\eta = 0$.

relevant coefficient value. We also show the average relevant coefficient between the ground truth (GT) and two methods (100 runs), as seen in Table 1.

In Fig. 3, we analyze the components of $Z = \frac{e^W}{e^H} = \frac{1}{e^H} \sum_{l=1}^{\infty} \frac{W^l}{l!}$. We draw normalized value of $|| \frac{1}{e^H} \frac{W^l}{l!} ||_2$ with respect to $l = 1, \ldots, 100$.

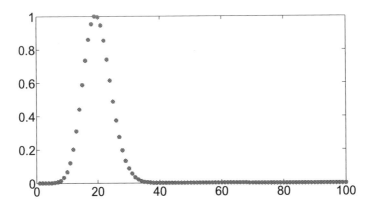

Fig. 3. Draw normalized value of $||\frac{1}{e^{l!}}\frac{W^l}{l!}||_2, l = 1, \ldots, 100$. $||\frac{1}{e^{l!}}\frac{W^l}{l!}||_2$ has largest value when $l = 19$ $(K = 20)$. The x-axis denotes l the y-axis denotes the value of $||\frac{1}{e^{l!}}\frac{W^l}{l!}||_2$. Parameters: $N = 400$, $K = 20$, size of ground $L = 7$, the absolute value of velocity $||v|| = 0.03$, the interaction radius $r = 1$ and $\eta = 0$.

Table 1. Average relevant coefficient between the ground truth (GT) and two methods (100 runs, 100 frames in each run). Fix $K = 20$, and $||v|| = 0.03$, $r = 1$, $L = 7$ and $\eta = 0$.

Relevant coefficient	GT vs. [11]	GT vs. proposed
$N = 200$	0.86	**0.90**
$N = 400$	0.81	**0.88**
$N = 500$	0.83	**0.92**

3.3   Crowd Collectiveness

We adopt the collective motion database collecting by [11] to analysis the collectiveness of crowd scene. The collective motion database in [11] consists of 413 video clips from 62 crowded scenes. The generalized KLT (gKLT) tracker method [20] is used as the feature extraction method. K-nn graph is adopt as the affinity matrix W and notice that $W \geq 0$. In [11], the collectiveness of 413 video clips are estimated by the following method: 10 subjects independently rate the level of collective motions in all clips from three options: low (0 score), medium (1 score), and high (2 score). We compare the proposed method with the state-of-the-art method in [11] in computing crowd collectiveness.

For each video clip, its collectiveness is the average value of collectiveness of all frames. Each clip are rated by ten objects, then the range of collective scores is [0, 20]. The collectiveness of clips can be defined as: 1. Low collectiveness (214 clips): $0 \leq \text{score} \leq 5$. 2. Medium collectiveness (105 clips): $0 < \text{score} < 5$. 3. High collectiveness (94 clips): $15 \leq \text{score} \leq 20$. Due to the number of different level collectiveness clips are unbalanced, we use the average binary classification accuracy of high and low, high and medium, and medium and low categories as the evaluation criteria. Note n_h as

Table 2. Average binary classification accuracy with different K.

		High-low	High-medium	Medium-low
$K = 20$	The proposed method	**91.9 %**	**82.1 %**	71.4 %
	Method in [11]	91.7 %	81.4 %	**74.1 %**
$K = 15$	The proposed method	**92.1 %**	**81.9 %**	73.4 %
	Method in [11]	91.2 %	81.4 %	**73.7 %**
$K = 10$	The proposed method	**92.0 %**	80.9 %	**73.3 %**
	Method in [11]	89.3 %	80.9 %	72.0 %
$K = 5$	The proposed method	**92.0 %**	**81.2 %**	**73.8 %**
	Method in [11]	87.6 %	79.9 %	72.6 %
$K = 3$	The proposed method	**90.3 %**	**80.6 %**	**73.5 %**
	Method in [11]	87.2 %	79.2 %	71.6 %

the number of high collectiveness clips and n_l as the number of low collectiveness clips. Note c_h as the number of recognized high collectiveness clips by algorithm and n_l as the number of recognized low collectiveness clips by algorithm. Then the average binary classification accuracy of high and low categories is

$$AC_{h-l} = \left(\frac{c_h}{n_h} + \frac{c_l}{n_l}\right)/2. \tag{13}$$

The average binary classification accuracy of high and medium, and medium and low categories is similar to (13). As seen in Table 2, the proposed method performs well, especially on the average binary classification of high-low categories. In addition, the proposed method is not sensitive to the parameter K.

We analysis the collective motion in clip by the threshold method. We can get the clusters of collective motion patterns as the connected components by setting a threshold τ and thresholding the values on Φ. Figure 4 illustrates the performance of extracting collective motions. According to [11], we set the parameters of method in [11] and set $\tau = 10^{-5}$ in the proposed method. The parameter K is set to 20. We can find that the proposed method has advantage in judging the number of collective motions.

There are two parameters in the proposed method, we use the finite grid method [21] to select parameters, as seen in Table 3.

Fig. 4. Performance of collective motion extraction at frame 10. Different collective motions are illustrated by different colors. (Color figure online)

Table 3. A finite grid of parameter values.

Parameter	Value
K	3, 5, 10, 15, 20, 30
τ	10^{-2}, 10^{-3}, 10^{-4}, 10^{-5}, 10^{-6}, 10^{-7}

4   Conclusion

In this paper, we define the path integral descriptor of an edge, a node and a set by the exponent generating function on the basis of the path integral and [1], respectively. Several fine properties of the path integral descriptor have been shown. We regard the path integral descriptor as the collectiveness measurement of a moving system, as same in [11]. Experiment on SDP model and crowd scene database show the good performance of the proposed method in measuring collectiveness. In future work, we will compute the collectiveness of various types of crowd systems, and provides useful information in crowd monitoring.

References

1. Zhang, W., Zhao, D., Wang, X.: Agglomerative clustering via maximum incremental path integral. Pattern Recogn. **46**(11), 3056–3065 (2013)
2. Moussaid, M., Garnier, S., Theraulaz, G., Helbing, D.: Collective information processing and pattern formation in swarms, flocks, and crowds. Topics Cogn. Sci. **1**, 469–497 (2009)
3. Couzin, I.: Collective cognition in animal groups. Trends Cogn. Sci. **13**, 36–43 (2009)
4. Ballerini, M., et al.: Empirical investigation of starling flocks: a benchmark study in collective animal behaviour. Anim. Behav. **76**, 201–215 (2008)
5. Zhou, B., Wang, X., Tang, X.: Understanding collective crowd behaviors: learning a mixture model of dynamic pedestrian-agents. In: Proceedings of IEEE Conference Computer Vision and Pattern Recognition (CVPR 2012) (2012)
6. Lin, D., Grimson, E., Fisher, J.: Learning visual flows: a lie algebraic approach. In: Proceedings of IEEE Conference Computer Vision and Pattern Recognition (CVPR) (2009)
7. Zhang, H., Ber, A., Florin, E., Swinney, H.: Collective motion and density fluctuations in bacterial colonies. Proc. Natl. Acad. Sci. **107**, 13626–13630 (2010)
8. Feynman, R.P.: Space-time approach to non-relativistic quantum mechanics. Rev. Mod. Phys. **20**, 367–387 (1948)
9. Kleinert, H.: Path Integrals in Quantum Mechanics, Statistics, Polymer Physics, and Financial Markets, 3rd edn. World Scientific, Singapore (2004)
10. Rudnick, J., Gaspari, G.: Elements of the Random Walk: An Introduction for Advanced Students and Researchers. Cambridge University Press, Cambridge (2004)
11. Zhou, B., Tang, X., Zhang, H., et al.: Measuring crowd collectiveness. IEEE Trans. Pattern Anal. Mach. Intell. **36**(8), 1586–1599 (2014)
12. Zhang, W., Wang, X., Zhao, D., Tang, X.: Graph degree linkage: agglomerative clustering on a directed graph. In: Fitzgibbon, A., Lazebnik, S., Perona, P., Sato, Y., Schmid, C. (eds.) ECCV 2012. LNCS, vol. 7578, pp. 428–441. Springer, Heidelberg (2012). doi:10.1007/978-3-642-33718-5_31
13. Biggs, N.: Algebraic Graph Theory. Cambridge University Press, Cambridge (1993)

14. Knuth, D.E.: The Art of Computer Programming, Volume 1 Fundamental Algorithms, 3rd edn. Addison-Wesley, Reading (1997)
15. Vicsek, T., Czirók, A., Ben-Jacob, E., Cohen, I., Shochet, O.: Novel type of phase transition in a system of self-driven particles. Phys. Rev. Lett. **75**, 1226–1229 (1995)
16. Buhl, J., Sumpter, D., Couzin, I., Hale, J., Despland, E., Miller, E., Simpson, S.: From disorder to order in marching locusts. Science **312**, 1402–1406 (2006)
17. Raafat, R.M., Chater, N., Frith, C.: Herding in humans. Trends Cogn. Sci. **13**, 420–428 (2009)
18. Zhao, D., Tang, X.: Cyclizing clusters via zeta function of a graph. In: Advances in Neural Information Processing Systems, pp. 1953–1960 (2009)
19. Liou, M.L.: A novel method of evaluating transient response. Proc. IEEE **54**(1), 20–23 (1966)
20. Tomasi, C., Kanade, T.: Detection and tracking of point features. Int. J. Comput. Vis. (1991)
21. Chapelle, O., Zien, A.: Semi-supervised classification by low density separation. In: Proceedings of the 10th International Workshop on Artificial Intelligence and Statistics, pp. 57–64 (2005)

Robust Face Frontalization in Unconstrained Images

Yuhan Zhang[✉], Jianjun Qian, and Jian Yang

School of Computer Science and Engineering,
Nanjing University of Science and Technology, Nanjing, China
chlwll9910502@126.com, {csjqian, csjyang}@njust.edu.cn

Abstract. The goal of face frontalization is to recover frontal facing views of faces appearing in single unconstrained images. The previous works mainly focus on how to achieve the frontal facing views effectively. However, they ignore the influences of the face images with occlusion. To overcome the problem, this paper presents a novel but simple scheme for robust face frontalization with only a single 3D model. We employ the same scheme with T. Hassner's work to render the non-frontal facing view to the frontal facing view and estimate the invisible (self-occlusion) region. Subsequently, we compute the differences of the local patches around each fixed facial feature points between the average face (male average face or female average face) and test images for occlusion detection. Finally, we combine the proposed local face symmetry strategy and the Poisson image editing to fill the invisible region and occlusion region. Experimental results demonstrate advantages of the proposed method over the previous work.

Keywords: Face frontalization · 3D model · Poisson image editing · Occlusion

1 Introduction

Face recognition has attracted a lot attention due to its potential value in real-world applications. Indeed, many researchers have developed numerous works in the field of face recognition during the past forty years and made a lot of achievements. However, face recognition is still a challenge problem since the face image remains significantly affected by the variations of illumination, expression, age and pose in the wild. Particularly, pose has been an important challenge because it can enhance the difference between homogeneous faces.

To address above problem, many promising works have been developed. Morphable model based methods [1–4] mainly use the linear combinations of a 3D face exemplar set to represent each face. The shape and texture parameters of 3D morphable model can be achieved in a iterative update scheme by using the morphable model to approximate the input image. In 2002, Blanz et al. firstly proposed the synthesis of 3D face based morphable model and then they applied it to face recognition [2, 3]. In [15], the authors introduced a novel method for 3D shape reconstruction of faces from a single image by using only a single reference model. They combined the shading information and generic shape information to exploits the global similarity.

© Springer Nature Singapore Pte Ltd. 2016
T. Tan et al. (Eds.): CCPR 2016, Part I, CCIS 662, pp. 225–233, 2016.
DOI: 10.1007/978-981-10-3002-4_19

Subsequently, some researchers attempted to adjust 3D reference face to match the texture of the query face in order to preserve natural appearances [4, 6, 14]. However, these methods relied on the landmark detection extremely and cost heavy computation time. Zhu et al. proposed a new deep learning framework to recover the canonical view of face images and achieve remarkable results [8]. Different from other face reconstruction methods, it directly learned the transformation from the face images with a complex set of variations to their canonical views. However, deep learning based method required substantial training since the characteristic of deep learning. Sagonas et al. believed that the frontal facial image has the minimum rank compared with all different poses. And they introduced a unified method for joint face frontalization (pose correction), landmark localization, and pose-invariant face recognition with a few frontal images [7]. It's a pity that the computation time of above method is expensive. In [9], Hassner et al. presented a fast and effective face frontalization method. They computed the correspondence between the input face and the reference 3D model to render the non-frontal face to frontal face. Thus, the invisible region is estimated and filled according to the global symmetry of facial image. However, there are three drawbacks of T. Hassner's work. (1) The global symmetry is used to fill the invisible region that leads to unnatural visual effect and face information loss; (2) copying the mirrored pixel intensity directly results in the obvious boundary; (3) occlusion (such as hands, microphone, ball etc.) will degrade the performance.

To overcome these problems, this paper proposes a novel method for robust face frontalization. Specifically, it is composed of feature points detection, location correspondence estimation, visibility estimation and occlusion detection. They leverage the local symmetry and Poisson image editing to fill the invisible region and occlusion region at the same time. Figure 1 shows the pipeline of the proposed scheme.

2 Frontalization Method

2.1 Frontal Facing Synthesis

In this part, we briefly review the process of how to generate the frontal facing view. Given a single 3D model, it just contains the shape information and ignores the texture information. We also provide the coordinate of the facial feature points of the 3D model. For query image, the same landmarks can be achieved via the facial feature points detection method. Subsequently, we can estimate the 3×4 projection matrix according to the facial feature points in the query image and the points on the surface of our 3D face model.

For facial feature detection, a lot of highly effective methods are proposed to detect facial feature points in unconstrained images [10–12]. In general, any on-the-shelf feature points detection method can be used in our scheme. Here, we choose SDM (supervised descent method) [10] to determine the position of landmarks in query image. SDM not only learns in a supervised manner generic descent directions, but also it is able to overcome many drawbacks of second order optimization schemes. Moreover, SDM can achieve better performance with little computation load.

For frontal view synthesis, given a single 3D face model, our goal is to compute the projection matrix, which reveals the correspondences between points in the query image and the points on the surface of the 3D face model.

$$p \sim C_Q P. \tag{1}$$

where $p = (x, y)^T$ are facial feature points in the query image (as shown in Fig. 1). C_Q is the 3×4 project matrix, $P = (X, Y, Z)^T$ are the feature points in our 3D face model.

Here, we use the same method like the reference [2, 3] to estimate the projection matrix C_Q. Based on this, we can generate the frontal face view by employing the Bi-linear interpolation algorithm to sample the intensities of the query image at p (as shown Fig. 1).

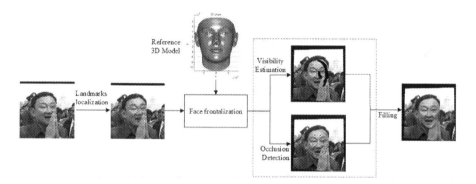

Fig. 1. The flow chart of the proposed method

2.2 Visibility Estimation

As well known, it's hard to estimate visibility directly. In general, most researchers prefer to use two or more views to estimate 3D geometry to estimate visibility. Borrow the idea of [9], we also use the simple but effective approach to estimate invisible region caused by self-occlusion. Specifically, all the points in 3D face model are mapped to the query image. In this way, most points in the query image may have one or more corresponding 3D points. Thus, we can count the number of the corresponding 3D points in each position of query image. Based on this, we can use a threshold value to determine the invisible region.

2.3 Occlusion Detection

Occlusion is still a challenge problem in the field of face recognition. Obviously, it's hard to complete the face recognition task when there are occlusions on frontal facing view. In this section, we mainly introduce our method to remove the occlusions. Our basic idea is that we wonder to design a simple method to estimate the occlusion regions as fast as possible.

Specifically, given 22 fixed facial feature points including boundary, forehead, eye corner, cheek and mouth corner (as shown in Fig. 1). Meanwhile, we provide two average face images (male and female) are computed by 200 frontal face images outside LFW dataset. These two average faces are used as reference standard template. Based on this, we can determine the local patches around the facial feature points with occlusion or not by employing the differences of the corresponding LBP feature between target face image and average image (male average face or female average face). Subsequently, an appropriate threshold can help to give the decision.

So, we should determine the gender property of the input image. In this study, LBP and linear SVM are used to estimate the gender of the test images. In this step, we firstly choose 739 images, including 445 males and 294 females, for training the gender classification model.

2.4   Occlusion Region Filling

In fact, invisible region also can be considered as the self-occlusion region. In other words, invisible region is a special case of occlusion regions. So, we mainly discuss the occlusion region filling here. Intuitively, a good occlusion region filling scheme will be beneficial to improve the face recognition rate. In [9], T. Hassner etc. actually utilize the half of face in visible region to recover the other half of face (as shown in Fig. 2 (b)). In other words, they just use half of face to complete the face recognition task. This scheme can be considered as a global face symmetry method. It's obviously that this strategy ignores some meaningful information and the frontal facing view is not reality.

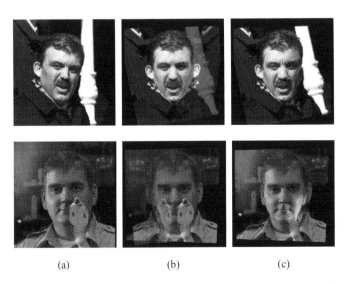

(a)   (b)   (c)

Fig. 2. Face frontalization results. (a) An input image. (b) A completely symmetrical face with face symmetry. (c) A frontal view using Poisson image editing.

To address above problems, we introduce an effective and simple local face symmetry method in conjunction with Poisson image editing for occlusion region filling. Here, we recover occlusion region by using its corresponding symmetry region according to the character of facial symmetry. However, face image is not complete symmetry due to the different illumination. So, we just copy the intensities directly will lead to non-smoothness and remarkable edge. Motivated by the work [13], we introduce Poisson image editing to fill the occlusion region. Because the Poisson image editing not only retains the gradient information of source image, but also fuses source image and target image's background perfectly. Figure 2(c) shows the final results with local facial symmetry in conjunction with Possion image editing.

3 Experiments

In this section, we will evaluate the robustness of the proposed method compared with previous work by experimenting on LFW face data set. Here, we employ the 'calib' function [9] to estimate project matrix. The SDM is leveraged to detect facial feature points. The source code of SDM[1] is from the authors' web. Note that we mainly aim to see how much is face recognition performance improved by using the proposed face frontalization method. The processing time is about 1.5 s per image due to the whole procedure is composed of some classical methods and do not need to train model.

3.1 Face Verification

LFW benchmark is the most commonly used database for unconstrained face recognition. It contains 13233 images of 5729 subjects. Among them, 4069 subjects just contain only one image and 1680 subjects contain two or more images. These images have a large degree of facial expression, occlusions, pose and illuminations since all of them are taken from the real world. Here, we simply crop the face image to remove the background, leaving a 100×100 face image.

In this experiment, we follow the standard evaluation protocol as specified in [9] and mainly compare the proposed method with T. Hassner's work for face verification in the unsupervised setting where we don't use any training samples. Specifically, we firstly achieve the face frontal view via the proposed method. Thus, LBP, TPLBP, FPLBP and SIFT are used for face representation and L_2 distance is employed to measure the similarity.

Table 1 lists the verification rates on the LFW benchmark. From Table 1, we can see that the proposed method achieves the leading results among all the competed methods. LFW means the original images without any preprocessing. LFW-a means all the images aligned with commercial alignment software. LFW3D represents the images generated by face frontalization method of T. Hassner. LFWRF represents the images generated by our proposed method. For SIFT, the performance of the proposed method is 2.5 % higher than that of LFW3D. For LBP, TPLBP and FPLBP, our method also

---

[1] http://www.humansensing.cs.cmu.edu/intraface.

gives the slight improvement. Compared with LFW3D, we can provide more natural synthesis face image by employing the poisson image editing to fill invisible region. Additionally, we also give the ROC curves to show the performances of competed methods in conjunction with SIFT and LBP as shown in Fig. 3. We can obtain the consistent results with Table 1.

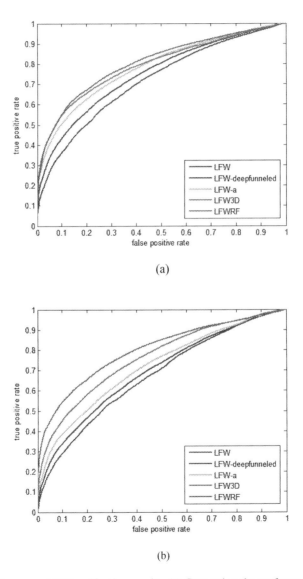

(a)

(b)

Fig. 3. ROC Curves on LFW verification results. (a) Comparing the performance of SIFT on LFW images, LFW-a, LFW-deepfunneled, LFW3D and LFWRF; (b) Comparing the performance of LBP on LFW images, LFW-a, LFW-deepfunneled, LFW3D and LFWRF.

Table 1. Face verification rates (%) on databases.

Method	Database				
	LFW	LFW-deepfunneled	LFW-a	LFW3D	LFWRF
LBP	62.15	65.37	65.63	69.50	73.75
TPLBP	65.28	67.05	67.42	67.40	69.52
FPLBP	64.93	67.43	67.60	68.08	69.77
SIFT	65.95	68.78	71.62	72.88	75.25

3.2 Gender Estimation

In this section, we evaluated the competed methods on LFW dataset for gender estimation. Here, we just select a subset of LFW, which contains 150 subjects. Each subject has one or more images. In total, there are 1054 images. We manually give the image labels 'male' or 'female'.

In this experiment, 739 images from 332 subjects were selected as training images. And the remaining images for testing. The proposed method is firstly used to achieve the frontal facing view of all the face images. LBP, TPLBP, FPLBP and SIFT are used for face representation and linear SVM is employed for classification.

Table 2 lists the performance of LBP, TPLBP, FPLBP and SIFT on LFW images, LFW-a, LFW3D and LFWRF. From Table 2, we see clearly the proposed method gives best performance than other competed methods. Our frontalized faces provide a performance boost of over 2 % compared with LFW3D.

Table 2. Gender estimation (%) on LFW dataset.

Method	Database				
	LFW	LFW-deepfuneled	LFW-a	LFW3D	LFWRF
LBP	89.16	91.11	90.25	91.85	93.65
TPLBP	88.57	90.48	89.52	92.75	94.29
FPLBP	86.51	88.69	87.62	89.21	92.05
SIFT	90.33	91.76	91.43	92.75	94.08

3.3 Face Verification with Occlusion

In this experiment, we mainly examine the robustness of our method when face images suffer different occlusions. We also follow the standard evaluation protocol of LFW dataset. The main difference is that we put different occlusions on the test images. Again, the location of occlusion is randomly chosen for each image and is unknown to the computer.

The performances of LBP, TPLBP, FPLBP and SIFT on LFW datasets, including LFW images with occlusion (LFW-O), LFW images with occlusion are frontalized via T. Hassner's method (LFW3D-O) and LFW images with occlusion are frontalized via our method (LFWRF-O), are listed in Table 3. From Table 3, we can see that our methods significantly outperform the other competed methods. It's pity that the

Table 3. Occlusion verification rate (%) on databases.

Method	Database			
	LFW3D-O	LFW-O	LFWRF-O (without occlusion detection)	LFWRF-O
LBP	65.57	61.82	68.47	71.84
TPLBP	64.07	61.53	64.37	68.35
FPLBP	64.28	62.83	64.75	68.64
SIFT	67.31	64.35	69.17	72.33

performance of each method decreased when the face images with occlusion. However, our frontalized faces provide better performance than other methods. Experimental results demonstrate the effectiveness and robustness of our method.

4   Conclusion

This paper proposes a robust face frontalization method to further improve the face recognition performance. Motivated by T. Hassner's work, we also only use a single reference 3D face model to synthesize frontal facing view. A simple occlusion localization procedure, which just uses the differences between average face and target faces in the fixed facial feature points, is introduced. Subsequently, local face symmetry scheme is employed to fill the self-occlusion region and occlusion regions at the same time. Additionally, we also acknowledge that the performance of the proposed method depends on the robustness of SDM. It's difficult to obtain the facial feature points for SDM when facing extreme profile faces. In this case, our method will fail to complete the face frontalization. In future, we will investigate how to address this problem.

Acknowledgment. This work was partially supported by the National Science Fund of China under Grant Nos. 91420201, 61472187, 61502235, 61233011 and 61373063, the Key Project of Chinese Ministry of Education under Grant No. 313030, the 973 Program No. 2014CB349303, and Program for Changjiang Scholars and Innovative Research Team in University.

References

1. Ding, C.: Multi-task pose-invariant face recognition. IEEE Trans. Image Process. **24**(3), 980–993 (2015)
2. Blanz, V., Vetter, T.: Face recognition based on fitting a 3D morphable model. IEEE Trans. Pattern Anal. Mach. Intell. **25**(9), 1063–1074 (2003)
3. Blanz, V., Vetter, T.: A morphable model for the synthesis of 3D faces. In: Computer Graphics Proceedings of SIGGRAPH 2002, pp. 187–194 (2002)
4. Taigman, Y., Yang, M., Ranzato, M., Wolf, L.: Deepface : closing the gap to human-level performance in face verification. In: IEEE Conference on Computer Vision and Pattern Recognition (CVPR), pp. 1701–1708 (2013)
5. Sun, Z.-l., Lam, K.-M.: Depth estimation of face images using the nonlinear least-squares model. IEEE Trans. Image Process. **22**(1), 895–898 (2013)

6. Zhu, X., Lei, Z., Yan, J., Yi, D.: High-fidelity pose and expression normalization for face recognition in the wild. In: 2015 IEEE Conference on Computer Vision and Pattern Recognition (CVPR), pp. 787–796 (2015)
7. Sagonas, C., Panagakis, Y., Zafeiriou, S., Pantic, M.: Robust statistical face frontalization. In: 2015 IEEE International Conference on Computer Vision (ICCV), pp. 4321–4329 (2015)
8. Zhu, Z., Luo, P., Wang, X., Tang, X.: Recover canonical view faces in the wild with deep neural networks. arXiv preprint arXiv:1404.3543 (2014)
9. Hassner, T., Harel, S., Paz, E., Enbar, R.: Effective face frontalization in unconstrained images. In: 2015 IEEE Conference on Computer Vision and Pattern Recognition (CVPR), pp. 4295–4304 (2015)
10. Xiong, X., De la Torre, F.: Supervised descent method and its application to face alignment. In: 2013 IEEE Conference on Computer Vision and Pattern Recognition (CVPR), pp. 532–539 (2013)
11. Zhu, X., Ramanan, D.: Face detection, pose estimation and landmark localization in the wild. In: Computer Vision and Pattern Recognition (CVPR) Providence, Rhode Island, June 2012
12. Kazemi, V., Sullivan, J.: One millisecond face alignment with an ensemble of regression trees. In: Computer Vision and Pattern Recognition (CVPR), Columbus, Ohio, June, 2014
13. Perez, P., Gangnet, M., Blake, A.: Poisson image editing. ACM Trans. Graph. (TOG) **22**, 313–318 (2003)
14. Hassner, T.: Viewing real-world faces in 3D. In: 2013 IEEE International Conference on Computer Vision (ICCV), pp. 3607–3614 (2013)
15. Kemelmacher-Shlizerman, I., Basri, R.: 3D face reconstruction from a single image using a single reference face shape. IEEE Trans. Pattern Anal. Mach. Intell. **33**(2), 394–405 (2011)

Research on the Stability of Plantar Pressure Under Normal Walking Condition

Ding Han[⊠], Tang Yunqi, and Guo Wei[⊠]

Department of Criminal Science and Technology,
People's Public Security University of China, Beijing 100038, China
dinghande@163.com, gd928@sina.com

Abstract. The plantar pressure of identity recognition in pattern recognition field has a certain application. Only each of the plantar pressure is stable, personalized, plantar pressure identification is scientific, but the stability problem of the plantar pressure has not been verified. In the public security practice footprints can be were personal identification, but there is no underlying scientific principles as the theoretical support. To process the collected data by computer programming, so as to prove the data of plantar pressure of normal person is basically stable. Use MATLAB software to conduct the dynamic plantar pressure's peak data table that acquired by FOOTSCAN plantar pressure measurement and analysis system, and use statistical correlation algorithm to deal with, so as to analysis the stability of plantar pressure quantitatively. Each one's feature point variance of plantar pressure is basically stable. The stability of plantar pressure data is better in normal human walking condition, so it can be used to study and apply by the plantar pressure data.

Keywords: Plantar pressure · Peak value · Stability

1 Introduction

With the gradually popularization of kinect, pressure insole and other equipment about plantar pressure, in public security, medical science and other fields, using plantar pressure identification has become a new research direction [1] by more and more attention. Compared to the biometric fingerprint recognition [2], face recognition [3], iris recognition [4], voiceprint recognition [5], the plantar pressure of personal identification has characteristics of concealment, non intrusive and so on. However, plantar pressure stability problem has not been resolved. If the plantar pressure is not stable, the plantar pressure of identity recognition has no scientific basis. Therefore, it is necessary to research on the problem of plantar pressure stability.

In the 60's of the last century, Chinese public security expert Ma Yulin found on the basis of long-term practice: because of human physiological structure and gait habits and other factors, the footprint of different pedestrians are generally different. Using this property, in 1959, Ma Yulin summed up the a gait tracking technology of "from printing to the shadow, from shadow to the printing, and to the printing and the shadow". Gait tracking technology for Chinese public security organs to solve the investigation and create a unique shortcut, provided an important basis for the

© Springer Nature Singapore Pte Ltd. 2016
T. Tan et al. (Eds.): CCPR 2016, Part I, CCIS 662, pp. 234–242, 2016.
DOI: 10.1007/978-981-10-3002-4_20

prosecution, the trial of criminal suspects. This magical detection method developed from the practice of public security and proved the stability of the plantar pressure on the experience level.

In this paper, we learn from the experience of the experts of public security footprint, combined with FOOTSCAN plantar pressure measurement and analysis system, and proposed a variance analysis algorithm of plantar pressure characteristic points to explore the stability of plantar pressure under the condition of normal walking, it provides a scientific basis for the use of plantar pressure for human identification, which can be used in criminal investigation and other fields better. The method is to use the plantar pressure peak data table, divided the plantar to three regions: heel area, paws area and toe area and extract the plantar pressure center points above three regions according to gravity formula as plantar pressure features. Finally, the stability of plantar pressure is judged by the formula of variance.

This paper is mainly about the introduction, related work, research on the stability based on the analysis of the variance of the plantar pressure features, the experiment and analysis and conclusion of five parts.

2 Related Work

Gait tactile information mainly refers to the interaction between foot and ground or other plane areas (collectively known as the bearing marks) during the normal walking. Gait tactile information includes the macroscopic gait feature information and microscopic gait feature information, and the feature of plantar pressure distribution belongs to the microscopic gait feature category. Affected by the factors of human muscle, bones, living environment and daily habits, gait tactile information can be used to reflect individual's physiological characteristics and behavior characteristics, which are considered to have specific features with relative stability.

At present, it is still at the initial stage to use gait tactile information for personal identification. Compared with gait recognition based on computer vision method, the gait recognition based on gait tactile information is not affected by the background, angle, clothes or other factors. Equipment used to measure gait tactile information mainly includes the trail gait analysis system (Fig. 1) and insole gait analysis system (Fig. 2).

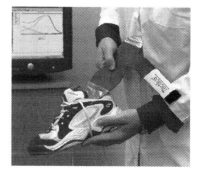

Fig. 1. Trail gait analysis system   **Fig. 2.** Insole gait analysis system

Gait recognition based on tactile information has a relatively late start. However, there are still lots of researchers who have made great contributions to this field. Mackey et al. [6] conducted analysis on the plantar pressure matrix from those patients suffered from diabetes to assess the action features between their foot and ground. Julie et al. [7] also carried out analysis on the application of characteristics of plantar pressure in clinical medicine. Kong et al. [8] developed a plantar pressure measurement footpath which can be used to measure the successive gait and the abnormal gait at the same time. Queen et al. [9], implemented analysis on the plantar pressure characteristics for those 50 healthy testers with standard and flat feet respectively, so as to make comparison to the influence caused by the different foot characteristics to the human gait. Nakajima et al. conducted normalization to the plantar pressure image to make the plantar pressure images of each tester is basically overlapped, then, extract the euclidean distance and angle at the relative position between the left and right foot of each tester as the gait feature to analysis for personal identification. Lemaire et al. [10], conducted experiments to obtain the center track of plantar pressure, the center position of plantar pressure, the maximum weight point and other parameters to study the stability of gait. Muniz [11] adopted principal component analysis method in plantar pressure, and this method enables to identify the normal gait and abnormal gait. Thus, good results have been achieved. Researchers from the University of Muenster in Germany and the University of Calgary in Canada [12] used Kistler three-dimensional dynamometry board and four Motion Analysis of video analysis system to obtain the testers' plantar pressure data, ankle joint angle and angular velocity, ankle joint and knee joint's resultant moment in vertical plane, coronal plane and transverse plane, and these information was also used for gait recognition for those females with high-heeled shoes. Begg et al. [13] used 3D motion capture system to obtain the gait spatial and temporal parameters and kinematic parameters; besides, SVM classification algorithm was used to identify two kinds of gait patterns by young people and the elderly, and the good recognition results have been obtained. Researchers from the Chinese Academy of Sciences Institute of Automation [14] used wavelet analysis method and the wavelet packet coefficients of plantar pressure as the gait feature for personal recognition. For database with 34 people's gait data included, the recognition rate has shown a maximum of 98.2 %.

3 Stability Analysis Based on the Analysis of the Variance of Plantar Pressure Characteristics

Based on the peak data of the 6 groups of plantar pressure of each person, the method of variance analysis of the characteristic points of the plantar pressure is designed, and the algorithm is as follows:

(1) Collect the data from the 50 testers and six groups of data is collected from a single person. Thus, a total of $50 \times 6 = 300$ groups of plantar pressure peak data table is obtained.
(2) Use MATLAB software to conduct pretreatment to the data, including the removal of all zero row/column of data table to obtain the normalized treatment of plantar pressure minimum matrix and plantar pressure data.

(3) Divide all the plantar pressure peak data into the heel area, palm area and toe area, and then the local center point of pressure for each group of data is calculated to get individual's plantar pressure features:

$$F = (P_1, P_2, P_3, P_4, P_5, P_6)$$

The mathematical formula of the pressure center point is expressed as:

$$Xc = \sum_{i=1}^{n}(Xi \times Pi)/ \sum_{i=1}^{n} Pi \ (1), \quad Yc = \sum_{i=1}^{n}(Yi \times Pi)/ \sum_{i=1}^{n} Pi \ (2)$$

(4) Use the 6 groups of data of each person to conduct variance respectively to the same feature. The variance formula is expressed as follows:

$$S^2 = \frac{1}{n}[(X_1 - X)^2 + (X_2 - X)^2 + \cdots + (X_n - X)^2]$$

Finally, the variance data of the 6 characteristics of each tester are obtained for analysis.

4 Experiment and Analysis

4.1 The Establishment of Gait Database

4.1.1 Laboratory Equipment

FOOTSCAN gait analysis system (flat type) and supporting software, MATLAB R2013a software.

FOOTSCAN gait analysis system is developed by the Belgian RSscan company, has the world advanced level and representation of the three-dimensional dynamic plantar pressure gait analysis system, plantar pressure distribution measurement and quantitative analysis using the method of mechatronics and image. The principle part of the system is the measuring force plate which size is 60 cm x 40 cm and the supporting software components. Among them, more than sixteen thousand sensors are uniformly distributed on the force plate, density per square centimeter 4, sampling frequency up to 500 Hz.

4.1.2 Data Acquisition

The data made in this experiment is from the 50 university students aged from 18 to 20 years old with normal weight and no disease to affect the normal walking. During the data acquisition, it is required to keep testers relaxed with normal walking posture, and every step a foot drops on the pressure testing plate (Fig. 3), plantar pressure and related data is recorded. The acquisition of the plantar pressure data of the left and right foot is called as a group of plantar pressure data, and six groups of data are collected from each person for analysis purpose (Fig. 4).

0.3	0	0	0	0	0
0	0	0	0	0	0
0	0.3	0.3	0.3	0.3	0.3
1.3	1.3	1	1	0.7	0.7
6.6	6.3	4.3	4.6	3.6	1.3
5.9	6.6	6.9	15.5	14.2	3.6
1.6	2.6	5.3	10.9	8.2	2.3
0.7	1	2	3	1.6	0.7
0	0.3	0.3	0.3	0	0.3
0.3	0	0	0	0	0.7
0.7	0.3	0.7	0.3	0.7	2.3
0.3	0.7	1	0.7	0.3	1.6
0	0.7	1	0.3	0	0.3
0	0	0.3	0	0	0
0	0	0	0	0	0
0	0	0	0	0	0

Fig. 3. Acquisition of plantar pressure data   **Fig. 4.** Plantar pressure peak data

4.2   Pretreatment of Plantar Pressure Data

As shown in Fig. 3, plantar pressure peak data matrix obtained from FOOTSCAN has multiple zero value, which forms all zero row/column that can be removed to merely get the minimum matrix of the real plantar pressure distribution to improve the extraction accuracy. After the data is obtained and for the purpose of making the different groups of data of a same person be extracted the stable features under the same coordinate system, preprocessing for plantar pressure data is needed to make the direction of the footprint equivalent. According to people's feet structure and the statistical analysis, the center of pressure of people's feet are, under normal circumstances, at the position where the center line and the feet joint. Then, it is needed to measure the farthest Euclidean distant point from the whole pressure points of a foot to the overall center point, that is, the position of the big toe's pressure points. Thus, the overall center of pressure and the big toe's pressure points constitute a line, which has a certain angle with horizontal line. With this angle as the center, plantar pressure matrix is rotated to the horizontal direction, and then the footprint direction is in a horizontal position.

4.3   Operating Results of the Analysis of Variance of the Plantar Pressure Feature Point

The variance analysis is conducted to the plantar pressure peak data of the 50 testers, and the results are shown in Table 1.

Combine with the center point features of a group of plantar pressure area of the same foot of the first 30 testers and obtain the 30 groups of data for comparison. The variance result show as below in Table 2.

By comparison from the experimental results, it is showed that the center point of the plantar pressure region of the same person is very stable. There are total 50 testers, and each of them has 6 groups of plantar pressure regional center point features. Thus, a total of $50 \times 6 = 300$ feature vectors of variance result are obtained. With the number of five as the unit, the variance structure of the characteristic points of the plantar pressure for the testers is statistically conducted and the result is shown in Table 3.

Table 1. Variance result for the plantar pressure area's feature point of all testers

NUM.	Variance result for the plantar pressure area's feature point					
	P_1	P_2	P_3	P_4	P_5	P_6
1	0.7796	0.0400	0.2992	0.3661	0.3078	0.2054
2	0.8583	0.5937	0.4268	0.7487	4.2301	0.8259
3	0.2738	0.1758	0.8551	0.9838	0.9263	0.4742
4	0.1301	0.1595	0.2647	0.1559	0.0621	0.4464
5	2.8088	70.7371	3.4701	4.0714	15.6077	1.2041
6	0.6427	1.1908	0.2478	0.2669	0.1074	1.0610
7	0.5394	0.2519	0.4090	0.9678	0.2851	0.3652
8	28.5180	33.4096	37.9234	26.0836	31.9421	19.2478
9	6.0130	2.5078	2.9075	3.0721	6.4239	83.8224
10	3.3576	88.2330	4.8229	5.6469	8.1646	2.2990
11	0.9225	0.4835	0.6647	0.0723	0.4408	0.7532
12	0.5489	0.6689	2.1100	0.1455	0.7902	0.3149
13	2.7457	0.9640	3.3570	0.2210	2.8900	1.2437
14	0.7915	0.1966	0.1181	0.6310	0.0556	0.4799
15	0.4144	1.2049	0.4621	0.7515	0.4370	0.5957
16	1.8841	0.2183	0.2232	0.5707	1.3394	0.1618
17	0.4071	0.2485	0.0698	0.3488	0.4337	0.1446
18	0.1391	0.5003	0.2936	0.1012	1.2745	0.1131
19	0.0899	1.0608	0.5832	0.2044	0.1910	0.2558
20	1.4865	3.8360	6.1047	2.8458	2.1037	66.3093
21	1.7182	1.1531	9.2088	4.2179	1.7385	63.4645
22	4.0573	81.9417	6.6365	8.4183	6.8691	5.4754
23	10.1372	10.5870	18.9838	55.2820	21.8981	54.5480
24	2.7085	3.3218	85.1183	23.1925	2.2618	2.1585
25	14.5032	2.7828	3.0041	3.1667	3.3705	75.1508
26	3.0640	3.5932	0.1920	1.9649	1.9433	0.6389
27	30.2884	45.7405	26.6926	28.7702	35.8827	35.8632
28	0.8450	0.7731	0.3284	0.1382	0.2750	1.5635
29	1.5686	0.9275	1.7057	0.4556	1.9429	8.0024
30	0.1666	0.1153	1.7314	0.0979	0.5293	0.4943
31	0.5972	0.1746	0.3116	0.1621	0.2626	0.2537
32	5.4886	2.0498	62.0011	3.6378	1.5106	2.5675
33	0.2815	2.3176	0.3156	1.4076	1.2199	1.6435
34	0.3282	0.2707	0.6294	0.1899	0.8490	0.9833
35	1.6804	2.5497	0.4003	0.3420	1.9793	0.7824
36	6.8856	2.4894	2.8130	3.3154	4.9935	86.5907
37	2.8690	6.3433	80.5211	3.7083	2.9886	3.4151
38	0.1235	0.6213	0.1570	0.3616	1.0659	0.3647
39	0.1835	0.3118	0.0362	0.0610	0.1028	0.7519
40	0.4229	0.1007	1.2343	0.1366	0.8876	0.2262
41	17.8336	14.2557	6.5526	61.4761	18.5563	43.0643
42	0.6658	0.1738	0.1821	0.3907	0.9582	0.1407
43	0.9509	2.0877	0.9563	0.2079	0.1184	0.3198
44	4.0479	2.1854	4.4633	71.5941	1.6129	4.5763

(*continued*)

Table 1. (*continued*)

NUM.	Variance result for the plantar pressure area's feature point					
	P_1	P_2	P_3	P_4	P_5	P_6
45	0.7999	0.0933	0.3489	0.1142	0.1691	0.3302
46	0.3046	0.2310	0.3148	0.2721	0.6070	0.1748
47	0.1105	0.2032	0.3437	0.1985	0.5613	0.3673
48	2.7555	3.9797	77.1419	3.6448	2.2185	5.7450
49	0.3976	1.1222	0.2192	0.9026	0.4062	0.1079
50	1.1538	0.3584	0.3129	0.3385	0.2707	0.6143

Table 2. Data from comparison

NUM.	Variance result for the plantar pressure area's feature point					
	P_1	P_2	P_3	P_4	P_5	P_6
1	0.9316	2.4183	4.6024	1.4236	15.5875	3.7228
2	4.8766	0.5324	3.8769	0.7273	0.5161	4.0096
3	0.7408	5.8255	0.5274	1.2284	0.6886	0.2753
4	1.6120	73.5256	5.2862	4.1381	13.2272	4.7745
5	0.8996	0.6697	3.8666	2.3496	1.4927	1.2617
6	4.8914	69.4325	3.7451	7.3978	8.7906	2.9836
7	2.4409	78.1356	2.8885	8.8164	5.1898	18.1872
8	4.5690	4.5656	86.2712	19.0774	6.8723	3.6184
9	5.1618	0.2790	1.9770	1.6320	12.1418	10.1454
10	7.5401	5.5396	30.1796	4.5405	7.0974	93.0097
11	28.0907	63.1666	8.6744	10.1120	16.1504	35.2632
12	6.8366	13.5550	5.0045	1.5718	1.8449	1.5822
13	55.2865	4.6295	2.4756	2.5533	2.4134	7.7407
14	2.8592	67.6229	4.3790	6.0337	7.4061	3.1147
15	3.7942	4.1606	58.5385	2.9548	11.9186	3.3712
16	1.0006	5.1817	6.0018	59.4184	4.2268	6.5687
17	6.6760	5.2033	5.0365	4.2479	73.1440	3.3580
18	3.4038	5.7907	4.7342	4.6146	4.2131	75.8688
19	13.0036	4.7422	1.5361	10.8270	11.1521	0.9804
20	1.9630	8.2406	5.6966	16.9368	14.4304	71.6772
21	1.2722	1.6639	12.3654	62.2752	3.3417	20.8584
22	30.2017	19.3809	54.5691	14.9187	7.3474	42.9010
23	9.5008	41.2810	36.0433	10.0175	16.7383	54.7221
24	24.1065	35.9366	49.8334	8.4949	8.2192	16.1984
25	11.4270	4.8734	3.5931	8.6532	11.1802	1.5520
26	5.2053	2.5067	1.2406	2.4943	5.2545	2.3197
27	3.7631	5.5315	0.6206	6.1978	2.8943	9.3411
28	8.3034	21.4177	2.9547	6.8593	2.7473	88.6352
29	42.5614	6.3811	12.5788	23.1730	66.5284	43.7674
30	0.5972	0.1746	0.3116	0.1621	0.2626	0.2537

Table 3. Statistical result of variance quantities of the center point features of the plantar pressure region for the testers

Variance interval	Variance quantities	Proportion
0–5 (include 5)	243	81.00 %
5–10 (include 10)	17	5.67 %
10–15 (include 15)	5	1.67 %
15–20 (include 20)	5	1.67 %
20–25 (include 25)	2	0.67 %
25–30 (include 30)	4	1.33 %
30–35 (include 35)	6	2.00 %
35–40 (include 40)	0	0
40–45 (include 45)	1	0.33 %
45–50 (include 50)	1	0.33 %
>50	16	5.33 %

Variance is the sum of the squares of the difference between each data and the delta of its mean. In Probability and Mathematical Statistics, variance is used to measure the deviation level between the random variables and mathematical expectation (also known as mean value). Variance is adopted by the fluctuations of the size of samples to estimate the overall size of the fluctuations. Smaller the variance is, less the deviation is, and better the stability is. On the contrary, worse the stability is. Through the horizontal comparison made to the six plantar pressure center point feature's variance of each person, it is found that, under the most cases, the variance is relatively smaller, and each person's characteristic variance shows a little volatility. A few cases show that every six variance characteristics of each person has a greater fluctuation which is obviously abnormal compared with the rest five. This may due to the emergence of some abnormal factors which affect the stability of the plantar pressure during the data acquisition. These potential abnormal factors may need to be analyzed and proved through monitoring video and other means in the experiment to obtain other factors to influence the stability of plantar pressure. Through longitudinal comparison made to the variance of the six plantar pressure center points features of each person, it is found that, within all the 300 data samples, 243 samples' variance is ranged within 0 to 5, accounting for 81 % of the total; only 16 samples' variance is greater than 50, accounting for only 5.33 % of the total. This indicates that the characteristics of people's plantar pressure are still relatively stable in general.

5 Conclusion

In this paper, the preprocessing to the peak data table of dynamic plantar pressure is conducted, together with the extraction to the plantar pressure regional feature points, aiming to obtain each tester's features of plantar pressure. By using the multiple groups of data, the statistical analysis is conducted to each feature to explore the stability of plantar pressure. This also helps to lay a foundation for the future application of plantar pressure for personal identification. The deficiency of the experiment is that the stability of the plantar pressure under normal walking is only verified, and a comprehensive analysis is still needed for those abnormal working or walking steps with various speed. For those unstable data obtained from the statistical analysis, further research and analysis are still essential for the monitor and the video during the data acquisition, aiming to find out some other factors to influence the stability of the plantar pressure. Besides, it can be combined with mechanism of plantar pressure's formation to have a more in-depth understanding on the plantar pressure stability.

References

1. Rozado, D., Moreno, T., San Agustin, J., et al.: Controlling a smartphone using gaze gestures as the input mechanism. Hum.-Comput. Interact. **30**(1), 34–63 (2015)
2. Amjad, A., Nasir, S.: GLCM—based fingerprint recognition algorithm. In: Proceedings of 2011 4th IEEE International Conference on Broadband Network and Multimedia Technology, vol. 4, pp. 207—211 (2011)
3. Frucci, M., Nappi, M., Riccio, D.: GSD baja. Pattern Recogn. **52**, 148–159 (2015)
4. Malcangi, M., Maroulis, G.:.Robust speaker authentication based on combined speech and voiceprint recognition. In: AIP Conference Proceedings-American Institute of Physics, vol. 48, no. 1, pp. 872–877 (2009)
5. Nakajima, K., Mizukami, Y., Tanaka, K., et al.: Footprint-based personal recognition. IEEE Trans. Biomed. Eng. **47**(11), 1534–1537 (2000)
6. Mackey, J., Davis, B.: Simultaneous shear and pressure sensor array for assessing pressure and shear at foot/ground interface. J. Biomech. **39**(15), 2893–2897 (2006)
7. Julie, E.P., James, O.H., Brian, L.D.: Simultaneous measurement of plantar pressure and shear forces in diabetic individuals. Gait Posture **15**, 101–107 (2002)
8. Kong, K., Tomizuka, M.: Smooth and continuous human gait phase detection based on foot pressure patterns. In: Proceedings of the 2008 IEEE International Conference on Robotics and Automation, pp. 3678–3683 (2008)
9. Robin, M.Q., Alicia, N.A., Johannes, I.W., et al.: Effect of shoe type on plantar pressure: a gender comparison. Gait Posture **31**, 18–22 (2010)
10. Lemaire, E.D., Biswas, A., Kofman, J.: Plantar pressure parameters for dynamic gait stability analysis. In: The 28th Annual International Conference of the IEEE Engineering in Medicine and Biology Society, pp. 4465–4468 (2006)
11. Muniz, A.M.S., Manfio, E.F., Andrade, M.C., Nadal, J.: Principal component analysis of vertical ground reaction force: a powerful method to discriminate normal and abnormal gait and assess treatment. In: Proceedings of the the 28th Annual International Conference of the IEEE Engineering in Medicine and Biology Society, New York, 30 August 2003, vol. 1–15, pp. 2294–2297. IEEE (2006)
12. Schollhorn, W., Nigg, B., Stefanyshyn, D., Liu, W.: Identification of individual walking patterns using time discrete and time continuous: uses in assessing rater reliability. Psychol. Bull. **86**(2), 420–428 (2002)
13. Begg, R., Kamruzzaman, J.: A machine learning approach for automatad recognition of movement patterns using basic, kinetic and kinematic gait data. J. Biomech. **38**(3), 401–408 (2005)
14. Xu, S., Zhou, X., Sun, Y.: A novel platform system for gait analysis. In: Proceedings of the 2008 International Conference on Human System Inerractions, pp. 1045–1049 (2008)

Convolutional Neural Networks with Neural Cascade Classifier for Pedestrian Detection

Bei Tong[✉], Bin Fan, and Fuchao Wu

National Laboratory of Pattern Recognition, Institute of Automation,
Chinese Academy of Sciences, Beijing, China
{bei.tong,bfan,fcwu}@nlpr.ia.ac.cn

Abstract. The combination of traditional methods (e.g., ACF) and Convolutional Neural Networks (CNNs) has achieved great success in pedestrian detection. Despite effectiveness, design of this method is intricate. In this paper, we present an end-to-end network based on Faster R-CNN and neural cascade classifier for pedestrian detection. Different from Faster R-CNN that only makes use of the last convolutional layer, we utilize features from multiple layers and feed them to a neural cascade classifier. Such an architecture favors more low-level features and implements a hard negative mining process in the network. Both of these two factors are important in pedestrian detection. The neural cascade classifier is jointly trained with the Faster R-CNN in our unifying network. The proposed network achieves comparable performance to the state-of-the-art on Caltech pedestrian dataset with a more concise framework and faster processing speed. Meanwhile, the detection result obtained by our method is tighter and more accurate.

Keywords: Convolutional Neural Network · Cascade classifier · Faster R-CNN · Pedestrian detection

1 Introduction

Object detection is an enduring topic in the field of computer vision. As a typical issue of object detection, pedestrian detection attracts increasing attention in the field of surveillance, autonomous driving and robotics applications. Since the robust real-time face detection method [30] was proposed, it was widely applied to pedestrian detection. Histogram of Oriented Gradient (HOG) [10] accelerated the development of pedestrian detection and led to the formation of the framework of features and classifier. Dollár et al. [11–14,20] proposed ten channel features and a cascade AdaBoost classifier based on [30] and Felzenszwalb et al. [16–18] presented the usage of Deformable Part-based Model (DPM) in pedestrian detection. Both of them had a great influence on the later methods.

As a fundamental component of pedestrian detection, feature extraction is vital to the subsequent processes. Various hand-crafted features had been proposed such as CSS [31], InformedHarr [34], Motion [23], Cross Channel feature

© Springer Nature Singapore Pte Ltd. 2016
T. Tan et al. (Eds.): CCPR 2016, Part I, CCIS 662, pp. 243–257, 2016.
DOI: 10.1007/978-981-10-3002-4_21

[33], Checkboard filters [36]. Researchers try to design more discriminative features since it has been proven that the combination of more features can bring better performance to a certain extent. However, more features also require more computations. The performance improvement is achieved with a sacrifice of the model's efficiency. What's more, although this kind of methods is effective, hand-crafted features are difficult to design and may be too task-specific to extend on more diverse datasets and general object detection tasks.

Recently, using CNNs to automatically learn features has become a tendency. VGG16 [26] was applied to extract features and a cascade AdaBoost classifier was trained based on these features [7,32]. Their good performance testified that CNNs have a strong power of extracting general and representative features without the need of human interference. However, these methods were always based on rectangle window of a fixed size, so they had to firstly get proposals provided by other methods such as ACF [11], stixel [2,3], Edge Boxes [37], BING [9], selective search [29], Objectness [1] and CPMC [8]. Moreover, their methods were not end-to-end and several stages of processes had to be gone through before giving the final results. Despite they had achieved good performance, sophisticated operations limit their practical use and may be time-consuming as well.

In this paper, we propose an end-to-end neural network based on Faster R-CNN [24] and neural cascade classifier for pedestrian detection. Mainly derived from the architecture of Faster R-CNN, our network is free of external proposal extraction methods and hand-crafted features. Moreover, motivated by the idea that low-level feature maps carry local information of the image [7], we utilize features from multiple convolutional layers rather than the last one to incorporate more information. These features are fed to a neural cascade classifier for hard negative mining and pedestrian detection. The neural cascade classifier consists of multiple softmax classifiers and helps to filter out negative samples in each classification stage. By integrating Faster R-CNN with neural cascade classifier, we build a unifying neural network whose inputs are images and outputs are the corresponding bounding boxes (bboxes) with the confidences of including a person. The proposed network makes a step closer to the real-time pedestrian detection since it can process an image in about 0.7 s. What's more, it achieves comparable performance to the state-of-the-art on Caltech pedestrian dataset with a more concise framework and can be extended to diverse object detection tasks.

Overall, the contributions of this paper and the merits of the proposed network can be summarized as follows:

(1) We propose an end-to-end neural network based on the Faster R-CNN and cascade neural classifier. The network does not resort to any hand-crafted features. All features are learnt by the network automatically.
(2) We utilize both low-level and high-level features and build a neural cascade classifier for hard negative mining. The cascade mechanism not only boosts the classification performance of the network, but also accelerates the processing procedure by quickly rejecting the majority of negatives.

(3) The network performs well on the Caltech dataset with the new annotations [35] and runs very fast. Besides, our network is more elegant than the frameworks proposed by the previous methods.

The remainder of this paper is organized as follows. Firstly, we give a review of related works about pedestrian detection in Sect. 2. In Sect. 3, we introduce our end-to-end neural network and implementation details. Experimental results are shown and discussed in Sect. 4. Finally, we conclude this paper in Sect. 5.

2   Related Work

Although excellent performance has been achieved on the Caltech reasonable subset, pedestrian detection still has a long way to go for the following reasons. (1) The false positive rate and false negative rate of the detection results are not satisfactory, let alone the speed of the model. (2) The original evaluation protocol [11] is not enough to describe the model's performance and there is still a large gap between the state-of-the-art results and the human baselines according to [35]. (3) Occlusion is impossible to neglect since nearly 70 % of the pedestrians captured in street scenes are occluded in at least one video frame according to [15], while many state-of-the-art methods do not consider the occlusion problem.

Recent pedestrian detection methods can be divided into three categories: hand-crafted feature based methods, CNNs based methods and combination (i.e., mixture of traditional methods and CNNs) based methods. The first category contains two mainstream branches, decision forest [4,11,20,34,36] and DPM variants [16–18] according to [5]. Both are based on hand-crafted features and traditional classifiers such as AdaBoost, SVM, etc. Most of these methods need to construct multilayer pyramid models and test by sliding window. Although the training and testing speeds of these methods are considerable at an early stage, they become slower when more features are used for improving performance. The second category becomes popular with the upsurge of deep learning in pedestrian detection. In the beginning, researchers tend to design and train their own networks for a certain task. Sermanet et al. [25] utilized an unsupervised method based on convolutional sparse coding to pre-train the filters at each stage and then trained a pedestrian detection model with multi-stage features. Wang et al. [19,21] designed their unique network structure as well. These shallow networks do make effects, but the performance is not very well. With the advent of VGG, GoogLeNet and other very deep CNNs, researchers find that for a certain task, fine-tuning these CNNs pre-trained with massive general object categories brings much better performance. Thus in the later period, most CNN based methods fine-tune these famous CNNs with a specific task. The third category is born because the combination of the previous two families [7,35] makes sense. It can integrate their advantages and be effective in detection and classification, but the tedious processes limit its possibility of application.

Overall, it is hard to judge which kind of method is obviously better than others since the state-of-the-art methods come from all three categories. However, there is no doubt that deep learning methods used in pedestrian detection form a tendency. Unfortunately, most of exiting methods are complicated and inelegant. Tian et al. [27] trained 45 different part models based on the proposals provided by LDCF [20] and selected 6 models for testing in order to save time. Each model was an independent VGG16 network and a SVM classifier was applied for combining these models to give the final result. The framework was time-consuming and not an end-to-end network. In [6,35], traditional methods were firstly applied to output a moderate good result and the best results were achieved after the binary classification via VGG16. For these methods, the traditional part requires much time to grasp, let alone the additional cost caused by VGG16. By contrast, the method proposed in this paper is more concise and its result is comparable to the state-of-the-art results. No hand-crafted features are introduced into the model and a single network is learnt automatically. Namely, the network is end-to-end without other operations.

3   Models

3.1   Datasets

Our model is trained on the Caltech10× pedestrian dataset relabeled by [35] since the old annotations have many wrong labels or labels shifting away from the real objects. Figure 1 shows two samples of the Caltech training dataset. The red bboxes are the old annotations and the green ones are the new annotations. It is obvious that the new annotations are more accurate than the old ones. However, the new labelled annotations also have some problems, such as missing

(a)   (b)

Fig. 1. Samples of Caltech train set. Red bboxes present the old annotations and the green ones are the new annotations. The left image shows that the old annotations have several pixels offset away from the objects and some missing bboxes, which may result in regarding the true positive detection bboxes as false positive if the Intersection Over Union (IOU) threshold is set high. The right image indicates that the new annotations still have some problems of the missing and shifting labels and need further revision. However, in the current stage, we choose relatively accurate dataset (i.e., new annotations provided by [35]) for training and testing. (Color figure online)

or shifting labels, which can be seen from Fig. 1(b). More details can refer to [35]. As our model is trained image by image, namely batch size of 1, wrong labels in one image may hurt the network's learning process and lead to a suboptimal converge path. Thus, the new and more accurate annotations are chosen for training and evaluation.

3.2   Architecture

CNNs have great potential in feature extraction, which can be seen in [19, 21, 22, 25, 27, 28, 32]. To avoid sophisticated hand-crafted features, our model utilizes convolutional layers to obtain features as well. The proposed network integrates a pedestrian proposal network and a neural cascade classifier in a unified framework. The pedestrian proposal network aims to predict the location of pedestrian and provides the confidence of the predicted rectangle boxes simultaneously. The neural cascade classifier attempts to give more accurate results through classification of several stages and enables the network to filter out negatives in an elegant way. Both structures share all the convolutional layers, which could obtain general characteristics and avoid too many parameters. The network takes the whole image as the input, and directly outputs the detection results with corresponding bboxes by no means of any external operations.

The basic architecture of our network is shown in Fig. 2. It consists of two parts: proposal extraction network (i.e., Part1) and neural cascade network (i.e., Part2). Since the best performance is achieved with a cascade classifier of two stages that is demonstrated in Sect. 4, the cascade classifier we refer to contains two-stage classification if not specified.

The input of the network is the whole image whose size is 800×600, which is resized from the Caltech train image whose size is 640×480. The resize operation enlarges the size of pedestrian and meanwhile maintains the aspect ratio

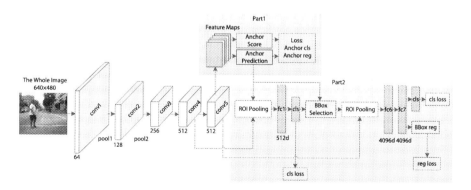

Fig. 2. Architecture of our model with two-stage classification. It can be divided into two parts, proposal extraction network and neural cascade network. The parameters in all the convolutional layers are shared by the two parts. The first part aims to predict the bbox's location and the second part is used to classify the proposals provided by the first one.

Table 1. The size of the feature map of each convolutional layer. The original image is resized before feeding to the network. Only the top two pooling operations of VGG16 are preserved since the RPN can gain more foreground bboxes for training.

Layer	L0	L1	L2	L3	L4	L5
Name	Original image	VGG16				
		conv1	conv2	conv3	conv4	conv5
Channels	3	64	128	256	512	512
Size (width × height)	640 × 480	800 × 600	400 × 300	200 × 150	200 × 150	200 × 150

of the whole image, which enables the network to deal with larger pedestrians since small size pedestrians are very difficult to detect. Then, several convolutional layers are applied to extract features, which are the same as Faster R-CNN except the absence of *pool3* and *pool4*. This is because that pedestrian in Caltech dataset is much smaller than the object in PASCAL VOC. To keep the pedestrian area not too small and guarantee adequate foreground anchors for *Region Proposal Network* (RPN) training, *pool3* and *pool4* operations are removed. The sizes of feature maps are listed in Table 1. After the shared convolutional layers, the proposal extraction network is firstly applied to predict the location of the pedestrian and meanwhile outputs two scores of each bbox. The neural cascade network is then used to filter out the negatives from the top 2,000 bboxes sorted by scores provided by the Part1. It can be divided into two stages and a bbox selection mechanism is employed between the two stages. The inputs of the first stage are the predicted bboxes and the corresponding region features from conv4_3. Then the combination of 512 dimensional fully connection layer and binary softmax classifier is adopted to eliminate negative bboxes. Subsequently, the bboxes regarded as foreground bboxes by the first stage are fed to the next stage through a bbox selection mechanism. In the network training phase, random sampling in predicted bboxes is employed when the number of foreground bboxes is less than 64. The second stage is followed with two 4,096 dimensional fully connection layers and one binary softmax classifier. Its inputs are the region features from conv5_3 and the outputs are the detection results.

The proposed neural cascade structure utilizes multilayer features and forms a strong classifier. It can quickly reject the majority of negative bboxes in the early stages and relief the burden of computation for the following stages, which is much helpful for accelerating processing speed and improving the output accuracy.

3.3   Aspect Ratio and Scale

Traditional proposal extraction methods such as ACF [11] use one aspect ratio of all detection bboxes if one model is employed. Several models have to be trained if different aspect ratio bboxes are need. Thanks to the characteristic of the Caltech test set whose pedestrian's aspect ratio is around 0.41, one aspect ratio has little effect on the model's final performance when evaluating on this dataset because its groundtruth bboxes are resized to guarantee a constant aspect ratio.

However, it is unreasonable to restrict the aspect ratio to a constant value in practical use due to the fact that the aspect ratios of pedestrians captured in street scenes distribute in a wide range. As a consequence, the proposed network should have the ability to deal with objects of different aspect ratios. There are usually two ways to solve this problem. One is resizing all bboxes to a fixed size and the other is building several models of different aspect ratios. RPN can deal with several different aspect ratios with one model by producing anchors when different aspect ratios and scales are set. The proposal extraction network of our model is based on RPN.

In RPN, more different aspect ratios and scales mean more different anchors, and theoretically should have more bboxes with high IOU values to the groundtruth. However, there is a trade-off between the number of anchors and the computation time. As a result, several relatively better aspect ratios and scales are chosen for training and testing according to the experimental results. Table 2 shows the results of using different configurations of aspect ratios and scales when three aspect ratios and scales have to be selected. The third group of configurations are chosen for an overall high performance.

Table 2. Results of configurations of different aspect ratios and scales. Three sets of one thousand images are randomly sampled from Caltech training dataset since it is time-consuming to take all training data into consideration. Aspect ratio and scale are generated randomly in a reasonable range. Only 9 different anchors (3×3) are used to compromise between time and precision. Four indicators including total_gt (i.e., total number of groundtruth bboxes), total_iou_fg (i.e., total number of bboxes whose IOU value is larger than 0.7 with any groundtruth bbox), total_fg (i.e., total number of bboxes regarded as foreground bbox) and zero_num (i.e., total number of images whose IOU value of generated bboxes with any groundtruth bbox is smaller than 0.7) are shown. The table lists only top ten results among 1,200 different random configurations sorted by total_iou_fg. A configuration should result in high total_iou_fg, total_fg and low zero_num with a smaller total_gt.

	Aspect ratio	Scale	total_gt	total_iou_fg	total_fg	zero_num
1	2.2, 2.3, 2.4	3.7, 7, 4.3	2,049	36,713	71,874	436
2	3, 2.3, 2.2	3.4, 3.6, 6.8	2,077	36,444	71,903	448
3	2.4, 2.9, 2.8	4.4, 2.9, 10.6	2,132	34,896	67,054	239
4	1.7, 2.9, 2.7	2.8, 3.8, 3.7	2,132	34,299	66,739	283
5	1.9, 2.8, 2.2	6.8, 3.8, 4.3	2,077	33,957	70,660	469
6	2.8, 1.9, 2.3	3.4, 4.2, 8.4	2,049	33,444	65,826	414
7	2.7, 2.9, 1.8	3, 7.1, 3.5	2,049	32,868	62,700	277
8	2.1, 2.6, 2.9	5.1, 3.7, 6	2,132	32,328	69,107	406
9	2.2, 1.5, 2.9	4, 6.6, 3.9	2,077	32,277	66,982	456
10	2.8, 3.6, 2.5	3.2, 6.5, 3.4	2,077	32,057	69,834	386

3.4   Optimization

Since the foreground bboxes of the top 2,000 bboxes predicted by the Part1 are limited at most 32, ordinary softmax loss is no longer applicable. All bboxes will be classified as negatives by the fully connection layers because of the huge difference between the number of background and foreground bboxes when the simple softmax loss is applied in back propagation. As a consequence, different loss weights are employed to negatives and positives. According to the approximate ratio of background and foreground bboxes, the loss is multiplied by 0.5 if the bbox is negative and 19 if otherwise.

4   Experiments

The Caltech train set and reasonable test set are employed for training and evaluation respectively. Due to the reasons explained in Sect. 3.1, the new annotations of Caltech10× dataset [35] are used to train our network with the initial VGG16 weights pre-trained on the ImageNet dataset. Since far or occluded person is not considered in the training data, the total number of qualified images for training is only 15,678, including mirrored images. Reasonable test set is a subset of the test dataset in which the pedestrian's height is larger than 49 pixels and the percentage of visual part is larger than 65 %. It is the most frequently used test set and evaluation on it is considered more representative than evaluated on the whole test set.

First of all, we give an overview of the training and testing processes. Figure 3 presents the pipelines of training and testing processes respectively. The main processes are similar and the input of both is the whole image and the output is the detection result. Besides using the groundtruth bboxes to compute the loss for back propagation and training RPN, another difference between training and testing processes is the bbox selection after each stage. In the training process, there are at least 64 bboxes to be judged in each stage, while in the testing process, any bbox which is classified as negative bbox in the previous stage will not appear in the next stage. Both processes are end-to-end and do not need any additional operations.

To verify how many stages make the best performance in terms of precision and time, a series of experiments have been conducted with one to four stages. The input batch size of the cascade with only one stage is restricted to 128 since one stage structure can not filter out part of bboxes in advance and too many bboxes fed to the next two 4,096 dimensional fully connection will sharply increase the burden of computation. The other structures with more than one stage do not need this additional operation because they can reject part of bboxes by a smaller fully connection layer beforehand. This results in the training time of one stage structure is the least among all the tested structures, which is 15 h for training with 10,000 iterations. The corresponding time cost by structures with two to four stages are 22 h, 19 h and 21 h respectively. Generally, more stages guarantee less time to train since the fewer bboxes need to be considered by the larger fully connection behind. However, at least 64 bboxes are remained

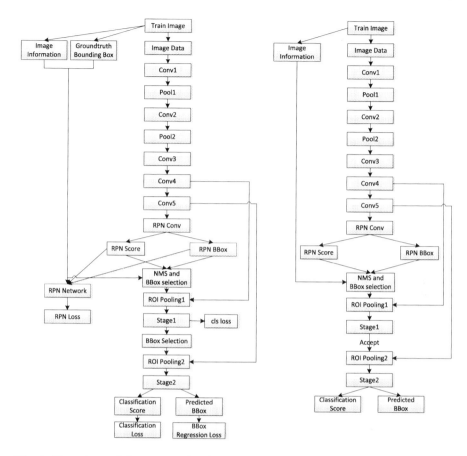

Fig. 3. Flow chart of the training (left) and testing (right) procedures. Conv1 to Conv5 present the five large convolutional layers of VGG16. RPN is used to extract proposals. More details can refer to [24, 26].

to guarantee the minimal batch size for the fully connection of each stage. This is the reason why the training time of three-stage structure is less than that of the four-stage structure.

Table 3 gives the miss rate of our network on the test set with varying number of classification stages. All parameters are the same except the stages of classification. The experimental results show that cascade softmax classifier works well in eliminating negative bboxes and two-stage classification already performs well. The output of the network is evaluated by the Caltech evaluation program [11] after Non-Maximum Suppression (NMS) operations with an IOU value of 0.65, which is the same as the output of ACF [11].

It can be known from Table 3 that classifier with more stages does not guarantee better performance. We think it is due to the small size of train set since the network tends to overfit with more fully connection layers. In addition, more stages mean a stronger classifier, which may cause some bboxes of low score

Table 3. Evaluation results of classifier with different stages. The miss rate is given by the Caltech MATLAB evaluation program [11] evaluated on the Caltech reasonable test set with the new and more accurate annotations [35]. The *Time* is the approximate average time of testing on one image and the *BBox* is the number of bboxes outputted by the model with the lowest miss rate among the 80,000 iterations detected on the test set after NMS operation.

Train iterations	One stage		Two stages		Three stages		Four stages	
	Miss rate	Time	Miss rate	Time	Miss rate	Time	Miss rate	Time
10,000	23.56 %	0.75 s	13.78 %	1.0 s	13.49 %	0.68 s	14.19 %	0.7 s
20,000	21.53 %	0.75 s	11.72 %	0.7 s	11.35 %	0.88 s	13.85 %	0.9 s
30,000	20.71 %	0.81 s	11.28 %	0.9 s	12.59 %	0.72 s	14.79 %	0.69 s
40,000	19.88 %	0.76 s	11.44 %	0.7 s	12.45 %	0.76 s	14.88 %	0.88 s
50,000	19.75 %	0.71 s	11.35 %	1.2 s	12.33 %	0.7 s	14.65 %	0.71 s
60,000	19.58 %	0.74 s	11.06 %	0.7 s	12.52 %	0.69 s	15.03 %	1.21 s
70,000	19.55 %	1.5 s	11.10 %	0.71 s	12.29 %	0.69 s	14.88 %	0.7 s
80,000	19.58 %	0.85 s	11.15 %	0.68 s	12.37 %	1.1 s	14.98 %	0.68 s
BBox	47,758		15,927		5,755		5,356	

such as cyclist, occluded person to be rejected by the classifier. The network can converge to a good result around 20,000 iterations. Although more iterations could bring better training performance, it leads to less testing accuracy due to overfitting. The network could potentially perform better if there are more different training images since the Caltech train set is sampled from a video with around 2,300 unique pedestrians and its diversity is far from rich. For instance, the low confidence of cyclist may be due to the biased train set which contains few cyclists.

Since all the experiments are carried out on the server cluster, the statistics of testing time are not very stable due to the discrepancy in load capacity. However, the majority of testing times of classifier with different stages are close. Nearly 0.7 s are cost for testing each image and 11.06 % miss rate is achieved on the test set. For a compromise of performance and time, two-stage classification is applied and the best result we achieved is used for comparing with the state-of-the-art methods.

Figure 4 compares our model with recent state-of-the-art methods evaluated on the same test set. Average miss rates over the FPPI range of $[10^{-2}, 10^0]$ (MR_{-2}) are shown. In the brackets, we also show the average miss rates over the range of $[10^{-4}, 10^0]$ (MR_{-4}). Figure 4(a) presents the results with IOU being 0.5, which means a detection bbox is regarded as a true positive when its IOU with any groundtruth in the same image is larger than 0.5. It can be seen that our model achieves a second lower MR_{-2} as 11.06 %, which is 5.44 %, 3.41 %, 0.32 % lower than those of TA-CNN, Checkerboards and DeepParts respectively. Similar results could be observed for MR_{-4}. Figure 4(b) shows the results of IOU being 0.65. The higher the IOU threshold is, the more accurate and tighter the detection results are. From this figure, we can find that our model significantly

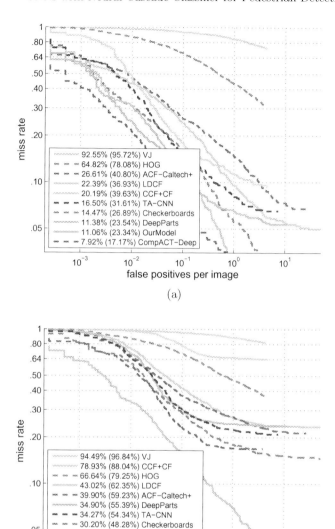

(a)

(b)

Fig. 4. Performance comparison to the state-of-the-art methods. (a) and (b) draw the ROC curves with IOU being 0.5 and 0.65 respectively. The results show that our model performs well and is comparable to the state-of-the-art.

outperforms other methods, with a decrease of 18.62 %, 14.55 % and 11.14 % compared to TA-CNN, Checkerboards and CompACT-Deep respectively. Average miss rates over the FPPI range of $[10^{-2}, 10^{0}]$ with other IOU settings are listed in Table 4. This result demonstrates that the detection results of our model are tighter and closer to the groundtruth.

Table 4. Average miss rates over the FPPI range of $[10^{-2}, 10^0]$ with different IOUs.

IOU	TA-CNN	Checkerboards	DeepParts	CompACT-Deep	OurModel
0.5	16.50 %	14.47 %	11.38 %	**7.92 %**	11.06 %
0.55	20.59 %	17.22 %	15.18 %	**11.48 %**	11.86 %
0.6	26.02 %	22.91 %	22.55 %	16.66 %	**12.75 %**
0.65	34.27 %	30.20 %	34.90 %	26.79 %	**15.65 %**
0.7	45.01 %	42.55 %	53.02 %	39.16 %	**19.21 %**
0.75	61.33 %	58.30 %	68.54 %	59.14 %	**27.16 %**
0.8	80.11 %	76.01 %	80.64 %	76.58 %	**42.05 %**
0.85	91.33 %	88.63 %	91.31 %	89.23 %	**64.13 %**
0.9	97.28 %	96.14 %	97.56 %	96.36 %	**86.17 %**

The number of bboxes detected on the test set of different methods are summarized in Table 5. It can be seen that our model has the fewest bboxes with an excellent performance. What's more, the three-stage structure can reduce the number to about one third of that of the two-stage structure with a 0.29 % increase of miss rate over the FPPI range of $[10^{-2}, 10^0]$. It indicates that our model produces smaller number of false positives, which is extremely important for practical use since too many false positives will increase the processor's burden.

Since training the model is time consuming, limited by our computational resources, our hyperparameter tuning strategy is relatively coarse-grained. More fine-grained tuning should bring better performance. Although our model's performance does not exceed [6,35], it is better than them if no additional VGG16 is applied to further classify the results of them. Overall, the model has following merits. (1) Our model is more concise and convenient for training and testing instead of dividing the train and test processes into several parts. (2) Although it is hard to compare all models' runtime, our model is at the leading level considering both performance and time according to the summarization of [5–7,11]. (3) The detection results of our model are tighter which can be seen from Table 4. (4) With a tight IOU threshold, our model achieves the best results.

Table 5. The number of bboxes of different methods.

Method	VJ	HOG	ACF-Caltech+	LDCF	CCF+CF
BBox	190,867	33,508	106,855	225,702	19,706
Method	TA-CNN	Checkerboards	DeepParts	CompACT-Deep	OurModel
BBox	31,676	1,487,711	46,684	16,337	15,927

5   Conclusion

In this paper, we have proposed an end-to-end neural network based on Faster R-CNN and neural cascade classifier. It uses multiple convolutional layers to learn rich features, which are shared by the RPN for predicting the locations of bboxes and the muti-stage softmax cascade classifier used for pedestrian classification. The model is concise and achieves comparable miss rate to the state-of-the-art with more accurate detections and being faster to process. The network can run on Tesla K80 by around 0.7 s for each image. Future work will be on improving the model's capability to deal with occlusion since most pedestrians captured in the street are occluded.

Acknowledgement. This work was supported in part by the Projects of the National Natural Science Foundation of China (Grant No. 61375043, 61403375, 61272394) and the Beijing Natural Science Foundation (Grant No. 4142057).

References

1. Alexe, B., Deselaers, T., Ferrari, V.: Measuring the objectness of image windows. IEEE Trans. Pattern Anal. Mach. Intell. **34**(11), 2189–2202 (2012)
2. Benenson, R., Mathias, M., Timofte, R., Van Gool, L.: Fast stixel computation for fast pedestrian detection. In: Fusiello, A., Murino, V., Cucchiara, R. (eds.) ECCV 2012 Ws/Demos, Part III. LNCS, vol. 7585, pp. 11–20. Springer, Heidelberg (2012)
3. Benenson, R., Mathias, M., Timofte, R., Van Gool, L.: Pedestrian detection at 100 frames per second. In: 2012 IEEE Conference on Computer Vision and Pattern Recognition (CVPR), pp. 2903–2910. IEEE (2012)
4. Benenson, R., Mathias, M., Tuytelaars, T., Gool, L.: Seeking the strongest rigid detector. In: Proceedings of the IEEE Conference on Computer Vision and Pattern Recognition, pp. 3666–3673 (2013)
5. Benenson, R., Omran, M., Hosang, J., Schiele, B.: Ten years of pedestrian detection, what have we learned? In: Agapito, L., Bronstein, M.M., Rother, C. (eds.) ECCV 2014 Workshops. LNCS, vol. 8926, pp. 613–627. Springer, Heidelberg (2015)
6. Cai, Z., Saberian, M., Vasconcelos, N.: Learning complexity-aware cascades for deep pedestrian detection. In: Proceedings of the IEEE International Conference on Computer Vision, pp. 3361–3369 (2015)
7. Cao, J., Pang, Y., Li, X.: Learning multilayer channel features for pedestrian detection (2016). arXiv preprint: arXiv:1603.00124
8. Carreira, J., Sminchisescu, C.: CPMC: automatic object segmentation using constrained parametric min-cuts. IEEE Trans. Pattern Anal. Mach. Intell. **34**(7), 1312–1328 (2012)
9. Cheng, M.M., Zhang, Z., Lin, W.Y., Torr, P.: BING: binarized normed gradients for objectness estimation at 300fps. In: Proceedings of the IEEE Conference on Computer Vision and Pattern Recognition, pp. 3286–3293 (2014)
10. Dalal, N., Triggs, B.: Histograms of oriented gradients for human detection. In: IEEE Computer Society Conference on Computer Vision and Pattern Recognition, CVPR 2005, vol. 1, pp. 886–893. IEEE (2005)
11. Dollár, P., Appel, R., Belongie, S., Perona, P.: Fast feature pyramids for object detection. IEEE Trans. Pattern Anal. Mach. Intell. **36**(8), 1532–1545 (2014)

12. Dollár, P., Appel, R., Kienzle, W.: Crosstalk cascades for frame-rate pedestrian detection. In: Fitzgibbon, A., Lazebnik, S., Perona, P., Sato, Y., Schmid, C. (eds.) ECCV 2012, Part II. LNCS, vol. 7573, pp. 645–659. Springer, Heidelberg (2012)

13. Dollár, P., Belongie, S., Perona, P.: The fastest pedestrian detector in the west. In: BMVC, vol. 2, p. 7. Citeseer (2010)

14. Dollár, P., Tu, Z., Perona, P., Belongie, S.J.: Integral channel features. In: BMVC, pp. 1–11. British Machine Vision Association (2009)

15. Dollar, P., Wojek, C., Schiele, B., Perona, P.: Pedestrian detection: an evaluation of the state of the art. IEEE Trans. Pattern Anal. Mach. Intell. **34**(4), 743–761 (2012)

16. Felzenszwalb, P., McAllester, D., Ramanan, D.: A discriminatively trained, multiscale, deformable part model. In: IEEE Conference on Computer Vision and Pattern Recognition, CVPR 2008, pp. 1–8. IEEE (2008)

17. Felzenszwalb, P.F., Girshick, R.B., McAllester, D.: Cascade object detection with deformable part models. In: 2010 IEEE Conference on Computer Vision and Pattern Recognition (CVPR), pp. 2241–2248. IEEE (2010)

18. Felzenszwalb, P.F., Girshick, R.B., McAllester, D., Ramanan, D.: Object detection with discriminatively trained part-based models. IEEE Trans. Pattern Anal. Mach. Intell. **32**(9), 1627–1645 (2010)

19. Luo, P., Tian, Y., Wang, X., Tang, X.: Switchable deep network for pedestrian detection. In: Proceedings of the IEEE Conference on Computer Vision and Pattern Recognition, pp. 899–906 (2014)

20. Nam, W., Dollár, P., Han, J.H.: Local decorrelation for improved pedestrian detection. In: Advances in Neural Information Processing Systems, pp. 424–432 (2014)

21. Ouyang, W., Wang, X.: Joint deep learning for pedestrian detection. In: Proceedings of the IEEE International Conference on Computer Vision, pp. 2056–2063 (2013)

22. Ouyang, W., Wang, X., Zeng, X., Qiu, S., Luo, P., Tian, Y., Li, H., Yang, S., Wang, Z., Loy, C.C., et al.: Deepid-net: deformable deep convolutional neural networks for object detection. In: Proceedings of the IEEE Conference on Computer Vision and Pattern Recognition, pp. 2403–2412 (2015)

23. Park, D., Zitnick, C., Ramanan, D., Dollár, P.: Exploring weak stabilization for motion feature extraction. In: Proceedings of the IEEE Conference on Computer Vision and Pattern Recognition, pp. 2882–2889 (2013)

24. Ren, S., He, K., Girshick, R., Sun, J.: Faster R-CNN: towards real-time object detection with region proposal networks. In: Advances in Neural Information Processing Systems, pp. 91–99 (2015)

25. Sermanet, P., Kavukcuoglu, K., Chintala, S., LeCun, Y.: Pedestrian detection with unsupervised multi-stage feature learning. In: Proceedings of the IEEE Conference on Computer Vision and Pattern Recognition, pp. 3626–3633 (2013)

26. Simonyan, K., Zisserman, A.: Very deep convolutional networks for large-scale image recognition (2014). arXiv preprint: arXiv:1409.1556

27. Tian, Y., Luo, P., Wang, X., Tang, X.: Deep learning strong parts for pedestrian detection. In: Proceedings of the IEEE International Conference on Computer Vision, pp. 1904–1912 (2015)

28. Tian, Y., Luo, P., Wang, X., Tang, X.: Pedestrian detection aided by deep learning semantic tasks. In: Proceedings of the IEEE Conference on Computer Vision and Pattern Recognition, pp. 5079–5087 (2015)

29. Uijlings, J.R., van de Sande, K.E., Gevers, T., Smeulders, A.W.: Selective search for object recognition. Int. J. Comput. Vis. **104**(2), 154–171 (2013)

30. Viola, P., Jones, M.J.: Robust real-time face detection. Int. J. Comput. Vis. **57**(2), 137–154 (2004)
31. Walk, S., Majer, N., Schindler, K., Schiele, B.: New features and insights for pedestrian detection. In: 2010 IEEE Conference on Computer Vision and Pattern Recognition (CVPR), pp. 1030–1037. IEEE (2010)
32. Yang, B., Yan, J., Lei, Z., Li, S.Z.: Convolutional channel features. In: 2015 IEEE International Conference on Computer Vision, ICCV 2015, Santiago, Chile, 7–13 December, pp. 82–90 (2015)
33. Yang, Y., Wang, Z., Wu, F.: Exploring prior knowledge for pedestrian detection. In: Proceedings of the British Machine Vision Conference, BMVC 2015, Swansea, UK, 7–10 September, pp. 176.1–176.12 (2015)
34. Zhang, S., Bauckhage, C., Cremers, A.: Informed haar-like features improve pedestrian detection. In: Proceedings of the IEEE Conference on Computer Vision and Pattern Recognition, pp. 947–954 (2014)
35. Zhang, S., Benenson, R., Omran, M., Hosang, J., Schiele, B.: How far are we from solving pedestrian detection? (2016). arXiv preprint: arXiv:1602.01237
36. Zhang, S., Benenson, R., Schiele, B.: Filtered feature channels for pedestrian detection. In: Proceedings of the IEEE Conference on Computer Vision and Pattern Recognition, pp. 1751–1760 (2015)
37. Zitnick, C.L., Dollár, P.: Edge boxes: locating object proposals from edges. In: Fleet, D., Pajdla, T., Schiele, B., Tuytelaars, T. (eds.) ECCV 2014, Part V. LNCS, vol. 8693, pp. 391–405. Springer, Heidelberg (2014)

Adaptive Multi-Metric Fusion for Person Re-identification

Penglin Li[1], Mengxue Liu[1], Yun Gu[2], Lixiu Yao[1], and Jie Yang[1(✉)]

[1] Institute of Image Processing and Pattern Recognition,
Shanghai Jiao Tong University, 800, Dongchuan Road, Shanghai 200240, China
{lipenglin,liumengxue,lxyao,jieyang}@sjtu.edu.cn
[2] School of Biomedical Engineering, Shanghai Jiao Tong University,
800, Dongchuan Road, Shanghai 200240, China
geron762@sjtu.edu.cn

Abstract. Person re-identification, which aims at recognizing a person of interest across spatially disjoint camera views, is still a challenging task. Plenty of approaches emerge in recent years and some of them achieve good matching results. Given a probe image, we observe that the ranking results generated by different approaches differ from each other. Considering these conventional methods are reasonable, we propose an Adaptive Multi-Metric Fusion (AMMF) method which fuses the existing ranking results with query-specific weights. Experiments on two challenging databases, VIPeR and ETHZ, demonstrate that the proposed method achieves further performance improvement.

Keywords: Person re-identification · Fusion · Multi-metric · Re-ranking

1 Introduction

The fundamental task of person re-identification is to recognise and associate a person over a distributed surveillance network. In general, we simplify this task as matching each *probe* image with a *gallery* data set composed of numerous candidates, which can be viewed as a special case of image retrieval. This problem is challenging due to camera variations, low image resolutions, changing poses, viewpoints and illumination conditions [1–4].

To tackle such challenges, current research efforts mainly focus on two aspects: feature representation methods and metric learning methods. Feature representation methods concentrate on designing discriminative descriptors which are robust to changes in viewpoint, illumination and background clutters. Some effective person image descriptors include kBiCov [5], salience match [6], SCNCD [7], query-adaptive fused feature [8] and LOMO [9]. Metric learning methods aim to learn a distance metric which measures the similarity between person images. Under the learned optimal metric, the intra-class distances are minimized while the inter-class distances are maximized. Representative learning methods include Information Theoretic Metric Learning (ITML) [10], Logistic

© Springer Nature Singapore Pte Ltd. 2016
T. Tan et al. (Eds.): CCPR 2016, Part I, CCIS 662, pp. 258–267, 2016.
DOI: 10.1007/978-981-10-3002-4_22

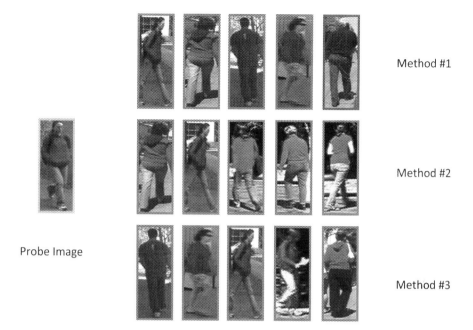

Fig. 1. Given a probe image x_p^i, its correct matching gallery image x_g^j is ranked at different positions (marked with green border) by 3 different re-identification methods. Obviously, Method #1 is more discriminative for gallery image x_g^j. (Color figure online)

Discriminant Metric Learning (LDML) [11], Large Margin Nearest Neighbors (LMNN) [12] and Keep It Simple and Straightforward (KISSME) [13], etc.

In this paper, we investigate the ranking results obtained from previous methods, finding that different methods get different ranking results for a specific probe image as shown in Fig. 1. What is more, the single decision taken by one particular method may not be enough to achieve a reliable re-identification. This motivates us to investigate how to fuse the ordered retrieval sets given by multiple methods, to further enhance the retrieval precision [8,14]. The idea is inspired by the feature fusion approach which has been demonstrated as effective in image retrieval. Typically, it is assumed that the to-be-fused methods work well by themselves for the probe. Therefore, we propose an Adaptive Multi-Metric Fusion (AMMF) method to assign different weights to the candidate images. These weights from the fused methods reflect different similarity metrics to the current image pair, which can lead to more reliable and improved re-ranking retrieval results.

The remainder of this paper is organized as follows: In Sect. 2, we present some basic concepts. Adaptive Multi-Metric Fusion (AMMF) method is illustrated in Sect. 3. In Sect. 4, we perform numerous experiments to compare the performance of our AMMF framework with its fused methods and other fusion methods. The conclusion is presented in Sect. 5.

2   Notations and Problem Statement

To clarify the person re-identification problem, we assume that the clipped person images from two camera views form the *probe* set and the *gallery* set respectively. The *probe* set is denoted as $X_P = \{x_p^1, x_p^2, ..., x_p^n\}$, where x_p^i is the i-th probe image feature vector. Correspondingly, we denote the *gallery* set as $X_G = \{x_g^1, x_g^2, ..., x_g^m\}$, in which x_g^j represents the j-th gallery image feature vector. Given a probe image, we expect a ranking list described as $(r_1 \succ r_2 \succ \cdots \succ r_m)$ from the *gallery* set, where the symbol $\succ$ indicates the order. Every ranking list results from the similarity scores of gallery images to the current probe image. The more similar a candidate is to the given probe image, the higher it can be ranked in the list.

3   Methodology

In this section, we detail the proposed AMMF method. Given a specific person image, the ranking results of the existing methods differ from each other. Considering these results are reasonable, we aim to fuse these existing methods to obtain a promoted re-ranking result. The main framework of our method is illustrated in Fig. 2.

3.1   Mapping Rank to Relevance Score

Given N conventional methods $M = \{\mathcal{M}_k\}_{k=1}^N$, the output of each method's decision function is a distance measure or a similarity score. Then, these methods have produced reasonable results described by $\mathcal{R} = \{R_{\mathcal{M}_k}\}_{k=1}^N$, in which $R_{\mathcal{M}_k}$ denotes the ranking result by method $\mathcal{M}_k$.

In this paper, we assume that: (1) these ranking lists are reasonable; (2) given a specific probe image, its similar candidates are ranked at the top of the list according to the relevance scores described as $(s_1 > s_2 > \cdots > s_m)$. Therefore, for a

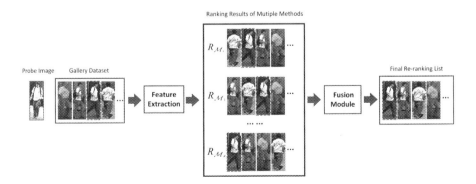

Fig. 2. Main framework of AMMF. For a specific probe image, we can get N different ranking lists by multiple methods. In the *Fusion Module*, these ranking results are mapped to top-heavy relevance scores before the final fusion and re-ranking.

given probe image x_p^i, we get a ranking list $R_{\mathcal{M}_k} := (r_1 \succ r_2 \succ \cdots \succ r_m)_{\mathcal{M}_k}$ in terms of method $\mathcal{M}_k$. While in the real-world scenarios, we are more interested in the top-ranked candidates which can be assigned higher scores by a nonlinear function:

$$\tilde{s}_j = f(r_j) = 1 - \frac{1}{1 + K^{(-r_j + R)}} \tag{1}$$

where $j = 1, 2, \cdots, m$ represents the index of a gallery image. K is set to be greater than 1. R controls the ranking position where the corresponding relevance score is 0.5 when r_j equals to R. As shown in Fig. 3(a), different gallery images are assigned different relevance scores in the range $(0, 1)$ according to their ranking orders. The top ranked gallery images are assigned higher scores.

Given a probe image $x_p^i \in X_P$, we get N ranking lists from the existing N methods which are illustrated in Fig. 3(b). For one gallery image $x_g^j \in X_G$ to the current probe image x_p^i, its ranking positions generated by N different methods differ from each other. The fused re-ranking relevance score can be defined as

$$\tilde{s}_{ij} = \sum_{k=1}^{N} \tilde{s}_{ij}^k \tag{2}$$

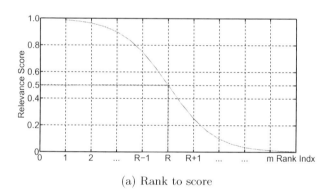

(a) Rank to score

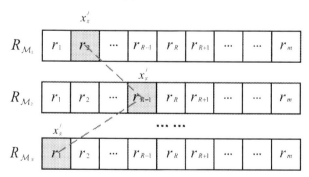

(b) N ranking lists

Fig. 3. Given a probe image x_p^i, one of its gallery image x_g^j is ranked at different positions (marked with gray background) by N different person re-identification methods.

where $\tilde{s}_{ij}^k$ is the k-th relevance score through method $\mathcal{M}_k$ generated by Eq. (1). $\tilde{s}_{ij}$ is the renewed similarity score for gallery image j to the current probe image i.

3.2   Adaptive Weighting Fusion

In fact, the retrieval results of multiple methods are not always of the same quality. Simply adding their relevance scores together without discriminating the relative importance of each method is not enough. Considering this, we give a larger weight to the relevance score of the method that achieves a high quality retrieval result, while a smaller weight is assigned to the one that has a low quality identification result. Given a probe image x_p^i and a gallery image x_g^j, the weight is defined as

$$w_{ij}^k = \frac{s_{ij}^k}{\sum_{j=1}^m s_{ij}^k} \tag{3}$$

where s_{ij}^k is the decision function output of method $\mathcal{M}_k$.

For the given probe image x_p^i, from Eq. (3) and Fig. 1, we can see that if method $\mathcal{M}_k$ assigns higher relevance score to x_g^j, it means this method is more confident of the retrieval result. In other words, the correct matching probability, regarded as w_{ij}^k, is much higher.

Consequently, we get the final renewed score

$$\tilde{s}_{ij} = \sum_{k=1}^N w_{ij}^k \tilde{s}_{ij}^k \tag{4}$$

Then we re-rank the updated relevance score set $\widetilde{S}_i = \{\tilde{s}_{ij}\}_{j=1}^m$ in a descending order with respect to the probe image x_p^i.

4   Experiments

In this section, we conduct the experiments on two widely used person re-identification datasets, the VIPeR dataset [1] and the ETHZ dataset [2]. Three learning based methods are selected to validate our approach including ITML [10], LMNN [12] and KISSME [13]. For evaluation, we use Cumulative Matching Characteristic (CMC) curves [15] to measure the performance of our approach, which represents the probability of finding the correct match over the top-r ranks.

4.1   Experiments on VIPeR Dataset

This dataset contains 1264 images of 632 pedestrians taken from two different cameras. 632 image pairs are randomly split into two equal sets with 316 image pairs each, one for training and another for testing. After extracting LOMO [9] features, PCA is applied to reduce the feature dimensionality to 100. In experiments, we find that the best result is obtained when we set $K = 3$ and $R = 10$ in Eq. (1). The experiments are performed for 10 times to get the average performance.

The re-ranking results have been illustrated in Fig. 4. By fusing multiple methods, our method gets an enhanced recognition performance. The experiments demonstrate that every final re-ranking result outperforms its fused methods in all ranking positions. For example, the re-ranking illustrated in Fig. 4(a) by fusing ITML and LMNN outperforms both ITML and LMNN. Table 1 shows several top ranked CMC results.

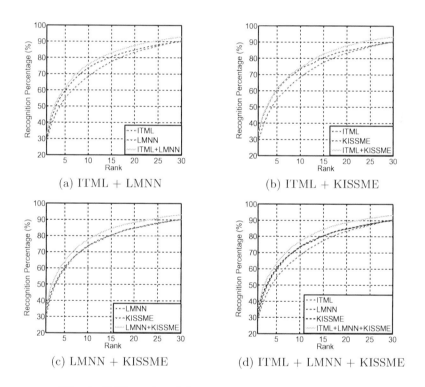

(a) ITML + LMNN

(b) ITML + KISSME

(c) LMNN + KISSME

(d) ITML + LMNN + KISSME

Fig. 4. CMC results of the top-30 ranks on VIPeR dataset.

As indicated in Table 1, fusing the three methods leads to a better performance than others. However, it is worth noticing that fusing LMNN and KISSME can get comparative results as well. It is due to the fact the ITML performance is poorer compared to LMNN and KISSME. Merging a poor method can not enhance the final performance observably. What is even worse is that the merged poorer method causes a degradation of the performance, which will be discussed in Subsect. 4.2.

Table 1. Comparison of top matching rate (%) on the VIPeR dataset. Best results for each rank are in bold font.

Rank-r	$r = 1$	$r = 5$	$r = 10$	$r = 15$	$r = 20$	$r = 25$	$r = 30$
ITML [10]	28.67	55.03	68.45	77.56	83.45	87.34	90.38
LMNN [12]	29.46	59.62	73.70	80.54	84.75	88.10	90.03
KISSME [13]	33.07	60.32	73.64	80.32	84.94	87.78	90.25
ITML + LMNN	32.50	61.39	75.79	82.91	87.44	90.60	92.94
ITML + KISSME	34.53	61.11	74.24	82.75	87.85	91.23	93.29
LMNN + KISSME	35.47	63.96	**77.47**	84.53	88.35	91.01	93.16
ITML + LMNN + KISSME	**35.60**	**63.99**	76.99	**85.30**	**88.54**	**91.46**	**93.32**
Max voting fusion	33.64	60.92	74.75	82.56	87.12	90.19	92.53
Query adaptive fusion [8]	30.89	51.95	62.96	70.25	75.05	78.47	80.55

4.2   Experiments on ETHZ Dataset

This dataset is captured by two moving cameras from crowded street scenes and structured as follows: SEQ.#1 contains 4857 images of 83 pedestrians; SEQ.#2 contains 1936 images of 35 pedestrians; SEQ.#3 contains 1762 images of 28 pedestrians. There are multiple images of the same individual with a range of

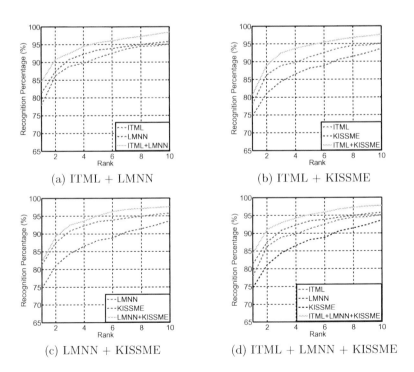

(a) ITML + LMNN

(b) ITML + KISSME

(c) LMNN + KISSME

(d) ITML + LMNN + KISSME

Fig. 5. CMC results of the top-10 ranks on ETHZ (SEQ.#1)

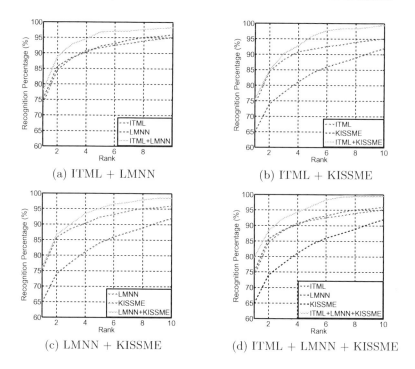

Fig. 6. CMC results of the top-10 ranks on ETHZ (SEQ.#2).

variations in appearances. All images are normalized to 64×32 pixels before extracting LOMO [9] features.

For this dataset, we randomly sample two images per person to build a training pair, and another two images to build the testing pair. The images of the testing pairs are then divided into the probe set and the gallery set. The feature dimensionality of SEQ.#1, SEQ.#2 and SEQ.#3 are reduced to 60, 30 and 25 by PCA respectively. To get the best fusion result, parameters K and R are set to 3 and 5. Its average performance is evaluated in the same way as the VIPeR's. The CMC curves are illustrated in Figs. 5, 6 and 7.

Table 2 has shown that the improved performance is achieved by fusing multiple methods. Our AMMF method outperforms its fused methods on all the 3 sequences. In addition, the rank 1 matching rates of fusing ITML and LMNN have visible increase than that of Max Voting Fusion and Query Adaptive Fusion [8]. Noticeably, fusing all methods (ITML, LMNN and KISSME) gets a poorer performance than fusing ITML and LMNN. Take SEQ.#2 for example, ITML and LMNN get comparative ranking results while the performance of KISSME is relatively poorer. Their rank 1 matching rates are 73.71 %, 75.14 % and 64.86 % respectively. Merging KISSME method into our AMMF framework causes a degradation of the final performance. As more and more state-of-the-art re-identification algorithms emerge in the future, fusing them by AMMF frame can achieve more reliable and enhanced results.

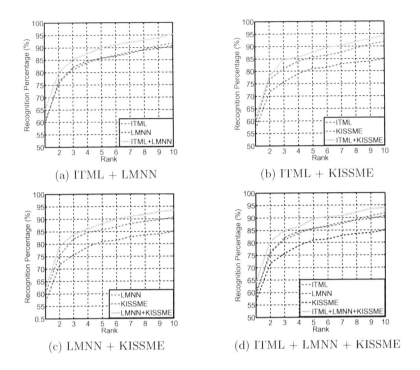

(a) ITML + LMNN

(b) ITML + KISSME

(c) LMNN + KISSME

(d) ITML + LMNN + KISSME

Fig. 7. CMC results of the top-10 ranks on ETHZ (SEQ.#3).

Table 2. Comparison of top matching rate (%) on the ETHZ dataset. Best results for each rank are in bold font.

Rank-r	SEQ.#1			SEQ.#2			SEQ.#3		
	$r=1$	$r=5$	$r=10$	$r=1$	$r=5$	$r=10$	$r=1$	$r=5$	$r=10$
ITML [10]	77.95	91.33	95.06	73.71	91.71	95.14	59.29	85.71	91.79
LMNN [12]	81.33	93.49	95.78	75.14	92.29	96.00	60.36	85.71	90.71
KISSME [13]	74.70	88.19	93.61	64.86	84.00	92.00	56.79	81.07	85.00
ITML + LMNN	**84.58**	**95.54**	**98.43**	**78.00**	**96.86**	98.29	**65.00**	**90.00**	**95.36**
ITML + KISSME	80.60	94.58	97.59	75.43	96.00	**99.43**	61.07	87.50	92.86
LMNN + KISSME	82.17	95.18	97.59	78.71	96.29	**99.43**	62.14	88.21	93.57
Fuse all *	84.22	95.30	97.71	76.00	95.14	98.57	64.29	89.64	93.93
Max voting fusion*	80.00	93.41	95.66	74.43	91.43	95.29	60.36	85.57	91.43
Query adaptive fusion	79.25	88.29	95.43	74.29	90.79	95.57	60.78	84.60	90.43

*Fuse ALL means the fusion of ITML, LMNN and KISSME.

5   Conclusion

In this paper, an Adaptive Multi-Metric Fusion (AMMF) framework is proposed. This framework fuses different ranking results of conventional methods to obtain a boosted performance. The idea is simple but effective. Experiments on two challenging person re-identification databases, VIPeR and ETHZ, validate our proposed approach. What is more, other person re-identification methods can exploit our framework to obtain superior performance.

References

1. Gray, D., Brennan, S., Tao, H.: Evaluating appearance models for recognition, reacquisition, and tracking. In: Proceedings of IEEE International Workshop on Performance Evaluation for Tracking and Surveillance, vol. 3(5) (2007)
2. Ess, A., Leibe, B., Gool, L.V.: Depth and appearance for mobile scene analysis. In: IEEE 11th International Conference on Computer Vision, pp. 1–8 (2007)
3. Alipanahi, B., Biggs, M., Ghodsi, A.: Distance metric learning vs. Fisher discriminant analysis. In: Proceedings of the 23rd International Conference on Artificial Intelligence, vol. 2, pp. 598–603 (2008)
4. Cheng, D.S., Cristani, M., Stoppa, M., Bazzani, L., Murino, V.: Custom pictorial structures for re-identification. Br. Mach. Vis. Conf. 1(2), 6–16 (2011)
5. Ma, B., Su, Y., Jurie, F.: Covariance descriptor based on bio-inspired features for person re-identification and face verification. Image Vis. Comput. 32(6), 379–390 (2014)
6. Zhao, R., Ouyang, W., Wang, X.: Unsupervised salience learning for person re-identification. In: Proceedings of the IEEE Conference on Computer Vision and Pattern Recognition, pp. 3586–3593 (2013)
7. Yang, Y., Yang, J., Yan, J., Liao, S., Yi, D., Li, S.Z.: Salient color names for person re-identification. In: European Conference on Computer Vision, pp. 536–551 (2014)
8. Zheng, L., Wang, S., Tian, L., He, F., Liu, Z., Tian, Q.: Query-adaptive late fusion for image search and person re-identification. In: Proceedings of the IEEE Conference on Computer Vision and Pattern Recognition, pp. 1741–1750 (2015)
9. Liao, S., Hu, Y., Zhu, X., Li, S.Z.: Person re-identification by local maximal occurrence representation and metric learning. In: Proceedings of the IEEE Conference on Computer Vision and Pattern Recognition, pp. 2197–2206 (2015)
10. Davis, J.V., Kulis, B., Jain, P., Sra, S., Dhillon, I.S.: Information-theoretic metric learning. In: Proceedings of the 24th International Conference on Machine Learning, pp. 209–216 (2007)
11. Guillaumin, M., Verbeek, J., Schmid, C.: Is that you? Metric learning approaches for face identification. In: IEEE 12th International Conference on Computer Vision, pp. 498–505 (2009)
12. Dikmen, M., Akbas, E., Huang, T.S., Ahuja, N.: Pedestrian recognition with a learned metric. In: Asian Conference on Computer Vision, pp. 501–512 (2010)
13. Koestinger, M., Hirzer, M., Wohlhart, P., Roth, P.M., Bischof, H.: Large scale metric learning from equivalence constraints. In: Proceedings of the IEEE Conference on Computer Vision and Pattern Recognition, pp. 2288–2295 (2012)
14. Zhang, S., Yang, M., Cour, T., Yu, K., Metaxas, D.N.: Query specific fusion for image retrieval. In: European Conference on Computer Vision, pp. 660–673 (2012)
15. Wang, X., Doretto, G., Sebastian, T., Rittscher, J., Tu, P.: Shape and appearance context modeling. In: IEEE 11th International Conference on Computer Vision, pp. 1–8 (2007)

Face Detection Using Hierarchical Fully Convolutional Networks

Jiang-Jing Lv[✉], You-Ji Feng, Xiang-Dong Zhou, and Xi Zhou

Intelligent Multimedia Technique Research Center,
Chongqing Institute of Green and Intelligent Technology,
Chinese Academy of Sciences, Chongqing 400714, People's Republic of China
lvjiangjing@cigit.ac.cn

Abstract. Face detection in unconstrained environment is a challenge problem. Recent studies show that deep convolutional networks (DCNs) have achieved outstanding performance on this task, but most of them have multiple stages (e.g., region proposal, classification), which are complex and time-consuming in practice. In this paper, we propose a fully convolutional network (FCN) framework which can be trained straightforward in an end-to-end manner. In our network, hierarchical feature layers with different resolutions are used to detect different scale faces. For each hierarchical layer, a specific default boxes set with different aspect ratios and scales is associated with each map cell. At prediction time, the network generates confidence scores for the default boxes and produces offsets of default boxes to get better bounding boxes of faces. The predictions of each hierarchical layer are combined into final detection result. Experimental results on the AFW and FDDB datasets confirm the effectiveness of our method.

Keywords: Face detection · Fully convolutional network · End-to-end

1 Introduction

Face detection is a core problem in face recognition systems. The face detection processes usually consist of region proposal generator and classifier. Region proposals can be generated by densely scanning an image in a sliding window fashion [1,2], or by using proposal mechanisms such as MCG [3], Selective Search [4], and then the classifier is used to classify each proposal. Currently, Deep Convolutional Neural Networks (DCNNs) are widely acted as a classifier to improve classification performance [5–7]. Even though some of them have significantly improved detection accuracy, they detect faces in multi-stage strategies which are slow and inconvenient. For instance, Yang *et al.* [6] introduced a Convolutional Channel Features (CCF) method for face detection. It transfers low-level features from pre-trained CNN models to feed the boosting forest model. However, it use sliding window mechanism as region proposal. A large number of ROIs need to be processed for each image, which is time-consuming. In addition,

© Springer Nature Singapore Pte Ltd. 2016
T. Tan et al. (Eds.): CCPR 2016, Part I, CCIS 662, pp. 268–277, 2016.
DOI: 10.1007/978-981-10-3002-4_23

it fail in the presence of strong occlusions. Therefore, a facial parts responses method [5] was proposed for severe occlusion and pose variation face detection. First, a set of attribute-aware deep networks were designed to generate responses maps. Then a set of face proposals were obtained by calculating the degree of face likeliness through facial parts' spatial arrangement in responses maps. Finally, a multi-task deep convolutional network was used to refine the face locations and confidences. However, it has a multi-stage training and different sub-nets are needed.

As most previous methods are run either in sliding windows fashion over the whole image or on some proposal regions [3,4] in an image, it is time-consuming and cannot meet the real-time face detection. In addition, for most proposed networks, the input image size of a CNN model is usually fixed. The resize of input image to the fixed size may result in unwanted geometric distortion. Furthermore, due to the size of feature maps is reduced with deeper layers, fine details of image will be lost which lead to the difficulty in detecting different scale objects or objects with overlap. Even though some methods [6,8] suggest resizing the image to different size and processing each scale image individually, it requires large extra computational expense.

For the sake of efficiency, we propose a new framework to integrate region proposal and face detection into a single network. It is an absolutely an end-to-end training mechanism for face detection and easy to train. Our network is a kind of FCN [9], which can take arbitrary-sized images as inputs. In addition, in order to efficiently detect multi-scale faces in an image, we design a hierarchical mechanism for detecting different scale faces. For each hierarchical layer, a specific default boxes set with different aspect ratios and scales is associated with each map cell. During prediction, the network generates confidence scores and offsets of each default box to identify and bound faces. According to multiple feature layers with different resolutions, faces with various sizes are automatically detected. Experimental results show that our method is able to detect face in arbitrarily size with close proximity.

2   Related Work

In this section, we discuss some previous methods closely related to this work.

Fully Convolutional Network. FCN has been widely exploited in recent years. Sermanet et al. [8] applied sliding windows on the top convolutional layer for object detection. Eigen et al. [10] used FCN to restore images form noises. However, FCN has become real prevalent, since Long et al. [9] first successfully used it for semantic segmentation. They adopted contemporary classification networks into FCN and de-convolution layers were used for up-sampling the feature maps into image size. In up-sampling layers, pixel-wise prediction was made for semantic segmentation. Besides, for further precision, former layers were combined with final layer to predict finer details. Inspired form this property, we use the former layer to predict smaller faces and later layers to predict larger faces. In addition, FCN supports arbitrary-sized inputs. According to this character, Fast R-CNN [11] first use FCN to extract semantic feature of image, then a region of interest (RoI) pooling layer extracts a fixed-length feature for each proposal.

Region proposal. Region proposal is a fundamental step for object detection. Various methods have been proposed in recently literature, including MCG [3], Selective Search [4], and sliding windows [1,2]. Traditionally, region proposals is thought as an independent part of object detection. Stage-wise training are adopted by most object detection methods [8,11,12]. This stereotypes have been broken since Ren *et al.* [13] proposed Region Proposal Networks (RPN), which shares full-image convolutional features with detection network and is trained end-to-end to generate high-quality region proposals. Even though RPN step is nearly cost-free and boost the speed and performance of object detection, RPN and detection network are trained in alternative manner. Liu *et al.* [14] combined these two network into a single deep net neural network and implemented an end-to-end training mechanism for object detection. It is much faster than previous methods and achieves much better accuracy. We also adopted this property for fast face detection.

3   Approach

For deep neural networks, spatial resolution is gradually reduced with the pooling operators in network, while the size of receptive field in top layers become larger. Therefore, the higher layers of DCNNs may have more semantic abstraction of the image, but they might lose more spatial details which are fundamental to precise object localization. As for object detection tasks, semantics are critical for classification and spatial details are critical for object localization. In order to balance these two properties, we design a hierarchical structure for face detection, as shown in Fig. 1. Different scales of faces are detected according to different convolutional layers.

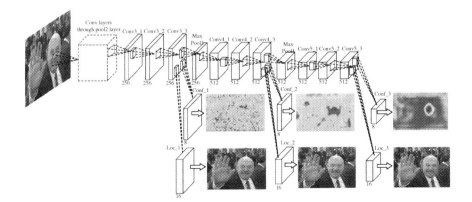

Fig. 1. The pipeline of hierarchical face detection.

In our method, we pick the VGG 16-layer network [15] and discard layers after *conv5_3*.

Hierarchical Feature Layers. The layers before *conv3_3* are used to encode semantic information, and the *conv3_3*, *conv4_3* and *conv5_3* layers are used for face detection. Two additional convolutional layers, which are used to predicting confidence scores and box offsets, are added to these layers respectively. These face detection layers decrease in resolution and increase the size of receptive field progressively. The former layers have much fine-grained details which is beneficial for small face detection, while the later layers have larger receptive fields which is useful for large face detection. According to this structure, faces with different scales are sequently detected.

Bounding Boxes Regression and Classification. Similar to the anchor boxes used in Faster R-CNN [13], we also associate a set of default bounding boxes with each feature map cell. During face detection, 3×3 filter kernels are applied to each face detection convolutional layer (*conv3_3*, *conv4_3* and *conv5_3*) and produce confidence scores of faces and shape offsets relative to the default boxes.

4   Implementation Details

4.1   Scales and Aspect Ratios for Default Boxes

As shown in [16], feature maps from different levels within a network have different empirical receptive field size. In our network, each hierarchical feature layer is designed to detect particular scales of faces. Table 1 lists the scales of default boxes in each face detection layer. We use three aspect ratios of 1:1, 2:1 and 1:2 for each scale.

Table 1. The scales and aspect ratios of default boxes using the VGG net.

Layer	*conv3_3*		*conv4_3*		*conv5_3*	
Scale	16^2	32^2	64^2	128^2	256^2	512^2
Proposal	16×16	32×32	64×64	128×128	256×256	512×512
	23×11	45×23	91×45	181×90	362×181	724×362
	11×23	23×45	45×91	90×181	181×362	362×724

4.2   Loss Function

For each hierarchical layer, we have two sibling output layers. The first output confidences for each default box, and the second later outputs bounding-box regression offsets for each default box. Inspired by [14], we select default boxes match ground truth with jaccard overlap higher than a threshold (e.g., 0.5) for bounding-box regression, which significantly simplifies the learning problem. The localization loss is computed as:

$$Loss_1 = \frac{1}{N_{reg}} \sum_i L_{reg}\left(t_i, t_i^*\right) \qquad (1)$$

where N_{reg} is the number of selected default boxes, t_i is predicted tuple off-sets and t_i^* is regression target. Similar to Fast R-CNN [11], the bounding-box regression loss is the Smooth L1 loss.

As for classification loss, there are large number of default boxes in a layer, but most of them are negatives. This leads to imbalance between the positive and negative training examples, which impacts optimization and stable training. In order to address this problem, we use the hard negative mining method introduced in [14]. We sort the default boxes with confidence and select top ones as negatives. The ratio between the negatives and positives is 3:1. The classification loss is defined as follows:

$$Loss_2 = \frac{1}{N_{cls}} \sum_i L_{cls}\left(p_i, p_i^*\right) \qquad (2)$$

where N_{cls} is the number of negatives and positives, p_i is the predicted probability of default box i and p_i^* is the ground-truth label whose value is 1 for positives and 0 otherwise. L_{cls} is the softmax loss over faces or background.

The combined loss function for each hierarchical layer is defined as follows:

$$Loss = Loss_1 + \lambda Loss_2 \qquad (3)$$

The two terms are weighted by λ. In our experiment, λ default set to 1.

4.3   Data Augmentation

As in face detection, faces may be under low resolution and serious blur. In order to make the model robust to various environment, some common data augmentation methods are randomly added to the training images:

- Rescale the input image with ratio between 0.3 and 1.2.
- Gaussian blur.
- Contrast transformation.
- Noises (e.g., random, salt).

5   Experiment

5.1   Experiment Setup

In this section, we evaluate the effectiveness of our method on Face in the wild (AFW) [17] and Face Detection Data Set and Benchmark (FDDB) [18]. We employ CelebFaces [19] and AFLW [20] databases to train our face detection networks. The details of each dataset are introduced as follows.

- AFW dataset is built from Flicker images, which has 205 images of 473 annotated faces.
- FDDB dataset is widely used face detection benchmark, which contains 5,171 annotated faces of 2,845 images.
- CelebFaces dataset contains 202,599 images and each image contains only one face. It covers large pose variations and background clutter.
- ALFW dataset contains 21,080 images with 24,386 annotated faces.

For training datasets, faces are annotated by ellipses to better fit face regions rather than traditional rectangles. The ellipses are generated according to the principles of [18]. We directly train our face detection network from scratch. SGD is adopted for updating weights with learning rate 0.001, 0.9 momentum, 0.0005 weight decay and iter size of 8. During training, we firstly use the 0.001 learning rate for 360k iterations, then we decay it to 0.0001 and continue for another 240k iterations.

During testing, a large number of boxes are predicted in a test image and have overlap with each other. To reduce redundancy, we use the confidence threshold of 0.01 to filter out most boxes, then non-maximum suppression (NMS) with jaccard overlap of 0.3 is applied to keep top 300 detections per image.

5.2   Experiment Results in AFW

We compare our method against some previous methods in the AFW. The average precision of each method is shown in Fig. 2. Our method outperforms all previous methods [5,21,22]. We further analysis the detection results of our method. As AFW omits annotations of many faces with a small scale, we not only almost detected all faces annotated in AFW but also detected faces without labeled which are shown in Fig. 3(a). Figure 3(b) shows some detection errors of our method.

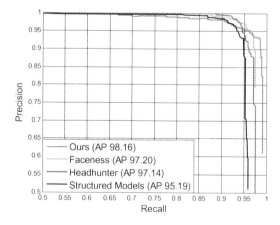

Fig. 2. Precision-recall curves on AFW. AP = average precision.

(a)   (b)

Fig. 3. Face detection results on AFW, where red boxes are our predicted results and blue boxes are the annotated faces. (a) Detect all faces in the image, (b) some detection errors (Color figure online)

5.3 Experiment Results on FDDB

FDDB is widely used face detection benchmark. In order to evaluate the performance of our method, we compare our method against previous state-of-the-art methods [5,6,13,23,24] on this database. We train the Faster RCNN model on the same training dataset as ours. As shown in Fig. 4, our method perform much better than most previous methods. Hoverer, contrast to most previous methods which have multiple stages and time-consuming to detect faces in an image, our method is end-to-end and more efficient. Figure 5 also shows some detection results of our method.

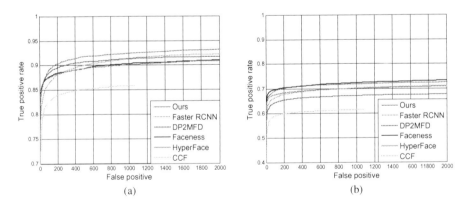

(a)   (b)

Fig. 4. Face detection ROCs on FDDB. (a) Discrete score, (b) continuous score.

5.4 Discussion

In contrast to previous face detection methods, our method is an end-to-end mechanism which is easy to train. Hierarchical layers help to detect different scales of faces. In addition, our network is an FCN, which can take arbitrary-sized

(a)   (b)

Fig. 5. Face detection results on FDDB. (a) Detect all faces in the image, (b) some detection errors. (Color figure online)

image as input and fast in face detection. For image size of 500×500, it takes 20 fps on GPU (Nvidia Titan Black). It can be faster for smaller images.

6   Conclusion

In this paper, we introduce a FCN based framework for fast face detection. A key feature of our method is detecting face in a hierarchical way. This allows us to efficiently detect different scale faces in an image. Besides, in contrast to multi-stage procedure, our network is an end-to-end structure, which is easy to train. Experimental results demonstrate that our method achieves promising performance on several challenging face detection benchmarks (e.g., AFW, FDDB). In addition, since our network only contains convolutional layers, face detection can be implemented in real time.

Acknowledgments. This work was partially supported by the National Natural Science Foundations of China (Grant nos. 61472386 and 61502444) and the Strategic Priority Research Program of the Chinese Academy of Sciences (Grant XDA 06040103).

References

1. Alexe, B., Deselaers, T., Ferrari, V.: Measuring the objectness of image windows. IEEE Trans. Pattern Anal. Mach. Intell. **34**(11), 2189–2202 (2012)
2. Zitnick, C.L., Dollár, P.: Edge boxes: locating object proposals from edges. In: Fleet, D., Pajdla, T., Schiele, B., Tuytelaars, T. (eds.) ECCV 2014, Part V. LNCS, vol. 8693, pp. 391–405. Springer, Heidelberg (2014)
3. Arbeláez, P., Pont-Tuset, J., Barron, J., Marques, F., Malik, J.: Multiscale combinatorial grouping. In: IEEE Conference on Computer Vision and Pattern Recognition, pp. 328–335 (2014)
4. Uijlings, J.R., van de Sande, K.E., Gevers, T., Smeulders, A.W.: Selective search for object recognition. Int. J. Comput. Vis. **104**(2), 154–171 (2013)

5. Yang, S., Luo, P., Loy, C.-C., Tang, X.: From facial parts responses to face detection: a deep learning approach. In: IEEE International Conference on Computer Vision, pp. 3676–3684 (2015)
6. Yang, B., Yan, J., Lei, Z., Li, S.Z.: Convolutional channel features. In: IEEE International Conference on Computer Vision, pp. 82–90 (2015)
7. Li, H., Lin, Z., Shen, X., Brandt, J., Hua, G.: A convolutional neural network cascade for face detection. In: IEEE Conference on Computer Vision and Pattern Recognition, pp. 5325–5334 (2015)
8. Sermanet, P., Eigen, D., Zhang, X., Mathieu, M., Fergus, R., LeCun, Y.: Overfeat: integrated recognition, localization and detection using convolutional networks. arXiv preprint arXiv:1312.6229
9. Long, J., Shelhamer, E., Darrell, T.: Fully convolutional networks for semantic segmentation. In: IEEE Conference on Computer Vision and Pattern Recognition, pp. 3431–3440 (2015)
10. Eigen, D., Krishnan, D., Fergus, R.: Restoring an image taken through a window covered with dirt or rain. In: IEEE International Conference on Computer Vision, pp. 633–640 (2013)
11. Girshick, R.: Fast R-CNN. In: IEEE International Conference on Computer Vision, pp. 1440–1448 (2015)
12. Girshick, R., Donahue, J., Darrell, T., Malik, J.: Rich feature hierarchies for accurate object detection and semantic segmentation. In: IEEE Conference on Computer Vision and Pattern Recognition, pp. 580–587 (2014)
13. Ren, S., He, K., Girshick, R., Sun, J.: Faster R-CNN: towards real-time object detection with region proposal networks. In: Advances in Neural Information Processing Systems, pp. 91–99 (2015)
14. Liu, W., Anguelov, D., Erhan, D., Szegedy, C., Reed, S.: SSD: single shot multibox detector. arXiv preprint arXiv:1512.02325
15. Simonyan, K., Zisserman, A.: Very deep convolutional networks for large-scale image recognition. arXiv preprint arXiv:1409.1556
16. Liu, W., Rabinovich, A., Berg, A.C.: Parsenet: looking wider to see better. arXiv preprint arXiv:1506.04579
17. Zhu, X., Ramanan, D.: Face detection, pose estimation, and landmark localization in the wild. In: IEEE Conference on Computer Vision and Pattern Recognition, pp. 2879–2886. IEEE (2012)
18. Jain, V., Learned-Miller, E.G.: FDDB: a benchmark for face detection in unconstrained settings. UMass Amherst Technical report
19. Liu, Z., Luo, P., Wang, X., Tang, X.: Deep learning face attributes in the wild. In: IEEE International Conference on Computer Vision, pp. 3730–3738 (2015)
20. Köstinger, M., Wohlhart, P., Roth, P.M., Bischof, H.: Annotated facial landmarks in the wild: a large-scale, real-world database for facial landmark localization. In: IEEE International Conference on Computer Vision Workshops, pp. 2144–2151. IEEE (2011)
21. Mathias, M., Benenson, R., Pedersoli, M., Van Gool, L.: Face detection without bells and whistles. In: Fleet, D., Pajdla, T., Schiele, B., Tuytelaars, T. (eds.) ECCV 2014, Part IV. LNCS, vol. 8692, pp. 720–735. Springer, Heidelberg (2014)
22. Yan, J., Zhang, X., Lei, Z., Li, S.Z.: Face detection by structural models. Image Vis. Comput. **32**(10), 790–799 (2014)

23. Ranjan, R., Patel, V.M., Chellappa, R.: A deep pyramid deformable part model for face detection. In: International Conference on Biometrics Theory, Applications and Systems, pp. 1–8. IEEE (2015)
24. Ranjan, R., Patel, V.M., Chellappa, R.: Hyperface: a deep multi-task learning framework for face detection, landmark localization, pose estimation, and gender recognition. arXiv preprint arXiv:1603.01249

Depth Supporting Semantic Segmentation via Deep Neural Markov Random Field

Wen Su[1,2,3(✉)] and Zengfu Wang[1,2,3]

[1] Institute of Intelligent Machines, Hefei Institutes of Physical Sciences,
Chinese Academy of Sciences, Hefei, Anhui, China
`wensu@mail.ustc.edu.cn`
[2] University of Science and Technology of China, Hefei, Anhui, China
[3] National Engineering Laboratory for Speech and Language Information Processing,
Hefei, Anhui, China

Abstract. Semantic segmentation is of great importance to various vision applications. Depth information plays an important role in human visual system to help people obtain meaningful segmentation results, but it is not well considered by most existing segmentation methods. In this paper, we address the problem of semantic segmentation by incorporating depth information via deep neural Markov Random Field. In our method, the color image and its corresponding depth map are first fed to a convolutional neural network. Then, a deconvolution approach is performed on the network output to obtain the pixelwise prediction in terms of the probability of labels assigned to pixels. Finally, the dense prediction is used to design unary term and pairwise term, which are determined by pixels coordinate, color and depth. Experiments are conducted on several public datasets to illustrate the effectiveness of the proposed method. On the PASCAL VOC 2011 test dataset, experimental results show that our method can get accurate results when compared with the ground truth. On the PASCAL VOC 2012 dataset and NYUDv2 dataset, the proposed method can obtain competitive results.

Keywords: Semantic segmentation · Depth · CNNs · MRF

1 Introduction

Semantic segmentation is an important research area in computer vision. Many sophisticated researches, such as image retrieval and object recognition, have benefited from it. The computer vision community has published a wide range of semantic segmentation algorithms so far. Those algorithms can be grouped by the kind of data they deal with or the method with which they produce the segmentation [1].

From the perspective of data category, semantic segmentation can be implemented by color images, depth maps or incorporating both of them. With the development of acquiring techniques for depth information, the depth map of a scene is easily accessed. The incorporation of depth maps shows a new insight

© Springer Nature Singapore Pte Ltd. 2016
T. Tan et al. (Eds.): CCPR 2016, Part I, CCIS 662, pp. 278–289, 2016.
DOI: 10.1007/978-981-10-3002-4_24

in semantic segmentation. Microsoft kinect sensor [2] makes it possible to collect color images along with their corresponding depth maps for indoor scenes. Silberman et al. [3] intergraded the dataset and dense label coverage to form a uniform dataset. Because the serious noise introduced into the sensor in outdoor scenes, they inclined to solve indoor scene segmentation. In the work of Zhang et al. [4], five view-independent 3D features were extracted from dense depth maps to train a conventional classifier using randomized decision forest technique. Obviously, their method relies on the hand-crafted features which are effort consuming.

From the viewpoint of core technique applied, semantic segmentation methods can be classified into two categories, namely, traditional methods and deep learning based methods. Recent advances have demonstrated the potential of deep learning architectures in semantic segmentation. Samuel et al. [5] used neural decision forests which adopt multi-layer perceptron in the split nodes to produce segmentation. Chen et al. [6] combined deep convolutional neural networks (CNNs) and fully connected conditional random fields (CRFs) to address the task of pixel-level classification. The combination of few strongly labeled and many weakly annotated images had been utilized to train a deep convolutional neural network for segmentation with weakly and semi-supervised learning [7]. However, all of the aforementioned methods ignore analyzing the pixels coordinate, color and depth relations in modelling the color images.

This paper proposes a novel depth supporting semantic segmentation via deep neural Markov Random Field (MRF). We feed depth maps and RGB images to CNN. The output prediction of CNN is then modeled by MRF: The output of de-convolutional layers is used to model depth supporting unary term; The designed layers following the CNN are used to model our depth supporting smoothing term. As the training of CNN need a large number of images with strong pixel-level annotations, we hope to find a dataset which provides a huge amount of images, corresponding depth maps and labels at the same time. Unfortunately, there exist no such a dataset to the best of our knowledge. However, large efforts have been put into estimating depth maps from single monocular images. Recent proposed methods in [8–10] exploit geometric priors, additional sources of information and hand-crafted features to estimate depth. However, their estimation is not quite accurate. Sexena et al. [11] proposed to formulate the depth estimation as a deep continuous CRF learning problem with a deep convolutional neural field model. Their experiments on both indoor and outdoor scene datasets demonstrate that the proposed method outperforms state-of-the-art depth estimation approaches. Thus, we adopt their method to estimate depth maps in our study. As depicted in [12], CNN is employed to get the dense prediction by changing the last three layers of classical models into fully convolution layers and feeding the output of fully convolutional layers to de-convolutional layers. Distincting from their work, we model a MRF on the dense prediction instead of feeding the output of de-convolutional layer to a $soft - max - log$ for producing a probability distribution over labels. It is somewhat like the method of [13]. Their method addresses semantic image segmentation by incorporating rich information into MRF, including high-order relations and mixture of label contexts. They solved MRF by proposing a CNN which extends a contemporary CNN architecture to model unary term. Additional layers were designed to approximate the mean field algorithm for pairwise term. Despite they claimed that their model yields a new state-of-the-art accuracy, they ignore the depth maps and just upsample the outputs of CNNs with bilinear interpolation simply when estimating the unary term. In our work, we estimate depth maps from single monocular images with the method mentioned in [11] so that our training dataset can be extended to any semantic segmentation dataset. We feed the output of CNN into de-convolutional layers for dense prediction. Images incorporating depth maps are used to model MRF based on the dense prediction. We design unary term and pairwise term in position, color and depth space simultaneously. Then we use the method similar to [13] to carefully devise follow-up layers to approximate the mean field algorithm (MF) for pairwise term.

Our contributions are summarized as below. (1) We use depth maps incorporating images for semantic segmentation. The depth maps are estimated from single monocular images. (2) The dense prediction is obtained by feeding the output of CNN to de-convolutional layers instead of using bilinear interpolation simply. (3) We design unary term and pairwise term in position space, color space and depth space simultaneously. Especially the rational application of the depth maps is explained carefully. The overview of the proposed depth supporting semantic segmentation via deep neural MRF framework is presented in Fig. 1.

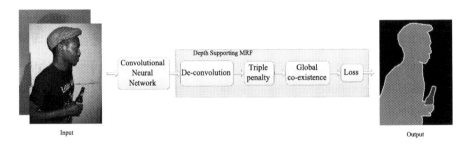

Fig. 1. Overview for depth supporting semantic segmentation via deep neural MRF. The input is depth maps and images. The depth maps are estimated from the corresponding images. We feed the output of the CNN to our depth supporting MRF to get the final segmentation.

2   Our Approach

The input of CNN comprise images and their corresponding depth maps. The depth maps are estimated from the single monocular images with the method aforementioned. Then we feed the input to CNN to get the dense prediction which also is regarded as unary term. The local context information in images and depth maps is utilized to model MRF by extending fully convolutional networks (FCN). We design unary term and pairwise term with pixel' coordinate, RGB and especially depth information simultaneously.

2.1   Methods

MRF [14] is a general approach to model image in computer vision. It is a undirected graph where each node represents the pixel in the image and every edge connecting two nodes represents some relations which we define between two pixels. Each node is associated with a indicator y_p^l. It is equal to 1 or 0, indicating that the pixel p has the label l or not. We have $l \in L = \{1, 2, 3, ..., C\}$, representing a set of C labels. The energy function of MRF is consisted of two terms: unary term and the pairwise term or smoothing term.

$$E(y) = \sum_{\forall p \in \mathcal{N}} \Phi(y_p^l) + \sum_{\forall p, q \in \mathcal{E}} \Psi(y_p^{l_1}, y_q^{l_2}). \tag{1}$$

where y, $\mathcal{N}$, $\mathcal{E}$ denote a set of variables, nodes and edges, respectively. $\Phi(y_p^l)$ is unary term, measuring the cost of assigning label l to $p-th$ pixel. For instance, if pixel p belongs to the a category instead of b, we should have $\Phi(y_p^a) < \Phi(y_p^b)$. Intuitively, the unary term represent per-pixel classification. The unary term is typically defined as

$$\Phi(y_p^l) = -\ln(p(y_p^l|I)). \tag{2}$$

where $p(y_p^l|I)$ represents the possibility of $p-th$ pixel has label l if given image I. Moreover, $\Psi(y_p^{l_1}, y_q^{l_2})$ is pairwise term, measuring the penalty of assigning labels l_1, l_2 to pixels p, q respectively. The pairwise term represents a set of smoothness constraints. It is typically defined as

$$\Psi(y_p^{l_1}, y_q^{l_2}) = f(\|I_p - I_q\|, \|p - q\|). \tag{3}$$

here I_p and I_q indicate feature vectors such as RGB values extracted from $p-th$ and $q-th$ pixel. p and q denote coordinates of pixels' positions, and f is the function which we define according to our smoothness constraints. Eq. (3) implies that if two pixels are close and look similar, they are encouraged to have labels that are compatible. It has been adopted by most of the recent deep models [6,15,16] for semantic image segmentation.

We redesign the unary term and the pairwise term with CNN incorporating depth maps.

Unary Term. Conventional unary terms construct on pixel' color features. It means that if a pixel is assigned a determinate label, the probability calculated from color space will support it. However, we find that the favour coming from color space is not enough. For instance, thinking about a person who wears a cloth with different color compared to his skin, segmentation based on probability will separate body into several different parts according to color information.

That semantic segmentation is wrong obviously. From depth view, the depth of hole body is almost same. So it is not difficult to infer that semantic segmentation based on probability coming from depth will be accurate. We design our unary term as

$$\Phi(y_p^l) = \max\{-\ln(P(y_p^l = 1|I)), -\ln(P(y_p^l = 1|D))\} \tag{4}$$

where D is depth maps. We model this unary term by FCN with changing some layers. It will be detailed in follow-up implementation part.

Pairwise Term. As described in [13], The pairwise term (also can be called as smoothness term) can be formulated as

$$\Psi(y_p^{l_1}, y_q^{l_2}) = \mu(l_1, l_2)d(p, q) \tag{5}$$

where the first term learns the penalty of global co-occurrence between any pair of labels, e.g. the output value of $\mu(l_1, l_2)$ is large if l_1 and l_2 should not coexist, while the second term calculates the distance between pixels. Similarly, we define our pairwise term as

$$\Psi(y_p^{l_1}, y_q^{l_2}) = \sum_{k=1}^{K} \lambda_k \mu_k(p, l_1, d_1, q, l_2, d_2) \sum_{\forall n \in N_q} d(q, n)p(y_q^{l_2} = 1|I, D) \tag{6}$$

where $\mu_k(p, l_1, d_1, q, l_2, d_2)$ learns the penalty of global co-existence between pairs from position, color and depth information, penalizing label assignment in a local region. It outputs labeling cost between (p, l_1) and (q, l_2) with respect to their relative positions and relative depth. For instance, the probability in which a person is under a table is so low that it can be neglected from position and color view. However if the table and person are not in the same depth level, from the two-dimensionality, the person is likely to be under the table. Let l_1 and l_2 represent table and person, the learned penalties of positions q that are at the bottom of center p should be large if the depth difference between p and q is samll. In the same situation, if the depth difference is large then the learned penalties should be small. $d(q, n)$ is the distance between $p-th$ and $n-th$ pixels. It can be formulated as

$$d(q, n) = \omega_1 \|I_q - I_n\|_2 + \omega_2 \|D_q - D_n\|_2 + \omega_3 \|(x_q, y_q) - (x_n, y_n)\|_2 \tag{7}$$

where I_q and I_n is the RGB value of $p - th$ and $n - th$ pixels. D_q and D_n is the depth value of $p - th$ and $n - th$ pixels, respectively. (x_q, y_q) and (x_n, y_n) are the coordinates of pixels' positions. It basically models a triple penalty, which involves pixels p, q and q' neighbors, implying that if (p, l_1) and (q, l_2) are compatible in color and depth, then (p, l_1) should be also compatible with q' nearby pixels (n, l_2). An intuitive illustration is given in Fig. 2(a). For overview, our energy function can be settled as

$$E(y) = \Phi(y_p^l) + \sum_{k=1}^{K} \lambda_k \mu_k(p, l_1, d_1, q, l_2, d_2) \sum_{\forall n \in N_q} d(q, n)p(y_q^{l_2} = 1|I, D) \tag{8}$$

Minimize difference between y and labels is equal to learning the parameters of CNN and the penalty of global co-existence between pairs which minimize the energy function. We need to infer Eq. (8) by the MF algorithm [17]. For detailed inference please refer to [13]. Finally we have

$$\alpha_p^{l_1} \propto \exp\{-\Phi_p^{l_1} - \sum_{k=1}^{K} \lambda_k \sum_{\forall l_2 \in L} \sum_{\forall q \in N_P} \mu_k(p, l_1, d_1, q, l_2, d_2) \sum_{\forall n \in N_q} d(q, n) \alpha_q^{l_2} \alpha_n^{l_2}$$

$$(9)$$

where $\alpha_p^{l_1}$ indicates the predicted probability of assigning lable l_1 to the $p - th$ pixel.

2.2 Implementation

We initialize our network based on the FCN. Tables 1 and 2 give a brief illustration on the structure of it. The "*layer*" means the name of the layer. The "*input*" means the input of the layer and the "*output*" indicates the output of the layer. "$t - p$" is short for "triple-penalty" layer. "$g - c$" is short for "global-coexistence" layer. "pd" is short for prediction. We put the input to the layer and use a filter whose size, stride, and channel are shown in "$filter - stride$" and "*channel*" respectively. The output of the filter is activated by "*activation*" and then is pooled by "*pooling*".

Table 1. Network structure 2-1

	1	2	3	4	5	6	7	8	9
Layer	2-conv	2-conv	3-conv	3-conv	3-conv	fc6	fc7	fc8	deconv1
Filter-stride	3-1	3-1	3-1	3-1	3-1	7-1	1-1	1-1	4-2
Channel	64	128	256	512	512	4096	4096	21	21
Activation	Relu	Relu	Relu	Relu	Relu	Relu	Relu	Relu	–
Size	384	192	96	48	24	12	12	12	24
Input	x0	x1	x2	x3	x4	x5	x6	x7	x8
Pooling	Max	Max	Max	Max	Max	–	–	–	–
Output	x1	x2	x3	x4	x5	x6	x7	x8	x9

We feed the images and depth maps to the CNN which is the "$x0$" in Table. The size of it is $384 \times 384 \times 4$. Then we upsample the output of the forth pooling layer with a de-convolutional layer. The output of this layer is summed with the output of the fully convolutional layer. The sum is regarded as one of the input to the next sum layer. Another input of this sum layer is the upsample result of the fifth pooling layer which is obtained by using de-convolutional layer again. Then we can get a dense prediction based on both color information and depth information. We regard this dense prediction as $p(y_q^{l_1} = 1|I, D)$ and calculate our unary term as Eq. (4).

Table 2. Network structure 2-2

	10	11	12	13	14	15	16	17	18
Layer	Skip1	Sum1	Deconv2	Skip2	Sum2	Deconv3	t-p	g-c	Objective
Filter-stride	1-1	–	4-2	1-1	–	16-8	–	9-1	–
Channel	21	21	21	21	21	21	–	21	–
Activation	–	–	–	–	–	Sigmoid	Linear	Linear	Loss
Size	24	24	48	48	48	384	–	384	–
Input	x4	(x9,x10)	x11	x3	(x12,x13)	x14	pd	x15	(x16,label)
Pooling	–	–	–	–	–	–	–	–	–
Output	x10	x11	x12	x13	x14	pd	x15	x16	objective

As mentioned in Eq. (9), we initialize $\alpha_p^{l_1}$ as $p(y_q^{l_1} = 1|I, D)$. Then $d(q, n)$ can be regarded as a "fixed" filter. "Fixed" means that the parameters in the filter are not learned from the training of the CNN. They are calculated from Eq. (7). Recall that each output feature map of "sum2" indicates a probabilistic label map of a specific object appearing in the image and depth maps. As a result, Eq. (9) suggests that the probability of object l_2 presented at position q is updated by weighted averaging over the probabilities at its nearby positions. In fact, the output of the "trip-penalty" is $\sum_{\forall n \in N_q} d(q, n)\alpha_q^{l_2}\alpha_n^{l_2}$. An intuitive illustration is given in Fig. 2(b).

As shown in Fig. 2(c). In order to learning the penalty of global co-existence between pairs from position, color as well as depth information, We learn five different filters of size $5 \times 5 \times 21$ for each feature channel and pooling the minimal value of 105 channels. In other words, "global-coexistence" learns a filter for each category to penalize the probabilistic label maps of "trip-penalty", corresponding to the second term in Eq. (9). The pooling operator get the minimal penalty for co-existence.

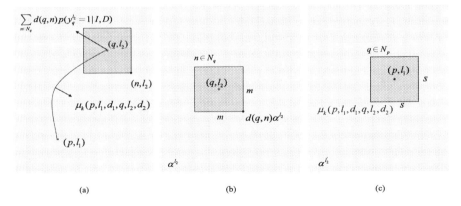

(a)   (b)   (c)

Fig. 2. (a) Illustration of pairwise term. (b) Explanation of triple-penalty.(c) Explanation of global-coexistence.

The last layer is a combination of the unary term and pairwise term. We use the $soft-max-log$ loss to minimize the difference between the final probabilistic prediction and ground-truth.

3 Experiments

We train our model on the PASCAL VOC 2011 segmentation challenge training set and BSDS500 additionally. There are totally 11287 images for training and 736 images for validation. The inputs of the network are shown in Fig. 3. All models are trained and tested with matconvnet [18] on a single NVIDIA GeForce GTX Titan Black.

The fundamental operation to learn the network is back propagation. We train it by stochastic gradient descent computing the derivative of the training loss with respect to the network parameters. We initialize the parameters of the front twelve layers from the parameters of FCN. The parameters of the rest layers are initialized randomly within 0 to 1. Firstly, we fine tune the net without the "triple-penalty" layers and "global-coexistence" layers. In other words, we just get the probabilistic prediction from unary term. The training results and the validation results are shown in Fig. 4(a). Then, we add the "triple-penalty" layers and "global-coexistence" layers to the net. As the parameters of the "triple-penalty" layers is not learned from the training, the learning rate of this layers is set to zero and the gradient do not be back propagated. The training results and the validation results are shown in Fig. 4(b).

Fig. 5 shows improvements in the fine structure of the outputs. In the relatively simple scene, our architecture is able to segment semantic object exactly. We determine the general outline of the objects and the correct labels of the

Fig. 3. Inputs of the CNN. The first row is the images. The second row is depth maps estimated from corresponding images.

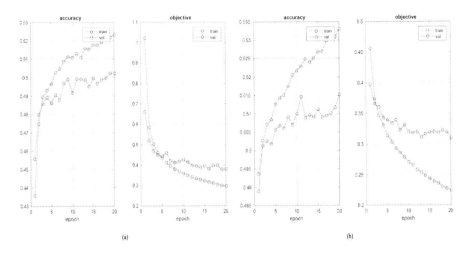

Fig. 4. (a) The training curve of CNN with unary term. (b) The training curve of *CNN* with unary term and pairwise term.

objects even in the failure. Notice the forth column fine structures are recovered. It is robustness to the situation which the color of objects is similar to the background (seventh row). We are able to separate closely interacting objects in the eighth row. The ninth row shows a failure case: the net do not see the correct labels of the semantic object and the complete boundary.

For comparing with other method's results, we report three metrics from common semantic segmentation and scene parsing evaluations that are variations on pixel accuracy and region intersection over union (IU). They are pixels

Table 3. Three quantitative indicators: On the PASCAL VOC 2012, FCN-8s and DPN [13] segment images without depth information. For the work of Gupta et al. please refer to [19]. On the NYUDv2, all of the shown methods segment images with depth information.

Dataset	Method	Pixel accuracy	Mean accuracy	Mean IU
PASCAL VOC 2012	FCN-8s	–	–	62.20
	DPN	–	–	**77.50**
	Our method	82.70	76.5	66.5
NYUDv2	Gupta et al.	60.30	–	28.60
	FCN-32s RGB	60.00	42.20	29.20
	FCN-32s RGBD	61.50	42.40	30.50
	FCN-32s HHA	57.10	35.20	24.20
	FCN-32s RGB-HHA	64.30	44.90	32.80
	FCN-16s RGB-HHA	65.40	**46.10**	49.50
	Our method	**67.50**	45.90	**49.70**

Fig. 5. (a) Images. (b) Depth maps. (c) The results from unary term. (d) Our results. (e) The ground truth.

accuracy, mean accuracy and mean IU respectively. For computation please refer to [12]. The measurement results of three quantitative indicators on two different datasets demonstrate that our nets produce competitive performance, as shown in Table 3. The proposed method can obtain competitive results.

4   Conclusion

We propose a novel framework where depth maps and images are fed into the CNN. CNN is used to model MRF. The output of CNN is obtained through deconvolutional layers instead of using bilinear upsample simply. Then we regard the dense prediction as the unary term of MRF. We construct the energy function of MRF and design the pairwise term carefully using depth information. The design of the depth supporting pairwise term is then implement in the form of net layers. The experiments show that our depth supporting framework is effective for semantic segmentation. We hope to generalize our framework to more varieties of computer vision research.

Acknowledgement. This work is supported by National Natural Science Foundation of China No. 61472393.

References

1. Thoma, M.: A survey of semantic segmentation. arXiv preprint arXiv:1602.06541 (2016)
2. Zhang, Z.: Microsoft kinect sensor and its effect. MultiMedia IEEE **19**(2), 4–10 (2012)
3. Silberman, N., Fergus, R.: Indoor scene segmentation using a structured light sensor. In: 2011 IEEE International Conference Computer Vision Workshops (ICCV Workshops), pp. 601–608. IEEE, November 2011
4. Zhang, H., Xiao, J., Quan, L., Zhang, C., Wang, L., Yang, R.: Semantic segmentation of urban scenes using dense depth maps. In: Daniilidis, K., Maragos, P., Paragios, N. (eds.) ECCV 2010, Part IV. LNCS, vol. 6314, pp. 708–721. Springer, Heidelberg (2010)
5. Bulo, S., Kontschieder, P.: Neural decision forests for semantic image labelling. In: Proceedings of the IEEE Conference on Computer Vision and Pattern Recognition, pp. 81–88 (2014)
6. Chen, L.C., Papandreou, G., Kokkinos, I., Murphy, K., Yuille, A.L.: Semantic image segmentation with deep convolutional nets and fully connected CRFs. arXiv preprint arXiv:1412.7062 (2014)
7. Papandreou, G., Chen, L.C., Murphy, K.P., Yuille, A.L.: Weakly-and semi-supervised learning of a deep convolutional network for semantic image segmentation. In: Proceedings of the IEEE International Conference on Computer Vision, pp. 1742–1750 (2015)
8. Gupta, A., Hebert, M., Kanade, T., Blei, D.M.: Estimating spatial layout of rooms using volumetric reasoning about objects and surfaces. In: Advances in Neural Information Processing Systems, pp. 1288–1296 (2010)

9. Gupta, A., Efros, A.A., Hebert, M.: Blocks world revisited: image understanding using qualitative geometry and mechanics. In: Daniilidis, K., Maragos, P., Paragios, N. (eds.) ECCV 2010, Part IV. LNCS, vol. 6314, pp. 482–496. Springer, Heidelberg (2010)

10. Ladicky, L., Shi, J., Pollefeys, M.: Pulling things out of perspective. In: Proceedings of the IEEE Conference on Computer Vision and Pattern Recognition, pp. 89–96 (2014)

11. Saxena, A., Chung, S.H., Ng, A.Y.: Learning depth from single monocular images. In: Advances in Neural Information Processing Systems, pp. 1161–1168 (2005)

12. Long, J., Shelhamer, E., Darrell, T.: Fully convolutional networks for semantic segmentation. In: Proceedings of the IEEE Conference on Computer Vision and Pattern Recognition, pp. 3431–3440 (2015)

13. Liu, Z., Li, X., Luo, P., Loy, C.C., Tang, X.: Semantic image segmentation via deep parsing network. In: Proceedings of the IEEE International Conference on Computer Vision, pp. 1377–1385 (2015)

14. Freeman, W.T., Pasztor, E.C., Carmichael, O.T.: Learning low-level vision. Int. J. Comput. Vis. **40**(1), 25–47 (2000)

15. Zheng, S., Jayasumana, S., Romera-Paredes, B., Vineet, V., Su, Z., Du, D., Torr, P.H.: Conditional random fields as recurrent neural networks. In: Proceedings of the IEEE International Conference on Computer Vision, pp. 1529–1537 (2015)

16. Schwing, A.G., Urtasun, R.: Fully connected deep structured networks. arXiv preprint arXiv:1503.02351 (2015)

17. Opper, M., Winther, O.: From naive mean field theory to the TAP equations (2001)

18. Vedaldi, A., Lenc, K.: MatConvNet: convolutional neural networks for matlab. In: Proceedings of the 23rd Annual ACM Conference on Multimedia Conference, pp. 689–692. ACM, October 2015

19. Gupta, S., Girshick, R., Arbeláez, P., Malik, J.: Learning rich features from RGB-D images for object detection and segmentation. In: Fleet, D., Pajdla, T., Schiele, B., Tuytelaars, T. (eds.) ECCV 2014, Part VII. LNCS, vol. 8695, pp. 345–360. Springer, Heidelberg (2014)

Omega-Shape Feature Learning for Robust Human Detection

Pengfei Liu, Xue Zhou$^{(\boxtimes)}$, and Shibin Cai

School of Automation Engineering,
University of Electronic Science and Technology of China, Chengdu, China
zhouxue@uestc.edu.cn

Abstract. For most top view surveillance scenes, due to having little pose variations and being robust to partial occlusion, people's head-shoulder Omega-like shapes are proven to be good cues for human detection. In this paper, we focus on learning Omega-shape features with improved discriminative ability in human detection. Orthogonal non-negative matrix factorization (ONMF) is introduced to model the local semantic parts of Omega-like shapes, which is much more robust to noise corruption and local occlusion. The properties only allowing additive, not subtractive combinations of ONMF can well suppress some background clutter. Furthermore, we introduce metric learning into SVM decision framework where object/nonobject classification is performed within a learned feasible metric space. Experimental results on a number of challenging datasets demonstrate the effectiveness and robustness of the proposed human detection method.

Keywords: Omega-shape · Human detection · Non-negative Matrix Factorization (NMF) · Metric learning

1 Introduction

As a fundamental and challenging problem in computer vision, human detection has a wide range of applications, including visual surveillance, human-computer interaction, self-driving, crowd counting, etc. However, due to variations in illumination, partial occlusions, articulated structure, complexity of the backgrounds, a lot of problems still remain in human detection.

Following the general detection framework which is composed of feature extraction and classifier design, most human detection methods mainly concentrate on improving the performance of the above two components. Oren et al. [1] first introduce the idea of machine learning and propose to adopt wavelet combined with Support Vector Machine (SVM) for human detection. VJ model [2] combines haar wavelet feature with a cascade architecture detector to improve the detection efficiency. Other features, like Local binary Pattern (LBP) [3], edgelet [4] and Histograms of Oriented Gradients (HOG) [5] are also chosen as feature descriptors to model the global property of human body. Recently, a

© Springer Nature Singapore Pte Ltd. 2016
T. Tan et al. (Eds.): CCPR 2016, Part I, CCIS 662, pp. 290–303, 2016.
DOI: 10.1007/978-981-10-3002-4_25

real-time human detector, named C^4 [6,7] is proposed, which adopts CENTRIST (CENsus TRansform hISTogram) visual feature [8] to capture the global contour information of human body. The aforementioned methods have something in common, they all capture global features for detection. Although these features have been verified with good performance in some cases, when confronted with large pose variations and partial occlusion, the global features may lead to unsuccessful detections.

Subsequently, part-based feature models are getting more and more popular. Because they are more imperative for handling partial object/inter-occlusions and are flexible in modeling shape articulations in human detection. A pose-invariant descriptor for human detection is proposed through constructing a part-template tree model [9]. Felzenszwalb et al. [10,11] propose a human detection method based on deformable part models (DPM) which are capable of coping with object pose changes. Different parts of object can be represented by several higher resolution part templates in conjunction with their spatial layout. The above part-template based models are often used for full-body human detection, which easily suffers from occlusions among individuals and scenes in which people are not necessarily standing [12]. Hence, different from the full-body human detection, lots of researchers instead focus on the upper part of human body. Especially for top view surveillance scenes, head-shoulder might be the last part being occluded and is not as flexible as entire human body. Omega-like shape has been proved to be a salient feature of head-shoulder [12–16]. Li et al. [12,13] propose an effective head-shoulder detection method based on boosting local HOG features. Experimental results from [13] indicate HOG feature performs better than Haar feature [2] and SIFT descriptor. Another edge-based feature similar to HOG, named Oriented Integration of Gradients (OIG), is introduced to describe subparts of human head-shoulder in [14]. Julio et al. [15,16] propose a graph-based segmentation model to estimate the head-shoulder contour.

Inspired by the above analysis, in our method we also adopt human head-shoulder Omega-like contour not the whole human body as shape cue for human detection. A matrix decomposition technique is introduced to obtain the meaningful local curvelets of Omega-like shape. Different from the above introduced methods lacking of the further feature learning and analysis, based on the HOG feature, we further perform the task of part-based Omega shape modeling by constructing a part-based shape model that captures the intrinsic semantic local shape variations. The problem of part-based Omega-shape feature learning is converted to that of Nonnegative Matrix Factorization (NMF)-based local feature learning. Especially for NMF method, only additive, not subtractive combinations are allowed and some background clutter can be well suppressed or downweighted.

Moreover, under different view angles, e.g., front view and side view, the intra-class variations of human head-shoulder contours still exist. In order to effectively measure the similarities of omega-shape samples with multimodal distribution, we introduce distance metric learning into support vector machine (SVM) classification. With learned distance metric, the local neighborhood property that examples within the same class are preserved while examples from different classes are separated by a large margin. Both intra-class compactness and inter-class separability are improved.

Therefore, in this paper, we focus on learning the Omega-like shape features for human detection. Specifically, the contributions of this work lie in the following two aspects:

1. We convert the local Omega shape feature learning problem to the part-based semantic shape representation problem. Orthogonal Nonnegative Matrix Factorization (ONMF) is introduced to encode the shape dictionary with each word representing the semantic part of Omega-like shape and reduce the feature dimensionality simultaneously. The learned features are robust to partial occlusion and background clutter.
2. In order to cope with Omega shape intra-class multimodal problem, we introduce distance metric learning into SVM classification where object/nonobject classification is performed in a learned Mahalanobis distance metric space. The intra-class compactness and inter-class separability are guaranteed.

2   The Proposed Human Detection Method

We give the overview of our method at first. During the off-line training stage, positive training samples are used to learn the part-based Omega shape bases by ONMF and then negative training samples are included to learn a Mahalanobis distance metric with neighborhood constraint. For the on-line detection stage, image patches with different scales are generated firstly. Secondly, after HOG feature extraction, part-based features are represented based on the learned local shape bases. Thirdly, SVM classifier is used to evaluate each input feature within a feasible metric space. Figure 1 is the framework of our proposed human detection method.

In the following subsections, we firstly introduce the construction of ONMF-based Omega shape feature descriptor, then describe SVM classifier training with learned distance metric, and finally the on-line detection process is presented.

2.1   ONMF-based Omega Shape Feature Learning

HOG feature [5] has been verified to be effective for capturing the omage-shape contour information [13]. In our method, we also adopt HOG to describe the

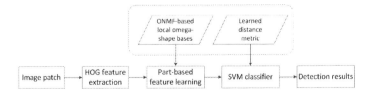

Fig. 1. Illustration of our proposed Omega shape feature learning-based human detection method

global Omega-shape feature. However, due to its cell-based calculation mechanism, it is inevitable for HOG to include some background noise and has a high dimensionality with some redundant information. Moreover, HOG features are weak in local Omega-shape variations modeling.

Therefore, to alleviate the above problems, we need to do further feature learning and dimension reduction based on HOG features. We present a simple but efficient Nonnegative Matrix Factorization (NMF)-based Omega shape feature learning method. NMF is a powerful tool of learning local features [18]. In this model, the construction of shape variations is viewed as a shape dictionary learning problem with each shape word encoding a local semantic part of shapes. Generally, the NMF problem is associated with the following constrained least-square optimization problem:

$$\min_{W,H} \|V - WH\|_2 \ \text{ s.t. } W \geq 0, H \geq 0 \qquad (1)$$

in our method, $V = (V_{x_1}, V_{x_2}, ... V_{x_n})$ is a nonnegative $m \times n$ Omega shape vector matrix, in which each column $\{V_{x_i}\}_{i=1}^n$ represents a HOG feature vector corresponding to a positive training sample, and W is an $m \times p$ matrix with each column representing a basis image. Each column of $H(p \times n)$ consists of the coefficients by which a sample is represented with a linear combination of basis images.

After W is learnt according to Eq. (1), given an input test sample V_{x_t}, we adopt its recovery coefficients as the corresponding NMF-based shape descriptor. By solving the following nonnegative least square (NNLS) problem, shape descriptor f_{x_t} can be obtained:

$$\min_{\mathbf{f}_{x_t}} \|V_{x_t} - W \cdot \mathbf{f}_{x_t}\|_2 \ \ \text{ s.t. } \mathbf{f}_{x_t} \geq 0 \qquad (2)$$

In order to verify the advantage of using NMF method to represent the local Omega features, i.e., the ability to selectively encode only the foreground of regions of interest, hence effectively rejecting unwanted background clutter or noise. We use NMF bases only additively to reconstruct test samples with clutter or partial occlusion according to their NNLS coefficients, i.e., $W \cdot \mathbf{f}_{x_t}$. Figure 2(a) shows the illustration of decomposed Omega shape basis images. Our proposed NMF-based Omega shape model is capable of representing variations of shapes with a set of basis images denoting the meaningful local curvelet features (shown in the 5×10 montages, highlighted by darker grey).

Some recovery results of input samples with background clutter and partial occlusion are shown in Fig. 2(b) and (c). For a better visualization, before applying NMF, we perform edge extraction on the gray images. From the recovery results, we can find the recovered silhouette images preserve the Omega contour information. The background clutter (Fig. 2(b)) can be clearly suppressed and partial occlusions (Fig. 2(c)) are well handled.

NMF-based shape descriptor f_{x_t} is often obtained by recursively solving a NNLS problem in Eq. (2), which is time-consuming. Another solution is to adopt a generalized NNLS method: $H = (W^T W)^{-1} W^T V$, which is not effective for

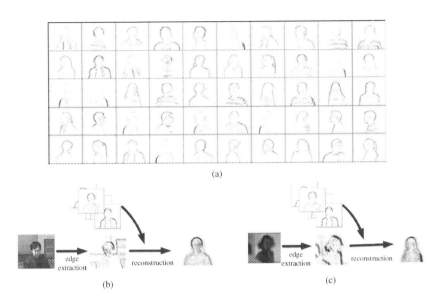

(a)

(b)   (c)

Fig. 2. Illustration of obtained Omega shape basis images and the recovery results of input samples with background clutter and partial occlusion. (a) Several examples of Omega shape basis images based on NMF; (b) reconstruction result of input sample with background clutter; (c) reconstruction result of input sample with partial occlusion.

high dimensional data. Thus, lots of improved NMF-based methods are proposed. In our method, we adopt an orthogonal NMF (ONMF) algorithm which constrains the basis matrix should be an orthogonal matrix, Eq. (1) can be re-written as:

$$\min_{W,H} \|V - WH\|_2 \tag{3}$$
$$\text{s.t. } W, H \geq 0, W^T W = I$$

The above optimization problem can be effectively solved based on the canonical metric of the Stiefel manifolds [19]. The iterative equations for W and H are directly given by [19]:

$$H = H \odot \frac{[W^T V]}{[W^T W H]} \tag{4}$$

$$W = W \odot \frac{[V H^T]}{[W H V W]} \tag{5}$$

where $\odot$ denotes the Hadamard product and $\frac{[\cdot]}{[\cdot]}$ represents an element-wise division. Similar to [20], a hierarchies of ONMF training mechanism is adopted in our method, allowing to represent parts of Omega shapes on different scales for a more robust classification. In the hierarchical structure, the value for W_1 and H_1 in the first layer are randomly initialized where the number of bases is set to be b_1. The initial value for the basis matrix of the next layer is obtained by copying s times of W_1, i.e., $\tilde{W}_2 = [W_1, ..., W_1]$. Using $\tilde{W}_2$ and a randomly initialized encoding $\tilde{H}_2$ both matrices are updated according to Eqs. (4) and (5), obtaining a basis W_2 and an encoding H_2. The hierarchical tree is growing continuously until the number of the basis matrix satisfies the convergence condition. The final representation of W is estimated by: $W = [W_1, ..., W_L]$, wherein L denotes the number of layers in the ONMF hierarchies. Overall, with respect to a test sample V_{x_t}, the shape descriptor f_{x_t} can be obtained by a linear transformation and no iterative procedure is required, i.e., $f_{x_t} = W^T V_{x_t}$.

2.2   Improved SVM Training with Learned Distance Metric

After the local Omega-shape descriptors are obtained, the next step is to train a SVM classifier using the label information. Due to its strong performance, SVM [21] has been widely adopted in classification and pattern recognition tasks. For a better separation, SVM usually projects the input data into a higher dimensional feature space through a kernel function, most of which measure the similarity between pairs of features using their Euclidean inner product or Euclidean distance. However, Euclidean measure treats different feature dimensions equally and ignores the correlation information between feature dimensions. As a result, the Euclidean distance measure is incapable of well reflecting the intrinsic affinity relationships between samples [22].

To address the above issue, distance metric learning has emerged as a useful tool in recent years [23–25]. Motivated by this observation, we introduce distance metric learning into the SVM classifier training and therefore adopt such a metric learning method called large margin nearest neighbor (LMNN) [25]. Different from the other distance metric learning methods which require the samples with the same labels are close to each other, LMNN aims to guarantee only the k nearest neighbors always belong to the same class while samples from different classes are separated by a large margin. Thus, LMNN is capable of effectively handling the metric learning problem with the training samples of multi-modal distributions. There exists intra-class variations about Omega-shapes, which has been shown in Fig. 3(a) with each row corresponding to a kind of Omega-shape. Due to different view angles (different rows in Fig. 3(a)), even if all of them are from Omega-shape class, there are obvious differences between different shapes generated from different view angles. Therefore, the goal of distance metric learning is to require in the learned feature space shapes from the same row move closer, and shapes from different rows move away, which is quite reasonable in real applications. This objective coincides with the key idea of learning a distance metric by LMNN, which is illustrated in Fig. 3(b). From the figure we can find the distance metric is optimized so that: (1) its $k = 3$ nearest neighbors lie within a smaller radius after training; (2) differently labeled inputs lie outside this smaller radius by some finite margin.

After metric learning operated by LMNN [25], we obtain a discriminative distance metric denoted as $\mathbf{M} = \mathbf{L}^T \mathbf{L}$. As a result, with respect to two feature vectors, we have a Mahalanobis distance measure:

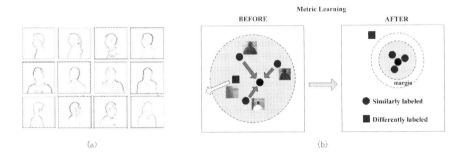

(a)   (b)

Fig. 3. Illustration of Omega shape intra-class variations and schematic explanation of LMNN [25]. (a) The learned ONMF-based Omega shape basis images are shown in the 3×4 montages with different rows corresponding to different view angles; (b) schematic illustration of one input's neighborhood before training (left) versus after training (right) in LMNN. The example images are placed next to their corresponding nodes for better illustration.

$$D_M(f_{x_i}, f_{x_j}) = \|\mathbf{L}(f_{x_i} - f_{x_j})\|_2^2 = (f_{x_i} - f_{x_j})^T \mathbf{M}(f_{x_i} - f_{x_j}) \tag{6}$$

where f_{x_i} and f_{x_j} denote the ONMF-based Omega shape descriptors.

Assume a set of training samples $\{f_{x_i}\}_{i=1}^n$ and their labels $y_i \in \{-1, 1\}$ are given. Training a SVM classifier is to find an optimal hyperplane that maximizes the margin between two classes. The corresponding Lagrangian dual problem is formulated by [21]:

$$\max_{\alpha} \sum_{i=1}^n \alpha_i - \frac{1}{2} \sum_{i,j=1}^n \alpha_i \alpha_j y_i y_j (\varphi(f_{x_i}) \cdot \varphi(f_{x_j})) \tag{7}$$

$$\text{s.t.} \sum_{i=1}^n y_i \alpha_i = 0, \qquad 0 \le \alpha_i \le C, i = 1, ..., n$$

where φ is a kernel feature mapping to a higher dimensional feature space, C is the regularization parameter and $(\cdot)$ denotes an inner product operator.

Without explicitly computing the feature mapping φ, the kernel function $k(f_{x_i}, f_{x_j}) = \varphi(f_{x_i}) \cdot \varphi(f_{x_j})$ is introduced to handle nonlinearly separable cases. One of the most common used kernel is the exponential radial basis function (RBF) kernel:

$$k_{rbf}(f_{x_i}, f_{x_j}) = \exp(-\frac{D^2(f_{x_i}, f_{x_j})}{\sigma^2}) \tag{8}$$

where σ is the bandwidth parameter of RBF kernel and $D(\cdot)$ is a distance measure.

A standard distance measure is the Euclidean distance such that $D_E(f_{x_i}, f_{x_j}) = (f_{x_i} - f_{x_j})^T (f_{x_i} - f_{x_j})$. In our method, we make use of the statistical regularities estimated from the training data and introduce a learned Mahalanobis metric. In this way, we directly replace $D(\cdot)$ with D_M (referred to Eq. (6)) in the kernel function in Eq. (8). The optimization problem in Eq. (7) is solved using quadratic programming.

2.3   On-Line Human Detection

For an input image, we adopt a variable-scale searching mechanism to obtain the test image patches, and ONMF-based Omega shape descriptor on each image patch is extracted. Then SVM classification is applied within a learned distance metric space. Detected candidate results are fused to generate the final detection results. Specifically, the detection procedure is given as follows:

step 1 **Generating Test Image Patches**.
In our method, we fix the size of sliding window, and change the scale of input image from 0.8 to 1.2 with interval 0.1. Test image patches with different scales are generated in the first step.

step 2 **Extracting ONMF-based Omega Shape Descriptors**.
With respect to each test image patch x_t, HOG feature vector V_{x_t} is obtained firstly. Then a linear transformation is employed to get a local part-based shape descriptor with reduced dimensionality, i.e., $f_{x_t} = W^T V_{x_t}$, where W is the trained ONMF-based shape basis matrix.

step 3 **Applying SVM Classification**.
The feature descriptor f_{x_t} obtained in the last step is classified according to the output value of SVM classifier in the learned Mahalanobis distance metric space: $y(f_{x_t}) = \text{sign}(\sum_{i=1}^{n} \alpha_i y_i k(f_{x_i}, f_{x_t}) + b)$, where $y(\cdot)$ is the decision function, α_i is the supporting vector coefficients, $k(\cdot)$ is the kernel function measured in the learned metric space and b is a constant item.

step 4 **Fusing Detection Results**.
Due to different scales being detected, there may exist overlap, inclusion and intersection among detected bounding boxes for one object. Therefore, the subsequent fusion process is needed. Obtained candidates from step 3 are grouped based on their location and size similarities. The bounding boxes belonging to the same class are fused to one and the others not satisfying the threshold condition are discarded.

3   Experimental Results

3.1   Experimental Setups

In order to evaluate the proposed human detection method, we have conducted a set of experiments on several challenging clips from CAVIAR dataset [26]. These clips include people walking along corridor browsing, going inside and coming out of stores in a shopping center and contain 1500 frames on average. Some challenging factors are considered: partial occlusion, background disturbance, pose variation, etc. Some sample images from CAVIAR dataset are shown in Fig. 4. Silhouettes of 600 human head-shoulders under different view angles with clean background are used to train ONMF-based Omega shape basis matrix. For SVM classifier training, we randomly choose one clip of CAVIAR dataset to generate training samples which are composed of 6085 images. In the 3085 positive images, human head-shoulders are automatically cropped based on the

Fig. 4. Some sample images from CAVIAR dataset.

ground truth benchmark and other 3000 negative samples are human free. All training samples are divided into 10 groups for cross validation. During each training, 9 groups of samples are used as training set while the left one as testing set.

The proposed detection algorithm is implemented in MATLAB on a workstation with Intel Core 3.6 GHz CPU and 4 GB RAM. The parameter configuration of our experiments is listed as follows. We extract 2592D HOG feature. The size of sliding window is 54×60. The number of bases (b_1) in the first layer of ONMF hierarchies, parameter s, level L parameters are set as 10, 2 and 4, respectively, leading to a 150D ONMF Omega shape descriptor. Three nearest neighbors are considered in the LMNN distance metric learning algorithm. A linear kernel function is adopted for SVM classifier. In order to demonstrate the effectiveness of our proposed method, we compare it with other representative

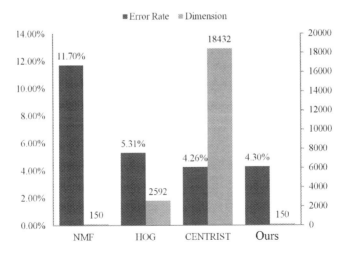

Fig. 5. Performance comparison of four feature extraction methods. The horizontal axis displays four different feature extraction methods, from left to right, namely, NMF, HOG, CENTRIST and Ours. The left and right vertical axes denote detection error rate and feature dimensions, respectively.

Our Method	HOG-SVM	ONMF150	C4	LMNN-R

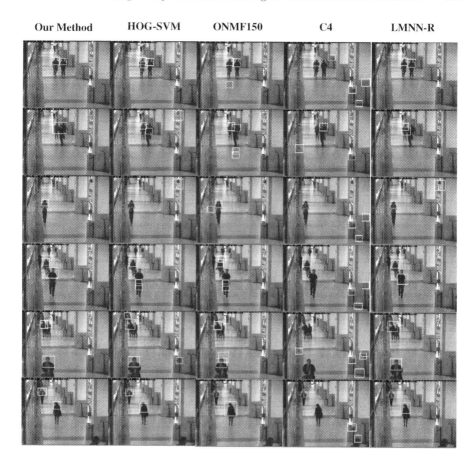

Fig. 6. Example detection results on CAVIAR dataset. The results by ours, HOG-SVM, ONMF150, C4 and LMNN-R are listed in column. Each row represents each case on CAVIAR dataset.

methods. They are referred to as ONMF150 [20], HOG-SVM [5], C4 [7] and LMNN-R [27]. To quantitatively evaluate these methods, we adopt FPPI (False Positive Per Image)-MR (Miss Rate) curve as the evaluation criterion.

3.2   Empirical Results

In this subsection, we present the effect of key components in our method as well as some comparisons with other state-of-art methods.

In order to demonstrate the effectiveness of our proposed ONMF-based Omega shape feature learning method, we compare it with other three feature extraction methods. They are referred to as NMF, HOG and CENTRIST [8]. Figure 5 shows the performance comparison among these four feature extraction methods. Cross-validation is adopted to obtain the average error rate (miss

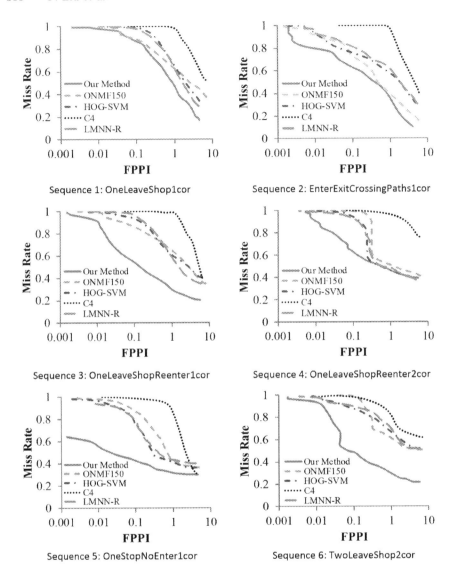

Fig. 7. Quantitative comparison of ONMF, HOG-SVM, C4, LMNN-R and our method on six typical sequences from CAVIAR dataset.

rate). Training set is randomly divided into five groups with four groups used for training and one left used for testing. A linear SVM model is selected as the classifier with four different kinds of input. From the results we can find although the error rate of our method is a little higher than CENTRIST method, however, CENTRIST gets a remarkable high dimensional feature which is not suitable for classification. Thus, taking error rate and feature dimensions these two factors into consideration simultaneously, our feature learning method outperforms the other feature learning methods.

We compare our proposed method with other representative methods including ONMF150 [20], HOG-SVM [5], C4 [7] and LMNN-R [27]. Figure 6 lists some example detection results on CAVIAR dataset. The detected human is represented by a white bounding box. From the results we can find with the other four methods they all have some miss-classifications. Even with the partial occlusion case, our method is capable of detecting people accurately. Figure 7 illustrates the quantitative comparison results of the five methods over six typical sequences from CAVIAR dataset. We observe that our detection method achieves the best detection performance with a lower FPPI and missing rate on most video sequences.

4 Conclusion

In this paper, we have proposed a human detection method based on part-based human Omega shape features. We have empirically proved that the HOG-ONMF based shape feature descriptor can capture the certain details which is robust to partial occlusion and background corruption. The improved SVM classifier is applied in a learned feasible metric space, both intraclass compactness and interclass separability of training samples are improved. We compare our proposed method with four competing methods on six challenging sequences on CAVIAR dataset. Both qualitative and quantitative experimental results have verified the effectiveness and robustness of our method.

Acknowledgments. This work was supported by the National Natural Science Foundation of China under Grant NSFC 61472063.

References

1. Oren, M., Papageorgiou, C., Sinha, P., et al.: Pedestrian detection using wavelet templates. In: IEEE Conference on Computer Vision and Pattern Recognition, pp. 193–199 (1997)
2. Viola, P., Jones, M.J., Snow, D.: Detecting pedestrians using patterns of motion and appearance. In: International Conference on Computer Vision, pp. 734–741 (2003)
3. Mu, Y., Yan, S., Liu, Y., et al.: Discriminative local binary patterns for human detection in personal album. In: IEEE Conference on Computer Vision and Pattern Recognition, pp. 1–8 (2008)
4. Wu, B., Nevatia, R.: Detection of multiple, partially occluded humans in a single image by bayesian combination of edgelet part detectors. In: IEEE International Conference on Computer Vision, pp. 90–97 (2005)
5. Dalal, N., Triggs, B.: Histograms of oriented gradients for human detection. In: IEEE International Conference on Computer Vision and Pattern Recognition, pp. 886–893 (2005)
6. Wu, J.X., Geyer, C., Rehg, J.M.: Real-time human detection using contour cues. In: IEEE International Conference on Robotics and Automation, pp. 860–867 (2011)

7. Wu, J.X., Liu, N.N., Geyer, C., Rehg, J.M.: C4: a real-time object detection framework. IEEE Trans. Image Process. **22**(10), 4096–4107 (2013)

8. Wu, J.X., Rehg, J.M.: CENTRIST: a visual descriptor for scene categorization. IEEE Trans. Pattern Anal. Mach. Intell. **33**, 1489–1501 (2011)

9. Lin, Z., Davis, L.S.: A pose-invariant descriptor for human detection and segmentation. In: Forsyth, D., Torr, P., Zisserman, A. (eds.) ECCV 2008. LNCS, vol. 5305, pp. 423–436. Springer, Heidelberg (2008). doi:10.1007/978-3-540-88693-8_31

10. Felzenszwalb, P.F., McAllester, D., Ramanan, D.: A discriminatively trained, multiscale, deformable part model. In: IEEE International Conference on Computer Vision and Pattern Recognition, pp. 1–8 (2008)

11. Felzenszwalb, P.F., Girshick, R.B., McAllester, D., et al.: Object detection with discriminatively trained part-based models. IEEE Trans. Pattern Anal. Mach. Intell. **32**(9), 1627–1645 (2010)

12. Li, M., Zhang, Z., Huang, K., et al.: Rapid and robust human detection and tracking based on omega-shape features. In: IEEE International Conference on Image Processing, pp. 2545–2548 (2009)

13. Li, M., Zhang, Z., Huang, K., et al.: Estimating the number of people in crowded scenes by mid-based foreground segmentation and head-shoulder detection. In: International Conference on Pattern Recognition, pp. 1–4 (2008)

14. He, F., Li, Y.L., Wang, S.J., Ding, X.Q.: A novel hierarchical framework for human head-shoulder detection. In: International Congress on Image and Signal processing, pp. 1485–1489 (2011)

15. Julio, C.S., Jung, C.R., Soraia, R.M.: Head-shoulder human contour estimation in still images. In: International Conference on Image Processing, pp. 278–282 (2014)

16. Julio, C.S., Soraia, R.M.: Improved head-shoulder human contour estimation through clusters of learned shape models. In: SIGBRAPI Conference on Graphics, Patterns and Images, pp. 329–336 (2015)

17. Lowe, D.G.: Distinctive image features from scale-invariant keypoints. Int. J. Comput. Vis. **60**(2), 91–110 (2004)

18. Lee, D.D., Seung, H.S.: Learning the parts of objects by nonnegative matrix factorization. Nature **401**, 788–791 (1999)

19. Yoo, J., Choi, S.: Orthogonal nonnegative matrix factorization: multiplicative updates on stiefel manifolds. In: Fyfe, C., Kim, D., Lee, S.-Y., Yin, H. (eds.) IDEAL 2008. LNCS, vol. 5326, pp. 140–147. Springer, Heidelberg (2008). doi:10.1007/978-3-540-88906-9_18

20. Mauthner, T., Kluckner, S., Roth, P.M., Bischof, H.: Efficient object detection using orthogonal NMF descriptor hierarchies. In: Goesele, M., Roth, S., Kuijper, A., Schiele, B., Schindler, K. (eds.) DAGM 2010. LNCS, vol. 6376, pp. 212–221. Springer, Heidelberg (2010). doi:10.1007/978-3-642-15986-2_22

21. Vapnik, V.N.: The Nature of Statistical Learning Theory. Springer, Heidelberg (2000)

22. Liu, Y., Caselles, V.: Improved support vector machines with distance metric learning. In: Blanc-Talon, J., Kleihorst, R., Philips, W., Popescu, D., Scheunders, P. (eds.) ACIVS 2011. LNCS, vol. 6915, pp. 82–91. Springer, Heidelberg (2011). doi:10.1007/978-3-642-23687-7_8

23. Davis, J.V., Kulis, B., Jain, P., Sra, S., Dhillon, I.S.: Information-theoretic metric learning. In: Proceedings of the International Conference on Machine Learning, pp. 209–216 (2007)

24. Globerson, A., Roweis, S.: Metric learning by collapsing classes. In: Proceedings of Advances in Neural Information Processing Systems, pp. 451–458 (2005)

25. Weinberger, K.Q., Saul, L.K.: Distance metric learning for large margin nearest neighbor classification. J. Mach. Learn. Res. **10**, 207–244 (2009)
26. http://homepages.inf.ed.ac.uk/rbf/CAVIARDATA1/
27. Dikmen, M., Akbas, E., Huang, T.S., Ahuja, N.: Pedestrian recognition with a learned metric. In: Kimmel, R., Klette, R., Sugimoto, A. (eds.) ACCV 2010. LNCS, vol. 6495, pp. 501–512. Springer, Heidelberg (2011). doi:10.1007/978-3-642-19282-1_40

Video Object Detection and Segmentation Based on Proposal Boxes

Xiaodi Zhang, Zhiguo Cao$^{(\boxtimes)}$, Yang Xiao, and Furong Zhao

National Key Laboratory of Science and Technology on Multi-Spectral Information
Processing, School of Automation,
Huazhong University of Science and Technology, Wuhan, China
{xiaodizhang,zgcao,Yang_Xiao,zhaofr639}@hust.edu.cn

Abstract. In this paper, we propose a new method to detect and segment foreground object in video automatically. Given a video sequence, our method begins by generating proposal bounding boxes in each frame, according to both static and motion cues. The boxes are used to detect the primary object in the sequence. We measure each box with its likelihood of containing a foreground object, connect boxes in adjacent frames and calculate the similarity between them. A layered Directed Acyclic Graph is constructed to select object box in each frame. With the help of the object boxes, we model the motion and appearance of the object. Motion cues and appearance cues are combined into an energy minimization framework to obtain the coherent foreground object segmentation in the whole video. Our method reports comparable results with state-of-the-art works on challenging benchmark dataset.

Keywords: Video analysis · Video object segmentation · Object proposals

1 Introduction

Video object segmentation is one of the fundamental problems in video analysis. It aims at separating the foreground object from its background in a video sequence. This technique is beneficial to a variety of applications, such as video summarization, video retrieval and action recognition.

Humans have the talent to distinguish an object from its background in static images or videos. However, it is difficult for a computer to accomplish such task. Finding or segmenting objects in images is one of the core problems in the field of computer vision. Many exciting approaches have been explored for this type of task, such as figure-ground segmentation, object proposal techniques, saliency detections and object discovery. When we process video data, the task of object discovery and segmentation changes a lot since continuous frames provide motion information. Therefore, both appearance and motion cues should be considered to find out the primary object.

Our approach aims at video object segmentation, while we also give out category independent object detection results in the form of bounding boxes.

© Springer Nature Singapore Pte Ltd. 2016
T. Tan et al. (Eds.): CCPR 2016, Part I, CCIS 662, pp. 304–317, 2016.
DOI: 10.1007/978-981-10-3002-4_26

Similar to some methods [1–3], we utilize an object proposal method to obtain the initial locations of the object. However, the former methods need to pre-segment each frame and measure the likelihood of each segment to be an object. This process costs a lot of time (more than 1 min per frame). Different from these methods, we generate proposal bounding boxes instead of object-like segments, and it is much more efficient (about 1 s per frame). Both static and motion cues are under consideration for the boxes generation. We attempt to measure the proposal bounding boxes with two scores: objectness score and motion score. Objectness score reports the likelihood of containing an object, and motion score estimates the motion difference between the box and its surrounding area. Since the object always moves differently from its background, the box which has the object inside it oughts to own high objectness score and motion score. In most situations, object moves smoothly across frames in a video, its location and appearance vary slowly. Therefore, the proper object boxes in the consecutive frames are also coherent in location and size. We connect two boxes in consecutive frames and measure the similarity between them. A layered Directed Acyclic Graph (DAG) is constructed to formulate the bounding boxes in the whole sequence, and the problem of selecting boxes is transformed into finding out the path with highest score in the layered DAG. When the boxes are determined, the foreground object is detected. With the help of the selected boxes, we locate the object in each frame, and model the motion and appearance of the object according to the regions inside the boxes. The final segmentation is performed in an energy minimization framework.

The rest of this paper is organized as follows. Section 2 of this paper reviews the related researches on the task of video object segmentation. In Sect. 3, our approach is introduced and discussed in detail. The experimental results and analysis are reported in Sect. 4. The paper is concluded in Sect. 5.

2 Related Work

Lots of methods have been explored to fulfill the task of video object segmentation. Divided by the need for manual annotation, the methods are summarized as semi-automatic manner and fully automatic manner. Semi-automatic methods require manual annotations of object position in some key frames for initialization, while the latter scenario doesn't need any human intervention. Without any priori knowledge of the foreground object, the fully automatic methods have to firstly answer the questions of what and where the object is. Different strategies have been developed for the questions. Trajectory analysis and object proposal techniques are often used to discover the object in videos.

Semi-automatic Methods. Some semi-automatic methods [4–7] require annotation of precious object segments in key frames. The segments are propagated to other frames under the constrains of motion and appearance. Other semi-automatic methods [8] require object location in the first frame as initialization.

These methods track the object regions in the rest frames. Semi-automatic methods usually get better result than fully automatic methods as they obtain prior knowledge about object annotated by interaction. However, since labor cost is expensive, these methods are unsuitable for large-scale video data processing.

Trajectory Based Methods. The main characteristic of trajectory based methods [9–11] is that they analyze long term motion over several frames rather than forward/backward optical flow. These methods assume that trajectories of moving object are similar with each other and different from background. Brox et al. [9] defined a distance between trajectories as the maximum difference of their motion over time. Given the distance between trajectories, an affinity matrix was built for the whole sequence and the trajectories were clustered based on the matrix. Lezama et al. [11] combined local appearance and motion measurements with long range motion cues in the form of grouped point trajectories. Fragkiadaki et al. [10] proposed an embedding discontinuity detector for localizing object boundaries in trajectory spectral embeddings. Trajectory clustering was replaced by discontinuities detection. These methods suffered from some problems such as model selection of clustering and no-rigid/articulated motion.

Proposal Based Methods. Object proposals are regarded as regions or windows likely to be an object in an image. Object proposal techniques [1–3] try to find out an object in the image based on bottom-up segmentation. These techniques are beneficial for video object segmentation task as they segment frames in advance. One of the most important steps in these methods is to distinguish which proposals are the regions of the object. Proposal based methods generate hundreds of proposals and measure the probabilities to be a foreground object using appearance and motion cues. After that video object segmentation transforms into a proposal selection problem. Lee et al. [12] preformed spectral clustering on proposals to discover the reliable proposals in the key frames. The proposal cluster with highest average score was corresponding to the primary foreground object. These proposals were used to generate object segments in rest frames. Although this method outperformed some semi-automatic methods, the main drawback was that clustering abandoned temporal connections of the proposals and the object-regions were only obtained in some key frames. Zhang et al. [13] designed a layered Directed Acyclic Graph to solve proposal selection problem. The layered DAG considers temporal relationship of proposals in consecutive frames and the problem of proposals selection transformed to get the longest weighted path in the DAG. When the path was determined, the most suitable proposal was selected in each frame. Our method is inspired by their brilliant idea. Perazzi et al. [14] formulated proposals in a fully connected manner and trained a classifier to measure the proposal regions. Proposal based methods report state-of-art result in this task. They also handle no-rigid and articulated motion well since a prior segmentation is performed in each frame without the influence of motion. However they face the problem of high computational complexity for generating proposal regions. It costs several minutes to get hundreds of proposal segments in a standard scaled image. The unacceptable time cost limits practical value of these methods.

Apart from the aforementioned paradigms, several other formulations have been explored for this task. Papazoglou *et al.* [15] proposed a novel solution, which combined motion boundaries and point-in-polygon problem for efficient initial foreground estimation. The method reported comparable results to proposal based methods while being orders of magnitude faster. Wang *et al.* [16] proposed a saliency-based method for video object segmentation. This method firstly generated framewise spatiotemporal saliency maps using geodesic distance. The object owned high saliency value in the maps and was easy to be segmented.

3   Our Approach

Our proposed approach is explained detailedly in this section. There are three main stages in the approach: (1) Proposal bounding boxes generation in each frame; (2) Layered DAG construction and bounding box selection; (3) Modeling motion and appearance for the foreground object and obtaining the final segmentation in an energy minimization framework. The forward optical flow between the t-th and $(t + 1)$-th frame is computed using [17] previously.

3.1   Proposal Bounding Boxes Generation

We utilize the efficient object proposal technique Edge Boxes [18] to generate proposal bounding boxes in each frame. Edge Boxes [18], as its name implies, measures edges in an image, returns hundreds to thousands bounding boxes along with their objectness scores. In order to obtain reliable bounding boxes of the foreground object, we extract two types of edges, one is image edges extracted by structured random forests [19], the other one is motion boundary extracted by [20] which improves the structured random forests [19] and makes it work for motion boundary detection. Both frame image and its forward optical flow are used to extract the motion boundary. Utilizing the edges, Edge Boxes [18] generates image bounding boxes and motion bounding boxes and their objectness scores. Notice that the object may be motionless and the optical flow may be inaccurate in some frames, which leads to false motion boundary and bounding boxes. However, we can always get high-quality image bounding boxes only if there are enough image boundaries. Figure 1 demonstrates the process of bounding boxes generation. As the illustration shows, in this case, the motion bounding boxes in (f) are more concentrated around the girl than the image bounding boxes in (c), due to the clear motion boundary in (e). But when the optical flow is inaccurate, motion bounding boxes turn to be unreliable. Figure 2 demonstrates a failing case of optical flow. The motion boundary fails following the inaccurate optical flow, and then the motion bounding boxes are outside of the target. However, the image bounding boxes are not influenced. We choose 100 bounding boxes with the highest scores from each type, and merge them to be the candidate regions of the object.

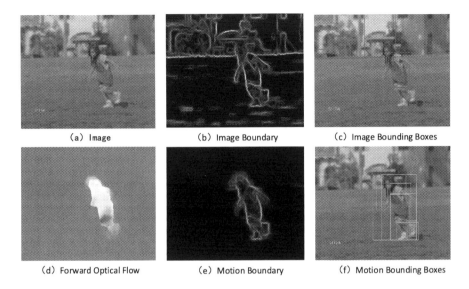

Fig. 1. (a) One of the input frames. (b) Image boundary extracted by structured random forests [19]. (c) Some top ranked bounding boxes using image boundary. (d) Forward optical flow of the image. (e) Motion boundary of the image extracted by [20]. The origin image and the forward optical flow are used to extract the motion boundary. (f) Some top ranked motion bounding boxes.

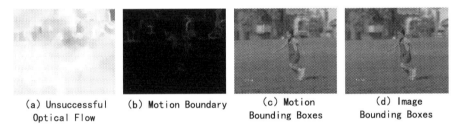

Fig. 2. (a) A failing case of optical flow. (b) Motion boundary. The motion boundary fails following the opical flow. (c) Some top ranked motion bounding boxes. The motion bouding boxes are outside of the target. (d) Some top ranked image bounding boxes.

3.2 Layered DAG for Bounding Box Selection

We have obtained hundreds of bounding boxes for each frame. It is difficult to determine which bounding box is best for the object in a single frame since the bounding boxes provide candidate regions of the object. Therefore, we consider the consistency of the boxes in the sequence. As the object moves smoothly across the frames, the box associated with it also moves. We want to obtain the boxes which have high probabilities to contain an object tightly and move coherently across the frames. A layered DAG is constructed to formulate the motion of the boxes. The problem of selecting the best box for each frame transforms into

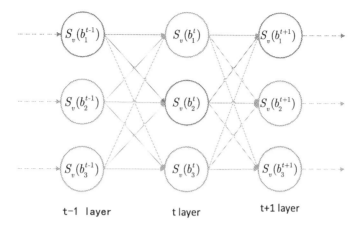

t−1 layer   t layer   t+1 layer

Fig. 3. Structure of layered DAG. Circles represent vertices of the graph. The vertices in consecutive frames are connected by straight lines in pairs. The straight lines represent edges of the graph. The path with maximum weight is selected by dynamic programming method. Red circles and lines are parts of the selected path. (Color figure online)

finding the longest path in the layered DAG. Figure 3 demonstrates the structure of the layered DAG. Each frame is represented by a layer in the graph, and each bounding box in the frame turns to be a vertex of the corresponding layer. Boxes in adjacent frames are directly connected in pairs by the edges of the DAG. All the vertices and the edges are weighted, which will be reviewing later on.

Vertices. We measure every bounding box with its potential of containing the foreground object. The scores of all bounding boxes are used as the vertices. $S_v(b_n^t)$ represents score of the n-th bounding box in the t-th frame. As both appearance and motion cues are useful for determining moving object in video, $S_v(b_n^t)$ is made of two parts as:

$$S_v(b_n^t) = A_v(b_n^t) + M_v(b_n^t),\tag{1}$$

where $A_v(b_n^t)$ is the objectness score of b_n^t obtained by Edge Boxes [18], and $M_v(b_n^t)$ is the motion score. $M_v(b_n^t)$ measures the moving difference between b_n^t and its surrounding area. $M_v(b_n^t)$ is defined according to the optical flow histogram:

$$M_v(b_n^t) = 1 - \exp(-\chi_{flow}^2(b_n^t, \overline{b_n^t})),\tag{2}$$

where $\chi_{flow}^2(b_n^t, \overline{b_n^t})$ is the χ^2-distance between L_1-normalized optical flow histograms, $\overline{b_n^t}$ is a set of pixels within a larger box and around b_n^t.

Edges. Bounding boxes in consecutive frames are connected in pairs to form the edges of the layered DAG. Edges measures the similarity between two connected boxes. As object moves smoothly across frames, object bounding box also changes coherently in location and size. Therefore, we measure the similarity score of the boxes in two consecutive frames by their locations, sizes and overlap ratio. The similarity score of the boxes is used as edges in the layered DAG and is defined as:

$$S_e(b_t, b_{t+1}) = \lambda * S_g(b_t, b_{t+1}) * S_o(b_t, b_{t+1}), \tag{3}$$

where b_t and b_{t+1} are two boxes from t-th frame and $(t+1)$-th frame, $S_g(b_t, b_{t+1})$ is location and size similarity, and $S_o(b_t, b_{t+1})$ is overlap similarity. λ is a balance factor between S_e and S_v. $S_g(b_t, b_{t+1})$ and $S_o(b_t, b_{t+1})$ are defined as:

$$S_g(b_t, b_{t+1}) = \exp(-\frac{\|g_t - g_{t+1}\|_2}{h_t + w_t}), \tag{4}$$

$$S_o(b_t, b_{t+1}) = \frac{|b_{t+1} \cap warp(b_t)|}{|b_{t+1} \cup warp(b_t)|}. \tag{5}$$

In Eq. 4, $g_t = [x, y, w, h]$ is location and size of box b_t, where $[x, y]$ is centroid coordinate of box and $[w, h]$ is width and hight respectively. In Eq. 5, $warp(b_t)$ is the warped region from b_t to frame $t+1$ by the optical flow.

Box Selection. A proper object bounding box in a certain frame is considered to own high objectness score and motion score, while proper boxes in the consecutive frames are close to each other, or high similarity score S_e in other words. Once the layered DAG is constructed, the path with maximum total score in the graph represents the most suitable boxes in the frames. This problem can be solved by dynamic programming in linear complexity. The vertices in the maximum weighted path represent object boxes in each frame. After box selection, we obtain a set of object boxes and only one box is left in each frame.

3.3   Object Segmentation

We oversegment frames into superpixels by the algorithm SLIC [21] for less computational complexity. After all these, video object segmentation is formulated as a superpixel labeling problem with two labels (foreground/background). Each superpixel in the video sequence takes a label l_i^t from $\mathbf{L} = \{0, 1\}$ where 0 represents background and 1 represents foreground. Similar to the related works [12,13,15,16], we define an energy function for the labeling problem:

$$E(l) = \sum_{t,i} A(l_i^t) + \alpha_1 \sum_{t,i} M(l_i^t) + \alpha_2 \sum_{(i,j,t) \in N_s} V(l_i^t, l_j^t) + \alpha_3 \sum_{(i,j,t) \in N_t} W(l_i^t, l_j^{t+1}), \tag{6}$$

where $A(l_i^t)$ is an appearance unary term and $M(l_i^t)$ is a motion unary term associated with a superpixel. $V(l_i^t, l_j^t)$ and $W(l_i^t, l_j^{t+1})$ are pairwise terms associated with spatial and temporal consistency respectively. N_s is a set of spatial neighborhoods of a superpixel and N_t is a set of temporal neighborhoods. A superpixel is warped to the next frame by forward optical flow, and the superpixels in the next frame which is overlapped with the warped region are considered as its temporal neighborhoods. i, j are indexes of superpixels. Unary terms try to determine labels of superpixels by appearance and motion cures, while pairwise terms insure spatial and temporal coherence of the segmentation. α_1, α_2 and α_3 are balance coefficients. Equation 6 is optimized via graph-cuts model [22] efficiently.

Motion Term. Saliency detection is always used as a technique to discover salient object in images. But saliency detection in video sequence always fails to get a satisfying result, as the primary object in video may not be salient in visualization. However, since object moves differently from its background, it will be salient in optical flow fields. We turn optical flow to RGB image by visualizing it, and use saliency detection method [23] to get motion saliency map. [23] provides a robust and efficient method for visual saliency detection. Figure 4 compares image saliency map with its motion saliency map. The image saliency map Fig. 4(c) fails to get the object (parachute), while it is outstanding in the motion saliency map (d). The motion term in Eq. 6 is defined as:

$$M(l_i^t) = \begin{cases} -\log(1 - S^t(x_i^t)) & l_i^t = 0; \\ -\log(S^t(x_i^t)) & l_i^t = 1. \end{cases} \tag{7}$$

where $S^t(x_i^t)$ is the motion saliency value of superpixel x_i^t.

Appearance Term. Two Gaussian Mixture Models (GMM) in RGB color space are estimated to model the appearance of foreground and background respectively. We select a part of superpixels and separate them into two sets of fg/bg. We set two conditions for a superpixel to be a member of foreground: (1) inside the selected object box; (2) its motion saliency value is larger than mean value of the frame. Superpixels outside the box with lower motion saliency values are regarded as background superpixels. After that we obtain two sets of super-pixels from the whole sequence. Mean RGB colors of the superpixels are used to estimate GMMs for both foreground and background. The appearance term $A(l_i^t)$ is the negative log-probability of x_i^k to take label l_i^t under the associated GMM.

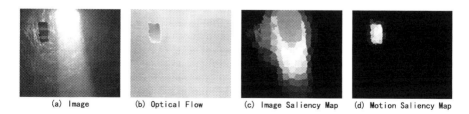

(a) Image   (b) Optical Flow   (c) Image Saliency Map   (d) Motion Saliency Map

Fig. 4. (a) One frame in sequence *parachute*. (b) Forward optical flow of (a). (c) Image saliency map of (a). (d) Motion saliency map of (a) and (b).

Pairwise Terms. $V(l_i^t, l_j^t)$ and $W(l_i^t, l_j^{t+1})$ are standard contrast-modulated Potts potentials, and follow the definition in [24]:

$$V(l_i^t, l_j^t) = dist(x_i^t, x_j^t)^{-1}[l_i^t \neq l_j^t] \exp(-\beta_1 col(x_i^t, x_j^t)^2), \qquad (8)$$

$$W(l_i^t, l_j^{t+1}) = \varphi(x_i^t, x_j^{t+1})[l_i^t \neq l_j^{t+1}] \exp(-\beta_2 col(x_i^t, x_j^{t+1})^2), \qquad (9)$$

where $dist(x_i^t, x_j^t)$ and $col(x_i^t, x_j^t)$ are the Euclidean distances between the average positions and average RGB colors of the two superpixels respectively, $[\bullet]$ is an indicator function, $\varphi(x_i^t, x_j^{t+1})$ is the overlap ratio of the warped region of x_i^t and x_j^{t+1}. The pairwise terms encourage superpixels with close RGB colors to get the same label if they are spatially or temporally connected.

4   Experimental Results

We evaluate our approach on SegTrack dataset [7] and FBMS dataset [9]. SegTrack dataset contains 6 videos and pixel-level ground-truth of foreground object in every frame. Following some related works [12,13,16], we abandon the sequence *penguin* since there are many penguins moving in the sequence and it is hard to determine which of them is the foreground object. The sequences in this dataset are quite challenging for tracking and video segmentation. *birdfall* has similar colors in fg and bg. There are large camera motion and large shape deformation in *cheetha* and *monkeydog*. *girl* suffers from articulated motion. Seg-Track is a benchmark for video object segmentation task. The dataset doesn't supply the object bounding boxes, therefore we manually annotate the box in each frame as ground-truth. We designate the object bounding box as the minimum rectangle that contains the whole object.

There are some parameters in our methods. In Eq. 6, the balance factor λ is fixed as 0.5. For the energy function Eq. 6, we set $\alpha_1 = 0.4$ and $\alpha_2 = \alpha_3 = 20$. In the pairwise terms Eqs. 8 and 9, we set $\beta_1 = \beta_2 = 1/100$. The parameters are kept fixed in all experiments.

Firstly, the effectiveness of $S_v(b_n^t)$ is evaluated. As mentioned previously, $S_v(b_n^t)$ is designed to measure the probability of a box to be the object bounding box. We rank proposal bounding boxes by descending order of $S_v(b_n^t)$, take out the top 100 boxes in each frame and calculate mean IoU of every 10 boxes. Figure 5 reports the ranked proposals-mean IoU results. As the graph shows, in most sequences high rank proposal bounding boxes get high mean IoU. Among the sequences, *girl* reports low and uniform mean IoU value among all the ranks. This is due to the object is large in this sequence, and it is possible to get many boxes that have high overlap ratio with ground-truth. On the other hand, the articulated parts such as legs and hands obtain large motion scores, and corresponding proposal boxes get high rank. Since these boxes lack of consistency, they are not selected as the object boxes. Table 1 reports the mean IoU of the selected boxes in each sequence. *cheetah* gets the lowest mean IoU score because in the beginning frames the box covers two moving objects (cheetah and antelope), since they are close to each other. Although *girl* reports bad results in

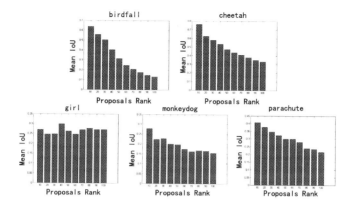

Fig. 5. The ranked proposal bounding boxes in different sequences and their mean IoU compared with ground-truth bounding boxes.

Fig. 5, it successes to get proper boxes and obtains a high mean IoU in Table 1, due to the consistency of the selected boxes.

Table 1. Mean IoU of sequences in SegTrack dataset.

Sequence	Birdfall	Cheetah	Girl	Monkeydog	Parachute
meanIoU	0.688	0.527	0.723	0.627	0.907

Figure 6 demonstrates some results of foreground object detection and the final segmentation on SegTrack dataset [7]. The green rectangles in Fig. 6 are the ground-truth bounding boxes annotated by us, while the red rectangles are the selected bounding boxes. As the illustration shows, most selected boxes catch object tightly and very close to the ground-truth bounding boxes, especially in sequence *parachute* and *birdfall*. In sequence *cheetah*, the antelope is regraded as the primary object in the dataset, while the cheetah also appears and moves in the beginning frames. Our method selects a big bounding box that contains both the antelope and the cheetah at first. When the cheetah disappears, the selected bounding box turns to only contain the antelope. In sequence *monkeydog*, when the monkey is close to boundary of the image, the selected box hasn't followed it. However, when the monkey returns to the middle of the image, the selected box catches it again. The regions in green boundaries in Fig. 6 are the segmented foreground objects. We notice that in *girl*, the foots are usually missed due to the heavy motion blur. In *cheetha*, the cheetah is segmented as the object at the beginning frames since both cheetah and antelope are inside the selected boxes. Table 2 reports quantitative results and comparison with some related works on SegTrack dataset. The results in Table 2 are the average number of mislabelled pixels pre frame compared to the ground-truth. The definition in

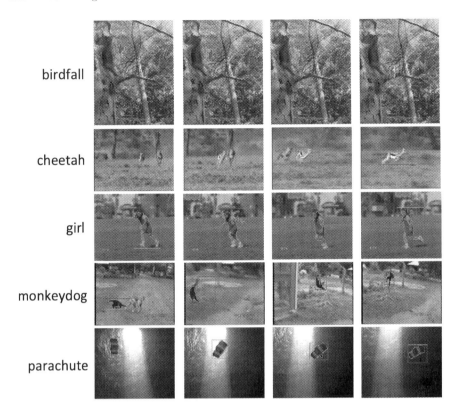

Fig. 6. Results of object detection and segmentation on SegTrack Dataset. The red rectangles are the selected bounding boxes, and the green rectangels are the manually annotated bounding boxes. The regions within green boundaries are the segmented foreground objects. (Color figure online)

[7] is $error = \frac{XOR(S,GT)}{F}$, where S is the segmentation result, GT is the ground-truth labeling and F is the frame number of the sequence. As Table 2 reports, our method gets comparably result with state-of-the-art works.

Table 2. Quantitative results and comparison with state-of-the-art works on SegTrack dataset.

Methods	Ours	[13]	[16]	[12]	[15]	[9]	[7]	[25]
Birdfall	183	155	209	288	217	458	252	189
Cheetah	849	633	796	905	890	1968	1142	806
Girl	1943	1488	1040	1785	3859	7595	1304	1698
Monkeydog	501	365	562	521	284	1434	563	472
Parachute	337	220	207	201	855	1113	235	221

Fig. 7. Some segmentation results on FBMS dataset. The regions within green boundaries are the segmented foreground objects. (Color figure online)

We also test our approach on FMBS dataset qualitatively. Figure 7 demonstrates some results in this dataset. We notice when the object moves in an articulated manner, our approach may select part instead of the whole object. In *marple1* and *marple3*, pedestrian's head is detected as the foreground object.

5 Conclusion

In this paper, we propose a new approach for the task of video object segmentation. Compared with the former works in this task, our approach works based on proposal bounding boxes and avoids heavy computation for generating segments of proposal regions. The boxes are integrated into a layered DAG, and the problem of box selection can be easily and efficiently solved. With the selected boxes, the foreground object is detected in each frame. The final segmentation is performed based on both motion cues and appearance cues. The experimental results on SegTrack dataset and FBMS dataset testifies the effectiveness of our approach.

Acknowledgements. This work was supported by the National High-Tech R&D Program of China (863 Program) under Grant 2015AA015904.

References

1. Alexe, B., Deselaers, T., Ferrari, V.: What is an object? In: Proceedings of the IEEE Conference on Computer Vision and Pattern Recognition (CVPR), pp. 73–80. IEEE (2010)

2. Endres, I., Hoiem, D.: Category independent object proposals. In: Daniilidis, K., Maragos, P., Paragios, N. (eds.) ECCV 2010. LNCS, vol. 6315, pp. 575–588. Springer, Heidelberg (2010). doi:10.1007/978-3-642-15555-0_42

3. Manen, S., Guillaumin, M., Gool, L.: Prime object proposals with randomized prim's algorithm. In: Proceedings of the IEEE International Conference on Computer Vision (ICCV), pp. 2536–2543. IEEE (2013)

4. Bai, X., Wang, J., Simons, D., Sapiro, G.: Video snapcut: robust video object cutout using localized classifiers. ACM Trans. Grap. (TOG) **28**, 70 (2009)

5. Price, B.L., Morse, B.S., Cohen, S.: Livecut: learning-based interactive video segmentation by evaluation of multiple propagated cues. In: Proceedings of the IEEE International Conference on Computer Vision (ICCV), pp. 779–786. IEEE (2009)

6. Yuen, J., Russell, B., Liu, C., Torralba, A.: Labelme video: building a video database with human annotations. In: Proceedings of the IEEE International Conference on Computer Vision (ICCV), pp. 1451–1458. IEEE (2009)

7. Tsai, D., Flagg, M., Nakazawa, A., Rehg, J.M.: Motion coherent tracking using multi-label MRF optimization. Int. J. Comput. Vis. **100**, 190–202 (2012)

8. Ren, X., Malik, J.: Tracking as repeated figure/ground segmentation. In: Proceedings of the IEEE Conference on Computer Vision and Pattern Recognition (CVPR), pp. 1–8. IEEE (2007)

9. Brox, T., Malik, J.: Object segmentation by long term analysis of point trajectories. In: Daniilidis, K., Maragos, P., Paragios, N. (eds.) ECCV 2010. LNCS, vol. 6315, pp. 282–295. Springer, Heidelberg (2010). doi:10.1007/978-3-642-15555-0_21

10. Fragkiadaki, K., Zhang, G., Shi, J.: Video segmentation by tracing discontinuities in a trajectory embedding. In: Proceedings of the IEEE Conference on Computer Vision and Pattern Recognition (CVPR), pp. 1846–1853. IEEE (2012)

11. Lezama, J., Alahari, K., Sivic, J., Laptev, I.: Track to the future: spatio-temporal video segmentation with long-range motion cues. In: Proceedings of the IEEE Conference on Computer Vision and Pattern Recognition (CVPR). IEEE (2011)

12. Lee, Y.J., Kim, J., Grauman, K.: Key-segments for video object segmentation. In: Proceedings of the IEEE International Conference on Computer Vision (ICCV), pp. 1995–2002. IEEE (2011)

13. Zhang, D., Javed, O., Shah, M.: Video object segmentation through spatially accurate and temporally dense extraction of primary object regions. In: Proceedings of the IEEE Conference on Computer Vision and Pattern Recognition (CVPR), pp. 628–635. IEEE (2013)

14. Perazzi, F., Wang, O., Gross, M., Sorkine-Hornung, A.: Fully connected object proposals for video segmentation. In: Proceedings of the IEEE International Conference on Computer Vision (ICCV), pp. 3227–3234. IEEE (2015)

15. Papazoglou, A., Ferrari, V.: Fast object segmentation in unconstrained video. In: Proceedings of the IEEE International Conference on Computer Vision (ICCV), pp. 1777–1784. IEEE (2013)

16. Wang, W., Shen, J., Porikli, F.: Saliency-aware geodesic video object segmentation. In: Proceedings of the IEEE Conference on Computer Vision and Pattern Recognition (CVPR), pp. 3395–3402. IEEE (2015)

17. Sundaram, N., Brox, T., Keutzer, K.: Dense point trajectories by GPU-accelerated large displacement optical flow. In: Daniilidis, K., Maragos, P., Paragios, N. (eds.) ECCV 2010. LNCS, vol. 6311, pp. 438–451. Springer, Heidelberg (2010). doi:10.1007/978-3-642-15549-9_32

18. Zitnick, C.L., Dollár, P.: Edge boxes: locating object proposals from edges. In: Fleet, D., Pajdla, T., Schiele, B., Tuytelaars, T. (eds.) ECCV 2014. LNCS, vol. 8693, pp. 391–405. Springer, Heidelberg (2014). doi:10.1007/978-3-319-10602-1_26

19. Dollár, P., Zitnick, C.: Structured forests for fast edge detection. In: Proceedings of the IEEE International Conference on Computer Vision (ICCV), pp. 1841–1848. IEEE (2013)
20. Weinzaepfel, P., Revaud, J., Harchaoui, Z., Schmid, C.: Learning to detect motion boundaries. In: Proceedings of the IEEE Conference on Computer Vision and Pattern Recognition (CVPR), pp. 2578–2586. IEEE (2015)
21. Achanta, R., Shaji, A., Smith, K., Lucchi, A., Fua, P., Susstrunk, S.: SLIC superpixels compared to state-of-the-art superpixel methods. IEEE Trans. Pattern Anal. Mach. Intell. **34**, 2274–2282 (2012)
22. Boykov, Y., Veksler, O., Zabih, R.: Fast approximate energy minimization via graph cuts. IEEE Trans. Pattern Anal. Mach. Intell. **23**, 1222–1239 (2001)
23. Zhu, W., Liang, S., Wei, Y., Sun, J.: Saliency optimization from robust background detection. In: Proceedings of the IEEE Conference on Computer Vision and Pattern Recognition (CVPR), pp. 2814–2821. IEEE (2014)
24. Rother, C., Kolmogorov, V., Blake, A.: Grabcut: interactive foreground extraction using iterated graph cuts. ACM Trans. Graph. (TOG) **23**, 309–314 (2004)
25. Ma, T., Latecki, L.J.: Maximum weight cliques with mutex constraints for video object segmentation. In: Proceedings of the IEEE Conference on Computer Vision and Pattern Recognition (CVPR), pp. 670–677. IEEE (2012)

Landmark Selecting on 2D Shapes
for Constructing Point Distribution Model

Xianghu Ji[1,2(✉)], Lili Wang[1,2(✉)], Pan Tao[1,2(✉)],
and Zhongliang Fu[1(✉)]

[1] Chengdu Institute of Computer Application, No 9, Section 4,
South Renmin Road, Chengdu, China
jxh.ucas@gmail.com, wanglili8773@163.com,
taopanpan@gmail.com, Fzliang@netease.com
[2] University of Chinese Academy of Sciences, 19 A Yuquan Road,
Shijingshan District, Beijing, China

Abstract. A new method of selecting landmarks on 2D shapes which are represented by Centripetal Catmull-Rom spline is proposed in this paper. Firstly, a mean shape is generated from training set and landmarks on mean shape are extracted based on curvature and arc-length information. Then the corresponding landmarks on each shape can be obtained by projecting the mean shape back to each sample using non-rigid registration method Coherent Point Drift. Experiments showed that landmarks auto-generated are more accurate than landmarks manual annotated when used in segmentation.

Keywords: Corresponding landmarks · Point distribution model · Active shape model · Non-rigid registration · Centripetal Catmull-Rom spline

1 Introduction

Segmentation algorithms based on point distribution model (PDM) [1], such as Active Shape Model (ASM) and Active Appearance Model (AAM) are widely used in human face tracking and segmentation [2, 3], and also in medical and biological image segmentation [4, 5].

PDM can be constructed using corresponding points. These points are called landmarks which represent the correspondence of shapes [6]. For example, Fig. 1 shows a training set of 6 hand shapes, each shape has 40 landmarks. If the sixth landmark represents thumb tip, then the sixth landmark should on the thumb tip on every shape in the training set.

Early landmark positioning is done manually, which is time-consuming and error prone. Compared with face and hand which have salient features like corners and T-junctions, biological objects such as left ventricle and diatom are very difficult to annotate because they do not have salient features as references. Moreover, even shapes that belong to the same class are vary in shape, size and rotation. To solve these problems, we need an automatic landmark selecting method.

The automatic method commonly used for selecting landmarks is MDL [7], non-rigid registration [8], shape descriptors [9], etc. Hill et al. [10] extract landmarks by

© Springer Nature Singapore Pte Ltd. 2016
T. Tan et al. (Eds.): CCPR 2016, Part I, CCIS 662, pp. 318–331, 2016.
DOI: 10.1007/978-981-10-3002-4_27

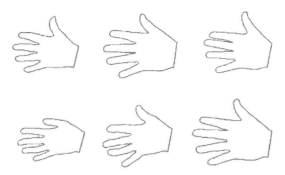

Fig. 1. A training set consists of six hand shapes

employing a binary tree of corresponded pairs of shapes to generate mean shape, then extract landmarks on the mean shape using Critical Point Detection (CPD) method and project these landmarks back to the training set.

Hicks et al. [6] build a template of typical curvature extrema of the training set using morphological scale space. Landmarks on each sample are extracted by matching extrema of the sample curvature to the extrema on the template.

Angelopoulou et al. [11] select landmarks of ventricles on 2D MRI brain images automatically using growing neural gas (GNG). This method can map low-dimensional GNG network to high-dimensional contours without requiring a priori knowledge and reference shape, and remain topological structure of the contours at the same time.

Souza and Udupa [12] also use a method of extracting landmarks by propagating a mean shape to each sample. The mean shape is calculated using signed distance function, and landmarks on the mean shape are extracted using Douglas-Peucker diagonalization. However, signed distance function can only be used to represent closed shape, and the method they used projecting mean shape to each sample is too simple, so the landmarks extracted can't fully express the shape information. Rueda et al. [13] propose a method of local curvature scale to select landmarks on mean shape automatically, and then map the mean shape to samples.

In this paper, we propose a method of selecting landmarks on shapes represented by Centripetal Catmull-Rom (CCR) spline. We densely select same number of points of equal arc-length on each sample, and then calculate mean shape of this point sets. The landmarks on the mean shape are extracted according to curvature extrema and arc-length. Finally, the mean shape is projected back to each sample using method of Coherent Point Drift (CPD) which can give the position of corresponding landmarks on each sample.

2 Point Distribution Model

Point distribution model, also called statistical shape model, is proposed by Cootes et al. [1]. The model is built on a set of similar shapes, and can be used to represent other shapes of the same class.

Given a training set of m shapes, each shape $x_i(i = 1, 2, \ldots, m)$ is a vector of $2n$ elements.

$$x_i = (x_{i1}, y_{i1}, x_{i2}, y_{i2}, \ldots, x_{in}, y_{in}) \tag{1}$$

Where $(x_{ij}, y_{ij})(j = 1, 2, \ldots, n)$ is the coordinate of the jth landmark. The shapes are aligned using Procrustes Analysis method, then a PDM model can be built using Principal Component Analysis (PCA):

$$\mathbf{x} = \bar{\mathbf{x}} + \mathbf{pb} \tag{2}$$

$$\bar{\mathbf{x}} = \frac{1}{m} \sum_{i=1}^{m} x_i \tag{3}$$

Here, $\bar{\mathbf{x}}$ is the mean shape. $\mathbf{b}$ is a vector of weights, $\mathbf{P}$ is a matrix of the first t eigenvectors of covariance matrix $\mathbf{S}$.

$$S = \frac{1}{1 - m} \sum_{i=1}^{m} (\mathbf{x_i} - \bar{x})^{\mathrm{T}} (x_i - \bar{\mathbf{x}}) \tag{4}$$

Assuming that $\lambda_k (k = 1, 2, \ldots 2n, \lambda_k \geq \lambda_{k+1})$ is the kth eigenvalue of $\mathbf{S}$, the proportion of the first t eigenvalues should be greater than a given threshold ϕ (for example, 95 %):

$$\frac{\sum_{i=1}^{t} \lambda_i}{\sum_{i=1}^{2n} \lambda_i} > \phi \tag{5}$$

3 Centripetal Catmull-Rom (CCR) Spline

In order to fully express the information of a shape, landmarks should be connected by lines. The simplest way is using straight line, but it is not accurate, therefore polynomial interpolation, B-spline, Bezier spline, etc. are used by some researchers. However, neither B-spline nor Bezier spline goes through control points, which is inconvenient when editing.

In this paper, we use CCR spline to represent shapes. It goes through control points, so the landmarks can be used as control points, and when editing is necessary, changing position of landmarks is all we need. For example, Fig. 2 shows a partial sample of the left ventricle's endocardium delineated by CCR spline. CCR spline is widely used in graphics because of three important features [14]: (1) The curve is smooth. (2) The position change of each control point only affects a small neighborhood on the curve. (3) CCR curves have an explicit piecewise polynomial representation, allowing them to be easily converted to other bases and manipulated computationally.

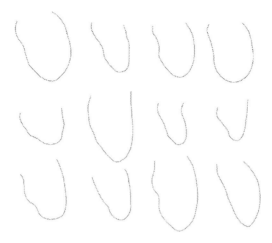

Fig. 2. Shapes of left ventricle delineated by CCR spline

Let $P = (x, y)^T$ denotes a control point. P_{i-1}, P_i, P_{i+1}, P_{i+2} can be used to define a curve segment Q_i. t_{i-1}, t_i, t_{i+1}, t_{i+2} are knots on the curve segment. The CCR curve segment can be plotted as:

$$Q_i = \frac{t_{i+1} - t}{t_{i+1} - t_i} L_{012} + \frac{t_i - t}{t_{i+1} - t_i} L_{123} \tag{6}$$

Here,

$$L_{012} = \frac{t_{i+1} - t}{t_{i+1} - t_{i-1}} L_{01} + \frac{t - t_{i-1}}{t_{i+1} - t_{i-1}} L_{12} \tag{7}$$

$$L_{123} = \frac{t_{i+2} - t}{t_{i+2} - t_i} L_{12} + \frac{t - t_i}{t_{i+2} - t_i} L_{23} \tag{8}$$

$$L_{01} = \frac{t_i - t}{t_i - t_{i-1}} P_{i-1} + \frac{t - t_{i-1}}{t_i - t_{i-1}} P_i \tag{9}$$

$$L_{12} = \frac{t_{i+1} - t}{t_{i+1} - t_i} P_i + \frac{t - t_i}{t_{i+1} - t_i} P_{i+1} \tag{10}$$

$$L_{23} = \frac{t_{i+2} - t}{t_{i+2} - t_{i+1}} P_{i+1} + \frac{t - t_{i+1}}{t_{i+2} - t_{i+1}} P_{i+2} \tag{11}$$

$$t_{i+1} = |P_{i+1} - P_i|^\alpha + t_i \tag{12}$$

α ranges between 0 and 1. If α equals 0, the curve is called Uniform Catmull-Rom spline; if α equals 1, then it is called Chordal Catmull-Rom spline; if α equals 0.5, it is called Centripetal Catmull-Rom spline. Uniform and Chordal Catmull-Rom spline will

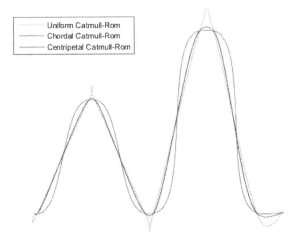

Fig. 3. Catmull-Rom curves

produce cusp and self-intersection (Fig. 3), and experiments showed that only Centripetal Catmull-Rom guarantees no cusp or self-intersections within a curve segment [14, 15].

4 Mean Shape

Mean shape can be calculated from the training set represented by CCR. Because CCR curve is C^1, not C^2 [14, 16, 17], its second derivative is discontinuous. In order to get the mean shape, we first extract same number of points with equal arc-length densely on each shape. Then the mean shape can be obtained according to formula (3) after aligning the discrete point sets.

4.1   Curvature of Mean Shape

After obtaining the mean shape which is represented by discrete points, we can delineate the mean shape using CCR spline, and discretize the curve by densely extracted points which divide the contour into subintervals with equal arc-length. Then the curvature on each point can be calculated using method below.

The curvature k of a parametric curve $\gamma(t) = (x(t), y(t))$ can be written as [18]:

$$k = \frac{\begin{vmatrix} x' & y' \\ x'' & y'' \end{vmatrix}}{(x'^2 + y'^2)^{3/2}} \tag{13}$$

A digital curve at point p_i can be approximated by second order polynomials using pixels p_{i-k_b} and p_{i+k_b}. The approximating polynomial $\gamma(t) = (x(t), y(t))$ is defined by

$$\begin{cases} x(t) = a_2 t^2 + a_1 t + a_0 \\ y(t) = b_2 t^2 + b_1 t + b_0 \end{cases} \tag{14}$$

with $t \in [-1, 1]$. Let $t = -1$ denotes p_{i-k_b}, $t = 0$ denotes p_i, and $t = 1$ defines p_{i+k_b}. In this case, Eq. (13) takes the following form:

$$k = \frac{2(a_1 b_2 - a_2 b_1)}{(a_1^2 + b_1^2)^{3/2}} \tag{15}$$

at point p_i, with

$$\begin{cases} a_2 - a_1 + a_0 = x_{i-k_b} \\ a_0 = x_i \\ a_2 + a_1 + a_0 = x_{i+k_b} \end{cases} \tag{16}$$

for $x(t)$. Analogously, for $y(t)$, we obtain

$$\begin{cases} a_1 = \dfrac{x_{i+k} - x_{i-k}}{2} \\ a_2 = \dfrac{x_{i+k} + x_{i-k}}{2} - x_i \\ b_1 = \dfrac{y_{i+k} - y_{i-k}}{2} \\ b_2 = \dfrac{y_{i+k} + y_{i-k}}{2} - y_i \end{cases} \tag{17}$$

Figure 4 shows the mean shape of left ventricle's endocardium and its curvature.

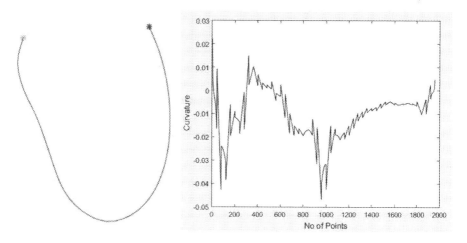

Fig. 4. Mean shape (left) and curvature (right) of left ventricle's endocardium. The green point of mean shape is start point, and the blue point is end point. In the right figure, the horizontal axis represents the number of points, and the vertical axis represents curvature. (Color figure online)

4.2   Curvature Extrema on Mean Shape

Landmarks on a shape are usually points that have extreme curvature, such as corner points. We use the following method to find the local curvature extrema: a local maximum point must be the highest point between two consecutive valleys, it is higher than the points around it. The local minima can be obtained in the similar way.

Figure 5 shows the curvature extrema and their distribution on mean shape. From the picture, we can find that some extremums are too close and some too far away, and mean shape can't be represented by these extremums. Therefore, the arc-length information was introduced in this paper. Assuming that $ArcDis$ is the arc-length between two neighboring landmarks, and $Tresh_{min}$ is the minimum acceptance threshold, $Thresh_{max}$ is the maximum acceptance threshold. If $ArcDis < Thresh_{min}$, the two landmarks are too close, then one of them is removed. If $ArcDis > Thresh_{max}$, new points are added between the two landmarks which are far away from each other. The number of points added can be given as:

$$l = round(ArcDis/Thresh_{max}) \tag{18}$$

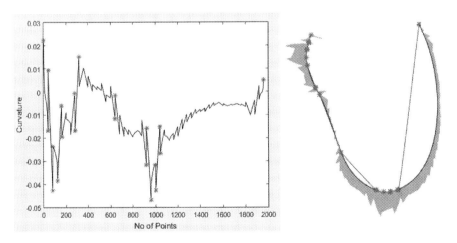

Fig. 5. Curvature extremums (left) and their distribution on the mean shape (right). The Red points represent curvature extrema. The blue line in the right figure denotes mean shape and the red line is mean shape redrawn by CCR spline with curvature extremums as control points. (Color figure online)

Here, $round()$ is a function that rounds a number to the nearest integer. The landmarks extracted based on curvature and arc-length are shown in Fig. 6.

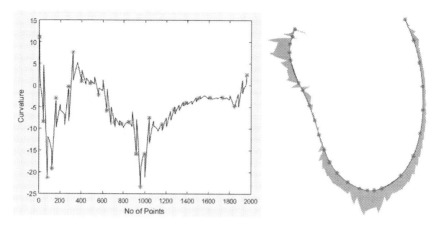

Fig. 6. Landmarks (red points) extracted based on curvature and arc-length (left) and their distribution on mean shape (right). (Color figure online)

4.3 Propagating Mean Shape to Training Set

After landmarks on mean shape are obtained, the mean shape is propagated to each shape in the training set to acquire the corresponding landmarks. Here, we use non-rigid registration method Coherent Point Drift (CPD) proposed by Myronenko etc. [19].

CPD considers the alignment of two point sets as a probability density estimation problem where one point set represents the Gaussian Mixture Model (GMM) centroids and the other represents the data points. The GMM centroids are moved coherently as a group, which preserves the topological structure of the point sets. CPD can be used in rigid, affine and non-rigid registration cases by using different transformation matrix. We will briefly introduce CPD algorithm in this paper.

Given target point set $X_{N \times D} = (x_1, \ldots, x_N)^{\mathrm{T}}$ and GMM centroids $Y_{M \times D} = (y_1, \ldots, y_M)^{\mathrm{T}}$, and N, M are the number of points and D is the dimension of the points sets. After adding a uniform distribution to account for the effects of noise and outliers, the final mixture model takes the form:

$$p(x) = \omega \frac{1}{N} + (1 - \omega) \sum_{m=1}^{M} \frac{1}{M} p(x|m) \tag{19}$$

Here, $p(x|m) = \frac{1}{(2\pi\sigma^2)^{D/2}} \exp(-\frac{\|x - y_m\|^2}{2\sigma^2})$, $\omega(0 \leq \omega \leq 1)$ is a weight parameter that reflects the assumptions about the amount of noise and outliers in the point sets.

The locations of GMM centroids are determined by a set of transformation parameters θ that can be estimated by minimizing the negative log-likelihood function

$$E(\theta, \sigma^2) = -\sum_{n=1}^{N} \log \sum_{m=1}^{M} p(m) p(x_n|m) \tag{20}$$

The Expectation Maximization (EM) algorithm is used to find θ and σ^2. The E-step constructs the objective function Q by computing a posteriori probability distributions $p^{old}(m|x_n)$ of mixture components based on Bayes' theorem. The M-step updates the parameters by maximizing Q. Ignoring the constants independent of θ and σ^2, Q can be written as

$$Q(\theta, \sigma^2) = \frac{1}{2\sigma^2} \sum_{n=1}^{N} \sum_{m=1}^{M} p^{old}(m|x_n)\|x_n - T(y_m, \theta)\|^2 + \frac{N_p D}{2} \log \sigma^2 \qquad (21)$$

Where $N_p = \sum_{n=1}^{N} \sum_{m=1}^{M} p^{old}(m|x_n)$, and

$$p^{old}(m|x_n) = \frac{\exp\left(-\frac{1}{2}\left\|\frac{x_n - T(y_m, \theta^{old})}{\sigma^{old}}\right\|^2\right)}{\sum_{k=1}^{M} \exp\left(-\frac{1}{2}\left\|\frac{x_n - T(y_m, \theta^{old})}{\sigma^{old}}\right\|^2\right) + (2\pi\sigma^2)^{D/2} \frac{\omega}{1-\omega} \frac{M}{N}}$$

The transformation T can be specified for rigid, affine and non-rigid point set registration cased separately.

Figure 7 shows landmarks of some samples extracted using CPD. The shapes delineated by CCR (red lines) coincide with the original shapes (blue lines), which means that little shape information is missing.

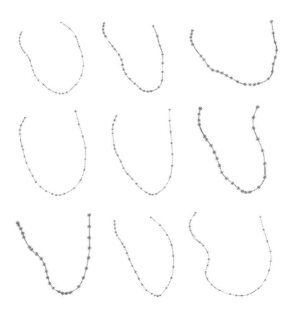

Fig. 7. Landmarks of some samples extracted using CPD. The red points are landmarks, blue lines are shapes annotated by experts, and the red lines are shapes delineated by CCR using landmarks as control points. (Color figure online)

5 Experimental Results

5.1 Evaluation Criteria

The corresponding landmarks we extracted are used to construct point distribution model. However, how to evaluate the performance of the model still remains a problem [20]. Actually, comparing the original shape with the shape redrawn by CCR can also reflect the performance of landmark extracting algorithms.

In his paper [21], Davies proposed three measures (compactness, specificity and generality) for evaluating shape-correspondence performance in terms of the PDM construction. However, considering that the point distribution model is used for segmentation, we care more about the accuracy of segmentation rather than the locations of landmarks. Here, two metrics proposed by Heimann et al. [22] are used to evaluate the performance of the segmentation, e.g.

1. Hausdorff Distance (HD)

    This metric measures the maximum distance between two shapes. If $d(x, y)$ is the distance between two points, the Hausdorff Distance between two shapes $\mathbf{x}$ and $\mathbf{y}$ is defines as:

$$D_{HD} = \max(h(\mathbf{x}, \mathbf{y}), h(\mathbf{y}, \mathbf{x})) \tag{22}$$

    With $h(\mathbf{x}, \mathbf{y}) = \max\limits_{x \in \mathbf{x}} \min\limits_{y \in \mathbf{y}} d(x, y)$.

1. Average Square Distance (ASD)

    Since Hausdorff Distance is very sensitive to outliers, an alternative is to average the squared distance over both shapes:

$$D_{ASD}(\mathbf{x}, \mathbf{y}) = \frac{1}{2} \left(\sum_{x \in \mathbf{x}} \min_{y \in \mathbf{y}} d^2(x, y) + \sum_{y \in \mathbf{y}} \min_{x \in \mathbf{x}} d^2(y, x) \right) \tag{23}$$

5.2 Experiments and Results

Four-chamber view transesophageal echocardiographic images are used in our experiments with 72 images as training set and 24 images as test set. The annotation of endocardium was finished by experts of West China Hospital, Sichuan Province. Figure 8 shows some images annotated in the training set.

In order to test the performance of the proposed automatic landmark selecting scheme, two point sets (one is landmarks annotated by experts and the other is auto-generated) are used to build two point distribution models. These two models are used to segment left ventricle under the same conditions of segmentation which use Active Shape Model (ASM) [24] as segmentation algorithm. Each ASM model is set to the same initial position and have four level image pyramids. Some segmentation results are shown in Fig. 9.

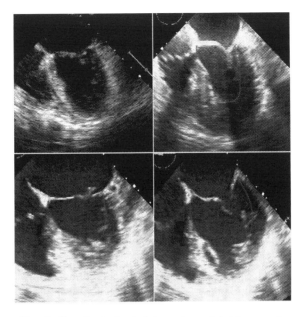

Fig. 8. Samples in the training set annotated by experts

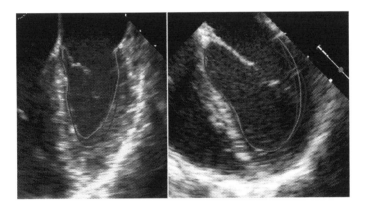

Fig. 9. Two segmentation results using manual annotated landmarks (blue line) and auto-generated landmarks (red line). The green line is annotated by experts. (Color figure online)

In our experiments, we got 24 segmentation results each for manual and auto-generated datasets. Their HD distance and ASD distance with ground-truth are shown in Figs. 10 and 11. The average HD distance and average ASD distance are shown in Table 1. From Figs. 10 and 11, we can find that auto-generated scheme are closer to ground truth and this can also be found in Table 1.

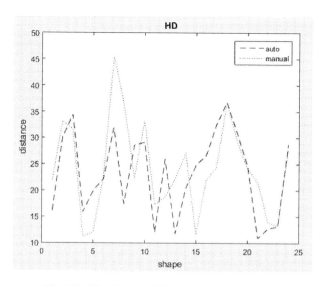

Fig. 10. The Hausdorff distance for both datasets.

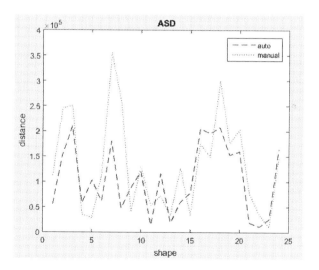

Fig. 11. The average square distance for both datasets

Table 1. Average HD and ASD for two datasets

	Mean HD	Mean ASD
Auto-generated landmarks	**23.2265**	**10423**
Manual landmarks	24.1701	13147

6 Conclusion

In this paper, a new method of selecting landmarks automatically is proposed. It combines the information of curvature and arc-length of mean shape. Firstly, a mean shape is generated from training set, then landmarks on the mean shape are extracted based on curvature and arc-length. Finally, the landmarks on each sample are extracted by projecting mean shape back to each sample using Coherent Point Drift. This avoids the disadvantages of manual annotation, improves the efficiency and accuracy of annotating.

However, in the method proposed, only shape information is used. Considering that the texture information on images is also valuable when extracting landmarks, how to combine information of shape and texture is the direction of our further research.

References

1. Cootes, T.F., Taylor, C.J., Cooper, D.H., Graham, J.: Active shape models-their training and application. Comput. Vis. Image Underst. **61**(1), 38–59 (1995)
2. Milborrow, S., Nicolls, F.: Active shape models with SIFT descriptors and MARS. In: VISAPP, no. 2, pp. 380–387 (2014)
3. Kossai, J., Tzimiropoulos, G., Pantic, M.: Fast and exact bi-directional fitting of active appearance models. In: 2015 IEEE International Conference on Image Processing (ICIP), pp. 1135–1139. IEEE (2015)
4. Santiago, C., Nascimento, J.C., Marques, J.S.: 2D segmentation using a robust active shape model with the EM algorithm. IEEE Trans. Image Process. **24**(8), 2592–2601 (2015)
5. Chen, X., Udupa, J.K., Bagci, U., Zhuge, Y., Yao, J.: Medical image segmentation by combining graph cuts and oriented active appearance models. IEEE Trans. Image Process. **21**(4), 2035–2046 (2012)
6. Hicks, I., Marshall, D., Martin, R.R., Rosin, P.L., Bayer, M.M., Mann, D.G.: Automatic landmarking for building biological shape models. In: 2002 Proceedings of International Conference on Image Processing, vol. 2, pp. II–801. IEEE (2002)
7. Davies, R.H., Twining, C.J., Cootes, T.F., Waterton, J.C., Taylor, C.J.: A minimum description length approach to statistical shape modeling. IEEE Trans. Med. Imaging **21**(5), 525–537 (2002)
8. Heitz, G., Rohlfing, T., Maurer Jr., C.R.: Statistical shape model generation using nonrigid deformation of a template mesh. In: Medical Imaging, pp. 1411–1421. International Society for Optics and Photonics (2005)
9. Rueda, S., Udupa, J.K., Bai, L.: A new method of automatic landmark tagging for shape model construction via local curvature scale. In: Medical Imaging, p. 69180 N. International Society for Optics and Photonics (2008)
10. Hill, A., Taylor, C.J., Brett, A.D.: A framework for automatic landmark identification using a new method of nonrigid correspondence. IEEE Trans. Pattern Anal. Mach. Intell. **22**(3), 241–251 (2000)
11. Angelopoulou, A., Psarrou, A., Rodriguez, J.G., Revett, K.R.: Automatic landmarking of 2D medical shapes using the growing neural gas network. In: Liu, Y., Jiang, T.-Z., Zhang, C. (eds.) CVBIA 2005. LNCS, vol. 3765, pp. 210–219. Springer, Heidelberg (2005)

12. Souza, A., Udupa, J.K.: Automatic landmark selection for active shape models. In: Medical Imaging, pp. 1377–1383. International Society for Optics and Photonics (2005)

13. Rueda, S., Udupa, J.K., Bai, L.: Shape modeling via local curvature scale. Pattern Recogn. Lett. **31**(4), 324–336 (2010)

14. Yuksel, C., Schaefer, S., Keyser, J.: Parameterization and applications of Catmull-Rom curves. Comput.-Aided Des. **43**(7), 747–755 (2011)

15. Dyn, N., Floater, M.S., Hormann, K.: Four-point curve subdivision based on iterated chordal and centripetal parameterizations. Comput. Aided Geom. Des. **26**(3), 279–286 (2009)

16. Yuksel, C., Schaefer, S., Keyser, J.: On the parameterization of Catmull-Rom curves. In: 2009 SIAM/ACM Joint Conference on Geometric and Physical Modeling, pp. 47–53. ACM (2009)

17. Kindlmann, G., Whitaker, R., Tasdizen, T., Moller, T.: Curvature-based transfer functions for direct volume rendering: methods and applications. In: Visualization, VIS 2003, pp. 513–520. IEEE (2003)

18. Hermann, S., Klette, R.: A comparative study on 2D curvature estimators. Technical report, CITR, The University of Auckland, New Zealand (2006)

19. Myronenko, A., Song, X.: Point set registration: coherent point drift. IEEE Trans. Pattern Anal. Mach. Intell. **32**(12), 2262–2275 (2010)

20. Munsell, B.C., Dalal, P., Wang, S.: Evaluating shape correspondence for statistical shape analysis: a benchmark study. IEEE Trans. Pattern Anal. Mach. Intell. **30**(11), 2023–2039 (2008)

21. Davies, R.H.: Learning Shape: Optimal Models for Analysing Natural Variability. University of Manchester, Manchester (2002)

22. Heimann, T., Wolf, I., Meinzer, H.P.: Optimal landmark distributions for statistical shape model construction. In: Proceedings of SPIE Medical Imaging: Image Processing, vol. 6144, pp. 518–528 (2006)

23. Niessen, W.J., Bouma, C.J., Vincken, K.L., Viergever, M.A.: Error metrics for quantitative evaluation of medical image segmentation. In: Klette, R., Stiehl, H.S., Viergever, M.A., Vincken, K.L. (eds.) Performance Characterization in Computer Vision, vol. 17, pp. 275–284. Springer, Heidelberg (2000)

24. Cootes, T., Baldock, E., Graham, J.: An introduction to active shape models. In: Image Processing Analysis, pp. 223–248 (2000)

Dual Camera Based Feature for Face Spoofing Detection

Xudong Sun[✉], Lei Huang, and Changping Liu

Institute of Automation, Chinese Academy of Sciences,
95 Zhongguancun East Road, Beijing 100190, China
{sunxudong2013,lei.huang,changping.liu}@ia.ac.cn

Abstract. This paper presents a fused feature using dual cameras for
face spoofing detection. The feature takes full advantage of input image
pairs in terms of texture and depth. It consists of two parts: 2D component and 3D component. For the former, we propose an algorithm based
on image similarity to combine every pair of input images into one gray-
level image, from which the 2D feature is extracted. For the latter, based
on point feature histograms (PFH) method, we describe the point cloud
obtained by stereo reconstruction algorithms. The concatenation of 2D
and 3D features above is used to represent the input image pair. Experiments on self collected dataset demonstrate the competitive performance
and potential of the proposed feature.

Keywords: Face spoofing detection · Dual cameras · Feature fusion ·
Similarity measurement

1 Introduction

Various face recognition systems have been deployed in our daily life, however,
traditional face recognition techniques are vulnerable to face spoofing attacks.
With digital cameras and cell phones becoming increasingly popular, it is much
easier to conduct photo and video spoofing attacks; even 3D face models are
used for spoofing. Face spoofing detection, which is aimed to judge the genuine
person from fake replicas, plays an important role in security systems and has
drawn much attention.

Many anti-spoofing methods have been proposed, which can be roughly classified into four categories: motion based methods [3,9,13,22], texture based
methods [2,8,14,18,20], 3D structure based methods [5–7,10,19], and fusion
methods [20,22].

Motion based methods mainly deal with the human physiological responses or
human physical motions, such as eye blinking [13,22], head rotation [3,9], and
mouth movement [9]. They can utilize relative features across several frames
and they are expected to achieve better results than some other methods, such
as texture based features. For some methods, users should strictly follow the
instructions. It usually takes a relatively long time to capture image sequences

© Springer Nature Singapore Pte Ltd. 2016
T. Tan et al. (Eds.): CCPR 2016, Part I, CCIS 662, pp. 332–344, 2016.
DOI: 10.1007/978-981-10-3002-4_28

for spoofing detection, and this kind of methods may be confused or fooled by background motions or replayed video attacks.

Texture based methods analyze skin, reflectance, and other texture properties to classify genuine and fake faces, given that many spoofing faces in photos or videos are different from genuine ones in terms of quality, blurriness, and light condition. Up to now, reflectance information [18,20], imaging banding effects [22], spectral information [14], multi-scale feature [2,23], and many other different kinds of texture based descriptors [7,8,20] are proposed. They can achieve satisfying results, but methods using texture features often do not take adjacent frame information into consideration and may be vulnerable to spoofing attacks made by high quality photos or videos.

3D structure based methods use depth information to distinguish between genuine and spoofing faces. Obviously printed photos or recorded videos show quite different depth structure. Choudhury et al. [5] mentioned that depth information can be used to detect planar faces spoofing attacks, but no further experiments were conducted. In [10], based on a binocular framework, the percentage of coplanar facial points was calculated for spoofing detection. It is said to be effective but the framework can not deal with warped spoofing images. Marsico et al. [6] presented a system exploiting 3D projective invariant. The system is quite effective and efficient, but the result highly depends on landmark precision. Erdogmus et al. [7] gave analyses on various LBP-based anti-spoofing methods using color and depth images obtained from Kinect, which relied on the depth sensors. Wang et al. [19] used normal cameras and recovered the sparse 3D shape of faces from image sequences. The recovered depth of landmarks are concatenated to form the final feature, and SVM is used as a classier. The method is able to work on different devices and achieves good performances. However, only depth information of the input images is employed, and other information, such as texture, is ignored.

Nowadays, some researchers are focusing their attentions on fusion methods, trying to make full use of input images. Erdogmus et al. [7] used both color and depth images obtained by Kinect in their countermeasures. Yan et al. [22] used three scenic clues, including non-rigid motion, face-background consistency and imaging banding effect. In [20], a concatenated feature of four different components (specula reflection, blurriness, chromatic moment, and color diversity) was extracted. Most of the fused clues are among texture and motion features, and 3D structure features are seldom utilized, which may restrain the performance of fusion methods.

With the help of various electronic devices, it is not a hard job to obtain high quality printed photos or deliberately recorded videos, which may even fool the human eyes, so, it may be not an effective way only to use texture descriptor from one image, or only to exploit human responses and motions. In this paper, we use a binocular camera system. Setting up dual cameras and obtaining two images at the same time can provide additional texture and structure information which makes face spoofing detection system more robust and effective. We propose a fused feature which combines both texture and 3D structure clues based on dual cameras. Compared with existing texture based methods, we use binocular images which can provide relative texture clues, and no specific depth sensors are needed in our method. The proposed feature can deal with both warped high quality spoofing images and replayed spoofing videos. Experiments show the promising performance of our proposed feature.

2   System Overview

Motivated by the simple fact that fooling two cameras can not be done simply with captured images or recorded videos any longer, we use dual camera systems to perform spoofing detection, which can capture two images at the same time, to increase the difficulty of face spoofing. Dual cameras are used in proposed framework. We use two cameras of the same type and mount them side by side. The two cameras are in the same direction and they are several centimeters apart. Stereo camera calibration [24] is performed before data collection process.

Our framework is illustrated in Fig. 1. The proposed feature consists of two main parts. At first, we combine left and right images and obtain one 2D gray-level image using similarity measures, and Gabor feature [11] is extracted from regions of interest in the similarity image. Then, using reconstructed sparse 3D facial structure by stereo reconstruction algorithms [24], a simplified PFH method based on PFH [17] and FPFH [16] is used to extract point cloud features. Finally, 2D texture features and 3D depth features are both normalized before concatenated into one vector feature, and we use this fused feature to represent the input image pair.

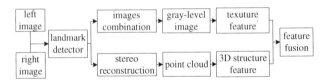

Fig. 1. Proposed feature overview

3   Proposed Feature

As we all know, the depth of genuine faces is quite sophisticated: eyes are sunken; noses are close to cameras, and ears are relatively far. So, the distances do not change smoothly especially around organs such as eyes and noses. The irregular surface contributes to complicated occlusion in face images and face images vary as view direction changes. In contrast, the depth is almost flat for printed or replayed spoofing faces; it changes smoothly even if the photo is deliberately warped. The gradually changed surface only leads to simple occlusion.

With the help of binocular camera images, we can easily analyze detailed texture differences caused by occlusion in face areas. The genuine faces are likely to have different appearances, especially where surface alters much, while spoofing faces may not contain such characteristics. We utilize this clue to construct proposed 2D-texture feature.

It should be noted that genuine and spoofing faces have quite various depth information. To make proposed methods robust and make full use of input image pairs, we also employ 3D-depth features based on depth values of the landmarks.

We have binocular framework, so we can obtain more than merely texture feature or depth feature. As discussed earlier, every single feature has its own drawbacks and the fusion feature may work better, so, it may be a wise idea to utilize both features to perform spoofing detection. As shown in our experiment in Sect. 4, the fusion method achieves better results indeed. Hence, in our system, both 2D and 3D features are exploited together to conduct final classification.

3.1   2D-Texture Feature

To extract the detailed occlusion differences in binocular images, we first adopt a similarity descriptor and combine binocular images into one gray-level image; then, we choose Gabor wavelet filter [11] to extract the differences.

Our similarity descriptor approach is inspired by BRIEF [4] and ORB [15] feature which both show that image patches can be used to extract effective and efficient local features. BRIEF [4] is an efficient feature point descriptor, which uses binary strings features and Hamming distance, and ORB is one of popular descriptors based on BRIEF. However, both descriptors are initially designed for key points only, and for two adjacent pixels, their descriptors may be similar and can not reflect possible diversities. So we propose a modified BRIEF feature descriptor M-BRIEF with Hamming distance to describe similarities for every corresponding pixel, which will be shown later.

First of all, we preprocess input images. A pre-trained supervised descent method [21] is used to locate landmarks on input faces. Then, input images are registered using eye coordinates. The normalized face images are cropped to the same size. To make the descriptor noise-insensitive, we use a Gaussian filter to smooth the normalized images.

Next, the modified BRIEF descriptors of all pixels are supposed to be calculated. For every point $\mathbf{p} = (u, v)^T$ in left normalized image, we create a patch $\mathbf{P}$ with the size $S \times S$ around it, and $\mathbf{x}$ is one of the points in the patch. At the same time, we can find a corresponding point $\mathbf{p}'$ with the same coordinate $(u, v)^T$ in right normalized image. A patch $\mathbf{P}$ can also be created and $\mathbf{y}$ is a point in the patch. Now, just like BRIEF descriptor [4], we define a similar test function τ on patch $\mathbf{P}$ as:

$$\tau(\mathbf{P}; \mathbf{x}, \mathbf{y}) := \begin{cases} 1 & \text{if } \text{sum}(\mathbf{x}) < \text{sum}(\mathbf{y}) \\ 0 & \text{otherwise} \end{cases}. \tag{1}$$

where sum($\mathbf{x}$) is the sum of intensities of pixel $\mathbf{x}$ and its eight surrounding pixels. Here we calculate the sum of surrounding pixels instead of only one pixel $\mathbf{x}$ in [4] and [15] to make it robust for potential errors in locating landmarks and registering faces. Now we randomly choose one set of n_d ($\mathbf{x}, \mathbf{y}$)-location pairs,

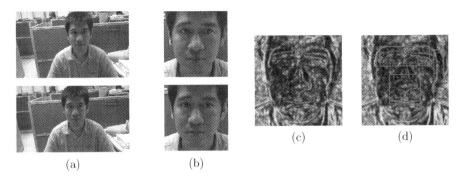

(a)   (b)   (c)   (d)

Fig. 2. Image combination process. (a) shows input left and right images, (b) shows normalized face images, (c) is combined similarity image, (d) illustrates the chosen regions

where $\mathbf{x}$ is in the left image patch and $\mathbf{y}$ is in the left image patch. We use the same set of n_d location pairs for patches of all pixeles.

In this way we uniquely define a set of binary tests. Thus the modified BRIEF descriptor of point $\mathbf{p} = (u, v)^T$ becomes one n_d-dimensional bit string:

$$\text{M-BRIEF}(\mathbf{p}) := \sum_{1 \leq i \leq n_d} 2^{i-1} \tau(\mathbf{P}; \mathbf{x}_i, \mathbf{y}_i). \tag{2}$$

We use the same parameters as mentioned in [4], where Gaussian kernel is set to 2, patch size $S \times S$ is set to 9×9 and n_d is set to 256.

After calculating M-BRIEF descriptor of all pixels, we use Hamming distance as similarity metrics between every pair of corresponding pixels with the same coordinate (u, v) in binocular images. Then, the distance is normalized to $[0, 1]$, denoted as $\mathrm{D}(u, v) \in [0, 1]$. Right now, we can simply get gray-level similarity image by $\mathrm{S}(u, v) = \text{floor}(\mathrm{D}(u, v) \times 255)$. The whole process of getting combined similarity image is illustrated in Fig. 2.

Pixels in combined gray-level images represent similarity distances between corresponding pixels of binocular images. The whiter the pixels are, the larger the differences exist in binocular images. As analyzed before, texture differences of binocular images exist especially where the depth changes dramatically, and the white pixels represent potential texture differences. For genuine faces, there are certainly more changes in depth, so, these bright white pixels may probably represent potential genuine faces.

Figure 3 shows more examples of constructed similarity images. It can be inferred that, in face areas from top row images, similarity images of spoofing ones contain relatively less bright white pixels, and streaks and organ outlines are also simpler and thinner, especially those around noses and eyes. This satisfies our hypotheses, but it is still quite hard for humans to make the correct judgements directly from similarity images. So, we adopt further feature extraction and construct our 2D-texture feature.

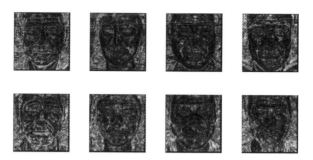

Fig. 3. Examples of constructed similarity images of spoofing faces (top row) and genuine faces (bottom row). However, in reality it is quite hard to tell them apart by humans directly.

From combined similarity images, we manually choose several regions of interest, which are considered to contain some major differences between genuine face images and spoofing ones. In this paper, based on our analyses and experiments, we choose 4 simple regions according to face landmarks, which roughly include regions around noses, mouths, eyes, and ears. We mainly choose those regions in which depth of genuine persons may alter suddenly, and regions where depths varies gradually, such as cheeks, are not used. The chosen regions are illustrated in red rectangles in Fig. 2(d).

Finally, we use Gabor wavelet filter [11] in eight directions and five scales, to extract feature vectors in each region. Downsampling process is also adopted in our system and downsampling factor is set to 64 as in [11] to perform feature reduction. Downsampled feature vectors of all chosen regions are concatenated together forming a feature vector, but Gabor texture feature is stll too long even after downsampling process. So, we employ traditional Principal Component Analysis (PCA) for dimension reduction, and only main energy is preserved, forming our final 2D-texture feature. For images in testing process, we use the same PCA parameters as those in training process.

3.2 3D-Depth Feature

Before obtaining proposed depth feature, we utilize calibrated cameras and stereo construction algorithms [24] to get depth information. In our proposed method, only depths and normal vectors of some landmarks are used, forming a sparse point cloud. The following depth features are extracted from the sparse ponit cloud we obtain.

Point feature histograms (PFH) [17] is a powerful 3D point feature, and it is used in many areas such as point cloud matching. We are inspired by PFH and FPFH [16] algorithms, and modify and simplify PFH algorithm into extracting features of sparse point cloud in our case. The main modifications of PFH [17] and our main algorithmic processes are listed as follows:

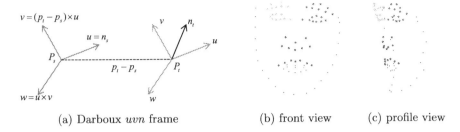

(a) Darboux *uvn* frame   (b) front view   (c) profile view

Fig. 4. (a) is illustration of Darboux *uvn* frame, (b) and (c) are front and profile views of 3D landmark structure used in proposed method. Different colors means different sets, and small points in black are not used. (Color figure online)

1. Note that we have a sparse point cloud, rather than using k-neighborhood of a specific point [17], we group landmarks manually into several sets, as illustrated in Fig. 4(b) and (c). To make our methods less noise sensitive, we omit the points on the face contour which may not be so precise, and mainly construct three sets around eyes and nose. Our 3D feature is extracted from each point set respectively.
   For point pairs $<p_s, p_t>$ in each set, their estimated surface normal vectors n_s, n_t and the Darboux *uvn* frame are defined exactly the same way as in [17] ($u = n_s, v = (p_t - p_s) \times u, w = u \times v$), which are illustrated in Fig. 4(a). It should be noted that given two sequential points $<p_s, p_t>$ and their corresponding normal vectors, the *uvn* frame will be uniquely defined.
2. The scale of faces does not change so much under our circumstance, we just omit the Euclidean distance component in original PFH method. In every set of points, we extract three primary features from point pairs, and we can get bin histogram as our 3D-feature, where each bin at index idx contains the *percentage* of the point pairs defined by idx:

$$\left.\begin{array}{l} f_1 = v \cdot n_t \\ f_2 = w \cdot n_t \\ f_3 = \frac{(u \cdot (p_t - p_s))}{|p_t - p_s|} \end{array}\right\} \quad idx = \sum_{i=1}^{i \leq 3} \text{step}(thre_i, f_i). \tag{3}$$

where $\text{step}(thre_i, f_i)$ is defined as 1 if $thre_i > f_i$ and 0 otherwise.

By setting an appropriate threshold $thre_i$, every primary feature can be turned into two categories and in this way we can obtain $2^3 = 8$ histogram bins for each depth feature. In every set of our sparse points, we calculate the histogram of all point pairs, and we can get an eight-dimension histogram feature. The final 3D-feature can be gained by concatenating histogram bins in all manually grouped sets. We have 3 sets of points in total as illustrated in Fig. 4(b), so the final dimension of proposed 3D-feature is $3 \times 8 = 24$.

4   Experiments

4.1   Dataset Description

As far as we know, there are no public binocular datasets for spoofing detection. So, we use dual cameras to collect data ourselves. The two cameras are both calibrated respectively, and stereo calibration [24] is performed. There are 35 users taking part and we allow them to move or rotate heads as they like. For each user, we use his or her high quality color photos to do the same thing imitating face spoofing attacks, and printed photos can be rotated or warped freely. We also record videos for all users and use these videos to perform tablet replay attacks. For those wearing glasses, the same goes for their printed photos and recorded videos. In our dataset, one image pair consists of one image from left camera and one image from right camera. The numbers of image pairs are shown in Table 1, and some examples are shown in Fig. 5.

Table 1. The number of image pairs in our dataset.

	Genuine	Photo spoofing	Tablet spoofing
Number of pairs	4869	4758	5674

Fig. 5. Some illustrations of our dataset. Column 1 and column 3 are from left view, and column 2 and column 4 are from right view respectively.

4.2   Baseline Method

The original BRIEF and PFH descriptor are not designed for the situations like ours as mentioned before, so, we have to choose our baseline methods.

There are only a few methods dealing with dual cameras for face spoofing detection, and spoofing photos may be warped in our dataset. We choose depth method mentioned in [19] as one of baseline methods. They proposed an algorithm that recovered the 3D sparse facial structure, and they used all 3D coordinates as a feature vector. We just calculate depth by stereo reconstruction algorithms and use the same number of landmarks as that in [19].

According to recent research [8], although many LBP based methods are proposed relatively a long time before, they are still powerful local texture descriptors in terms of face spoofing detection, such as CoA-LBP [12]. Many of the methods have public access code [8], which makes it easier and more convincing to make comparisons. Besides CoA-LBP, multiscale local binary pattern (MsLBP) and its related methods draw much attention in previous works such as [2,23]. In this paper, we find MsLBP can achieve a bit better results on our collected dataset than CoA-LBP with a linear SVM classifier, so we utilize MsLBP from [23] on every single image from our dataset as another baseline method.

4.3   Results and Discussions

As used in [8], we also employ a simple linear Support Vector Machine (SVM) as final classifier, and leave-one-out configuration is adopted. More specifically, we use images of 34 persons to train SVM model; then, we test the model on images of remaining person.

We adopt traditional False Acceptance Rate (FAR), False Rejection Rate (FRR), Equal Error Rate (EER), and Half Total Error Rate (HTER) to evaluate these methods. Besides, based on the metric developed in ISO/IEC 30107-3 [1], we also report Attack Presentation Classification Error Rate (APCER), Normal Presentation Classification Error Rate (NPCER), and Average Classification Error Rate (ACER). In our experiment tables, FAR, FRR, APCER, and NPCER are the exact values when the highest SVM classification accuracy is reached, and all the evaluation metrics are mean values according to leave-one-out configuration.

The results on both photo and tablet spoofing images between proposed method and baseline methods are shown in Tables 2 and 3 especially. We not only evaluate the results of our proposed methods, but also examine our proposed 2D and 3D feature alone.

Table 2. Results on photo spoofing images (%)

Methods	FAR	FRR	HTER	APCER	NPCER	ACER	EER
Proposed method	0.99	2.00	1.50	1.13	1.76	1.44	0.76
Proposed texture	6.10	4.19	5.15	7.21	3.53	5.37	3.52
Proposed depth	1.83	2.85	2.34	2.09	2.50	2.29	1.63
MsLBP [23]	14.78	8.91	11.85	18.56	6.93	12.74	6.71
3D coordinates [19]	1.75	5.59	3.67	1.95	5.05	3.50	2.33

Table 3. Results on tablet spoofing images (%)

Methods	FAR	FRR	HTER	APCER	NPCER	ACER	EER
Proposed method	0.78	1.21	0.99	0.66	1.42	1.04	0.63
Proposed texture	4.65	3.32	3.99	5.45	2.82	4.14	3.34
Proposed depth	1.21	1.47	1.34	1.03	1.73	1.38	0.73
MsLBP [23]	7.72	8.20	7.96	6.48	9.74	8.11	6.06
3D coordinates [19]	2.30	2.38	2.34	2.66	2.06	2.36	1.49

From both tables, we can see multiscale local binary pattern does not perform as desired. The main reason is that we use high quality images and videos; using texture alone methods such as MsLBP may not be a good idea. Besides, MsLBP can only perform on one single image, and binocular information in our dataset is not utilized at all, which is one of the disadvantages of traditional texture based spoofing detection methods.

The 3D depth based methods, which utilize binocular images to recover the depth information, perform much better than MsLBP. However, we do not get as excellent results as that in [19], in which all the images are classified correctly. It may be explained that by following reasons: we reconstruct depths only based on one left and one right image, so stereo correspondence algorithm may easily be affected by some factors, such as the glasses, bangs hair, and pale faces. These factors all leads to noisy or even mistaken depth values, which worsen the results directly. We have checked the reconstructed results of our dataset manually, and we have found for a minority of persons, especially for some female persons wearing glasses and bangs hair, the recovered depth is not so ideal. For examples, we find that results of depth based methods on persons among the first two rows in Fig. 5 are much worse than average values, which are not listed here though. So using depth based feature alone may not produce very good results and it might be a wise choice to use the fused feature instead.

Comparing results between Tables 2 and 3, it can be easily concluded that almost all methods perform better on photo spoofing images than tablet spoofing ones. The spoofing media contributes a lot to such results. Images displayed by tablet screen is always flat and screens can not be warped at all, which benefits depth based methods. In addition, the screen emits lights itself so there are no interferences by environment lighting such as shadows, which makes texture methods relatively easier.

We also evaluate methods above on all collected images, imitating the situation in which we have no idea about how the spoofing detection systems will be challenged. The experiment results are shown in Table 4, and DET curves are illustrated in Fig. 6. It should be noted that although both proposed 2D-texture and 3D-depth methods do not have overwhelming advantages over depth based baseline method [19], the fusion method obtains quite satisfied results with EER of 0.68 % and ACER of 1.40 %.

Table 4. Results on all collected images (%)

Methods	FAR	FRR	HTER	APCER	NPCER	ACER	EER
Proposed method	1.06	1.11	1.09	0.51	2.28	1.40	0.58
Proposed texture	5.62	3.90	4.76	6.62	3.30	4.96	3.54
Proposed depth	2.04	2.62	2.33	0.97	5.42	3.20	1.83
MsLBP [23]	18.63	9.49	14.06	9.07	19.41	14.24	10.52
3D coordinates [19]	2.21	1.93	2.07	2.06	2.37	2.22	1.75

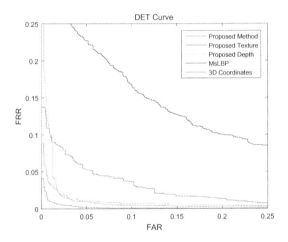

Fig. 6. Results obtained on all collected images.

5 Conclusion

In this paper, we propose a novel feature for face spoofing detection using dual cameras. On one hand, we investigate image texture details caused by different surface situations between genuine and spoofing faces. We combine a pair of binocular images into one gray-level image, on which texture features are extracted. On the other hand, we extract depth feature by using a sparse point cloud with limited number of points. Experiment results demonstrate the promising performance of the fused feature. In the future, we will focus on dealing with different kinds of spoofing attacks, such using 3D masks or 3D head models.

References

1. ISO/IEC 30107–3 Biometric presentation attack detection - part 3: testing and reporting. International Organization for Standardization (2015)
2. Arashloo, S.R., Kittler, J., Christmas, W.: Face spoofing detection based on multiple descriptor fusion using multiscale dynamic binarized statistical image features. IEEE Trans. Inf. Forensics Secur. **10**, 2396–2407 (2015)

3. Bao, W., Li, H., Li, N., Jiang, W.: A liveness detection method for face recognition based on optical flow field. In: International Conference on Image Analysis and Signal Processing, pp. 233–236. IEEE (2009)

4. Calonder, M., Lepetit, V., Strecha, C., Fua, P.: BRIEF: binary robust independent elementary features. In: Maragos, P., Paragios, N., Daniilidis, K. (eds.) ECCV 2010, Part IV. LNCS, vol. 6314, pp. 778–792. Springer, Heidelberg (2010)

5. Choudhury, T., Clarkson, B., Jebara, T., Pentland, A.: Multimodal person recognition using unconstrained audio and video. In: International Conference on Audio- and Video-Based Person Authentication, pp. 176–181. Citeseer (1999)

6. De Marsico, M., Nappi, M., Riccio, D., Dugelay, J.L.: Moving face spoofing detection via 3D projective invariants. In: IAPR International Conference on Biometrics, pp. 73–78. IEEE (2012)

7. Erdogmus, N., Marcel, S.: Spoofing in 2D face recognition with 3D masks and anti-spoofing with kinect. In: International Conference on Biometrics: Theory, Applications and Systems, pp. 1–6. IEEE (2013)

8. Gragnaniello, D., Poggi, G., Sansone, C., Verdoliva, L.: An investigation of local descriptors for biometric spoofing detection. IEEE Trans. Inf. Forensics Secur. **10**(4), 849–863 (2015)

9. Kollreider, K., Fronthaler, H., Bigun, J.: Evaluating liveness by face images and the structure tensor. In: Workshop on Automatic Identification Advanced Technologies, pp. 75–80. IEEE (2005)

10. Li, Q., Xia, Z., Xing, G.: A binocular framework for face liveness verification under unconstrained localization. In: International Conference on Machine Learning and Applications, pp. 204–207. IEEE (2010)

11. Liu, C., Wechsler, H.: Gabor feature based classification using the enhanced fisher linear discriminant model for face recognition. IEEE Trans. Image Process. **11**, 467–476 (2002)

12. Nosaka, R., Ohkawa, Y., Fukui, K.: Feature extraction based on co-occurrence of adjacent local binary patterns. In: Ho, Y.-S. (ed.) PSIVT 2011, Part II. LNCS, vol. 7088, pp. 82–91. Springer, Heidelberg (2011)

13. Pan, G., Sun, L., Wu, Z., Wang, Y.: Monocular camera-based face liveness detection by combining eyeblink and scene context. Telecommun. Syst. **47**, 215–225 (2011)

14. Pinto, A., Pedrini, H., Robson Schwartz, W., Rocha, A.: Face spoofing detection through visual codebooks of spectral temporal cubes. IEEE Trans. Image Process. **24**(12), 4726–4740 (2015)

15. Rublee, E., Rabaud, V., Konolige, K., Bradski, G.: ORB: an efficient alternative to SIFT or SURF. In: International Conference on Computer Vision, pp. 2564–2571. IEEE (2011)

16. Rusu, R.B., Blodow, N., Beetz, M.: Fast point feature histograms (FPFH) for 3D registration. In: International Conference on Robotics and Automation, pp. 3212–3217. IEEE (2009)

17. Rusu, R.B., Blodow, N., Marton, Z.C., Beetz, M.: Aligning point cloud views using persistent feature histograms. In: International Conference on Intelligent Robots and Systems, pp. 3384–3391. IEEE (2008)

18. Tan, X., Li, Y., Liu, J., Jiang, L.: Face liveness detection from a single image with sparse low rank bilinear discriminative model. In: Daniilidis, K., Maragos, P., Paragios, N. (eds.) ECCV 2010, Part VI. LNCS, vol. 6316, pp. 504–517. Springer, Heidelberg (2010)

19. Wang, T., Yang, J., Lei, Z., Liao, S., Li, S.Z.: Face liveness detection using 3D structure recovered from a single camera. In: International Conference on Biometrics, pp. 1–6. IEEE (2013)

20. Wen, D., Han, H., Jain, A.K.: Face spoof detection with image distortion analysis. IEEE Trans. Inf. Forensics Secur. **10**, 746–761 (2015)
21. Xiong, X., De la Torre, F.: Supervised descent method and its applications to face alignment. In: Computer Vision and Pattern Recognition, pp. 532–539. IEEE (2013)
22. Yan, J., Zhang, Z., Lei, Z., Yi, D., Li, S.Z.: Face liveness detection by exploring multiple scenic clues. In: International Conference on Control Automation Robotics and Vision, pp. 188–193. IEEE (2012)
23. Yang, J., Lei, Z., Liao, S., Li, S.Z.: Face liveness detection with component dependent descriptor. In: 2013 International Conference on Biometrics (ICB), pp. 1–6. IEEE (2013)
24. Zhang, Z.: A flexible new technique for camera calibration. IEEE Trans. Pattern Anal. Mach. Intell. **22**, 1330–1334 (2000)

Structured Degradation Model for Object Tracking in Non-uniform Degraded Videos

Yuan Feng, Sheng Liu$^{(\boxtimes)}$, and ShaoBo Zhang

College of Computer Science and Technology,
Zhejiang University of Technology,
Hangzhou 310023, Zhejiang, People's Republic of China
edliu@zjut.edu.cn

Abstract. Structure information is a hot spot currently in the domain of computer vision. As many people had applied structure information to their method, few people employed degradation information to their algorithm. However the degradation itself contains some important information. In this paper, we introduce a Structured Degradation Model with degradation assessment of the target to solve the tracking problem. To track the target in non-uniform degraded video, autocorrelation is used to generate the direction map and Tenengrad is used to extract the degradation degree of each target part. In our Structured Degradation Model, an undirected graph of the target is generated to track the target. The nodes of the graph are the target parts and the edges are the interactions between the parts. Experimental result shows that our method performs well especially for object tracking in degraded video.

Keywords: Super-pixel · Non-uniform degradation · Degradation assessment · Object tracking

1 Introduction

Object tracking plays an important role in computer vision, it has lots of applications in human behaviour analysis, intelligent surveillance system, military object detection, etc. Certainly, object tracking is quite a challenging task, to accomplish this task, researchers usually try to increase the accuracy and robustness in three common ways: feature, usually colours [1,2], and textures [3,4], representation model, like Support vector machine [5], and sparse representations [6,7], and structure information [8–10]. Since structure information has been widely employed in tracking. However, few people take the degradation information into account. In this paper, we introduce a structured degradation model to achieve the robustness of our algorithm for various degradation. Such as motion degradation and defocus degradation. We generate the structure information though dividing the target into several non-uniform parts with

This work was supported by Zhejiang Provincial Natural Science Foundation of China under Grant number LY15F020031 and LQ16F030007, National Natural Science Foundation of China (NSFC) under Grant numbers 11302195 and 61401397.

T. Tan et al. (Eds.): CCPR 2016, Part I, CCIS 662, pp. 345–355, 2016.
DOI: 10.1007/978-981-10-3002-4_29

Fig. 1. Some results of our method

over-segmentation and obtaining the relations among the neighbouring parts. The target is then represented by an undirected graph which the nodes is the super-pixel and the edges is the relation between the super-pixels. The degradation information is generated by finding the motion blur level and direction of the target. The information of degradation is combined in the graph construction and the matching of the graph. When tracking the undirected graph, the degradation information is used to eliminate the negative influence caused by large movements and serious defocus (Fig. 1).

Our work can be summarized as following aspects: Firstly, we introduced a new method to assess the degradation of the super-pixel parts. We assign each super-pixel part a direction by autocorrelation and a degree of degradation by Tenengrad. Secondly, we combine the degradation information into graph matching and the generating of the potential graph. Therefore, we can accurately catch the position of moving object and achieve accurate object tracking (Fig. 2).

2   Related Works

The general tracking algorithm [3,4,11] represents target as a bounding box template, structural information is not included. An on-line incremental space in the [1] model is robust to represent the target.

Some other trackers [3–5] using support vector machine or improve the effect between different target and background of a classification model. Similarly, ignorance of structural information also leads to poor performance of structural deformation and occlusion. [6,7] the objective of the model as a sparse representation of the structure of the dictionary is not sensitive to the appearance characteristics of the historical part, which is not sensitive to the change of the structure. Part-based model has been applied widely in target recognition and object detection [12,13]. There is also a part of the tracking field generated by the basis of the model [5,8,10,14,15].

At 2011, Han [16] did some researches about an alternative appearance model formulation with sparse representation, in their research the projection of the trace is used to find the sparse representation of the sub-picture feature set, the features are sampled near the tracking object. The results of tracking are associated with the most resemblances of coefficient distribution of the tracked

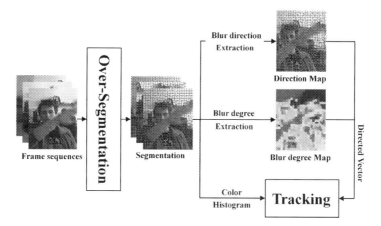

Fig. 2. Overview of our algorithm (Firstly, the frame sequence will be over-segmented to super-pixels. Secondly, we extract degradation informations such as degree and direction) from the super-pixels, then we combine the degradation information in to the procedure of segmentation of candidate parts to enhance the tracking

target. The experiments goes well in the case of occlusion. Yet, it is not sure if such a technique can track target successfully when they experience noteworthy light and stance varieties.

Despite the fact that the objective is proportional to a physically labelled part [15], just the restricted structure data is not redesigned for a company and the different parts of the template. In [9], a fixed number of parts are generated without updating and the relative positions of the parts are fixed. Won et al. [8] a portion of the method is described, and the method is very unstable, and the tracking result usually includes a bad tracking part. The former tracking [2] only calculates the probability that a part of the foreground is not subject to any structural constraint, and it is easy to distance from the tracking results of similar objects in other colours. The tracker [10] demonstrate low-level super relating pixel and chronic renal failure during the time spent figure/ground division, and high complex level of chronic renal failure(CRF) limits the further application in visual tracking.

Another similar work is [17], in Yang's work, Data mining techniques were used to help objects, and connected with their goals, to convert to a star topology graph model for robust visual tracking. While, our segmentation is combined to collect the part of candidate targets, so segmentations rather than pixel level is sufficient for the super-pixel in here to fit our work.

3 Structured Degradation Model

3.1 Degradation Assessment

We give every super-pixel a directed vector to describe the motion degradation. The direction assessment of a single super-pixel is obtained by a local auto-correlation [14] function, and the blur degree is calculated by Tenengrad [18]

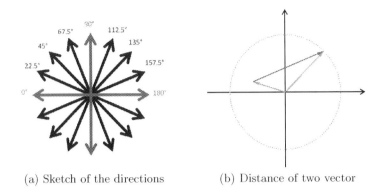

(a) Sketch of the directions   (b) Distance of two vector

Fig. 3. (a) is the sketch of the 8 normalized directions and (b) is the distance of two directed vector in different super-pixel parts

function. Then we construct a directed vector to represent the motion of single super-pixel. In this way, the direction map of the image is generated.

Direction Assessment. The local autocorrelation function for (x, y) in the local part p which centred the point:

$$f(x, y) = \sum_{(x_i, y_i) \in P} [I(x_i, y_i) - I(x_i + \Delta x, y_i + \Delta y)]^2, \tag{1}$$

where $I(x_i, y_i)$ means the gradient of $I(x_i, y_i)$ in part $p(3 \times 3)$, Δx and Δy is the deviation in the direction of x and y. The formula can be shown generally as

$$f(x, y) \cong \sum_{(x_i, y_i) \in P} [I_x(x_i, y_i) - I_y(x_i, y_i)] \begin{bmatrix} \Delta x \\ \Delta y \end{bmatrix}^2, \tag{2}$$

$$= [\Delta x, \Delta y] M \begin{bmatrix} \Delta x \\ \Delta y \end{bmatrix}^2, \tag{3}$$

where

$$M = \sum_{(x_i, y_i) \in P} \begin{bmatrix} I_x^2(x_i, y_i) & I_x(x_i, y_i) I_y(x_i, y_i) \\ I_x(x_i, y_i) I_y(x_i, y_i) & I_y^2(x_i, y_i) \end{bmatrix}, \tag{4}$$

Though the calculated matrix $M(2 \times 2)$ of pixel, there are two eigenvalue of the matrix M are obtained. The eigenvalue of the two eigenvalues which is smaller shows the direction of the pixel (direction of the motion). We take these values of each pixel to $[0, 180)$, so each value indicates the two directions in one same line. The image of motion direction I_m is generated as Fig. 4.

Then we normalize the directions into eight directions $\{0, 22.5, 45, 67.5, 90, 112.5, 135, \text{ and } 157.5\}$ as Fig. 3(a), to divide it equally and take the Maximum number of the direction as the number of direction of the super-pixel, so the degradation direction of the image part is obtained.

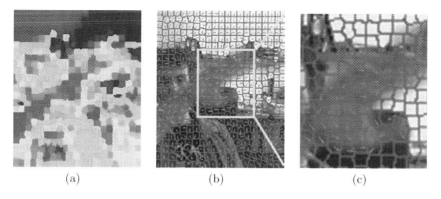

(a)   (b)   (c)

Fig. 4. (a) shows the degradation degree map of the input image (blue means the part is clear and red means part is fuzzy), (b) is the degradation direction map of the image (the direction of the green arrow is the degradation direction and the length of the arrow is the degradation degree of each part), (c) shows the enlarged result of the image in the yellow box in (b). (Color figure online)

Blur Level Assessment. Tenengrad is one of the most fast way to obtain the degradation degree information. This method is a gradient based degradation measure which has been widely used in Camera technology. It can be expressed by [18]:

$$F_{ten} = \sum_{i=1}^{M-2} \sum_{j=1}^{N-2} \left[S_x^2(i,j) + S_y^2(i,j) \right], \tag{5}$$

where $s_x(i,j)$ and $s_y(i,j)$ are respectively the convolution results of the image $f(i,j)$ with Sobel operators on the horizontal and vertical directions; M and N are respectively the image height and width. The degradation degree is given by dividing the Tenengrad result to the maximum of the Tenengrad result, it can be viewed as

$$F_p = \frac{F_{ten}}{max(F)}, \tag{6}$$

where F_{ten} is the Tenengrad result, and $max(F)$ is the maximum of the Tenengrad result. In this way, the Tenengrad result can be normalized into $(0,1]$. As a matter of fact, when the image is degraded the results of Tenengrad are usually close to zero.

The degradation vector of super-pixel is defined as:

$$\boldsymbol{d_p} = \{ F_p \boldsymbol{i_k} |\ k \in K \} \tag{7}$$

where F_p is the degradation degree $\boldsymbol{i_k}$ is the degradation direction and K is the direction number of the direction vector.

To assess the degradation, the length of the directed vector is determined by the result of Degradation degree assessment, and the direction of the vector is determined by the result of Degradation direction assessment.

3.2 Structured Model

Firstly, the tracking window of the image in our method is segmented by SLIC [19] into a set of super-pixels. Apparently, by segmenting the image into super pixels, each part have similar properties. The Dynamic Graph Tracker (DGT) proposed in [20] is adopted to track the target though colour information, and the results combined with degradation information are shown in Fig. 4.

When a set of super-pixels T_p is given, we will obtain the candidate target super-pixels set T_i^P and generate the candidate graph $G(V, E)$. Pairwise Markov Random Field (MRF) is used to take apart the candidate foreground super-pixels from background. In order to represent the appearance of the target more robustly, a generative colour histogram and a discriminative SVM classifier are concurrently combined to calculate the unary potentials. The MRF energy is respectively optimized as:

$$E(\mathbf{B}) = \sum_{p \in S} D_p(b_p) + \sum_{p,q \in N} V_{p,q}(b_p, b_q), \tag{8}$$

where $\mathbf{B} = b_p | b_p \in 0, 1, p \in S$ is the labelling of super-pixels set T_p, bp an indicator function for part T_p ($b_p = 1$ if T_p belongs to foreground and $b_p = 0$ otherwise), $D_p(b_p)$ a unary potential connected with super-pixel T_p, and $V_{p,q}(b_p, b_q)$ a pairwise potential for connected super-pixels T_p and T_q. S is the set of super-pixels in the tracking search window, and N the set of pairs of connected super-pixels with same edges.

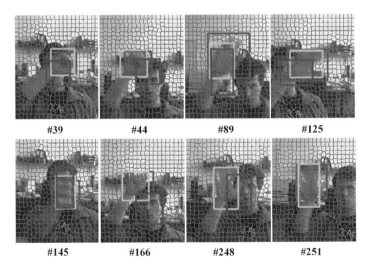

Fig. 5. The Tracking result (The blue box is ground truth, red box is the tracking result of DGT and green box is our result and the number under each frames is the frame number) (Color figure online)

The Unary Potential.

$$D_p(b_p) = \alpha D_p^g(b_p) + D_p^d(b_p), \tag{9}$$

$D_p(b_p)$ is a weighted combination of a generative colour histogram potential $D_p^g(b_p)$ and a discriminant Support Vector Machine classifier potential $D_p^d(b_p)$. $\alpha = 0.1$ is a constant to rebalance the effect of the two potential terms. The generative potential is of the form:

$$D_p^g(b_p) = \begin{cases} -\frac{1}{N_p} \sum_{i=1}^{N_p} \log P(C_i|\mathcal{H}^b) & b_p = 0, \\ -\frac{1}{N_p} \sum_{i=1}^{N_p} \log P(C_i|\mathcal{H}^f) & b_p = 1, \end{cases} \tag{10}$$

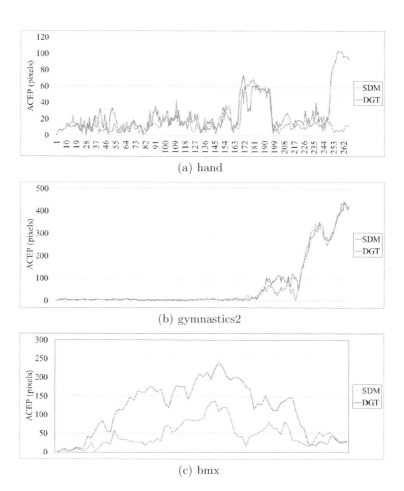

(a) hand

(b) gymnastics2

(c) bmx

Fig. 6. ACEP (Average Center Error in Pixels) result of each frame in VOT1015 sequence hand (a), gymnastics2 (b) and bmx(c). The blue line is our method and the red is DGT (Color figure online)

where $\mathcal{H}^f$ and $\mathcal{H}^b$ are normalized colour histograms of the target and the background, correspondingly, C_i is the RGB information of pixel I, and N_p the quantity of pixels in super-pixel Tp. $P(C_i|\mathcal{H})$ is the probability of C_i within histogram $\mathcal{H}$. The discriminant potential is the classification result of an on-line Support Vector Machine classifier trained from colour features obtained from the object and the background parts,

$$D_p^g(b_p) = \begin{cases} \lambda\hat{y}(f_p) & \hat{y}(f_p) \geq 0, b_p = 1, \\ 1 - \lambda\hat{y}(f_p) & \hat{y}(f_p) \geq 0, b_p = 0, \\ \lambda\hat{y}(f_p) & \hat{y}(f_p) < 0, b_p = 1, \\ 1 - \lambda\hat{y}(f_p) & \hat{y}(f_p) < 0, b_p = 0, \end{cases} \tag{11}$$

where $y(f_p) = w \cdot \Phi(f_p) + b$ is the Support Vector Machine discriminant, and f_p is the colour information of T_p. $= 15$ is the constant factor that strengthen the factor of Support Vector Machine classifier when it classifies T_p as foreground, however we want to keep real foreground super-pixels try where possible. $V_{p,q}(b_p, b_q)$ captures the discontinuity between two neighbouring super-pixels which is considered as

Smoothness Term.

$$V_{p,q}(b_p, b_q) = exp\{-D(f_p, f_q)\} - \beta\theta(\boldsymbol{d_p}, \boldsymbol{d_q}), \tag{12}$$

$$\theta(\boldsymbol{d_p}, \boldsymbol{d_q}) = |\boldsymbol{d_p} - \boldsymbol{d_q}| \tag{13}$$

where $D(f_p, f_q)$ is the X^2 distance between colour feature. d_p is the degrade feature vector. $\theta(\boldsymbol{d_p}, \boldsymbol{d_q})$ is the degrade features distance. $\beta = 0.15$ is a constant to balance the influences of the degradation vector distance, β is determined by our tests.

4   Experiments

We used two Intel(R) Xeon(R) CPU E5-2670 @ 2.60 GHz machine with Windows 10 to run our algorithm and DGT. In order to evaluate the superiority of our method, we test it on the most challenging sequences VOT2015. Lots of the challenges of Video tracking are included in the VOT2015, such as: severe occlusion, complex movement, large scale changes, abrupt movement and inner structure deformation.

4.1   Evaluation Criteria

All the methods are evaluated under two criteria: Average Center Error in Pixels (ACEP) and Success Rate. Tracking should be considered successful when its Pascal score is more than 0.5. The Pascal score is defined in the PASCAL VOC [12] where the overlaping ratio between the tracked bounding-box B_{tr} and ground truth bounding-box B_{gt} : $(area(B_{tr} \bigcap B_{gt}))/(area(B_{tr} \bigcup B_{gt}))$ A lower ACEP and a higher Success Rate are indicative of better tracking performance.

Table 1. Average center error in pixels (ACEP)

Sequences	DGT	Ours
hand	24.92	16.76
gymnastics2	66.93	62.81
singer1	73.63	22.78

Table 2. Success rate

Sequences	DGT	Ours
hand	38.20 %	57.67 %
gymnastics2	59.16 %	71.66 %
singer1	47.13 %	74.46 %

4.2 Experiment Analysis

Our tracker performs better when large appearance and structure variations occurs, as shown in Fig. 5. We mainly contrast our tracking results to Dynamic Graph Tracker (DGT) and other state-of-the-art trackers under different challenges, and we will give detailed analysis in the fol-lowing to explain the advantage of our method.

Severe movements, such as acute motion (hand, gymnatics2), sudden motion (hand, bmx), and fast rotations (bmx), are usually grim challenge for object tracking, these abnormal movements usually do not follow the movement hypothesis.

For example, in hand, most trackers failed to track the target when it starts to accelerate. Also, DGT suddenly lose the target in hand when the man abruptly change direction which make the image of hand very fuzzy.

Lots of the trackers are fragile when facing the fast moving athlete in gymnastics2. The degradation assessment of the object, the integration of appearance based segmentation and geometric constraints makes our method more sensitive to these types of abnormal movements (Fig. 6).

As what is shown in Tables 1 and 2, our method performs better in serious degraded videos, that is because we have apply degradation information in to the tracking procedure. These sequences are also tested by common tracking methods (like struck, MIL, TLD, CT, KCF), due to the abrupt motion, they both failed when the degradation occurs.

5   Conclusion

A part-based degradation structured model is proposed in this paper, in which the parts semantically generated by over-segmentation and the interaction between parts are modelled as a dynamic graph.

To assess the degradation of the video, we adopt Autocorrelation to obtain the degradation direction and Tenengrad to extract the degradation degree. The spatial prior with MRF helps us to separate the candidate target parts from the background and form the candidate graph. By calculating the distance of degradation information between super-pixels, we optimized the track result greatly.

In future work, we plan to accelerate our tracker with GPU. That would further improve the overall performance of our approach. The speed of our algorithm is approximatively 3 frames per second. We plan to accelerate it in the future work.

References

1. Lim, J., Ross, D.A., Lin, R.S., Yang, M.H.: Incremental learning for visual tracking. In: Advances in neural information processing systems, pp. 793–800 (2004)
2. Wang, S., Lu, H., Yang, F., Yang, M.H.: Superpixel tracking. In: 2011 International Conference on Computer Vision, pp. 1323–1330. IEEE (2011)
3. Grabner, H., Bischof, H.: On-line boosting and vision. In: 2006 IEEE Computer Society Conference on Computer Vision and Pattern Recognition (CVPR 2006), vol. 1, pp. 260–267. IEEE (2006)
4. Babenko, B., Yang, M.H., Belongie, S.: Robust object tracking with online multiple instance learning. IEEE Trans. Pattern Anal. Mach. Intell. **33**, 1619–1632 (2011)
5. Tian, M., Zhang, W., Liu, F.: On-line ensemble SVM for robust object tracking. In: Yagi, Y., Kang, S.B., Kweon, I.S., Zha, H. (eds.) ACCV 2007. LNCS, vol. 4843, pp. 355–364. Springer, Heidelberg (2007). doi:10.1007/978-3-540-76386-4_33
6. Mei, X., Ling, H.: Robust visual tracking using 1 minimization. In: 2009 IEEE 12th International Conference on Computer Vision, pp. 1436–1443 (2009)
7. Liu, B., Huang, J., Yang, L., Kulikowsk, C.: Robust tracking using local sparse appearance model and k-selection. In: IEEE Conference on Computer Vision and Pattern Recognition, pp. 1313–1320 (2011)
8. Kwon, J., Lee, K.M.: Tracking of a non-rigid object via patch-based dynamic appearance modeling and adaptive basin hopping monte carlo sampling. In: IEEE Conference on Computer Vision and Pattern Recognition 2009, CVPR 2009, pp. 1208–1215 (2009)
9. Adam, A., Rivlin, E., Shimshoni, I.: Robust fragments-based tracking using the integral histogram. In: IEEE Computer Society Conference on Computer Vision and Pattern Recognition, pp. 798–805 (2006)
10. Ren, X., Malik, J.: Tracking as repeated figure/ground segmentation. In: IEEE Conference on Computer Vision and Pattern Recognition, pp. 1–8 (2007)
11. Kalal, Z., Matas, J., Mikolajczyk, K.: P-N learning: bootstrapping binary classifiers by structural constraints. In: CVPR, IEEE Computer Society Conference on Computer Vision and Pattern Recognition, pp. 49–56. IEEE Computer Society Conference on Computer Vision and Pattern Recognition (2010)
12. Felzenszwalb, P.F., Girshick, R.B., Mcallester, D., Ramanan, D.: Object detection with discriminatively trained part-based models. IEEE Trans. Softw. Eng. **32**, 1627–1645 (2010)
13. Quattoni, A., Collins, M., Darrell, T.: Conditional random fields for object recognition. In: Advances in Neural Information Processing Systems, pp. 1097–1104 (2004)

14. Wang, J., Liu, S., Zhang, S.: A novel saliency-based object segmentation method for seriously degenerated images. In: IEEE International Conference on Information and Automation (2015)

15. Nejhum, S.M.S., Ho, J., Yang, M.H.: Online visual tracking with histograms and articulating blocks. Comput. Vis. Image Underst. **114**, 901–914 (2010)

16. Han, Z., Jiao, J., Zhang, B., Ye, Q., Liu, J.: Visual object tracking via sample-based adaptive sparse representation (AdaSR). Pattern Recogn. **44**, 2170–2183 (2011)

17. Tsai, D., Flagg, M., Rehg, J.M.: Motion coherent tracking with multi-label MRF optimization. Int. J. Comput. Vis. **100**, 190–202 (2010)

18. Tenenbaum, J.M.: Accommodation in Computer Vision (1970)

19. Achanta, R., Shaji, A., Smith, K., Lucchi, A., Fua, P., Süsstrunk, S.: SLIC super-pixels. EPFL (2010)

20. Cai, Z., Wen, L., Lei, Z., Vasconcelos, N., Li, S.Z.: Robust deformable and occluded object tracking with dynamic graph. IEEE Trans. Image Process. Publ. IEEE Signal Process. Soc. **23**, 5497–5509 (2014)

An Improved Background Subtraction Method Based on ViBe

Botao He[1(✉)] and Shaohua Yu[2]

[1] School of Optical and Electronic Information,
Huazhong University of Science and Technology, Wuhan, China
hebotao@hust.edu.cn
[2] Wuhan Research Institute of Posts and Telecommunications, Wuhan, China

Abstract. The classic ViBe method has shortcoming that it may detect the "Ghosting" area, when the initial frame contains a moving target or a target moves from a stationary position. In this paper, the Ghosting phenomenon was investigated, and an improved background subtraction method based on ViBe was proposed. The proposed method provided an enhanced pixel classification mechanism and background update mechanism, a significantly better Ghosting melting speed was obtained in the proposed method as compared to the classic ViBe method. The experimental results found that the proposed method had a good performance in static background scenes, and a low computational cost, that the proposed method can be used in real-time supervisory control system.

Keywords: Object detection · Background subtraction · Background modeling · Ghost elimination

1 Introduction

Moving object detection provides the basis for intelligent video surveillance and extracts motion change information from video image sequences. The target detection effect influences the performance and application of intelligent monitoring systems. In recent years, scholars have made significant advancements in the field of target detection, such as, the point detection method [16,31], image segmentation method [3,4,19], inter-frame difference method [10,28,32], optical flow method [7,18,20], the background subtraction method [11,13,17,23], motion vector field method [33] and clustering analysis method [14,15,26]. The inter-frame difference method, optical flow method and the background subtraction method are among the most commonly used methods.

The inter-frame difference method [10,28,32] is based on the strong correlation between adjacent frames and uses a differential method to detect moving targets. The principle is simple and has low computational complexity, thus it can be used in real-time video surveillance systems. However, when the time interval between two frames is very short, the method cannot detect areas of overlap and the detected target can appear larger than the real object [12].

© Springer Nature Singapore Pte Ltd. 2016
T. Tan et al. (Eds.): CCPR 2016, Part I, CCIS 662, pp. 356–368, 2016.
DOI: 10.1007/978-981-10-3002-4_30

Horn and Schunch [7] linked the two-dimensional velocity field with the pixel and establish optical flow constraint equation, and then applied the optical flow field to moving target detection. [7] calculated the dense optical flow, while [20] calculated the sparse optical flow. The application of the optical flow method does not require any prior knowledge of scene information and has good adaptability. However, the method has higher computational complexity, that the optical flow velocity estimation is an iterative process.

The background subtraction method uses the difference between the current frame and background image to detect the moving object. However, there is a challenge to maintain and update the background. There has been widespread attempts to improve the existing background subtraction method for different application, like the Mixture of Gaussians [2,22,29]. These methods can effectively suppress some disturbances in the background; thus the Mixture of Gaussians, although that has higher computational complexity, have better reliability and accuracy.

A general background subtraction method, known as ViBe (Visual Background Extractor) is presented in [1], that applied time sub-sampling and a neighborhood propagation mechanism to background modeling and updating. That the ViBe method has no specific requirements for scene content and color space, with a small amount of calculation, the method can be used for real time systems and embedded systems [24]. When the first frame contains moving objects, the classic ViBe method has high risk to detect the Ghosting area, because the method use foreground pixels to initialize the background model. Also, that the ViBe method uses a conservative update strategy, some undesirable phenomena may be observed in the experiment, such as slow elimination of the Ghosting area, failure to detect some stationary targets or some shadows and incomplete phenomenon in moving object detection. Some modifications [6,8,9,25,27,30,35] have been proposed to enhance the behavior of ViBe for specific problems.

In this paper, the performance of ViBe method is presented, an improved background subtraction method based on ViBe. The proposed method presents an improved pixel classification mechanism and background update mechanism. The subsequent chapters demonstrate the robustness of the proposed method.

2   An Improved Background Subtraction Method Based on ViBe

The background subtraction technique is proposed mainly to address the following: (1) how should a background model be created? (2) how should a background model be updated? These problems will be addressed in this section.

Firstly, the ViBe method initializes a background model for each pixel in the image, that is, the background sample values set (history storage values set). The size of the background sample values set is dependent on the complexity of the scene [24]. Secondly, a comparison between the current value of the pixel with the value of the background sample set can determine the need to update. If the pixel is a background pixel, then the value of the background sample set can be randomly updated by the value. At the same time, the value of the neighboring pixel background sample can be randomly updated by the value. The following sections will describe the proposed method in detail.

2.1   Pixel Model and Initialization

The classic ViBe method uses one frame to complete the model initialization. More specifically, the algorithm randomly samples from eight-neighboring of each pixel in the first frame of the video to create the background model. However, one frame alone cannot distinguish whether pixels belong to the foreground or background. Therefore, video sequences should be adopted to initialize the background model. In this paper, the classic ViBe background modeling method is improved using the first n frames of the video sequence to complete the initialization of the background model. Pixel (i, j) of the image is used as an example to illustrate the improved initialization method. Where, $v^t(i, j)$ is the value of pixel (i, j) at frame t, and $N_G^t(i, j)$ is the set of eight neighboring pixels of the t-th frame at pixel (i, j). The background model of pixel (i, j) is

$$m(i, j) = \{v_0(i, j), v_1(i, j), v_2(i, j), \ldots, v_N(i, j)\} \tag{1}$$

where, $v_0(i, j)$ is the mean value of the first n frames at pixel (i, j), $v_0(i, j) = \sum_{t=1}^{n} v^t(i, j)/n$. It not only plays the role of a filter, but also smoothly represents the background signals. Therefore, pixels without background characteristics can be determined. The values $v_1(i, j), v_2(i, j), \ldots, v_N(i, j)$ are randomly sampled from the eight-neighbors of pixels (i, j) of former n frames. If the width and height of the frame are w and h, the background model of the frame is as follows:

$$M^t = \begin{pmatrix} m(1,1) \ldots m(1,j) \ldots m(1,w) \\ \vdots \quad \ddots \quad \vdots \quad \ddots \quad \vdots \\ m(i,1) \ldots m(i,j) \ldots m(i,w) \\ \vdots \quad \ddots \quad \vdots \quad \ddots \quad \vdots \\ m(h,1) \ldots m(h,j) \ldots m(h,w) \end{pmatrix}_{w \times h} \tag{2}$$

For subsequent narrative convenience, x is used instead of (i, j), and the model (1) is simplified to the following:

$$m(x) = \{v_0(x), v_1(x), v_2(x), \ldots, v_N(x)\} \tag{3}$$

2.2   Pixel Classification

In order to classify the value of pixel x, the classic ViBe method is used to create a sphere $S_R(v^t(x))$ of radius R centered on $v^t(x)$ in two-dimensional Euclidean color space, and the number of pixels U is counted within the sphere $S_R(v^t(x))$.

The mathematical expression is as follows, which denotes the number of samples within the sphere $S_R(v^t(x))$.

$$U = |S_R(v^t(x)) \cap \{v_1(x), v_2(x), \ldots, v_N(x)\}| \qquad (4)$$

At the same time, the threshold U_{min} is set. When U is no less than U_{min}, $v^t(x)$ is classified as a background pixel, and otherwise is classified as a foreground pixel. When using this classification mechanism, the instance that $v^t(x)$ is assumed to be a foreground pixel and there are U_{min} sample values in the sphere $S_R(v^t(x))$ would lead to the mistaken assumption that $v^t(x)$ is a background pixel. In order to reduce the likelihood of such mistakes, this paper aims to make some improvements to the classification mechanism of pixels.

Over a short period of time, changes in background pixels can be seen as a relatively smooth process, and the mean can be regarded as a benchmark. The real background pixels change around the benchmark. However, the foreground pixels have no such features, and changes in the foreground pixels may thus be more intense. According to this policy, when the sphere $S_T(v_0(x))$ of radius T centered on $v_0(x)$ is defined, with U no less than U_{min} and the pixel within the sphere $S_T(v_0(x))$, the pixel will be classified as a background pixel. The pixel classification method is:

$$ifn(x) = \begin{cases} 0, & \text{if } U \geq U_{min} \text{ and } v^t(x) \in S_T(v_0(x)); \\ 255, & \text{else.} \end{cases} \qquad (5)$$

The sphere $S_T(v_0(x))$ has the same function as the sphere $S_R(v^t(x))$ in order to ensure consistency between parameter T and parameter R. The value of T is defined as follows:

$$T = \begin{cases} R, & \text{if } \sigma \leq \beta R; \\ \sigma/\beta, & \text{else.} \end{cases}, \sigma = \sqrt{\frac{1}{n} \sum_{t=1}^{n} (v^t(x) - v_0(x))^2} \qquad (6)$$

where, σ reflects the historical scene change intensity, and smaller σ indicates slow historical scene changes. In order to ensure the accuracy of the algorithm when the value of σ is very large, β is recommended to take an integer value larger than 2.

The classification process that $N = 7$ and $U_{min} = 2$ is illustrated in Fig. 1. (C_1, C_2) is a 2-D Euclidean color space. Based on (3), (4), and (5), it is known that $v^t(x)$ is classified as a background pixel value.

2.3   Update of the Model

The model of t-th frame at pixel x is as: $m^t(x) = \{v_0^t(x), v_1(x), v_2(x), \ldots, v_N(x)\}$. The update method incorporates the following components:

(1) Memory-less update policy. The classic ViBe method uses a memory-less update policy to update the background model. That is, if the value $v^t(x)$

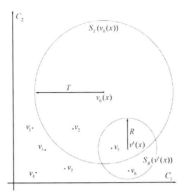

Fig. 1. The classification process in a 2-D Euclidean color space (C_1, C_2).

of the pixel x is classified as a background pixel value, and a value from the background model $v_i(x)(i = 1, 2, \ldots, N)$ is randomly selected; using $v^t(x)$ instead of $v_i(x)$. The application of a memory-less update policy can effectively reduce the probability of discarded background pixels. As a result, the proposed method follows this policy.

(2) Memory update policy. The classic ViBe method uses a memory-less update policy that ignores the linkage between frames. In this paper, the method proposed also uses a memory update policy, and the specific update method is as follows:

$$v_0^t(x) = \sum_{i=1}^{n} v^{t-i+1}(x)/n = v_0^{t-1}(x) + \frac{1}{n}v^t(x) - \frac{1}{n}v_0^{t-n+1}(x) \tag{7}$$

$$T = \begin{cases} R, & \text{if } \sigma \leq \beta R; \\ \sigma/\beta, & \text{else.} \end{cases}, \sigma = \sqrt{\frac{1}{n}\sum_{i=1}^{n}(v^{t-i+1}(x) - v_0(x))^2} \tag{8}$$

The memory update policy links to the history values of pixel x and inhibits random noise. The policy can limit the background model update frequency, although it cant only effectively use history values of the pixel.

(3) Spatial propagation. The classic ViBe method assumes that the neighboring background pixels share a similar temporal distribution. As a result, a new background sample of a pixel should also update the models of neighboring pixels. In this paper, the proposed method follows this rule with a propagation method that considers spatial consistency and allows for spatial diffusion.

(4) Foreground pixels counting policy. The random update policy ensures exponential decay in the probability of a sample value remaining inside the background sample values set. However, the classic ViBe method does not intervene with pixels classified as foreground pixels. The proposed method

uses the foreground pixels counting method to change the long term foreground pixels into background pixels or to quickly eliminate Ghosting pixels. A pixel detected as a foreground pixel for continuous M times will be updated to a background pixel through the foreground pixels counting mechanism. In order to reduce the amount of calculations, a pixel x with continuous M detected as a foreground pixel will use the first m results ($M \geq m$) in the following statistical method:

$$A_t(x) = if n_{t-m}(x) \cap if n_{t-m+1}(x) \cap \cdots \cap if n_{t-1}(x) \qquad (9)$$

Binary image $A_t(x)$ is the collection of pixels continuous m times detected as foreground pixels. An edge corrosion method is used to detect the edge of foreground pixels and the corrosion result of $A_t(x)$ is $E(A_t(x))$.

$$D_t(x) = A_t(x) - E(A_t(x)) \qquad (10)$$

$D_t(x)$ is the edge image of the binary image $A_t(x)$, and a pixel y of $N_G^t(x)$ is randomly selected. When $D_t(y) = 0$, $v^t(x)$ is classified according to (5). If $v^t(x)$ is classified as a background pixel, the background model $m^t(x)$ is updated. Otherwise, no action is taken.

The classic ViBe method uses a random time sub-sampling policy to reduce the frequency of the background update, which can improve computing speed and efficiency. In this paper, the proposed method uses sphere and a foreground pixels counting policy to limit the frequency of updates.

3   Experimental Results

In this section, the optimal values for the parameters of the proposed method will be determined. The proposed method will also be compared with the classic ViBe method and its improved methods using the public dataset2012 provided on the http://www.changedetection.net website. In this section, all the used sequences are available on the website. The optimal values of the classic ViBe method parameters is also used in the proposed method.

3.1   Determination of Parameters

From previous discussion, it is all known that the following parameters needed to be determined for the proposed method: the initialization frames number n, the factor β, the statistic values M and m. As shown in the analysis of the second chapter, the factor β is influential for the background model update frequency, and the value of β is dependent on the scene change speed. In the experience, the use of a factor $\beta = 3$ leads to excellent results in every situation.

The initialization and update method of $v_0(x)$ is similar to the method of the average background model. With the increased of frames n, the background model will be produced to closer the real background. Zheng and Fan [34] proposed that the number n of the average background model initialization frames is generally set between 30 and 1000. Where, the proposed method needed to calculate the standard deviation of consecutive n frames in the model update process, the number n was set to 20 in the follow-up experiment in order to guarantee the efficiency of the method.

M and m not only determine the update frequency of the background model, but also relate to the fusion rate of the detected foreground and detected background. The Ghosting area is used as an example to illustrate the impact of M and m. In a Ghosting area where other parameter values are unchanged, a larger M value will have Ghosting with a longer life time and a slower melting rate, which also means that a larger M will have a static object with a longer life time. Meanwhile, the background model update frequency decreases with the increase of M. If the other parameter values are unchanged, the value of m has a large influence on computing efficiency. Where $M \geq m$, to increase m value will correlate with a slower background model update. In this paper, the computing veracity of the proposed method is assured through setting the parameter $m = 10$. The proposed method computes the evolution of the Percentage of Wrong Classifications(PWC) of the "intermittent object motion" and "thermal" categories for M ranging from 30 to 100 to determine an optimal value for M. The terms of the metric PWC [5] as follow:

$$PWC = 100 * (TP + TN)/(TP + TN + FP + FN) \tag{11}$$

where, (a) True positive (TP): the number of foreground pixels are correctly detected. (b) False positive (FP): the number of background pixels are classified as a foreground pixel number. (c) True negative (TN): the number of background pixels are correctly detected. (d) False negative (FN): the number of foreground pixels is classified as a background pixel number. The method hopes the PWC to be as small as possible according to (11). The "intermittent object motion" category contain 6 sequences: abandoned Box, parking, sofa, street Light, tram stop, winter Driveway; while, the "thermal" category contain 5 sequences: corridor, din-

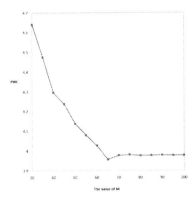

Fig. 2. PWC for M ranging from 30 to 100. The other parameters of proposed method were to $R = 20, N = 20, U_{min} = 2, m = 10, \beta = 3, n = 20$.

ing Room, lake Side, library, park. Values of the average PWC of these sequences is shown in Fig. 2, which shows that the minimum PWC is obtained for $M = 65$.

3.2   Ghost Phenomenon

The so-called "Ghosting", according to Ref. [24] is where "a series of pixels are detected as moving targets, but these pixels represent are not really moving objects". The causes of a Ghosting area can be divided into two categories: (1) the background model initialization contains moving objects; or (2) the update rate of the background and the change rate of the scene cannot match and a false target is detected when a target moves from a stationary position. To address these, the video sequence "winter Driveway" was used to compare the effect of the classic ViBe method and the proposed method. In the results, the white and black represent foreground and background pixels.

In Fig. 3(a), the car began to move from the 1811th frame, so, the background model initialization contained the target region. If the update rate of the background and the change rate of the scene cannot match, the region where the car initially located presented as a Ghosting region. For the classic ViBe uses spatial propagation which has low Ghosting melting speed, whereas, the proposed method has the improved pixel classification mechanism and background update mechanism that can eliminates the Ghosting area at a faster rate.

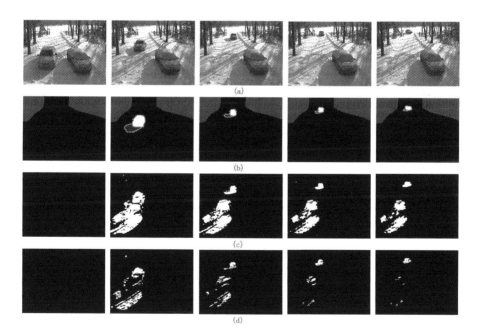

Fig. 3. A target moves from a stationary position (winter Driveway). (a) the 1500th, 1900th, 2000th, 2100th, 2200th frame; (b) the real background corresponding with (a); (c) The detection results of ViBe; (d) The detection results of the proposed method.

3.3   Comparison of the Accuracy of Segmentation

This section will compare the proposed method with the classic ViBe method and its improved methods under different scenarios using the public dataset2012. The dataset2012 contains 31 video sequences, and has be grouped in 6 categories: baseline, camera jitter (cam.jitter), dynamic background (dyn.bg), intermittent object motion (int.obj.mot), shadow, and thermal. The proposed method donot detail their content as the names of the categories are quite explicit. The official metrics are Recall (Re), Specificity (Sp), False Positive Rate (FPR), False Negative Rate (FNR), Percentage of Wrong Classifications (PWC), Precision (Pr) and F-Measure (FM). The method proposed many modifications to the classic ViBe method. From a practical point of view, it is very difficult to isolate the effects of each change separately either because the changes interact or because their behavior depends on the video sequence. Therefore, in this paper only presents the global results. For these experiments, the proposed method uses an unique set of parameters $R = 20, N = 20, U_{min} = 2, M = 65, m = 10, \beta = 3, n = 20$. All these video sequences are processed and then binary masks (where a 0 value represents the background) are compared to groundtruth masks.

In Table 1 results using the seven proposed performance measures are shown for all six scenarios. In the results, the overall performance in the baseline, shadow and thermal categories is very good, and that the Recall and F-Measure metrics of the three categories are higher. The "camera jitter" and the "intermittent object motion" categories pose bigger challenge which can be seen from the value of PWC. The proposed method has better performance in static background scenes which can be seen from the measures of "overall-s", that the values denote the average measure of the all categories apart from the "dynamic background" category and the "camera jitter" category.

Table 1. Results of proposed method.

Scenarios	Re	Sp	FPR	FNR	PWC	Pr	FM
baseline	0.862	0.998	0.002	0.138	0.701	0.930	0.895
cam.jitter	0.471	0.976	0.024	0.529	4.420	0.553	0.492
dyn.bg	0.458	0.995	0.005	0.542	1.069	0.635	0.475
int.obj.mot	0.575	0.987	0.013	0.425	4.596	0.764	0.641
shadow	0.952	0.990	0.010	0.048	1.136	0.837	0.889
thermal	0.847	0.991	0.009	0.153	1.623	0.809	0.825
overall	0.694	0.989	0.011	0.306	2.257	0.755	0.703
overall-s	0.809	0.992	0.008	0.191	2.014	0.835	0.812

The results show in Tables 2 and 3 how the proposed method compares to ViBe, ViBe+ [25], PBAS [6] and SuBSENSE [21] for each category. These tables have displayed that the proposed method has best performance for the "shadow" category compared to the other methods. When the area of the dynamic background is small in the scene, such as the waving trees of the "baseline" category, the proposed method has a good performance (Fig. 4). When the area of the dynamic background is big in the scene and the area of the dynamic background is much larger than the area of the moving target, such as the sparkling lake, the fountain, the falling leaves of the "dynamic background" category, that the proposed method is slightly less efficient than the other four methods. Also, the proposed method is less efficient for the "camera jitter" category. Not to surprise as modifications that produces primarily to enhance the behavior of ViBe for Ghosting melting. Meanwhile, the proposed method has better performance than the ViBe and ViBe+ methods which can be seen from the measures of "overalls". Although the proposed method is slightly less efficient than the PBAS and SuBSENSE methods, the proposed method has lower computational cost that is 34 ms per frame size 320*240 (3.30 GHz, Core(TM) i3-3220 CPU, 4 GB of RAM) than PBAS (105 ms) and SuBSENSE (239 ms), which can be used in real-time supervisory control system.

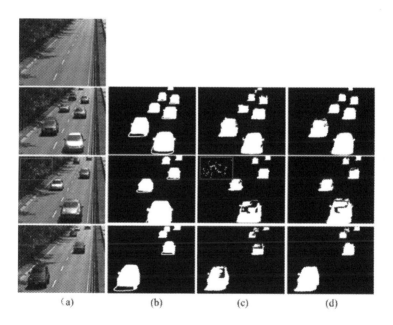

(a)   (b)   (c)   (d)

Fig. 4. The compared performance of ViBe and the proposed method when the area of the dynamic background is small in the scene (highway). (a) the 480th, 827th, 1068th 1366th frame, red rectangle highlight the waving trees; (b) the real background corresponding with (a); (c) The detection results of ViBe, red rectangle highlight the Ghosting area; (d) The detection results of the proposed method. (Color figure online)

Table 2. Comparison of the proposed method to the others using Re.

Scenarios	ViBe	ViBe+	PBAS	SuBSENSE	Proposed
baseline	0.820	0.828	0.959	0.952	0.862
cam.jitter	0.711	0.729	0.737	0.824	0.471
dyn.bg	0.722	0.762	0.696	0.777	0.458
int.obj.mot	0.512	0.473	0.670	0.658	0.575
shadow	0.784	0.811	0.913	0.942	0.952
thermal	0.543	0.541	0.728	0.816	0.847
overall	0.682	0.691	0.784	0.828	0.694
overall-s	0.665	0.663	0.818	0.842	0.809

Table 3. Comparison of the proposed method to the others using FM.

Scenarios	ViBe	ViBe+	PBAS	SuBSENSE	Proposed
baseline	0.870	0.872	0.924	0.950	0.895
cam.jitter	0.599	0.754	0.722	0.815	0.492
dyn.bg	0.565	0.720	0.683	0.818	0.475
int.obj.mot	0.507	0.509	0.575	0.657	0.641
shadow	0.803	0.815	0.860	0.899	0.889
thermal	0.665	0.665	0.756	0.817	0.825
overall	0.668	0.722	0.753	0.826	0.703
overall-s	0.711	0.715	0.779	0.831	0.812

4   Conclusions

In object detection, the presence of a moving target in the initial frame can lead to issues in distinguishing whether a pixel belongs to the foreground or background. To compare the classic ViBe method, the proposed method has the improved pixel classification mechanism and background update mechanism that can eliminates the Ghosting area at a faster rate. The experimental results have showed that the proposed method has a good performance in static background scenes. Meanwhile, the proposed method has a low computational cost which can be used in real-time supervisory control system.

References

1. Barnich, O., Van Droogenbroeck, M.: Vibe: a powerful random technique to estimate the background in video sequences. In: IEEE International Conference on Acoustics, Speech and Signal Processing, ICASSP 2009, pp. 945–948, April 2009
2. Bilmes, J.: A gentle tutorial of the EM algorithm and its application toparameter estimation for Gaussian mixture and hidden Markov models. Technical report, International Computer Science Institute, Berkley, CA, April 1998

3. Caselles, V., Kimmel, R., Sapiro, G.: Geodesic active contours. In: Fifth International Conference on Computer Vision, Proceedings, pp. 694–699, June 1995
4. Comaniciu, D., Meer, P.: Mean shift analysis and applications. In: The Proceedings of the Seventh IEEE International Conference on Computer Vision, vol. 2, pp. 1197–1203 (1999)
5. Goyette, N., Jodoin, P., Porikli, F., Konrad, J., Ishwar, P.: Changedetection.net: a new change detection benchmark dataset. In: 2012 IEEE Computer Society Conference on Computer Vision and Pattern Recognition Workshops (CVPRW), pp. 1–8, June 2012
6. Hofmann, M., Tiefenbacher, P., Rigoll, G.: Background segmentation with feedback: the pixel-based adaptive segmenter. In: 2012 IEEE Computer Society Conference on Computer Vision and Pattern Recognition Workshops (CVPRW), pp. 38–43, June 2012
7. Horn, B.K., Schunck, B.G.: Determining optical flow. Artif. Intell. 17(1–3), 185–203 (1981)
8. Kumar, K., Agarwal, S.: A hybrid background subtraction approach for moving object detection. In: Confluence 2013: The Next Generation Information Technology Summit (4th International Conference), pp. 392–398, September 2013
9. Leng, B., He, Q., Xiao, H., Li, B., Wang, H., Hu, Y., Wu, W., Guan, G., Zou, H., Liang, L.: An improved pedestrians detection algorithm using HOG and ViBe. In: 2013 IEEE International Conference on Robotics and Biomimetics (ROBIO), pp. 240–244, December 2013
10. Lipton, A., Fujiyoshi, H., Patil, R.: Moving target classification and tracking from real-time video. In: Fourth IEEE Workshop on Applications of Computer Vision, WACV 1998, Proceedings, pp. 8–14, October 1998
11. Liu, C.: Gabor-based kernel PCA with fractional power polynomial models for face recognition. IEEE Trans. Pattern Anal. Mach. Intell. 26(5), 572–581 (2004)
12. Liu, G., Ning, S., You, Y., Wen, G., Zheng, S.: An improved moving objects detection algorithm. In: 2013 International Conference on Wavelet Analysis and Pattern Recognition (ICWAPR), pp. 96–102, July 2013
13. Monnet, A., Mittal, A., Paragios, N., Ramesh, V.: Background modeling and subtraction of dynamic scenes. In: Ninth IEEE International Conference on Computer Vision, Proceedings, vol. 2, pp. 1305–1312, October 2003
14. Papageorgiou, C., Oren, M., Poggio, T.: A general framework for object detection. In: Sixth International Conference on Computer Vision, pp. 555–562, January 1998
15. Rowley, H., Baluja, S., Kanade, T.: Neural network-based face detection. IEEE Trans. Pattern Anal. Mach. Intell. 20(1), 23–38 (1998)
16. Ryu, J.B., Park, H.H.: Log-log scaled harris corner detector. Electron. Lett. 46(24), 1602–1604 (2010)
17. Saleemi, I., Shafique, K., Shah, M.: Probabilistic modeling of scene dynamics for applications in visual surveillance. IEEE Trans. Pattern Anal. Mach. Intell. 31(8), 1472–1485 (2009)
18. Shen, X., Wu, Y.: Exploiting sparsity in dense optical flow. In: 2010 17th IEEE International Conference on Image Processing (ICIP), pp. 741–744, September 2010
19. Shi, J., Malik, J.: Normalized cuts and image segmentation. IEEE Trans. Pattern Anal. Mach. Intell. 22(8), 888–905 (2000)
20. Spruyt, V., Ledda, A., Philips, W.: Sparse optical flow regularization for real-time visual tracking. In: 2013 IEEE International Conference on Multimedia and Expo (ICME), pp. 1–6, July 2013

21. St-Charles, P.L., Bilodeau, G.A., Bergevin, R.: Subsense: a universal change detection method with local adaptive sensitivity. IEEE Trans. Image Process. **24**(1), 359–373 (2015)
22. Stauffer, C., Grimson, W.: Adaptive background mixture models for real-time tracking. In: IEEE Computer Society Conference on Computer Vision and Pattern Recognition, vol. 2, pp. 246–252 (1999)
23. Stauffer, C., Grimson, W.: Learning patterns of activity using real-time tracking. IEEE Trans. Pattern Anal. Mach. Intell. **22**(8), 747–757 (2000)
24. Van Droogenbroeck, M., Barnich, O.: ViBe: a universal background subtraction algorithm for video sequences. IEEE Trans. Image Process. **20**(6), 1709–1724 (2011)
25. Van Droogenbroeck, M., Paquot, O.: Background subtraction: experiments and improvements for ViBe. In: 2012 IEEE Computer Society Conference on Computer Vision and Pattern Recognition Workshops (CVPRW), pp. 32–37, June 2012
26. Viola, P., Jones, M., Snow, D.: Detecting pedestrians using patterns of motion and appearance. In: Ninth IEEE International Conference on Computer Vision, Proceedings, no. 2, pp. 734–741, October 2003
27. Wang, H., Suter, D.: Background subtraction based on a robust consensus method. In: 18th International Conference on Pattern Recognition, ICPR 2006, vol. 1, pp. 223–226 (2006)
28. Weng, M., Huang, G., Da, X.: A new interframe difference algorithm for moving target detection. In: 2010 3rd International Congress on Image and Signal Processing (CISP), vol. 1, pp. 285–289, October 2010
29. White, B., Shah, M.: Automatically tuning background subtraction parameters using particle swarm optimization. In: 2007 IEEE International Conference on Multimedia and Expo, pp. 1826–1829, July 2007
30. Xu, H., Yu, F.: Improved compressive tracking in surveillance scenes. In: 2013 Seventh International Conference on Image and Graphics (ICIG), pp. 869–873, July 2013
31. Xu, W., Huang, X., Li, X., Zhang, Y., Zhang, J., Zhang, W.: An affine invariant interest point and region detector based on gabor filters. In: 2010 11th International Conference on Control Automation Robotics Vision (ICARCV), pp. 878–883, December 2010
32. Yin, Z., Collins, R.: Moving object localization in thermal imagery by forward-backward MHI. In: Conference on Computer Vision and Pattern Recognition Workshop, CVPRW 2006, pp. 133–140, June 2006
33. Zhao, Y., Fan, X., Liu, S.: Fast motion region segmentation based on motion vector field. In: 2012 International Conference on Wavelet Active Media Technology and Information Processing (ICWAMTIP), pp. 153–156, December 2012
34. Zheng, Y., Fan, L.: Moving object detection based on running average background and temporal difference. In: 2010 International Conference on Intelligent Systems and Knowledge Engineering (ISKE), pp. 270–272, November 2010
35. Zhu, F., Jiang, P., Wang, Z.: ViBeExt: the extension of the universal background subtraction algorithm for distributed smart camera. In: 2012 International Symposium on Instrumentation Measurement, Sensor Network and Automation (IMSNA), vol. 1, pp. 164–168, August 2012

Novel Face Hallucination Through Patch Position Based Multiple Regressors Fusion

Changkai Jiao, Zongliang Gan$^{(\boxtimes)}$, Lina Qi, Changhong Chen, and Feng Liu

Jiangsu Provincial Key Lab of Image Processing and Image Communication, Nanjing University of Posts and Telecommunications, Nanjing 210003, China
{1014010410,ganzl,qiln,chenchh,liuf}@njupt.edu.cn

Abstract. The task of face hallucination is to estimate one high-resolution (HR) face image from the given low-resolution (LR) one through the learning based approach. In this paper, a novel local regression learning based face hallucination is proposed. The proposed framework has two phases. In the training phase, after the training samples is separated into several clusters at each face position, the Partial Least Squares (PLS) method is used to project the original space onto a uniform manifold feature space and multiple linear regression are learned in each cluster. In the prediction phase, once the cluster of the LR patch is gotten, the corresponding learned regression function can be used to estimate HR patch. Furthermore, a multi-regressors fusion model and HR induced clustering strategy are proposed to further improve the reconstruction quality. Experiment results show that the proposed method has a very competitive performance compared with other leading algorithm with low complexity.

Keywords: Face hallucination · Local regression · Partial Least Squares · Multi-regressors fusion · HR induced clustering

1 Introduction

Face hallucination [1, 2] aims to infer the HR face image from the LR one based on machine learning approach, in which the relationship between the LR and HR patch pairs can be gotten by the face image database. As known, the face image resolution is usually low in surveillance video, but the facial features are very important to face recognition application. Thus, the face hallucination methods can be used to get a HR face image from the LR ones to improve the recognition results. Ma *et al.* [3] firstly proposed a face hallucination method by position-patch, where the training LR and HR samples are well alignment. The method predicts the HR patch using the same position patches of each training samples, and the training samples position-patches

This research was supported in part by the National Nature Science Foundation, P.R. China (No. 61071166, 6172118, 61071091, 61471201), Jiangsu Province Universities Natural Science Research Key Grant Project (No. 13KJA510004), and the "1311" Talent Plan of NUPT.

© Springer Nature Singapore Pte Ltd. 2016
T. Tan et al. (Eds.): CCPR 2016, Part I, CCIS 662, pp. 369–382, 2016.
DOI: 10.1007/978-981-10-3002-4_31

weights are calculated. After, the same weights are used to estimate the HR patches. And, many face hallucination methods [7, 8, 10, 18, 19] follow this idea. In this paper, we mainly focus on patch position based method for face hallucination.

Generally, several of face hallucination methods have been proposed and there are three kinds of face hallucination method. Including, Maximum a Posteriori (MAP) based method [4, 5], manifold consistency learning method [6, 7], regression based method [9–11]. For MAP method, the maximum a posteriori method is used in face hallucination framework. Capel and Zisserman [4] firstly use principal component analysis (PCA) method to decompose the training samples, and combine the MAP estimation to generate a HR face image from principal feature space. Liu et al. [5] proposed a two-step method to integrate a global linear model and a local nonparametric model in Maximum a posteriori solution frame for face hallucination. For manifold consistency learning, the face hallucination methods are conducted in the unified feature space. Li et al. [6] perform alignment by learning two mapping which project the HR and LR pairs from the origin manifold into the common manifold. Hao et al. [7] employ Easy-Partial Least Squares as a local constraint to learn two mappings through project LR and HR samples onto a common feature space. The reconstruct result in the subspace is satisfactory. There also has been some other method on face super-resolution using Partial Least Squares [8].

For regression based method, the relationship between LR and HR samples can be directly learned instead of consider the manifold consistent assumption. The regression-based methods have been already used in different frameworks for face hallucination. Ni and Nguyen [9] proposed a face hallucination algorithm via support vector regression. Jiang et al. [10] introduced a face hallucination algorithm, which learning a mapping function between the local support LR and the HR patches, the local support can be acquired from the HR or the LR manifolds. Moreover, [11] is also based on this regression method. In the recent years, some face hallucination methods [12–14] based regularization constraint and Convolutional Neural Network (CNN) also been proposed.

In this paper, we proposed a patch position based method to learn the multiple regression models in the feature subspace. There are two steps on our method. During training, the main process includes following: Firstly, the training patches are divided into several clusters at each position, then projection the origin space onto a common manifold space by PLS method and learning multiple regressors for each cluster; Secondly, we applied a cross validation method to iteration learning the training samples partition and regression models until convergence. During prediction, once the most nearest cluster of the input LR patch is obtained, the corresponding appropriate regressor model is used to predict the HR patch. Moreover, a multi-regressors model and HR induced clustering are introduced to refine improvements the reconstruction quality. The optimal K nearest neighbor (K-NN) patches are unknown before reconstruction, so we should guarantee the numbers of nearest patches are adequate for reconstruction. Therefore, we can search several nearest clusters and use a multi-regressors strategy to improve the result. HR manifold space is more credible than LR space, which is not influenced by degradation. Thus, this information can be used to guide K-means clustering. Experiments result shown that the HR induced clustering is better than LR induced clustering method. Compared to other method, our method not needs to search the KNN, so the computational complexity is low.

2 Related Works

2.1 Regression-Based Face Hallucination

Give a set of LR-HR face sample pairs, which have been well-aligned by the position of two eyes. All the samples are divided into patches utilizing the same strategy according to the position. We can to learn a mapping function among the K nearest neighbor patch pairs at the same position to predict the target HR patch given an input LR patch. The neighbor local manifold geometric structure information is applied to search the K nearest neighbor patches. Thus the locally map from LR patch manifold to HR patch manifold maintain the manifold geometrical structure of LR-HR training samples and the data fidelity. Integrating all the obtained HR patches according to the position to generate the final HR face image. This way is shown in Fig. 1.

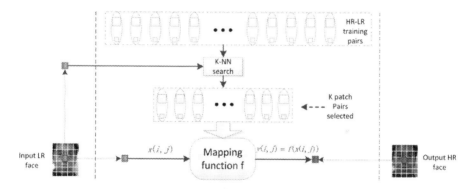

Fig. 1. Flowchart of regression-based face hallucination

The motivation of our regression-based method to generate HR images is derived from the spirit of [10]. Jiang *et al.* [10] proposed a locality-induced support regression face hallucination algorithm, which utilized the local geometrical structure to model the relationship between the LR training patches and the HR training patches. The local geometrical structure preservation of manifold learning concept was applied in this article. When an input LR patch $x(i,j)$ at position (i,j) was given, the K-NN patches of $x(i,j)$ is acquired by the Euclidean distances between $x(i,j)$, the LR training patch matrix $X = \{x_1(i,j), x_2(i,j), \cdots, x_M(i,j)\} \in R^{J \times M}$:

$$\text{dist}_K = sort(\|x(i,j) - x_t(i,j)\|_2^2)|_K, \quad 1 \le t \le M \tag{1}$$

where $dist_K$ is a set of indexes of the K nearest neighbor LR patches with the smallest Euclidean distance, $x_t(i,j)$ is the LR training patch which have same position with $x(i,j)$ LR patch, M is the number of training samples at current position. Thus, we can get the corresponding HR patches from HR training patch matrix $Y = \{y_1(i,j), y_2(i,j), \cdots, y_M(i,j)\} \in R^{l \times M}$ according to $dist_K$. Suppose $X_K \in R^{J \times K}$ is the matrix of K nearest neighbor LR patches, $Y_K \in R^{l \times K}$ is the matrix of K nearest neighbor HR patches, the regression function obtained:

$$F = argmin_F \|Y_K - FX_K\|^2 \tag{2}$$

where F is the regression function. So the HR patch $y(i,j)$ can be predicted by:

$$y(i,j) = F(x(i,j)) \tag{3}$$

Traverse all patches according to the position to acquire all HR patches, and integrate them to obtain the target HR face image.

Despite that the method [10] to conduct face hallucination with a satisfying results, there still exist drawbacks. The local regression model learning is directly in the pixel space, and it is not clear whether the space is optimal for the learned local regression model. To overcome this drawback, in this article we propose a local regression model learning in a compact feature space. We use Partial Least Squares to analyze the common latent variables present in the LR-HR patch pairs and project the pixel space onto a compact feature space to learn the local regression models. Experimental results show that the reconstruction images in the feature space are performed results in the pixel space.

2.2   Partial Least Squares Latent Variables Analysis

Partial least squares is to find the shared information present in data that collected from the same of observations, which obtain two sets of latent variables, there are liner combinations of variables of data, and latent variables are required to have maximal covariance.

Partial least squares analyze the information common to two data matrix. Suppose the one matrix is $X \in R^{J \times M}$, another is $Y \in R^{l \times M}$. The column of $X(Y)$ is an observation sample. Z_X is normalization of X, Z_Y is normalization of Y. The correlation matrix denoted C is computed as:

$$C = (Z_Y)^T Z_X \tag{4}$$

Decomposes C use SVD method as follows:

$$C = U \Delta V^T \tag{5}$$

The goal of PLS is to find two latent vectors $L_{X,i}$ and $L_{Y,i}$ with maximal covariance. So, we want to find $L_{X,i} = Z_X V_i$ and $L_{Y,i} = Z_Y U_i$ satisfy the following term:

$$cov(L_{X,i}, L_{Y,i}) \propto L_{X,i}^T L_{Y,i}^T = V_i^T Z_X^T Z_Y U_i = max \tag{6}$$

subject to

$$U_i^T U_i = V_i^T V_i = \mathbf{I} \tag{7}$$

Then, iterative method is repeated till the wanted dimensions of feature have been obtained. The iterative steps are as follows:

$$Z_{X,i+1}^T = Z_{X,i}^T - Z_{X,i}^T V_i V_i^T \tag{8}$$

$$Z_{Y,i+1}^T = Z_{Y,i}^T - Z_{Y,i}^T U_i U_i^T \tag{9}$$

So the two projection matrices $V \in R^{M \times q}$ and $U \in R^{M \times q}$, q is less than or equal to the minimal of J and I.

3   Proposed Method

3.1   Overview of Our Algorithm

In general learning based methods, given an input LR image, the learning based methods firstly find its k-NN patches in the LR patch set. And then obtain the k-NN HR patches according to the index of LR patch. So we can learn the local regression model in the LR-HR patch pairs. From above, we can find that the reconstruction time computational complexity is relatively high and hence limits its use in application where available at limited time. In order to solve this problem, we proposed a clustering method to partitioning the training samples into local clusters at each position and learning a local regression model for each cluster. For each input LR patch, the most appropriate regression model is selected to generate HR patch. Figure 2 shows an overview of our method.

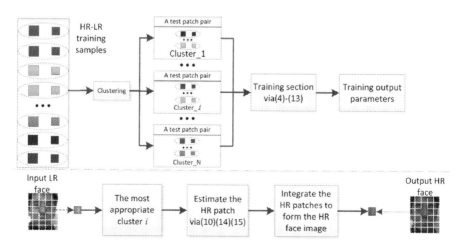

Fig. 2. The overview of proposed method: the part above the dotted lines is training phase, the other part is testing phase.

Training: Suppose $\mathbf{\Phi} = \{(X_1, Y_1), (X_2, X_2), \cdots, (X_M, Y_M)\}$ contain the corresponding LR-HR patch pairs located at the same position; our goal is learning a local regression model for each cluster. We initialize the cluster with a pre-determined cluster number by K-means clustering. But a challenge is that the partition of training samples is being ascertained before learned local regression model, and it is unknown if the samples partition is optimal for local regression model learning. If we get the optimal nearest neighbor patches for the input LR patch, we can compute the optimal regression function. However, the practical matter is that the input LR patch is unknown on the training phase. Thus, we propose an adaptive iteratively strategy rise to this challenge. We learning local regression model for each initialization cluster, and a leave-one-out cross validation procedure was performed at clusters, the nuts and bolts will be described in the following passage.

When clusters are obtained, we can get two projection matrices U and V by Partial Least Squares for each cluster, and the two latent matrices of LR and HR patch pairs can be computed as following:

$$L_{X,i} = \left(V_q\right)^T (X_q^i - \textbf{\textit{Xmean}})$$
(10)

$$L_{Y,i} = \left(U_q\right)^T (Y_q^i - \textbf{\textit{Ymean}})$$
(11)

The value of $\textbf{\textit{Xmean}}$ or $\textbf{\textit{Ymean}}$ is the mean value of current cluster training samples. Suppose the matrices L_X and L_Y have been obtained, the mapping function between L_X and L_Y can be learned based on the Least Square regression. The mapping function denoted F is obtained to minimize the following regularized cost function:

$$\min\{\left\|\sum_p^c \left(FL_X^p - L_Y^p\right)\right\|^2 + \alpha\|F\|^2\}$$
(12)

where c is the number of current cluster at p position, $\|F\|^2$ is a regularized term, α is a regularized parameter.

We take the derivative of Eq. (12) with respect to F utilizing some matrix properties and make the equation to zero, so the solution of Eq. (12) is like this:

$$F = L_Y^p (L_X^p)^T (L_X^p (L_X^p)^T + \alpha I)^{-1}$$
(13)

After we have learned a mapping function for every cluster at one position, and also an LR-HR pair sample patch have been left for each cluster. So given a LR patch of current cluster, the mapping function of every cluster are used to predict the HR patch. Obviously, the result of generate HR patch should be better on which cluster belongs to than others if the regression is optimal. But when the error between the original HR patch and the reconstruction HR patch on its cluster is more than the reconstruction error of other clusters, the samples which belong to this cluster will be partition to the cluster with a minimum reconstruction error. This process is repeated until to convergence. And for a new cluster, the training process is necessary. The training algorithm of our algorithm is summarized in Algorithm 1:

Algorithm 1: Training algorithm of the proposed method.

---

1. **Input:** A set of LR-HR patch pairs; number of clusters N , patchsize, overlap and the regularization parameter α.
2. Initialize all training patch pairs to N clusters by K-means, leave a test patch pair for each cluster; then utilize the partial least squares to get the two projection matrices $\boldsymbol{U}, \boldsymbol{V}$ via (4)-(9), and initialize the mapping function $\boldsymbol{F}$ via (10)-(13) for each cluster.
3. **while** not converged **do**
4.   Calculate the reconstruction error by (14)-(15) for each test pair using $\boldsymbol{F}$ of all clusters, and merging clusters according to reconstruction error;
5.   Update the center of all clusters and test patch pair; update the projection matrices by (4)-(9) and the mapping function by (10)-(13) ;
6. **end while**
7. Re-calculate the two projection matrix, the mapping function, the cluster mean value $\boldsymbol{Xmean}$ and $\boldsymbol{Ymean}$ for each final cluster.
8. **Output:** the center of cluster $\boldsymbol{Cla}$, projection matrices $\boldsymbol{U}$ and $\boldsymbol{V}$, the cluster mean value $\boldsymbol{Xmean}$ and $\boldsymbol{Ymean}$, the mapping function of cluster $\boldsymbol{F}$.

Testing: On the test phase, for the LR test patch, picking out the nearest cluster according to the minimum Euclidean distance between LR patch and the center of clusters. Suppose x is LR patch, we transform x into the feature space use Eq. (10), denoted l_x. So, the estimation of l_x can be obtained:

$$l_y = Fl_x \tag{14}$$

Then, projection l_y back to the origin image space use deformation of Eq. (11):

$$y = (U^T)^{-1}l_y + \boldsymbol{Ymean} \tag{15}$$

Iterate over all patches and merges the HR patch according to the position. Since overlapping regions on the patch, averaging the pixel values to generated the last result HR image. The testing description of our proposed method is summarized as the following:

Algorithm 2: Testing algorithm of the proposed method.

---

9. **Input:** projection matrices U and V, training set mean value $\boldsymbol{Xmean}$ and $\boldsymbol{Ymean}$, the center of cluster $\boldsymbol{Cla}$, the mapping function of cluster $\boldsymbol{F}$, a LR test image X_L, patchsize, overlap.
10. Compute the patch number of every column W, the patch number of every row R.
11. Divide LR test image into WR patches according the same location face, $X_L: \{x(i,j) | 1 \le i \le W, 1 \le j \le R\}$.
12. **for** i = 1 to W **do**
13.   **for** j = 1 to R **do**
14.   dist $= ||x - Cla||_2^2$
15. $S_L = support(dist)$
16. $l_x = V_{S_L}^T(x - Xmean_{S_L})$
17. $l_y = F_{S_L}l_x$
18. $y = (U_{S_L}^T)^{-1}l_y + Ymean_{S_L}$
19.   **end for**
20. **end for**
21. Integrate all the reconstruction HR patches according to the position and generated by averaging pixel values in the overlapping regions.
22. **Output:** HR face hallucination image Y_H.

Table 3 in Sect. 4 demonstrate that the clustering method saved a lot of calculation time at reconstruction.

3.2 Multi-regressors Fusion and HR Induced Clustering

Different with the traditional method, we cannot ensure the number of one cluster equal to the nearest neighbor number of traditional method. So In order to prevent the nearest neighbor numbers are too small to degrade the reconstruction result, more than one cluster are selected to generate the result. Suppose t is the number of nearest cluster, HR_i is the result of the $i - th$ cluster via reconstruction. So the fusion HR reconstruction will be generated:

$$HR = \sum_{i=1}^{t} \beta_i HR_i \qquad (16)$$

In the above method, the clustering process only considered on the LR samples, but a readily understood truth is that to clustering the HR samples will be more precise. So we think that the HR manifold information should be considered on clustering implement, and also training the projection matrix as above.

Firstly, we can obtain a reconstruction result from the LR induced clustering. Secondly, the reconstruction result is utilized to choose the nearest cluster on the next clustering based on HR samples. Lastly, we can get a result from the second step.

Just like the work in [15], the weighting β is defined on the distance between the input patch and the center of cluster, which are computed as:

$$\beta = \exp(-\frac{\|x - Cla\|_2^2}{\sigma})$$

(17)

where *Cla* is the center of cluster, *x* is the input patch, σ is parameter represent the width of Gaussian function, $\sum_i \beta_i = 1$.

4 Experiments and Results

We performed our face hallucination method on the FEI Face Database [16] with 400 images from 200 subjects (100 men and 100 women). In the face database, we randomly select 360 images (180 subjects) for training and the rest of 40 images (20 subjects) for testing. All the face images are cropped to the size 120×100 and well-aligned by two eyes (some training samples are shown in Fig. 3). The LR images were smoothed with averaging filter of size 4×4 and down-sampled with factor of 4, thus the LR images are 30×25. We set the LR patch to 8×8 pixels and the overlapped pixels are 4.

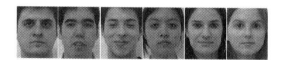

Fig. 3. Some training samples in the FEI face database

The next experiments mainly consist of four parts. The first is that under the condition of traditional experiment (no clustering), we compare the hallucination quality of our method with the state-of-art methods with NE [17], LSR [18], LcR [19], Hao's method [7], LiSR [10]. The next one is that under the clustering experiment condition, we compare the time complexity and super resolution quality with the first experiment of our method. In the third part, we analyze how the number of cluster to reconstruction affect the performance of the method, is it necessary for HR manifold induced clustering. In the last part, we compare the reconstruct time with other method.

4.1 Comparison the Hallucination Quality with Others Approaches

In this part, we compare our method with several representative learning-based super-resolution methods, including the neighborhood embedding algorithm (NE) [17], LSR [18], LcR [19], Hao's [7] and LiSR [10] method. In order to get the best result, we adjust the parameters according to the origin paper for each comparative method. For our method, we set the value of regularized parameter is 10^{-5}. In Table 1, we have given the result for different methods. From the table, we can see that our method over the state-of-art methods on the peak signal-to-noise ratio (PSNR) and the structural similarity index (SSIM). This is because the NE [17] hallucination only uses the LR local structure but also not considers locality constraints to the model. Even though the LcR [19] method impose a locality constraint onto the reconstruction function, but the manifold consistent assumption cannot be fully satisfied. Hao [7] employs Easy-Partial Least Squares (EZ-PLS) method to project the original images onto a common feature space, which guarantees the consistent relation of manifold. However, our method learning the regression model on the projection features space. So the proposed method outperforms other methods. We randomly choose some results reconstructed by our method and the comparative face Hallucination methods are shown in Fig. 4. For the Fig. 4, we can find that our face super-resolution images have more details than the next-best method, like the mouth, nose, and the eyes structures are more similar with ground images.

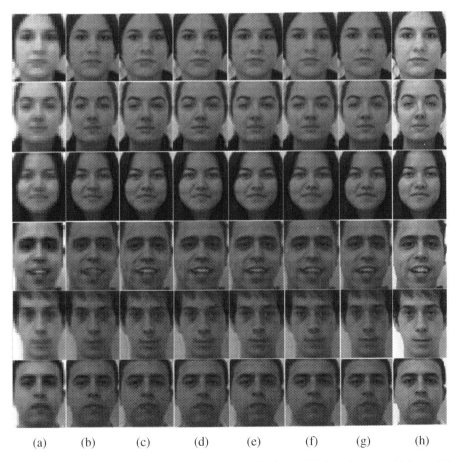

(a)   (b)   (c)   (d)   (e)   (f)   (g)   (h)

Fig. 4. Comparison of results based on different methods on FEI face database. (a) Input LR faces, (b) NE [17], (c) LSR [18], (d) LcR [19], (e) Hao's [7], (f) LiSR [10], (g) Our, (h) The origin HR faces.

4.2   Clustering Guided by LR Samples

From above, we can see that our method has outperforms the state-of-art methods, but the time complexity increased a lot because the LR-HR training patch pairs and K nearest neighbor patch pairs selected are all on reconstruction phase. For improving the computational efficiency, we partition the training samples space into a set of clusters and learn a local regression model for each cluster.

We directly use the experiment results of Table 1. We set the parameter of initial cluster number is 16, the value of regularized parameter is 10^{-5}. All results have been given in Table 2, and the time cost comparison is shown in Table 3.

Table 1. Average PSNR and SSIM comparison of different methods on FEI face database

Method	PSNR (dB)	SSIM
NE [17]	30.5196	0.8712
LSR [18]	31.9000	0.9031
LcR [19]	33.1044	0.9207
Hao's [7]	32.8500	0.9168
LiSR [10]	32.6158	0.9133
Our (no clustering)	**33.8494**	**0.9300**

Table 2. Average PSNR and SSIM comparison of different methods on FEI face database

Method	PSNR (dB)	SSIM
NE [17]	30.5196	0.8712
LSR [18]	31.9000	0.9031
LcR [19]	33.1044	0.9207
Hao's [7]	32.8500	0.9168
LiSR [10]	32.6158	0.9133
Our (non-clustering)	33.8494	0.9300
Our (clustering)	**33.4021**	**0.9240**

Table 3. The comparison results of time consumption

Method	Total time (s)	Average time (s)
Hao's [7]	948.3604	23.7090
Our (non-clustering)	1180.2169	29.5054
Our (clustering)	**167.2434**	**4.1811**
HR induced clustering	394.9910	9.8748

4.3   Re-improvement Using Multi-cluster Model and HR Induced Clustering

When a cluster is selected, it can not confirm that the number of neighbor patches is equal to its inherent neighbor patches number, so the face hallucination result has a low quality than the original result. To make up the quality losses as limited neighbor training samples, a multi-cluster fusion model is proposed to improve the reconstruction result. We chose the optimal weighted strategy based on Euclidean distance using the Eq. (17). The results are shown in Fig. 6.

As shown by Fig. 6, muti-cluster fusion result is better than one cluster, but the time consumption is expensive with the increase of neighbor cluster. So we give a cluster number, which is a trade-off between time-consuming and super-resolution reconstruction result. The cluster number is set to 2 and the σ is set to 0.07. Figure 5 is shows the average PSNR and SSIM values of all 40 test samples for different values of the cluster number.

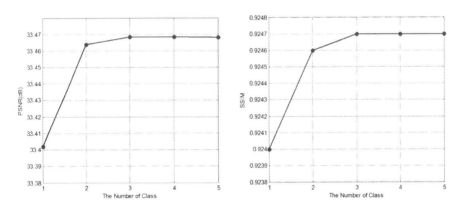

Fig. 5. Influence of the number of nearest cluster for average PSNR and SSIM.

However, the LR image patch space cannot describe the underlying manifold space since the LR image via a degraded process, so the structure of HR image space is much more feasible. Therefore, the HR samples should be used to guide clustering. The LR induced clustering reconstruction result can be used to guide the nearest cluster choose in the HR induced clustering reconstruction process. After this, a new HR image will be generated. The parameters are the same as LR induced clustering. The Fig. 6 have shown that our proposals are correct.

4.4   Time Consumption

In this section, we compare the reconstruction time on 4.1 method, 4.2 method, HR induced clustering method and Hao's [7] for 40 test images. Here, we only comparison with the Hao's [7], because this method also using the PLS method. The result is shown as Table 3. From Table 3, we can see that the time cost has descended a lot than

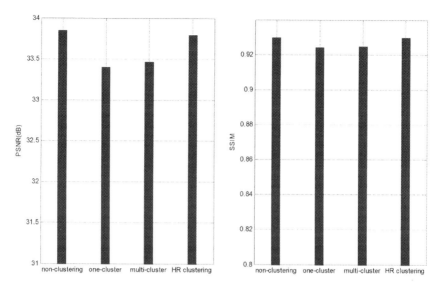

Fig. 6. The improvement of average PSNR and SSIM by multi-cluster and HR induced clustering

non-clustering method. But the HR induced clustering method corresponds to reconstruction two times, so the time consumption is equal to the sum of the two times cost. At the same time, our method has a competitive performance on reconstruction results.

5   Conclusion

In this paper, we propose a clustering-based local regression method in the feature space for face hallucination by iteration learning of the training samples partition and local regression models. We project the original image space into a feature space by Partial Least Squares for learning local regression models, whose result is better than original space. We iterate partitioning the training samples to learn a local regression model for each cluster, and to reconstruct each test patch by its most appropriate regression model, which has a faster reconstruction time than non-clustering methods. Moreover, a multi-regressors fusion model and HR induced clustering have been utilized to re-improve the recovery quality performance. Experiments results on the FEI face database demonstrated that our proposed method has outperformed competing methods.

References

1. Liu, C., Shum, H., Freeman, W.: Face hallucination: theory and practice. Int. J. Comput. Vis. **75**, 115–134 (2007)
2. Wang, N., Tao, D., Gao, X., Li, X., Li, J.: A comprehensive survey to face hallucination. Int. J. Comput. Vis. **106**, 9–30 (2014)

3. Ma, X., Zhang, J., Qi, C.: Hallucination face by position-patch. Pattern Recogn. **43**(6), 2224–2236 (2010)
4. Capel, D., Zisserman, A.: Super-resolution from multiple views using learnt image models. In: Proceedings of the IEEE Conference Computer Vision and Pattern Recognition, vol. 2, pp. II–627 (2001)
5. Liu, C., Shum, H., Zhang, C.: A two-step approach to hallucinating faces: global parametric model and local nonparametric model. In: Proceedings of the IEEE Conference Computer Vision and Pattern Recognition, vol. 1, pp. I–192 (2001)
6. Li, B., Chang, H., Shan, S.G., et al.: Aligning coupled manifolds for face hallucination. Proc. IEEE Sig. Process. Lett. **16**(11), 957–960 (2009)
7. Hao, Y., Qi, C.: Face hallucination based on modified neighbor embedding and global smoothness constraint. Proc. IEEE Sig. Process. Lett. **21**(10), 1187–1191 (2014)
8. Wu, W., Liu, Z.: Learning-based super resolution using kernel partial least squares. Image Vis. Comput. **29**(6), 394–406 (2011)
9. Ni, K., Nguyen, T.: Image super resolution using support vector regression. IEEE Trans. Image Process. **16**(6), 1596–1610 (2007)
10. Jiang, J., Hu, R., Liang, C., Han, Z., Zhang, C.: Face image super-resolution through locality-induced support regression. Sig. Process. **103**, 168–183 (2014)
11. Huang, H., Wu, N.: Fast facial image super-resolution via local linear transformations for resource-limited applications. IEEE Trans. Image Process. **21**(10), 1363–1377 (2011)
12. Hao, Y., Qi, C.: A unified regularization framework for virtual frontal face image synthesis. IEEE Sig. Process. Lett. **22**(5), 559–563 (2015)
13. Jiang, J., Hu, R., Wang, Z., Han, Z., Ma, J.: Facial image hallucination through coupled-layer neighbor embedding. IEEE Trans. Circ. Syst. Video Technol. **PP**(99), 1 (2015)
14. Zhou, E., Fan, H., Cao, Z., Jiang, Y., Yin, Q.: Learning face hallucination in the wild. In: National Conference on Artificial Intelligence (2015)
15. He, X., Yan, S., Hu, Y., Niyogi, P., Zhang, H.-J.: Face recognition using laplacianfaces. IEEE Trans. Pattern Anal. Mach. Intell. **27**(3), 328–340 (2005)
16. FEI Face Database. http://fei.edu.br/∼cet/facedatabase.html
17. Chang, H., Yeung, D., Xiong, Y.: Super-resolution through neighbor embedding. In: Proceedings of the IEEE Conference Computer Vision and Pattern Recognition, pp. 275–282 (2004)
18. Ma, X., Zhang, J., Qi, C.: Position-based face hallucination method. In: Proceedings of the IEEE Conference Multimedia and Expo, pp. 290–293 (2009)
19. Jiang, J., Hu, R., Wang, Z., Han, Z.: Noise robust face hallucination via locality-constrained representation. IEEE Trans. Multimedia **16**(5), 1268–1281 (2014)

Gait Retrieval: A Deep Hashing Method for People Retrieval in Video

Muhammad Rauf[1,2(✉)], Yongzhen Huang[1], and Liang Wang[1]

[1] National Lab of Pattern Recognition, Institute of Automation, CAS, Beijing, China
rauf@nlpr.ia.ac.cn
[2] University of Chinese Academy of Sciences, Beijing, China

Abstract. Automated surveillance systems are required for the state of the art security. Everyday networks of cameras generate a very-large set of data, which makes recognition and identification tasks harder. In this paper, we present a new problem called Gait Retrieval in order to address the challenge of large-scale surveillance data. We have an interest in retrieving similar videos based on the human gait. Gait is the most important biometric for long distance human identification. We also propose a solution for the Gait Retrieval problem by using gait biometrics. The solution is based on the deep hashing technique to learn a hash function that preserves the similarities between the same labeled images. Deep hash function with convolutional neural network learns features and maps them to hash codes. Images with similar appearance should have similar hash codes. Training samples are arranged in a batch of triplets. Our proposed method outperforms traditional methods with good margin.

Keywords: Gait retrieval · Gait Energy Image (GEI) · Deep hashing · Surrogate loss · Convolutional Neural Network (CNN)

1 Introduction

Efficient retrieval of specific persons in videos is the key for the state of the art surveillance systems. In this paper our focus is on people's retrieval from large-scale surveillance videos. The most common strategy used by surveillance systems is person re-identification. Re-identification systems are mostly based on images and ignore motion information, which is however vital in uncontrolled environments. In the last few years many models have been studied for person re-identification [1], and several feature representation techniques are used, for example, salience match [2], mid-level filter [3], the ensemble of local features (ELF) [4], SDALF [5] and fisher vectors (LDFV) [6]. Besides these features, metric learning has been widely applied for person re-identification systems [7–9].

For motion information gait recognition is the classical problem. Mainly two different approaches are used for gait recognition: model based and model-free approaches. In a model based approach, human body structure is used [10–12],

© Springer Nature Singapore Pte Ltd. 2016
T. Tan et al. (Eds.): CCPR 2016, Part I, CCIS 662, pp. 383–391, 2016.
DOI: 10.1007/978-981-10-3002-4_32

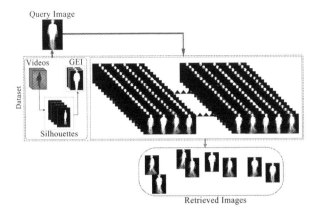

Fig. 1. Overview of gait retrieval

and in a model free approach, body motion patterns are studied [13–15]. Even though, gait recognition does not directly face the retrieval problem and thus cannot be well used for effective person retrieval in surveillance videos. In this paper, we propose a new problem called Gait Retrieval. It is originally motivated by the retrieval problem which is addressing people retrieval in video surveillance systems. In particular, we present a solution to this problem, based on the deep hash function learning to incorporate Gait Energy Image (GEI) [16]. The block diagram of our proposed method is shown in Fig. 1. We use a three-channel convolutional Neural Network to learn a hash function. Training data is provided in the form of triplets, two images with the same label and one with different label. This method generates hash codes by learning similarities and dissimilarities between images. Our contribution to this work can be summarized as follows.

– For the first time we introduce a new problem, i.e., gait based image retrieval, in the computer vision research.
– We provide one solution based on deep hash learning, which is demonstrated to be good for person retrieval in video surveillance.

The rest of paper is organized as follows. The next part will cover our proposed method. In Sect. 3, experiments and results are given. Section 4 concludes the paper.

2   Proposed Method

In this section, we introduce our gait retrieval framework in detail. The detailed diagram of our proposed method is shown in Fig. 2. In our work, a deep architecture of CNN is used to learn the hash function and generate K-bit hash codes. The proposed deep hash model is a three-channel deep network, and each channel architecture is the same. For deep feature representation, five convolutional

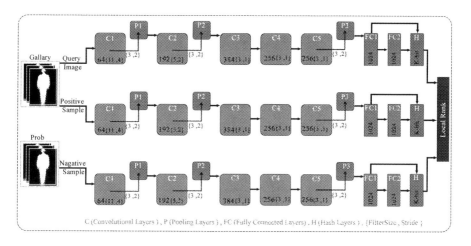

Fig. 2. Detailed framework of gait retrieval. In hash Layers K is 16, 32, 64 or 128-bit

layers and two fully connected layers are used, and then the hash layer generates a compact binary hash code.

Our objective is to learn a set of hash functions $h(x) = [h_1(x), \ldots, h_K(x)]$ to generate a K-bit hash code. The $h(x)$ hash function projects gait based p-dimensional feature vector onto q-dimensional binary hash code $h \in \{-1, 1\}$. In general we can describe hash function $h : R^D \rightarrow \{-1, 1\}$, where $D = \{x\}_{n=1}^{N}$ and $x \in R$ is a real valued data point associated with label. According to [17], the first FC layer provides more precise semantic distinction, and the second FC layer is class dependent. In this model, the FC layers have a direct connection with the hash layer. We can define the hash function as:

$$h(x;) = sign(w^T[f_a(x); f_b(x)]) \tag{1}$$

where f_a and f_b are the first and the second FC layers and w is the weights in the hash layer. We train the hash function by learning similarities and dissimilarities between images. A pair of points from D, labeled same if they are from the same class or different if they belong to different classes. We use gait energy images [16] as the gait feature to train a deep hash function and arrange batches of triplet. Each triplet contains three images, two of them with the same label and one with a different label. Channels one and two of the network receive images with the same label and channel three receives images with the different label, and all these images are selected randomly. If the samples are labeled with multi labels, then the similarity depends on how many labels two data points are sharing. To preserve the association of one point with other points in the dataset, an association ranking list is computed based on the number of sharing labels. Ranking loss function for the set of triplet can be defined as:

$$L(h(q), \{h(x_i)\}_{i=1}^{M}) = \sum_{i=1}^{M} \sum_{r_j < r_i} [d_H(h(q), h(x_i), h(x_j)) + \rho]_+ \tag{2}$$

where M is the length of the ranking list, ρ is the margin between a pair of images, $[.] = max(0,.)$, $d_H(h, h_1, h_2) = d_H(h, h_1) - d_H(h, h_2), ..$, is the hamming distance, and r_i, r_j is the similarity levels of ith and j^{th} data points. This kind of surrogate loss treats all triplets equally, so for more accurate ranking, adoptive weight [18] is used. On the basis of surrogate loss, adoptive weight and Eq. 1, the objective function is defined as:

$$F(W) = \sum_{i=1}^{M} \sum_{r_j < r_i} \frac{2^{r_i} - 2^{r_j}}{Z} [d_H(h(q), h(x_i), h(x_j)) + \rho]$$

$$+ \frac{\alpha}{2} \|mean_q(h(q;))\|_2^2 + \frac{\beta}{2} \|W\|_2^2$$

$$(3)$$

where the first term is the surrogate loss with adoptive weight on the set of triplets, the second term is the balance penalty and the third term is L_2 weight decay. For a detailed understanding of deep hash function and the CNN model, please refer to [17,19].

3   Experiments

3.1   Dataset

To show the performance of gait retrieval, we use the CASIA-B Gait dataset developed by Institute of Automation, Chinese Academy of Science (CASIA). The CASIA-B dataset consists of 124 individuals with three different conditions and 11 distinct viewing angles. The three different conditions are 'bag(bg)', which means individuals appear in a video with a bag, 'cloths(cl)', which means individual wearing a coat when they appear in the video and 'normal(nm)', which means having any kind of clothes. The angles' range is from $0°$ to $180°$ and distributed with $18°$ variation. These angles are $0°, 18°, 36°, 54°, 72°, 90°, 108°, 126°, 144°, 162°$ and $180°$. We use GEI as a biometric feature for the gait retrieval system. The mini-batch size is set to 128 and dropout is 0.5 on the fully connected layers to avoid over fitting. The regularization setting is the same as the original model. Note training and testing sets are equally divided for all experiments, and we use four preset code sizes, i.e., 16, 32, 64 and 128 bits [22].

3.2   Single Condition

In single condition (SC), we use three original conditions separately. That means we divide the dataset into three separate parts according to the wearing conditions. This type of problem is considered as a short-time retrieval problem. Peoples' images are captured by the camera for a very short period of time, and they cannot change their appearance. The original three conditions are normal, bag and clothes.

3.3   Mixed Condition

In the second kind of experiments, we perform mixed condition. Mixed conditions are considered as a long-time retrieval problem. Individuals are captured by the cameras for a long period of time and their appearance can change. In other words, they were captured by cameras on different time with different appearances. This kind of gait retrieval is very important in real-world applications. Further we divide these experiments into two different parts.

View Independent. In View Independent Mixed Condition (VIMC), we use multiple label representations for every person. This is called view independent, because an angle attribute is not considered as a label. Only person's ID and their wearing condition are used as two different labels.

View Dependent. In View Dependent Mixed Condition (VDMC), we add the angle attribute as a label. As we mentioned, the CASIA-B dataset contains 11 different viewing angles. For these experiments, we categorize these 11 angles into three groups. The first group contains those angles which are closer to the front view, the second are closer to side view and the last are closer to back view. According to this categorization angles $0°–36°$ are considered as front view (FV), $54°–126°$ as side view (SV) and $144°–180°$ as back view (BV).

3.4   Comparison Methods

We provide two different kinds of comparisons. First, we compare the performance of gait retrieval if we use different bits of hash code. Our intent is to check the effects on the performance of gait retrieval if we increase or decrease the code size. As we mentioned earlier, the gait retrieval problem is not addressed in the past as per our knowledge, that's why no experimental work available for a comparative study. So for performance evaluation of our proposed method we use three traditional methods and compare their results with our method. The first method we implemented for comparison is GEI + Appearance based feature. Zhao et al. [20] implemented appearance based feature on Re-Identification, these appearances based feature consists on Hue-Saturation-Value (HSV) and Gabor filter [23]. The second comparison method is based on GEI + Globality Locality Preserving Projection (LPP) proposed by Huang et al. [21]. The last comparison method is also a subspace learning method, Local Fisher Discriminant Analysis (LFDA) [12]. This method improved performance of view-invariant gait recognition in [12]. Comparison results are provided against these methods in Table 1.

3.5   Evaluation Criteria

The Normalized Discounted Cumulative Gain (NDCG) score is one of the most popular evaluation methods in information retrieval. In our experiments, we use

NDCG to evaluate the performance of our method. We calculate the NDCG score for the top-100 relevant items.

$$NDCG@p = \frac{1}{Z} \sum_{i=1}^{p} \frac{2^{r_i} - 1}{log(1 + i)} \tag{4}$$

where p is the truncate value of position in a ranking list, Z is a constant to ensure correct NDCG [24] scores of rank one and r is the similarity level of the ith data point.

3.6   Bit Size Comparison

In Fig. 3 we show results on four different code sizes with single condition. These results illustrate that the best performance of hash function is achieved on 64-bit for single condition and start dropping as we increase or decrease the bit size. In single condition we divide the dataset into three parts (normal, bag and clothes), and images are labeled with only single attribute.

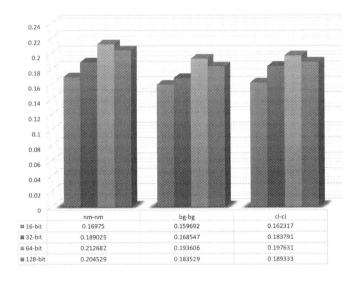

	nm-nm	bg-bg	cl-cl
▩ 16-bit	0.16975	0.159692	0.162317
▩ 32-bit	0.189025	0.168547	0.183791
▩ 64-bit	0.212682	0.193606	0.197631
▩ 128-bit	0.204529	0.183529	0.189333

Fig. 3. Different bit comparison on single condition

In, Figs. 4 and 5 we provide comparison on VDMC and VIMC. In mixed condition, 128-bit code performs better than 16, 32 and 64-bit. In mixed condition experiments, the dataset is the collection of normal, bag and clothes images. More over, the images have multi-labels. There are 2 reasons for explaining different optimal sizes: one is the size of dataset and the other is the number of attributes. So more labels will help to learn hash function better if we increase the code size.

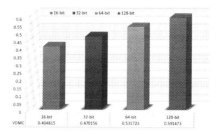

Fig. 4. Different bit comparison on view dependent mixed conditions

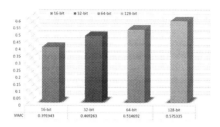

Fig. 5. Different bit comparison on view independent mixed conditions

3.7   Comparison with Traditional Methods

Table 1, shows the collective results of SC, VIMC and VDMC. We compare our method with the above mentioned comparison methods in Sect. 3.4 and our proposed method outperforms the compared methods with a large score. Deep hash method has the ability to preserve more semantic information and is more accurate in the gait retrieval problem than other compared methods. The code size for deep hashing method is set to 64-bit. Besides the best performance of our method, we can observe that subspace learning methods GEI+LFDA and GEI+LPP are close candidates for gait retrieval.

Table 1. Comparison of deep hashing methods with traditional methods

	nm-nm	cl-cl	bg-bg	VIMC	VDMC
Deep hashing	0.2126	0.1976	0.1936	0.5753	0.5914
GEI+ABF [20]	0.0971	0.0362	0.0332	0.1031	0.2517
GEI+LFDA [12]	0.2016	0.1703	0.1801	0.4499	0.5299
GEI+LPP [21]	0.1937	0.1634	0.163	0.5059	0.5403

4 Conclusion

In this paper, we have discussed a new research problem called Gait Retrieval. In Gait Retrieval, we want to retrieve images of a certain person and track their movements across camera networks based on the gait biometric. We have also provided one solution to this problem. It is based on deep hashing methods to learn similarities and dissimilarities between images and generate a k-bit hash code. Our proposed method outperforms all traditional methods with good margin. We also study the performance of deep hashing with different sizes of hash code. Gait retrieval will help to improve the performance of the automated surveillance system. In the future, we are interested in exploring other techniques that can help to improve the performance of gait retrieval. We are also interested in investigating the effect of gait retrieval in the re-identification system.

Acknowledgmets. This work is jointly supported by National Basic Research Program of China (2012CB316300), National Natural Science Foundation of China (61420106015, 61525306). The authors would also like to acknowledge the support by Strategic Perority Research Program of the CAS (Grant XDB02070001), Youth Innovation Promotion Association CAS (200612), and SAMSUNG GRO Program.

References

1. Vezzani, R., Baltieri, D., Cucchiara, R.: People re-identification in surveillance and forensics: a survey. ACM Comput. Surv. (CSUR) **46**(2), 29 (2013)
2. Zhao, R., Ouyang, W., Wang, X.: Unsupervised salience learning for person re-identification. In: IEEE Conference on Computer Vision and and Pattern Recognition (CVPR), pp. 3586–3593. IEEE (2013)
3. Zhao, R., Ouyang, W., Wang, X.: Learning mid-level filters for person re-identification. In: IEEE Conference on Computer Vision and Pattern Recognition, pp. 144–151 (2014)
4. Gray, D., Tao, H.: Viewpoint invariant pedestrian recognition with an ensemble of localized features. In: European Conference on Computer Vision, pp. 262–275, October 2008
5. Bazzani, L., Cristani, M., Murino, V.: Symmetry-driven accumulation of local features for human characterization and re-identification. Comput. Vis. Image Under Standing **117**(2), 130–144 (2013)
6. Ma, B., Su, Y., Jurie, F.: Local descriptors encoded by fisher vectors for personre-identification. In: European Conference on Computer Vision Workshops, pp. 413–422, October 2012
7. Davis, J.V., Kulis, B., Jain, P., Sra, S., Dhillon, I.S.: Information-theoretic metric learning. In: Proceedings of the 24th International Conference on Machine Learning, pp. 209–216, June 2007
8. Weinberger, K., Blitzer, J., Saul, L.: Distance metric learning for large margin nearest neighbor classification. In: Advances in Neural Information Processing Systems, pp. 1473–1480 (2005)
9. Li, Z., Chang, S., Liang, F., Huang, T.S., Cao, L., Smith J.R.: Learning locally-adaptive decision functions for person verification. In: IEEE Conference on Computer Vision and Pattern Recognition, pp. 3610–3617 (2013)

10. Bashir, K., Xiang, T., Gong, S.: Gait recognition without subject cooperation. Pattern Recogn. Lett. **13**(13), 2052–2060 (2010)
11. Xu, D., Huang, Y., Zeng, Z., Xu, X.: Human gait recognition using patch distribution feature and locality-constrained group sparse representation. IEEE Trans. Image Process. **21**(1), 316–326 (2012)
12. Hu, M., Wang, Y., Zhang, Z., Zhang, D., Little, J.J.: Incremental learning for video-based gait recognition with LBP flow. IEEE Trans. Cybern. **43**(1), 77–89 (2013)
13. Lee, L., Grimson, W.E.L.: Gait analysis for recognition and classification. In: Proceeding 5th IEEE Internationa Conference AFGR, pp. 148–155, May 2002
14. Bhanur, B., Han, J.: Human Recognition at a Distance in Video. Springer Science and Business Media, Berlin (2011)
15. Gu, J., Ding, X., Wang, S., Wu, Y.: Action and gait recognition from recovered 3-D human joints. IEEE Trans. Syst. Man Cybern. **40**(4), 1021–1033 (2010)
16. Han, J., Bhanu, B.: Individual recognition using gait energy image. IEEE Trans. Pattern Anal. Mach. Intell. **28**(2), 316–322 (2006)
17. Fang, Z., Huang, Y., Wang, L., Tan, T.: Deep semantic ranking based hashing for multi-label image retrieval. In: Proceedings of the IEEE Conference on Computer Vision and Pattern Recognition (2015)
18. Cao, Y., Xu, J., Liu, T., Li, H., Huang, Y., Hon, H.: Adapting ranking SVM to document retrieval. In: Proceeding Annual ACM SIGIR Conference, pp.186–193, August 2006
19. Krizhevsky, A.: One weird trick for parallelizing convolutional neural networks. arXiv preprint arXiv:1404.5997, March 2014
20. Zheng, L., Zhang, Z., Wu, Q., Wang, Y.: Enhancing person re-identification by integrating gait biometric. Neurocomputing **168**, 1144–1156 (2015)
21. Huang, S., Elgammal, A., Lu, J., Yang, D.: Cross-speed gait recognition using speed-invariant gait templates and globality locality preserving projections. IEEE Trans. Inf. Forensics Secur. **10**(10), 2071–2083 (2015)
22. Li, X., Lin, G., Shen, C., van den Hengel, A., Dick, A.R.: Learning hash functions using column generation, arXiv preprint arXiv, March 2013
23. Liu, C., Wechsler, H.: Gabor feature based classification using the enhanced fisher linear discriminant model for face recognition. IEEE Trans. Image Process **11**(4), 467–476 (2002)
24. Jarvelin, K., Kekalainen, J.: IR evaluation methods for retrieving highly relevant documents. In: Proceedings of the 23rd Annual International ACM SIGIR Conference on Research and Development in Information Retrieval. ACM (2000)

Background Subtraction Based on Superpixels Under Multi-scale in Complex Scenes

Chenqiu Zhao[✉], Tingting Zhang, Qianying Huang, Xiaohong Zhang, Dan Yang, Yinq Qu, and Sheng Huang

Laboratory of Intelligent Services and Software Engineering,
School of Software Engineering, Chongqing University, Chongqing 401331, China
zhaochenqiu@gmail.com, dyang@cqu.edu.cn

Abstract. Background subtraction in complex scenes is a challenging problem of computer vision. Most existing algorithms analyze the variation in pixels or regions for background subtraction. Unfortunately, these works ignoring the neighborhood information or similarity among pixels and do not work well in complex scenes. To solve this problem, a novel background subtraction method based on SuperPixels under Multi-Scale (SPMS) is proposed. In SPMS, the foreground consists of superpixels with foreground or background label, which decided by the statistic of its variation. The variation in superpixels is robust to noise and environmental changes, which endows the SPMS with the ability to work in extreme environment such as adverse weather and dynamic scenes. Finally, the summary of foregrounds under multiple scales improve the accuracy of the proposed approach. The experiments on standard benchmarks demonstrate encouraging performance of the proposed approach in comparison with several state-of-the-art algorithms.

Keywords: Background subtraction · Motion detection · Superpixels · Neighborhood information · Multi-scale

1 Introduction

Background subtraction is a fundamental issue of computer vision, which has a wide range of application [2,28]. The main challenge in background subtraction comes from the complexity of nature scenes, such as waving tree or rippling water. Recently, large numbers of works used pixels' neighborhood information to improve the robustness of background subtraction. Unfortunately, most of those algorithms were based on pixels or regions and ignored the similarity between pixels themselves. And those pixels or regions based algorithms utilize the information of all pixels' neighborhoods. However, since the similarity between a pixel and their neighborhoods is different, not all pixels' neighborhoods contribute positively to improve the robustness of algorithms. In this work, the superpixel contains the pixels' neighborhood information as well as the similarity among them is utilized to detect moving objects. And a novel background subtraction based on superpixels under multi-scale is proposed to background subtraction in complex scenes.

© Springer Nature Singapore Pte Ltd. 2016
T. Tan et al. (Eds.): CCPR 2016, Part I, CCIS 662, pp. 392–403, 2016.
DOI: 10.1007/978-981-10-3002-4_33

Background subtraction is the classification of pixels according its variation. Since the complexity of natural scenes, several pixels with different labels have the similar variation. For example, as the left part of Fig. 1 shows, the pixels labeled as foreground or background are mixing together in the variation of RGB space. However, a pixel is not independent [7] but related with its neighborhoods, especially the similar one. Since the superpixels contain the information on pixels and their similar neighborhoods, the variation in the superpixels is easier to be classified, such as the right part of Fig. 1 shows. Unlike previous work, we analyze the variation in superpixels instead of pixels for background subtraction. Moreover, since the segmentation of superpixels does not refer to motion information, not all the pixels belong to a particular superpixel are shared the same label of foreground or background. To solve this brought issue, the labelled superpixels are obtained under multiple scales to compose different foreground in each scale. And the final foreground of our SPMS is captured by the summary of several foregrounds consisted of superpixels under different scales.

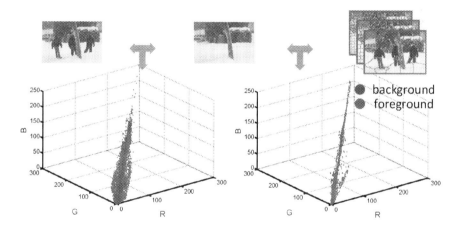

Fig. 1. The variation in pixels and superpixels with foreground or background labels.

Our SPMS method can be divided into three step. The first step of SPMS is learning a background via the statistical function such as Gaussian function [3]. Meanwhile, the current frame is segmented into several superpixels by a *K-means* based superpixel segmentation algorithm [1]. Then, we subtract the current frame and background to capture the variation of each pixel. And the means of pixels' variation in a particular superpixel is used as the variation in it. The variation is classified by a threshold to decide whether it is foreground or background. And the foreground of SMPS is consisting of these labeled superpixels in a particular scale. Finally, the segmentation of foreground in SPMS is done under multiple scales. We modify the size and zone of superpixels in each scale to capture different foregrounds. And the summary of these foregrounds under different scales is used as our final foreground result. The contributions of this paper are summarized as follows:

1. Our SPMS utilizes superpixels instead of pixels or regions for background sub-traction. Since the superpixel reveals the neighborhood information and the similarity among pixels, proposed approach achieves encouraging robustness in complex scenes, such as adverse weather and dynamic scene.
2. The final foreground of SPMS is captured by the summary of foregrounds captured in different scales. Benefited from the difference about superpix-els' zone in multiple scales, the accuracy of our SPMS background model is improved by the summary process.

2   Related Work

In the early works about background subtraction, pixels are assumed independent of each other. The MoG [3,4] and the CodeBook [5,6] algorithms are two popular approaches to segment foreground. However, since those works ignored the neighborhood information, both the MoG [4] and CodeBook [6] are working hard in complex scenes, like waving tree and rippling water.

Recently, larger numbers of works improved the robustness of background subtraction by neighborhood information (e.g. [7–9,12,13,15,17,22]). Among those works, several algorithms are based on features with neighborhood information. Heikkila et al. [9] modeling the background by LBP [9] texture descriptor, which is captured from pixels and its neighborhoods. Since the LBP is robust to illumination change and noise, the algorithm in [9] work well in the shade regions. Similarly, Liao et al. [12] improved the LBP into SILTP and segmented the foreground by Kernel Density Estimation. Hence, Lin et al. [13] detected moving objects in different levels to improve the efficiency of their algorithm. And Zaharescu et al. [17] extended the CodeBook algorithm by multiple scales for the accuracy of foreground.

Besides the texture features, the spatio-temporal features (e.g. [15,20]) also reveals the neighborhood information. Lin et al. [15] detect moving objects by pursuing dynamic spatio-temporal model. They employed the auto regressive moving average model to pursue the subspace. And model joints the appearance consistency and temporal coherence of the observations. Lu et al. [20] collected a set of samples on different spatial scales for each location in a dynamic scene. And the foreground is segmented by a non-parametric model. Barnich et al. proposed the ViBe [7] and ViBe+ [8] algorithm, which is an excellent method to utilize neighborhood information. The ViBe algorithm maintains a sample set, which contains the pixels and their neighborhoods, for each pixels. And the pixel is label as background if it is matched with any entries of the sample set. In addition, the histogram is another technique of the utilization about neighborhood information. Kita et al. [10] segmented foreground by analysing the intensity histogram during video sequence. And a similar method also based on histogram is proposed in [11].

Furthermore, these are also several works detected moving objects by regions instead of pixels. Riahi et al. [14] extended the MoG with rectangle region. Since the variation in the regions' intensity is robust to environment change, the algorithm in [14] achieves well performance in complex scenes. Varadarajan et al. [23] proposed a region-based MoG algorithm. In particular, the spatio-temporal features is extracted from the regions and as the input of MoG to segment foreground. The main difference between our work and regions based algorithms is the consideration of similarity between pixels themselves. The pixels in a superpixel have stronger connection with each others compared with pixels in a region. And the stronger connection prevents the accuracy of our superpixel-based approach.

One should note, the superpixels also utilized for background subtraction in several algorithms (e.g. [24–26]). In the [25], Patwardhan et al. [25] cluster pixels into several pixel layers. And the foregrounds are segmented from different pixel layers. Moreover, Wu et al. [26] detect moving objects by the trajectory, and reshape the foreground by superpixels. In addition, proposed approach is closely related with the work in [24], which is a saliency detection algorithm. However, besides the difference in purpose, the similarity measure also different between us. The algorithm in [24] just threshold the gray-scale of current frames. And in this work, the similarity of superpixels is captured by the analysing of its variation during time sequence.

3   Modeling Background by Superpixels

In this section, the details of SPMS model are explained and its flow chart is shown in Fig. 2. In the Sect. 3.1, we introduce the segmentation of foreground via analysing the variation of superpixels' intensity under multi-scale. Simultaneously, the process of capturing background, which is used to compare with current frame for extracting superpixels' variation, is explained in Sect. 3.2.

3.1   Foreground Segmentation via Superpixels Under Multi-Scale

In our SPMS model, the foreground consists of superpixels with foreground or background label. The label of each superpixel is decided by its variation. And the variation in a superpixel is actually the means of their pixels' variation compared with the corresponding pixels in the background. In particular, the process of capturing background will be explained later. Moreover, in order to improve the accuracy of proposed approach, we segment several foregrounds under multiple scales, and the summary of all these foreground information is used as the final result of proposed approach.

In this work, the superpixel are obtained from current frame by the SLIC algorithm [1]. Since the SLIC [1] is based on *K-Means*, it is convenient to control the size of superpixels according to the numbers. And the convenience is the main reason we choose SLIC algorithm [1]. Let's denote the superpixels segmented under multi-scale are:

$$\{\mathcal{SP}_1, \mathcal{SP}_2, \ldots \mathcal{SP}_s\} = \{Sp_{i,j} | i \in [1, s], j \in [1, N_i]\}, \tag{1}$$

where i is the index of scale and N_i is the number of superpixels under scale i.

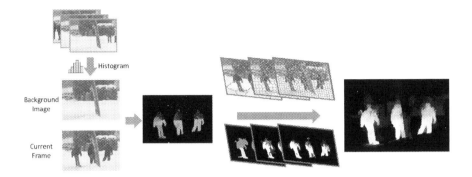

Fig. 2. The pipeline of proposed approach.

As we mentioned before, the final foreground is captured by the summary of foregrounds segmented under multiple scales. And the foreground in each scale consists of superpixels with foreground or background segmented under this scale. In the superpixel $Sp_{i,j}$, all their pixels are subtract from pixels of background $I_b(x,y)$ in the corresponding position. Moreover, in the subtracting process, we utilized a region searching method to improve the robustness of the subtraction result. In the region searching method, for each a pixel, we find the most similar pixels of background in a small region R on respective position of it. And the definition of region searching method $G(x,y)$ is shown as follows:

$$G(x,y) = \mathrm{argmin}\|I_t(m,n) - I_b(x,y)\|_1 \quad m,n \in x,y \pm R, \tag{2}$$

where x and y is the location of pixels. And I_t is the current frame, where t is the time index.

Then, the score of this superpixel labelled as background is captured from the means of all the subtraction results on it, which is shown as follows:

$$P(Sp_{i,j}) = \frac{\sum\limits_{x,y \in Sp_{i,j}} G(x,y)}{|Sp_{i,j}|}, \tag{3}$$

where N_i is the number of pixels belong to superpixel $Sp_{i,j}$, and $G(x,y)$ is the similarity function of position (x,y). The $P(Sp_{i,j})$ is compared with a threshold T_{sp} to decide whether the superpixel $Sp_{i,j}$ is foreground or background. And foreground of our SPMS under a particular scale i consists of all the labelled superpixels. And the process is shown as follows:

$$Fg_i(x,y) = g(Sp_{i,j}(x,y), T_{sp}), \quad I_t(x,y) \in Sp_{i,j}, \tag{4}$$

where $g(x,y)$ is the piecewise function to decide whether the pixel with location (x,y) is foreground or background, and the definition is shown as follows:

$$g(x,y) = \begin{cases} 1, & x < y \\ 0, & otherwise \end{cases}. \tag{5}$$

The superpixel is the set of similar pixels and does not consider motion information. Background subtraction based on superpixel achieves well robustness but suffer from accuracy problem. We captured foregrounds under different scales. The summary of all these foregrounds improve the accuracy of proposed approach, which is defined as:

$$Fg(x,y) = \frac{\sum_{i=0}^{s} Fg_i(x,y)}{s} \tag{6}$$

where $Fg(x,y)$ is the proximity image of SPMS and is utilized to compare with threshold T for our final foreground segmentation.

3.2 Background Updating

The purpose of background updating is to capture a background without moving objects. The updating procedures are different to the pixels labeled as foreground and background. If a pixel is labelled as background in the procedure of foreground segmentation. The respective pixel in background is replaced by this pixel. When a pixel is labeled as foreground, the process shown in Fig. 3 is done to capture the background intensity. And the background intensity is used to update the pixel with the corresponding location in the background.

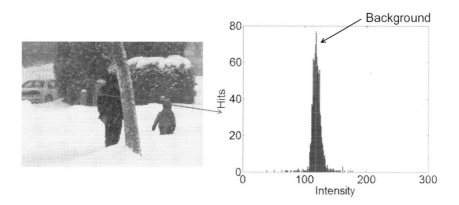

Fig. 3. The process of updating foreground is analysing the variation of pixels' intensity.

The procedure shown in Fig. 3 is a histogram of the pixel's variation during time sequence. Let's denote the observation of a pixel labeled as foreground in time sequence as $P = p_1, p_2, \ldots, p_t$. The variation of this pixel is analyzed by the histogram. and the intensity with the max peak value is used as the background intensity. Mathematically, the background intensity is defined as:

$$p_{bk} = \underset{p}{\arg\max} \sum_{i=1}^{t} p_i \cap p \quad p = \{p_1, p_2, \ldots, p_t\}, \tag{7}$$

where the p is pixel's intensity. Next, the P_{bk} is used to update the corresponding pixel in background.

4   Experiment

In this section, we have conducted the comprehensive experiments to analyze our SPMS. We compare proposed approach to four state-of-the-art algorithms with or without pixels' neighborhood information. All the experiments are ran in several video sequences of ChangeDetection.Net [19] (CDN) benchmark.

In this work, The Recall (Re), Precision (Pr) and F-measure (Fm) metrics are utilized to compare proposed approach with other algorithms. Re and Pr means the measures of completeness and accurateness respectively. The Fm is a combination of Re and Pr. The definition of these metrics is shown as follows:

$$Re = \frac{TP}{TP + FN}, Pr = \frac{TP}{TP + FN}, Fm = \frac{2 \times Pr \times Re}{Pr + Re}, \tag{8}$$

where the TP and FP is the True Positive and False Positive. In detail, the positive represents foreground, and negative represents background. True means the result of this detection is right, and Negative means the result of this detection is wrong. So the TP means the result of detection is foreground as well as the groundtruth.

The comparison algorithms include MoG [27], ViBe+ [8], RMoG [21] and MST [20]. In particular, the MoG algorithm is based on the pixels and without using neighborhood information. The ViBe and ViBe+ are pixel-based algorithms and utilizing neighborhood information for foreground segmentation. And RMoG is a region-based algorithm, MST is algorithm based on feature contains neighborhood information. In contrast, proposed algorithms is based on superpixels and improve the robustness of foreground by neighborhood information.

The CDN [19] benchmark is almost the largest dataset for Change Detection, which contains 11 categories video sequences. And in this work, the Baseline, Dynamic Background, Camera Jitter and Bad Weather categories include with thousand frames are chosen for evaluating our SPMS. The detection result of comparison state-of-the-art algorithms is available in CDN [19] benchmark. And for SPMS, we used two different parameters, which is shown in Table 2, in our $SPMS_1$ and $SPMS_2$. In Table 2, T is the threshold value for foreground segmentation, and (N_b, N_a, N_s) are utilized to control the number of superpixels in multiple scales. N_a is the number of scales. N_b is the number of superpixels in first scale, and N_s is the increment of superpixels' number in different scales.

Table 1 reports the performances of SPMS and other comparison algorithms on the video sequence belonging to Baseline, Dynamic Background and Bad Weather scenes under the metrics of Re. Pr and Fm. Moreover, due to the length of paper, Fig. 4 only shows the foreground of SPMS and other algorithms in Dynamic Background and Bad Weather scenes. In addition, since the result of ViBe+ [8] in Bad Weather scenes is not published, there is a empty on the foreground of ViBe+ in Fig. 4.

Table 1. The performance comparison of SoAF and four state-of-the-art algorithms on the video sequences belong to baseline, dynamic background, camera jitter and bad weather scenes under the metrics of Re, Pr and Fm (from left to right in each cell).

Videos	Baseline				Camera jitter
	Highway	Office	Pedestrians	PETS2006	Badminton
MoG [27]	0.92 0.93 0.92	0.49 0.75 0.59	**0.99** 0.92 0.95	**0.88** 0.79 **0.83**	0.76 0.64 0.69
RMoG [21]	0.79 0.95 0.87	0.43 0.92 0.59	0.91 0.96 0.94	0.70 0.81 0.75	0.72 0.87 0.79
MST [20]	0.83 0.92 0.87	0.70 0.94 0.80	0.95 0.96 0.95	0.78 0.73 0.75	0.81 0.45 0.58
ViBe+ [8]	**0.93** 0.93 **0.93**	0.70 0.92 0.80	0.95 0.96 **0.96**	0.73 **0.89** 0.80	0.85 **0.93 0.89**
SPMS$_1$	0.70 **1.00** 0.82	0.75 **0.92** 0.83	0.91 **0.98** 0.94	0.59 0.79 0.67	0.88 0.74 0.80
SPMS$_2$	0.76 0.97 0.85	**0.88** 0.87 **0.87**	0.97 0.89 0.93	0.78 0.72 0.75	**0.97** 0.39 0.56

Videos	Camera jitter			Dynamic background	
	Boulevard	Sidewalk	Traffic	Boats	Canoe
MoG [27]	0.83 0.40 0.54	0.58 0.43 0.49	0.76 0.59 0.66	0.76 0.70 0.73	0.87 0.90 0.88
RMoG [21]	0.60 0.45 0.51	0.62 **0.94 0.75**	0.72 **0.77** 0.75	0.82 0.85 0.83	0.90 **0.97** 0.94
MST [20]	0.77 0.53 0.63	0.45 0.27 0.34	0.83 0.34 0.48	0.51 0.45 0.48	0.91 0.86 0.89
ViBe+ [8]	0.76 **0.80 0.78**	0.43 0.77 0.55	0.87 0.73 **0.79**	0.43 0.69 0.53	0.92 **0.97** 0.94
SPMS$_1$	0.92 0.38 0.54	0.45 0.48 0.46	0.95 0.45 0.62	0.74 **0.96** 0.83	0.94 0.96 **0.95**
SPMS$_2$	**0.97** 0.33 0.49	**0.69** 0.22 0.33	**0.99** 0.30 0.46	**0.93** 0.83 **0.88**	**0.97** 0.72 0.83

Videos	Dynamic background				Camera jitter
	Fall	Fountain01	Fountain02	Overpass	Blizzard
MoG [27]	0.88 0.29 0.44	**0.80** 0.04 0.08	0.87 0.75 0.80	0.83 0.92 0.87	**0.81** 0.68 0.74
RMoG [21]	0.72 0.64 0.67	0.55 0.12 0.20	**0.91** 0.83 0.87	0.84 0.97 0.90	0.43 0.90 0.58
MST [20]	0.85 0.27 0.41	0.49 0.08 0.14	0.85 0.79 0.82	0.82 0.85 0.84	0.56 **0.99** 0.71
ViBe+ [8]	0.89 **0.71 0.79**	0.62 **0.19 0.30**	0.87 0.88 0.87	0.84 0.93 0.88	-
SPMS$_1$	0.91 0.39 0.55	0.23 0.06 0.09	0.64 **0.98** 0.78	0.85 **0.99** 0.92	0.65 0.97 0.78
SPMS$_2$	**0.97** 0.15 0.26	0.44 0.02 0.03	0.86 0.91 **0.89**	**0.94** 0.96 **0.95**	0.71 0.93 **0.80**

Videos	Bad weather		
	Skating	SnowFall	WetSnow
MoG [27]	0.80 0.97 0.88	0.72 0.74 0.73	0.54 0.69 0.61
RMoG [21]	0.67 0.98 0.79	0.66 0.89 0.76	0.47 **0.82** 0.60
MST [20]	0.81 0.46 0.59	0.58 0.92 0.71	0.43 0.70 0.53
SPMS$_1$	0.89 **0.99** 0.94	0.80 **0.96** 0.87	0.56 0.81 0.67
SPMS$_2$	**0.95** 0.95 **0.95**	**0.90** 0.93 **0.91**	**0.74** 0.63 **0.68**

In the Baseline scenes, there are four video sequences and SPMS achieves significant improvement in office video sequence under Fm and Pr metric. The foreground of our SPMS consists of superpixels which contributed the completeness of proposed approach. With the same reason, the Re score of proposed approach also better than other algorithms in pedestrians video sequence. Moreover, since the neighborhood information reserved in superpixels contributed the robustness of our SPMS model, proposed approach also achieves well result under Pr metric. However, the SPMS does not work well in the highway and PETS2006 video sequences. The reason of this situation is the shade of cars and pedestrians. And the similarity measures in SPMS can not handle the moving objects' shadow.

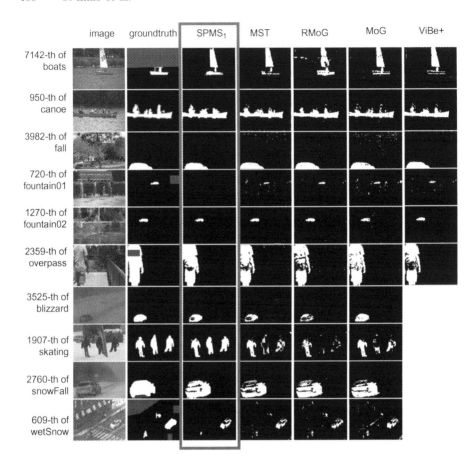

Fig. 4. The qualitative evaluation of SPMS model, all the qualitative result is followed in the CDN [19] Benchmark.

In the Dynamic Background category, proposed approach work well in the boats canoe, fountain02 and overpass video sequences. In this kind of category, the main challenges are come from the dynamic background, such as waving tree and rippling water. Although the intensity of pixels in the background is dynamic changing, the variation in the superpixels more invariant. In addition, the SPMS algorithm utilizes the range search method to against the variation of pixels' position. And the robustness of superpixels also improves the efficiency of our SPMS. Both these methods contributed the ability of SPMS to adapt the environment change and handle the noise. However, the SPMS does not work well in fall and fountain01 video sequences, which may result from the parameter we used in SPMS.

In the Bad Weather scene, our SPMS method achieves good performance in all four video sequences included blizzard, skating, snowFall and wetSnow. The main challenges of Bad Weather scene result from the raindrop and snowflake

Table 2. The Parameter value of SPMS model.

Method	R	(N_b, N_a, N_s)	T_{sp}	T
$SPMS_1$	2	$(20, 16, 20)$	0.3	0.45
$SPMS_1$	2	$(20, 16, 20)$	0.3	0.25

which can be recognized as kinds of noise. In our SPMS, the robustness of super-pixels' variation improves the efficiency of proposed approach. And the accuracy of proposed approach is prevented by the integration of multiple scales. And SPMS work well in the Bad Weather scene.

Unfortunately, the SPMS does not perform well in the Camera Jitter scene. Besides the effect from the arguments we utilized. The similarity measure we utilized is another reason that our SPMS does not work well in Camera Jitter scene.

5   Conclusion

In this paper, we proposed the SPMS background model to detect moving in complex scenes via the integration of superpixels with foreground or background label under multi-scale. Our SPMS analysing the variation of the superpixel to label if it is foreground or background. And the superpixel revered neighborhood information contributed the robustness of proposed approach. Moreover, we captured different superpixels under multiple scales and integrating these superpixels to improve the accuracy of proposed approach. Both the utilization of neighborhood information and multi-scale contributed the performance of our SPMS. And proposed approach achieves encouraging performance in complex scenes included Dynamic Background and Bad Weather. The comprehensive experiment demonstrate that the proposed approach achieves encouraging performance in comparison with some state-of-the-art approaches.

References

1. Achanta, R., Shaji, A., Smith, K., Lucchi, A., Fua, P., Susstrunk, S.: SLIC super-pixels compared to state-of-the-art superpixel methods. IEEE Trans. Pattern Anal. Mach. Intell. **34**(11), 2274–2282 (2012)
2. Bouwmans, T.: Traditional and recent approaches in background modeling for foreground detection: an overview. Comput. Sci. Rev. **1112**, 31–66 (2014)
3. Wren, C., Azarbayejani, A., Darrell, T., Pentland, A.: Pfinder: realtime tracking of the human body. IEEE Trans. Pattern Anal. Mach. Intell. **19**(7), 780–785 (1997)
4. Stauffer, C., Grimson, W.: Adaptive background mixture models for real-time tracking. Proc. IEEE Conf. Comput. Vis. Pattern Recognit. **2**, 252 (1999)
5. Kim, K., Chalidabhongse, T., Harwood, D., Davis, L.: Background modeling and subtraction by codebook construction. In: Proceedings of International Conference on Image Processing, vol. 5, pp. 3061–3064, October 2004

6. Kim, K., Chalidabhongse, T.H., Harwood, D., Davis, L.: Real-time foreground-background segmentation using codebook model. Real-Time Imaging **11**, 172–185 (2005)
7. Barnich, O., Van Droogenbroeck, M.: ViBe: a powerful random technique to estimate the background in video sequences. In: Proceedings of IEEE International Conference on Acoustics Speech Signal Processing, pp. 945–948, April 2009
8. Barnich, O., Droogenbroeck, M.V.: ViBe: a universal background subtraction algorithm for video sequences. IEEE Trans. Image Process. **20**(6), 1709–1724 (2011)
9. Heikkila, M., Pietikainen, M.: A texture-based method for modeling the background and detecting moving objects. IEEE Trans. Pattern Anal. Mach. Intell. **28**(4), 657–662 (2006)
10. Kita, Y.: Background modeling by combining joint intensity histogram with time-sequential data. In: Proceedings of International Conference on Pattern Recognition, pp. 991–994, August 2010
11. Kuo, C.M., Chang, W.H., Wang, S.B., Liu, C.S.: An efficient histogram-based method for background modeling. In: Proceedings of IEEE International Conference on Innovation Computing, Information and Control, pp. 480–483 (2009)
12. Liao, S., Zhao, G., Kellokumpu, V., et al.: Modeling pixel process with scale invariant local patterns for background subtraction in complex scenes. In: 2010 IEEE Conference on Computer Vision and Pattern Recognition (CVPR), pp. 1301-1306 (2010)
13. Yeh, C.H., Lin, C.Y., Muchtar, K., Kang, L.W.: Real-time background modeling based on a multi-level texture description. Inf. Sci. **269**, 106–127 (2014)
14. Riahi, D., St-Onge, P., Bilodeau, G.: RECTGAUSS-Tex: blockbased background subtraction. In: Proce Dept. genie Inform. genie logiciel, Ecole Polytechn. Montr. Montr. QC, Canada, pp. 1–9 (2012)
15. Lin, L., Xu, Y., Liang, X., Lai, J.: Complex background subtraction by pursuing dynamic spatio-temporal models. IEEE Trans. Image Process. **23**(7), 3191–3202 (2014)
16. Chiranjeevi, P., Sengupta, S.: Neighborhoodhood supported model level fuzzy aggregation for moving object segmentation. IEEE Trans. Image Process. **2**, 645–657 (2014)
17. Zaharescu, A., Jamieson, M.: Multi-scale multi-feature codebookbased background subtraction. In: Proceedings of IEEE International Conference on Computer Vision Work, pp. 1753–1760 (2011)
18. Yao, J.Y.J., Odobez, J.-M.: Multi-layer background subtraction based on color and texture. In: Proceedings of IEEE Conference on Computer Vision Pattern Recognition, pp. 1–8 (2007)
19. Wang, Y., Jodoin, P.-M., Porikli, F., Konrad, J., Benezeth, Y., Ishwar, P.: CDnet 2014: an expanded change detection benchmark dataset. In: Proceedings of IEEE Workshop on Change Detection (CDW-2014) at CVPR-2014, pp. 387-394 (2014)
20. Lu, X.: A multiscale spatio-temporal background model for motion detection. In: Proceedings of IEEE International Conference on Image Processing, pp. 3268–3271 (2014)
21. Varadarajan, S., Miller, P., Zhou, H.: Spatial mixture of Gaussians for dynamic background modelling. In: Proceedings IEEE International Conference on Advance Video Signal Based Surveillance, pp. 63–68 (2013)
22. Van Droogenbroeck, M., Paquot, O.: Background subtraction: experiments and improvements for ViBe. In: IEEE Conference on Computer Vision and Pattern Recognition Work, pp. 32–37 (2012)

23. Varadarajan, S., Miller, P., Zhou, H.: Spatial mixture of Gaussians for dynamic background modelling. In: 2013 10th IEEE International Conference on Advanced Video and Signal Based Surveillance (AVSS), pp. 63–68, 27–30 August 2013

24. Zhu, W., Liang, S., Wei, Y., Sun, J.: Saliency optimization from robust background detection. In: IEEE Conference on Computer Vision and Pattern Recognition (CVPR), pp. 2814–2821, June 2014

25. Patwardhan, K., Sapiro, G., Morellas, V.: Robust foreground detection in video using pixel layers. IEEE Trans. Pattern Anal. Mach. Intell. **30**(4), 746–751 (2008)

26. Wu, Y., He, X., Nguyen, T.: Moving objects detection with freely moving camera via background motion subtraction. IEEE Trans. Circuits Syst. Video Technol. **PP**(99), 1–1 (2015)

27. Zivkovic, Z.: Improved adaptive Gaussian mixture model for background subtraction. In: Proceedings of International Conference on Pattern Recognition, vol. 2 (2004)

28. Wu, L., Zhang, Z., Wang, Y., Liu, Q.: A segmentation based change detection method for high resolution remote sensing image. In: Li, S., Liu, C., Wang, Y. (eds.) CCPR 2014, Part I. CCIS, vol. 483, pp. 314–324. Springer, Heidelberg (2014)

Direct Discriminant Analysis Using Volterra Kernels for Face Recognition

Guang Feng[1], Hengjian Li[1(⋈)], Jiwen Dong[1], and Jiashu Zhang[2]

[1] Shandong Provincial Key Laboratory of Network Based Intelligent
Computing, University of Jinan, Ji'nan, China
ise_lihj@ujn.edu.cn
[2] Sichuan Key Laboratory of Signal and Information Processing,
Southwest Jiaotong University, Chengdu 610031, China

Abstract. Based on non-linear Volterra kernels mapping and direct discrimination analysis(DD-Volterra), a novel face recognition algorithm is proposed. Firstly, the original image is segmented into specific sub blocks and seeks functional mapping using truncated Volterra kernels. Next, simultaneous diagonalization obtain Volterra kernel optimal projection matrix. This matrix can discard useless information that exist in the null space of the inter-class. Also, it can reserve discriminative information that exist in the null space of the intra-class. Finally, in the test, each block of the test image is classified separately, voting strategy and nearest neighbor classifier algorithm are used for classification. Experiments show that the proposed DD-Volterra method has better performance for it is more effective than Volterrafaces during the extracting facial feature stage.

Keywords: Face recognition · Feature extraction · Volterra kernels · Direct discriminant analysis

1 Introduction

Face recognition is a front subject in the field of pattern recognition and computer vision, because it has the advantages of non-contact, concealment, easy to understand and low cost of image acquisition equipment, has been increasingly used in safety monitoring, human interaction, artificial intelligence and e-commerce security.

In the past 20 years, researchers have proposed many methods of face recognition. For example, a Principal Component Analysis (PCA) [1] method is proposed, which is based on the location of the sample points in the space, the maximum variance direction of the sample points in the multidimensional space is used as the judgment vector to achieve data compression and feature extraction. Linear Discriminant Analysis(LDA) [2], by looking for a set of optimal discriminant projection vector which can project the data sample from the high dimensional space to the low dimensional space, to make the between-class scatter among the projection samples is the largest, and the within-class scatter is the smallest. In the literature [3] a method of Locality Preserving Projections (LPP) is proposed, it is a linear approximation of the nonlinear Laplacian Eigenmaps [4]. In order to avoid the problem of singular value, a direct LDA algorithm

© Springer Nature Singapore Pte Ltd. 2016
T. Tan et al. (Eds.): CCPR 2016, Part I, CCIS 662, pp. 404–412, 2016.
DOI: 10.1007/978-981-10-3002-4_34

is proposed in [5], through the simultaneous diagonalization find a set of discriminant vectors. This kind of direct discriminant analysis methods can be popularly employed as they have been proved to perform well under several similar situations. For example, in [6], a Gabor based optimized discriminant locality preserving projections (ODLPP) algorithm which can directly optimize discriminant locality preserving is proposed. An adaptive slow feature discriminant analysis method may be found in [7]. Furthermore, Kumar [8] presented face classification method based on Volterra kernels, the method using the truncated form of Volterra kernels and construct a fitness function, which can minimize the intra-class distance and maximize the inter-class distance. Volterra kernels use the convolution operation as their foundation. Convolution is an important operation in image analysis. In the field of image classification, convolution has traditionally been employed at the feature extraction stage and the convolution feature is fixed. However, Volterra classifiers try to find the best filter that groups elements of the same class together and drives elements of different classes apart.

In this paper, based on volterra kernel, a novel method, direct discriminant Volterra kernels (DD-Volterra) is proposed to improve the effectiveness of feature extraction. Instead of matrix inversion operation, the projection matrix is obtained through the two eigenvalue decomposition, and the identification information of the data is extracted to the maximum extent. The experiments prove that the proposed method in this paper is better than many popular contemporary algorithms. The remainder of this paper is organized as follows. DD-Volterra is proposed in Sect. 2. In Sect. 3, the specific implementation methods and classification issues are described in detail. The experimental results are described in Sects. 4, 5 makes the conclusion.

2 Volterra Kernels

Mathematically, Volterra series is essentially a non-linear time-invariant systems functional series expansion, it also can be seen as the promotion of a one-dimensional linear convolution system on a multidimensional space. Therefore, according to convolution operation in the linear system and its related concepts as a reference, may well help people better understand the Volterra series theory. This chapter briefly describes the Volterra series and its application in face recognition.

2.1 Volterra Series

For linear systems, the linear relation between the input and output of the system in time domain can be expressed as the convolution:

$$y(t) = \int_{-\infty}^{+\infty} h(t - \tau)x(\tau)d\tau \tag{1}$$

Where $x(t)$ is the input signal, $y(t)$ is the input signal, $h(t)$ is the system impulse response function.

Correspondingly, any continuous nonlinear time invariant system, in the zero initial condition, if the system input signal $x(t)$ energy is limited, the system response is available Volterra series representation [9]. Volterra series is an expanded form of the Eq. (1). It is an infinite series, the following formula:

$$\begin{cases} y(t) = y_0 + \sum_{n=1}^{\infty} y_n(t) \\ y_n(t) = \int_{-\infty}^{+\infty} \cdots \int_{-\infty}^{+\infty} h_n(\tau_1, \ldots, \tau_n) \prod_{i=1}^{n} x(t - \tau_i) d\tau_1 \ldots d\tau_n \end{cases} \tag{2}$$

Here, $h_1(\tau), h_2(\tau_1, \tau_2), \cdots, h_n(\tau_1, \cdots \tau_n)$ are called the Volterra kernel function. It is a generalization of one dimensional impulse response function of linear system in high dimensional space. In addition, the general assumption is that the system move near zero equilibrium that $y_0(t) = 0$, it can avoid the mutual coupling between the input frequency caused by constant output. It can be seen that when the second-nuclear Volterra is above are zero, the nonlinear system is degenerates to a linear system. For nonlinear discrete time-invariant system, using Volterra series can be expressed as:

$$y(k) = y_0 + \sum_{n=1}^{\infty} \sum_{m_1}^{\infty} \cdots \sum_{m_n}^{\infty} h_n(m_1, \cdots m_n) x(k - m_1) \cdots x(k - m_n) \tag{3}$$

Where $u(k), y(k) \in R$ respectively represent the input and output of the system. $h_n(m_1, \cdots, m_n)$ is the n-th order kernel function of discrete Volterra series. It needs to be pointed out that the Volterra kernel function is a prominent feature that satisfies the symmetry. In practice, the use of infinite nuclear will undoubtedly increase the computational complexity, it can not be better applied to reality. Moreover, in the face of practical problems to be solved, most of the time we just need to reach a certain accuracy can be a good solution to the problem. So, more of the time we use the truncated form of Volterra Series.

The literature [10] describes this truncated form. Define the truncated form of Volterra series:

$$y_p(m) = \sum_{n=1}^{p} x(m) \otimes_n h(m) \tag{4}$$

Where p mean truncated order, $h(m)$ represents n-th order truncated Volterra kernel.

2.2   Construction of Adaptive Functions

Our goal is to classify a set of images $I = \{g_i\}$ into a set of class $C = c_k$, where $i = \{1, 2, \cdots, N_i\}, k = \{1, 2, \cdots, N_c\}$. To do this, we can get to build a fitness functional among the smallest class distance and maximizing class distance.

$$O_k(I) = \frac{\sum_{c_k \in C} \sum_{i,j \in c_k} \left\| I_i \otimes_p K - I_j \otimes_p K \right\|^2}{\sum_{c_k \in C} \sum_{m \in c_k, n \notin c_k} \left\| I_n \otimes_p K - I_m \otimes_p K \right\|^2} \tag{5}$$

Where K is convolution mask, like $h(m)$ in Eq. (4), is a placeholder for all the different order convolution kernels. I_i like $x(m)$ in Eq. (4), the mask K is unknown, our goal is to seek the mapping K to minimize the value of the function.

Furthermore, training images from each class are stacked up and divided into equal sized patches (overlapping or non-overlapping). Corresponding patches from each class are respectively used to learn Volterra kernels by minimizing intraclass distance over interclass distance.

By using the linear property of convolution, the convolution operation is transformed into multiplication operation.

$$I_i \otimes_p K = A_i^p \bar{K} \tag{6}$$

Where $\bar{K}$ is the vector form by two dimensional mask K transformation, A_i^p depends on the order of the convolutions. I_i is the sub-block of the input image.

Through this form, the formula (7) can be further written as

$$O(I) = \frac{\sum_{c_k \in C} \sum_{i,j \in c_k} \left\| A_i^p \cdot \bar{K} - A_j^p \cdot \bar{K} \right\|^2}{\sum_{c_k \in C} \sum_{m \in c_k, n \notin c_k} \left\| A_n^p \cdot \bar{K} - A_n^p \cdot \bar{K} \right\|^2} \tag{7}$$

This can be written as

$$O(I) = \frac{\bar{K}^T S_W \bar{K}}{\bar{K}^T S_B \bar{K}} \tag{8}$$

Where $S_W = \sum_{c_k \subset C} \sum_{i,j \in c_k} (A_i^p - A_j^p)^T (A_i^p - A_j^p)$ and $S_B = \sum_{c_k \subset C} \sum_{m \in c_k, n \notin c_k} (A_i^p - A_j^p)^T (A_i^p - A_j^p)$.

Thus, the face recognition problem using Volterra kernels is essentially an optimization problem.

3   Computational Issue and Result Classification

Li [11] suggested a two-dimensional direct discriminant LPP algorithm to optimize the projection matrix. Reference to the idea of direct discriminant analysis, in which the null space of S_W and the column space of S_B cotains the most discriminative information. Inspired by the direct discriminant analysis, the direct discriminant algorithm using Volterra Kernels is proposed.

The direct algorithm is described as follows:

Step 1. Diagonalize S_B, Suppose there is a matrix V, it satisfies $V^T S_B V = \Lambda$, where $V^T V = I$. Λ is a diagonal matrix which sorted in decreasing order. Since the matrix S_B might be singular, some of its eigenvalues will be zero or close to zero. Furthermore, the optimal discriminant vectors exist in the column space of S_B, as projection directions with a total scatter of zeros do not carry any discriminative power at all, so we can discard the eigenvectors corresponding to the zero eigenvalues.

The rank of the matrix S_B is assumed to p, so the number of nonzero eigenvalues is p. Let Y be the first p columns of V, then

$$Y^T S_B Y = D_B > 0 \tag{9}$$

Where D_B is the p × p sub-matrix of Λ

Step 2. Define a matrix Z, let $Z = YD_B^{-\frac{1}{2}}$. So

$$Z^T S_B Z = \left(YD_B^{-\frac{1}{2}}\right)^T S_b \left(YD_B^{-\frac{1}{2}}\right) = I_p \tag{10}$$

Step 3. Factorize $Z^T S_W Z$, solving the matrix U, to make it meet $U^T(Z^T S_W Z)U = D_W$ and $U^T U = I$, where D_W is in ascending order matrix. We can get the matrix D_W and U by the method of eigenvalue decomposition of matrix $Z^T S_W Z$.

Step 4. Let matrix W and is defined as $W = ZU$, obviously W can simultaneous diagonalization S_B and S_W, W is a matrix which is constructed by the corresponding characteristic vector of the matrix $S_B^{-1} S_W$ nonzero eigenvalues. W is the solution of the proposed direct discriminant projection algorithm. Through this method we can get the mapping $W(i)$ of each sub-block.

The classification algorithm is described as follows:

For a test image, it is first divided into specific sub-blocks (the same operation during the training). Each block is converted to the A_i^p matrix form. Through the former d dimension of the matrix $W(i)$, we can get the corresponding characteristic matrix $Y_d(i) = A_i * W_d(i)$, Where i represents the i-th block is segmented image, its value is $1 - N$. Comparing the 2-norm of the corresponding sub-block feature matrix $Y_d(i)$ in the test image with all the training images. The 2-norm minimum value represents the same subject in the test block and the training block. The same operation is performed on each of the sub blocks of the test image, and ultimately get the most sub blocks of a class, the test image is considered to belong to this class.

4   Experimental Result

In order to verify the effectiveness of our method, experiments were done on the Yale A[1], Extented Yale B[2] and CMU PIE [21] face databases. To widely evaluate the performance of our proposed DD-Volterra algorithm, we also enumerate comparative results for various methods that have been applied to face recognition.

4.1   Databases

The Yale A face database composed of 15 distinct subjects, 11 pictures with different expression and illumination per subject. The databases consist of 165 different images. The image size of the Yale A database used in this experiment is 64 × 64 pixels. Yale

---

[1] http://cvc.yale.edu/projects/yalefaces/yalefaces.html.
[2] http://www.cad.zju.edu.cn/home/dengcai/Data/FaceData.html.

database of facial images mainly light and facial expression changes. The Extended Yale B face database contains 16128 images of 38 human subjects under 9 poses and 64 illumination conditions. In this experiment, we chose 64 different illumination conditions under the frontal pose, so we get 64 images for each subject. All the face images are manually aligned and cropped. The size of each cropped image is 32×32 pixels. For the CMU PIE database, we used all 170 images (except for a few corrupted images) from 5 near frontal poses (C05, C07, C09, C27, C29) for each of the 68 subjects. The size of each cropped image is 32×32 pixels.

4.2   Face Recognition Result

The results (average recognition error rates/standard deviation) on the Yale A, Extented Yale B and CMU PIE databases are presents in Tables 1, 2 and 3. The recognition performance under different training set size is tested. Each experiment was repeated 10 times for 10 random choices of the train set. The results are compared to other stare-of-the-art methods. The specific parameter settings are attached to the table below.

Linear kernel size was 5×5 and quadratic kernels size was 3×3. For both cases, overlapping patches of size 12×12 were used. The linear Volterra kernels mapping matrix $W_d(i)$ dimension is 3. The quadratic Volterra kernels mapping matrix $W_d(i)$ dimension is 5. Images of size 64×64 were used. For other methods (except Volterrafaces), best results as reported in the respective papers are used. All values have been given as mean/standard deviation of recognition error percentages.

All Volterrafaces methods used kernels of size 3×3 and for all cases non-overlapping patches of size 8×8 were used. The linear Volterra kernels mapping matrix $W_d(i)$ dimension is 7. The quadratic Volterra kernels mapping matrix $W_d(i)$ dimension is 9. Images of size 32×32 were used. For other methods (except Volterrafaces), best results as reported in the respective papers are used. All values have been given as mean/standard deviation of recognition error percentages.

Table 1. YaleA recognition error rates

Train set size	2	3	4	5	6	7
UVF [13]	27.11	17.38	11.71	8.16	6.27	5.07
TANMM [14]	44.69	29.57	18.44	-	-	-
OLAP [12]	44.3	29.9	22.7	17.9	-	-
Eigenfaces [12]	56.5	51.1	47.8	45.2	-	-
Fisherfaces [12]	54.3	35.5	27.3	22.5	-	-
Laplacianfaces [12]	43.5	31.5	25.4	21.7	-	-
Volterrafaces(linear) [8]	18.59	11.56	9.27	8.11	6.40	5.33
(Standard deviation)	3.96	3.19	2.17	3.57	2.37	4.57
Volterrafaces(quadratic) [8]	24.53	16.08	12.86	12.44	11.49	10.92
(Standard deviation)	7.61	4.17	4.71	8.35	9.70	11.04
DD- Volterra(linear)	17.24	10.17	7.95	6.50	5.58	4.08
(Standard deviation)	3.84	2.38	0.81	1.46	2.07	2.65
DD- Volterra(quadratic)	28.37	17.50	12.77	9.57	9.07	7.42
(Standard deviation)	7.97	2.19	2.76	5.70	7.67	6.02

Table 2. Extended Yale B recognition error rates

Train set size	2	3	4	5	10	20	30	40
ORO [15]	-	-	-	-	-	-	9.0	-
SR [16]	-	-	-	-	12.0	4.7	2.0	1.0
RDA [16]	-	-	-	-	11.6	4.2	1.8	0.9
KLPPSI [17]	-	-	-	24.74	9.93	3.15	1.39	-
KRR [18]	-	-	-	23.9	11.04	3.67	1.43	-
CTA [19]	-	-	-	16.99	7.60	4.96	2.94	-
MALSSO [20]	58.0	54.0	50.0	-	-	-	-	-
Eigenfaces [19]	-	-	-	54.73	36.06	31.22	27.71	-
Fisherfaces [19]	-	-	-	37.56	18.91	16.87	14.94	-
Laplacianfaces [19]	-	-	-	34.08	18.03	30.26	20.20	-
Volterrafaces(linear) [8]	24.82	18.00	10.97	7.60	2.55	0.84	0.59	0.38
(Standard deviation)	7.56	8.54	3.68	2.39	1.32	0.22	0.21	0.25
Volterrafaces(quadratic) [8] (Standard deviation)	34.99	24.87	17.90	12.21	4.69	1.43	0.99	0.55
	8.13	7.16	4.92	3.12	1.32	0.64	0.38	0.38
DD- Volterra(linear)	13.38	8.43	5.63	5.48	1.12	0.44	0.36	0.27
(Standard deviation)	4.62	3.35	2.38	1.49	0.62	0.14	0.16	0.16
DD- Volterra(quadratic)	30.78	20.99	15.34	8.77	5.50	0.63	0.40	0.24
(Standard deviation)	5.73	6.87	5.78	3.00	1.22	0.51	0.25	0.17

Table 3. CMU PIE recognition error rates

Train set size	2	3	4	5	10	20	30	40
KLPPSI [17]	-	-	-	27.88	12.32	5.48	3.62	-
KRR [18]	-	-	-	26.4	13.1	5.97	4.02	-
ORO [15]	-	-	-	-	-	-	6.4	-
TANMM [14]	-	-	-	26.98	17.22	5.68	-	-
SR [16]	-	-	-	-	-	-	6.1	5.2
OLAP [12]	-	-	-	21.4	11.4	6.51	4.83	-
MLASSO [20]	54.0	43.0	34.0	-	-	-	-	-
Eigenfaces [12]	-	-	-	69.9	55.7	38.1	27.9	-
Fisherfaces [12]	-	-	-	31.5	22.4	15.4	7.77	-
Laplacianfaces [12]	-	-	-	30.8	21.1	14.1	7.13	-
Volterrafaces(linear) [8]	42.79	31.44	25.73	21.44	11.63	5.79	4.22	3.16
(Standard deviation)	6.50	5.63	4.61	4.52	2.60	1.82	1.32	1.09
Volterrafaces(quadratic) [8] (Standard deviation)	61.12	47.14	39.72	34.59	19.01	8.55	5.31	4.13
	9.35	8.47	9.47	6.91	4.27	2.01	1.71	1.49
DD- Volterra(linear)	39.57	29.53	23.67	18.94	9.42	4.41	3.21	2.43
(Standard deviation)	6.82	6.48	5.21	4.31	2.27	1.46	1.19	0.95
DD- Volterra(quadratic)	50.40	38.53	32.97	27.48	14.59	6.88	4.19	3.19
(Standard deviation)	6.61	7.49	9.31	5.31	3.21	1.75	1.51	1.18

Linear kernel size was 5×5 and quadratic kernels size was 3×3, Linear cases used overlapping patches of size 12×12 and quadratic cases used non-overlapping patches of size 8×8. The linear Volterra kernels mapping matrix $W_d(i)$ dimension is 4. The quadratic Volterra kernels mapping matrix $W_d(i)$ dimension is 5. Images of size 32×32 were used. For other methods (except Volterrafaces), best results as reported in the respective papers are used. All values have been given as mean/standard deviation of recognition error percentages.

Through the comparison of these experimental data, it is obvious that the results of our algorithm are better than the Volterrafaces and other state-of-the-art methods. Especially in the Extended Yale B face databases, it is more obvious. This shows that our algorithm has better robustness in the face of illumination changes. Of course, in other factors, our method also has a better recognition performance than some other methods.

In addition, our method not only has a lower error rate but also has a higher recognition rates than other algorithm when the number of the training set is small. It is effective to solve the small sample problem. The three different databases used here have different characteristics, it also shows that the applicability of the DD-Volterra method is excellent.

5 Conclusion

Direct discrimination using Volterra kernels (DD-Volterra) functional is proposed for face recognition in this paper. Our method directly optimizes the goodness function via simultaneous diagonalization, it captures more useful information about the spatial structure of the image, especially enhances the ability of the nonlinear description. Also, the proposed DD-Volterra utilizes discriminant analysis to minimum-in-cluster-distance and maximum-between-cluster-distance. Experimental results on the three benchmark face images databases, that is, Yale A, Extended Yale B and CMU PIE, are compared to traditional as well as the state-of-the-art techniques in discriminant analysis for faces. From the results presented in this paper it can be concluded that the DD-Volterra is more accurate than some other algorithms.

Acknowledgements. This work is supported by grants by National Natural Science Foundation of China (Grant No. 61303199).

References

1. Turk, M., Pentland, A.: Eigenfaces for recognition. J. Cogn. Neurosci. **3**(1), 71–86 (1991)
2. Belhumeur, P., Hespanha, J., Kriegman, D.: Face recognition: eigenfaces vs. fisherfaces: recognition using class specific projection. IEEE Trans. Pattern Anal. Mach. Intell. **19**(7), 711–720 (1997)
3. He, X., Yan, S., Hu, Y., et al.: Face recognition using laplacianfaces. IEEE Trans. Pattern Anal. Mach. Intell. **27**(3), 328–340 (2005)

4. Belkin, M., Niyogi, P.: Laplacian eigenmaps and spectral techniques for embedding and clustering. Adv. Neural Inf. Process. Syst. **14**(6), 585–591 (2002)

5. Hua, Y., Jie, Y.: A direct LDA algorithm for high-dimensional data with application to face recognition. Pattern Recogn. **34**(10), 2067–2070 (2001)

6. Chen, X., Zhang, J.: Optimized discriminant locality preserving projection of gabor feature for biometric recognition. Int. J. Secur. Its Appl. **6**(2), 321–328 (2012)

7. Gu, X., Liu, C., Wang, S., et al.: Feature extraction using adaptive slow feature discriminant analysis. Neurocomputing **154**, 139–148 (2015)

8. Kumar, R., Banerjee, A., Vemuri, B.C.: Volterrafaces: discriminant analysis using volterra kernels. In: IEEE Computer Society Conference on Computer Vision and Pattern Recognition, pp. 150–155 (2009)

9. Orcioni, S.: Improving the approximation ability of volterra series identified with a cross-correlation method. Nonlinear Dyn. **78**(4), 2861–2869 (2014)

10. Kumar, R., Banerjee, A., Vemuri, B.C., et al.: Trainable convolution filters and their application to face recognition. IEEE Trans. Pattern Anal. Mach. Intell. **34**(7), 1423–1436 (2012)

11. Li, H., Dong, J., Li, J.P.: Two-dimensional direct discriminant locality preserving projection analysis for face recognition. In: International Symposium on Biometrics and Security Technologies, pp. 18–23 (2014)

12. Cai, D., He, X., Han, J., et al.: Orthogonal laplacianfaces for face recognition. IEEE Trans. Image Process. **15**(11), 3608–3614 (2006)

13. Shan, H., Cottrell, G.W.: Looking around the backyard helps to recognize faces and digits. In: IEEE Conference on Computer Vision and Pattern Recognition, pp. 1–8 (2008)

14. Wang, F., Zhang, C.: Feature extraction by maximizing the average neighborhood margin, pp. 1–8 (2007)

15. Hua, G., Viola, P.A., Drucker, S.M.: Face recognition using discriminatively trained orthogonal rank one tensor projections. In: IEEE Conference on Computer Vision and Pattern Recognition, pp. 1–8 (2007)

16. Cai, D., He, X., Han, J.: Spectral regression for efficient regularized subspace. In: IEEE Conference on Computer Vision and Pattern Recognition, pp. 1–8 (2007)

17. An, S., Liu, W., Venkatesh, S.: Exploiting side information in locality preserving projection. In: IEEE Conference on Computer Vision and Pattern Recognition, pp. 1–8 (2008)

18. An, S., Liu, W., Venkatesh, S.: Face recognition using kernel ridge regression. In: IEEE Conference on Computer Vision and Pattern Recognition, pp. 1–7 (2007)

19. Fu, Y., Huang, T.S.: Image classification using correlation tensor analysis. IEEE Trans. Image Process. **17**(2), 226–234 (2008)

20. Pham, D.S., Venkatesh, S.: Robust learning of discriminative projection for multicategory classification on the Stiefel manifold. In: IEEE Conference on Computer Vision and Pattern Recognition, pp. 1–7 (2008)

21. Sim, T., Baker, S., Bsat, M.: The CMU pose, illumination, and expression (PIE) database. In: IEEE International Conference on Automatic Face and Gesture Recognition, Proc. IEEE **25**(12), 46–51 (2002)

An Improved MEEM Tracker via Adaptive Binary Feature Encoding

Yang Liu[1], Yuehuan Wang[1,2(✉)], and Jun Wang[1]

[1] School of Automation, Huazhong University of Science and Technology,
Wuhan, People's Republic of China
yuehwang@hust.edu.cn
[2] National Key Laboratory of Science and Technology on Multi-spectral
Information Processing, Wuhan, People's Republic of China

Abstract. We propose an adaptive binary feature encoding method to improve the tracking performance of the MEEM-tracker on real surveillance videos by enhancing the distinguished ability between the target object and the background. The adaptive binary feature encoding method transfers the source image data into binary features by calculating the online encoding parameters such as quantization number and quantization thresholds of each feature channel according to the current image data. The quantization number is calculated based on the dissimilarity between the target region and the surrounding region, and the quantization thresholds are decided by the feature clusters of each channel using the K-Means method. Our improved MEEM-tracker (IPMEEM) restores the online encoding parameters for producing distinguishing binary feature vectors in the current training and tracking procedure. In the experiments, our tracker achieves better overall performance on a surveillance dataset which has 12 new collected and labeled sequences under challenging scenes like "low contrast" and "low resolution". We show that our tracker is more robust for real surveillance videos.

Keywords: Visual tracking · Bianry feature encoding · MEEM-tracker · Surveillance videos

1 Introduction

Visual object tracking is one of the core problems of computer vision as it is widely applied in the automatic object identification, human-computer interaction, vehicle navigation and many others. A visual tracking algorithm which performs well should be designed to overcome many cases such as occlusions, illumination changes and scale changes. Numerous state-of-the-art tracking methods have been proposed in the past few years.

Online AdaBoost feature selection algorithm [1] raises the capability of online training. Therefore object appearance changes are handled quite naturally. Moreover, the features selected are the most discriminating ones because of the dependence on background. In order to handle ambiguous labeled training examples which are provided by the tracker itself, Babenko et al. [2] use Multiple Instance

© Springer Nature Singapore Pte Ltd. 2016
T. Tan et al. (Eds.): CCPR 2016, Part I, CCIS 662, pp. 413–425, 2016.
DOI: 10.1007/978-981-10-3002-4_35

Learning (MIL) which puts all ambiguous positive and negative samples into bags to learn a discriminative model for tracking. Hare et al. [3] use an online structured output SVM learning framework, making it easy to incorporate image features and kernels. Zhang et al. [4] model particles as linear combinations of dictionary templates that are updated dynamically, learning the representation of each particle is considered a single task in Multi-Task Tracking (MTT). Henriques et al. [7] use kernelized correlation filter (KCF) for tracking and perform convolution via Fast Fourier Transform. This method raises the tracking speed greatly. Zhang et al. [15] use Structural Sparse Tracking (SST) algorithm which not only exploits the intrinsic relationship among target candidates and their local patches to learn their sparse representations jointly, but also preserves the spatial layout structure among the local patches inside each target candidate. Zhang et al. [16] also propose a consistent low-rank sparse tracker (CLRST) exploiting temporal consistency, which adaptively prunes and selects candidate particles. By using linear sparse combinations of dictionary templates, the proposed method learns the sparse representations of image regions corresponding to candidate particles jointly by exploiting the underlying low-rank constraints. Zhang et al. [8] propose a multi-expert tracking framework, where the base tracker can evolve backwards to correct undesirable effects of bad model updates using an entropy-regularized restoration scheme. The base tracker exploits an online linear SVM algorithm, which uses a prototype set to manage the training samples, and an explicit feature encoding technique for efficient model update.

Despite great progresses, many challenges still remain when dealing with the real surveillance videos. On one hand, low image resolution in real surveillance sequences degrades the performance of the trackers. On the other hand, surveillance scene is much more complicated, and the background clutter is more serious in surveillance videos. Because of the limitation of the employed object visual representation, the state-of-the-art trackers are hard to overcome the mentioned problems and cannot distinguish the object from the background. In this paper, we propose an improved MEEM-Tracker with a novel flexible online feature binary encoding method to improve the tracking performance in real surveillance scenes.

In the stage of object visual representation, MEEM-Tracker [8] employs a fixed feature binary encoding method with manual thresholds to transfer the multiple-channels source image to binary features which are efficient for online SVM training. That is to say, the source images of multiple channels are quantized respectively with a fixed number of thresholds that are set from 0 to 255 with an equal step to produce a few of binary feature vectors. However this binary feature encoding method cannot distinguish the tracked object from the background in difficult situations of the real surveillance videos.

The main contribution of this paper is that we propose a flexible adaptive feature binary encoding method which is appropriate for the MEEM-tracker to represent the target object, this feature representation method can be described as two steps:

Firstly, it's reasonable to calculate the adaptive feature mapping dimensions according to the dissimilarity between the target region and the surrounding region. The bigger the dissimilarity is, the lower the dimensions of feature binary encoding method are. We divide the surrounding region into several patches which are separately compared with the target region aiming at finding the minimal feature disparity. This minimal feature disparity indicates the dissimilarity between the target region and its surrounding region. Secondly, in order to determine the binary mapping thresholds of dimension online, we obtain several clusters from image data of the region of interest by using the K-Means method.

The proposed adaptive feature binary encoding method improves the performance of MEEM-tracker in real complicated surveillance scenes by representing the target object with distinguishable binary features.

2 Proposed Binary Feature Encoding Method

Binary feature encoding technique transfers the source image data to binary features. This feature encoding technique is helpful for efficiently approximating the min kernel SVM and obtaining nonlinear decision boundaries with linear SVM [9].

2.1 Binary Feature Encoding Method

Reference [8] uses the feature mapping technique proposed in [9] to approximate the min kernel SVM in order to obtain the binary features for SVM training.

The following steps for preprocessing are taken in this feature binary encoding method. First, current frames are transformed into CIE Lab color space in order to acquire a wide perceptron of colors. Then on the L channel of the image, a non-parametric local rank transform [10] is applied to produce a feature map which is invariant to monotonically increasing transformations of pixel intensities. This feature map and the Lab channels constitute a 4-channel source image $I_c, c \in \{1, 2, 3, 4\}$ of the current frame. After preprocessing steps, a quantization is applied to transfer the 4-channel source image to binary features.

Let $K = \{K_1, K_2, K_3, K_4\}$ be the feature quantization number of each channel l_c. Suppose that each component a of a feature vector $v = [a_i]$ is in the range $[0, 1]$, and a is discretize $[0, 1]$ into K levels using K thresholds which are uniform distributed. Then the mapping is defined as

$$\phi(a) = \Omega(\Re(Ka)) \qquad (1)$$

where $\Re(.)$ is a rounding function and $\Omega(.)$ is a unary representation transformation. For example, when $K = 5$, $\phi(0.4) = \Omega(2) = [1, 1, 0, 0, 0]$. Then $\phi(v) = [\phi(a_i)]$ is used to the SVM classifier for training and inference. In MEEM-tracker, for color frames $K = 4$ and for gray scale frames $K = 8$.

Nevertheless, using this method we cannot obtain feature mapping results which vary according to the specific situation in real surveillance. For example, in color sequence *soldiers*, the feature vectors of 4 source image channels are quantized respectively with 4 thresholds produces 16 binary feature vectors. The feature vectors of the L channel are displayed as binary images for convenient.

Figure 1 shows us an example image from the sequence *soldiers*. Figure 2 gives us an example of the L channel feature vectors of Fig. 1 using this method. As our

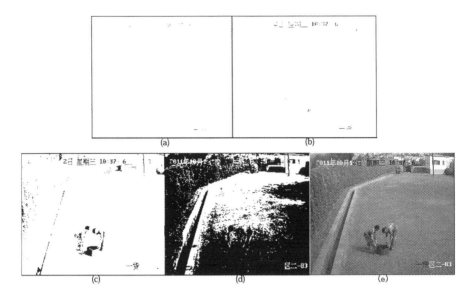

Fig. 1. Binary feature encoding results using fixed binary feature encoding. (a)–(d) represent the binary feature images of *soldiers*' L channel using fixed binary feature encoding. (e) represents the source image of *soldiers*.

observation, this method provides less valid binary features than we thought it should be because of the great similarity between the object and background. Since Almost all the pixel values lie in a small scale, it's supposed to set the thresholds in this scale as well.

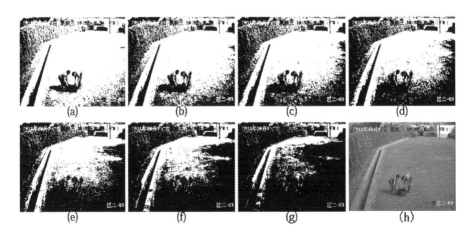

Fig. 2. Binary feature encoding results using adaptive binary feature encoding. (a)–(g) represent the binary feature images of *soldiers*' L channel using adaptive binary feature encoding. (h) represents the source image of *soldiers*.

2.2 Adaptive Binary Feature Encoding Method

The fixed quantization number K and uniform distributed thresholds provides an inflexible and limited ability to distinguish the target region and background region. Thus we need to introduce a more flexible feature encoding technique where our quantization parameters are decided dynamically at every channel of frames. In order to conduct an adaptive flexible feature binary encoding, the previous method can be improved by two steps:

At the beginning, we need to adaptively calculate the feature quantization number K of Lab channel which indicates the quantization precision. In this paper, the feature quantization number K of each image channel can be defined in the first frame I_0 as

$$K = \exp(-C * \min_{B \in B_0} |A(T_0) - A(B_0)|), \tag{2}$$

where $A(.)$ is an average function outputting the average feature value, we use $C = 7$ as an empirically set value. Given the target object region T_0 and the background region $B_0 = \{B_0^1, B_0^2, \ldots, B_0^n\}$ around the object of the frame I_0, if T_0 and B_0 is similar, the value of K would be set greater for efficiently separating the object and the background. For more details, we divide B_0 into 4 patches $\{B_0^1, B_0^2, \ldots, B_0^4\}$ to compare with T_0 respectively, and use the minimal distance to measure the similarity between the object and the surrounding background region in order to get a more accurate measurement.

Next, for every frame, we use the K-Means method to get $K + 1$ cluster centers. Considering that image data from different clusters should be discretized to different binary value, we define a function f to get appropriate thresholds:

$$f(B_t) = g(Kmeans(B_t)) \tag{3}$$

where $Kmeans(.)$ is the K-Means method which returns the cluster centers $centers = \{cent_0, cent_1, \ldots, cent_K\}$ if we use the feature value in frame I_t as input, and $g(.)$ calculates the thresholds which separate $centers$ at the most proper locations. The $g(.)$ can be described according to Eq. (4). Since the feature values $F = \{F_0, \ldots, F_K\}$ are labeled with cluster numbers from 0 to K after using the K-Means method, we are able to get the thresholds and separate the values that have different cluster number labeled. To be simple, we use (4) which returns the needed thresholds.

$$thr_i = (\max(F_{i-1}) + \min(F_i))/2, i \in \{1, \ldots K\} \tag{4}$$

After the above discussion, $\phi(a) = \Omega(\Re(Ka))$ is transformed into $\phi(a) = \Omega'(\Re(Ka))$. Apparently, the result of $\phi(a)$ is likely to change when $\Omega(.)$ is transformed into $\Omega'(.)$ because the thresholds are not uniformed distribution anymore. For example, $\phi(0.4) = \Omega(2)$ may equals $[1, 0, 0, 0, 0]$ or $[1, 1, 1, 0, 0]$. And when i = 1 we use $K_1 = 8$, because larger K is more reasonable for surveillance sequences.

By using this method which dynamically figuring the quantized number K_i of every channel, we can translate the source image to a more distinguishable binary feature. For color sequence $soldiers$, the feature dimension vector of the four channels is $[8, 7, 7, 7]$, and thus the 7 binary images of the L channel are as shown in Fig. 3. Obviously the features are more efficient for SVM training because the distinguishability for object and background.

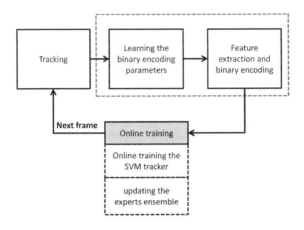

Fig. 3. The tracking framework of IPMEEM

3   Improved MEEM Tracker (IPMEEM)

The MEEM Tracker [8] adopts the multi-expert tracking framework using the minimum entropy criterion. $\mathbf{E} := \{\mathcal{T}, \mathcal{S}_{t1}, \mathcal{S}_{t2}, \ldots\}$ is an experts ensemble, $\mathcal{S}_t$ denotes a snapshot of the classifier $\mathcal{T}$ at time t. Let E denotes an expert in the ensemble. Each expert E is assigned a loss ε_E^t at each step t, and the best expert is determined by its cumulative loss within a recent temporal window $[t - \Delta, t]$:

$$E^* = \arg \min_{E \in \mathbf{E}} \sum_{k \in [t-\Delta, t]} \varepsilon_E^k \tag{5}$$

If a disagreement among the experts is detected, the best expert is selected via (5). Given the instance bag $x = \{x^1, \ldots, x^n\}$ and the possible label set $z = \{y_1, \ldots, y_n\}$ where the ground truth label y must be contained in, the base loss function ε_E is designed as a n entropy-regularized form:

$$\varepsilon_E(x, z) = -L(\theta_E; x, z) + \lambda H(y \mid x, z; \theta_E) \tag{6}$$

In this criterion, $L(\theta_E; x, z)$ is the log likelihood of the current expert model parameterized by θ_E, $H(y \mid x, z; \theta_E)$ is the entropy of class labels conditioned on the training data and the possible label sets. The scalar λ controls the tradeoff between the likelihood and the prior. The criterion (6) is inspired by the semi-supervised partial-label learning (PLL) problem. In [12], learning is based on partially labeled training samples $\ell = \{(x_i, z_i)\}$, where z_i represents a possible label set that contains the true label y_i of instance x_i.

The online linear SVM tracker is employed as the base tracker which contains a compact prototype set to gain an improved approximation to the offline SVM. This tracker is reformulated using the online SVM algorithm [13]. To summarize the previous training data, the prototype set $\mathcal{Q} = \{\zeta_i = (\phi(q_i), q_i, \omega_i, s_i)\}_1^B$ is used where $\phi(q_i)$ is the feature vector of an image patch q_i, ω_i is the binary label, and s_i is a counting number indicating how many support vectors are represented by this instance. Training the SVM classifier at each frame using the prototype set $\mathcal{Q}$ and the new current data $\mathcal{L} = \{(x_i, y_i)\}_1^J$ is realized by minimizing the following objective function:

$$\min_{w,b} \frac{1}{2}\|w\|^2 + C\left\{\sum_{i=1}^{B}\frac{s_i}{N_{\omega_i}}L_h(\omega_i, q_i; w) + \sum_{i=1}^{J}\frac{1}{N_{y_i}}L_h(y_i, x_i; w)\right\}, \qquad (7)$$

where L_h is the hinge loss, $N_l = \sum_{\omega_i=l} s_i + \sum_{y_i=l} 1$, $l \in \{+1, -1\}$ is used to equalize the total weight of the positive samples and that of the negative samples. After the above online training, support vectors from the new training data are added to the prototype set with counting number 1.

The framework of our improved MEEM tracker (IPMEEM) is depicted as Fig. 3. Firstly, we learn the adaptive encoding parameters which are used for feature extraction and binary feature encoding. Then we use the obtained binary feature vector for online training and tracking the object in the next frame. The added adaptive binary feature encoding contributes to achieving more separable features in the feature space, which are more efficient for online SVM learning. The online learning and tracking phase are exactly same to the original MEEM tracker, whose main context is described above. Because of the adaptive binary feature encoding, IPMEEM tracker has stronger ability in handling the typical challenges like low contrast and low resolution in surveillance videos.

4 Experiments

4.1 Trackers and Dataset

To evaluate our IPMEEM Tracker, we collect and label a set of 12 challenging surveillance sequences that possess about 7500 frames in total. These sequences are recorded by surveillance cameras in real scenarios. The frame numbers, quality and fps of each video are not usually same. These sequences cover typical challenging attributes of surveillance videos including illumination variation, background clutter, low resolution, low contrast, scale variation and occlusion. One sequence may cover several attributes. Figure 4 shows the initial frames and Table 1 shows the detail information of the 12 sequences.

Here we compare our IPMEEM tracker against 5 other representative state-of-the-art visual trackers denoted as: OAB [1], Struck [3], MEEM [8], TLD [6], L1PGA [5], and CSK [14]. It's conventional to run the trackers on a test sequences with initialization from the ground truth in the first frame and report the evaluation results. This is referred as one-pass evaluation (OPE). Two evaluation methods are used: precision plot and success plot. The precision plot thresholds the center location error and the success plot thresholds the intersection over union

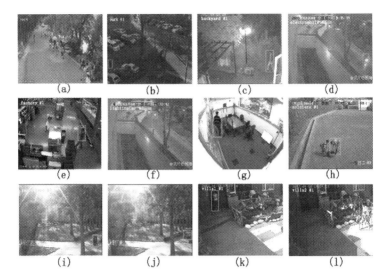

Fig. 4. Initial frames of the 12 sequences in the dataset

Table 1. Information of our dataset. IV: illumination variation, BC: background clutter, LR: low resolution, LC: low contrast, SV: scale variation, OCC: occlusion.

seqName	Frame number	Attributes
(a) road	599	BC, SV, OCC
(b) park	385	IV, BC, LC, OCC
(c) backyard	1766	BC, LR, LC, OCC
(d) electrombile	129	IV, BC, LR, LC, SV, OCC
(e) factory	1573	BC, LC, SV, OCC
(f) lightingCar	243	IV, BC, LR, SV, OCC
(g) room	206	BC, SV, OCC
(h) soldiers	431	BC, LR, LC
(i) sunnyCar	226	IV, BC, LR, SV, OCC
(j) twoWalkers	358	IV, BC, LR, OCC
(k) villa1	271	IV, BC
(l) villa2	1139	IV, BC, SV, OCC

(IOU) metric. As discussed in [11], the precision plot and the success plot are more informative than some widely used metrics, e.g. the success rate and the average center location error.

4.2 Overall Comparisons

Overall precision and success plots on the testing dataset are shown in Fig. 5. It can be seen that IPMEEM outperforms the other trackers by a significant margin. In the

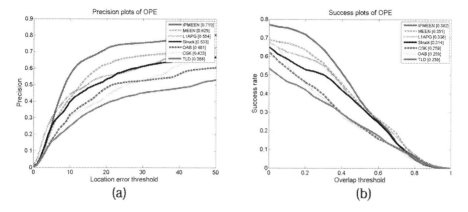

Fig. 5. Precision plot and success plot of OPE

precision plot, the ranking score of IPMEEM is 0.09 point higher than the second best one, which is a performance gain of over 15 %. In success plot, the ranking score is 0.03 higher than the second best one, which is a performance gain of over 8 %.

4.3 Attributes-Based Comparisons

Each of the 12 sequences is annotated with attributes that indicate the challenging situations we face in surveillance sequences. Figures 6 and 7 show the AUC and

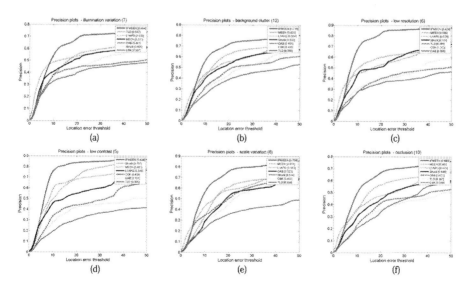

Fig. 6. Precision plots on each attributes

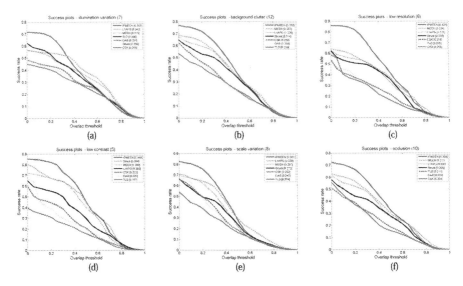

Fig. 7. Success plots on each attributes

threshold based ranking scores. The IPMEEM substantially outperforms other state-of-the-art-trackers on every attribute groups. Each plot correspond to an attribute, such as "illumination variation", "background clutter", "low resolution", and "low contrast", "occlusion" and "scale variation". The number after each attribute name is the number of sequences that have this attribute. Tables 2 and 3 report the ranking scores of all trackers in precision plot and success plot. It can be seen that in precision plots and success plot, IPMEEM substantially outperforms the other state-of-the-art

Table 2. Ranking scores in precision plot. The score of the best tracker on each attribute is noted by red, the score of the second best tracker is noted by dark blue.

Precision Plot	average	IV(7)	BC(12)	LR(6)	LC(5)	SV(8)	OCC(10)
IPMEEM	**0.719**	**0.684**	**0.719**	**0.826**	**0.836**	**0.756**	**0.665**
MEEM	0.625	0.513	0.625	0.599	0.683	0.611	0.563
L1APG	0.554	0.535	0.554	0.530	0.548	0.583	0.473
Struck	0.533	0.409	0.533	0.507	0.707	0.504	0.445
OAB	0.481	0.441	0.481	0.356	0.421	0.523	0.411
CSK	0.433	0.397	0.433	0.383	0.439	0.455	0.336
TLD	0.388	0.547	0.388	0.400	0.305	0.356	0.357

Table 3. Ranking scores in success plot. The score of the best tracker on each attribute is noted by red, the score of the second best tracker is noted by dark blue.

Success Plot	average	IV(7)	BC(12)	LR(6)	LC(5)	SV(8)	OCC(10)
IPMEEM	**0.382**	**0.356**	**0.382**	**0.428**	**0.446**	**0.361**	**0.350**
MEEM	0.351	0.315	0.351	0.334	0.390	0.297	0.311
L1APG	0.338	0.342	0.338	0.321	0.303	0.339	0.300
Struck	0.314	0.256	0.314	0.302	0.394	0.272	0.265
OAB	0.259	0.257	0.259	0.205	0.225	0.247	0.209
CSK	0.259	0.244	0.259	0.218	0.233	0.272	0.204
TLD	0.239	0.306	0.239	0.205	0.167	0.204	0.211

trackers on all 6 groups with different typical challenge attributes of surveillance videos. By adopting the proposed adaptive binary feature encoding method, our tracker is more suitable for such scenes in surveillance videos and achieves outstanding performance.

More intuitive tracking results can be observed in Fig. 8. As we can see, tracking failure happens on state-of-the-art trackers while our tracker performs well facing the most challenging problems in surveillance videos. To be more specific, IPMEEM tracker is better at handing sequences like *electrombile*, *soldiers*, *lightingCar*, and *backyard* where image contrast and resolution are low. For example, the target object is so similar with the background in sequence *soldiers* due to the low contrast that most trackers choose an image patch from background as wrong output. But our tracker successfully distinguishes the target object by using the adaptive binary feature encoding method which improves the distinguishing ability. And in sequence *electrombile*, although the target object appearance has significant features, the low resolution of this sequence provides little image information content, which also causes the other state-of-the-art trackers' tracking failure. Similar situations can be observed on sequence *backyard* and *factory*.

Furthermore, in the situations of handling illumination variations and occlusions, the proposed IPMEEM still achieves accurate performance, which can be seen in the tracking results of sequence *lightingCar* and *sunnyCar*. Only L1APG and IPMEEM successfully handle the tree occlusion in *sunnyCar*, and finally IPMEEM obtains more accurate tracking results. The car light changes significantly in *lightingCar*, IPMEEM and TLD both can tackle with this difficult illumination variation challenge, while the other trackers fail. In a word, because of the great ability of distinguishing the object from the background, our tracker is better at handing surveillance videos than others.

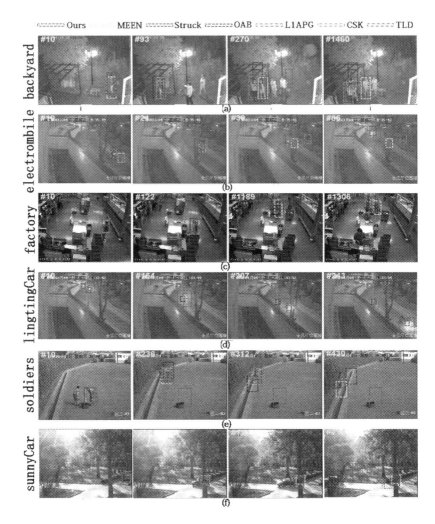

Fig. 8. Intuitive tracking results of typical sequences from the test dataset

5   Conclusions

In this paper, we propose an improved MEEM-Tracker with a flexible adaptive feature binary encoding method which improves the tracking performance in real surveillance scenes. The feature encoding method of our IPMEEM tracker can flexibly distinguish the tracked object from the background in the feature space when facing the challenging situations in surveillance videos. The experiment results show that under challenging circumstance, the performance of our tracker on real-world sequences is robust and efficient.

References

1. Grabner, H., Grabner, M., Bischof, H.: On-line boosting and vision. In: IEEE Conference on Computer Vision and Pattern Recognition, pp. 260–267 (2006)
2. Babenko, B., Yang, M.-H., Belongie, S.: Visual tracking with on-line multiple instance learning. In: IEEE Conference on Computer Vision and Pattern Recognition, pp. 983–990 (2009)
3. Hare, S., Saffari, A., Torr, P.H.S.: Struck: structured output tracking with kernels. In: IEEE International Conference on Computer Vision, pp. 263–270 (2011)
4. Zhang, T., Ghanem, B., Liu, S., Ahuja, N.: Robust visual tracking via multi-task sparse learning. Int. J. Comput. Vis. **101**(2), 367–383 (2013)
5. Bao, B., Wu, Y., Ling, H., Ji, H.: Real time robust L1 tracker using accelerated proximal gradient approach. In: IEEE Conference on Computer Vision and Pattern Recognition, pp. 1830–1837 (2012)
6. Kalal, Z., Matas, J., Mikolajczyk, K.: P-N learning: bootstrapping binary classifiers by structural constraints. In: IEEE Conference on Computer Vision and Pattern Recognition, pp. 188–203 (2010)
7. Henriques, J.F., Caseiro, R., Martins, P.: High-speed tracking with kernelized correlation filters. IEEE Trans. Pattern Anal. Mach. Intell. **37**(3), 583–596 (2015)
8. Zhang, J., Ma, S., Sclaroff, S.: MEEM: robust tracking via multiple experts using entropy minimization. In: European Conference on Computer Vision, pp. 188–203 (2014)
9. Maji, S., Berg, A.C.: Max-margin additive classifiers for detection. In: IEEE Conference on Computer Vision and Pattern Recognition, pp. 40–47 (2009)
10. Zabih, R., Woodfill, J.: Non-parametric local transforms for computing visual correspondence. In: European Conference on Computer Vision, pp. 151–158 (1994)
11. Wu, Y., Lim, J., Yang, M.H.: Online object tracking: a benchmark. In: IEEE Conference on Computer Vision and Pattern Recognition, pp. 2411–2418 (2013)
12. Grandvalet, Y., Bengio, Y.: Semi-supervised learning by entropy minimization. In: NIPS (2005)
13. Wang, Z., Vucetic, S.: Online training on a budget of support vector machines using twin prototypes. Stat. Anal. Data Min. **3**(3), 149–169 (2010)
14. Henriques, J.F., Caseiro, R., Martins, P., Batista, J.: Exploiting the circulant structure of tracking-by-detection with kernels. In: Fitzgibbon, A., Lazebnik, S., Perona, P., Sato, Y., Schmid, C. (eds.) ECCV 2012. LNCS, vol. 7578, pp. 702–715. Springer, Heidelberg (2012). doi:10.1007/978-3-642-33765-9_50
15. Zhang, T., Liu, S., Xu, C., Yan, S., Ahuja, N., Yang, M.: IEEE Conference on Computer Vision and Pattern Recognition, pp. 150–158 (2015)
16. Zhang, T., Liu, S., Ahuja, N., Ghanem, B.: Robust visual tracking via consistent low-rank sparse learning. Int. J. Comput. Vis. **111**(2), 171–190 (2015)

Real-Time Object Tracking Using Dynamic Measurement Matrix

Jing Li, Hong Cheng$^{(\boxtimes)}$, Runzhou Wang, and Lu Yang

Center for Robotics, University of Electronic Science and Technology of China,
Sichuan 611731, China
hcheng@uestc.edu.cn

Abstract. Object tracking has attracted a lot of attention over the past decades. Features represent the main and primary information of object, however, fixed and invariable feature extraction methods would make the features losing their representation. In this paper, we propose a novel robust single object tracking approach using Dynamic Measurement Matrix to extract dynamic features. In particular, we employ the dynamic measurement matrix to adaptively extract features for discriminating object and background so that features have better and clear representativeness. In additional, our approach is a tracking-by-detection approach via a Naive Bayes Classifier with online updating. Compared to traditional approaches, we not only utilize a Naive Bayes Classifier to classify samples but exploit the nature of this classifier to weigh each compressive feature unit, which would be used to update the measurement matrix. The proposed approach runs in real-time and is robust to pose variation, illumination change and occlusion. Furthermore, both quantitative and qualitative experiments results show that our approach has more stable and superior performance.

Keywords: Object tracking · Dynamic measurement matrix · Compressive sensing · Dynamic feature

1 Introduction

Object tracking is a longstanding problem in computer vision and numerous approaches have been proposed. However, object tracking remains challenging due to the appearance change caused by occlusion and motion and the low resolution.

Comaniciu *et al.* [10] proposed MST (Mean Shift Tracking) approach in tracking, which utilized RGB channels' histogram to represent and match the target. Briechl *et al.* [8] used sliding window to search for the target around the object location, which comes from the previous frame. KLT (Kanade-Lucas Tracker) [7] utilized optical flow to find the best area that matches with the affine transformation of the target, and then regarded this area as the object location. Above tracking approaches can track target well in certain situations. However,

© Springer Nature Singapore Pte Ltd. 2016
T. Tan et al. (Eds.): CCPR 2016, Part I, CCIS 662, pp. 426–436, 2016.
DOI: 10.1007/978-981-10-3002-4_36

these approaches still have disadvantages if the object has obvious deformation. These approaches overlook and waste background information, which can improve the efficiency and accuracy of the tracking. Generally, most of these tracking approaches assume the target without obvious and severe deformation during the process of tracking, which can cause tracking drift.

Yu Xiang et al. [22] utilized Markov Decision Processing (MDPs) to model the lifetime of an object, and then formulate the problem of online Multi-Object Tracking (MOT) as decision making in MDPs. Xinge You et al. [23] proposed a local metric learning approach to well handle exemplar-based object detection while few exemplars are available. Kiran Kale et al. [19] combined optical flow with motion vector estimation to realize real-time object tracking even in dynamic environment. Zhang et al. [29] made use of the information of camera and laser range finder, and visual recognition and laser detection were combined together resulted in object can still be detected when out of camera's view. In real life, the object could have severe transformations, and then Yang et al. [17] formulated this task as proposal selection task, in which they build a rich candidate set for predicting the object location.

Discriminative model considered the current area of object as an only positive sample, and labeled areas around the object location as negative samples. In the case that the target location is not accurate, these approaches would utilize the nearest samples to update classifier instead of positive samples. These approaches also cause tracking drift and reduce the credibility of model as time goes on. Grabner et al. [14] proposed an online semi-supervised boosting approach where labeled examples come from the first frame and unlabeled samples come from second frame or latter frames. Babenko et al. [6] introduced multiple instances learning into online tracking where the appearance model is updated with a set of image patches, even though it is not known which patch of the image precisely captures the object of interest. Compressive tracking (CT) [24, 25] utilized a very sparse matrix to compress the image feature space and can greatly simplify the process of feature extraction. However, this random measurement matrix remains unchanged and stable once it is generated and initialized at the beginning of the object tracking. Hence, it would overlook some important potential information of the object during the process of tracking.

Recently, sparse representation has been used in object tracking. [26] regarded particles as linear combinations of dictionary templates that are updated dynamically. In [27], the proposed approach used underlying low-rank constraints to learn the sparse representations of candidate particles, and candidate particles were pruned and selected by exploiting temporal consistency. Zhang et al. [28] proposed Structural Sparse Tracking (SST) to preserves the spatial layout structure among the local patches what always is ignored.

Deep convolutional neural networks has been used in tracking and shows amazing performance. A three-layer convolutional neural network (CNN) is trained [20] to tracking object. [15] using deep features extracted from pretrained deep convolution networks to predict saliency maps. To reduce feature

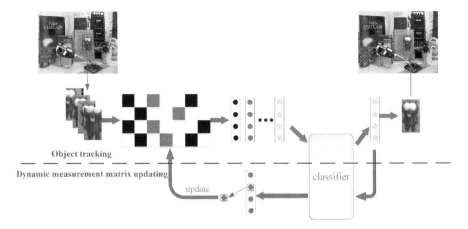

Fig. 1. The overall process of our approach. We update the random measurement matrix to extract better features, which would be a excellent and self-adaptive representation of object. Surprisingly, the classifier is used to provide key and useful information for updating the random measurement matrix.

correlation and avoid over-fitting, Wang *et al.* [21] regard online training CNN as learning group with different loss criterions.

In this paper, we present a novel object tracking approach using dynamic measurement matrix. The overall process of our proposed method is shown Fig. 1. We employ a random measurement matrix that is similar to CT [24], but our measurement matrix is dynamic and can adaptively extract features during the tracking. We also employ Naive Bayes Classifier in our approach, which not only can classify samples, but can provide the information for dynamic measurement matrix to update in every frame except the first frame. After determining the object location in the current frame and getting its compressive features, classifier-based metric function is defined to measure the ability of each compressive feature unit to distinguish the object from the background. And the row vector in dynamic measurement matrix, which corresponds to the compressive feature unit with worst discrimination capacity, will be replaced by a new random row vector. Hence, the random measurement matrix can be adaptively updated to improve the discrimination ability of the compressive features.

The organization of rest of this paper as follows. Section 2 presents our proposed dynamic measurement matrix and classifier-based metric function in details. Section 3 shows experiment results and comparison with CT [24]. Finally, Sect. 4 shows a conclusion and future work of this paper.

2   Object Tracking Approach

In this paper, we propose an adaptive features extracting approach for object tracking. Features of object are extracted using dynamic measurement matrix.

Then we use Naive Bayes Classifier to classify compressive features. Lastly, the compressive feature unit with high score is employed to update the dynamic measurement matrix. So our object tracking approach using dynamic measurement matrix is called DMM.

2.1   Dynamic Measurement Matrix

Features are the main and primary representative information of object, and high dimension features cannot be directly applied in the real-time object tracking problem for complex and massive calculation. So we need to reduce feature dimension to satisfy the real-time object tracking overall speed requirement. To compress features, CT [24] uses compressive sensing theory [12] to generate a random measurement matrix R, whose rows have project information from the high-dimensional image space to a low-dimensional space.

But this random measurement matrix R, meeting the Restricted Isometry Property (RIP) [9], is randomly generated in the process of initialization and remains unchanged during the tracking. However, the appearance of object would change during the tracking, due to motion, illumination and occlusion. Moreover, the object area may become similar to background that caused by motion, illumination and occlusion. If some compressive features have poorer ability to distinguish the object from the background, the generated object features may cause the tracking drift and failure. In these situations, those features would lose their previously distinctive information, So it is hard for this random measurement matrix R to keep the distance information, which represents the projection from high-dimensional space to the subspace. In other words, this random measurement matrix R can not be adjusted according to the object appearance change, which would cause tracking drift and missing tracking object. And our main contribution is to use a appropriate approach to replace the compressive feature with poorer ability.

To change this fixed way of features extraction and utilize the current tracked object information, we propose a dynamic measurement matrix $R(t) \in \mathbb{R}^{n \times m}$ ($n << m$), where t is image sequence frame number. Namely, this dynamic measurement matrix $R(t)$ varies with the image sequence frames number t. According to random projection, $R(t)$ rows have the project information, which is regarded as generalized haar-like feature. So in this paper, we regard each row of $R(t)$ as a compressive feature unit. $R(t)$ is generated and initialized using the same random way in CT [24], so we have $v = R(t)x$, where $x \in \mathbb{R}^m$ is high-dimensional image space, and $v \in \mathbb{R}^n$ is the corresponding low-dimensional space. And $R(t)$ update strategy in our paper is depending on feedback information of classifier.

To utilize the feedback information of classifier, we define a $error(\cdot)$ function to measure which compressive feature unit lose distinctive information, which represents the capacity to distinguish object from background. The specific form of $error(\cdot)$ function we defined depends on the classifier we used. The greater function value means that the distinctive information of this compressive feature unit is worse, and we can determine the compressive feature unit with the worst capacity, which corresponds to a row vector in dynamic measurement matrix. Then, the index of the row vector in $R(t)$ with the worst capacity is marked as idx.

$$idx = \arg \max_{1 \leq i \leq n} error(v_i), \tag{1}$$

where v_i is i-th dimension compressive feature unit of tracked object.

In order to keep the randomness of $R(t)$, we also use the same random way following CT [24] to get $R^* \in \mathbb{R}^{1 \times m}$. Simultaneously, we utilize this row vectors R^* to replace the compressive feature unit in $R(t)$ with index of idx. Specifically, the dynamic measurement matrix $R(t)$ is updated by

$$r_{i,j}(t) = \begin{cases} r_{i,j}(t-1), & i \neq idx \\ r_{1,j}^*, & i = idx, \end{cases} \tag{2}$$

where the idx is the index of compressive feature unit, which has vague and ambiguous distinctive information.

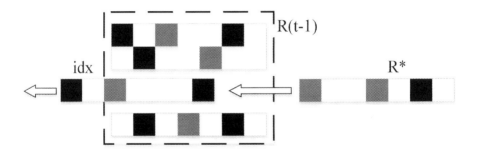

Fig. 2. The process for updating dynamic measurement matrix.

As shown in Fig. 2, after complete t-th frame target tracking, we can determine the target location. We calculate the tracked object's compressive feature in which the compressive feature unit is classified as positive sample with the minimum possibility, is called "bad feature". The "bad feature" we care about is a dimension feature which is classified as positive sample with the minimum possibility. In other words, the "bad feature" with the idx as the index has the poorest ability to separate positive sample from negative samples, and it could play a key role in the process of positive sample was classified as negative sample. We use R^* to replace the line of idx in the $R(t-1)$, and then we get the new matrix $R(t)$ in t-th frame.

2.2 Classifier-Based Metric Function

The classifier is related to dynamic measurement matrix updating strategy, and affects the efficiency of our approach. To update the dynamic measurement matrix, we propose two requirements for classifier:

1. The error of each generalized haar-like feature on the classification results can be measured and the $error(\cdot)$ function in Eq. (1) can be expressed in an explicitly way.
2. The classifier is online learned, and training time complexity is low.

There are several classifiers satisfying the two necessary conditions that mentioned above. And Naive Bayes Classifier (NBC) [18] also satisfies the above two conditions. In our approach, NBC is used to Establish the probability model of each feature unit respectively,

$$H(v) = \sum_{i=1}^{n} \log \left(\frac{p(v_i|y=1)}{p(v_i|y=0)} \right) = \sum_{i=1}^{n} \log \left(\frac{p(v_i|y=1)}{p(v_i|y=0)} \right), \tag{3}$$

where $y \in \{0,1\}$ represents the sample label, and we assume that $p(y=1) = p(y=0)$.

Diaconis and Freedman [11] had proved that the random projection vector of high dimension matrix is always Gaussian. If background and object were viewed as two Gaussian Mixture Model (GMM) respectively, each feature unit extracted from background or object can represented by a Gaussian. We utilize NBC to model each feature unit, which makes classification result is the linear superposition of each dimension classification result. so

$$H(v) = \sum_{i=1}^{n} h_i(v); \quad h_i(v) = \log \left(\frac{p(v_i|y=1)}{p(v_i|y=0)} \right), \tag{4}$$

where $h_i(v)$ is the classification result of i-th dimension feature unit.

In traditional approaches, NBC is always just used to classify samples. In contrast to traditional approaches, we not only employ NBC to classify samples, but also utilize the nature of NBC to measure compressive feature units capacity to distinguish information from object and background.

After getting the compressive features, we learn models for each compressive feature unit separately by NBC. And the classifier can be described with the linear combination of each compressive feature unit

$$v^* = \arg \max_{v \in V^\gamma} H(v) = \arg \max_{v \in V^\gamma} \sum_{i=1}^{n} h_i(v), \tag{5}$$

where V^γ is the feature space extracted from the tracked object.

We utilize NBC classifier to classify samples, and then we take the sample with the highest classification score as the object. The v^* represents the sample with highest score, and then we can get the object location in current frame. As shown in Eq. 3, we are clear that the final classification result of classifier is a linear superposition of each compressive feature unit classification results. Namely, $error(\cdot)$ function, weighing the separating capacity of each compressive feature unit, can be explicitly described. Simultaneously, the classification results can be measured by each generalized haar-like feature as showed in Eq. (5), so we have

$$error(v_i) = -h_i(v). \tag{6}$$

3 Experiment Results

We implement experiment in six sequences to validate the accuracy and effectiveness of our approach. Object in sequences has obvious appearance changes caused by illumination, motion and occlusion, which are challenges for object tracking.

3.1 Experiment Settings

After determining object location in current frame, the radius of extracting positive samples around the object is set to 4, and the radius of extracting negative samples is set 25. We would extract 45 positive samples and 50 negative samples to train Naive Bayesian Classifier. For a new frame, we set the radius of search object new location to 20, which would around object location determined in previous frame. Subsequently, there will be generated 1100 samples. The dimension of each compressive feature is set to 50, namely, the row vector number of dynamic measurement matrix is 50. Commonly, we set learning rate to 0.85 based on experience.

We proposed approach is based on the CT. However, we improve the way of features extraction for making features have better adaptability. Furthermore, we propose two requirements for the classifier, which make contributions to extraction adaptive features. Therefore, to validate the accuracy and effectiveness of our approach, we initialize our approach (DMM) and CT [24] parameters with same initial values, because both DMM and CT utilize Random measurement matrix, which contributes to the tracking results. The dynamic measurement matrix would play a key role in the situation where two methods have same initial values and CT cannot track object steadily.

3.2 Evaluation Metrics

Object tracking problem has two common evaluation metrics: success rate (SR) and center location error (CLE). In this paper, we also use those two common evaluation metrics to quantitative comparison of the experimental results.

We use percentage overlap to measure whether a object is tracked successfully or unsuccessfully. The percentage overlap is defined as $score = \frac{area(ROI_T \bigcap ROI_G)}{area(ROI_T \bigcup ROI_G)}$, where ROI_T is the area of tracked object rectangle box, ROI_G is the area of real object rectangle box. If $score > 50\%$, we think that the object is tracked successfully in the current frame, and then we can calculate the SR. Furthermore, we should care about the tracking accuracy during tracking, which can be measured by the CLE. And the CLE is defined as $CLE = \sum_{i=1}^{N} \frac{\|l_T(i) - l_G(i)\|}{N}$, where $l_T(i)$ is the tracked target location, $l_G(i)$ is the real target location, N is the real object location.

3.3 Results and Analysis

The quantitative experiments results are summarized in Table 1, which shows comparison of 2 trackers on six sequence in terms of success rate and center location error, respectively. And the qualitative experiments are compared as in Fig. 3, where green box is the tracking result of our approach DMM, and red box is the tracking result of CT.

Table 1. Success Rate (SR) and Center Location Error (CLE). **Bold** number indicates the best performance. N represents frame number of each sequence.

Sequence	N	Approach	SR	CLE
Lemming	900	CT [24]	0.2444	12.9452
		DMM	**0.8467**	**10.8761**
Liquar	487	CT [24]	0.3860	24.6831
		DMM	**0.9363**	**14.6539**
Panda	900	CT [24]	0.4156	48.7070
		DMM	**0.72**	**7.9236**
Sylvestr	1100	CT [24]	0.5914	15.2502
		DMM	**0.7516**	**10.9707**
Girl	500	CT [24]	0.5398	35.2328
		DMM	**0.6693**	**31.3075**
Occluded face	885	CT [24]	**0.7054**	30.7030
		DMM	0.6806	**25.4447**

Lemming: In *Lemming* sequence, which comes from [1], the moving lemming has significant appearance changes, illumination, and distracting background, which provide interference information for trackers. Our approach can track moving lemming successfully with higher success rate and lower center location error, while CT loses the object when it undergoes illumination variation. Some examples of tracking results are shown in Fig. 3(a).

Liquar: In *liquar* sequence, the task is to track a bottle in indoor scenes. This sequence comes from [1]. There are several interference information caused by illumination and similar backgrounds during bottle motion. Experiment shows that our approach has robust performance to track the bottle even the object has similar background during moving. But CT loses the object when the bottle pass a similar background. See Fig. 3(b).

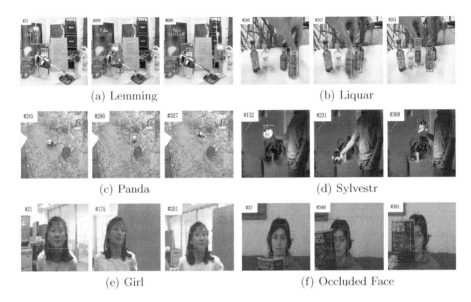

Fig. 3. Tracking results of CT [24] (red) and our approach DMM (green) in six sequences. Frame number are shown in the top left of each figures. Each row shows tracking comparison result in two different sequences. (Color figure online)

Panda: The *panda* sequence, which is proposed in [2], illustrates a problem that panda has significant appearance changes, occlusion, and similar background during moving in outdoor scenes. Our approach can track moving panda successfully and accurately, while CT loses the object when it undergoes similar background. Some examples of tracking results are shown in Fig. 3(c).

Sylvestr: In *Sylvestr* sequence, the task is to track moving dolls in indoor. This sequence shows challenges of lighting, scale, and pose changes, which are always challenges for trackers. This sequence comes from [3] and is used in several tracking paper [13,16]. Experiment shows that our approach has better performance. See Fig. 3(d).

Girl: The *Girl* sequence comes from [4]. The task in this sequence is to track a girl face under the situations that girl face has scale changes, motion, and angel variation caused by girl motion. The tracking task would become a challenge because of rotation, which would cause girl facial feature do not apply to the back of the head. Experiment shows our approach has robust tracking performance. Some examples of tracking results are shown in Fig. 3(e).

Occluded Face: In *Occluded Face* sequence, which comes from the author of [5], the task is to track human face under the situations that human face has scale changes, motion, occlusion and angel variation. Although our approach has slightly lower success rate than CT, our approach has lower center location error, which shows that our approach has better accuracy and precision during tracking. See Fig. 3(f).

4   Conclusion and Future Work

In this paper, we propose object tracking approach by using dynamic measurement matrix, which can compress samples features and keep the distinctive information between object and background. Naive Bayes Classifier is adopted not only as a classifier, but to evaluate the distinctive information in each compressive feature unit, which will help to update the dynamic measurement matrix. Furthermore, the feature without distinctive information is determined by Naive Bayes Classifier, and this feature will be replaced by a random feature. This approach improves the precision and robustness during the object tracking. In our future work, we would like to further refine the measure matrix by learned Haar-like features.

Acknowledgments. The authors would like to thank all the reviewers for their insightful comments. This work was supported by the National Natural Science Foundation of China (Grant No.61305033, 61273256 and 6157021026), Fundamental Research Funds for the Central Universities (ZYGX2014Z009).

References

1. http://gpu4vision.icg.tugraz.at/index.php?content=subsites/prost/prost.php
2. http://info.ee.surrey.ac.uk/Personal/Z.Kalal/
3. http://www.cs.toronto.edu/~dross/ivt/
4. http://vision.ucsd.edu/~bbabenko/miltrack.shtml
5. Adam, A., Rivlin, E., Shimshoni, I.: Robust fragments-based tracking using the integral histogram. In: Computer Vision and Pattern Recognition (2006)
6. Babenko, B., Yang, M.H., Belongie, S.: Robust object tracking with online multiple instance learning. Pattern Anal. Mach. Intell. **33**(8), 1619–1632 (2011)
7. Baker, S., Matthews, I.: Lucas-Kanade 20 years on: a unifying framework. Int. J. Comput. Vision **56**(3), 221–255 (2004)
8. Briechle, K., Hanebeck, U.D.: Template matching using fast normalized cross correlation. In: Aerospace/Defense Sensing, Simulation, and Controls. International Society for Optics and Photonics (2001)
9. Candes, E.J., Tao, T.: Decoding by linear programming. IEEE Trans. Inf. Theory **51**(12), 4203–4215 (2005)
10. Comaniciu, D., Ramesh, V., Meer, P.: Real-time tracking of non-rigid objects using mean shift. In: Proceedings of the IEEE Computer Vision and Pattern Recognition (2000)
11. Diaconis, P., Freedman, D.: Asymptotics of graphical projection pursuit. Ann. Stat. **12**(3), 793–815 (1984)
12. Donoho, D.L.: Compressed sensing. IEEE Trans. Inf. Theory **52**(4), 1289–1306 (2006)

13. Gao, J., Xing, J., Hu, W., Maybank, S.: Discriminant tracking using tensor representation with semi-supervised improvement. In: Proceedings of the IEEE International Conference on Computer Vision (2013)

14. Grabner, H., Leistner, C., Bischof, H.: Semi-supervised on-line boosting for robust tracking. In: European Conference on Computer Vision (2008)

15. Hong, S., You, T., Kwak, S., Han, B.: Online tracking by learning discriminative saliency map with convolutional neural network. arXiv preprint arXiv:1502.06796 (2015)

16. Hong, Z., Mei, X., Prokhorov, D., Tao, D.: Tracking via robust multi-task multi-view joint sparse representation. In: Proceedings of the IEEE International Conference on Computer Vision (2013)

17. Hua, Y., Alahari, K., Schmid, C.: Online object tracking with proposal selection. In: Proceedings of the IEEE International Conference on Computer Vision (2015)

18. Jordan, A.: On discriminative vs. generative classifiers: a comparison of logistic regression and naive bayes. In: Annual Conference on Neural Information Processing Systems (2002)

19. Kale, K., Pawar, S., Dhulekar, P.: Moving object tracking using optical flow and motion vector estimation. In: International Conference on Reliability, Infocom Technologies and Optimization. IEEE (2015)

20. Li, H., Li, Y., Porikli, F.: Robust online visual tracking with a single convolutional neural network. In: Cremers, D., Reid, I., Saito, H., Yang, M.-H. (eds.) ACCV 2014. LNCS, vol. 9007, pp. 194–209. Springer, Heidelberg (2015). doi:10.1007/978-3-319-16814-2_13

21. Wang, L., Ouyang, W., Wang, X., Lu, H.: STCT: sequentially training convolutional networks for visual tracking. In: Computer Vision and Pattern Recognition (2016)

22. Xiang, Y., Alahi, A., Savarese, S.: Learning to track: online multi-object tracking by decision making. In: Proceedings of the IEEE International Conference on Computer Vision (2015)

23. You, X., Li, Q., Tao, D., Ou, W., Gong, M.: Local metric learning for exemplar-based object detection. IEEE Trans. Circuits Syst. Video Technol. **24**(8), 1265–1276 (2014)

24. Zhang, K., Zhang, L., Yang, M.H.: Real-time compressive tracking. In: European Conference on Computer Vision (2012)

25. Zhang, K., Zhang, L., Yang, M.H.: Fast compressive tracking. Pattern Anal. Mach. Intell. **36**(10), 2002–2015 (2014)

26. Zhang, T., Ghanem, B., Liu, S., Ahuja, N.: Robust visual tracking via multi-task sparse learning. In: Computer Vision and Pattern Recognition. IEEE (2012)

27. Zhang, T., Liu, S., Ahuja, N., Yang, M.H., Ghanem, B.: Robust visual tracking via consistent low-rank sparse learning. Int. J. Comput. Vision **111**(2), 171–190 (2015)

28. Zhang, T., Liu, S., Xu, C., Yan, S., Ghanem, B., Ahuja, N., Yang, M.H.: Structural sparse tracking. In: Proceedings of the IEEE conference on computer vision and pattern recognition (2015)

29. Zhao, Z., Chen, W., Wu, X., Wang, J.: Object tracking based on multi information fusion. In: Control and Decision Conference. IEEE (2015)

A Novel Discriminative Weighted Pooling Feature for Multi-view Face Detection

Shiwei Shi[1], Jifeng Shen[1(✉)], Xin Zuo[2], and Wankou Yang[3]

[1] School of Electrical and Information Engineering,
Jiangsu University, Zhenjiang, Jiangsu, China
shenjifeng@ujs.edu.cn
[2] School of Computer Science and Engineering,
Jiangsu University of Science and Technology, Zhenjiang, Jiangsu, China
[3] School of Automation, Southeast University, Nanjing, Jiangsu, China

Abstract. Finding discriminative feature is crucial for building a high-performance object detection system, which has an effect on the detection speed and accuracy. In this paper, we propose a novel discriminative weighted pooling feature based on the multiple channel maps for multi-view face detection. The color and shape statistics of face structure can be utilized to enhance the discriminative ability of the box filter, which is generalized from the square channel filter. The discriminative information can be obtained with LDA and imbalance embedding LDA method, which is superior to the baseline box filter. The experimental result on the FDDB dataset shows that our proposed method has some advantages in accuracy or speed when compared with many other state-of-the-art methods.

Keywords: Discriminative weighted pooling feature · Multiple channel maps · Face detection · Color and shape statistics

1 Introduction

Face detection plays an important role in computer vision field due to its widely used applications such as face tracking, face recognition and human-machine interaction, etc. The method proposed by Viola and Jones [1] (V&J) is regarded as a milestone work, which realizes the first real-time face detection.

Based on the V&J' work, many researchers have proposed some improved methods from two aspects. The first one is related to more complicated feature. Though Haar wavelet [1] is the most famous feature, it only reflects change of intensity in horizontal, vertical and diagonal. Based on it, some improved filters are proposed. Extended Haar-like feature [2] can decrease the false positives at the same recall rate due to its bigger feature pool. Disjoint Haar-like feature [3] can handle multi-view face detection, which is because it can break up the connected sub-regions. The LBP-based feature [4] is another famous feature that can improve training efficiency due to its small feature pool. The classifier construction is another aspect that can be improved. Though the cascade structure [1] achieves the first real-time object detection, it sets parameters empirically and uses thousands of features to remove a large number of the negative

© Springer Nature Singapore Pte Ltd. 2016
T. Tan et al. (Eds.): CCPR 2016, Part I, CCIS 662, pp. 437–448, 2016.
DOI: 10.1007/978-981-10-3002-4_37

samples, which is imprecise and time-consuming. Therefore, some advanced classifiers which set parameters adaptively and can improve the detection speed are proposed, such as boosting chain [5], waldboost [6], soft cascade [7], dynamic cascade [8], etc.

There is a family of channel feature which has achieved better performance in another domain of pedestrian detection [9, 10]. It computes multiple channel maps of the original image like color, texture and histograms of oriented gradients. Moreover, it also has a good performance in face detection due to the rich representation capacity and fast calculation [11].

As mentioned above, though all of these methods are proved to be quite effective in face detection, there still has room of improvement for the detection speed or accuracy. In this paper, we propose a novel discriminative weighted pooling feature (DWPF) based on aggregate channel features (ACF) [10] for multi-view face detection. The Gentle AdaBoost and soft-cascade are utilized to learn the final detector. The experiment results indicate that our method has a faster speed and a strong discriminative ability in detection.

The remaining parts of this paper are organized as follows. Section 2 discusses the baseline square channel filter. Section 3 describes our proposed novel feature in detail. Section 4 introduces the feature learning algorithms. The experimental results and analysis are given in Sect. 5 and we conclude the paper in Sect. 6.

2 Squares Channel Feature (SCF)

Recently, a novel Squares Channel Feature (SCF) based on the multiple channel maps framework is proposed by Benenson, which utilizes the square box as filter and achieves good performance for face detection [12].

Figure 1 shows some different channel images can be generated from the original image. Although there exists many channel types, only 10 channel maps are combined as the basic framework in SCF due to its effective calculation and strong representative capacity. The aforementioned ten channels including 3 color space channels (LUV), 1 gradient magnitude channel and 6 angle gradient histogram channels.

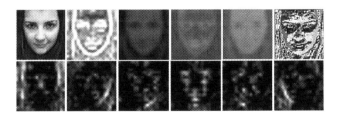

Fig. 1. Different channel maps of the face image (First row corresponds to the original, magnitude, LUV, LBP; Second row corresponds to gradient angle(6)) (Color figure online)

The feature pool is used to extract the discriminative feature, which is a key step during the training procedure of the traditional Integral Channel Features (ICF) [9]. There is no doubt that an appropriate feature pool is more effective for detector training.

As we know, using too many features will consume a large amount of memory during the training process or even reaches the machine limits, while using too few filters may impoverish the capacity of the classifier and leads to a bad detection result. Unfortunately, the feature pool of random size rectangle inside a 80×80 face model is very large, which is a shortcoming of ICF. Based on this point, Benenson proposes SCF which utilizes the squares of all size inside the same face model as ICF, positioned regularly each 4 pixels [12], which is tested to be a better method in face detection. The procedure of SCF computation is shown in Fig. 2.

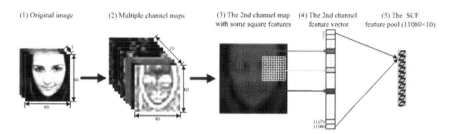

Fig. 2. The procedure of SCF computation

Both SCF and the ACF are based on the multiple channel maps, which are proved to be a good method in face detection [11, 12]. The only difference between them is the step 3 in Fig. 2, which the ACF shrinks the multiple channel images with a factor of 0.25 and uses a pixel information instead of the original square information with size 4×4 to obtain the feature pool, which is shown in Fig. 3. The shrinking procedure has an effect of anti-interference ability to noise, which is the reason why we utilize ACF instead of ICF as the framework of our novel feature in Sect. 3.

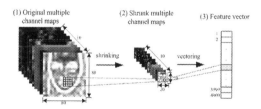

Fig. 3. The key steps of ACF computation

3   Discriminative Weighted Pooling Feature (DWPF)

We can learn from Sect. 2 that the filter plays an important role in the detector training procedure, and a discriminative filter can achieve higher detection accuracy. Recent study [13] shows that the feature embedded with discriminative information obtained from the training data can improve the performance for object detection. The motivation of this method aims to learn a filter which makes use of the discriminative color and shape statistic information of face structure from the training data. A novel DWPF is proposed which learns the discriminative information with LDA and imbalance embedding LDA method, which will be explained in Sect. 4.

The average face image of ten channels for six different views are shown in Fig. 4. We can see that each of them has notable structure information for human face, especially in the fourth and eighth channel. The variety of brightness can be captured by rectangle filter with different size and position which is widely used in the V&J' work in the early studies.

Fig. 4. The average face image of ten channels

The square channel feature is a box filter pooling the information in a local area with identical weights. A sample box filter in the average face template is illustrated in Fig. 5. In Fig. 5(a), the size of average face template is 20×20. The sample template is overlapped with two face parts (Fig. 5(c)), area above the chin (blue rectangle) and the chin (red rectangle).

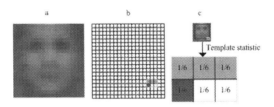

Fig. 5. Average frontal face template and an original box filter (Color figure online)

There are two reasons why we design the bounding box in Fig. 5(b). Firstly, as mentioned in Sect. 2, an appropriate feature pool is computationally efficient for training. The number of DWPF in feature pool is 41480 which is a subset of that without box. Moreover, from the average face template in Fig. 5(a) we can see that the poor pixel information in the area outside of the bounding box is less discriminative for the structure information of face.

In order to get some intuition of why the proposed method works, the distribution of the first selected SCF and DWPF value of positive and negative samples are illustrated in Fig. 6(a) and (b) respectively. The horizontal axis indicates the feature value of samples while the vertical axis shows the number of samples corresponding to the values. The red bins correspond to the positive data while the blue bins represent the negative data. Both of shape of the distribution is similar to the Gaussian

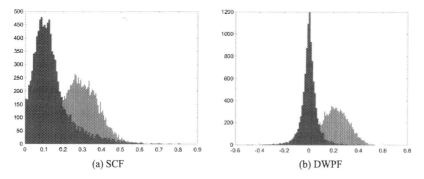

(a) SCF   (b) DWPF

Fig. 6. The distribution of filtering results on the training data (Color figure online)

distribution. It is easy to observe that, the overlapping area of DWPF is smaller than SCF, which indicates the error rate of the former one is much smaller. It is also can be conclude that DWPF is more discriminative than SCF to separate faces from non-faces in the training data.

The SCF is essentially an average pooling operation on the channel maps, whereas DWPF is a weighted pooling operation which can distill discriminative information in the specified location. The key step of DWPF is to find an optimized weight for box filter which produces a hyperplane to separate the faces and non-faces in the feature space. It is easy to see that SCF can be formulated as a dot product between channel feature vector and pooling weight vector, which is shown in Eq. (1).

$$f = w^T a \tag{1}$$

$$w = (w_1, w_2, \ldots, w_k)^T, \quad a = (a_1, a_2, \ldots, a_k)^T$$

where k is the number of the pixels of the local area of channel map, e.g. $k = 6$ in the Fig. 7. When $w = \left[\frac{1}{6}, \frac{1}{6}, \frac{1}{6}, \frac{1}{6}, \frac{1}{6}, \frac{1}{6} \right]^T$, Eq. (1) is identical to the average

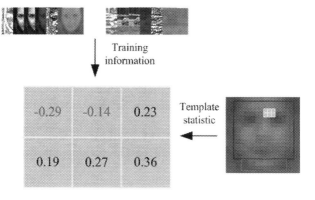

Fig. 7. The procedure of producing DWPF

pooling operation (Fig. 5(c)). It can also be explained that DWPF can pool the local area with statistical information of the face and non-face (Fig. 7), which is a generalized version of SCF. The numbers in block in Fig. 7 means the optimized weight for this box filter which plays the key role in our method. It can be learned by utilize the training information and template statistic which will be explained in detail in Sect. 4.

The procedure of learning a DWPF based face detector is shown in Fig. 8. Firstly, the channel maps for each training sample are calculated. Then shrink the channel maps with a factor of $1/4$. The next step is to learn the optimized weight of each DWPF. The Gentle AdaBoost algorithm is used to select discriminative feature as weak classifier and the final strong detector will be trained multiple times which bootstrap hard negatives from hard negative pools.

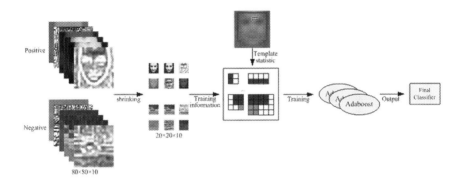

Fig. 8. The procedure of learning a DWPF based detector

4   Feature Learning Algorithm

4.1   Linear Discriminative Analysis (LDA)

It is obvious to see from Eq. (1) that the weight w is very important for the discriminative feature. So we decide to embed the discriminative information from the training samples into the feature, such that the face samples have a low intra-class variance and the inter-class variance between the face samples and non-face samples is high. The idea is similar to the fisher discriminative criterion, which is shown in Eq. (2)

$$w_f = \arg\max_{w} \frac{w^T S_b w}{w^T S_w w} \tag{2}$$

where S_b is the between-class scatter matrix and S_w is the within-class scatter matrix. The closed solution of Eq. (2) can be obtained, $w^* = S_w^{-1}(m_1 - m_2)$. We have named the filter learned with LDA as DWPF-LDA feature.

4.2 Imbalance Embedding LDA (LDA-IB)

It is a common phenomenon that the number of positive samples is far less than the number of negative samples in object detection field. The separating hyperplane ling in the feature space will be affected due to the imbalance number of positive and negative samples. There is a solution that embeds the imbalance information into Eq. (2) which may solve this problem. It is a binary classification problem that the sample can be classified to face or non-face. So we can use $\mathbf{X}^c = \{\mathbf{x}_i^c\}_{i=1}^{N_c}$ to represent the training data, where $\mathbf{x} \in R^k$, $c = 1, 2$, denotes the face or non-face respectively, and i indicates the index of sample. The bold variable denotes vector and matrix while the italic type denotes the scalar. Then a mean vector $\mathbf{m}^c$ and a covariance matrix $\mathbf{S}^c$ are used for two types of the training data, which can be shown in Eqs. (3) and (4).

$$\mathbf{m}^c = \frac{1}{N_c}\sum_{i=1}^{N_c} \mathbf{x}_i^c, \quad c = 1, 2 \tag{3}$$

$$\mathbf{S}^c = \sum_{i=1}^{N_c} (\mathbf{x}_i^c - \mathbf{m}^c), \quad c = 1, 2 \tag{4}$$

The fisher discriminative rule measures the coherence of between-class and within-class value in the case of the k dimensional data project into one dimensional real value. Based on it, we project the mean vector of two types into one dimensional space, which is denoted in Eqs. (5) and (6). Those equations below them are used to calculate the answer.

$$Y^c = \{y_i^c\}_i^{N_c}, \quad c = 1, 2, \quad y \in R, y = \mathbf{w}^T \mathbf{x} \tag{5}$$

$$m^c = \mathbf{w}^T \mathbf{m}^c, \quad c = 1, 2 \tag{6}$$

$$s_w = s_1^2 + s_2^2 = \mathbf{w}^T \mathbf{s}_w \mathbf{w}, \quad \mathbf{S}_w = \mathbf{S}^1 + \mathbf{S}^2 \tag{7}$$

$$s_b = (m^1 - m^2)^2 = \mathbf{w}^T \mathbf{s}_b \mathbf{w}, \quad \mathbf{S}_b = (\mathbf{m}^1 - \mathbf{m}^2)(\mathbf{m}^1 - \mathbf{m}^2)^T \tag{8}$$

$$J_F(\mathbf{w}) = \frac{(m^1 - m^2)^2}{s_1^2 + s_2^2} = \frac{\mathbf{w}^T \mathbf{s}_b \mathbf{w}}{\mathbf{w}^T \mathbf{s}_w \mathbf{w}} \tag{9}$$

$$\mathbf{w}^* = \arg \max J_F(\mathbf{w}) = \mathbf{s}_w^{-1}(\mathbf{m}^1 - \mathbf{m}^2) \tag{10}$$

We can infer from the above equations that the $\mathbf{s}_w$ is the only matrix that can be affected by the imbalance number of face and non-face data, while the means of face and non-face data can be thought as a fixed parameter. The N_1 and N_2 means the number of face and non-face samples respectively, there is no doubt that N_1 far less than N_2. Therefore, s^2 will accounted for the main part of $\mathbf{s}_w$, which has a bad effect on the weight optimization. In order to solve this problem, we utilize a balance factor λ to balance the information from face and non-face samples, which can be formulated in Eq. (11)

$$\mathbf{S}_w = \lambda \mathbf{S}^1 + \mathbf{S}^2, \quad \lambda \geq 1 \tag{11}$$

It is easy to see from the Eq. (11) that the balance factor λ will has different results with different value. When $\lambda = 1$, the computation of $J_F(\mathbf{w})$ is similar to the Eq. (2). When $1 < \lambda < N_2/N_1$, it means there is a imbalance between face and non-face samples, and the former has little status in the process of computation. When $\lambda = N_2/N_1$, the face and non-face samples achieve a balanced state that the number of the former is equal to the later. The last situation is $\lambda > N_2/N_1$, which denotes the face samples will outweigh the non-face samples and the decision hyperplane will move towards to the former. The final feature learned with LDA-IB is called DWPF-LDA-IB.

5   Experiment

5.1   Experiment Setting

The training samples are cropped with size 80×80 from the AFLW dataset [14]. The face views have divided into six angles as $[-90, -60]$, $[-60, -30]$, $[-30, 0]$, $[0, -30]$, $[30, 60]$, $[60, 90]$ degree by yaw while limiting the angle of pitch and roll in $[-22.5, 22.5]$ degree. The number of face samples for each view are 3726, 4024, 4636, 5069, 4024 and 3726 respectively. The training procedure is conducted by four rounds (32, 128, 512, 2048), each with 5000 hard negatives which is bootstrapped from a large negative pools. The total number of DWPF is 41480 which only consider the scenario of overlapping with average face template which is shown in Fig. 5. The piotr's toolbox [15] is used to compute the ACF, and the evaluation code [15] which is public available is also utilized. We use 1.5 as the imbalance factor in imbalance feature weight learning. The experiment is conducted on the T7610 DELL server (dual CPU 16 core, 2.6 GHz, 64G memory). We use MATLAB + VS2012 to implement the final face detector.

In order to verify the effectiveness of our proposed method, the experiment is tested in the FDDB dataset which comprised of 2845 images with 5171 faces. It is the most challenging dataset, with a variety of complex environmental factors, such as low resolution, blur, occlusion and light transformation, etc.

5.2   Experimental Results and Analysis

Comparison of SCF and DWPF-*. The DWPF-* is used to represent the DWPF-LDA and DWPF-LDA-IB features in our experiment. The comparison between ACF, SCF and DWPF-* is shown in Fig. 9, where the ACF and SCF are calculated using our training data. It is worth to mention that our method is nearly 1.7 % higher detection rate than ACF-ours, which means the effectiveness of using high-order statistics in box filter. The same as ACF-ours, SCF-ours has the similar gap in detection rate with DWPF-*, which indicates the weight optimization process of a box filter is effective to improve the performance. Furthermore, DWPF-LDA-IB is slightly better than DWPF-LDA in general due to the embedding of the imbalance factor.

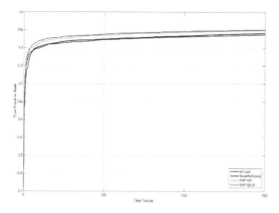

Fig. 9. Comparison among ACF, SCF and DWPF-* features

Top 3 Templates Learned on the FDDB Dataset. The top 3 templates of optimized box filter selected by Gentle AdaBoost are shown in Fig. 10, where the six columns correspond to the six views of face. There are three features selected as the first weak classifier in each view. It is obvious that the gradient orientation channels of 60 and 120 degress are discriminative, which is because the edge of face is a discriminative area where exists a incline about nearly 60 and 120 degress. Moreover, the most discriminative area focuses on the forehead in gradient map due to its strong edge information. The eye is another discriminative area, which has obvious structure information when compared with the cheek.

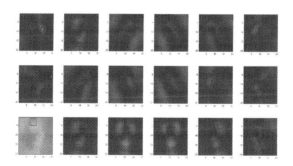

Fig. 10. First template of optimized box filter selected by the boosting algorithm (the six columns correspond to the six views of face)

Comparison with State-of-the-Art Algorithms. The comparison between our proposed method and some other state-of-the-art methods can be shown in Fig. 11, where (a) is the comparison on continuous data while (b) is on discrete data. The evaluation code is downloaded from piotr's toolbox [16] which is available online. It can be seen from Fig. 11(a) that the performance of ACF-ours is far better than original Haar-like feature (Viola-Jones), which means the effectiveness of the multiple channel maps

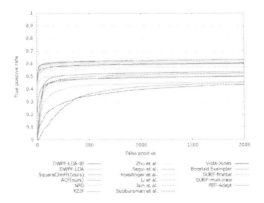

(a) The ROC curve of continuous data

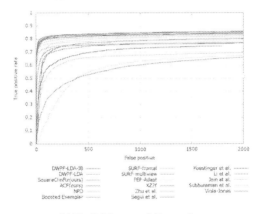

(b) The ROC curve of discrete data

Fig. 11. Experiment comparison on the FDDB dataset

framework. It is easy to notice that DWPF-* is about 2 % higher detection rate than the famous NPD feature [17], and the performance of some other methods such as XZJY [18], Boosted Exemplar [19], SURF-frontal/multi-view [20] and PEP-Adapt [21] are worse than our method obviously. Furthermore, the ROC curve of discrete data in Fig. 11(b) shows that our feature also has the best detection performance when the number of the false positive less than 240.

Comparison of Detection Speed. The comparison of running speed of different methods for frontal face detection is shown in Table 1. It is obvious to see that the

Table 1. Running speed comparison for 640 × 480 images.

Name	Running speed(fps)
ACF	30
SCF	18
DWPF	20

proposed method is superior to the baseline SCF method. It can also be observed that our method is slower than the ACF method.

5.3   Experimental Error Analysis

It can be known from Sect. 5.2 that our method has a good performance in face detection. Figure 12 shows some detection results of our method on the FDDB dataset, where the ellipse shows the ground truth while the red box is our result. It can be seen that a small amount of the missed and false detection are still exist. The occurrence of the former one is due to the large plane rotation angle of face, occlusion, low resolution, the influence of shooting angle and other factors. The false detection mainly occurs in the area overlapping with the human face, which is because there are no face image in the negative training samples.

Fig. 12. Some detection results on the FDDB dataset

6   Conclusion

In this paper, a novel DWPF is proposed based on the multiple channel maps which make use of the distribution information from training data and average face template. This method takes advantage of the color and shape statistics of face structure, which enhances the discriminative ability of feature. The experimental result shows that our method can achieve a high detection rate with low false positive rate, which surpasses many other state-of-the-art methods. Furthermore, the method has good generalization ability which can also be used in the detection of rigid body or semi rigid object, such as pedestrian detection, vehicle detection and other fields.

Acknowledgments. This project is supported by the NSF of China (61305058, 61473086), the NSF of Jiangsu Province (Grants No. BK20140566, BK20150470, BK20130471), the Fundamental Research Funds for the Jiangsu University (13JDG093), the NSF of the Jiangsu Higher Education Institutes of China (15KJB520008), and China Postdoctoral science Foundation (2014M561586).

References

1. Viola, P., Jones, M.J.: Robust real-time face detection. IJCV **57**(2), 137–154 (2004)
2. Lienhart, R., Maydt, J.: An extended set of Haar-like features for rapid object detection. In: International Conference on Image Processing, vol. 1, pp. 900–903 (2001)
3. Stan, Z.L., Long, Z., Zhenqiu, Z., Blake, A., Hongjiang, Z., Shum, H.: Statistical learning of multi-view face detection. In: Proceedings of the 7th European Conference on Computer Vision—Part IV, pp. 67–81 (2002)
4. Ojala, T., Pietikäinen, M., Maenpaa, T.: Multiresolution gray-scale and rotation invariant texture classification with local binary patterns. IEEE Trans. Pattern Anal. Mach. Intell. **24** (7), 971–987 (2002)
5. Rong, X., Long, Z., Hongjiang, Z.: Boosting chain learning for object detection. In: IEEE International Conference on Computer Vision, vol. 701, pp. 709–715 (2003)
6. Sochman, J., Matas, J.: WaldBoost—learning for time constrained sequential detection. In: IEEE Conference on Computer Vision and Pattern Recognition, vol. 152, pp. 150–156 (2005)
7. Bourdev, L., Brandt, J.: Robust object detection via soft cascade. In: IEEE Conference on Computer Vision and Pattern Recognition, vol. 232, pp. 236–243 (2005)
8. Rong, X., Huaiyi, Z., He, S., Xiaoou, T.: Dynamic cascades for face detection. In: IEEE International Conference on Computer Vision, pp. 1–8 (2007)
9. Dollár, P., Tu, Z.W., Perona, P., Belongie, S.: Integral channel features. In: BMVC (2009)
10. Dollár, P., Appel, R., Belongie, S., Perona, P.: Fast feature pyramids for object detection. TPAMI **36**(8), 1532–1545 (2014)
11. Bin, Y., Junjie, Y., Zhen, L., Stan, Z.L.: Aggregate channel features for multi-view face detection. In: Proceedings of International Joint Conference on Biometrics, pp. 194–201 (2014)
12. Benenson, R., Mathias, M., Tuytelaars, T., Gool, L.V.: Seeking the strongest rigid detector. In: CVPR 2013, pp. 1, 2, 6, 7 (2013)
13. Jifeng, S., Changyin, S., Wankou, Y.: A novel distribution-based feature for rapid object detection. Neurocomputing **74**, 2767–2779 (2011)
14. Kostinger, M., Wohlanrt, P., Roth, P.M., Bischof, H.: Annotated facial landmarks in the wild: a large-scale, real-world database for facial landmark localization. In: 2011 IEEE International Conference on Computer Vision Workshops (ICCV Workshops) (2011)
15. Dollár, P., Wojek, C., Schiele, B., Perona, P.: Pedestrian detection: an evaluation of the state of the art. IEEE Trans. Pattern Anal. Mach. Intell. **34**(4), 743–761 (2012)
16. Dollár, P., Appel, R., Belongie, S., Perona, P.: Fast feature pyramids for object detection. IEEE Trans. Pattern Anal. Mach. Intell. **36**(8), 1532–1545 (2014)
17. Shengcai, L., Jain, A.K., Stan, Z.L.: A fast and accurate unconstrained face detector. IEEE Trans. Pattern Anal. Mach. Intell. **38**(2), 211–223 (2016)
18. Xiaohui, S., Zhe, L., Jonathan, B., Ying, W.: Detecting and aligning faces by image retrieval. In: CVPR (2013)
19. Haoxiang, L., Zhe, L., Jonathan, B., Xiaohui, S., Gang, H.: Efficient boosted exemplar-based face detection. In: 2014 IEEE Conference on Computer Vision and Pattern Recognition (2014)
20. Jianguo, L., Yimin, Z.: Learning SURF cascade for fast and accurate object detection. In: IEEE Conference on Computer Vision and Pattern Recognition (CVPR) (2013)
21. Haoxiang, L., Gang, H., Zhe, L., Jonathan, B., et al.: Probabilistic elastic part model for unsupervised face detector adaptation. In: Proceedings of the 2013 IEEE International Conference on Computer Vision, pp. 793–800. IEEE Computer Society (2013)

Non-rigid 3D Model Retrieval Based on Weighted Bags-of-Phrases and LDA

Hui Zeng[1(✉)], Huijuan Wang[1], Siqi Li[1], and Wei Zeng[2]

[1] School of Automation and Electrical Engineering,
University of Science and Technology Beijing, Beijing 100083, China
hzeng@ustb.edu.cn
[2] CAS Key Lab of Network Data Science and Technology,
Institute of Computing Technology, Chinese Academy of Sciences,
Beijing 100190, China

Abstract. This paper presents an improved BOP model, called weighted bags-of-phrases (W-BOP), and its application in non-rigid 3D shape retrieval. The W-BOP model uses the Gaussian weighted function and the 3D points' ring based neighborhood to construct the spatial arrangement model of visual words. Compared with BOP model, it can describe the 3D model more detailedly. Compared with the SS-BOW model and the BOFG model, it needn't perform feature detection step and has higher computation efficiency. To further improve the retrieval performance, the LDA algorithm is used to reduce the dimension of the W-BOP descriptor. Extensive experiments have validated the effectiveness of the designed W-BOP model and LDA based non-rigid 3D model retrieval method.

Keywords: Non-rigid 3D model shape retrieval · SI-HKS · Ring based neighborhood · Gaussian weighted · LDA

1 Introduction

With the development of computer technology and multimedia technology, 3D model plays a more and more important role in many fields, such as biometric recognition, virtual reality, medical diagnostics, intelligent robots, and so on. Because of the explosive growth of the number of 3D models in recent years, 3D model retrieval has become one of the key technologies for effective 3D model analysis and understanding [1–3]. Generally, the 3D model can be divided into two categories: rigid model and non-rigid model. As the non-rigid 3D objects exist widely in the real world and the virtual world, it is necessary to study the non-rigid 3D model retrieval method. In addition to the translation, rotation and scale transformations, there are many non-rigid deformations between the same kind of non-rigid 3D models. So the non-rigid 3D model retrieval technology is more difficult than rigid 3D model retrieval technology, and 3D non-rigid shape analysis and retrieval have attracted more and more researchers' attentions in recent years.

According to the description methods, the existing non-rigid 3D model shape analysis and retrieval methods can be divided into two categories: local feature based

© Springer Nature Singapore Pte Ltd. 2016
T. Tan et al. (Eds.): CCPR 2016, Part I, CCIS 662, pp. 449–460, 2016.
DOI: 10.1007/978-981-10-3002-4_38

methods and global feature based methods [4]. Compared with the global feature based method, the local feature based methods needn't transform the non-rigid 3D model into the canonical form and needn't perform the pose normalization step. Furthermore, they have better robustness to the occlusions and missings. So they have been widely used in the non-rigid 3D model retrieval systems. Similar to the field of local feature based image retrieval, the most popular local feature based 3D model retrieval methods is BOW model based methods. The BOW (Bag of Words) model was originally proposed for text classification and identification of documents. Because of its simple and effectiveness, the BOW model has been widely used in the field of text retrieval, image retrieval and 3D model retrieval. The BOW model based 3D model retrieval method mainly includes the following four steps: (1) for each 3D model, extract the local features according to certain rules; (2) generate the codebook using local features via clustering method (such as K-means algorithm), and the centers of the clusters are named as the visual words; (3) map each local feature of the 3D model to one or several visual words according to certain rules (nearest neighbor, soft allocation etc.); (4) each 3D model can be represented as a word histogram and design classifier to complete retrieval. This kind of method can represent the 3D model with different number of local features as a histogram vector with the same length, which has certain robustness to the resolution and pose changes of 3D model. But the BOW based method doesn't consider the spatial position information of the local features, which may make some effective identification information lost.

Up to now, most of the existing 3D model retrieval methods use the BOW model. Bronstein et al. proposed a BOW based shape retrieval method which uses heat kernel signature (HKS) as local feature descriptor [5]. They performed matching using the similarity distance of the distribution vectors of the visual words. Later, they developed a scale-invariant version of HKS, named as scale invariant heat kernel signature (SI-HKS) [6], and used it to design the BOW model based 3D model retrieval system. The experimental results have shown that the SI-HKS descriptor has better stability to the scale transformation and has better discriminative ability. But both above methods consider only the distribution of the words and lose the spatial position information of the local features. To overcome this problem, the spatially sensitive bag of words (SS-BOW) was proposed [7]. Instead of using the frequency of individual words, the SS-BOW method uses the frequencies of words pairs, thus not only the frequency but also the spatial relations between words are considered. However, the computation complexity of this paper is so big that it's necessary to reduce the number of points through feature detection algorithms. Hou et al. advocated a new paradigm for non-rigid 3D model retrieval, called bag-of-feature-graphs (BOFG), which represents a shape by constructing graphs among its features [8]. Given a vocabulary of geometric words, it firstly builds a graph that records spatial information of features, weighted by their similarities to this word. Then the multi-dimensional scaling (MDS) method is used for reduce the dimension of the BOFG matrices. Although the SS-BOW and the BOFG method record spatial information among features, they have to do feature detection at first to decrease the computation complexity. Then a reduced number of points may not be sufficient to faithfully represent the shape, which likely results in information loss.

In the field of the image retrieval, bag-of-phrases (BOP) model is proposed as an improved version of the BOW model [9]. Compared with the BOW model, the BOP model uses the relative position information between the visual words and achieves better results. Inspired by the idea of the BOP model for 2D image, Li extended it to 3D model representation and proposed a BOP model construction method hat is suitable for 3D model [10]. This method turns the visual words into visual phrase by combining the characteristics of each point in the model and its first order neighborhood point into a local feature phrase. Meanwhile, this method uses the distances from the point to its neighborhood points as correlation weights. The smaller the distance is, the larger the correlation weight is. Compared with the SS-BOW model and the BOFG model, the BOP model not only considers the relative position information between the visual words but also has higher computation efficiency. Although the BOP model contains the "word" space position relationship, the experimental results have shown that the performance of the BOP based 3D retrieval method is worse than the BOW based method.

In this paper, we proposed an improved BOP model for non-rigid 3D model retrieval, which is called weighted bags-of-phrases (W-BOP) model. Similar to the BOP model, the W-BOP model uses the 3D points' "1-ring" neighborhood to construct the spatial arrangement model of visual words. So the computation efficiency of the W-BOP is higher than the SS-BOW model and the BOFG model. To improve the retrieval performance, we use the Gaussian weighted distance as the correlation weight between two points and the LDA method are used for reducing the dimension of the W-BOP feature matrix. The experimental results have shown that, for non-rigid 3D model retrieval, the W-BOP and LDA based method has better performance than the Bow method and the BOP method.

2   Related Work

2.1   HKS and SI-HKS

The heat kernel signatures (HKS) feature is an effective local shape descriptor for non-rigid 3D model retrieval [5]. On compact Riemann manifolds, the heat kernel can be presented as:

$$k_t^X(x, y) = \sum_{i=0} e^{-\lambda_i t} \phi_i(x)\phi_i(y) \qquad (1)$$

where $\lambda_0, \lambda_1 \ldots \geq 0$ are the eigenvalues of the Laplace-Beltrami operator, and $\phi_0, \phi_1, \ldots, \phi_n$ are the eigenfunctions of Laplace-Beltrami operator, which satisfies $\Delta_X \phi_i = \lambda_i \phi_i$. $t \in R^+$ is diffusion time, x and y are two 3D points. Then the HKS feature can be expressed as

$$h(x, t) = k_t^X(x, x) = \sum_{i=0} e^{-\lambda_i t} \phi_i^2(x) \qquad (2)$$

The HKS feature has isometric invariance, and has good robustness to small distortion of non-rigid 3D model. But it is sensitive to the scale change of the 3D model. To solve this problem, Bronstein and Kokkinos [6] proposed scale invariant heat kernel signature (SI-HKS). First of all, compute the HKS feature of each point in the model X. Then the diffusion time is processed by logarithm, and replace t with α^τ:

$$h_\tau = h(x, \alpha^\tau) \tag{3}$$

Let $s = 2\log_\alpha \beta$, and we can obtain the following equation:

$$h'_\tau = \beta^2 h_{\tau+s} \tag{4}$$

Then we remove the multiplicative constant β^2 by taking the logarithm of h. Assume $\dot{h}_\tau = \log h_{\tau+1} - \log h_\tau$, and the discrete derivative w.r.t. to τ:

$$\dot{h}'_\tau = \dot{h}_{\tau+s} \tag{5}$$

Finally, compute the discrete-time Fourier transform of $\dot{h}_\tau$:

$$H'(\omega) = H(\omega)e^{2\pi\omega s}, \omega \in [0, 2\pi] \tag{6}$$

where H and H' denote the Fourier transform of h and $\dot{h}$. Next, the Fourier transform modulus is used for eliminating the phase:

$$\left|H'(\omega)\right| = |H(\omega)| \tag{7}$$

The SI-HKS feature uses low frequency components of $|H(\omega)|$ to construct the scale-invariant feature descriptor. To a certain degree, the SI-HKS feature overcomes the shortcomings of the HKS feature. It not only retains the good performances of the HKS feature but also has better robustness to scale. So the SI-HKS based 3D retrieval method has higher accuracy than HKS based 3D retrieval method. In this paper, we adopt the SI-HKS feature to describe local 3D shapes.

2.2   BOW Model

The BOW (Bag-of-words) is a popular method in the field of the text retrieval and image retrieval, which has recently been introduced into 3D model retrieval. It uses the visual words distribution to describe the shape for 3D shape retrieval. Given a 3D model X, the feature descriptor $p(x)$ of each point $x \in X$ is firstly computed. Then the vocabulary $P = \{p_1, p_2, \ldots, p_V\}$ with size V is obtained by the clustering algorithm, and it is used to represent feature vectors in the descriptor space. Finally the feature distribution $\theta(x) = (\theta_1(x), \theta_2(x), \ldots \theta_V(x))^T$ is computed using the Eq. (8). The similarity of x and word p_i is given by:

$$\theta_i(x) = c(x) \exp\left(-\frac{\|p(x) - p_i\|_2^2}{2\sigma^2} \right) \tag{8}$$

where σ is a parameter computed in advance, and the constant $c(x)$ is selected so that $\|\theta(x)\|_1 = 1$. Then the BOW descriptor of shape X is computed using a $V \times 1$ vector by integrating the feature distribution over the entire shape:

$$f(X) = \int_X \theta(x) d\alpha(x) \tag{9}$$

where $\alpha(x)$ presents the neighboring surface area of the point x.

2.3   SS-BOW Model

Although the BOW model can represent a 3D model using a vector, the BOW model considers only the distribution of the words and loses the relations between them. To overcome this problem, Bronstein et al. introduced an improved BOW model called the Spatially Sensitive Bag of Words (SS-BOW). Instead of looking at the frequency of individual geometric words, they looked at the frequency of words pairs. So the SS-BOW model includes not only the frequency information but also the spatial relations between features. As described in [7], the amount of heat transferred from the point x to the y at time t can be represented as:

$$k_t^X(x, y) = \sum_{i=0}^{\infty} e^{-\lambda_i t} \phi_i(x) \phi_i(y) \tag{10}$$

where λ_i and ϕ_i are the i-th eigenvalue and eigenfunction of the Laplace-Beltrami operator. Then the SS-BOW descriptor can be obtained by

$$F(X) = \int_{X \times X} \theta(x) \theta^T(y) k_t^X(x, y) d\alpha(x) d\alpha(y) \tag{11}$$

From Eq. (4) we can see that, the SS-BOW descriptor is a $V \times V$ matrix representing the frequency of appearance of nearby geometric words.

2.4   BOFG Model

To reduce the computation burden of the SS-BOW model, Bronstein et al. reduced the number of points through feature detection algorithms. However, a reduced number of points may not be sufficient to faithfully represent the shape. To solve this problem, Hou et al. [8] proposed the bag-of-feature-graphs (BOFG) model for non-rigid 3D shape retrieval.

For the shape X with feature set F, the vocabulary is clustered and the feature distribution is computed using Eq. (1). Then construct a matrix G_i using the following equation:

$$G_i(x, y) = \theta_i(x)\theta_i(y)k_t^X(x, y) \tag{12}$$

where $(x, y) \in F \times F$. The heat kernel between the point x and y is used as the weight to the word p_i. Finally we can obtain the matrix set $G(X) = \{G_1, ..., G_V\}$ as the representation of the shape X. The matrices encode spatial relations between features by capturing global structures of graphs. Compared with the SS-BOW model, the BOFG model contains the geometric information in a multi-scale way, and it can faithfully represent the shape. But it also needs performing feature detection step to reduce computation cost.

2.5 BOP Model

The Bag-of-phrases (BOP) model is an improved version of the BOW model, which is designed to encode the spatial information among features. As described in [10], it combines each point's local feature with its first-order neighborhood points' local feature to generate a local feature phrase. Compared with the BOW model, it uses visual phrases instead of visual words. Compared with the SS-BOW model and the BOFG model, the BOP model uses the distance between the point and its neighborhood points as the weight instead of using the amount of heat transferred between two points. The BOP descriptor can be computed by:

$$H(X) = \iint_{X \times X} \theta(x)\theta^T(y)d\alpha(x)d\alpha(y)\big/dist(x, y) \tag{13}$$

where $(x, y) \in F \times F$. The heat kernel between the point x and y is used as the weight to the word p_i. Finally we can obtain the matrix set $H(X)$ as the representation of the shape X. The matrices encode spatial relations between features by capturing global structures of graphs. Compared with The SS-BOW model, the BOFG model contains the geometric information in a multi-scale way, and it can faithfully represent the shape. But it also needs performing feature detection step to reduce computation cost.

3   Weighted BOP Model Based Non-rigid 3D Model Retrieval

3.1   Weighted BOP Model

From the analysis of Sect. 2 we can conclude that: although the SS-BOW model, the BOFG model and the BOP model all include the spatial relation information of visual words, they all have some impactions. The computation burden of the SS-BOW model and the BOFG model is huge, and the feature detection steps need to be performed before construct the model descriptor. Unfortunately, the reduced number of feature points may not be sufficient to describe the shape. The BOP model has higher computational efficiency than the SS-BOW model and the BOFG model, but its retrieval performance is worst. To overcome the above problems, we proposed weighted bags-of-phrases (W-BOP) model. Instead of using the Euclidean distance between two points as weight, we use the Gaussian weighted distance to increases the retrieval accuracy. To further improve computational efficiency, we use the "ring" neighborhood to construct the spatial model of visual words [11]. As shown in Fig. 1, the green points are the 1-ring neighborhood points of the red point. Compared with the distance based local neighborhood points, the ring based local neighborhood points is much more easy to obtain and compute. In this paper, we use the 1-ring neighborhood points to compute the W-BOP model.

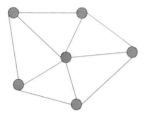

Fig. 1. The 1-ring neighborhood points (Color figure online)

The proposed W-BOP descriptor is computed by,

$$W(X) = \sum_{x \in X} \sum_{y \in Ne(x)} \theta(x)\theta^T(y)\alpha(x)\alpha(y)/G(x,y) \tag{14}$$

where $W(X)$ is a $V \times V$ matrix, and V is the dimension of the visual words. $Ne(x)$ denotes the 1-ring neighborhood point set of the point x, and $G(x, y)$ is the Gaussian distance between x and y. It is used as a relevance weight given by,

$$G(x,y) = \exp\left(-\frac{\|x - y\|_2^2}{2\sigma^2}\right) \tag{15}$$

where σ represent the average distance of the point x with its 1-ring neighborhood points. The smaller the $G(x, y)$ is, the bigger the relevance weight between x and y is. Compared with BOP model, the W-BOP model can describe the spatial relationship more detailedly.

3.2 Dimension Reduction of W-BOP Descriptor

If the W-BOP descriptor is directly used for non-rigid 3D model retrieval, we have to use a feature vector with dimension $V \times V$. However the high dimensional feature vectors are not suitable for 3D model retrieval. A commonly used strategy is to reduce the dimension of feature vectors. In this paper, we adopt the Linear Discriminant Analysis (LDA) algorithm to reduce the dimensionality of W-BOP descriptor [12–14].

Different from the Principal Component Analysis (PCA) algorithm, which is the most classical feature dimension reduction method in the field of pattern recognition, the LDA algorithm aims to find the best projection space to maximize the ratio of projection's inter-class scatter and intra-class scatter.

At first, we compute inter-class scatter matrix using the W-BOP feature vectors of the training 3D models. The intra-class scatter matrix S_W can be computed by:

$$S_W = \sum_{i=1}^{c} \sum_{x_j \in X_i} (x_j - m_i)(x_j - m_i)^T \tag{16}$$

where x_j is the W-BOP feature vector of the jth 3D model that belongs to class i, m_i is the mean of feature vectors of the class i, and c is the number of the classes. The inter-class scatter matrix S_B can be computed by:

$$S_B = \sum_{i=1}^{c} N_i(m_i - m)(m_i - m)^T \tag{17}$$

where m is the total mean of the feature vectors.

Then the projective matrix can be obtained by solving the following generalized eigenvalue problem:

$$S_B W = \lambda S_W W \tag{18}$$

where W is the projective matrix, and λ is the diagonal matrix whose diagonal elements are the eigenvalue of the matrix $S_B^{-1} S_W$.

Finally, for both training 3D models and testing 3D models, we compute the reduced feature vectors using the projective matrix W.

3.3   Non-rigid 3D Model Retrieval Algorithm

In this paper, the W-BOP based non-rigid 3D model retrieval algorithm can be summarized as follows:

(1) For each 3D model, compute the SI-HKS descriptor of each point.
(2) Perform k-means clustering algorithm to obtain the vocabulary $P = \{p_1, p_2, \ldots, p_V\}$ using SI-HKS descriptors of the training 3D models.
(3) For each point of each 3D model, compute its corresponding feature distribution $\theta(x) = (\theta_1(x), \theta_2(x), \ldots \theta_V(x))^T$ using Eq. (8).
(4) Compute the W-BOP descriptor of each 3D model using Eqs. (14) and (15).
(5) For the W-BOP feature vectors of the training 3D models, compute the projective matrix W based on the LDA algorithm using Eq. (18). Then reduce the dimension of the W-BOP feature vectors using the projective matrix W for both training 3D models and testing 3D models.
(6) The Euclidean distance between the reduced feature vectors of two 3D models is selected as the similarity measure for 3D model retrieval.

4   Experimental Results

To investigate the performance of the W-BOP model, we used the McGill 3D shape benchmark for non-rigid 3D model retrieval experiments [15]. The McGill 3D shape benchmark contains 255 non-rigid 3D models, which contains 10 different classes: ant, crab, spectacle, hand, human, octopus, plier, snake, spider and teddy-bear. Each class has 20 to 30 3D models. Among these models, there are rotational transformation, scale transformation and non-rigid deformation. In the 3D model retrieval experiments, 15 3D models are randomly selected for training and the others for testing. Some example 3D models are shown in Fig. 2.

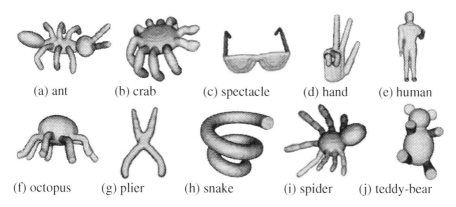

(a) ant   (b) crab   (c) spectacle   (d) hand   (e) human

(f) octopus   (g) plier   (h) snake   (i) spider   (j) teddy-bear

Fig. 2. Example 3D models of the McGill 3D shape benchmark

For the SI-HKS, we used the parameter settings given in the Ref. [6]. We used a logarithmic scale-space with base $\alpha = 2$ and τ ranging from 1 to 25 with increments of 1/16. Then the first 6 discrete lowest frequencies were used as the local SI-HKS descriptor. The k-means clustering algorithm was performed to obtain the vocabulary with 48 words. So the dimension of the BOW descriptor is 48, and the dimension of the W-BOP without dimension reduction is 2304 (48×48). The 3D model retrieval performance was quantified using the Precision-Recall (PR) curve, the mean average precision (MAP), Nearest Neighbor (NN), First-tier (FT), Second-tier (ST) and Discounted Cumulative Gain (DCG).

At first, we investigated the influences of the numbers of the dimension reduction to the 3D model retrieval. The LDA algorithm was used for dimension reduction, and the numbers of the dimension reduction were selected as 1000, 500, 400, 300, 250, 200 and 100 respectively. The PR curves were shown in Fig. 3, and four measures of the retrieval results were listed in Table 1. From Fig. 3 and Table 1 we can see that the retrieval performances are similar with different numbers of the dimension reduction. It has better performance while the numbers of the dimension reduction range from 100 to 400. Comprehensively analyzing different measures of the retrieval results, the

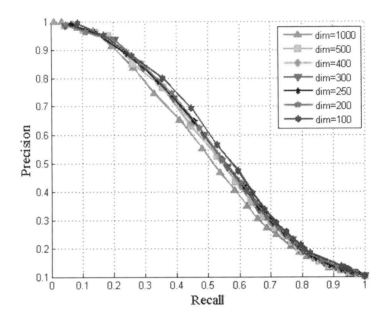

Fig. 3. PR curves of W-BOP based method with different numbers of the dimension reduction

Table 1. Retrieval results of W-BOP based method with different numbers of the dimension reduction

	dim = 1000	dim = 500	dim = 400	dim = 300	dim = 250	dim = 200	dim = 100
MAP	0.7322	0.7410	0.7394	0.7428	0.7399	0.7416	**0.7510**
NN	0.8190	0.8190	**0.8381**	**0.8381**	0.8190	0.8190	0.8095
FT	0.5219	0.5371	**0.5714**	0.5689	**0.5714**	0.5492	0.5606
ST	0.6952	0.7200	0.7441	**0.7454**	0.7435	0.7321	0.7397
DCG	0.8014	0.8100	0.8260	0.8269	**0.8271**	0.8222	0.8215

retrieval method seems to have best performance when the number of the dimension reduction is 300.

Then we evaluated the effectiveness of our proposed W-BOP and LDA based method. We compared our proposed method with BOW based method and BOP based method. The PR curves were shown in Fig. 4, and four measures of the retrieval results were listed in Table 2. As shown in Fig. 4 and Table 2, the W-BOP based method with LDA dimension reduction performs best. The performance of the BOP based method is slightly lower than that of the BOW based method. So Gaussian weighted model can describe the 3D model in more detail and the LDA based dimension reduction step is necessary for W-BOP based 3D model retrieval.

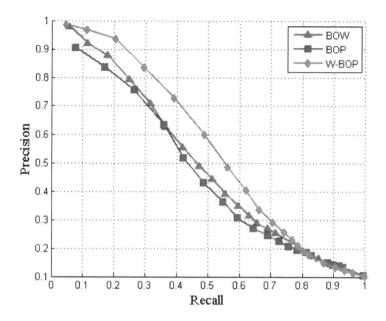

Fig. 4. PR curves of BOW based method and W-BOP based method.

Table 2. Retrieval performance indicators of BOW and W-BOP

	MAP	NN	First-tier	Second-tier	DCG
BOW	0.7276	0.7524	0.5060	0.6851	0.7885
BOP	0.7241	0.7714	0.5035	0.6660	0.7847
W-BOP	0.7428	0.8381	0.5689	0.7454	0.8269

5   Conclusion

This paper presents the W-BOP model and LDA based non-rigid 3D model retrieval method. The W-BOP model is an improved version based on the BOP model, and it uses the Gaussian function to weight the influence of the neighboring points and needn't compute the amount of heat transferred from one point to another. Compared with BOP model, it can describe the 3D model more detailedly. Compared with the SS-BOW model and the BOFG model, it has lower computation burden. In order to further improve the accuracy of 3D model retrieval, the LDA algorithm is used for reducing the dimension of the W-BOP feature matrix. Extensive experimental results have shown that our proposed W-BOP based method has better performance than the Bow model based method and the BOP model based method. In our future work, we will aim at improving the visual words spatial model based on machine learning method to describe the non-rigid 3D model more detailedly.

Acknowledgments. This article is supported by the National Natural Science Foundation of China (Grant No. 61375010 and No. 61005009).

References

1. Guo, Y., Bennamoun, M., Sohel, F., et al.: A comprehensive performance evaluation of 3D local feature descriptors. Int. J. Comput. Vis. **116**, 1–24 (2015)
2. López, G.L., Negrón, A.P.P., Jiménez, A.D.A., et al.: Comparative analysis of shape descriptors for 3D objects. Multimed. Tools Appl. 1–48 (2016)
3. Lu, K., He, N., Xue, J., et al.: Learning view-model joint relevance for 3D object retrieval. IEEE Trans. Image Process. **24**(5), 1449–1459 (2015)
4. Guo, Y., Bennamoun, M., Sohel, F., et al.: 3D object recognition in cluttered scenes with local surface features: a survey. IEEE Trans. Pattern Anal. Mach. Intell. **36**(11), 2270–2287 (2014)
5. Sun, J., Ovsjanikov, M., Guibas, L.: A concise and provably informative multi-scale signature based on heat diffusion. Comput. Graph. Forum **28**(5), 1383–1392 (2009)
6. Bronstein, M.M., Kokkinos, I.: Scale-invariant heat kernel signatures for non-rigid shape recognition. In: Proceedings of IEEE Conference on Computer Vision & Pattern Recognition, pp. 1704–1711 (2010)
7. Bronstein, A.M., Bronstein, M.M., Guibas, L.J., et al.: Shape google: geometric words and expressions for invariant shape retrieval. ACM Trans. Graph. **30**(1), 623–636 (2011)
8. Hou, T., Hou, X., Zhong, M., Qin, H.: Bag-of-feature-graphs: a new paradigm for non-rigid shape retrieval. In: International Conference on Pattern Recognition, pp. 1513–1516 (2012)
9. Zhang, L., Wang, C., Xiao, B., Shao, Y.: Image representation using bag-of-phrases. Acta Autom. Sinica **38**(1), 46–54 (2012)
10. Li, Y.: Non-rigid 3D model retrieval and tagging based on HKS. Master dissertation, Beijing Jiaotong University (2012)
11. Zeng, H., Zhang, R., Huang, M.: Improved 3D local feature descriptor based on rotational projection statistics and depth information. In: Zha, H., Chen, X., Wang, L., Miao, Q. (eds.) Communications in Computer and Information Science, vol. 547, pp. 139–147. Springer, Heidelberg (2015)
12. Belhumeur, P.N., Hespanha, J.P., Kriegman, D.J.: Eigenfaces vs. fisherfaces: recognition using class specific linear projection. IEEE Trans. Pattern Anal. Mach. Intell. **19**(7), 711–720 (2013)
13. Zhou, D., Yang, X., Peng, N.: A modified linear discriminant analysis and its application to face recognition. J. Shanghai Jiaotong Univ. **39**(4), 527–530 (2005)
14. Zhuang, Z., et al.: Based on an optimized LDA algorithm for face recognition. J. Electron. Inf. Technol. **29**(9), 2047–2049 (2007)
15. Siddiqi, K., Zhang, J., Macrini, D., Shokoufandeh, A., Bouix, S., Dickinson, S.: Retrieving articulated 3D models using medial surfaces. Mach. Vis. Appl. **19**(4), 261–274 (2008)

Adaptively Weighted Structure Preserved Projections for Face Recognition

Yugen Yi[1(✉)], Wei Zhou[2], Yanjiao Shi[3], Guoliang Luo[1],
and Jianzhong Wang[4(✉)]

[1] School of Software, Jiangxi Normal University, Nanchang, China
yiyg510@gmail.com
[2] College of Information Science and Engineering,
Northeastern University, Shenyang, China
[3] School of Computer Science and Information Engineering,
Shanghai Institute of Technology, Shanghai, China
[4] College of Computer Science and Information Technology,
Northeast Normal University, Changchun, China
wangjz019@nenu.edu.cn

Abstract. In this paper, a new algorithm named Adaptively Weighted Structure Preserved Projections (Aw-SPP) is proposed for face recognition. Firstly, the configural structure relationship of sub-images in each face image is preserved in Aw-SPP. Then, an adaptive non-negative weight vector is introduced to take different contributions of various sub-pattern sets into account, which combines the Laplacian matrices obtained by different sub-pattern sets. Simultaneously, a Laplacian penalty constraint is also incorporated to preserve the intrinsic 2D structure of each sub-image. Finally, the procedures of feature extraction and non-negative weight vector learning are integrated into a unified framework. Moreover, an efficient iterative algorithm is designed to optimize our objective function. To validate the feasibility and effectiveness of the proposed approach, extensive experiments are conducted on three face databases (Extended YaleB, CMU PIE and AR). Experimental results demonstrate that the proposed Aw-SPP outperforms some other state of the art algorithms.

Keywords: Face recognition · Sub-pattern · Adaptively weighted · Local matching

1 Introduction

Automatic face recognition from still images or video sequences has become an active research topic for its wide range of potential applications in information security, video surveillance, human-computer interface and more general image understanding [1]. However, the face recognition tasks are always complex and difficult in practice due to the following two problematic factors. One is the appearance variations including facial expression, pose, age and illumination changes, the other is the man-made variations, e.g., the noises from the cameras [4].

As a result, numerous methods have been developed for face recognition in recent decades [1–3]. According to some studies [1–3], face recognition methods can be

© Springer Nature Singapore Pte Ltd. 2016
T. Tan et al. (Eds.): CCPR 2016, Part I, CCIS 662, pp. 461–473, 2016.
DOI: 10.1007/978-981-10-3002-4_39

roughly divided into two categories: holistic matching based methods and local matching based methods. Holistic matching based methods, which take advantage of the whole face image region as raw input to the face recognition system, have been extensively studied. In these methods, a low-dimensional feature subspace is first learned by some dimensionality reduction techniques to avoid the curse of dimensionality problem. Then, the face images are mapped into the low-dimensional feature subspace to carry out the classification or recognition. Currently, the most representative and popular algorithms employed for holistic matching based face recognition are Principal Component Analysis (PCA) [5], Linear Discriminant Analysis (LDA) [6], Non-negative Matrix Factorization (NMF) [7], Local linear embedding (LLE) [8], Laplacian Eigenmap (LE) [9] and Locality Preserving Projections (LPP) [10].

Although the aforementioned holistic matching based methods yielded impressive results on some benchmark datasets, their recognition performances may be affected by some problematic factors (such as illumination, expression, disguises and pose) in real-world face images [11–18]. The main reason is that they employ the holistic information of face images for recognition. Based on the observation that some of the local facial features would not vary with pose, lighting, facial expression and disguise, some local matching based methods which extract facial features from different levels of locality have been proposed and shown more promising results in face recognition tasks. The general idea of these methods is to first partition the face images into several smaller sub-image patches, and then classify the face images by comparing and combining the corresponding local features. The most representative and popular algorithms include Modular PCA (ModPCA) [11], Sub-pattern based PCA (SpPCA) [12], Sub-pattern NMF (SpNMF) [13], Sub-pattern PCA (Aw-SpPCA) [14] and Adaptively Weighted Sub-pattern LPP (Aw-SpLPP) [15].

Recently, Structure Preserved Projection (SPP) [16] which considered the configural structure of sub-images in each face image was developed for face recognition. From the experimental results in [16], it has been shown that the SPP is superior to some holistic and local matching based face recognition algorithms. However, SPP algorithm still has two limitations which may weaken its recognition performances. Firstly, SPP algorithm only considers the configural structure of sub-patterns in each face image and the different contributions made by various sub-pattern sets were neglected. Secondly, in order to extract the low-dimension local facial features, SPP algorithm needs to transform 2D sub-image into 1D feature vector. However, this scheme fails to consider the spatial correlation between neighboring pixels, which leads to break the intrinsic 2D structure of the pixels in the sub-images and discard much useful information [17].

In this paper, a new feature extraction method called Adaptively Weighted Structure Preserved Projection (Aw-SPP) is proposed to overcome the shortcomings of the SPP. In order to take the different contributions of various sub-pattern sets into account, an adaptive non-negative weight vector is learned to combine the different Laplacian matrices obtained by various sub-pattern sets in our Aw-SPP. Then, a Laplacian penalty constraint is incorporated into the proposed Aw-SPP to take advantage of the spatial structure of pixels in the sub-images. Finally, the procedures of feature extraction and non-negative weight vector learning are integrated into a unified framework. Moreover, an efficient iterative algorithm is designed to optimize our objective function. Numerical experiments on three face databases (Extended YaleB, CMU PIE and AR) show that Aw-SPP can obtain better recognition performances than SPP and some other local matching based algorithms.

The remainder of the paper is organized as follows. In Sect. 2, the proposed Aw-SPP is introduced. Section 3 presents the experimental results for face recognition. Finally, a conclusion is given in Sect. 4.

2 The Proposed Method

In this section, we first introduce the objective function of the proposed Adaptively Weighted Structure Preserved Projection (Aw-SPP) algorithm. Then, a simple yet efficient iterative update algorithm is presented to optimize the objective function of Aw-SPP. Finally, the recognition criterion of our algorithm is given.

We suppose that $X = [x_1, x_2, \ldots, x_N]$ are N face images belonging to C classes, each class contains n_c ($c = 1, 2, \ldots, C$) face images, and the size of each image is $H_1 \times H_2$. In our study, each face image is first divided into M equally sub-images with the size of $n_1 \times n_2$ and concatenating these sub-images into corresponding column vectors with dimensionality of $d = n_1 \times n_2$. Then, we use $X_i = [x_i^1, x_i^2, \ldots, x_i^M] \in R^{d \times M}(i = 1, \ldots, N)$ to denote all sub-patterns from the i-th face image and $X^m = [x_1^m, x_2^m, \ldots, x_N^m] \in R^{d \times N}$ ($m = 1, \ldots, M$) to denote the sub-pattern set containing the m-th sub-pattern of all face images [17].

2.1 The Objective Function

Similar to SPP, the first aim of the proposed Aw-SPP is to preserve the configural structure of sub-images in the same face image. Therefore, the reconstruction coefficients are employed to characterize the configurations of each face image. Specifically, for each X_i ($i = 1, 2, \ldots, N$), the reconstruction coefficients can be obtained by minimizing the following objective function:

$$\min_a \sum_{m=1}^{M} ||x_i^m - \sum_{j \in \{-m\}} a_i^{mj} x_i^j||_2^2 \tag{1}$$

where a_i^{mj} represents the reconstruction coefficients of x_i^m and $\{-m\} = \{1, 2, \ldots, m-1, m+1, \ldots, M\}$.

Let $W = [w_1, w_2, \ldots, w_d] \in \Re^{D \times d}(d << D)$ denotes the projection matrix, $y_i^m = W^T x_i^m$ represents the low-dimensional representation of x_i^m. Consequently, we need to minimize the following cost function:

$$
\begin{aligned}
\min_y \sum_{i=1}^{N} \sum_{m=1}^{M} ||y_i^m - \sum_{j \in \{-m\}} a_i^{mj} y_i^j||_2^2 \\
= \min_W \sum_{i=1}^{N} \sum_{m=1}^{M} ||W^T x_i^m - \sum_{j \in \{-m\}} a_i^{mj} W^T x_i^j||_2^2 \\
= \min_W \sum_{i=1}^{N} tr(W^T X_i (I - A_i)^T (I - A_i) X_i^T W)
\end{aligned}
\tag{2}
$$

where $A_i \in R^{M \times M}$ denotes the reconstruction coefficient matrix of the i-th face image, $I \in R^{M \times M}$ is defined as an identity matrix and $tr()$ denotes the trace operation.

The second aim of our Aw-SPP is to improve the discriminative ability of the sub-patterns in low-dimensional space by employing both the discrimination structure information of sub-pattern set and the different contributions made by various sub-pattern sets. Therefore, a weighted undirected graph G^m is firstly constructed for each sub-pattern set X^m. In this study, we adopt the simple binary weighting function to characterize the neighborhood relationship between sub-patterns x_i^m and x_j^m in graph G^m as follows:

$$S_{ij}^m = \begin{cases} 1, & \text{if } l(x_i^m) = l(x_j^m) \\ 0, & \text{otherwise} \end{cases} \tag{3}$$

where $l(x_i^m)$ is the class label of x_i^m. In order to ensure that the connecting samples shared the same class label would be close to each other in the low-dimensional space, we need to minimize the following objective function:

$$\min_y \sum_{i=1}^N \sum_{j=1}^N ||y_i^m - y_j^m||_2^2 S_{ij}^m$$
$$= \min_W \sum_{i=1}^N \sum_{j=1}^N ||W^T x_i^m - W^T x_j^m||_2^2 S_{ij}^m \tag{4}$$

By simple algebra formulation, the Eq. (4) can be reduced to

$$\min_W \sum_{i=1}^N \sum_{j=1}^N ||W^T x_i^m - W^T x_j^m||_2^2 S_{ij}^m$$
$$= \min_W \sum_{i=1}^N \sum_{j=1}^N (W^T x_i^m - W^T x_j^m)(W^T x_i^m - W^T x_j^m)^T S_{ij}^m \tag{5}$$
$$= \min_W tr(W^T X^m L^m X^{m^T} W)$$

where D^m is the diagonal matrix of the m-th sub-pattern set, whose entries are column or row sums of S^m, $L^m = D^m - S^m$ is the Laplacian matrix of S^m. In order to capture different contributions of various sub-pattern sets, the different Laplacian matrices computed by various sub-pattern sets are integrated by employing an adaptive non-negative weight vector $\omega = [\omega_1, \omega_2, \ldots, \omega_M] \in R^{l \times M}$, where ω_m ($m = 1, 2, \ldots, M$) is the weight for the graph G^m constructed by the m-th sub-pattern set. Thus, Eq. (5) can be extended as:

$$\min_W \sum_{m=1}^M \omega^m tr(W^T X^m L^m X^{m^T} W) + \alpha ||\omega||_2^2$$
$$= \min_W W^T (\sum_{m=1}^M \omega^m tr(X^m L^m X^{m^T}))W + \alpha ||\omega||_2^2 \tag{6}$$
$$s.t. \sum_{m=1}^M \omega^m = 1, \omega \geq 0$$

where the l_2-norm regularization term $||\omega||_2^2$ is utilized to avoid the weight vector ω over-fitting to a single Laplacian matrix [19, 22] and $\alpha \geq 0$ is a tradeoff parameter that balances the two terms.

Finally, a Laplacian penalty constraint [20] is incorporated into the proposed Aw-SPP to take advantage of the spatial correlation of pixels in the sub-image. Suppose f is a function defined on a region of interest $\Omega \subset R^B$. The Laplacian penalty function which measures the smoothness of f over the region Ω is defined as:

$$J(f) = \int_\Omega [Lf]^2 dt \tag{7}$$

where L is the Laplacian operator defines as follows:

$$Lf(t) = \sum_{i=1}^{B} \{(\partial^2 f)/(\partial^2 t_i^2)\} \tag{8}$$

In this study, the region of interest Ω is the sub-images partitioned from the face images, which is a $n_1 \times n_2$ lattice. Thus, we choose $B = 2$ in Eq. (8). Let w_i be a projection vector of W in $R^{n1 \times n2}$. Without loss of generality, w_i can be considered as a function defined on a $R^{n1 \times n2}$ lattice. Therefore, we can define the discrete Laplacian penalty function as [20]:

$$\Delta = D_1 \otimes I_2 + D_2 \otimes I_1 \tag{9}$$

where I_1 and I_2 are identity matrices with the size of $n_1 \times n_1$ and $n_2 \times n_2$, $\otimes$ is the kronecker product, and $D_j (j = 1, 2)$ is the discrete approximations to $\partial^2/\partial t_j^2$ as:

$$D = n_j^2 \begin{pmatrix} -1 & 1 & & & & 0 \\ 1 & -2 & 1 & & & \\ & & \cdots & \cdots & \cdots & \\ & & & 1 & -2 & 1 \\ 0 & & & & 1 & -1 \end{pmatrix}_{n_j \times n_j} \tag{10}$$

For a $n_1 \times n_2$ dimensional vector w_i, $\|\Delta w_i\|^2$ is proportional to the sum of squared differences among nearby grid points in its matrix form. In other words, it is a measurement of smoothness of w_i on the $n_1 \times n_2$ lattice. Therefore, the spatially smooth constraint utilized in Aw-SPP is

$$\min\|\Delta W\|_2^2 = tr(W^T \Delta^T \Delta W) \tag{11}$$

Now, through combining the functions in Eqs. (2), (6) and (11), we can get the objective function of Aw-SPP as:

$$\min_W \sum_{i=1}^{N} \sum_{m=1}^{M} \|W^T x_i^m - \sum_{j \in \{-m\}} a_i^{mj} W^T x_i^j\|_2^2$$
$$+ (\sum_{m=1}^{M} \omega_m (\sum_{i=1}^{N} \sum_{j=1}^{N} \|W^T x_i^m - W^T x_j^m\|_2^2 S_{ij}^m) + \alpha\|\omega\|_2^2) + \beta\|\Delta W\|_2^2 \tag{12}$$
$$s.t. \sum_{m=1}^{M} \omega_m = 1, \omega \geq 0$$

where $\beta \geq 0$ is the regularization parameter. Let $P = \sum_{i=1}^{N} X_i(I - A_i)^T(I - A_i)X_i^T$ and $Q^m = X^m L^m X^{m^T}$, Eq. (12) can be reduced to

$$\min\ tr(W^T(P + \sum_{m=1}^{M} \omega^m Q^m + \beta \Delta^T \Delta)W) + \alpha||\omega||_2^2$$
$$s.t.\ \sum_{m=1}^{M} \omega_m = 1, \omega \geq 0 \tag{13}$$

In this work, a constraint $W^T ZZ^T W = I$ is imposed to remove arbitrary scaling factor in the projection, where $Z = [x_1^1, \dots x_N^1, x_1^2, \dots, x_N^2, \dots, x_1^M, \dots, x_N^M]$ is the set of all sub-patterns. As a result, the objective function of Aw-SPP can be rewritten as the following optimization problem:

$$\min\ tr(W^T(P + \sum_{m=1}^{M} \omega^m Q^m + \beta \Delta^T \Delta)W) + \alpha||\omega||_2^2$$
$$s.t.\ W^T ZZ^T W = I, \sum_{m=1}^{M} \omega_m = 1, \omega \geq 0 \tag{14}$$

2.2   Optimization Method

From Eq. (14), it can be obviously found that the objective function of our algorithm is not convex in W and ω jointly. Thus, it is unrealistic to expect an algorithm to find the global optimal solution of the proposed objective function. Fortunately, since Eq. (14) is respectively convex in each variable, an alternating scheme can be employed to optimize our objective function.

Firstly, we suppose that ω_m ($m = 1, 2, \dots, M$) is fixed, the optimization problem with respect to W in Eq. (14) can be reduced to

$$\min\ tr(W^T(P + Q + \beta \Delta^T \Delta)W)$$
$$s.t.\ W^T ZZ^T W = I \tag{15}$$

where $Q = \sum_{m=1}^{M} \omega_m Q^m$. Moreover, it is easy to find that $Q + P + \beta \Delta^T \Delta$ is a symmetric and positive semi-definite matrix. Thus, Eq. (15) can be further converted to an eigensolver problem as:

$$(P + Q + \beta \Delta^T \Delta)W = \lambda ZZ^T W \tag{16}$$

Suppose $\lambda_1, \lambda_2, \dots, \lambda_d$ are the first d smallest eigenvalues of Eq. (16), and $w_1, w_2, \dots, w_d$ are the corresponding eigenvectors. The projection matrix W that projects the high-dimensional data into low-dimensional subspace is given by

$$W = [w_1, w_2, \dots, w_d] \tag{17}$$

Next, we fix W to optimize the weight vector ω. After removing the irrelevant terms, the optimization problem with respect to ω in Eq. (14) can be reduced to

$$\min_{\omega} \; \rho(\omega) \; = q^T\omega + \alpha||\omega||_2^2$$

$$s.t. \quad \sum_{m=1}^{M} \omega_m = 1, \omega \geq 0 \tag{18}$$

where $q = (q_1, q_2, \ldots, q_M)^T$ and $q_m = tr(W^T X^m L^m X^{m^T} W)$. From Eq. (18), we can easily find that if $\alpha = 0$, the trivial solution will be

$$\omega_i = \begin{cases} 1 & \text{if } q_i = \min_{k=1,\ldots,M} q_k \\ 0 & \text{otherwise} \end{cases} \tag{19}$$

From Eq. (19), we can find that only the information of one sub-pattern set is considered and the latent complementary information of various sub-pattern sets is neglected in the proposed algorithm. Therefore, it is an extremely sparse and undesirable solution. Moreover, when α is set as a too large value (i.e., $\alpha \rightarrow +\infty$), all the sub-pattern sets are endowed with a uniform weight value (i.e., $\omega_1 = \omega_2 = \ldots = _M = 1/M$) and the different contributions of various sub-pattern sets is ignored. Therefore, selecting an appropriate value for parameter α is very necessary. In our experiment, we empirically set parameter α as 1000.

The function in Eq. (18) is a typical quadratic programming (QP) problem, which can be solved by the generic QP solver (e.g., CVX [21]). However, this solver is often time consuming and shows slow convergence for larger size data. Therefore, according to the suggestion in Ref. [19, 22], the Coordinate Descent Algorithm (CDA) is employed to solve Eq. (18) in our algorithm.

Therefore, we can solve Eq. (14) by updating W and ω alternately. At the t-th iteration, W_t is updated with ω_{t-1}, then ω_t is updated with W_t. This procedure is repeated until convergence.

2.3   Recognition Criterion

In this selection, the recognition criterion of our Aw-SPP method is introduced. Here, we utilize the similar manner as the other local matching algorithms [15–17] to recognize a testing face image U. Specifically, the testing face image U is first to partition into M equally sized sub-images as Sect. 2.1, which are denoted as $u_1, u_2, \ldots, u_M$. For each sub-image, the low-dimensional feature is calculated using

$$\tilde{u}_m = W^T u_m, \quad m = 1, 2, \ldots, M \tag{20}$$

where W is the projection matrix obtained by Aw-SPP algorithm. Next, we adopt the nearest neighbor classifier with Euclidean distance to compute the recognition result of each low-dimensional feature.

Then, we employ a majority voting strategy to decide the final recognition result of the testing face image based on the recognition results of all M sub-images. The probability of the testing image belonging to the j-th person can be computed by

$$p_j = \sum_{m=1}^{M} \omega_m q_m^j / M \tag{21}$$

where ω_m is a weight value for the m-th sub-pattern learned by Aw-SPP and q_m^j is defined as

$$q_m^j = \begin{cases} 1, & \text{if the } m\text{ - th sub - pattern is classified to the } j\text{ - th person} \\ 0, & \text{otherwise} \end{cases} \tag{22}$$

The final recognition result of U is defined as:

$$Identity\,(U) = \arg \max_j (p_j) \tag{23}$$

3   Experimental Results and Analysis

In this section, the performance of the proposed algorithm is tested and compared with other related algorithms such as ModPCA [11], SpPCA [12], Aw-SpPCA [13], SpNMF [14], Aw-SpLPP [15] and SPP in its supervised model (S-SPP) [16]. Here, three benchmark face databases including Extended YaleB [23], CMU PIE [24] and AR [25] are employed. All images are manually aligned, cropped, and then resized to the resolution of 64×64. For each database, we randomly choose p samples of each individual for training and the remaining samples are used for testing. The random training sample selection is repeated 10 times and the averaged recognition accuracies are reported. More detailed information of three face image databases can be found in Table 1 and some face images from them can be seen in Fig. 1.

In our Aw-SPP and other algorithms, there are some parameters which need to be set in advance. For Aw-LPP and S-SPP, we fix the size of neighborhood k as $p-1$ on all the databases [16, 17]. In order to fairly compare the performances of different local matching-based algorithms, we tune the parameters for all algorithms by a gird-search strategy from $\{10^{-3}, 10^{-2}, 10^{-1}, 1, 10^1, 10^2, 10^3\}$, and set the subspace dimensionality from 5 to 140 with the interval of 5. Moreover, in order to consider the relation between nearby sub-images, the over-lapping partition strategy which can connect the adjacent local regions is adopted. The size of sub-image is set as 16×16 for three databases and the overlap between adjacent sub-images is set as 4 pixels.

In the first experiment, the performances of the proposed Aw-SPP and other state-of-art algorithms on the three databases are tested. The best average recognition rates and standard deviations obtained by different algorithms on three databases are

Table 1. Summary of three face databases

Databases	Size	Classes	Number	Train	Test
Extended YaleB	64×64	38	64	10	54
CMU PIE	64×64	68	24	10	14
AR	64×64	100	14	5	9

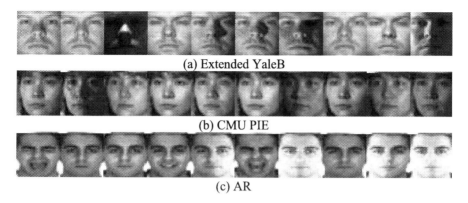

(a) Extended YaleB

(b) CMU PIE

(c) AR

Fig. 1. Some images of the face databases used in the experiment.

listed in Table 2. From these results, several interesting points can be revealed below. Firstly, we can see that the recognition rates of Aw-SpLPP, S-SPP and Aw-SPP are all higher than other local matching based algorithms. This is because that these three methods can preserve the geometry structure of the sub-pattern set during feature extraction. Secondly, we find that due to S-SPP and Aw-SPP can preserve the configural structure of sub-images from the same face image, their recognition performances are better than Aw-SpLPP. At last, since the proposed Aw-SPP can capture different contributions of various sub-patterns and take advantage of relationships between nearby pixels in the sub-images, it achieves the best recognition rates among all algorithms.

Moreover, the average recognition results versus feature dimensionality of the proposed Aw-SPP and other algorithms are shown in Fig. 2. From this figure, we can find that the performances of the proposed Aw-SPP under different subspace dimensions are consistently better than other algorithms on AR database. Although the performances obtained by our Aw-SPP algorithm on Extended YaleB and CMU PIE databases are worse than the S-SPP when the subspace dimensionality is relatively low. With the increase of subspace dimensionality, the recognition rates of Aw-SPP become

Table 2. The best average recognition rates (%) and standard deviations (%) of different algorithms on three face databases

Methods	Extended YaleB	CMU PIE	AR
SpPCA	70.26 ± 1.76(110)	85.56 ± 1.00(135)	89.61 ± 0.79(75)
Aw-SpPCA	75.08 ± 1.83(105)	86.32 ± 1.20(135)	90.28 ± 0.97(135)
ModPCA	69.65 ± 1.28(140)	84.47 ± 1.30(140)	89.33 ± 0.82(65)
SpNMF	78.27 ± 1.29(135)	88.86 ± 2.06(110)	86.43 ± 1.07(15)
Aw-SpLPP	79.15 ± 1.90(140)	89.88 ± 1.31(140)	90.09 ± 1.03(115)
S-SPP	92.35 ± 1.52(40)	91.94 ± 0.88(40)	91.83 ± 0.77(25)
Aw-SPP	94.34 ± 0.54(135)	93.10 ± 0.55(90)	94.63 ± 0.42(25)

Note: The numbers in parentheses are the corresponding subspace dimension with the best results.

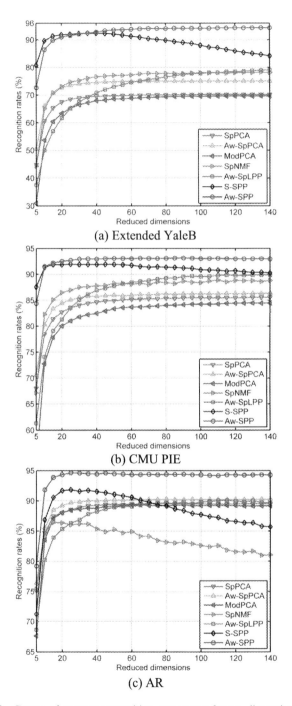

(a) Extended YaleB

(b) CMU PIE

(c) AR

Fig. 2. Curves of average recognition rate versus feature dimensionality.

Table 3. The best average recognition rates (%) and standard deviations (%) of Aw-SPP with varied β values on three face databases

Parameter β	Extended YaleB	CMU PIE	AR
0.001	93.78 ± 0.65(135)	92.56 ± 0.53(85)	94.41 ± 0.77(20)
0.01	94.02 ± 0.49(120)	92.66 ± 0.79(65)	94.48 ± 0.56(25)
0.1	94.34 ± 0.54(135)	92.89 ± 0.72(85)	94.63 ± 0.42(25)
1	94.14 ± 0.49(105)	93.03 ± 0.67(40)	94.58 ± 0.74(35)
10	93.21 ± 0.81(110)	93.10 ± 0.55(90)	94.58 ± 0.68(30)
100	91.22 ± 1.03(135)	92.63 ± 0.69(130)	94.50 ± 0.61(30)
1000	88.01 ± 1.24(115)	91.96 ± 0.58(140)	94.35 ± 0.61(45)

Note: The numbers in parentheses are the corresponding subspace dimension with the best results.

better than S-SPP and other algorithms. Furthermore, we can also see that after achieving its top performances, the recognition results of S-SPP begin to decrease with the increase in subspace dimensionality. This phenomenon makes it difficult to choose the proper subspace dimensionality for S-SPP. However, the above phenomenon can be alleviated in the proposed algorithm.

Next, the effect of β value on performance of the proposed Aw-SPP is tested. The best average recognition results under different parameter values obtained by our proposed method on the three face image databases are listed in Table 3. As we can see from these results, the performance of our proposed method first improves and then degrades with increase of the parameter β. Meanwhile, we can find that the parameter β has a relatively significant influence on the performance on Extended YaleB and CMU PIE databases, while has a small influence on AR database.

Finally, we utilize the one-tailed Wilcoxon rank sum to verify whether the performance of the proposed Aw-SPP significantly outperforms the other algorithms. In this test, the null hypothesis is defined as that our Aw-SPP algorithm makes no difference when compared to other algorithms, and the alternative hypothesis is defined as that Aw-SPP makes an improvement when compared to other algorithms. For example, if we need to compare the performance of our algorithm with that of SpPCA (e.g., Aw-SPP vs. SpPCA), the null and alternative hypotheses can be denoted as H_0: $M_{\text{Aw-SPP}} = M_{\text{SpPCA}}$ and H_1: $M_{\text{Aw-SPP}} > M_{\text{SpPCA}}$, where $M_{\text{Aw-SPP}}$ and M_{SpPCA} are the medians of the recognition rates obtained by Aw-SPP and SpPCA algorithms on all

Table 4. The p-values of the pairwise one-tailed Wilcoxon rank sum tests.

	Extended YaleB	CMU PIE	AR
Aw-SPP vs. SpPCA	9.1336e−05	8.8805e−05	8.9307e−05
Aw-SPP vs. Aw-SpPCA	9.1336e−05	8.8805e−05	8.9811e−05
Aw-SPP vs. ModPCA	9.0826e−05	8.9307e−05	8.9307e−05
Aw-SPP vs. SpNMF	9.0826e−05	8.7312e−05	8.9307e−05
Aw-SPP vs. Aw-SpLPP	9.1336e−05	8.9811e−05	8.9811e−05
Aw-SPP vs. S-SPP	5.7109e−05	4.4376e−03	8.9307e−05

face databases. In this experiment, the significance level value is set to 0.01. From the test results in Table 4, it can be found that the p-values obtained by all pairwise Wilcoxon rank sum tests are much less than the significance level value, which demonstrates that the null hypotheses are rejected in all pairwise tests and the proposed algorithm performs significantly better than the other algorithms.

4 Conclusions

In this paper, we propose a new supervised feature extraction algorithm named Adaptively Weighted Structure Preserved Projections (Aw-SPP), which not only takes the configural structure of sub-images in each face image and the different contributions of various sub-pattern sets into account, but also preserves the intrinsic 2D structure of each sub-image in low-dimension feature subspace. Meanwhile, an efficient iterative algorithm is also proposed to optimize the objective function of the proposed algorithm. Through extensive recognition experiments on three benchmark databases, it can be found that the proposed algorithm outperforms some state-of-the-art local matching based approaches.

Acknowledgment. This work is supported by National Natural Science Foundation of China (Nos. 61403078 and 61562044), Science and Technology Research Project of Liaoning Province Education Department (No. L2014450).

References

1. Li, S.Z., Jain, A.K.: Handbook of Face Recognition. Springer, Heidelberg (2011)
2. Subban, R, Mankame, D.P.: Human face recognition biometric techniques: analysis and review. Recent Adv. Intell. Inf. **235**, 455–463 (2014)
3. Zou, J., Ji, Q., Nagy, G.: A comparative study of local matching approach for face recognition. IEEE Trans. Image Process. **16**(10), 2617–2628 (2007)
4. Bi, C., Zhang, L., Qi, M., et al.: Supervised filter learning for representation based face recognition. PLoS One **11**(7), e0159084 (2016)
5. Jolliffe, I.: Principal Component Analysis. Wiley, Hoboken (2002)
6. Fukunaga, K.: Introduction to Statistical Pattern Recognition. Academic Press, Cambridge (2013)
7. Lee, D.D., Seung, H.S.: Algorithms for non-negative matrix factorization. In: Advances in Neural Information Processing Systems, pp. 556–562 (2001)
8. Roweis, S.T., Saul, L.K.: Nonlinear dimensionality reduction by locally linear embedding. Science **290**(5500), 2323–2326 (2000)
9. Belkin, M., Niyogi, P.: Laplacian eigenmaps and spectral techniques for embedding and clustering. In: Advances in Neural Information Processing Systems, vol. 14, pp. 585–591 (2001)
10. Niyogi, X.: Locality preserving projections. In: Advances in Neural Information Processing Systems, vol. 16, p. 153 (2004)
11. Gottumukkal, R., Asari, V.K.: An improved face recognition technique based on modular PCA approach. Pattern Recogn. Lett. **25**(4), 429–436 (2004)

12. Chen, S., Zhu, Y.: Subpattern-based principle component analysis. Pattern Recogn. **37**(5), 1081–1083 (2004)
13. Tan, K., Chen, S.: Adaptively weighted sub-pattern PCA for face recognition. Neurocomputing **64**, 505–511 (2005)
14. Zhu, Y.: Sub-pattern non-negative matrix factorization based on random subspace for face recognition. In: International Conference on Wavelet Analysis and Pattern Recognition, vol. 3, pp. 1356–1360 (2007)
15. Wang, J., Zhang, B., Wang, S., et al.: An adaptively weighted sub-pattern locality preserving projection for face recognition. J. Netw. Comput. Appl. **33**(3), 323–332 (2010)
16. Wang, J., Ma, Z., Zhang, B., et al.: A structure-preserved local matching approach for face recognition. Pattern Recogn. Lett. **32**(3), 494–504 (2011)
17. Yi, Y., Zhou, W., Wang, J., et al.: Face recognition using spatially smoothed discriminant structure-preserved projections. J. Electron. Imaging **23**(2), 023012 (2014)
18. Wang, J., Yi, Y., Zhou, W., et al.: Locality constrained joint dynamic sparse representation for local matching based face recognition. PLoS One **9**(11), e113198 (2014)
19. Li, P., Bu, J., Chen, C., et al.: Relational multimanifold coclustering. IEEE Trans. Cybern. **43**(6), 1871–1881 (2013)
20. Cai, D., He, X., Hu, Y., et al.: Learning a spatially smooth subspace for face recognition. In: IEEE Conference on Computer Vision and Pattern Recognition, pp. 1–7 (2007)
21. Boyd, S., Vandenberghe, L.: Convex Optimization. Cambridge University Press, New York (2009)
22. Yi, Y., Bi, C., Li, X., et al.: Semi-supervised local ridge regression for local matching based face recognition. Neurocomputing **167**, 132–146 (2015)
23. Lee, K., Ho, J., Kriegman, D.: Acquiring linear subspaces for face recognition under variable lighting. IEEE Trans. Pattern Anal. Mach. Intell. **27**(5), 684–698 (2005)
24. Terence, S., Simon, B., Maan, B.: The CMU pose, illumination, and expression (PIE) database. IEEE Trans. Pattern Anal. Mach. Intell. **25**(12), 1615–1618 (2003)
25. Martinez, AM.: The AR face database. CVC Technical report, 24 (1998)

A Color Model Based Fire Flame Detection System

Qing Lu[1,3], Jun Yu[1,3], and Zengfu Wang[1,2,3(✉)]

[1] Department of Automation, University of Science and Technology of China,
Hefei 230026, China
ustclq@mail.ustc.edu.cn,
{harryjun,zfwang}@ustc.edu.cn
[2] Institute of Intelligent Machines, Chinese Academy of Sciences,
Hefei 230031, China
[3] National Engineering Laboratory for Speech and Language Information
Processing, University of Science and Technology of China,
Hefei 230026, China

Abstract. Fire flame detection using color information is an important problem for public security and has many applications in computer vision and other domains. The color model based method used for fire flame detection has many advantages over conventional methods, such as simple, feasible and understandable. In order to improve the performance of fire flame detection based on video, we propose an effective color model based method for fire flame detection and build a corresponding fire flame detection system. Firstly, candidate fire flame regions are detected using the chromatic and dynamic measurements. Secondly, the fire flame regions are determined based on the area of the candidate regions. Finally, the fire flame detection system will give an alarm voice when the number of successive fire frames surpasses threshold. Experimental results show the effectiveness of our system on various fire-detection tasks in real-world environments.

Keywords: Computer vision · Color model · Fire flame detection system · Chromatic and dynamic measurements

1 Introduction

Fire flame detection is a challenging task in the field of public security and computer vision. To solve this problem, many techniques have been proposed and most of them are based on temperature sampling, particle sampling, relative humidity sampling and smoke analysis [1–3]. However, using the traditional methods, it is still difficult to detect the state of burning, e.g., fire flame size, location, direction of motion and so on. In a sensor-based fire flame detection method, coverage of wide areas in outdoor applications can not be achieved due to the requirement of reasonable distribution of sensors in close range [4–6]. Great success has been made over past years, especially in image processing used for fire flame detection, owing to the rapid development of computer vision, such as [7, 8].

© Springer Nature Singapore Pte Ltd. 2016
T. Tan et al. (Eds.): CCPR 2016, Part I, CCIS 662, pp. 474–485, 2016.
DOI: 10.1007/978-981-10-3002-4_40

In order to capture the feature of fire flame more accurately, color image processing technique used for fire flame detection is better than that of gray-sacle processing. Cappellini et al. [9] applied color video to recognize the fire flame from smoke, which has a good performance during the night. Some modified color image processing methods are proposed to achieve real-time detection performance of fire flame [10–15]. Nevertheless, the methods above can't provide the validation information of a real flame. Chen et al. [16] presented a set of rules to judge the flame pixels, using the relationship information among R, G and B color components. Töreyin et al. [17] used a mixture of Gaussians model in RGB space which is acquired from a training set of fire flame pixels. Later, Töreyin et al. [18] employed Chen's fire flame pixel detection rules along with motion information and Markov field modeling of the fire flame flicker process [6]. To reduce the false alarm rate and decrease the alarm reaction time, the fire flame detection system is required not only high reliability but also great flexibility [19–21]. Marbach et al. [19] provided YUV color model for fire flame detection, using luminance component Y to declare the candidate fire flame pixels and the chrominance components U and V to classify the candidate pixels to be in the fire sector or not [6]. From the viewpoint of the conventional fire flame detection methods, it is essential to propose a novel method for detecting fire more effectively.

In this paper, to tackle the aforementioned difficulties, we first detect each pixel of input image using a color model algorithm which combines RGB information and YCrCb information. Fire regions are then extracted by taking advantage of the movement information of flame. After this stage, the real regions of fire flame can be further determined. By combining the powerful decision-making mechanism which will be introduced subsequently with the excellent result of last step, our system achieves the leading performance on the experiments. The framework is illustrated in Fig. 1.

This paper is organized as follows: Sect. 2 describes fire flame pixels prejudging based on our new color model. In Sect. 3, we introduce the method used for extracting the regions of moving fire flame. The powerful decision-making mechanism and system implementation is introduced in Sect. 4. The experimental results are showed in Sect. 5. We conclude this paper in Sect. 6.

2   Candidate Fire-Pixel Detection

Each digital color image contains three color planes: red, green and blue (R, G and B) [6], and each of them have an arbitrary intensity. By combining the value R, G and B, we can get almost all the color that can be perceived by human eye. To detect fire flame, Chen et al. [16] proposed a fire detection method based on video processing, which combine the R channel threshold and saturation value.

For a fire flame region, the value of red channel tends to be larger than the green channel, and the value of green channel tends to be larger than the value of blue channel in the spatial location [6]. What is more, in all channels, red channel has highest saturation [16]. As shown in Fig. 2, the original RGB color fire flame images are listed in column (a), and the column (b), (c) and (d) show their R, G and B channels,

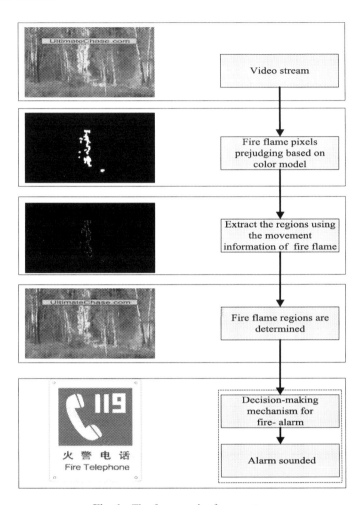

Fig. 1. The framework of our system

respectively. It is clear that the red channel is brighter than the other two in the region of fire flame, and the blue channel is darker than the others.

Although a suitable RGB color model may be found for fire flame detection, it has many drawbacks when the lighting conditions are unstable. So we try to find a model that is insensitive to illumination, that is to say, to build a model that can avoid the influence of illumination. In consideration of YCrCb color space which can separate the luminance signal Y with the chromaticity signal Cr and Cb, we use it to improve our model. The conversion from RGB to YCbCr color space is formulated as formula (1) [22]. As shown in Fig. 2, the column (e), column (f) and column (g) show the Y, Cr and Cb channels of their original images. It becomes obvious that the Y channel is brighter than the other two in the regions of fire flame, and the Cb channel is darker than the others.

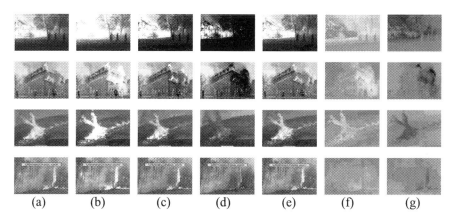

Fig. 2. Original RGB color images in column(a), *R*, *G*, and *B* channels in columns (b)–(d), respectively, *Y*, *Cb* and *Cr* channels in columns (e)–(g), respectively. (Color figure online)

$$\begin{bmatrix} Y \\ C_b \\ Cr \end{bmatrix} = \begin{bmatrix} 0.2568 & 0.5041 & 0.0979 \\ -0.1482 & -0.2910 & 0.4392 \\ 0.4392 & -0.3678 & -0.0714 \end{bmatrix} \begin{bmatrix} R \\ G \\ B \end{bmatrix} + \begin{bmatrix} 16 \\ 128 \\ 128 \end{bmatrix} \tag{1}$$

In order to obtain a better fire flame detection model combined RGB with YCrCb information, a large number of images which only contain fire flame regions are collected. After that, we calculate the mean value of *R*, *G*, *B*, *Y*, *Cr* and *Cb*, respectively. For a given region, we can calculate the mean value by formula (2). Table 1 shows the results of computation of Fig. 3.

Table 1. Mean values of *R*, *G*, *B*, *Y*, *Cb* and *Cr* planes of the fire flame regions of images given in Fig. 3.

The index in Fig. 3	R_{mean}	G_{mean}	B_{mean}	Y_{mean}	Cr_{mean}	Cb_{mean}
a	240	166	15	171	177	39
b	208	127	22	140	177	61
c	246	149	27	164	186	50
d	249	199	37	195	166	38
e	190	116	29	128	172	72
f	240	176	48	181	170	53
g	201	146	33	149	165	62
h	226	155	38	163	172	57
i	238	173	52	179	170	56
j	227	166	51	171	167	60
Mean	226	157	35	164	172	54

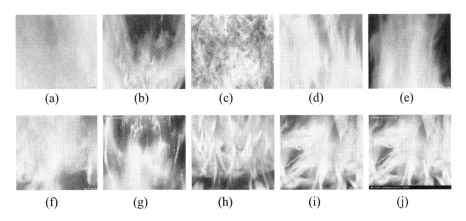

Fig. 3. Images that only contain fire flame region

$$X_{\text{mean}} = \frac{1}{n} \sum_{i=1}^{n} X(x_i, y_i) \tag{2}$$

where X can be replaced by *R, G, B, Y, Cr* and *Cb*, respectively, n is the total number of pixels in image, and (x_i, y_i) is the spatial location of the pixel.

From Table 1 and Fig. 4, it can be seen that our previous statement is reasonable. It is clear that the value of red channel is larger than the other two in the regions of fire flame, and the value of blue channel is smaller than the others. What's more, on the average, the fire flame pixels show the characteristics that their *Cb* color value is far less than the values of *Y* and *Cr* color. Deducing from the RGB fire flame color model [16] and the YCrCb fire flame color model [6], we get a new model for extracting fire flame pixels from an image, as described in the following:

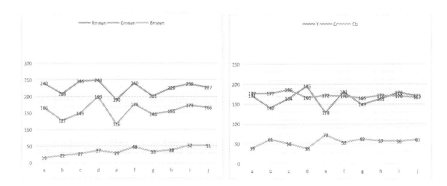

Fig. 4. Comparison between the numerical size of each component of Fig. 3.

rule 1: $R > R_T$

rule2: $R > g_T *G > b_T *B$

rule 3: $S > ((255-R)*S_T/G_T)$

rule 4: $Y > y_T *C_b$

rule 5: $C_r > c_{r_r} *C_b$

IF ((rule1)AND(rule2)AND(rule3)OR((rule4)AND(rule5)))==TRUE
THEN fire-pixel
ELSE not fire-pixel

In rule 1, R_T denotes the threshold of R component. In rule 2, the value of G component and B component are weighted by a certain coefficient g_T and b_T, respectively. In rule 3, S_T denotes the threshold of saturation when the threshold of R component is set to be R_T for the same pixel [16]. What's more, in order to highlight the value of Y component and C_r component, we set two coefficients y_T and c_{r_r} in rule 4 and rule 5, respectively. In the above decision rules, the parameters are defined according to a variety of experimental results. And the value of R_T and S_T range from 55 to 65 and 115 to 135, respectively [16], what's more, the value of g_T, b_T, y_T and c_{r_r} are set to be 1.2, 2, 1.5 and 1.5, respectively.

3   Fire Flame Regions Confirmation

If a background image without fire flame can be obtained, it will be relatively easy to get the regions of fire flame, just using the image to minus the background. However, in most cases, it is not possible to get the background image, such as when the scenarios are complex or the lighting conditions are not steadfast. Therefore, in order to solve this problem, a dynamic background image is necessary. It is a simple approach by taking the average of some adjacent fire images, but to do so, there are many drawbacks. First, you need to input a lot of images before calculating the background images. Second, in the process of calculating the background images, the regions of fire flame are not allowed to appear. Therefore, it is a relatively good way to establish the background image by calculating and updating it dynamically and real-timely. Gaussian mixture model [23] can meet this demand and will be explained as follows:

Each pixel in the image is constituted by a mixture of K Gaussian functions. The probability of a pixel as fire flame pixel can be defined as:

$$p(x) = \sum_{j=1}^{K} w_j \eta(x; \mu_j, \Sigma_j) \tag{3}$$

Where w_j is the mixture weight, $\eta(x; \mu_j, \Sigma_j)$ is the jth Gaussian distribution.

$$\eta(x; \mu_j, \Sigma_j) = \frac{1}{(2\pi)^{\frac{D}{2}} \left|\Sigma_j\right|^{\frac{1}{2}}} e^{-\frac{1}{2}(x-\mu_j)^T \Sigma_j^{-1}(x-\mu_j)} \tag{4}$$

Where μ_j and $\sum_j = \sigma_j^2 I$ are the mean and the covariance of the jth component, respectively.

The rank of K distributions is according to the fitness value w_k/σ_k. The first B distributions which satisfy (5) are chosen as the representative of the background.

$$B = \arg_b \min\left(\sum_{j=1}^{b} w_j > T_B\right) \tag{5}$$

where T_B is the threshold for the distributions.

The above parameters can be updated as follows [23]:

$$w_k^{N+1} = (1 - \alpha)w_k^N + \alpha p(w_k|x_{N+1}) \tag{6}$$

$$\mu_k^{N+1} = (1 - \alpha)\mu_k^N + \rho x_{N+1} \tag{7}$$

$$\sum_k^{N+1} = (1 - \alpha)\sum_k^N + \rho(x_{N+1} - \mu_k^{N+1})(x_{N+1} - \mu_k^{N+1})^T \tag{8}$$

$$\rho = \alpha \eta(x_{N+1}; \mu_k^N, \sum_k^N) \tag{9}$$

$$p(w_k|x_{N+1}) = \begin{cases} 1 & \text{if } w_k \text{ is the first match Gaussian component} \\ 0 & \text{otherwise} \end{cases} \tag{10}$$

Where w_k denotes the kth Gaussian component, and α is the learning rate for the weight. When the parameters K and α are set to be 5 and 0.01, respectively, our system shows good performance in extensive experiments.

4   Alarm Mechanism Design and System Implementation

After the above series of process, the fire flame regions can be accurately detected. In order to make the system more efficient for monitoring service, an alarm decision-making mechanism is designed. The algorithm is as follows:

Alarm decision-making algorithm

Step1: Structure some fire flame blocks using the fire pixel
Step2: Calculate the number of blocks(n) of each frame and their corresponding area
s_i , $i=1,2,...n$

Step3: If $s_i > s_{th}$, $i=1,2,...n$

      Then this frame is fire flame frame, and the flame accumulator T +1
Step4: If T> T_{th}

      Then alert sound , and clear the flame accumulator

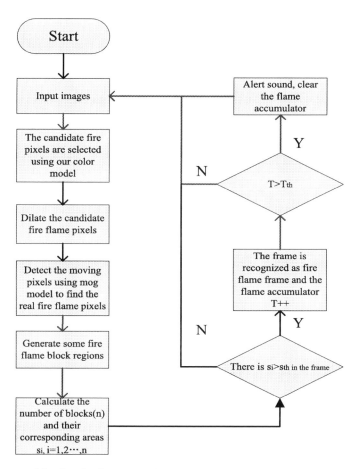

Fig. 5. The flow chart of our system for fire flame detection

Based on the descriptions in above sections, the proposed fire flame detection system can be summarized by a flowchart shown in Fig. 5.

5 Experiments

The system is constructed on a Intel Core(TM) i5-2430 M CPU @ 2.4 GHz, equipped with 4 GB of RAM. The color model can detect the location of each block accurately as shown in Fig. 6(d). The dimension of the image sequences is 400×256. We use our color model to search the fire flame pixels, shown in Fig. 6(b), in original images, shown in Fig. 6(a). In order to extract the fire flame pixels with motion feature, Gaussian mixture model is used for our system, and the results are shown in Fig. 6(c).

As can be seen from the column (b) of Fig. 6, the color model can approximately detect the fire flame pixels. The motion property of the flame can not only be used for detecting the contour of the fire flame regions, but also removing some noisy parts, which can be seen from the column (c) of Fig. 6. It can be easily observed from the last

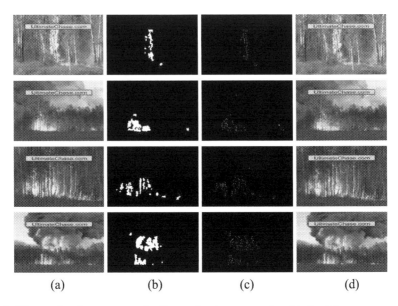

(a)   (b)   (c)   (d)

Fig. 6. The results of our system, the first to fourth rows are four kinds of fire scenes, column (a) is original images, column (b) and column (c) are the candidate fire flame pixels and the fire flame pixels with motion feature, respectively, the detected fire flame blocks are presented in column (d).

column of Fig. 6, our fire flame color model can accurately mark the flame regions, using some red bounding boxes.

In order to test the accuracy of our model, some representative flame videos are collected from the Internet. The results are shown in Table 2, N is the total number of fire frames in the video sequences, n is the number of flame frames which are detected accurately and the last column shows the mean processing time for per frame.

Table 2. The results of videos test

Video	N	n	Detection rate (%)	Running time per frame (ms)
a	260	257	98.85	76.55
b	246	245	99.59	122.63
c	208	206	99.04	193.98
d	200	199	99.50	128.54
e	245	244	99.59	131.09
f	219	217	99.09	115.58
g	218	217	99.54	121.07
sum	1596	1585	99.31	127.06

The fire flame detection is successful for the captured video sequences and 99.31 % correct detection rate is achieved. Owning to the limit of the hardware configuration, the mean processing time for per frame is not ideal, which is about 127.06 ms. In order to achieve the real-time performance, a GPU is adopted for acceleration.

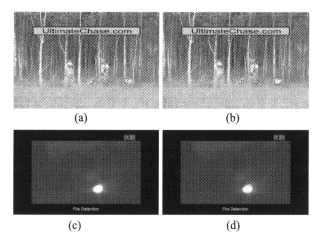

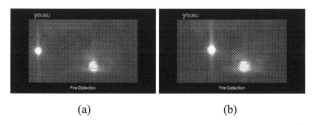

Fig. 7. Comparison to RGB color model [1]. (a) and (c): using RGB color model. (b) and (d): using our color model for fire flame detection. (Color figure online)

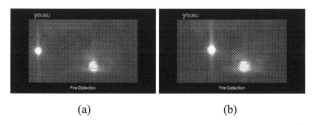

Fig. 8. Example results on fire flame images with pseudo-flame region. Highlight on the left side of each image is pseudo-flame region and the right is fire flame.

From Fig. 7, we can observe that the RGB color model [1] fails to detect the fire flame in several video sequences due to its inability to handle significant illumination variation caused by external environment, while our fire flame detection model can solve this problem easily.

Figure 8 presents the detection results on the fire flame images with pseudo-flame region. Notice that the highlight on the left side is pseudo-flame region generated by flashlight. Comparing the two images in Fig. 8, it is clear that our color model is desirable for distinguishing the real fire flame regions from images that contain moving pseudo-flame regions.

6   Conclusion

There are many methods to detect fire flame for reducing the damage of accidents. In this paper, we present a novel color model for fire flame pixels detection and apply it to build a system for fire flame detection. The color model used for fire flame pixels detection utilizes not only the specific relationship among the RGB components, but also the YCrCb color space for avoiding the influence of illumination. In order to make the system more efficient for monitoring service, an alarm decision-making mechanism is designed. Experimental results demonstrate that our color model is robust to detect fire flame pixels, and our system achieves ideal performances.

Acknowledgement. This work is supported by the National Natural Science Foundation of China (No. 61472393, No. 61572450 and No. 61303150), the Fundamental Research Funds for the Central Universities (WK2350000002).

References

1. Chen, T.H., Kao, C.L., Chang, S.M.: An intelligent real-time fire-detection method based on video processing. In: Proceedings of IEEE 37th Annual 2003 International Carnahan Conference on Security Technology, pp. 104–111 (2003)
2. Healey, G., Slater, D., Lin, T., Drda, B., Goedeke, A.D.: A system for real-time fire detection. In: Proceedings of IEEE Computer Society Conference on Computer Vision and Pattern Recognition, pp. 605–606 (1993)
3. Habiboğlu, Y.H., Günay, O., Çetin, A.E.: Covariance matrix-based fire and flame detection method in video. Mach. Vis. Appl. **23**(6), 1103–1113 (2012)
4. Töreyin, B.U., Cetin, A.E.: Online detection of fire in video. In: Proceedings of IEEE Conference on Computer Vision and Pattern Recognition, pp. 1–5 (2007)
5. Wang, D.C., Cui, X., Park, E., Jin, C., Kim, H.: Adaptive flame detection using randomness testing and robust features. Fire Saf. J. **55**, 116–125 (2013)
6. Celik, T., Demirel, H.: Fire detection in video sequences using a generic color model. Fire Saf. J. **44**(2), 147–158 (2009)
7. Jun, Yu., Wang, Z.: A video, text and speech driven realistic 3D virtual head for human-machine interface. IEEE Trans. Cybern. **45**(5), 977–988 (2015)
8. Yu, J., Wang, Z.: 3D facial motion tracking by combining online appearance model and cylinder head model in particle filtering. Sci. Chin. – Inf. Sci. **57**(2), 274–280 (2014)
9. Cappellini, V., Mattii, L., Mecocci, A.: An intelligent system fur automatic fire detection in forests. In: Proceedings of IEEE 3th International Conference on Image Processing and its Applications, pp. 563–570 (1989)
10. Phillins, W., et al.: Flame recognition in video. In: Proceedings of IEEE 5th Workshop on Applications of Computer vision, pp. 224–229 (2000)
11. Horng, W.B., Peng, J.W.: Real-time fire detection from video: a preliminary report. In: Proceedings of 14th IPPR, Computer Vision, Graphics and Image Processing, pp. 1–5 (2001)
12. Borges, P.V.K., Izquierdo, E: A probabilistic approach for vision-based fire detection in videos. In: Proceedings of IEEE Transactions on Circuits and Systems for Video Technology, pp. 721–731 (2010)
13. Rubio, E., Castillo, O.: Designing type-2 fuzzy systems using the interval type-2 fuzzy c-means algorithm. In: Castillo, O., Melin, P., Pedrycz, W., Kacprzyk, J. (eds.) Recent Advances on Hybrid Approaches for Designing Intelligent Systems, vol. 547, pp. 37–50. Springer, Berlin (2014)
14. Duong, H.D., Nguyen, D.D., Ngo, L.T., Tinh, D.T.: On approach to vision based fire detection based on type-2 fuzzy clustering. In: International Conference of Soft Computing and Pattern Recognition, pp. 51–56 (2011)

15. Ko, B.C., Cheong, K.H., Nam, J.Y.: Fire detection based on vision sensor and support vector machines. Fire Saf. J. **44**(3), 322–329 (2009)
16. Chen, T.H., Wu, P.H., Chiou, Y.C.: An early fire-detection method based on image processing. In: Proceedings of IEEE International Conference on Image Processing, vol. 3, pp. 1707–1710 (2004)
17. Töreyin, B.U., Dedeoglu, Y., Cetin, A.E.: Flame detection in video using hidden Markov models. In: Proceedings of IEEE International Conference on Image Processing, pp. 1230–1233 (2005)
18. Töreyin, B.U., Dedeoglu, Y., Güdükbay, U., Cetin, A.E.: Computer vision based method for real-time fire and flame detection. Pattern Recogn. Lett. **27**(1), 49–58 (2006)
19. Marbach, G., Loepfe, M., Brupbacher, T.: An image processing technique for fire detection in video images. Fire Saf. J. **41**(4), 285–289 (2006)
20. Den Breejen, E., Breuers, M., Cremer, F., Kemp, R., Roos, M., Schutte, K., De Vries, J.S.: Autonomous forest fire detection. In: Proceedings of 3rd International Conference on Forest Fire Research, pp. 2003–2012 (1998)
21. Cheong, K., Ko, B., Nam, J.: Vision sensor-based fire monitoring system for smart home. In: Proceedings of 1st International Conference of Ubiquitous Information Technology and Applications (2007)
22. Poynton, C.A.: A Technical Introduction to Digital Video. Wiley Inc, London (1996)
23. KaewTraKulPong, P., Bowden, R.: An improved adaptive background mixture model for real-time tracking with shadow detection. In: Remagnino, P., Jones, G.A., Paragios, N., Regazzoni, C.S. (eds.) Video-Based Surveillance Systems, pp. 135–144. Springer Science+ Business Media, New York (2002)

Robust Object Tracking Based on Collaborative Model via L2-Norm Minimization

Xiaokang Qiao, Kaile Su, Zhonglong Zheng[(✉)],
Huawen Liu, and Xiaowei He

Zhejiang Normal University, Jinhua, China
Zhonglong@zjnu.cn

Abstract. The computational cost of the tracking algorithms with sparse representation is relatively large, however, we proposed a robust object tracking algorithm based on a sparse collaborative model that exploits both holistic templates and local representation to account for drastic appearance model, properly solved by the l_2 norm minimization solution in a Bayesian inference framework, which is proved to be effective and efficient. In the process of the object tracking process, the positive template and negative template of the discriminant model together with the coefficient of the generative model are timely updated so as to have a strong adaptability and robust discrimination. In the discriminative module, we introduce an effective method to compute the confidence value that assigns more weights to the foreground than the background. In order to speed up the tracking algorithm, a variance particle filter algorithm is proposed to avoid the computational load of the particles with low similarity. Experiments on some challenge video sequences demonstrate that our proposed tracker is robust and effective to challenge issues such as illumination change, clutter background partial occlusion and so on and perform favorably against state-of-art algorithms.

Keywords: Object tracking · l_2 norm minimization · Discriminative model · Generative model · Variance particle filter algorithm · Sparse representation

1 Introduction

Visual tracking aims evaluate the states of a moving target in a video. It has been one of research hotspots in the field of computer vision with pervasive application such as military reconnaissance, medical imaging, robotics, and human computer interaction, to name a few. Despite numerous object tracking methods [7, 20, 27] having been proposed in recent years, it remains challenging to design a robust visual tracking algorithm for complex and dynamic scenes due to factors such as occlusion, background clutter, illumination, camera motion, varying viewpoints, and pose and scale changes.

Current tracking algorithm can be divided into generative or discriminative appearances. Discriminative method treat tracking as a question of boundary questions of classification which aims to find decision boundary for separation of target and background. Therefore, it is necessary to generate positive and negative training to

© Springer Nature Singapore Pte Ltd. 2016
T. Tan et al. (Eds.): CCPR 2016, Part I, CCIS 662, pp. 486–500, 2016.
DOI: 10.1007/978-981-10-3002-4_41

construct and update the classifier. For example, an adaptive method for selecting colors was proposed by Collins and Liu [1]. Avidan [7] combines a group of weak classifiers into a strong one to develop an ensemble tracking method. Kalal et al. [24] propose the P-N learning algorithm to exploit the underlying structure of positive and negative samples to learn availably classifiers for object tracking. Wang et al. [23] base the discriminative appearance model on super pixels, which facilitates the tracker to distinguish between the target and background. In contrast, generative tracking methods typically learn a model to represent the target object and then use it to search for the image region with minimal reconstruction error. For example, these methods are based on either templates [6, 16, 18] or subspaces model [5, 21]. The Frag tracker [8] addresses the partial occlusions of features. The IVT method [28] utilizes an incremental subspace model to adapt appearance changes. The mean shift tracking algorithm models a target with nonparametric distributions of features.

In recent years, numerous algorithms based on l_1 minimization have been proposed for visual tracking. Motivated by its popularity in face recognition, sparse coding techniques have recently migrated over to object tracking. Mei and Ling [14] apply the sparse representation to target tracking. In this case, the tracker represents each target candidate as a sparse linear combination of dictionary templates that can be dynamically updated to maintain an up-to-date target appearance model. This representation has been shown to be robust against partial occlusion, which leads to improve tracking performance. Nevertheless these formulations entail solving l_1 minimization problem, which is known to be time-consuming, Furthermore, since the target states are usually estimated in a particle filter framework, the computation cost grows linearly with the number of sampled particles.

For the above-mentioned problems, a large number of researchers use different approaches to the sparse representation for target tracking. Liu et al. [26] propose a tracking algorithm problem based on local sparse model which employs histograms of sparse coefficients and the mean-shift algorithm tracking. Zhang et al. [10] only used a multi-task sparse tracking to improve tracking algorithm. Si Liu and Zhang [27] propose the sparse representation model of a structure for visual tracking algorithm. Bao et al. [3] propose accelerate the approximate gradient method for solving l_1 norm minimization problem. Zhang [28] propose a new particle-filter based tracking algorithm that exploits the relationship between particles. During this period, many scholars apply sparse representation to face recognition and image classification. In the [29] are adopted l_2 norm minimization instead of l_1 norm minimum for face recognition experiment, we made the original algorithm have the same or even higher degree of recognition. Although the sparsity l_2 norm significantly is weaker than the l_1 norm, it solves simple and fast. However, these methods do not consider the background information and do not take effective occlusion processing mechanism, and hence may fail with more possibility when there is similar object or occlusion in the scene.

In this paper, we proposed a robust object tracking algorithm based on a sparse collaborative model that exploits both holistic templates and local representation to account for drastic appearance model, properly solved by the l_2 norm minimization solution in a Bayesian inference framework. On the one hand, the global template is used to distinguish the foreground and background; on the other hand, using local information to deal with occlusion makes full use of the advantages of these two models to improve the robustness of tracking. The proposed algorithm uses the l_2 norm to minimize the target appearance. The calculation process is simpler than that of the l_1 norm, and greatly reduced low computational complexity. In model updating, discriminative model several frames update the positive and negative template sets every several frames to obtain the latest accurate target and background information, which makes it having adaptability and discrimination. In the generative model, whether or not to update the template coefficient vector by set a threshold, thereby making it more effective and robust. In order to speed up the tracking algorithm, a variance particle filter algorithm is proposed to avoid the computational load of the particles with low similarity. Numerous experiments on various challenging sequences show that the proposed algorithm performs favorably against the state-of-the-art methods.

2   Particle Filter

Target tracking can be viewed as a Bayesian inference problem under Markoff's hidden state variables. It consists of essentially two steps: prediction and update. The predicting distribution of x_t given all available observations $y_{1:t-1} = \{y_1 y_2 \cdots y_{t-1}\}$ up to time t − 1, denoted by $p(x_t|y_{1:t-1})$, is recursively

$$p(x_t|y_{1:t-1}) = \int p(x_t|x_{t-1})p(x_{t-1}|y_{1:t-1})dx_{t-1} \qquad (1)$$

Where $p(x_t|x_{t-1})$ denotes the movement of two consecutive state model. $p(y_t|x_t)$ denotes the observation likelihood. Optimal tracking target state can be obtained to estimate the maximum a posteriori probability.

$$\hat{x}_t = argmaxp\left(y_t^i|x_t^i\right)p\left(x_t^i|x_{t-1}\right), \ i = 1, 2\ldots, N \qquad (2)$$

where x_t^i denotes the i-th sample of the state x_t, y_t^i denotes x_t corresponds to the observation vector.

2.1   State Transition Model

The state variable x_t is modeled by the six parameters of the affine transformation parameters $\{x_t, y_t, \theta_t, \omega_t, \alpha_t, \phi_t\}$, where $x_t, y_t, \theta_t, \omega_t, \alpha_t, \phi_t$ are target tracking in the x and y axis displacement rotation angle changes in scale aspect ratio and vergence direction respectively in the t-th frame. The deformation parameters in x_t are modeled independently by a Gaussian distribution, namely $p(x_t|x_{t-1}) = N(x_t; x_{t-1}, \psi)$, Ψ is diagonal covariance matrix.

2.2   Observation Model

The observation model $p(y_t|x_t)$ reflects the similarity between a target candidate and the target templates. Find the target to be tracked from a large number of candidate samples; combination based on the l_2 norm minimization objective apparent model; observation likelihood function:

$$p(y_t^i | x_t^i) = P \tag{3}$$

3 Objective Representation Based on L_2 Norm Minimization

In visual tracking, how to describe the objective is a very important problem. The appearance model based on discriminative model is used to distinguish the target and the background. The appearance model based on discriminative model is used to describe the goals, look for candidates with the do small reconstruction error. This paper combines the discriminant model and the generative model to described target; and using the l_2 norm to minimize the target apparent system the number of solutions; and achieved good results.

3.1 Discriminant Model Based on L_2 Norm Minimization

Given the image set of the training template $t = \mathbb{T}_1, \mathbb{T}_2, \mathbb{T}_3, \dots \mathbb{T}_{n+p} \in \mathcal{R}^{k \times (n+p)}$. At the time of the first frame by hand the choice of the location of the object, the results for the next frame tracking. $T^\varepsilon = \{zl(z) - l(t-1) < \varepsilon\}$ and $T^{\alpha,\beta} = \{z | \alpha < l(z) - l(t-1) < \beta\}$ according to a Gaussian distribution sampling n positive template $T_+ \in \mathcal{R}^{k \times m}$ and p negative template $T_- \in \mathcal{R}^{k \times p}$, ε represent circular radius threshold, α, β represent a boundary threshold. (See Fig. 1) Then avoid taking into account the scale of the impact and effectiveness of our unity as the sample size and stacked in the same one-dimensional vector, $(32 \times 32$ in our experiment) for efficiency. The negative training set contains both the background and images with parts of the target object, so that the part of the foreground of the sample can be defined as a negative sample to improve the location of the target location.

Fig. 1.

For each target candidate $z \in \mathcal{R}^k$ is an image patch in the current frame and normalized to have the same size as the template. Therefore, it approximately lies in the linear span of

$$z \approx \mathbb{T}c = c_1 \mathbb{T}_1 + c_2 \mathbb{T}_2 + \ldots + c_{n+p} \mathbb{T}_{n+p} \tag{4}$$

Where $c = (c_1, c_2, \cdots, c_{n+p})^T \in \mathfrak{R}^{(n+p)}$ is called a target coefficient vector. Due to c is weak sparsity; using this feature by l_2 minimum norm solution,

$$c = \|z - \mathbb{T}c\|_2^2 + \lambda_1 \|c\|_2^2, s.t. c \not> 0 \tag{5}$$

Where $\|\cdot\|_2^2$ denote the l_2 norm and λ_1 is a weight parameter.

The role of l_2-regularization term $\|c\|_2$ is two-folds: First, it makes the solution of c has certain sparsity, yet this sparsity is much weaker than that by l_1-regularization; Second, it minimizes the stability of the solution and l_2-norm minimization is easy to solve. Because the l_1-norm function is a convex, nonsmooth, global nondifferentiable function. The l_2-norm function is a convex, smooth, global, differentiable function.

To solve (5), $\|z - \mathbb{T}c\|_2^2 + \lambda_1 \|z\|_2^2$ ticked to zero, namely $\frac{d}{dz}\left(\|z - \mathbb{T}c\|_2^2\right) + \lambda_1 \|c\|_2^2 = 2\left(-\mathbb{T}^T\right)(z - \mathbb{T}c) + 2\lambda_1 c = 0$

$$c = \left(\mathbb{T}^T\mathbb{T} + \lambda_1 I\right)^{-1}\mathbb{T}^T z \tag{6}$$

Where $I \in \mathcal{R}^{k \times k}$ is an identity matrix. In Eq. 6 can be analytically derived as $c = Pz$, where $P = \left(\mathbb{T}^T\mathbb{T} + \lambda_1 I\right)^{-1}\mathbb{T}^T$, Clearly, P is independent of z so that it can be pre-calculated. If all the candidate samples are considered as a vector set Z, we can simply project Z onto P via PZ. This makes the coding very fast. The coefficients of all samples can be obtained at once to be a target object:

$$C = PZ \tag{7}$$

vice versa. Thus, we formulate the confidence value H_c of the candidate z by

$$H_c = 1 + \left(-\left(\mathcal{E}_f - \mathcal{E}_b\right)/\delta\right) \tag{8}$$

Where $\varepsilon_f = \|z - \mathbb{T}_+c_+\|_2^2$ is the reconstruction error of the sample z observed by the positive template set $\mathbb{T}_+$, and c_+ is the corresponding foreground sparse coefficient vector. Similarly, $\mathcal{E}_b = \|z - \mathbb{T}_-c_-\|_2^2$ is the reconstruction error of the sample z is observed by the negative template set $\mathbb{T}_-$, and c_- is the corresponding background sparse coefficient vector.δ is fixed to be a small constant that balances the weight of the discriminative classifier and the generative model.

3.2   Computational Complexity

This method is the most similar to l_1 tracking, $\mathbb{T} \in \mathbb{R}^{d \times k}$ denotes l_1 tracking using the template, $\mathrm{P} \in \mathbb{R}^{d \times k}$ denotes l_2 norm regularization projection matrix of the code. The computational complexity of the 2 tracking methods is shown in Table 1. The computationally most expensive operation for a PCG step is a matrix-vector product which has $\mathbb{O}(d(2d+n)) = \mathbb{O}(d^2 + dn)$ computing complexity. The most time consuming part of this paper is to use projection matrix to calculate the coefficient of representation. The most time consuming part of this paper is to use projection matrix to calculate the coefficient of representation, which has $\mathbb{O}(dk)$ computing complexity.

Table 1. Computer complexity and computing time

Method	Computer complexity			Computing time
l_1 tracking	$\mathbb{O}(d(2d+k))$	$\mathbb{O}(d^2+dk)$	$\mathbb{O}(dk)$	13.1
Our	$\mathbb{O}(dk)$			2.1

3.3   Generative Model Based on L_2 Norm Minimization

We present a robust observation function for object representation that takes local appearance information of patches and occlusions into consideration. Due to a large number of candidate target estimate to select the best candidate target; we apply a variance particle filtering algorithm, effective avoidance of particles with lower similarity calculation. In order to improve the robustness of tracking algorithm when the target is occluded, the image is divided into blocks of samples. We use overlapped sliding windows on the normalized images to obtain n patches and in the experimental section of each image block into a vector $y_i \in \mathbb{R}^{d \times 1}$, where d denotes the size of the patch. The l_2 norm minimization is used to solve the coefficient vector of each image block.

$$\alpha_i = \|y_i - D\alpha_i\|_2^2 + \lambda_2 \|\alpha_i\|_2^2 \qquad (9)$$

Where the dictionary $D \in \mathbb{R}^{d \times m}$ is generated from k-means cluster centers (m denotes the number of cluster centers)

$$\alpha = \left[\alpha_1^T, \alpha_2^T, \cdots, \alpha_n^T \right]^T \qquad (10)$$

Where $\alpha \in \mathbb{R}^{(m \times n) \times 1}$ is the proposed histogram for one candidate.

3.4   The Variance of Particle Filter Method

The method can quickly screen out the particle that is not highly similar to the target, and greatly reduces the computation of the target tracking algorithm. In the particle filter, it need to calculate the similarity between all particles and the target; it is not only difficult to find the ideal target feature and the appropriate measure of similarity, but also often requires a lot of computation in the case of the change of the object's shape and light and other factors. However, it can be used to find a relatively simple and fast method to find the particles with low similarity. From this point of view, this section studies a method of fast identification of particles with low target similarity.

The method can filter out a large number of particles with low similarity of target template and make the algorithm focus on the particle with higher similarity of target template. This method is referred to as variance particle filter.

Supposed as $y = y_1 y_2 \ldots \ldots y_n$ denotes candidate particle set in a particle filter, where $y \in \mathbb{R}^{d \times n}, D_{st}$ indicate target template. In order to distinguish other templates, it called the standard template. Divide each column vector of y with the D_{st} to divide the division operation, accessibility $U = [u_1 u_2 \ldots \ldots u_n]$

$$u_i = \frac{y_i / \sum_{j=1}^{d} (y_{i,j})}{D_{st} / \sum_{j=1}^{d} D_{st,j}} \tag{11}$$

Where $y_{i,j}$ represents the first j element in i-th candidate targets, $D_{i,j}$ represents the first j element in standard template. U in each column vector contains the similarity of the target and the standard template information, the purpose of particle filter is based on the information in the U to find a higher degree of similarity with the standard template to participate in the operation of the particle. Taking y_i as an example, if the similarity between the target and the standard template is high, the numerical value of each element in the u_i is small, and otherwise, the numerical value of each element in the u_i is large. Therefore, the lh_i can be used to determine the size of each element in the u_i to judge the similarity of the candidate target and template, and the volatility of each element in u_i can be used to measure the variance of each element:

$$lh_i = \sum_{j=1}^{d} (u_{i,j} - \mathbb{U}_i)^2 \tag{12}$$

where $\mathbb{U}_i$ represent mean value.

In order to deal with occlusions, the occluded image patch will affect relatively coefficient vector when describing the target object, because the occlusion image patch and the template of the difference is very large, this will cause the error to be larger than the error of the candidate sample itself. So in this paper, by setting a threshold value to determine whether the image patch is occluded or not. When the reconstruction error is relatively large, the image patch is considered to be covered; the weight of this image patch is set to 0. When the reconstruction error is relatively small, the image patch is not occluded; the weight of the image patch is set to 1. We modify the constructed histogram to excluded patches. Can be achieved by the following method

$$\delta_i = \begin{cases} 1, & \varepsilon_i < \varepsilon_0 \\ 0, & otherwise \end{cases} \tag{13}$$

Where $\varepsilon_i = \|y_i - D\alpha_i\|_2^2$ denotes the reconstruction error of patch y_i, ε_0 is a predefined threshold. Thus a weighted coefficient vector is generated by

$$\varepsilon = \alpha \odot \delta \tag{14}$$

Where $\odot$ denotes two vector corresponding multiplication. δ the weight coefficient vector of each element.

By comparing the similarity of the candidate sample coefficient vector and the template coefficient vector, defend

$$G = \sum_{k=1}^{m \times n} min\left(\varepsilon^j, \theta^j\right) \tag{15}$$

Where ε^j and θ^j are the coefficient vector for the candidate and the template. G denotes similarity of candidate samples and templates. The coefficient vector of the initial time template is adopted and the candidate in the first frame. The sample coefficient vector is obtained by the same method.

4   Collaborative Model

We propose a collaborative model combines the generative and the discriminative model modules within the particle filter framework. A new likelihood function expression is obtained:

$$P = GH \tag{16}$$

The multiplicative formula is more effective in our tracking scheme compared with the alternative additive operation. First, his likelihood function is able to distinguish between foreground and background. Compared with the background, the target area is given more weight. The confidence value of the discriminant method can be used as the weight of the likelihood function of the model. Therefore, the model can be overcome by using discriminant model, which has the ability to eliminate the interference of complex background.

At the same time, the likelihood function of the model is more similar when the likelihood function of the model is only a small amount of background degree value. Similarly, the likelihood function of the model can be used as the weight of the classification function of the discriminant model. The likelihood function of the target or the false background of the degree value is not obvious, and the likelihood function of the target is approximately the same. The likelihood function of the probability model can be distinguished by another model in the case of a similar situation. Therefore combine model is more flexible, strong and complementary to each other.

4.1   Templated Update

Tracking with fixed templated is prone to fail in dynamic scenes as it does not consider inevitable appearance changed due to factors such as illumination and pose changed. So the online update model is very important to improve the adaptive ability of the target tracker. We develop an updated scheme in which the generative and discriminative modules are updates independently.

For generative model, the dictionary D is fixed during the tracking process. Update according to the following:

$$\theta_i = \eta\theta_1 + (1 - \eta)\varepsilon_i O_i < O_0 \qquad (17)$$

The variable O_i denotes the occlusion condition of the tracking result in the new frame

$$O_i = \sum_{k=1}^{m \times n}(1 - \delta_i^k) \qquad (18)$$

Where η denotes update probability, θ_1 is the first frame of the coefficient vector. ε_i is the latest frame get coefficient vector. The update is performed as long as the occlusion condition O_i in this frame is smaller than a predefined constant O_0.

For discriminative model, we update the negative templates every several frames (5 in our experiments) from image regions away from the current tracking result. The positive templates remain the same in the entire sequence, as the discriminative model aims at distinguishing the foreground from the background, it must make sure that the positive templates and the negative templates are all correct and distinct.

5   Experimental Results

We implemented the proposed approach in MATLAB and evaluated the performance on numerous video sequences. In order to evaluate the performance of our tracker, we conduct experiments on eight challenging image sequences. The videos were recorded in indoor and outdoor environment in different formats where the targets underwent lighting and scale changes, out-of-plane rotation, and occlusion. For comparison, we run six state-of-the-art algorithms with the same initial position of the target.

In the experiment, the parameters are presented as follows, the numbers of positive templates $\mathbb{T}_+$ and negative templates $\mathbb{T}_-$ are 80 and 200 respectively. The variable λ in Eqs. 3 is fixed to be 0.001. The variable λ in Eqs. 5 and 10 is fixed to be 0.01. The row number m and column number n of dictionary D in Eq. 5 are 36 and 50. The threshold ε_i in Eq. 17 is 0.04. The update rate μ is set to be 0.9. The threshold η in O_0 Eq. 17 is 0.8.

5.1   Quantitative Comparison

In order to further analyze the performance of this algorithm, we draw the center point error curve and the overlap rate curve in the first 8 video sequences. The center point position error is the error between the center position of the tracking result and the standard center position of the frame. Given the tracked bounding box $\mathbb{R}_t$ and the ground truth bounding box $\mathbb{R}_g$. The overlap score is defined as $S = \text{area}(\mathbb{R}_t \cap \mathbb{R}_g) / \text{area}(\mathbb{R}_t \cup \mathbb{R}_g)$, where $\cap$ and $\cup$ represent the intersection and union of two regions, respectively. We count the number of successful frames whose overlap S is larger than the given threshold t_0 (Tables 2 and 3).

Table 2. Average enter location error (in pixel)

	Frag	MIL	L$_1$	VTD	TLD	IVT	Our
car4	18.4	60.1	4.1	13.2	18.4	2.9	4.8
board	45.4	66.7	184.0	105.0	140.7	160.7	35
car11	64.0	43.5	33.3	27.1	25.3	2.3	3.4
sylv	22.1	34.5	112.6	12.4	23.5	92.1	11.5
david	9.3	15.2	7.2	14.5	9.7	3.6	5.9
faceocc1	6.6	34.4	6.9	10.6	17.9	9.3	5.5
faceocc2	18.1	14.1	11.2	11.1	18.6	10.3	5.6
stone	8.2	32.8	19.2	31.5	8.2	2.2	6.5

Table 3. Average overlap rate

	Frag	MIL	L$_1$	VTD	TLD	IVT	Our
car4	0.63	0.43	0.84	0.72	0.64	0.92	0.86
board	0.65	0.46	0.12	0.32	0.19	0.18	0.71
car11	0.08	0.17	0.43	0.41	0.37	0.51	0.75
sylv	0.54	0.53	0.14	0.70	0.58	0.33	0.71
david	0.60	0.34	0.73	0.57	0.61	0.72	0.78
faceocc1	0.81	0.61	0.86	0.81	0.56	0.84	0.84
faceocc2	0.60	0.61	0.67	0.69	0.52	0.67	0.72
stone	0.41	0.36	0.32	0.43	0.43	0.70	0.61

5.2 Qualitative Comparison

Illumination Change: We select the car4 and car11 sequences used to evaluate the algorithm to cope with the challenges of light change. Video car4 is recorded during the day, the video of the vehicle in the tree and the tunnel through the trees and the tunnel will be a significant change in light. In the car11, due to the variety of automotive lighting and the background of the flashing lights have a low contrast, while the intensity of light changes will occur (Fig. 2).

Figure shows the tracking results of all the tracking methods when the object is subjected to light changes. Frag tracker in the above two video performance is not good, because the appearance of the Frag tracker based on the comparison of pixel values is directly and to take a fixed template, cannot be very good to adapt to changes in light. When the vehicle enters or comes out the shadow area, the vehicle has the intense illumination change causes the VTD tracker and MIL have the slight drift. Other tracking performance is good in car4 and car11 sequences, IVT tracker and the proposed tracker can in all sequences in the successful completion of target tracking, and other tracker can shift or the surrounding background as a target such as mil tracker in 180 frames and frame 269, VTD tracker, TLD tracker and, tracker at 269 frame and 300 frame. This can be attributed to the use of incremental subspace learning template update strategy so that the target tracker can be adaptive to changes in the light.

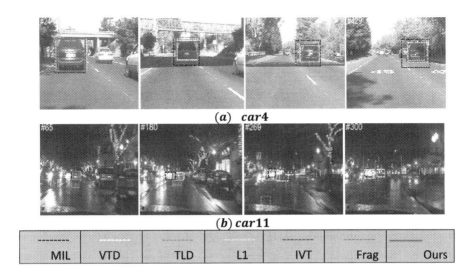

(a) car4

(b) car11

| MIL | VTD | TLD | L1 | IVT | Frag | Ours |

Fig. 2.

Rotation: We select the david and sylv sequences used to evaluate the algorithm to cope with the rotation and pose change challenge. In the video davidIndoor, David from the dark room to the light room (frame 1 and frame 396), while the scale and attitude of the target is also slowly changing (frame 148). In the sylv sequence, people operate the doll frequently and rotate and accompanied by changes in light. Figure shows the tracking results of all the tracking methods when the target is subjected to rotation and appearance changes. And some algorithms are not able to adapt to the scale changes brought about by plane rotation, such as MIL tracker, l_1 tracker. Although the proposed method, TLD tracker and VTD tracker can always track the face, the proposed algorithm in this paper is more accurate. Other trackers gradually drift f-rom the target area to the background area. In the video sylv, l_1 tracker slowly drifting away from the target (frame 220 and frame 728), l_1 tracker will lost in some frames (frame 270 and frame 728). Although our tracker, l_1 tracker, VTD tracker, TLD tracker, Frag tracker, MIL tracker can track the dolls, Frag tracker and MIL tracker can not adapt to the change of the doll's rotation at 1220, 270th frame (Fig. 3).

Heavy Occlusion: We select the faceoccl, faceocc2 sequences for each assessment algorithm to cope with the partial occlusion challenge. A woman in faceoccl from various directions seriously covered his face with a book. A man in faceocc2 occludes his appearance with a book or a hat, but also turns the head to change the attitude. Figure shows the tracking results of all the tracking methods when the target is partially occluded.

In faceoccl sequence, because the character's face is blocked by the book, there is no change in light and movement. Most of the tracker performs well except for MIL tracker appear a slight drift in frame 522. In the faceocc2 video sequence, when the character's face is almost completely blocked, most of the tracker began to drift, the positioning of the target is not accurate or when the face of severe occlusion when the

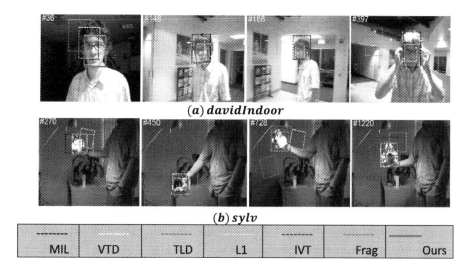

(a) *davidIndoor*

(b) *sylv*

MIL	VTD	TLD	L1	IVT	Frag	Ours

Fig. 3.

scale of the bad. In general, the Frag tracker,l_1 tracker and the proposed algorithm have achieved good results in these two sequences. Although Frag tracker can deal with the occlusion by using the histogram of the image block and the confidence map, it cannot deal with the appearance change and occlusion. The l_1 tracker based on the sparse representation of the appearance model and the trivial template against partial shading block. Because l_1 tracker a simple dictionary update strategy can easily lead to drift (Fig. 4).

Complex Background: We select the board and stone sequences used to evaluate the complexity of the algorithm to cope with the challenges of the background. There are a lot of stones in the shape and color of the object in the stone video, and there is a person's hand moving to interfere. Board sequence background color is rich and there is a variety of interference debris.

Figure 3(a) shows the tracking results of all tracking methods in complex background. In the stone sequence, Frag tracker, MIL tracker, VTD tracker will drift when the target occlusion, the IVT tracker and our tracker successfully track the target in this sequence. In frame 400 and 510, l_1 tracker is prone to drift in complex background. Frag tracker, MIL tracker and VTD tracker drift when the target is occluded. However, the IVT tracker and our tracker can track the target position successfully (frame 516 and 592). Although TLD tracker is able to capture the target when the tracker drifts from the target, there are high error rates and low success rates (frame 516 and frame 140). The board sequence is challenging as the background is cluttered and the target object experiences out-of-plane rotations. Most trackers fail as holistic representations inevitably include background pixels that may be considered as part of foreground object through straightforward update schemes. Our tracker performs well in this sequence. In addition, the update scheme uses the newly arrived negative templates that facilitate separation of the foreground object and the background (Fig. 5).

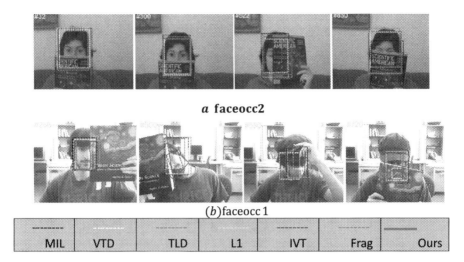

a **faceocc2**

(*b*)faceocc1

| MIL | VTD | TLD | L1 | IVT | Frag | Ours |

Fig. 4.

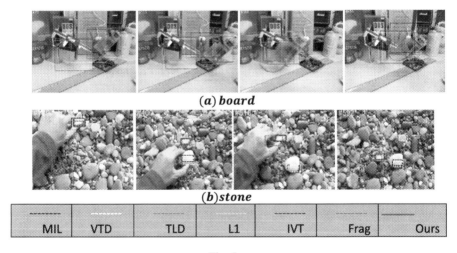

(*a*) *board*

(*b*)*stone*

| MIL | VTD | TLD | L1 | IVT | Frag | Ours |

Fig. 5.

Conclude: In this paper, we propose and demonstrate an effective and robust tracking method based on the collaboration of generative and discriminative modules. In our trackers, the occlusion processing mechanism used in this paper can effectively deal with occlusion. Moreover, the online update scheme reduces drifts and enhances the proposed method to adaptively account for appearance change in dynamic scenes. Quantitative and qualitative comparisons with six state-of-the-art algorithms on eight challenging image sequences demonstrate the robustness of our tracker.

Acknowledgement. The authors are appreciative to the anonymous reviewers for valuable comments. It is also acknowledged that the research was mainly supported by National Natural Science Foundation (Nos. 61170109, 61572443, 61572023), Zhejiang Provincial Natural Science Foundation (No. LY14F030008 and No. 2015C31095).

References

1. Wu, Y., Lim, J., Yang, M.-H.: Online object tracking: a benchmark. In: CVPR (2013)
2. Zhong, W., Lu, H., Yang, M.-H.: Robust object tracking via sparsity-based collaborative model. In: CVPR (2012)
3. Bao, C.L., Wu, Y., Ling, H.: Real time robust l_1 accelerated proximal gradient approach. In: CVPR. IEEE (2012)
4. Babenko, B., Belongie, S., Yang, M.-H.: Visual tracking with online multiple instance learning. In: CVPR (2009)
5. Ross, D., Lim, J., Lin, R.-S., Yang, M.-H.: Incremental learning for robust visual tracking. IJCV **77**, 125–141 (2008)
6. S. Avidan. Ensemble tracking. In CVPR, 2007
7. Adam, A., Rivlin, E., Shimshoni, I.: Robust fragments-based tracking using the integral histogram. In: CVPR (2006)
8. Zhang, T., Ghanem, B., Liu, S., Ahuja, N.: Robust visual tracking via multi-task sparse learning. In: CVPR (2012)
9. Bao, C., Wu, Y., Ling, H., Ji, H.: Real time robust l_1 tracker using accelerated proximal gradient approach. In: CVPR (2012)
10. Jia, X., Lu, H., Yang, M.-H.: Visual tracking via adaptive structural local sparse appearance model. In: CVPR (2012)
11. Zhang, K., Zhang, L., Yang, M.-H.: Real-Time Compressive Tracking. In: Fitzgibbon, A., Lazebnik, S., Perona, P., Sato, Y., Schmid, C. (eds.) ECCV 2012. LNCS, vol. 7578, pp. 864–877. Springer, Heidelberg (2012). doi:10.1007/978-3-642-33712-3_62
12. Avidan, S.: Support vector tracking. PAMI **26**, 1064–1072 (2004)
13. Li, X., Hu, W., Shen, C., Zhang, Z., Dick, A., Hengel, A.: A survey of appearance models in visual object tracking. TIST **4**, 58 (2013)
14. Mei, X., Ling, H.: Robust visual tracking using l_1 minimization. In: ICCV (2009)
15. Mei, X., Ling, H., Wu, Y., Blasch, E., Bai, L.: Minimum error bounded efficient L1 tracker with occlusion detection. In: CVPR (2011)
16. Mei, X., Ling, H., Wu, Y., Blasch, E., Bai, L.: Efficient minimum error bounded particle resampling L1 tracker with occlusion detection. TIP **22**, 2661–2675 (2013)
17. Liu, B., Huang, J., Yang, L., Kulikowsk, C.: Robust tracking using local sparse appearance model and K-selection. In: CVPR (2011)
18. Yu, Q., Dinh, T.B., Medioni, G.: Online tracking and reacquisition using co-trained generative and discriminative trackers. In: Forsyth, D., Torr, P., Zisserman, A. (eds.) ECCV 2008. LNCS, vol. 5305, pp. 678–691. Springer, Heidelberg (2008). doi:10.1007/978-3-540-88688-4_50
19. Wright, J., Ma, Y., Maral, J., Sapiro, G., Huang, T., Yan, S.: Sparse representation for computer vision and pattern recognition. Proc. IEEE **98**(6), 1031–1044 (2010)
20. Wang, S., Lu, H., Yang, F., Yang, M.-H.: Super pixel tracking. In: ICCV (2011)
21. Kalal, Z., Matas, J., Mikolajczyk, K.: P-N learning: bootstrapping binary classifiers by structural constraints. In: CVPR (2010)
22. Mei, X., Ling, H.: Robust visual tracking using L1 minimization. In: ICCV (2009)

23. Gao, S., Tsang, I.W.-H., Chia, L.-T., Zhao, P.: Local features are not lonely - Laplacian sparse coding for image classification. In: CVPR (2010)
24. Grabner, H., Bischof, H.: On-line boosting and vision. In: CVPR (2006)
25. Kwon, J., Lee, K.M.: Visual tracking decomposition. In: CVPR (2010)
26. Zhang, D., Yang, M., Feng, X.: Sparse representation or collaborative representation: which helps face recognition? In: Computer Vision (2011)
27. Zhang, T., Liu, S., Xu, C., Yan, S., Ahuja, N., Yang, M.-.H.: Structural sparse tracking. In: IEEE International Conference on Computer Vision and Pattern Recognition (CVPR) (2015)
28. Zhang, T., Liu, S., Ahuja, N., Yang, M.-H., Ghanem, B.: Robust visual tracking via consistent low-rank sparse learning. Int. J. Comput. Vis. (IJCV) **111**(2), 171–190 (2015)
29. Zhang, T., Jia, C., Xu, C., Ma, Y., Ahuja, N.: Partial occlusion handling for visual tracking via robust part matching. In: IEEE International Conference on Computer Vision and Pattern Recognition (CVPR) (2014)

Online Adaptive Multiple Appearances Model for Long-Term Tracking

Shuo Tang, Longfei Zhang$^{(\boxtimes)}$, Xiangwei Tan, Jiali Yan, and Gangyi Ding

Digital Performance and Simulation Key Laboratory, School of Software,
Beijing Institute of Technology, Beijing 100081, China
{shuo_tang,longfeizhang,dgy}@bit.edu.cn

Abstract. How to build a good appearance descriptor for tracking target is a basic challenge for long-term robust tracking. In recent research, many tracking methods pay much attention to build one online appearance model and updating by employing special visual features and learning methods. However, one appearance model is not enough to describe the appearance of the target with historical information for long-term tracking task. In this paper, we proposed an online adaptive multiple appearances model to improve the performance. Building appearance model sets, based on Dirichlet Process Mixture Model (DPMM), can make different appearance representations of the tracking target grouped dynamically and in an unsupervised way. Despite the DPMM's appealing properties, it characterized by computationally intensive inference procedures which often based on Gibbs samplers. However, Gibbs samplers are not suitable in tracking because of high time cost. We proposed an online Bayesian learning algorithm to reliably and efficiently learn a DPMM from scratch through sequential approximation in a streaming fashion to adapt new tracking targets. Experiments on multiple challenging benchmark public dataset demonstrate the proposed tracking algorithm performs 22 % better against the state-of-the-art.

Keywords: Object tracking · Multiple appearance model · Online Dirichlet process mixture model

1   Introduction

Object tracking plays an important role in numerous vision applications, such as motion analysis, activity recognition, visual surveillance and intelligent user interfaces. However, while much progress has been made in recent years, it is still a challenging problem to track a moving object in a long term in the real-world because of the variations of tracking environment such as view port exchanging, illuminance varying, and etc. For visual tracking problem, an appearance model is used to represent the target object and predicted the likely states of tracking target in future frame [1]. However, using one appearance model is not suitable to describe all the historical appearance information, especially for long term tracking task. So we mainly focused on building multiple appearance models

© Springer Nature Singapore Pte Ltd. 2016
T. Tan et al. (Eds.): CCPR 2016, Part I, CCIS 662, pp. 501–516, 2016.
DOI: 10.1007/978-981-10-3002-4_42

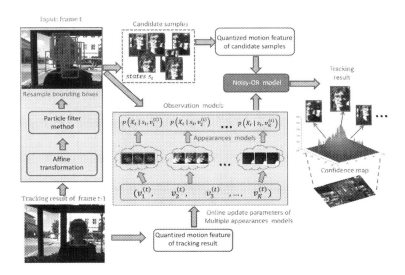

Fig. 1. The framework of online adaptive multiple appearances model based tracking.

dynamically in an unsupervised way in order to adapt the changing appearance of tracking target (Fig. 1).

In this paper, we used Bayesian non-parametric clustering method to cluster multiple different appearances dynamically. It can cover more aspects of the target appearance to make the proposed algorithm more robust to abrupt appearance changes, and the number of clusters can be inferred from the observation. Among the different probabilistic models, Bayesian non-parametric method has several properties which suit the object tracking application well. In particular, DPMM represents mixture distributions with an unbounded number of components where the complexity of the model adapts to the observed data. This property is important for building multiple appearance models dynamically. In general, the number of appearances is uncertain and varying over time.

However, despite the appealing properties of DPMM, it characterized by computationally intensive inference procedures, which often based on Gibbs samplers [2]. While Gibbs sampling can be an appropriate inference mechanism when execution time is not an issue. It is not applicable in visual tracking, as it needs more faster inference. In [3] a variational inference method which maximizes a lower bound to the true underlying distribution and after each iteration, the obtained parameters define a distribution which approximates the true one in a properly defined way. However, variational inference method is extremely vulnerable to local optima for non-convex unsupervised learning problems, and is frequently yielding poor solutions.

In visual tracking literature, the appearance model based on BNP is not applied as usual as the parametric methods [4]. The main strategy of the BNP tracking methodology is based on three aspects shown as follows: we need to solve (1) how to represent the observation of tracking target by Bayesian non-parametric models, (2) how to create multiple appearance models dynamically without knowledge of cluster numbers and model parameters to adapt in tracking environment variation, (3) how to update multiple appearance models effectively and reliably in tracking process.

Our proposed algorithm is mostly inspired by [5,6] which are online Bayesian learning algorithm to estimate DP mixture models. This method does not require random initialization like Gibbs samplers. Instead, it can reliably and efficiently learn a DPMM from scratch through sequential approximation in a single pass. The algorithm takes data in a streaming fashion, and thus can be easily adapted to new tracking target.

The rest of the paper is organized as follows: Sect. 2 reviews some of the related work. Section 3 reviews Bayesian non-parametric model on which our proposed model based. Section 4 introduces multiple appearances modeling and representation, and proposes related probabilistic distributions which can describe the generation process of tracking features. Section 5 introduces an online sequential Bayesian method to build multiple appearance models. Section 6 presents the framework of the proposed tracking algorithm. Section 7 reports the experimental results. Section 8 makes the conclusion of the paper.

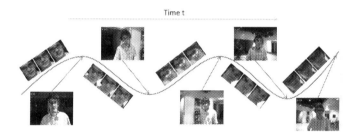

Fig. 2. The appearances when operating the online adaptive multiple appearances model tracking. (Color figure online)

2   Related Work

There is a rich literature in object tracking approatches [7]. As a main component in tracking algorithms, tracking target appearance modeling plays a key role in tracking performance. A good appearance representation should have strong description or discrimination power to distinguish the target from the background. In order to adapt to the appearance variations of the target during tracking, there are many adaptive appearance models have been proposed for object tracking including both generative and discriminative methods.

For generative appearance modeling methods, Jepson et al. [8] learn a Gaussian mixture model via an online expectation maximization algorithm to account for target appearance variations during tracking. Incremental subspace methods have also been used for online object representation [9]. This method uses target observations obtained online to learn a linear subspace for object representation. Since the appearance of a target in a long-time interval may be quite different, these generative models may not describe the appearance variations of the target well.

For discriminative appearance modeling methods, Avidan et al. [10] use online boosting method for tracking. They proposed an ensemble tracking framework to construct a strong classifier to distinguish the target from the background. Babenko et al. [11] use Multiple Instance Learning (MIL) instead of traditional supervised learning to avoid the inaccuracy accumulation problem caused by self-learning. In these methods, tracking is usually treated as a binary classification problem. In order to train and update the classifiers, samples usually needs to be correctly labeled, which may not be available in many real tracking applications.

The most related methods to our model is [12], which proposed the original Adaptive Multiple Appearance Model (AMAM) framework to maintain not only one appearance model as many other tracking methods but appearance model set to describe all historical appearances of the tracking target during a long term tracking task. This method employed DPMM to build multiple appearance models unsupervised to tackle drifting problem, and experiment in several public datasets shows that this tracker has high tracking performance compared with several other state-of-the-art. In order to infer the number of different appearances underlying tracking observations, this tracker resorts to Gibbs sampler [2] for approximate inference and also requires random initialization of components. However, as this sampler needs to maintain the entire configuration, the computational complexity of this tracker is quite high, which limits its applications in real-time scenarios.

Compared with the tracking methods as described above, our proposed method shows three mainly characteristics in dealing with appearance variations of the target. Firstly, our method can cluster multiple different appearances dynamically, and the number of clusters can be inferred from the tracking observation. Secondly, in our method, different kinds of tracking target appearances can be modeled by new model or constructed appearance models. It covered various target appearances, which made the proposed method more robust to abrupt appearance changes. Finally, our method begins with an empty model and progressively refines the models as tracking observation come in, adding new appearance models on the fly when needed.

3   Bayesian Non-parametric Model

The Dirichlet process (DP) introduced in [14], is a popular nonparametric stochastic process that defines a distribution over probability distributions. The DP is parameterized by a base distribution H which has corresponding density $h(\mu)$, and a positive scaling parameter $\alpha > 0$. We denote a DP as follows:

$$G|\{\alpha, H\} \sim DP(\alpha, H), G \triangleq \sum_{k=1}^{\infty} \omega_k \delta_{\phi_k},$$

$$v_k \sim Beta(1, \alpha), \omega_k = v_k \prod_{l=1}^{k-1}(1 - v_l)$$

The DP is most commonly used as a prior distribution on the parameters of a mixture model when the number of mixture components is uncertain. Such a model is called a Dirichlet process mixture model (DPMM) which can be specified as:

$$G \sim DP(\alpha, H), \theta_i|G \sim G, x_i|\theta_i \sim F(\theta_i), i \in \{1, \cdots, N\}$$

Let z_i indicate the subset, or cluster, associated with the ith observation, the DP mixture model can also be modeled by using the Chinese Restaurant Process (CRP) representation [17] of the DP, leading to the followings:

$$p(z_i = k|z_{-i}) \propto \sum_{j \neq i} 1_{z_j=k}, p(z_i = k_{new}|z_{-i}) \propto \alpha_0$$

So, a model equivalent to the DPMM using the CRP can be specified as:

$$z \sim CRP(\alpha), \theta_k|G \sim G, x_i|z_i \sim F(\theta_{z_i})$$

4   Multiple Appearance Modeling and Representation

In this section, our goal is to develop a probabilistic method to cluster multiple different appearances unsupervised, which can cover aspects of the target appearance. In order to do so, we represent motion features (e.g. HOG, color feature etc.) using histograms, and then, quantize motion feature values of tracking observations to 20 or more levels, which is a common practice for similar histogram-based descriptors, such as [13]. Thus, considering N tracking observations $X = \{X_i\}_{i=1}^{N}$, which can be clustered into K clusters or different appearances, and each $X_i = \{x_i\}_{i=1}^{D}$ represents a quantized D dimensional motion feature, and x_i is the corresponding histogram quantized bin counts, which is a quantized integer. With the new tracking observation arrival, the number of clusters became to be varies. Given the cluster assignment for i^{th} each observation X_i, its likelihood for that cluster is $F(X_i|\theta_k)$, while the $\theta_{1:K}$ are drawn from the base distribution of DPMM.

4.1   Exponential Family and Sufficient Statistics

In order to describe motion feature X which is the collection of small integers of histograms and the histogram bin counts, we adopt component distributions of which are members of the exponential family distributions. The base measure

---

Algorithm 1. The method of building multiple appearance models

---

Input: Given the concentration parameter α of $DPMM$, base measure
parameter $\mu = (\mu_1, \cdots \mu_D)$ of $H(\mu)$ and HOG features $\{X_i\}_{i=1}^N$ of tracking
observations, until frame N.
Output: multiple appearance model parameters $\zeta_k, (k = 1 : K)$
Let $K = 1, \rho_1(1) = 1, \omega_1 = \rho_1, \zeta_1 = \mu$
for $i = 2$ *to* N **do**
    Compute the marginal likelihood $f_k(X_i) = p(X_i|\zeta_k)$, for $k = 1 : K$
    Compute $f_k(X_i) = p(X_i|\lambda)$using Eq. (3), for $k = K + 1$
    Compute $\rho_i(k) = \omega_k f_k(X_i)/\sum_l \omega_l f_l(X_i)$ for $k = 1 : K + 1$, with $\omega_{K+1} = \alpha$
    see, Eq. (9)
    if $\rho_i(K + 1) > \varepsilon$ **then**
        for $k = 1 : K$ **do**
            $\omega_k = \omega_k + \rho_i(k)$,
            Update parameters according to posterior and
            $\rho_i(k) : \theta_k = \theta_k + X_i \rho_i(k)$ using Eq. (10)
        $\omega_{K+1} = \rho_i(K + 1)$, $\zeta_{K+1} = X_i \zeta_i(K + 1)$
        $K = K + 1$
    else
        Re-normalize ρ_i such that $\sum_{k=1}^K \rho_i(k) = 1$
        $\omega_k = \omega_k + \rho_i(k), \zeta_k = \zeta_k + X_i \rho_i(k)$ for $k = 1 : K$

---

of the DPMM will be the conjugate prior, because it has many well-known
properties, which can admit efficient inference algorithms. Thus, in this paper,
we will consider to describe the distributions as follows:

$$p(X_n|\theta_i) = l(X_n)exp(\theta_i^T X_n - a(\theta_i)),$$

where a is the log-partition function. We take H to be in the corresponding
conjugate family:

$$h(\theta|\lambda) = l(\theta)exp(\lambda_1^T \theta - \lambda_2 a(\theta) - a(\lambda)),$$

where the sufficient statistics are given by the vector $(\theta^T, -\alpha(\theta))$, and $\lambda = (\lambda_1^T, \lambda_2)$.

4.2   Model Representation

Specifically, we choose Multinomial distribution $F(X_i|\theta)$, which we denote
$Mult(\theta_k; n)\theta_k = (p_1, \cdots, p_D)$, is a discrete distribution over D dimensional non-
negative integer vectors $X_i = (x_1, x_2, \cdots x_D)$ where $\sum_{i=1}^D x_i = n$. The probability mass function is given as follows:

$$f(X_i; p_1, \cdots, p_D, n) = \frac{\Gamma(n + 1)}{\prod_{i=1}^D \Gamma(x_i + 1)} \prod_{i=1}^D p_i^{x_i} \tag{1}$$

The cluster prior $H(\theta|\lambda)$ is represented by a Dirichlet distribution which is conjugate to $F(X_i|\theta)$. We denote cluster prior $H(\theta|\lambda)$ as follows, which is a Dirichlet distribution and is conjugate to $F(X_i|\theta)$.

$$h(p_1, \cdots p_D; \lambda_1, \cdots \lambda_D, n) = \frac{1}{B(\lambda_1, \cdots \lambda_D)} \prod_{i=1}^{D} p_i^{\lambda_i - 1}, \tag{2}$$

where the normalizing constant is the multinomial Beta function. Because $H(\lambda)$ is conjugate to $F(\theta)$, then the marginal joint distribution can be obtained by integrating out $(p_1, \cdots, p_D)$ as follows:

$$p(X_i|\lambda_1, \cdots \lambda_D) = \frac{N!}{\prod_{i=1}^{D}(n_i!)} \frac{\Gamma(A)}{\Gamma(N+A)} \prod_{i=1}^{D} \frac{\Gamma(n_i + \lambda_i)}{\Gamma(\lambda_i)} \tag{3}$$

where $A = \Sigma_i \lambda_i$ and $N = \Sigma_i n_i$, and where $n_i = $ number of x_i's with value i.

4.3   Multiple Appearance Modeling

When the number of clusters K is estimated, the multiple appearance model can be built. Considering all of the model parameters, which is comprised of the model parameters $\theta_{1:K}$ and the cluster indicator $z_{1:N}$, the joint distribution of this Bayesian Non-parametric mixture model can be written as in Eq. (4).

$$p(\theta_{1:K}, z_{1:N}|X_{1:N}) \propto$$
$$p(z_{1:N})(\prod_{i=1}^{N} p(X_i|\theta_{z_i})) \prod_{k=1}^{K} p(\theta_k) \tag{4}$$

Here, $z_i \in \{1 \cdots K\}$ with $i \in \{1 \cdots N\}$ indicates the cluster label of the observation X_i and θ_k are the parameters for the k-th appearance model. The target of our proposed method is to infer the joint posterior distribution $p(\theta_{1:K}, z_{1:N}|X_{1:N})$ unsupervised and dynamical, then we can get the parameters $\theta_{1:K}$ of multiple appearance models.

5   Online Sequential Approximation

In order to infer the joint posterior distribution $p(\theta_{1:K}, z_{1:N}|X_{1:N})$, we can initialize the components randomly, and then resort to Gibbs sampler for approximate inference [12]. However, this method needs to maintain the entire configuration, so the computational complexity of this tracker is rather high, which limits its applications in real-time scenarios.

We improved it by using an online sequential variational approximation method to learn a DPMM from scratch through sequential approximation in a streaming, which is easily adapted to new observation.

By marginalizing out the cluster assignment $z_{1:N}$, we obtain the posterior distribution $p(\theta|X_{1:N})$:

$$p(\theta|X_{1:N}) = \sum_{z_{1:n}|X} p(z_{1:N}|X_{1:N})p(\theta|X_{1:N}, z_{1:N}) \tag{5}$$

In order to compute the distribution above, it requires enumerating all possible partitions $z_{1:N}$, which grows exponentially as n increases. To tackle this difficulty, we resort to variational approximation [6] to choose a tractable distribution to approximate $p(\theta|X)$ as follows:

$$q(\theta|\rho, \upsilon) = \sum_{z_{1:n}} \prod_{i=1}^{n} \rho_i(z_i)q_\upsilon^{(z)}(\theta|z_{1:n}) \tag{6}$$

We begin our tracker with one appearance model (i.e. $K = 1$) and progressively refine the model as samples come in, adding new appearance models on the fly when needed. Specifically, when we have $\rho = (\rho_1, \rho_2, \cdots, \rho_i)$ and $\upsilon^{(i)} = (\upsilon_1^{(i)}, \upsilon_2^{(i)}, \cdots, \upsilon_K^{(i)})$ after processing i frames. To determine X_{i+1}, we can use either of the K existing appearance models or generate a new model θ_{K+1}. Then the posterior distribution of $z_{i+1}, \theta_1, \cdots, \theta_{K+1}$ given $x_1, \cdots, x_{i+1}$ is

$$\begin{aligned} p(z_{i+1}, \theta_{1:K+1}|X_{1:i+1}) \propto \\ p(z_{i+1}, \theta_{1:K+1}|X_{1:i})p(X_{i+1}, z_{i+1}|\theta_{1:K+1}) \end{aligned} \tag{7}$$

Using the tractable distribution $q(\theta|\rho, \upsilon)$ to approximate the posterior $p(z_{i+1}, \theta_{1:K+1}|X_{1:i})$, we get the following:

$$p(z_{i+1}, \theta_{1:K+1}|X_{1:i+1}) \propto q(z_{i+1}|\rho_{1:i}, \upsilon^{(i)})p(X_{i+1}|z_{i+1}\theta_{1:K+1}) \tag{8}$$

Then, for our model, the optimal setting of q_{i+1} and $\upsilon^{(i+1)}$ minimizes the Kullback-Leibler divergence between $q(z_{i+1}, \theta_{1:K+1}|\rho_{1:i+1}, \upsilon^{(i+1)})$ and the approximate posterior in Eq. (8) are given as follows:

$$\rho_{i+1} \propto \begin{cases} \omega_k^{(i)}\int F(X_{i+1}|\theta)\upsilon_k^{(i)}(d\theta) & (k \leq K) \\ \alpha\int F(X_{i+1}|\theta)h(d\theta) & (k = K+1), \end{cases} \tag{9}$$

with $\omega_k^{(i)} = \sum_{j=1}^{i} \rho_j(k)$, and

$$\upsilon_k^{(i+1)}(\theta) \propto \begin{cases} h(\theta)\prod_{j=1}^{i+1} F(X_j|\theta)\rho_j(k) & (k \leq K) \\ h(\theta)F(X_{i+1}|\theta)\rho_{i+1}(k) & (k = K+1) \end{cases} \tag{10}$$

Algorithm 1 illustrates the basic flow of our algorithm. More details can be found in [5]. The implementation of this algorithm is under the circumstance where H and F are exponential family distributions that form a conjugate pair. In such cases, base measure h and posterior measures υ_k can be represented by natural parameter denoted by λ and ζ_k.

Algorithm 2. A Summary of the proposed tracking method

---

(1)$L_t(X) \in R_2$ denotes the location of sample X at the t-th frame. We have the object location $L_t(X)$ where we assume the corresponding sample is X_t representing the quantized HOG feature.

(2)We apply the affine transformation to $L_t(X)$ with six affine parameters to product candidate samples s_t.

(3)For each candidate samples s_t, we extract quantized HOG featureX_t, then use NOR model of Eq. 13 and each of the multiple appearance models $zeta^k$ to compute the likelihood of X_t.

(4)We select the state s_t which has maximum probability of X_t.

(5)Let X_t represents the quantized HOG feature of the target at frame t, and then use Algorithm 1 to update parameters of multiple appearance models online in a streaming fashion.

---

6   Proposed Tracking Algorithm

Given the observation set of the target $X_{1:t} = [X_1, \ldots, X_t]$ up to time t, where each X_t represents a quantized HOG target feature at time t, the target state s_t(motion parameter set) can be determined by the maximum a posteriori(MAP) estimation as follows:

$$\hat{s}_t = argmax \; p(s_t | X_{1:t}) \tag{11}$$

where $p(s_t | X_{1:t})$ can be inferred by the Bayesian theorem in a recursive manner (with Markov assumption)

$$p(s_t | X_{1:t}) \propto p(X_t | s_t) p(s_t | X_{1:t-1}) \tag{12}$$

where $p(s_t | X_{1:t-1}) = \int p(s_t | s_{t-1}) p(s_{t-1} | X_{1:t-1}) ds_{t-1}$. The tracking process is governed by a dynamic model, i.e. $p(s_t | s_{t-1})$, and an observation model, i.e. $p(X_t | s_t)$.

A particle filter method [15] is adopted here to estimate the target state. In the particle filter, $p(s_t | X_{1:t})$ is approximated by a finite set of samples with important weights. Let $s_t = [l_x, l_y, \theta, s, \alpha, \phi]$, where $l_x, l_y, \theta, s, \alpha, \phi$ denote x, y translations, rotation angle, scale, aspect ratio, and skew respectively. We approximate the motion of a target between two consecutive frames with affine transformation. The state transition is formulated as $p(s_t | s_{t-1}) = N(s_t; s_{t-1}, \sum)$ where $\sum$ is the covariance matrix of six affine parameters. The observation model $p(X_t | s_t)$ denotes the likelihood of the observation X_t at state s_t. The Noisy-OR (NOR) [17] model is adopted for doing this:

$$p(X_t | s_t) = 1 - \prod_k (1 - p(X_t | s_t, \zeta^k)) \tag{13}$$

where $\zeta^k, k \in (1, 2, \ldots, K)$ represents the multiple appearance model parameters learned from Algorithm 1. The equation above has the desired property that if one of the appearance models has a high probability, the resulting probability will be high as well. Algorithm 2 illustrates the basic flow of our tracking algorithm.

Figure 2 shows how the online adaptive multiple appearances model working. These small face images show the appearance instance belong to each appearance model and the historical instances while tracking. The red rectangle in main frame is the tracking result based on our proposed model, and the green one is the ground truth. With the new tracking observation arrival, the number of clusters became varies.

7   Experiments

To evaluate our tracker, we compared the proposed tracker with 10 latest algorithms using 10 challenging public tracking datasets introduced by [20]. When evaluating the trackers, there are several problems should be discussed. We followed the evaluation methods from [20]. As object tracking is a traditional problem in computer vision, these trackers have quite different frameworks, so that all of them have advantage and disadvantage when meeting different challenges like occlusion and etc. Table 1 shows all the trackers (including our proposed algorithm) and their features and models. Note that in our proposed algorithm the HOG feature can be replaced by other features.

Table 1. Compare trackers and their representations in our experiment [20]

Trackers	Features	Models
LOT [22]	C	L
IVT [9]	PCA	H
ASLA [23]	SR	L, GM
L1ANG [25]	SR	H, GM
MTT [28]	SR	H, GM
VTD [24]	SPCA	H, GM
OAB [26]	Haar	H, DM
MIL [21]	Haar	H, DM
TLD [28]	BP	L, DM
Struck [27]	Haar	H, DM
AMAM [12]	Optional	DPMM
OAMAM	Optional	H, DPMM

One thing to emphasis is that all the trackers are running with adjusted parameters or simply use the parameters given by their publication for fair evaluation.

As mentioned before, a tracker might face tons of problems listed below in a real usage. According to the [20,29], we divided these variation into six groups and analyzed some datasets by using this division. In the Table 2, we also add a short form of each challenge on each datasets. Here, the OCC stands for Occlusion, IV stands for Illumination Variation, R stands for Rotation which contains in-place rotation and out-of-place rotation, SV stands for Scale Variation while BC stands for Background Clutters.

One general problem for tracking is that the object may be occluded by other objects for several seconds. While in the dataset *Bolt*, the main object Bolt just kept the sportsman near him out in some of the frames and this will lead trackers to track on the sportsman near Bolt.

Table 2. Datasets and their problems

Dataset	Problems
CarDark	IV, BC
David2	R
Car4	IV, SV
Trellis	IV, SV, R, BC
Singer1	IV, SV, OCC, R, BC
Singer2	IV, R, BC
Bolt	OCC, R
Crossing	SV, R, BC
MountainBike	R, BC
Dog1	SV, R

This method we proposed didn't limited any certain kind of features for tracking task. Better features can get better tracking results. We simply applied HOG feature to implement.

It's common to use Center Location Error (CE [20]) and Overlap Score (OS [29]), to estimate the performance of the tracker. OS is calculated by the formula $score = \frac{area(ROI_T \cap ROI_G)}{area(ROI_T \cup ROI_G)}$. In the experiment, the $area(ROT_T)$ is the area of bounding box of tracking, and the $area(ROT_G)$ is the area of the ground truth. The CE is the Euclidean distance between the centers of tracking bounding box and the ground truth.

7.1 Online AMAM vs. Original AMAM [12]

In the previous sections, we compared our new proposed tracking method with AMAM tracking method [12]. As online method benefits the predicting speed on a long-term object, we compared these two methods in the time consumption. Figure 4(a) illustrates the DPMM time consumption of each frame in *Trellis* for both OAMAM method and AMAM method. It's obvious that AMAM method has a quite unaffordable time cost tracking for a long time while our online method performs relatively stable.

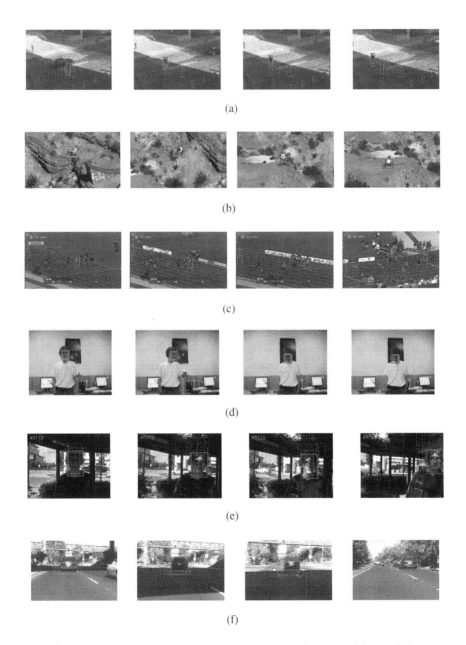

Fig. 3. All the images above are tracking results by trackers in Table 1 and dataset in Table 2, in which the bounding boxes in red are our results. (Color figure online)

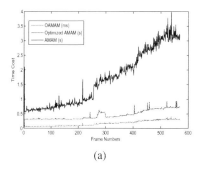 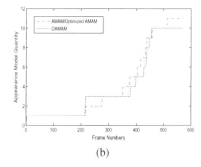

(a)   (b)

Fig. 4. (a) Is the DPMM time cost of each frame in *Trellis*, where OAMAM using *ms* and the other is *s*. (b) Is the quantity of appearance models of each frame when processing in *Trellis*.

Besides the time cost, they have a slightly difference in forming appearance models during the tracking task. Figure 4(b) shows the amount of appearance models for both methods in every frame in dataset *Trellis*.

7.2   Qualitative Comparison

Our tracker has a robust performance while solving different challenges in different video sequences. Typical background problem can be seen in *MountainBike*, *Crossing*. In the Fig. 3 in *Crossing*, when a car was passing by the pedestrian, they shared similar dark colors in the frame 31 and result in the ASLA, Struck, and TLD's failure in tracking. In the frame 73 to frame 85 the target pedestrian blurred himself with the dark shade and only Struck, MIL and our tracker catched the target successfully(even Struck failed to track the pedestrian in the frame 31). In *MountainBike*, our tracker still performed well while the target was on the grass or dark shade in frame 62(VTD lost the target entirely from this frame), frame 150, frame 199, and frame 225. During the whole period of these two video sequences, our tracker tracked the target perfectly and constantly performed better than other trackers.

At the same time, there are view port varying problem in *Bolt* and rotation challenge in *David*2. In *Bolt*, the view port of the camera varied three times. It firstly lied in frame 97, as shown in the Fig. 3 Bolt was running towards the camera. The Second variation lied in frame 137 while Bolt was running parallel to the camera. The third variation lied in frame 252 while Bolt running away from the camera. Most of the trackers lost the target at the first stage, except four were still catching the Bolt. Only three trackers tracked Bolt successfully at the second stage. At the last stage, only our tracker was still working. In *David*2, there were abundant in-plane rotations and out-of-plane rotations. During the out-of-plane rotation(from the frame 79 to 115), half of the trackers had high CE rate even they did not lost the target.

In *Trellis* and *Car*4, there are significant illumination variations. In *Trellis*, The illumination of target varied from all dark to half dark during frame 139 to frame 213, and changed to bright in the frame 230. All the bounding box of these frames is shown in Fig. 3. In the frame 282 we could clearly find that only two trackers (ours and MIL) succeed in tracking the target while others drifted away because of the dark background. In *Car*4, the video sequences undergo serious illumination changes when the vehicle ran through a tunnel or under trees. At the frame 182, most of trackers performed well except two trackers fail to track the vehicle. But in the frame 207, 6 trackers enlarged its bounding box and drift away in frame 233 while the vehicle ran outside the tunnel. After the frame 490 and passed several trees and billboards, only 4 trakers including our tracker, MIL, ASLA and VTD were succeed in tracking target, and only our tracker didn't falsely enlarge its bounding box comparing to the ground truth.

We employed the protocol above to finish a comparison and analyzed all the data after evaluating. By adopting the OPE evaluation matrics, we compared the performance of trackers in all testing datasets with the same testing result shown in Fig. 5. From Fig. 5, we found that our tracking method outperforms state-of-art in the OPE evaluation on these 10 datasets. In the plot we can also infer that our tracking method is approximately 22 % better than the second best tracking method.

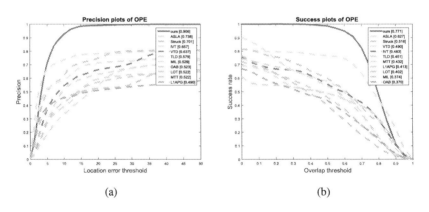

(a)   (b)

Fig. 5. Success rate and precision of eleven trackers versus different thresholds under different attributions on ten video sequences.

8   Conclusion

In this paper, we proposed a new online adaptive multiple appearances model for long-term tracking. This approach remained more historical information on appearances of tracking target to avoid the target drifting or lost during the tracking caused by varying illumination or pose changing. We employed HOG to build the basic appearance representation of the tracking target in our algorithm framework. Multiple appearances representations were grouped unsupervised and dynamically by an online sequential approximation BNP learning method. Tracking result can be selected from candidate targets, which were predicted by trackers based on those appearance models, by using Noisy-OR method. Experiments on public datasets show that, our tracker has low variation (less than 0.002), low time cost for real-time tracking, and high tracking performance (22 % better than other 10 trackers in average) when compared to the state-of-the-art.

References

1. Li, Y., Ai, H., Yamashita, T., Lao, S., Kawade, M.: Tracking in low frame rate video: a cascade particle filter with discriminative observers of different life spans. IEEE Trans. Pattern Anal. Mach. Intell. **30**(10), 1728–1740 (2008)
2. An, S., An, D.: Stochastic relaxation, Gibbs distributions, the Bayesian restoration of images. IEEE Trans. Pattern Anal. Mach. Intell. **6**, 721–741 (1984)
3. Blei, D.M., Jordan, M.I.: Variational inference for dirichlet process mixtures. Bayesian Anal. **1**(121–144), 1 (2005)
4. Stauffer, C., Grimson, W.E.L., Adaptive background mixture models for real-time tracking. In: IEEE Computer Society Conference on Computer Vision and Pattern Recognition, vol. 2, pp. 246–252. IEEE, Los Alamitos, August 1999
5. Lin, D.: Online Learning of nonparametric mixture models via sequential variational approximation. In: Proceedings of NIPS (2013)
6. Ulker, Y., Gunsel, B., Cemgil, A.T.: Sequential Monte Carlo samplers for Dirichlet process mixtures. In: AISTATS 2010 (2010)
7. Cannons, K.: A review of visual tracking. Department of Computer Science, York Univ., Toronto, ON, Canada, Technical report CSE-2008–07 (2008)
8. Jepson, A.D., Fleet, D.J., El-Maraghi, T.F.: Robust online appearance models for visual tracking. IEEE Trans. Pattern Anal. Mach. Intell. **25**(10), 1296–1311 (2003)
9. Ross, D., Lim, J., Lin, R.-S., Yang, M.-H.: Incremental learning for robust visual tracking. IJCV **77**(1–3), 125–141 (2008)
10. Avidan, S.: Ensemble tracking. PAMI **29**(2), 261–271 (2007)
11. Babenko, B., Yang, M.-H., Belongie, S.: Visual tracking with online multiple instance learning. In: CVPR, pp. 983–990 (2009)
12. Tang, S., Zhang, L., Chi, J., Wang, Z., Ding, G.: Adaptive multiple appearances model framework for long-term robust tracking. In: Ho, Y.-S., Sang, J., Ro, Y.M., Kim, J., Wu, F. (eds.) PCM 2015. LNCS, vol. 9314, pp. 160–170. Springer, Heidelberg (2015). doi:10.1007/978-3-319-24075-6_16
13. Lowe, D.G.: Object recognition from local scale-invariant features. In: Proceedings of ICCV, pp. 1150–1157 (1999)
14. Ferguson, T.: A Bayesian analysis of some nonparametric problems. Ann. Stat. **1**(2), 209–230 (1973)
15. Isard, M., Blake, A.: CONDENSATION-conditional density propagation for visual tracking. Int. J. Comput. Vis. **29**(1), 5–28 (1998)
16. Pitman, J.: Combinatorial Stochastic Processes. Lecture Notes in Mathematics, vol. 1875. Springer, Heidelberg (2006)
17. Viola, P., Platt, J.C., Zhang, C.: Multiple instance boosting for object detection. In: Proceeding of Neural Information Processing Systems, pp. 1417–1426 (2005)
18. Bernardo, J.M., Smith, A.F.M.: Bayesian Theory. Wiley, New York (1994)
19. Bissacco, A., Yang, M., Soatto, S.: Detecting humans via their pose. Adv. Neural Inf. Process. Syst. **19**, 169 (2007)

20. Wu, Y., Lim, J., Yang, M.: Online Object Tracking: a benchmark. In: Proceeding IEEE Conference on Computer Vision and Pattern Recognition (CVPR), pp. 2411–2418 (2013)
21. Babenko, B., Yang, M., Belongie, S.: Robust object tracking with online multiple instance learning. IEEE Trans. Pattern Anal. Mach. Intell. (TPAMI) **33**(8), 1619–1632 (2011)
22. Oron, S., Bar-Hillel, A., Levi, D., Avidan, S.: Locally orderless tracking. In: CVPR (2012)
23. Jia, X., Lu, H., Yang, M.: Visual tracking via adaptive structural local sparse appearance model. In: CVPR (2012)
24. Kwon, J., Lee, K.M.: Visual tracking decomposition. In: CVPR (2010)
25. Bao, C., Wu, Y., Ling, H., Ji, H.: Real time robust L1 tracker using accelerated proximal gradient approach. In: CVPR (2012)
26. Grabner, H., Grabner, M., Bischof, H.: Real-time tracking via online boosting. In: BMVC (2006)
27. Hare, S., Golodetz, S., Saffari, A., Vineet, V., Cheng, M.M., Hicks, S., Torr, P.: Struck: structured output tracking with kernels. In: ICCV (2011)
28. Zhang, T., Ghanem, B., Liu, S., Ahuja, N.: Robust visual tracking via multi-task sparse learning. In: CVPR (2012)
29. Everingham, M., Van Gool, L., Williams, C.K.I., et al.: The Pascal visual object classes (VOC) challenge. Int. J. Comput. Vis. **88**(2), 303–338 (2010)

Multi-stream Deep Networks for Person to Person Violence Detection in Videos

Zhihong Dong, Jie Qin[✉], and Yunhong Wang

Laboratory of Intelligent Recognition and Image Processing,
State Key Laboratory of Virtual Reality Technology and Systems,
School of Computer Science and Engineering, Beihang University, Beijing, China
dzhwinter@gmail.com, qinjiebuaa@gmail.com, yhwang@buaa.edu.cn

Abstract. Violence detection in videos has numerous applications, ranging from parental control and children protection to multimedia filtering and retrieval. A number of approaches have been proposed to detect vital clues for violent actions, among which most methods prefer employing trajectory based action recognition techniques. However, these methods can only model general characteristics of human actions, thus cannot well capture specific high order information of violent actions. Therefore, they are not suitable for detecting violence, which is typically intense and correlated with specific scenes. In this paper, we propose a novel framework, i.e., multi-stream deep convolutional neural networks, for person to person violence detection in videos. In addition to conventional spatial and temporal streams, we develop an acceleration stream to capture the important intense information usually involved in violent actions. Moreover, a simple and effective score-level fusion strategy is proposed to integrate multi-stream information. We demonstrate the effectiveness of our method on the typical violence dataset and extensive experimental results show its superiority over state-of-the-art methods.

Keywords: Violence detection · Acceleration feature · Convolutional neural networks · Long short-term memory

1 Introduction

With the rapid development of digital media, massive collections of video materials have become ubiquitous online. Detecting different types of human actions has a wide range of applications. Among various applications, for the reason of protecting children against offensive video contents and providing people the ability of content-based video filtering or retrieval, detecting violent actions in videos has recently received considerable attentions. Violence detection poses big challenges to the computer vision community. On one hand, because of the subjective nature, one may have an ambiguous concept of violence in definition. Here, we adopt the common definition from VSD [1], i.e., physical violence or accident resulting in human injury or pain. On the other hand, violence detection in surveillance videos always turns into the crowd scene analysis problem.

© Springer Nature Singapore Pte Ltd. 2016
T. Tan et al. (Eds.): CCPR 2016, Part I, CCIS 662, pp. 517–531, 2016.
DOI: 10.1007/978-981-10-3002-4_43

In this paper, we are specifically interested in content based person to person violence detection at a relatively short distance in videos.

To address the above problem, previous researchers prefer employing trajectory-based action recognition techniques [2,3,11]. Conventional approaches often follow the standard bag-of-words pipeline for representing general human actions. Specifically, they first extract several types of features of entire videos, then quantize features into histograms using k-means clustering, VLAD [29] or Fisher Vector [19]. The key step of these methods is extracting proper features to model human actions. For instance, improved dense trajectory [26] extracts Motion Boundary Histogram (MBH), Histogram of Oriented Gradients (HOG), and Histogram of Optical Flows (HOF) as feature descriptors for every trajectory. Trajectory-based methods have been proved to be effective for action recognition problems. Since violence detection is closely related to action recognition, a lot of recent violence detection methods [10,23] have been developed by using the trajectory-based framework. And those methods have been proved to be useful on some datasets for the violence detection task. Besides, some methods based on Optical Flow orientations can handle some specific fight scenes effectively, because they can capture the correlation between people. In addition to the above methods, others combine color and HOG features to discover more useful clues for violent actions.

However, these methods, which are based on traditional action recognition approaches, are not completely suitable for the task of violence detection and have the following limitations. Firstly, these methods lack high order discriminate information of neighbor frames, e.g., the velocity change over time (i.e. acceleration) has not been exploited yet. Secondly, some violence happens with high degree of confidence in specific scenes. For instance, supposing people holding iron bars are confronting with others in the street and hitting with each other, this violence is tightly correlated with scenes, such as cold arms. Their intense actions, such as kick and hit, contain the acceleration in formation, which contributes significantly to the final detection results. In addition, trajectory-based methods can only detect short-term information of actions, while semantic violent actions are usually long-term actions.

Recently, it is worth noting that learning from raw data with deep neural networks has been receiving more and more attentions. Convolutional neural networks (ConvNets) based approaches [6,7,13,21,24,31] have yielded the state of the art performance in many tasks, such as semantic segmentation, image caption, action classification and recognition. Specifically, there are two classic frameworks for action classification/recognition, namely two-stream ConvNets [21] and 3D CNN (C3D) [24]. They have been receiving massive attentions on action classification/recognition applications. In [24], C3D is proposed by replacing 2D convolution filters with 3D cube filters, and treating the temporal dimension in the same way as the spatial dimension. Thus, C3D can gain action information inherently over the time axis. The two-stream framework [21] extracts spatial stream and temporal stream respectively. The spatial stream captures still image information of video frames and the temporal stream extracts stacked

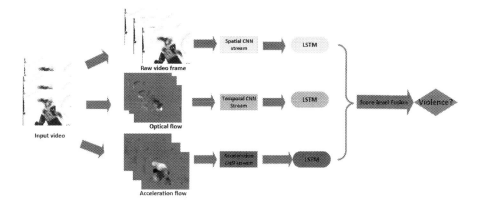

Fig. 1. The overall framework of the proposed multi-stream deep neural networks. Given a video sequence, by using different types of features or raw data as input, we aim to automatically judge whether the video is violent or not.

optical flow information to get short-term movement information. Based on the two-stream framework, several more recent works [30] utilize Long Short Term Memory (LSTM) network to model long-term action information.

The above deep frameworks have demonstrated their advantages over hand-crafted feature representations. However, they are specially developed for general action recognition problems, thus not appropriate for the particular task of violence detection. As aforementioned, two-stream ConvNets and C3D encode videos into deep features, which cannot capture the particular characteristic of violence without any specific side information. For instance, supposing there are two videos of football matches, in which one involves fight inside, and the other not. Based on the above analysis, it would be difficult to train such deep networks to learn specific discriminative action features for violence detection. Additionally, deep networks require massive training data to achieve satisfactory performance, which cannot be satisfied in terms of violence videos.

To overcome the above shortcomings, in this paper, we aim at enhancing conventional two-stream ConvNets with additional acceleration information to capture more violent clues. Specifically, we propose a three-stream ConvNets framework shown in Fig. 1, which integratedly incorporates spatial, temporal and dynamic information for person-person violence detection. A novel feature based on the acceleration of actions is first proposed to capture the dynamic and intense information, which is of vital importance for detecting violence. Subsequently, we combine spatial, temporal and dynamic streams together to generate the proposed multi-stream ConvNets. Finally, a score-level fusion strategy is employed for the final detection. The main contributions of this paper are listed as follows:

- We propose a novel feature descriptor based on the dynamic information, which extracts the acceleration feature map from raw video data, for detecting intense violent actions.

– We develop a three-stream deep neural networks framework for person to person violence detection, which adopts the Long Short Term Memory on top of three streams to model long-term temporal information. We show that all the three streams (i.e., spatial, temporal and acceleration streams) can work well with LSTM and are complementary to each other.
– Through extensive experiments on the typical violence dataset, we demonstrate that our proposed framework outperforms the state-of-the-art methods.

The remainder of the paper is organized as follows. We give a brief review of previous violence detection methods in Sect. 2. In Sect. 3, we introduce the proposed multi-stream ConvNets framework, including framework structure and components. Section 4 introduces the datasets and experimental setup, and demonstrates the experimental results. We finally draw our conclusions in Sect. 5.

2   Related Work

Since violent actions are essentially similar to highly intense human actions, conventional human action recognition approaches have been explored a lot for violence detection. The popular trajectory based method was employed by M. Sjoberg et al. [4] to model violent scenes. Specifically, spatial temporal trajectories were described by Histograms of Oriented Gradients and further improved with color histograms. Jiang et al. [11] combined Spatial Temporal Interest Points (STIP) and the SIFT feature to capture sparse interest points in videos, which were then quantized to represent the violent action. Dai et al. [2] proposed to integrate STIP with Part-level Attribute features. The Attribute features were learned from deformable part-based models, aiming to find certain objects and scenes in videos. Later, Dai et al. [3] improved their feature with Trajectory Shape [12] (i.e., HOF, HOG and MBH) to represent human actions in videos.

As aforementioned, these trajectory-based methods have some limitations, i.e., losing the scene and dynamic information of violence. Although Matt and Dai attempted to take the scene information into account, local feature descriptors restricted the ability for holistic representations. In addition, Derbas et al. explored the optical flow features as the average velocity to detect violent videos. Deniz et al. [17] proposed a SIFT variant called MoSIFT to detect violence actions, which compared SIFT corners between consecutive frames. Martin et al. [14] proposed a violence detector based on the dynamics of new multi-scale local binary histogram features.

3   The Proposed Method

We propose a novel architecture for deep convolutional neural networks, consisting of three individual streams, which can respectively extract different types of violence information from raw videos. Figure 2 shows the overall framework of the proposed deep neural networks. Specifically, three components/streams are developed for different purposes. The spatial stream captures correlations

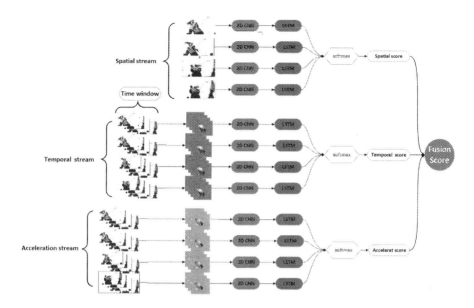

Fig. 2. The proposed framework of deep convolutional neural networks, which consists of three streams, namely spatial stream, temporal stream, and acceleration stream.

between violent actions and scenes; the temporal stream extracts short-term action information; the acceleration stream discovers intense clues of violence. Each of them goes through the same Long Short Term Memory network respectively to acquire the long-term temporal information and score-level fusion is adopted for integrating the three streams to predict the final confidence score for violence videos.

In the following, we will first elaborate each of the three streams respectively. Subsequently, the LSTM network is briefly introduced and multi-stream fusion strategy is present for final detection.

3.1 Acceleration Stream

Person to person violence always involves fights and leads to physical injuries. To capture the intense dynamic information of violent actions, we propose a novel feature descriptor called acceleration flow and form the acceleration flow feature map as input for the acceleration stream of the overall deep neural networks.

Acceleration Flow. When dealing with actions, two important elements are velocity and acceleration respectively. In terms of violence, we are especially interested in the acceleration, since acceleration is the changing rate of velocity over time and can well represent the intensity of violent actions.

Optical flow can successfully represent the velocity between any consecutive frames. It computes each pixel with forward and backward corresponding pixels, then generates optical flow feature maps for horizontal (x) and vertical (y) directions respectively. We assume that the frame order is identical to the time dimension. Obviously optical flow is the first-order derivatives of original images over the time axis, which models the velocity of movement.

We aim to keep tracking velocity variations for interest points over time in both directions. Given one point (x, y) at time t, the velocity field at this moment is denoted as $w = (u(x, y, t), v(x, y, t))$. We assume the movements in horizontal, vertical and temporal directions are denoted as Δx, Δy, and Δt, respectively. Then the increment of velocity field can be computed as $w = (u(x + \Delta x, y + \Delta y, t + \Delta t), v(x + \Delta x, y + \Delta y, t + \Delta t))$. Due to the small movement in every direction during Δt, we ignore the high order terms and the equation at point $p_t = (x, y, t)$ in time t can be expanded with Taylor series as follows:

$$u(p_{t+\Delta t}) = u(p_t) + \frac{\partial u}{\partial x}\Delta x + \frac{\partial u}{\partial y}\Delta y + \frac{\partial u}{\partial t}\Delta t \tag{1}$$

$$v(p_{t+\Delta t}) = v(p_t) + \frac{\partial v}{\partial x}\Delta x + \frac{\partial v}{\partial y}\Delta y + \frac{\partial v}{\partial t}\Delta t \tag{2}$$

To measure the change of velocity over time, we derive the following equations with regard to the time dimension:

$$\frac{du}{dt} = \frac{\partial u}{\partial x}\frac{dx}{dt} + \frac{\partial u}{\partial y}\frac{dy}{dt} \tag{3}$$

$$\frac{dv}{dt} = \frac{\partial v}{\partial x}\frac{dx}{dt} + \frac{\partial v}{\partial y}\frac{dy}{dt} \tag{4}$$

where $(\frac{dx}{dt}, \frac{dy}{dt})$ corresponds to (u, v), which represents the horizontal and vertical components of the velocity. Then, the acceleration vector $a = (a_x, a_y)$ in the velocity field can be calculated as:

$$a = (a_x, a_y) = (w\Delta u, w\Delta v). \tag{5}$$

We model the acceleration information using the second orders of original neighbor frames, which describe the velocity changing speed over time. The acceleration can well represent the action intensity, which is helpful for distinguishing violence from normal actions. As Fig. 3 shows, both optical flow feature map and acceleration information can capture interesting areas where violent actions occur. However, acceleration flow feature maps emphasize action intensity and generate a more accurate area. An acceleration flow map is computed based on three consecutive frames. In the whole framework, acceleration flow serves as an individual stream, which captures the action intensity information for detecting more clues of violence.

3.2 Spatial Stream

Obviously, key frames of videos have high correlations with its content. In most situations, human can recognize violent actions based one still images. Scene

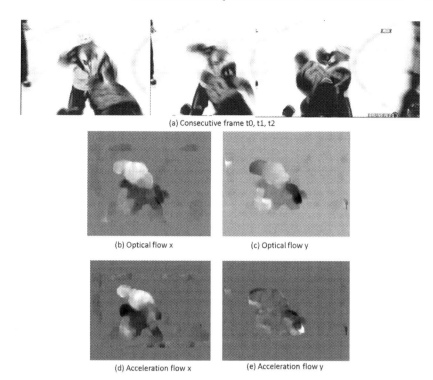

(a) Consecutive frame t0, t1, t2

(b) Optical flow x   (c) Optical flow y

(d) Acceleration flow x   (e) Acceleration flow y

Fig. 3. Example feature maps of optical flow and acceleration flow from $t0$ to $t2$ of the fight. x and y denote horizontal and vertical flows, respectively.

information, such as indoor or outdoor, object information and person-to-person distance could help us identify whether violence accidents happen or not. 2D CNN has revealed the significant abilities for feature extraction in many applications such as semantic segmentation [18], human tracking [9]. In this framework, we adopt 2D CNN in the spatial stream to extract the correlation between violence and scene.

In practice, we adopt the VGG19 model which is pre-trained on the ImageNet dataset and consists of 16 convolution-pool-norm layers and three fully connected layers. Taking an image as input, we adopt the feature map from the second last layer as the image feature representation. Specifically, the output of each video frame is a 4096-dimensional feature vector.

Note that the spatial stream has some limitations. For instance, when human body deformation is caused by movement or people are sheltering each other, it is hard to recognize the action in such scenarios. Besides, some actions such as punch or kick can only be recognized by scanning through the whole actions. In other words, we need information in the time dimension to classify human actions. Particularly, we denote the time dimension as T. Given a video containing T frames, 2D CNN generates feature maps with shape $(T, 4096)$ as the representation of a single video. Then the extracted maps go through the LSTM network.

3.3   Temporal Stream

In order to detect violence in videos, we model human actions to find more clues. Recently, two-stream ConvNets has yielded the state-of-the-art performance in action classification. As aforementioned, optical flow computes first order of original images with regard to the time axis, which has been used to model short motion information between consecutive frames. In the two-stream framework, optical flow images are stacked together to extract the temporal information for actions. In our temporal stream, we follow the idea of two-stream ConvNets. As shown in later experimental results in Sect. 5, it can model the temporal information for violent actions very well.

Similar to the spatial stream, we follow the pipeline of extracting video level representation. Specifically, given a video with F frames, we first compute optical flow in each consecutive frame and obtain horizontal (x) and vertical (y) optical flow values. In order to make different optical flow feature maps comparable between frames, we normalize the values into the range of $[0, 255]$. Subsequently, we stack flow-x, flow-y as one multi-channel image. Here we denote the stacked frames count as time window L. In this paper, we choose $L = 10$ in this temporal stream, which means F consecutive frames video will generate $T = F - L$ stacked feature map images, and every stacked multi-channel image has the same spatial size with video frame and its channel size is $2L$.

We take pre-trained TDD [27] motion networks to construct the temporal stream, which is pre-trained on UCF101 [22]. The network consists of 5 pairs of convolution-pool layers and 3 fully connected layers. We take stacked multi-channel optical flow images as input. By following the forward step in the network, we take the second last layer feature maps as final representations. Given an input video, temporal representation is extracted in the shape of $(T, 4096)$, then LSTM network follows up in the same way as the spatial stream.

3.4   Modeling Long-Term Information with LSTM

Neither the above three streams, i.e., spatial, temporal and acceleration streams, can well model the whole video-level action information. Stacked optical flow and acceleration feature maps can only model short-term movements in time window of certain length. As the framework shows, all the streams represent videos in different aspects. We then employ the Long Short Term Memory (LSTM) network to model long-term action information based on the outputs of three streams. Since most violent actions need long time windows to perform completely, LSTM can model the long-term action clues effectively.

Specifically, the LSTM architecture contains two layers. In our experiments, the first layer takes 1024 LSTM units and the second layer consists of 512 LSTM units. A softmax layer is then employed to compute the probability in terms of violence.

LSTM Network. Long Short Term Memory (LSTM) network is a variant of recurrent neural networks, which has been successfully employed in many tasks,

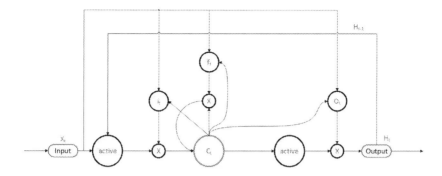

Fig. 4. A diagram of an LSTM memory cell.

such as image caption [25], speech recognition [20, 28], text and speech translation [15, 16]. Generally, LSTM is designed for capturing important information for sequential data. The structure of one LSTM unit is illustrated in Fig. 4. The middle controllable memory cell is stateful, which can choose to store predecessor state or clear its content. The controller consists of three gates, which are called input gate, output gate and forget gates, respectively. Stacking LSTM units from layer to layer will form an LSTM Model. Generally, LSTM recursively maps the input representations at current time step to output labels, and every LSTM unit memory cell may contain current content inside, which depends on the controller's state. The built-in memory cell stores the information over time to exploit long range dynamics. Furthermore, non-linear gate controller unit governs the information flows, which enhances the ability of modelling the non-linear feature of sequential data.

In our framework, we denote x_t as the feature representation of video frame, i.e., stacked optical flow, or stacked acceleration flow at time order of t-th step. Generally, an LSTM maps an input sequence of $(x_1, x_2, \ldots, x_T)$ into $(y_1, y_2, \ldots, y_T)$ following the frame or stacked image order from $t = 1$ to T. At time t, the output gate o_t, memory cell c_t, and hidden state h_t are computed as follows:

$$c_t = f_t \odot c_{t-1} + i_t \odot \tanh(W_{xc}x_t + W_{hc}h_{t-1} + b_c) \tag{6}$$

$$o_t = \sigma(W_{xo}x_t + W_{ho}h_{t-1} + W_{co}c_t + b_o) \tag{7}$$

$$h_t = o_t \odot \tanh(c_t) \tag{8}$$

In the above equations, W_{xc}, W_{hc}, W_{xo}, W_{ho}, W_{co} represent weight matrix connecting LSTM units; b_c and b_o denote the bias terms; $\odot$ denotes an element-wise product operator; σ is the sigmoid activation function; the following i_t and f_t are the input and forget gates respectively, which are calculated as follows:

$$i_t = \sigma(W_{xt}x_t + W_{hi}h_{t-1} + W_{ci}c_{t-1} + b_t) \tag{9}$$

$$f_t = \sigma(W_{xf}x_t + W_{hf}h_{t-1} + W_{cf}c_{t-1} + b_f) \tag{10}$$

The key component in the LSTM unit is memory cell content, which is passed to the hidden state h_t to influence the computation in the next time step.

In terms of the LSTM model, layer l takes the output of layer $l-1$ as input. On the top of stacked forward layers, a softmax layer is placed on the last LSTM layer, which will generate the posterior probability p_c of $c-th$ class with the following value:

$$p_c = soft \max(h_t^L) = \frac{\exp(u_c^T h_t^L + b_c)}{\sum_{c' \in C} \exp(u_c^T h_t^L + b_c)} \quad (11)$$

where p_c denotes the probability for $c-th$ category, and u_c and b_c represent the corresponding weight vector. The LSTM network is trained by the Back-Propagation Through Time (BPTT) algorithm [8], which unrolls the model into a feed forward network and adjusts the parameters in the network.

In our framework, we adopt the output of last layer as the final output of our LSTM model since it contains all the temporal information of the entire video sequence. At the training phase, three LSTM models are trained for each stream individually. Loss is computed with the video label separately, and we fine-tune the parameters to achieve better performance. At the test phase, we use score-level fusion scheme to generate the fused label for testing videos, which will be elaborated in the following.

3.5   Multi-stream Fusion

We propose the score-level fusion scheme for our framework. For each video, we generate three independent streams and put them through the LSTM network respectively. In this way, the model of every single stream is trained individually and can predict a degree of confidence in the range [0, 1]. Specifically, the max-pooling method is used when predicting scores for videos. The final score will be compared with the ground-truth video label. We also do complementary experiments based on average-pooling and fusion at the feature level. Our empirical results show that using individual model and max-pooling based fusion at the score level is a better choice. We show that different models focus on different clues of violence. For instance, in terms of the violent action 'two men boxing in the ring', the spatial stream may capture the information of ring and predict a high rate of violence; the temporal stream catches the information of large area movement and may consider this as a violent event; the acceleration stream models their long arm boxing actions and measures its intensity, and also identifies this as the violent scene. However, these three streams describe the same action from different perspectives and the fusion scheme can discover the complementary information among them and further lead to better performance.

4   Experiments and Results

4.1   Dataset and Experimental Setup

The violence scene varies from conventional scenes, and some may contain cold arms and bloody content. Due to the subjective nature of violence and the

(a) Normal action

(b) Violence action

Fig. 5. Sample videos from the Hockey Fights dataset: (a) normal action and (b) violence action.

difficulty of collecting and annotating adequate materials, there are only few publicly available datasets. In spite of this, the Violence Scene Detection (VSD) dataset [1] and Hockey Fights dataset [17] have been extensively discussed in the literature. The VSD dataset, originally designed for movies violence and targeting rating movie automatically, contains videos specially edited for film artistic expression, which is quite different from our targeting problem. On the contrary, the Hockey Fights dataset comes from National Hockey League matches and contains different viewpoints of person-to-person violence, which is exactly the focus of this paper. Therefore, we evaluate our framework on the Hockey Fights dataset.

Specifically, Hockey Fights contains 1,000 videos and the whole duration lasts approximately two hours, with half of them containing violence inside. Following previous works, we split the dataset into 5 fold for cross validation. Some sample videos are shown in Fig. 5. We can see that the video is very challenging, since it contains severe occlusions and large variations.

In terms of the experimental settings, we set time window L as 10 for stacked images. On the top of three streams, we set the LSTM model with learning rate as 0.01, momentum decay over time as 0.0001. We utilize Stochastic Gradient Descent (SGD) for optimization. We compare the proposed framework with several state-of-the-art methods on this dataset, i.e., STIP, MoSIFT, improved Dense Trajectory and 3D CNN. Accuracies and precision-recall curves are shown to demonstrate the performance of different methods.

4.2   Results and Analysis

We show the comparison results with the state-of-the-art methods in terms of accuracy in Table 1. From the table, we can see that the proposed method consistently outperforms other hand-crafted features, such as improved Dense

528   Z. Dong et al.

Table 1. Comparison results with the state of the art on the Hockey Fights dataset.

Methods	Accuracy
STIP(HOG) + Bag of visual words [17]	89.1 %
STIP(HOF) + Bag of visual words [17]	88.6 %
MoSIFT + Bag of visual words [17]	90.6 %
Improved dense trajectory + Bag of visual words [26]	91.0 %
3D CNN [5]	91.0 %
Three streams + LSTM (Proposed)	93.9 %

Trajectory or STIP. This is because those kinds of hand-crafted feature based methods mainly focus on modeling general actions, which are lack of capturing the special scene clues correlated with violent actions. Additionally, our multi-steam deep networks also have the advantage over 3D CNN, which shows our superiority over other deep networks based methods in terms of violence detection.

We also show the performance of different network frameworks containing different numbers of streams in Fig. 6. From the figure, we can see that even single-stream networks in our framework can outperform hand-crafted feature based methods, which shows the powerful representation ability of combination of 2D CNN and LSTM models. Our proposed acceleration stream performs much better than the other two single-stream frameworks, which shows the effectiveness of the proposed acceleration feature. This is mainly because the acceleration flow feature can explicitly represent violent actions in interesting areas. In addition, due to the ability of discovering the complementary scene information, there

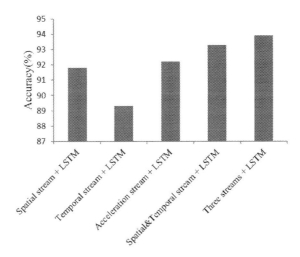

Fig. 6. Performance comparisons of single-stream and multi-stream networks.

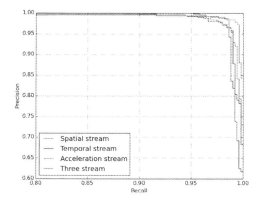

Fig. 7. Precision-recall curves of single-stream and multi-stream networks.

is a slight improvement of the single spatial stream, compared with the single temporal stream. It is interesting that the single temporal stream only obtains 89.3 % in accuracy, which is slightly worse than the improved dense trajectory method. This is partially because it is lack of human body information like Histogram of Oriented Gradients and may confuse people with moving objects. Note that fusing spatial and temporal streams can lead to better performance than each single stream, which demonstrates that the two streams are complementary to each other. Undoubtedly, the proposed three-stream framework achieves the best performance among all variants, proving the effectiveness of our fusion strategy.

Finally, we show the precision-recall curves of different single-stream frameworks, which are illustrated in Fig. 7. From the figure, it can be seen that our proposed acceleration flow feature achieves higher recall rates than the other two single-stream alternatives. It is reasonable that lack of action intensity may lead to high false alarm rates. Furthermore, combining the streams together could always achieve the best performance.

5   Conclusions

In this paper, we have proposed a novel multi-stream deep neural networks framework for person-to-person violence detection in videos. The proposed framework addresses the problem of detecting intensive violent actions. We have exploited the acceleration feature to generate more useful clues for violence. The proposed method has been systematically evaluated on the typical dataset and the results demonstrate the superiority of our method over state-of-the-art methods. In future work, it is worth considering exploring the integration of trajectory-based methods and deep neural networks to further enhance the performance.

Acknowledgment. This work was supported by the Hong Kong, Macao and Taiwan Science Technology Cooperation Program of China (No. L2015TGA9004), and the National Natural Science Foundation of China (No. 61573045).

References

1. Claire-Heilene, D.: VSD, a public dataset for the detection of violent scences in movies: design, annotation, ananlysis and evaluation. In: The Handbook of Brain Theory and Neural Networks, vol. 3361 (1995)
2. Dai, Q., Tu, J., Shi, Z., Jiang, Y.G., Xue, X.: Fudan at mediaeval 2013: violent scenes detection using motion features and part-level attributes. In: MediaEval (2013)
3. Dai, Q., Wu, Z., Jiang, Y.G., Xue, X., Tang, J.: Fudan-NJUST at mediaeval 2014: violent scenes detection using deep neural networks. In: MediaEval (2014)
4. Demarty, C.-H., Penet, C., Gravier, G., Soleymani, M.: A benchmarking campaign for the multimodal detection of violent scenes in movies. In: Fusiello, A., Murino, V., Cucchiara, R. (eds.) ECCV 2012. LNCS, vol. 7585, pp. 416–425. Springer, Heidelberg (2012). doi:10.1007/978-3-642-33885-4_42
5. Ding, C., Fan, S., Zhu, M., Feng, W., Jia, B.: Violence detection in video by using 3D convolutional neural networks. In: Bebis, G., et al. (eds.) ISVC 2014. LNCS, vol. 8888, pp. 551–558. Springer, Heidelberg (2014). doi:10.1007/978-3-319-14364-4_53
6. Donahue, J., Anne Hendricks, L., Guadarrama, S., Rohrbach, M., Venugopalan, S., Saenko, K., Darrell, T.: Long-term recurrent convolutional networks for visual recognition and description. In: Proceedings of the IEEE Conference on Computer Vision and Pattern Recognition, pp. 2625–2634 (2015)
7. Girshick, R., Donahue, J., Darrell, T., Malik, J.: Rich feature hierarchies for accurate object detection and semantic segmentation. In: Proceedings of the IEEE Conference on Computer Vision and Pattern Recognition, pp. 580–587 (2014)
8. Graves, A., Schmidhuber, J.: Framewise phoneme classification with bidirectional LSTM and other neural network architectures. Neural Netw. **18**(5), 602–610 (2005)
9. Hahn, M., Chen, S., Dehghan, A.: Deep tracking: visual tracking using deep convolutional networks (2015). arXiv preprint arXiv:1512.03993
10. Ionescu, B., Schlüter, J., Mironica, I., Schedl, M.: A naive mid-level concept-based fusion approach to violence detection in hollywood movies. In: Proceedings of the 3rd ACM International Conference on Multimedia Retrieval, pp. 215–222. ACM (2013)
11. Jiang, Y.G., Dai, Q., Tan, C.C., Xue, X., Ngo, C.W.: The Shanghai-Hongkong team at mediaeval 2012: violent scene detection using trajectory-based features. In: MediaEval (2012)
12. Jiang, Y.-G., Dai, Q., Xue, X., Liu, W., Ngo, C.-W.: Trajectory-based modeling of human actions with motion reference points. In: Fitzgibbon, A., Lazebnik, S., Perona, P., Sato, Y., Schmid, C. (eds.) ECCV 2012. LNCS, vol. 7576, pp. 425–438. Springer, Heidelberg (2012). doi:10.1007/978-3-642-33715-4_31
13. Karpathy, A., Toderici, G., Shetty, S., Leung, T., Sukthankar, R., Fei-Fei, L.: Large-scale video classification with convolutional neural networks. In: IEEE Conference on Computer Vision and Pattern Recognition (CVPR) (2014)
14. Martin, V., Glotin, H., Paris, S., Halkias, X., Prevot, J.M.: Violence detection in video by large scale multi-scale local binary pattern dynamics. In: MediaEval. Citeseer (2012)

15. Mohamed, A.r., Seide, F., Yu, D., Droppo, J., Stolcke, A., Zweig, G., Penn, G.: Deep bi-directional recurrent networks over spectral windows. In: ASRU (2015)

16. Na, S.H.: Deep learning for natural language processing and machine translation (2015)

17. Bermejo Nievas, E., Deniz Suarez, O., Bueno García, G., Sukthankar, R.: Violence detection in video using computer vision techniques. In: Real, P., Diaz-Pernil, D., Molina-Abril, H., Berciano, A., Kropatsch, W. (eds.) CAIP 2011. LNCS, vol. 6855, pp. 332–339. Springer, Heidelberg (2011). doi:10.1007/978-3-642-23678-5_39

18. Raj, A., Maturana, D., Scherer, S.: Multi-scale convolutional architecture for semantic segmentation (2015)

19. Sánchez, J., Perronnin, F., Mensink, T., Verbeek, J.: Image classification with the fisher vector: theory and practice. Int. J. Comput. Vis. **105**(3), 222–245 (2013)

20. Schmidhuber, J.: Deep learning in neural networks: an overview. Neural Netw. **61**, 85–117 (2015)

21. Simonyan, K., Zisserman, A.: Two-stream convolutional networks for action recognition in videos. In: Advances in Neural Information Processing Systems, pp. 568–576 (2014)

22. Soomro, K., Zamir, A.R., Shah, M.: Ucf101: A dataset of 101 human actions classes from videos in the wild (2012). arXiv preprint arXiv:1212.0402

23. de Souza, F.D., Chávez, G.C., do Valle, E.A., Araujo, A.D.: Violence detection in video using spatio-temporal features. In: 2010 23rd SIBGRAPI Conference on Graphics, Patterns and Images (SIBGRAPI), pp. 224–230. IEEE (2010)

24. Tran, D., Bourdev, L., Fergus, R., Torresani, L., Paluri, M.: Learning spatiotemporal features with 3D convolutional networks (2014). arXiv preprint arXiv:1412.0767

25. Vinyals, O., Toshev, A., Bengio, S., Erhan, D.: Show and tell: a neural image caption generator. In: Proceedings of the IEEE Conference on Computer Vision and Pattern Recognition, pp. 3156–3164 (2015)

26. Wang, H., Schmid, C.: Action recognition with improved trajectories. In: Proceedings of the IEEE International Conference on Computer Vision, pp. 3551–3558 (2013)

27. Wang, L., Qiao, Y., Tang, X.: Action recognition with trajectory-pooled deep-convolutional descriptors. In: Proceedings of the IEEE Conference on Computer Vision and Pattern Recognition, pp. 4305–4314 (2015)

28. Weninger, F., Bergmann, J., Schuller, B.: Introducing current: the Munich open-source CUDA recurrent neural network toolkit. J. Mach. Learn. Res. **16**(1), 547–551 (2015)

29. Wu, J., Zhang, Y., Lin, W.: Towards good practices for action video encoding. In: Proceedings of the IEEE Conference on Computer Vision and Pattern Recognition, pp. 2577–2584 (2014)

30. Wu, Z., Wang, X., Jiang, Y.G., Ye, H., Xue, X.: Modeling spatial-temporal clues in a hybrid deep learning framework for video classification. In: Proceedings of the 23rd Annual ACM Conference on Multimedia Conference, pp. 461–470. ACM (2015)

31. Yue-Hei Ng, J., Hausknecht, M., Vijayanarasimhan, S., Vinyals, O., Monga, R., Toderici, G.: Beyond short snippets: deep networks for video classification. In: Proceedings of the IEEE Conference on Computer Vision and Pattern Recognition, pp. 4694–4702 (2015)

Basic Theory of Pattern Recognition

1-Norm Projection Twin Support Vector Machine

Rui Yan, Qiaolin Ye[✉], Dong Zhang, Ning Ye, and Xiaoqian Li

School of Information Science and Technology, Nanjing Forestry University,
Nanjing, China
{1951657338,65620868}@qq.com

Abstract. In this paper, we propose a novel feature selection method which can suppress the input features automatically. We first introduce a Tikhonov regularization term to the objective function of projection twin support vector machine (PTSVM). Then we convert it to a linear programming (LP) problem by replacing all the 2-norm terms in the objective function with 1-norm ones. Then we construct an unconstrained convex programming problem according to the exterior penalty (EP) theory. Finally, we solve the EP problems by using a fast generalized Newton algorithm. In order to improve performance, we apply a recursive algorithm to generate multiple projection axes for each class. To disclose the feasibility and effectiveness of our method, we conduct some experiments on UCI and Binary Alphadigits data sets.

Keywords: Projection twin support vector machine · Twin support vector machine · Unconstrained convex programming · Suppression of input features

1 Introduction

The conventional support vector machine (SVM [1]) requires solving quadratic programming problems (QPPs), and the computational complexity is not more than O (N^3) (N is sample dimension). Solving the QPP is extremely time-consuming, which limits the application of conventional support vector machine to large-scale problems. In order to solve the problem of high computing cost of the conventional SVM. In 2001, Mangasafian et al. proposed the approximate support vector machine proximal (PSVM [2]). In order to solve the optimization problem of conventional SVM quickly, this algorithm fit samples with two parallel planes by solving the equations instead of the quadratic programming problems. In 2006, Mangasarian and Wild improved the performance of PSVM by relaxing its parallel constraint and proposed Multisurface proximal support vector machine (GEPSVM [3]), which solve the problem by calculating two generalized eigenvalues. However, GEPSVM cannot guarantee the excellent classification performance on solving singularity problem. That same year, Twin support vector machine (TWSVM [4]) was presented based on GEPSVM by Jayadeva et al. This algorithm does not solve the eigenvalue like GEPSVM, but convert the GEPSVM problem to two smaller SVM-type problems. Finally, TWSVM is as competitive as the conventional SVM in terms of performance whereas the former is around four times faster than the latter. TWSVM inherited the advantage of GEPSVM in

© Springer Nature Singapore Pte Ltd. 2016
T. Tan et al. (Eds.): CCPR 2016, Part I, CCIS 662, pp. 535–550, 2016.
DOI: 10.1007/978-981-10-3002-4_44

solving XOR problem, but still needs solving quadratic programming problem result in being slower than GEPSVM. Recently, many methods based on TWSVM were proposed to improve performance. In 2013, Qi et al. proposed Structural Twin Support Vector Machine (S-TWSVM) [13]. In 2016, Chen et al. proposed multi-label twin support vector machine (MLTSVM) [14].

To avoid solving the quadratic programming problem, the multi-weight vector projection support vector machine (MVSVM [5]) was proposed based on GEPSVM which is a novel algorithm to seek one weight vector instead of a hyper-plane for each class. The weight vectors of MVSVM can be found by solving a pair of eigenvalue problems, avoiding the complex solution of quadratic programming problem. In 2011, Chen et al. proposed the projection twin support vector machine (PTSVM [6]) inspired by MVSVM and TWSVM. PTSVM seeks the projection directions by solving a pair of SVM-type problems rather than eigenvalue problems. Some new methods based on PTSVM such as projection twin support vector machine with regularization term (RPTSVM) [15] and Least Squares Projection Twin Support Vector Machine (LSPTSVM) [16] were proposed. However, these new methods based PTSVM are still as slow as TWSVM on account of solving quadratic programming problems. Similar to GEPSVM, TWSVM and MVSVM, PTSVM also needs using all input features, which limits it to solve high-dimensional classification problems.

As we known, feature suppression is essential for some classification problem in high-dimensional spaces. Recently, many feature selection methods for SVM such as LPNewton [9], NELSVM [17], FLSTSVM [18] were proposed to suppress features by 1-norm. However, the performance of LPNewton which is based on traditional SVM is poor; NELSVM and FLSTSVM which are based on TWSVM cannot get satisfactory results in features suppression. Inspired by these methods, we replace 2-norm terms in PTSVM by 1-norm terms to suppress features. Traditionally, we solve the LP problem in 1-norm SVM by standard LP packages which have millions of constraints. Furthermore, it is infeasible to optimize 1-norm SVM problem due to its complex optimization structures. NELSVM achieves favorable results in features suppression by using the exterior penalty (EP) theory [8, 9].

In this paper, we propose a novel feature selection method based on PTSVM, termed as feature selection based on projection twin support vector machine (FPTSVM). At first, we introduce a Tikhonov regularization term [7] to the objective function of PTSVM. Due to the regularization term in the objective function, we can improve the generalization ability by avoiding the matrix singularity in PTSVM. Secondly, we give the linear programming form of PTSVM by measuring the distance of the point to the sample mean in 1-norm. Then we convert the LP problem to a completely unconstrained quadratic convex minimization by using the exterior penalty (EP) theory. Finally, we solve the EP problem by using a generalized Newton method, and prove that these optimization problems are solvable. Our method has two advantages: (1) It has better classification performance than PTSVM, and can suppress input features; (2) It has high speed due to the using of a generalized Newton method. These advantages are especially important in high dimensional classification.

2   Brief Reviews of TSVM and PTSVM

Suppose that we have a training data set of m points in n-dimensional input space $\mathbf{R}^n$, which consists of the two subsets $\left\{ (\mathbf{x}_j^{(i)}, y_j) \right\} \in \mathbf{R}^n \times \{-1, 1\}$, $i, j = 1, .., m_i$ where $\mathbf{x}_j^{(i)}$ denotes the jth training samples belonging to Class i and $m_1 + m_2 = m$. We further denote the m_1 samples of Class 1 by a $m_1 \times n$ matrix $\mathbf{A}$ and m_2 samples of Class 2 by a $m_2 \times n$ matrix $\mathbf{B}$.

2.1   Twin Support Vector Machine

The aim of TWSVM is to seek two nonparallel planes in n-dimensional input space, each plane is generated such that is closest to the samples of one class and be away from the other class as far as possible. We give the formulation of TWSVM1:

$$\mathbf{x}^T \mathbf{w}_1 + b_1 = 0, \ \mathbf{x}^T \mathbf{w}_2 + b_2 = 0 \tag{1}$$

For TWSVM, each nonparallel plane is generated by solving a SVM-type problem. The formulation of TWSVM can be written as follows:

$$\begin{aligned} \text{Min} \quad & 0.5 \| \mathbf{A} \mathbf{w}_1 + \mathbf{e}_1 b_1 \|^2 + C_1 \mathbf{e}_2^T \boldsymbol{\xi} \\ \text{s.t.} \quad & -(\mathbf{B} \mathbf{w}_1 + \mathbf{e}_2 b_1) + \boldsymbol{\xi} \geq \mathbf{e}_2, \ \boldsymbol{\xi} \geq 0. \end{aligned} \tag{2}$$

where $C_1 \geq 0$ is a penalty constant, where $\mathbf{e}_1$ and $\mathbf{e}_2$ are a column vector of ones of appropriate dimensions, superscript T denotes transposition. The objective function measure by 2-norm, and ensure that it is closest to the samples of Class 1. Optimization problem (2) is a typical SVM formulation. The conventional method is to convert it to a dual problem.

$$\begin{aligned} \text{Max} \quad & \mathbf{e}_2^T \mathbf{a} - 0.5 \mathbf{a}^T \mathbf{G} (\mathbf{H}^T \mathbf{H})^{-1} \mathbf{G}^T \mathbf{a} \\ \text{s.t.} \quad & 0 \leq \mathbf{a} \leq C_1 \end{aligned} \tag{3}$$

where $\mathbf{H} = [\mathbf{A} \ \mathbf{e}_1]$, $\mathbf{G} = [\mathbf{B} \ \mathbf{e}_2]$. Generally speaking $\mathbf{G}(\mathbf{H}^T \mathbf{H})^{-1} \mathbf{G}^T$ and $\mathbf{H}(\mathbf{G}^T \mathbf{G})^{-1} \mathbf{H}^T$ is positive semi-definite matrix $(m \gg n)$. Therefore, the TWSVM problem is not a strong convexity and needs to care about the matrix singularity. The TWSVM is infeasible to solve large-scale classification problem due to the QPPs.

2.2   Projection Twin Support Vector Machine

The goal of PTSVM is to seek a projection axis for each class, such that the projected samples are closest to own class meanwhile be away from the other class as far as possible. We give the formulation of PTSVM1:

$$\text{Min } 0.5 \sum_{i=1}^{m_1} (\boldsymbol{w}_1^T \boldsymbol{x}_i^{(1)} - \boldsymbol{w}_1^T 1/m_1 \sum_{j=1}^{m_1} \boldsymbol{x}_j^{(1)})^2 + C_1 \sum_{k=1}^{m_2} \xi_k$$
$$\text{s.t. } -(\boldsymbol{w}_1^T \boldsymbol{x}_k^{(2)} - \boldsymbol{w}_1^T 1/m_1 \sum_{j=1}^{m_1} \boldsymbol{x}_j^{(1)}) + \xi_k \geq 1, \xi_k \geq 0, k = 1, 2, \ldots, m_2 \tag{4}$$

where C_1 is a penalty constant, and ξ_k is a non-negative slack variable. Minimizing the first term in the objective function of Eq. (4) can ensure the projected samples of Class 1 be clustered around its mean. The second term reduces the hinge loss. The constraints in Eq. (4) aim at maximizing the separation between the two classes. Equation (4) can be written as follows:

$$\text{Min } 0.5 \| \boldsymbol{A}\boldsymbol{w}_1 - 1/m_1 \boldsymbol{e}_1 \boldsymbol{e}_1^T \boldsymbol{A}\boldsymbol{w}_1 \|^2 + C_1 \boldsymbol{e}_2^T \boldsymbol{\xi}$$
$$\text{s.t. } -(\boldsymbol{B}\boldsymbol{w}_1 - 1/m_1 \boldsymbol{e}_2 \boldsymbol{e}_1^T \boldsymbol{A}\boldsymbol{w}_1) + \boldsymbol{\xi} \geq \boldsymbol{e}_2, \boldsymbol{\xi} \geq 0 \tag{5}$$

write down the Lagrange function of the Eq. (5), then converted it to a dual problem:

$$\text{Max } \boldsymbol{\beta}^T \boldsymbol{e}_2 - 0.5 \boldsymbol{\beta}^T \left[\boldsymbol{G}(\boldsymbol{H}^T \boldsymbol{H})^{-1} \boldsymbol{G}^T \right] \boldsymbol{\beta} \tag{6}$$

where $\boldsymbol{H} = \boldsymbol{A} - 1/m_1 \boldsymbol{e}_1 \boldsymbol{e}_1^T \boldsymbol{A}$, $\boldsymbol{G} = \boldsymbol{B} - 1/m_1 \boldsymbol{e}_2 \boldsymbol{e}_1^T \boldsymbol{A}$. Similar to (3), $\boldsymbol{G}(\boldsymbol{H}^T \boldsymbol{H})^{-1} \boldsymbol{G}^T$ and $\boldsymbol{H}(\boldsymbol{G}^T \boldsymbol{G})^{-1} \boldsymbol{H}^T$ is a positive semi-definite matrix. This problem is still not strong convexity and needs to solve two QPPs. In this paper, we avoid the matrix singularity by introducing regularization term.

3 Feature Selection Based on PTSVM

Different from TWSVM, PTSVM seeks a projection direction rather than a hyper-plane for each class. It can achieve better performance on complex Xor problem than TWSVM. However, PTSVM still needs to solve two QPPs and cannot deal with the singularity problem.

3.1 Model of FPTSVM

In order to avoid the singularity problem in PTSVM, we first introduce a Tikhonov regularization term into the primal problems of PTSVM. Then we replace all the 2-norm terms with 1-norm ones to improve the training speed of PTSVM and suppress the features of input space. We give two formulations of FPTSVM as follows:

FPTSVM1:

$$\text{Min } \| \boldsymbol{A}\boldsymbol{w}_1 - 1/m_1 \boldsymbol{e}_1 \boldsymbol{e}_1^T \boldsymbol{A}\boldsymbol{w}_1 \|_1 + \varepsilon \|\boldsymbol{w}_1\|_1 + C \boldsymbol{e}_2^T \boldsymbol{\xi}$$
$$\text{s.t. } -(\boldsymbol{B}\boldsymbol{w}_1 - 1/m_1 \boldsymbol{e}_2 \boldsymbol{e}_1^T \boldsymbol{A}\boldsymbol{w}_1) + \boldsymbol{\xi} \geq \boldsymbol{e}_2, \boldsymbol{\xi} \geq 0 \tag{7}$$

FPTSVM2:

$$\text{Min} \quad \left\| Bw_2 - 1/m_2 e_2 e_2^T Bw_2 \right\|_1 + \varepsilon \|w_2\|_1 + Ce_1^T \xi$$
$$\text{s.t.} \quad Aw_2 - 1/m_2 e_1 e_2^T Bw_2 + \xi \geq e_1, \xi \geq 0 \tag{8}$$

It is essential for FPTSVM to introduce regularization term. When we remove the regularization term, FPTSVM becomes the 1-norm PTSVM problem which cannot ensure sparse solution due to the variable w_i is composed of the class i sample linearly. The two linear programming problems can be easily solved by using the exterior penalty theory (EP [8, 9]). Given a linear programming problem (LP):

$$\text{Min} \quad a^T x + d^T y$$
$$\text{s.t.} \quad Ax + By \geq b, \ Ex + Gy = f, \ x \geq 0, \tag{9}$$

where $a \in R^n$, $d \in R^l$, $A \in R^{m \times n}$, $B \in R^{m \times l}$, $E \in R^{k \times n}$, $G \in R^{k \times l}$, $b \in R^m$ and $f \in R^k$. The dual of (9) can be obtained as follows:

$$\text{Max} \qquad b^T u + f^T v$$
$$\text{s.t.} \quad A^T u + E^T v \leq a, B^T u + G^T v = d, u \geq 0, \tag{10}$$

The EP problem of the dual problem (10) can be reduced to be (11).

$$\text{Min} \ \mu(-b^T u - f^T v) + 0.5(\|(A^T u + E^T v - a)_+\|^2$$
$$+ \|B^T u + G^T v - d\|^2 + \|(-u)_+\|^2), \tag{11}$$

where $\| \cdot \|$ represent 2-norm, $z_+ = \max(0, z)$, $\mu > 0$. In [8, 9], we can obtain the following lemma.

Lemma 1: If the primal LP (9) is solvable, then the dual EP problem (11) is solved for all $\mu > 0$. For any $\mu \in (0, \overline{\mu}]$ ($\overline{\mu} > 0$), an exact solution with respect to any solution x and y to the primal LP (9) can be obtained as follows:

$$x = 1/\mu(A^T u + E^T v - a)_+ \quad y = 1/\mu(B^T u + G^T v - d)$$

The problem (11) is an unconstrained quadratic programming problem, then the solution given by Lemma 1 can be obtained by the generalized Newton method [10] with less computing cost. In combination with the exterior penalty theory, we can solve FPTSVM easily by using Lemma 1. We now begin with the analysis of FPTSVM. First, discuss the model of FPTSVM1:

$$H = A - \frac{1}{m_1} e_1 e_1^T A, \ G = B - \frac{1}{m_1} e_2 e_1^T A$$
$$w_1 = p_1 - q_1, Hw_1 = r_1 - s_1 \tag{12}$$

Substituting all the equations in (12) into (7) results in the following LP problem:

(FPTSVM1)

$$\min_{p_1,q_1,r_1,s_1,\xi} \quad e_1^T(r_1+s_1)+\varepsilon e^T(p_1+q_1)+Ce_2^T\xi$$
$$\text{s.t.} \quad -G(p_1-q_1)+\xi \geq e_2, \tag{13}$$
$$H(p_1-q_1)=r_1-s_1, p_1, q_1, r_1, s_1, \xi \geq 0.$$

where $e \in R^{n+1}$. However, for large-scale problems, it is not feasible to solve the problem (13) by using the traditional linear programming method. We can solve it by combining the exterior penalty theory. First of all, to solve dual form of problem (13), the corresponding Lagrange function is

$$L(r_1,s_1,p_1,q_1,\xi) = e_1^T(r_1+s_1)+\varepsilon e^T(p_1+q_1)+Ce_2^T\xi - \beta_1^T(-G(p_1-q_1)+\xi-e_2)$$
$$- \beta_2^T(H(p_1-q_1)-r_1+s_1)-\beta_3^T r_1 - \beta_4^T s_1 - \beta_5^T p_1 - \beta_6^T q_1 - \beta_7^T \xi \tag{14}$$

where $\beta_i, i=1,2,3,\ldots,7$ are non-negative vectors of the Lagrange multiplier. We differentiate the primal variables (r_1,s_1,p_1,q_1,ξ) and equate them to zero for (14) resulting in

$$e_1+\beta_2-\beta_3=0, e_1-\beta_2-\beta_4=0, Ce_2-\beta_1-\beta_7=0.$$
$$\varepsilon e+G^T\beta_1-H^T\beta_2-\beta_5=0, \varepsilon e-G^T\beta_1+H^T\beta_2-\beta_6=0, \tag{15}$$

Substituting all the equations in (15) into Lagrange function (14) results in the following Wolfe dual problem of primal LP (13):

$$\text{Max} \quad \beta_1^T e_2 \tag{16}$$
$$\text{s.t.} \quad -\varepsilon e \leq G^T\beta_1 - H^T\beta_2 \leq \varepsilon e, -e_1 \leq \beta_2 \leq e_1, 0 \leq \beta_1 \leq Ce_2.$$

Optimization problem (16) contains two Lagrange multipliers, is not easy to solve by using the exterior penalty theory directly. For convenience of computation, we combine β_1 and β_2 to solve the vector $u = \begin{bmatrix} \beta_1 \\ \beta_2 \end{bmatrix}$ as the target:

By defining

$$E = \begin{bmatrix} e_2 \\ O_2 \end{bmatrix}, L = \begin{bmatrix} G \\ -H \end{bmatrix}, S_1 = \begin{bmatrix} O_1 \\ e_1 \end{bmatrix}, S_2 = \begin{bmatrix} Ce_2 \\ e_1 \end{bmatrix} \tag{17}$$

where $O_2 \in R^{m_1}$, $O_1 \in R^{m_2}$, are two zero vectors. The dual problem (16) can be obtained as follows:

$$\text{Max} \quad E^T u \tag{18}$$
$$\text{s.t.} \quad -\varepsilon e \leq L^T u \leq \varepsilon e, -S_1 \leq u \leq S_2.$$

The EP problem of the dual problem (18) can be reduced to be (19).

$$
\text{Min } f(\boldsymbol{u}) = -\mu \boldsymbol{E}^{\mathrm{T}} \boldsymbol{u} + 0.5 \|(\boldsymbol{L}^{\mathrm{T}} \boldsymbol{u} - \varepsilon \boldsymbol{e})_{+}\|^2 + 0.5 \|(-\boldsymbol{L}^{\mathrm{T}} \boldsymbol{u} - \varepsilon \boldsymbol{e})_{+}\|^2
$$
$$
+ 0.5 \|(\boldsymbol{u} - \boldsymbol{S}_2)_{+}\|^2 + 0.5 \|(-\boldsymbol{u} - \boldsymbol{S}_1)_{+}\|^2 \tag{19}
$$

Inspired by Lemma 1, the solution of (19) can be determined.

Theorem 1: If the primal LP (16) is solvable, then the dual EP problem (19) is solved for all $\mu > 0$. For any $\mu \in (0, \bar{\mu}]$ $(\bar{\mu} > 0)$, an exact solution with respect to $\boldsymbol{p}, \boldsymbol{q}, \boldsymbol{r}$ and $\boldsymbol{s}$ to the primal LP (16) can be obtained as follows:

$$
\boldsymbol{p}_1 = 1/\mu(-\boldsymbol{L}^{\mathrm{T}} \boldsymbol{u} - \varepsilon \boldsymbol{e})_{+}, \; \boldsymbol{q}_1 = 1/\mu(\boldsymbol{L}^{\mathrm{T}} \boldsymbol{u} - \varepsilon \boldsymbol{e})_{+},
$$
$$
\boldsymbol{r}_1 = 1/\mu((-\boldsymbol{u} - \boldsymbol{S}_1)_{+})_{m_1}, \boldsymbol{s}_1 = (1/\mu(\boldsymbol{u} - \boldsymbol{S}_2)_{+})_{m_1}, \boldsymbol{\xi} = (1/\mu(\boldsymbol{u} - \boldsymbol{S}_2)_{+})_{m_2} \tag{20}
$$

where, $\boldsymbol{c}_{m_1} = (\boldsymbol{c})_{m_1}$ denotes the m_1 rows of c, and $\boldsymbol{c}_{m_2} = (\boldsymbol{c})_{m_2}$ denotes the m_2 rows of c. Further, we obtain the solution with respect to $\boldsymbol{w}_1$:

$$
\boldsymbol{w}_1 = 1/\mu(-\boldsymbol{L}^{\mathrm{T}} \boldsymbol{u} - \varepsilon \boldsymbol{e})_{+} - 1/\mu(\boldsymbol{L}^{\mathrm{T}} \boldsymbol{u} - \varepsilon \boldsymbol{e})_{+} \tag{21}
$$

Proof: The solutions of (20) are existent.

By defining

$$
\boldsymbol{\eta}_1 = (\boldsymbol{L}^{\mathrm{T}} \boldsymbol{u} - \varepsilon \boldsymbol{e})_{+}, \boldsymbol{\eta}_2 = (-\boldsymbol{L}^{\mathrm{T}} \boldsymbol{u} - \varepsilon \boldsymbol{e})_{+},
$$
$$
\boldsymbol{\eta}_3 = (\boldsymbol{u} - \boldsymbol{S}_2)_{+}, \boldsymbol{\eta}_4 = (-\boldsymbol{u} - \boldsymbol{S}_1)_{+}. \tag{22}
$$

then the EP problem of FPTSVM1 (19) is equivalent to:

$$
\text{Min} \quad -\mu \boldsymbol{E}^{\mathrm{T}} \boldsymbol{u} + 0.5(\|\boldsymbol{\eta}_1\|^2 + \|\boldsymbol{\eta}_2\|^2 + \|\boldsymbol{\eta}_3\|^2 + \|\boldsymbol{\eta}_4\|^2)
$$
$$
\text{s.t.} \quad -\boldsymbol{L}^{\mathrm{T}} \boldsymbol{u} + \varepsilon \boldsymbol{e} + \boldsymbol{\eta}_1 \geq 0, \; \boldsymbol{L}^{\mathrm{T}} \boldsymbol{u} + \varepsilon \boldsymbol{e} + \boldsymbol{\eta}_2 \geq 0, \tag{23}
$$
$$
-\boldsymbol{u} + \boldsymbol{S}_2 + \boldsymbol{\eta}_3 \geq 0, \; \boldsymbol{v} + \boldsymbol{S}_1 + \boldsymbol{\eta}_4 \geq 0.
$$

The Lagrange function of (23) as follows:

$$
L(\boldsymbol{u}, \boldsymbol{\eta}_1, \boldsymbol{\eta}_2, \boldsymbol{\eta}_3, \boldsymbol{\eta}_4) = -\mu \boldsymbol{E}^{\mathrm{T}} \boldsymbol{u} + 0.5(\|\boldsymbol{\eta}_1\|^2 + \|\boldsymbol{\eta}_2\|^2 + \|\boldsymbol{\eta}_3\|^2 + \|\boldsymbol{\eta}_4\|^2)
$$
$$
- \boldsymbol{\beta}_1^{\mathrm{T}}(-\boldsymbol{L}^{\mathrm{T}} \boldsymbol{u} + \varepsilon \boldsymbol{e} + \boldsymbol{\eta}_1) - \boldsymbol{\beta}_2^{\mathrm{T}}(\boldsymbol{L}^{\mathrm{T}} \boldsymbol{u} + \varepsilon \boldsymbol{e} + \boldsymbol{\eta}_2) \tag{24}
$$
$$
- \boldsymbol{\beta}_3^{\mathrm{T}}(-\boldsymbol{u} + \boldsymbol{S}_2 + \boldsymbol{\eta}_3) - \boldsymbol{\beta}_4^{\mathrm{T}}(\boldsymbol{u} + \boldsymbol{S}_1 + \boldsymbol{\eta}_4)
$$

Differentiate the primal variables $(\boldsymbol{u}, \boldsymbol{\eta}_1, \boldsymbol{\eta}_2, \boldsymbol{\eta}_3, \boldsymbol{\mu}_4)$ and equate them to zero gives the following K.K.T conditions:

$$
-\mu \boldsymbol{E} + \boldsymbol{L} \boldsymbol{\beta}_1 - \boldsymbol{L} \boldsymbol{\beta}_2 + \boldsymbol{\beta}_3 - \boldsymbol{\beta}_4 = 0,
$$
$$
\boldsymbol{\beta}_1 = \boldsymbol{\eta}_1, \boldsymbol{\beta}_2 = \boldsymbol{\eta}_2, \boldsymbol{\beta}_3 = \boldsymbol{\eta}_3, \boldsymbol{\beta}_4 = \boldsymbol{\eta}_4. \tag{25}
$$

Substituting all the equations in (25) into Lagrange function (24) results in the following Wolfe dual problem:

$$
\begin{aligned}
\text{Min} \quad & 0.5(\|\pmb{\eta}_1\|^2 + \|\pmb{\eta}_2\|^2 + \|\pmb{\eta}_3\|^2 + \|\pmb{\eta}_4\|^2) + \varepsilon e^{\mathrm{T}}\pmb{\beta}_1 + \varepsilon e^{\mathrm{T}}\pmb{\beta}_2 + S_2^{\mathrm{T}}\pmb{\beta}_3 + S_1^{\mathrm{T}}\pmb{\beta}_4 \\
\text{s.t.} \quad & -\mu E + L\pmb{\beta}_1 - L\pmb{\beta}_2 + \pmb{\beta}_3 - \pmb{\beta}_4 = 0, \\
& \pmb{\beta}_1 = \pmb{\eta}_1 \geq 0, \pmb{\beta}_2 = \pmb{\eta}_2 \geq 0, \pmb{\beta}_3 = \pmb{\eta}_3 \geq 0, \pmb{\beta}_4 = \pmb{\eta}_4 \geq 0.
\end{aligned} \tag{26}
$$

where

$$
\pmb{\beta}_3 = \pmb{\eta}_3 = (\pmb{u} - S_2)_+ = \begin{bmatrix} \pmb{\rho}_1 \\ \pmb{\rho}_2 \end{bmatrix}, \pmb{\rho}_1 = ((\pmb{u} - S_2)_+)_{m_2}, \pmb{\rho}_2 = ((\pmb{u} - S_2)_+)_{m_1},
$$

$$
\pmb{\beta}_4 = \pmb{\eta}_4 = (-\pmb{u} - S_1)_+ = \begin{bmatrix} \pmb{\rho}_3 \\ \pmb{\rho}_4 \end{bmatrix}, \pmb{\rho}_3 = ((-\pmb{u} - S_1)_+)_{m_2}, \pmb{\rho}_4 = ((-\pmb{u} - S_1)_+)_{m_1}. \tag{27}
$$

Expanding the first equation of (25), divide it by μ on both sides, then the following two linear equations can be obtained:

$$
-e_2 + 1/\mu(H\pmb{\eta}_1 - H\pmb{\eta}_2 + \pmb{\rho}_1) = 1/\mu\pmb{\rho}_3 \geq 0 \tag{28}
$$

$$
1/\mu(-G\pmb{\eta}_1 + G\pmb{\eta}_2 + \pmb{\rho}_2 - \pmb{\rho}_4) = 0 \tag{29}
$$

Also, the problem (26) can be written as follows:

$$
\begin{aligned}
\text{Min} \quad & 0.5(\|\pmb{\eta}_1\|^2 + \|\pmb{\eta}_2\|^2 + \pmb{\rho}_1^2 + \pmb{\rho}_2^2 + \pmb{\rho}_3^2 + \pmb{\rho}_4^2) + \\
& \varepsilon e^{\mathrm{T}}\pmb{\eta}_1 + \varepsilon e^{\mathrm{T}}\pmb{\eta}_2 + Ce_2^{T}\pmb{\rho}_1 + e_1^{T}\pmb{\rho}_2 + e_1^{T}\pmb{\rho}_4 \\
\text{s.t.} \quad & 1/\mu(H\pmb{\eta}_1 - H\pmb{\eta}_2 + \pmb{\rho}_1) \geq e_2, \pmb{\rho}_1, \pmb{\eta}_2, \pmb{\eta}_1 \geq 0, \\
& 1/\mu(-G\pmb{\eta}_1 + G\pmb{\eta}_2) = 1/\mu(\pmb{\rho}_2 - \pmb{\rho}_4), \ \pmb{\rho}_2, \pmb{\rho}_3, \pmb{\rho}_4, \geq 0.
\end{aligned} \tag{30}
$$

By defining

$$
\pmb{p}_1 = 1/\mu\pmb{\eta}_2, \pmb{q}_1 = 1/\mu\pmb{\eta}_1, \pmb{\xi} = 1/\mu\pmb{\rho}_1, \pmb{r}_1 = 1/\mu\pmb{\rho}_4, \pmb{s}_1 = 1/\mu\pmb{\rho}_2. \tag{31}
$$

Substituting all the equations in (31) into (30) leads to the following formulation:

$$
\begin{aligned}
\text{Min} \quad & \varepsilon e^{\mathrm{T}}(\pmb{p}_1 + \pmb{q}_1) + Ce_2^{T}\pmb{\xi} + e_1^{T}(\pmb{s}_1 + \pmb{r}_1) + 0.5\mu(\|\pmb{p}_1\|^2 + \|\pmb{q}_1\|^2 + \\
& \|\pmb{\xi}\|^2 + \|\pmb{s}_1\|^2 + \| - H(\pmb{p}_1 - \pmb{q}_1) + \pmb{\xi} - e_2\|^2 + \|\pmb{r}_1\|^2) \\
\text{s.t.} \quad & -H(\pmb{p}_1 - \pmb{q}_1) + \pmb{\xi} \geq e_2, \pmb{p}_1, \pmb{q}_1, \pmb{\xi} \geq 0, \\
& G(\pmb{p}_1 - \pmb{q}_1) = \pmb{r}_1 - \pmb{s}_1, \ \pmb{r}_1, \pmb{s}_1 \geq 0.
\end{aligned} \tag{32}
$$

Since the original problem (13) is feasible, the problem (32) is feasible, because the objective function is a strong convexity, so it is bounded, and (32) is solvable.

According to linear programming perturbation theory, the exact solution with respect to p, q, r and s to the primal LP (13) can be obtained.

Similar to FPTSVM1, we give the solution of FPTSVM2 as follows:

By defining

$$H = B - 1/m_2 e_2 e_2^T B, \; G = A - 1/m_2 e_1 e_2^T B$$
$$w_2 = p_2 - q_2, \; Hw_2 = r_2 - s_2$$

Similar to FPTSVM1, we can obtain the final solution $w_2 = 1/\mu(-L_2^T v - \varepsilon e)_+ - 1/\mu(L_2^T v - \varepsilon e)_+$, where

$$E_2 = \begin{bmatrix} e_1 \\ O_1 \end{bmatrix}, L_2 = \begin{bmatrix} H \\ G \end{bmatrix}, S_1 = \begin{bmatrix} O_2 \\ e_2 \end{bmatrix}, S_2 = \begin{bmatrix} Ce_1 \\ e_2 \end{bmatrix}, v = \begin{bmatrix} \beta_1 \\ \beta_2 \end{bmatrix}$$

Following [8, 9], we solve the EP problems (19) by using a generalized Newton method. Firstly, we denote the problem (19) by $f(u)$, and then give its gradient and Hessian:

$$\nabla f(u) = - E\mu - L(-L^T u - \varepsilon e)_+ + L(L^T u - \varepsilon e)_+ \qquad (33)$$
$$+ (u - S_2)_+ - (-u - S_1)_+$$

$$\partial^2 f(u) = L(\text{diag}((-L^T u - \varepsilon e)_* + (L^T u - \varepsilon e)_*)L^T \qquad (34)$$
$$+ \text{diag}((u - S_2)_* + (-u - S_1)_*)$$

where z_* is the gradient vector of $(z)_+$, which denotes the vector with members $(z_*)_i = 1$ if $z_i > 0$ or $(z_*)_i = 0$ if $z_i < 0$ or $(z_*)_i \in [0, 1]$ if $z_i = 0$.

Algorithm 1: Generalized Newton Method. Setting the parameter values ($tol = 10^{-3}$, $imax = 50$, C and ε can be obtained through cross-validation), The starting point is any $u^0 \in R^{m_1}$. $i = 0, 1, \ldots \ldots$

1. $u^{i+1} = u^i - \lambda_i (\partial^2 f(u_i) + \delta I)^{-1} \nabla f(u^i) = u^i + \lambda_i d^i$ where *Armijo* step length $\lambda_i = \max(1, 0.5, 0.25, \ldots)$, then

$$f(u^i) - f(u^i + \lambda_i d^i) \geq - 0.25\lambda_i \nabla f(u^i)^T d^i, \qquad (35)$$

where

$$d^i = -(\partial^2 f(u^i) + \delta I)^{-1} \nabla f(u^i), \qquad (36)$$

is the modified Newton direction.

2. If $\|u^i - u^{i+1}\| \leq tol$ or $i = imax$, then stop iterating. Otherwise $i = i + 1$, and go back to step 1.
3. And then comes to w_1.

For the $m \gg n$ classification problem, the Algorithm 1 has a $m \times m$ matrix inversion. Therefore, we only have a $(n+1) \times (n+1)$ matrix inversion by using Sherman-Morrison-Woodbury equation

$$
\begin{aligned}
Q^2 &= \mathrm{diag}((-L^\mathrm{T}u - \varepsilon e)_* + (L^\mathrm{T}u - \varepsilon e)_*, \\
F &= \mathrm{diag}((u - S_2)_* + (-u - S_1)_*) + \delta I, \ P = LQ.
\end{aligned}
\tag{37}
$$

Combined (35) can be obtained

$$
(\partial^2 f(u) + \delta I)^{-1} = F^{-1}(I - P(I + P^\mathrm{T}F^{-1}P)^{-1}P^\mathrm{T}F^{-1})
\tag{38}
$$

After the optimal projection axes are obtained by using **Generalized Newton Method**, the training stage of FPTSVM is completed. In testing stage, we define the label of a new sample x as

$$
label(x) = \arg\min_{i=1,2}\{d_i\} = \arg\min_{i=1,2}\left| w_i^T x - w_i^T 1/m_1 \sum_{j=1}^{m_i} x_j^{(i)} \right|
\tag{39}
$$

where $d_i = \left| w_i^T x - w_i^T 1/m_1 \sum_{j=1}^{m_i} x_j^{(i)} \right|$ is the projected distance between x and the mean of Class i.

3.2   FPTSVM for Multiple Orthogonal Projection Directions

Through the above steps, we can find a single projection axis for each class. To improve the performance of our method, we extend FPTSVM to seek multiple orthogonal directions. Similar to PTSVM, we apply a recursive algorithm to generate multiple projection axes for each class in FPTSVM. The recursive algorithm for FPTSVM works as follows:

Algorithm 2: Recursive algorithm for FPTSVM.

1. To initialize the iteration $t = 0$ and the training sets $S_1(t) = S_2(t) = \{x_i | i = 1, 2, \ldots, m.\}$
2. To obtain the optimal projection direction $w_1(t)$ and $w_2(t)$ by solving the pair of QPPs (14) and (15) on sets $S_1(t)$ and $S_2(t)$, respectively.
3. To normalize $w_1(t)$ and $w_2(t)$ by setting $w_1(t) = w_1(t)/\|w_1(t)\|, w_2(t) = w_2(t)/\|w_2(t)\|$.
4. To generate the following two data sets by projecting the samples into two sub-spaces that are orthogonal to $w_1(t)$ and $w_2(t)$:

$$
\begin{aligned}
S_1(t+1) &= \{x_i(t+1)|x_i(t+1) = x_i(t) - w_1^T x_i(t)w_1\} \\
S_2(t+1) &= \{x_i(t+1)|x_i(t+1) = x_i(t) - w_2^T x_i(t)w_2\}
\end{aligned}
$$

5. If the predefined criterion is satisfied then stop iterating. Otherwise $t = t + 1$, and go back to step 2.

4 Experimental Results

The FPTSVM, GEPSVM, TWSVM, MVSVM, PTSVM, LPNewton, NELSTSVM and
FLSTSVM methods were implemented by using MATLAB 2014a running on a PC
with an Intel(R) Core(TM)i5-3230M CPU (2.60 GHz), 4 GB RAM. The algorithms
were evaluated on UCI and Binary Alphadigits data sets. Classification accuracy of
each algorithm was estimated by the standard 10-fold cross-validation (CV) methodology. We search the optimal values of the regularization and penalty parameters for all
the above algorithms from the set $\{2^i|i = -7, -6, \ldots, +7\}$ by the standard 10-fold
cross-validation (CV) methodology. To avoid overfitting phenomenon, we choose the
parameter μ of FPTSVM from the set $\{i \times 10^j|i = 1, 4, 8, j = -6, -5, -4\}$ manually
before the 10-fold CV.

4.1 UCI Data Sets

We experimented with 10 UCI data sets [11], then report the results of the five algorithms on the selected data sets in Table 1. It notes that the number in brackets of
PTSVM and FPTSVM column show how many orthogonal projection axes were used
in this data set.

Comparing to GEPSVM, TWSVM, MVSVM and PTSVM, FPTSVM gains better
accuracy on 7 of 10 data sets. In addition, on Monk2, Monk3, Tic-Tac-Toe and Cancer,
FPTSVM achieves its best performance when there is a single projection for each class.
It is worth noting that FPTSVM also gains comparable accuracy on Cleve and
Housing. Actually, also it is evident that FPTSVM is comparable or better than
LPNewton, NELSTSVM and FLSTSVM in classification accuracy on 7 of 10 data sets.
Specially, FPTSVM is better than NELSTSVM on all data sets. In general, FPTSVM
gains the best performance among these eight methods on accuracy.

Meanwhile, FPTSVM attempts to suppress input features during the process of
classification. The number of features is the average value of the features used to
optimize nonparallel plane classifiers. For GEPSVM, TWSVM, MVSVM and
PTSVM, all features must be used to estimate two planes. As expected, the proposed
algorithm selects few features at the classification stage automatically. For instance,
FPTSVM on Wpbc requires only 8.57 features, whereas in the other SVMs like
GEPSVM, TWSVM, MVSVM, and PTSVM, they require 33 features. From the
Table 1, we can also see that FPTSVM uses a minimum number of input features on 5
of 10 data sets. It suggests that FPTSVM has a better ability to suppress features during
the process of classification.

In Table 1, we also show the computing time of all the algorithms. FPTSVM
requires a series of linear equation solvers and no QP package as TWSVM and
PTSVM. This is why that FPTSVM is considerably faster than TWSVM and PTSVM,
although slower than the other algorithms. When it requires more projection directions
to boost performance on Cleve, FPTSVM becomes slower than PTSVM. We also find
that FPTSVM slower than LPNewton and NELSTSVM but still has a comparable
speed to FLSTSVM.

Table 1. Classification accuracy (%) on UCI data sets.

Dataset	GEPSVM test % / features / time (s)	TWSVM test % / features / time (s)	MVSVM test % / features / time (s)	PTSVM test % (dim) / features / time (s)	LPNewton test % / features / time (s)	NELSTSVM test % / features / time (s)	FLSTSVM test % / features / time (s)	FPTSVM test % (dim) / features / time (s), μ
Spect (267 * 44)	62.17 ± 8.7706 / 44 / 0.0661	77.93 ± 7.5990 / 44 / 1.6418	73.87 ± 9.2162 / 44 / 0.0105	76.85 ± 8.4331(2) / 44 / 0.8029	78.32 ± 6.9570 / 12.90 / 0.0250	79.44 ± 8.0939 / 20.05 / 0.1458	**80.20 ± 8.5351** / **13.90** / **0.1233**	79.07 ± 6.5466(2) / 22.93 / 0.2691, $4 \times 1e{-}4$
Wpbc (194 * 33)	71.58 ± 10.9469 / 33 / 0.0651	78.37 ± 11.8332 / 33 / 1.7731	74.32 ± 6.4538 / 33 / 0.0096	76.74 ± 6.4053(2) / 33 / 0.6786	75.89 ± 11.2675 / 1.80 / 0.0186	76.42 ± 10.5120 / 11.80 / 0.1356	**79.50 ± 9.2199** / **13.25** / **0.1158**	78.47 ± 10.6407(3) / 8.57 / 0.2402, $1 \times 1e{-}4$
Pimadata (768 * 8)	75.91 ± 4.4794 / 8 / 0.0573	**77.87 ± 4.5977** / **8** / **4.8927**	66.68 ± 5.0913 / 8 / 0.0080	75.01 ± 7.2912(2) / 8 / 3.1998	77.22 ± 4.2597 / 5.7 / 0.0289	65.11 ± 3.8596 / 3.75 / 0.1303	76.95 ± 3.8474 / 7.95 / 0.0948	76.05 ± 6.8656(1) / 7.7 / 0.1775, $1 \times 1e{-}4$
Sonar (208 * 60)	75.07 ± 7.1445 / 60 / 0.0702	74.45 ± 10.2453 / 60 / 1.1355	76.98 ± 8.0799 / 60 / 0.0089	77.83 ± 9.2290(2) / 60 / 0.8498	77.52 ± 10.9665 / 15.7 / 0.0305	69.36 ± 12.8554 / 29.60 / 0.1249	75.50 ± 10.0444 / 17.05 / 0.1171	**78.90 ± 9.9427(8)** / 12.02 / **0.4978, 4 × 1e-4**
Cleve (297 * 13)	**86.54 ± 5.1454** / **13** / **0.0606**	83.20 ± 5.0785 / 13 / 2.2295	85.54 ± 5.3437 / 13 / 0.0072	83.85 ± 3.8609(2) / 13 / 0.1386	85.21 ± 5.7662 / 9.7 / 0.0192	79.14 ± 6.8267 / 10.10 / 0.0925	83.84 ± 4.9117 / 9.25 / 0.0963	86.20 ± 4.8302(4) / 4.25 / 0.2194, $4 \times 1e{-}4$
Monk2 (601 * 6)	69.24 ± 7.4586 / 6 / 0.0577	65.55 ± 5.4714 / 6 / 4.1038	65.56 ± 5.5096 / 6 / 0.0081	65.89 ± 6.3009(1) / 6 / 0.1213	65.72 ± 5.6768 / 4.5 / 0.0162	64.38 ± 8.7636 / 3.95 / 0.0746	65.72 ± 5.6768 / 2.25 / 0.0965	**72.22 ± 6.5565(1)** / **4.3** / **0.1131, 1 × 1e-6**
Monk3 (554 * 6)	79.41 ± 3.2096 / 6 / 0.0573	88.44 ± 3.2746 / 6 / 1.7051	80.32 ± 3.5866 / 6 / 0.0079	83.39 ± 3.9301(2) / 6 / 0.9547	80.86 ± 3.6481 / 2.3 / 0.0163	79.41 ± 8.2542 / 3.05 / 0.0821	86.63 ± 4.1732 / 5.80 / 0.1363	**88.80 ± 3.1378(1)** / **2** / **0.1813, 1 × 1e-6**
Tic-Tac-Toe (958 * 9)	65.87 ± 3.1758 / 9 / 0.0687	67.74 ± 4.0487 / 9 / 2.8359	58.12 ± 7.3175 / 9 / 0.1287	69.51 ± 3.2254(2) / 9 / 0.3266	65.97 ± 4.7934 / 2.7 / 0.0157	63.88 ± 4.6323 / 5.65 / 0.0971	67.43 ± 3.9298 / 9.00 / 0.1295	**70.78 ± 3.7924(1)** / **5.9** / **0.2369, 4 × 1e-4**
Housing (506 * 13)	79.64 ± 5.4994 / 13 / 0.0677	**86.58 ± 3.9712** / **13** / **4.0897**	71.53 ± 7.0447 / 13 / 0.0098	85.97 ± 3.5792(2) / 13 / 0.2007	84.20 ± 4.4686 / 5.4 / 0.0199	84.58 ± 4.3411 / 11.45 / 0.0996	85.97 ± 4.5106 / 11.35 / 0.1332	86.38 ± 4.4847(2) / 12.1 / 0.1976, $4 \times 1e{-}4$
Cancer (683 * 9)	94.74 ± 3.3415 / 9 / 0.0643	97.21 ± 2.0842 / 9 / 2.0359	96.19 ± 2.2962 / 9 / 0.0099	96.78 ± 1.8355(2) / 9 / 0.1616	96.33 ± 2.4871 / 8.2 / 0.0207	92.52 ± 10.5165 / 8.05 / 0.1085	96.77 ± 2.1625 / 8.50 / 0.1280	**97.22 ± 2.2262(1)** / **4** / **0.1521, 1 × 1e-6**

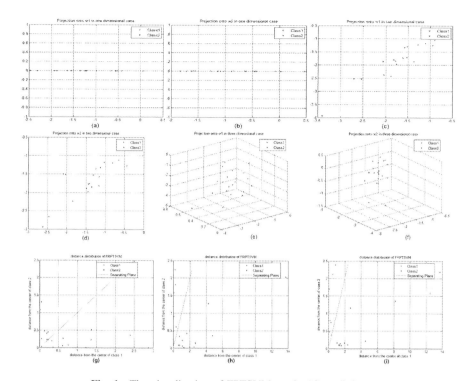

Fig. 1. The visualization of FPTSVM on the "Sonar" data set

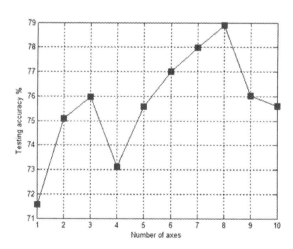

Fig. 2. The classification accuracy of FPTSVM on the "Sonar" data set

In order to verify the effectiveness of the algorithm, we take some experiments on the "Sonar" data set. The data set contains signals obtained from a variety of different aspect angles, spanning 90° for the cylinder and 180° for the rock. Firstly, we take

Table 2. Classification accuracy (%) on Binary Alphadigits data sets

Dataset	GEPSVM test % / Features / time (s)	TWSVM test % / Features / time (s)	MVSVM test % / Features / time (s)	PTSVM test % (dim) / Features / time (s)	LPNewton test % / Features / time (s)	NELSTSVM test % / Features / time (s)	FLSTSVM test % / Features / time (s)	FPTSVM test % (dim) / Features / time (s), μ
0 vs. 1	97.32 ± 5.3720 320 0.7273	97.32 ± 5.3720 320 3.7607	100 ± 0 320 0.1361	98.75 ± 3.7500(1) 320 0.1812	89.82 ± 17.5337 52.80 0.0878	94.29 ± 17.1429 15.90 0.3750	**100 ± 0** 15.70 **1.1147**	**100 ± 0(1)** 11.2500 **0.1761, 4e-05**
1 vs. 2	96.07 ± 6.0187 320 0.5819	**97.32 ± 5.3720** **320** **1.2843**	96.07 ± 6.0187 320 0.1417	**97.32 ± 5.3720(1)** **320** **0.1820**	84.82 ± 13.4404 51.70 0.0807	87.50 ± 23.0489 21.50 0.6168	96.07 ± 6.0187 28.35 1.1100	**97.32 ± 5.3720(2)** 24.225 **0.6787, 4e-04**
1 vs. 7	**98.75 ± 3.7500** **320** **0.4761**	**98.75 ± 3.7500** 319.55 1.3979	**98.75 ± 3.7500** **320** **0.1406**	98.57 ± 4.2857(1) 320 0.1833	93.57 ± 6.4484 26.70 0.0811	88.75 ± 18.9159 12.00 0.5222	95.00 ± 6.1237 8.75 0.8986	**98.75 ± 3.7500(3)** 13.67 **0.8799, 4e-04**
2 vs. 3	96.07 ± 6.0187 320 0.5605	96.25 ± 5.7282 320 1.7587	96.07 ± 6.0187 320 0.0998	92.5 ± 11.4564(1) 320 0.1872	70.54 ± 19.5656 44.40 0.3204	91.25 ± 9.7628 41.30 0.4695	90.00 ± 12.2474 24.30 1.3372	**97.32 ± 5.3720(2)** 31.32 **0.6198, 4e-04**
2 vs. 7	95 ± 11.4564 320 0.5067	98.75 ± 3.7500 320 1.8691	98.75 ± 3.7500 320 0.1459	97.5 ± 7.5000(1) 320 0.1813	84.46 ± 11.2387 46.10 0.1454	95.00 ± 8.2916 34.85 0.4032	98.75 ± 3.7500 37.40 1.1859	**100 ± 0(3)** 28.07 **0.5845, 4e-04**
3 vs. 8	87.5 ± 17.6777 320 0.6758	90 ± 12.2474 320 2.0595	**91.25 ± 12.5623** **320** **0.1370**	87.5 ± 15.8114(1) 320 0.1701	67.32 ± 19.5460 50.50 0.2317	91.25 ± 9.7628 39.20 0.5216	**91.25 ± 9.7628** **28.60** **0.9376**	**91.25 ± 12.5623(2)** 17.23 **0.3663, 4e-04**
4 vs. 7	**98.75 ± 3.7500** **320** **0.8752**	97.32 ± 5.3720 320 1.6736	**98.75 ± 3.7500** **320** **0.1320**	**98.75 ± 3.7500(1)** **320** **0.1878**	95.00 ± 8.2916 41.20 0.1106	96.25 ± 8.0039 23.65 0.4615	97.50 ± 5 7.45 1.2881	**98.75 ± 3.7500(3)** 11.93 **0.5136, 4e-04**
5 vs. 6	98.75 ± 3.7500 320 0.9191	**100 ± 0** **320** **1.2808**	98.57 ± 4.2857 320 0.1243	96.07 ± 8.2143(1) 320 0.1933	80.71 ± 15.7872 54.90 0.1974	94.64 ± 9.1752 26.05 0.5836	93.75 ± 10.0778 9.95 1.3997	97.5 ± 5.0000(1) 2.85 0.1762, 1 e-06

experiments on the first fold data set which divided into 188 training set and 20 testing set. Secondly, we obtain w_1, w_2 from the model constructed by 188 training samples. Then we plot the projections of test samples when there are one, two and three orthogonal projection axes for each class in Fig. 1(a)–(f), respectively. As the number of axes increased, the samples of one class are clustered around whereas the others scatter away. At last, we show the classification results of FPTSVM_1d, FPTSVM_2d and FPTSVM_3d on the testing set by plotting the two-dimensional scatter diagrams, respectively, in Fig. 1(g) and (i). We plot testing points with coordinates (d1, d2) under the definition (46). The separating planes of FPTSVM_1d, FPTSVM_2d and FPTSVM_3d are also shown in Fig. 1(g)–(i). A sample whose value of d1 is less than d2 is assigned to Class1 and vice versa. As we can see from Fig. 1(g) and (h), some samples of Class 1 is misclassified to Class 2 by FPTSVM_1d whereas it is correctly classified by FPTSVM_2d. Also in Fig. 1(h) and (i), one sample of Class 1 is misclassified to Class 2 by FPTSVM_2d whereas it is correctly classified by FPTSVM_3d.

The following experiments will further show the relationship between the number of axes and average 10-fold CV classification accuracy. We take some experiments when the number of axes varies from 1 to 10 and show the results in Fig. 2. It shows that the accuracy increases as the number of axes increase except 4, whereas it decreases when the number exceeds 8. In a word, better classification accuracy can be achieved if the number of axes is selected appropriately.

4.2 Binary Alphadigits Data Sets

We also conducted experiments on the Binary Alphadigits [12] which consists of binary digit images from 0 to 9. Each class has 39 samples which are 20×16 pixels with 2 gray level. We select eight pair digits of varying difficulty for odd vs. even digit recognition. For instance, "0 vs. 1" represents the classification between digit 1 and digit 2. The classification results reported in Table 2 indicate that FPTSVM is comparable or better than the other algorithms in terms of accuracy in most case. In Table 2, we also show the computing time and the feature number of all the algorithms. As we expected, FPTSVM uses only a few input features to finish the classification work and the result testified the effectiveness of the algorithm.

5   Conclusions

In this paper, we propose a novel feature selection method based on PTSVM. Without solving LP or QPP problems, this method achieves a fast convergence by using the exterior penalty (EP) theory. Similar to PTSVM, this method generates more than one projection axis for each class by a recursive algorithm. The experimental results on UCI and Binary Alphadigits data sets indicate the effectiveness of FPTSVM by comparing it with the other algorithms. As a result, this method can obtain comparable or better performance with faster computational time and better suppression of input features. Therefore, this method can be used for classifying large-scale data sets and solving high dimensional classification problems. In future, we will extend this method to the nonlinear case by the kernel trick.

Acknowledgement. This work was supported in part by the National Foundation for Distinguished Young Scientists under Grant 31125008, in part by the Scientific Research Foundation for Advanced Talents and Returned Overseas Scholars of Nanjing Forestry University, in part by the Natural Science Foundation of the Jiangsu Higher Education Institutions, China, under Grant 14KJB520018, in part by the Practice Innovation Training Program Projects for Jiangsu College Students under Grant 2015sjcx119, and in part by the National Science Foundation of China under Grant 61101197 and Grant 61401214.

References

1. Bradley, S., Mangasarian, O.L.: Massive data discrimination via linear support vector machines. Optim. Methods Softw. **13**, 1–10 (2000)
2. Fung, G.M., Mangasarian, O.L.: Multicategory proximal support vector machine classifiers. Mach. Learn. **59**(1–2), 77–97 (2005)
3. Mangasarian, O.L., Wild, E.: MultisurFace proximal support vector machine classification via generalized eigenvalues. IEEE Trans. Pattern Anal. Mach. Intell. **28**(1), 69–74 (2006)
4. Jayadeva, Khemchandani, R., Chandra, S.: Twin support vector machines for pattern classification. IEEE Trans. Pattern Anal. Mach. Intell. **29**(5), 905–910 (2007)
5. Ye, Q., Zhao, C., Ye, N., et al.: Multi-weight vector projection support vector machines. Pattern Recogn. Lett. **42**(13), 2006–2011 (2010)
6. Chen, X., Yang, J., Ye, Q., et al.: Recursive projection twin support vector machine via within-class variance minimization. Pattern Recogn. **44**(10–11), 2643–2655 (2011)
7. Tikhonov, A.N., Arsenin, V.Y.: Solutions of Ill-posed Problems. Translated from the Russian, Preface by translation John, F. (ed.), Scripta Series in Mathematics (1977)
8. Fung, G., Mangasarian, O.L.: A feature selection Newton method for support vector machine classification. Comput. Optim. Appl. **28**(2), 185–202 (2004)
9. Mangasarian, O.L.: Exact 1-norm support vector machines via unconstrained convex differentiable minimization. J. Mach. Learn. Res. **7**, 1517–1530 (2006)
10. Mangasarian, O.L.: A finite Newton method for classification problems. Optim. Methods Softw. **17**, 913–929 (2002)
11. http://archive.ics.uci.edu/ml/
12. http://www.cs.nyu.edu/~roweis/data.html
13. Qi, Z., Tian, Y., Shi, Y.: Structural twin support vector machine for classification. Knowl. Based Syst. **43**, 74–81 (2013)
14. Chen, W.J., Shao, Y.H., Li, C.N., et al.: MLTSVM: a novel twin support vector machine to multi-label learning. Pattern Recogn. **52**, 61–74 (2016)
15. Shao, Y.H., Wang, Z., Chen, W.J., et al.: A regularization for the projection twin support vector machine. Knowl. Based Syst. **37**, 203–210 (2013)
16. Guo, J., Yi, P., Wang, R., et al.: Feature selection for least squares projection twin support vector machine. Neurocomputing **144**, 174–183 (2014)
17. Gao, S., Ye, Q., Ye, N.: 1-Norm least squares twin support vector machines. Neurocomputing **74**(17), 3590–3597 (2011)
18. Ye, Q., Zhao, C., Ye, N., et al.: A feature selection method for nonparallel plane support vector machine classification. Optim. Methods Softw. **27**(3), 431–443 (2012)

Consistent Model Combination of Lasso via Regularization Path

Mei Wang[1,2], Yingqi Sun[1], Erlong Yang[3(✉)], and Kaoping Song[2,3]

[1] School of Computer and Information Technology, Northeast Petroleum University,
Daqing, China
wangmei@nepu.edu.cn
[2] Center for Post-Doctoral Studies of Beijing Deweijiaye Science
and Technology Ltd., Beijing, China
[3] Key Laboratory of Enhanced Oil Recovery in Ministry of Education,
Northeast Petroleum University, Daqing, China
erlongyang.dqpi@mail.com

Abstract. It is well-known that model combination can improve prediction performance of regression model. We investigate the model combination of Lasso with regularization path in this paper. We first define the prediction risk of Lasso estimator, and prove that Lasso regularization path contains at least one prediction consistent estimator. Then we establish the prediction consistency for convex combination of Lasso estimators, which gives the mathematical justification for model combination of Lasso on regularization path. With the inherent piecewise linearity of Lasso regularization path, we construct the initial candidate model set, then select the models for combination with Occam's Window method. Finally, we carry out the combination on the selected models using the Bayesian model averaging. Theoretical analysis and experimental results suggest the feasibility of the proposed method.

Keywords: Model combination · Lasso · Prediction consistency · Regularization path · Occam's Window

1 Introduction

In machine learning, it is very unlikely that only one model needs to be considered. When multiple plausible models are present, the common practice is to take a reasonable guiding principle, such as C_p, BIC, AIC and cross-validation, to find a single from which one makes the final estimation. Due to the underestimation of the uncertainty associated with the whole estimation procedure, even the selected model is optimized, it may give an overly optimistic or misleading answer [1,2].

Model combination is an alternative way to address the issue. Instead of selecting one model, model combination can integrate all useful information from the candidate models into the final hypothesis. This will tend to minimize the implicit risk in taking into account just one model to guarantee good

© Springer Nature Singapore Pte Ltd. 2016
T. Tan et al. (Eds.): CCPR 2016, Part I, CCIS 662, pp. 551–562, 2016.
DOI: 10.1007/978-981-10-3002-4_45

generalization performance. The problem of model combination has always been of interest to the regression community [3–8]. Many experimental works show that combining learning machines often leads to improved generalization performance. And, there are some theoretical studies to explain their efficiency. Lugosi and Vayatis [9] investigated the Bayes-risk consistency of regularized boosting methods. Steinwart [10] showed that various classifiers based on minimization of a regularized risk with universal kernels are universally consistent.

In this paper, we will construct consistent and selective model combination of Lasso (least absolute shrinkage and selection operator). Lasso, introduced by Tibshirant [11], has been an attractive technique for regularization and variable selection [12–14]. Some reasons for the popularity might be that the entire regularization path of the Lasso can be computed efficiently with the LARS algorithm [12]. Regularization path is the set of solutions for all values of the regularization parameter.

The important work in model combination is how to obtain the model set for combining [6,15,16]. In traditional model combination methods, each of the individual regression model is trained by one subset of the original data obtained by the data sampling method known as Stacking [17], Bagging [18] or Boosting [19]. However, the sampling process is more complex and time-consuming. More recently, Wang et al. have proposed model combination method for Lasso [20] and support vector machines on regularization path [7,8,21]. Different from traditional algorithms for learning base models, each individual SVR model is trained by using the same entire training dataset, but with different regularization parameter.

In this paper, we will establish the consistency of the model combination of Lasso and then construct model combination of Lasso with regularization path. The rest of the paper is organized as follows. In the next section, we show that Lasso regularization path is prediction consistent. In Sect. 3, we present the consistency result of model combination of Lasso in terms of prediction risk. Then we realize the selective Bayesian model combination of Lasso in Sect. 4. Finally, we illustrate our set of results on simulation examples in Sect. 5 and provide the conclusion in Sect. 6.

2   Consistent Lasso Regularization Path

2.1   Regularization Path of Lasso

Suppose we are given training data

$$T = \{(X_1, y_1), ..., (X_n, y_n)\} \in (\mathcal{X} \times \mathcal{Y})^n,$$

where the input $X_i = (x_{i1}, \ldots, x_{ip})^\top \in \mathbb{R}^p$ is a vector with p predictor variables, and the output $y_i \in \mathbb{R}$ denotes the response variables, $i = 1, \ldots, n$. We generally assume that the samples in T are independent and identically distributed (i.i.d) with respect to a distribution P on $X \times Y$, which is (almost) completely unknown. Assume the training data is generated by a linear regression model

$$\boldsymbol{Y} = \boldsymbol{X}\boldsymbol{\beta} + \varepsilon,$$

where $Y = (y_1, \ldots, y_n)^\top$, $X = ((X_1)^\top, \ldots, (X_n)^\top)^\top$, $\varepsilon = (\varepsilon_1, \ldots, \varepsilon_n)^\top$ is a vector of i.i.d random variables with mean 0 and variance σ^2, and $\boldsymbol{\beta} = (\beta_1, \ldots, \beta_p)^\top$ is the vector of model coefficients. Here $\beta_1, \ldots, \beta_p$ and σ are unknown constants.

The Lasso estimates $\hat{\boldsymbol{\beta}} = (\hat{\beta}_1, \ldots, \hat{\beta}_p)^\top$ are defined by

$$\hat{\boldsymbol{\beta}}(\lambda) = \arg\min_{\boldsymbol{\beta}} \; \sum_{i=1}^n (y_i - \boldsymbol{\beta}^\top X_i)^2 + \lambda\|\boldsymbol{\beta}\|_1, \tag{1}$$

where $\|\cdot\|_1$ stands for the L_1 norm of a vector. The parameter $\lambda \geq 0$ controls the amount of regularization applied to the estimate. In general, moderate values of λ will cause shrinkage of the solutions towards 0, and some coefficients may end up being exactly 0.

The LARS algorithm can compute the exact entire Lasso regularization path. The path

$$RP = \{\hat{\boldsymbol{\beta}}_\lambda, 0 \leq \lambda < \infty\}$$

ranges from the least regularized model to the most regularized model. The solution $\hat{\boldsymbol{\beta}}_\lambda$ is piecewise linear as a function of λ. We let the sequence

$$\infty > \lambda_1 > \cdots > \lambda_k > 0$$

denote the corresponding break points on the path.

2.2 Prediction Consistency of the Lasso Regularization Path

Suppose that $x_1, \ldots, x_p$ are random variables, the value of y is unknown and our task is to predict y using the values of $x_1, \ldots, x_p$. If the coefficient vector $\hat{\boldsymbol{\beta}} = (\hat{\beta}_1, \ldots, \hat{\beta}_p)^\top$ was known, then predictor of y based on $X = (x_1, \ldots, x_p)^\top$ would be the linear combination

$$\hat{y} = \hat{f}(\hat{\boldsymbol{\beta}}, X) = \hat{\boldsymbol{\beta}}^\top X. \tag{2}$$

Following the regularization path algorithm, we can obtain all possible estimator $\hat{\boldsymbol{\beta}}_\lambda (0 \leq \lambda < \infty)$. The "mean squared prediction error" of any estimator $\hat{\boldsymbol{\beta}}_\lambda$ is defined as the expected squared error in estimating y using $\hat{\boldsymbol{\beta}}_\lambda$, that is,

$$MSPE(\hat{\boldsymbol{\beta}}(\lambda)) := \mathbb{E}(y^* - \hat{y})^2, \tag{3}$$

where $y^* = \boldsymbol{\beta}^\top X$ and $\hat{y} = \hat{\boldsymbol{\beta}}^\top X$. Now, for any estimator $\hat{\boldsymbol{\beta}}$ if

$$\lim_{n \to \infty} MSPE(\hat{\boldsymbol{\beta}}) \to_P 0,$$

it is said to be *prediction consistent*.

It is of great importance to make sure that the regularization path indeed contains at least one "desirable" estimator. In our context, a estimator $\hat{\boldsymbol{\beta}}$ is considered desirable if it is prediction consistent. And, we call a Lasso regularization path is *prediction consistent* if it contains at least one prediction consistent estimator. Now, we establish prediction consistency of the Lasso regularization path. We consider the following assumptions:

(A1) $\lambda \to 0$ and $\lambda^2 n \to \infty$.
(A2) $\log(p^2)/n \to 0$.
(A3) $|X_{ij}| \leq M$, where $M \geq 0$ is some constant, $i = 1, \ldots, n$; $j = 1, \ldots, p$.

Proposition 1. *Assume (A1)–(A3). The Lasso regularization path is prediction consistent.*

Proof: As we know the Karush-Kuhn-Tucker (KKT) condition is the necessary and sufficient condition for optimal solution of optimization problem (1). The Lasso regularization path is computed on the basis of KKT theorem. Combining the result of Theorem 1 of [22], we easily achieve the proof.   □

3   Combination Consistency

We define $\mathcal{G}$ as a class of combined estimators $\bar{\boldsymbol{\beta}}$ obtained as convex combinations of the estimators

$$\mathcal{G} = \{\bar{\boldsymbol{\beta}} = \sum_{j=1}^{m} w_j \hat{\boldsymbol{\beta}}_{\lambda_j} : \hat{\boldsymbol{\beta}}_{\lambda_j} \in RP, \ w_1, \ldots, w_m \geq 0, \ \sum_{j=1}^{m} w_j = 1\}.$$

This is the weighted averaging on subset of the Lasso regularization path. In this section we establish prediction consistency of model combination of Lasso via regularization path. The only additional assumption for the class $\mathcal{G}$ is sufficiently rich to approximate the optimal combination. This assumption is easily achieved by assumption (A1) and the continuity of the regularization path.

Theorem 1. *Assume (A1)–(A3).*
Let $X = (x_1, \ldots, x_p)^\top \in \mathbb{R}^p$, $y^* = \boldsymbol{\beta}^\top X$, and $\bar{y} = \bar{f}(\bar{\boldsymbol{\beta}}, X) = \bar{\boldsymbol{\beta}}^\top X$. The Lasso combination is prediction consistent, that is,

$$MSPE(\bar{\boldsymbol{\beta}}) = \mathbb{E}(y^* - \bar{y})^2 \to_P 0$$

with $n \to \infty$.

The following proof follows [22] in a streamlined fashion and refer readers to the references mentioned above for technical details.

Proof: Firstly, the Lasso estimator can be defined as the minimizer of the optimisation problem

$$\min_{\boldsymbol{\beta}} \sum_{i=1}^{n} (y_i - \boldsymbol{\beta}^\top X_i)^2$$

$$s.t. \sum_{j=1}^{p} |\beta_j| \leq t, \tag{4}$$

for some $t > 0$. Optimisation Problems (1) and (4) are equivalent, that is, for any given λ, there exist a t such that the two problems share the same solution, and vice versa.

Let $X = (x_1, \ldots, x_p)^\top \in \mathbb{R}^p$, $y^* = \boldsymbol{\beta}^\top X$, and $\hat{y}_\lambda = \hat{f}_\lambda(\hat{\boldsymbol{\beta}}_\lambda, X) = \hat{\boldsymbol{\beta}}_\lambda^\top X$.
Following the Theorem 1 of [22], for any regularization parameter λ,

$$MSPE(\hat{\boldsymbol{\beta}}_\lambda) := \mathbb{E}(y^* - \hat{y}_\lambda)^2 \leq 2tM\sigma\sqrt{\frac{2\log(2p)}{n}} + 8t^2M^2\sqrt{\frac{2\log(2p^2)}{n}}.$$

Let $\gamma = 2tM\sigma\sqrt{\frac{2\log(2p)}{n}} + 8t^2M^2\sqrt{\frac{2\log(2p^2)}{n}}$.

Let $\hat{\boldsymbol{\beta}}_{\lambda_j} \in RP$ for any λ_j, $j = 1, \ldots, m$, and $\hat{f}_{\lambda_j}(\hat{\boldsymbol{\beta}}_{\lambda_j}, X) = \hat{\boldsymbol{\beta}}_{\lambda_j}^\top X$. The combination prediction is

$$\bar{y} = \bar{f}(\bar{\boldsymbol{\beta}}, X) = \bar{\boldsymbol{\beta}}^\top X$$
$$= (\sum_{j=1}^m w_j \hat{\boldsymbol{\beta}}_{\lambda_j})^\top X$$
$$= \sum_{j=1}^m w_j (\hat{\boldsymbol{\beta}}_{\lambda_j}^\top X)$$
$$= \sum_{j=1}^m w_j \hat{f}_{\lambda_j}(\hat{\boldsymbol{\beta}}_{\lambda_j}, X)$$
$$= \sum_{j=1}^m w_j \hat{y}_{\lambda_j},$$

where $w_1, \ldots, w_m \geq 0$, $\sum_{j=1}^m w_j = 1$. Therefore,

$$MSPE(\bar{\boldsymbol{\beta}}) := \mathbb{E}(y^* - \bar{y})^2$$
$$= \mathbb{E}(y^* - \sum_{j=1}^m w_j \hat{y}_{\lambda_j})^2$$
$$= \mathbb{E}(\sum_{j=1}^m w_j y^* - \sum_{j=1}^m w_j \hat{y}_{\lambda_j})^2$$
$$= \mathbb{E}(\sum_{j=1}^m w_j(y^* - \hat{y}_{\lambda_j}))^2$$
$$\leq \sum_{j=1}^m w_j \mathbb{E}(y^* - \hat{y}_{\lambda_j})^2$$
$$\leq \sum_{j=1}^m w_j \gamma$$
$$= \gamma.$$

Using assumptions (A1), (A2) and (A3), γ converges in probability to 0 as $n \to \infty$, which implies the theorem. $\square$

The consistency result, in theory at least, confirms us to construct consistent model combination of Lasso via regularization path.

4   Selective Model Combination

In this section, we will present how to construct the candidate model set according to the Lasso regularization path and how to combine the models.

4.1   Initial Model Set Based on Regularization Path

Generally speaking, the purpose of regularization is to control the complexity of the fitted model. The least regularized Lasso corresponds to Ordinary Least Squares; while the most regularized Lasso yields a constant fit. An informative measurement of Lasso model complexity is the effective degrees of freedom, which also plays an important role in estimating the prediction accuracy of the fitted model. Zou et al. [13] studied the degrees of freedom of the Lasso in the framework of Stein's unbiased risk estimation, and showed that the number of non-zero coefficients is an unbiased estimate for the degrees of freedom of the Lasso.

In the interior of any interval $(\lambda_{l+1}, \lambda_l)$, $0 < l < k - 1$, the models share the same degrees of freedom. The whole solution path $\boldsymbol{\beta}(\lambda)$ is piecewise linear. As long as the break points can be establish, all values in between can be found by simple linear interpolation. Figure 1 shows the paths of all the $\{\beta_j(\lambda) \mid 0 < \lambda < \infty\}$.

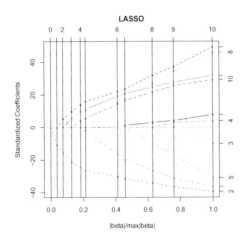

Fig. 1. The entire collection of piecewise liner paths $\beta_j(\lambda)$, $j = 1, \ldots, p$.

The piece-wise constant property of the degrees of freedom of the Lasso is basically determined by the piece-wise linearity of β. Therefore, all the regularization parameter in $(\lambda_{l+1}, \lambda_l)$ can lead to models with the same complexity. So we select βs and λs on the break points to obtain the initial model set.

We adopt the notation $f_\lambda(\hat{\boldsymbol{\beta}}_\lambda, X)$ (Eq. (2)) for a model with coefficient vector $\hat{\boldsymbol{\beta}}_\lambda$ and regularization parameter λ. It should be understood that the different models may be parameterized differently. Now, we use the notation f_λ for simplicity. The initial model set is denoted as

$$\mathcal{M}_{init} = \{f_{\lambda_1}, \ldots, f_{\lambda_k}\}.$$

4.2   Occam's Window

All possible models are involved in the initial model set $\mathcal{M}_{init}$, including the good performance ones and the poor performance ones. Madigan and Raftery [23] used the Occam's Window method for graphical model and showed combination on the selected models provided better inference performance than basing inference on a single model in each of the examples they considered. In this paper, we apply the Occam's Window method to eliminate the poor performance models. In the proposed method, posterior model probabilities are used as a metric to guide model selection. There are two basic principles underlying this approach.

First, if a model predicts the data far less well than the model which provides the best prediction, then it has been discredited and should no longer be considered. Thus the model not belonging to

$$\mathcal{M}' = \left\{f_{\lambda_j} : \frac{\max_l\{\Pr(f_{\lambda_l} \mid T)\}}{\Pr(f_{\lambda_j} \mid T)} \leq W\right\}$$

should be excluded from the combination candidate model set, where the posterior probability ratio W is usually a constant as in [23], and $\max_l\{\Pr(f_{\lambda_l} \mid T)\}$ denotes the model in initial candidate model set $\mathcal{M}_{init}$ with the highest posterior model probability.

Secondly, appealing to Occam's razor, we exclude models which receive less support from the data than any of their simpler submodels. Here we give the definition of submodel. If f_{λ_i} is a submodel of f_{λ_j}, we mean that all the support vectors involved in f_{λ_i} are also in f_{λ_j}. Thus we also exclude from models belonging to

$$\mathcal{M}'' = \left\{f_{\lambda_j} : \exists f_{\lambda_l} \in \mathcal{M}_{init}, f_{\lambda_l} \subset f_{\lambda_j}, \frac{\Pr(f_{\lambda_l} \mid T)}{\Pr(f_{\lambda_j} \mid T)} > 1\right\},$$

then we obtain the model set $\mathcal{M}_{com} = \mathcal{M}' \backslash \mathcal{M}'' \subseteq \mathcal{M}_{init}$.

The Occam's Window algorithm, as shown in the Algorithm 1, can greatly reduce the number of models in the candidate model set.

4.3   Bayesian Model Combination

The Bayesian model combination over $\mathcal{M}_{com}$ has the form

$$f_{bmc}(X) = \sum_{j=1}^{k} f_{\lambda_j}(\hat{\beta}_{\lambda_j}, X) \Pr(f_{\lambda_j} \mid T), \tag{5}$$

Algorithm 1. The Occam's Window algorithm.

Input: $\mathcal{M}_{init} = \{f_{\lambda_1}, \ldots, f_{\lambda_k}\}$, $\mathcal{P} = \{\mathrm{Pr}(f_{\lambda_1}), \ldots, \mathrm{Pr}(f_{\lambda_k})\}$, W
Output: $\mathcal{M}_{com}$
$MP \leftarrow \max(\mathcal{P})$;
$\mathcal{M}_{tmp} \leftarrow \emptyset$;
while $\mathcal{M}_{init} \neq \emptyset$ *And* $f \in \mathcal{M}_{init}$ **do**
    if $MP/\mathrm{Pr}(f) \leq W$ **then**
        $\mathcal{M}_{tmp} \leftarrow \mathcal{M}_{tmp} \cup \{f\}$;
        $\mathcal{M}_{init} \leftarrow \mathcal{M}_{init} \backslash \{f\}$;
    end
end
$\mathcal{M}_{com} \leftarrow \mathcal{M}_{init}$;
for $f \in \mathcal{M}_{ao}$ **do**
    for $f_1 \in \mathcal{M}_{com} \backslash \{f\}$ **do**
        if $(\mathrm{Pr}(f_1) > \mathrm{Pr}(f))$ *and* $(f_1 \subset f)$ **then**
            $\mathcal{M}_{com} \leftarrow \mathcal{M}_{cand} \backslash \{f\}$;
            break;
        end
    end
end

---

where X denotes any input vector, $\mathrm{Pr}(f_{\lambda_j} \mid T)$ is the posterior probability of model f_{λ_j}, $j = 1, \ldots, k$. Then this is the process of estimating the prediction under each model f_{λ_j} and then averaging the estimates according to how likely each model is.

In general, the posterior probability of model f_{λ_j} in Eq. (5) is given by

$$\mathrm{Pr}(f_{\lambda_j} \mid T) \propto \mathrm{Pr}(T \mid f_{\lambda_j})\mathrm{Pr}(f_{\lambda_j}), \tag{6}$$

where $\mathrm{Pr}(T \mid f_{\lambda_j})$ is the marginal likelihood of model f_{λ_j} and $\mathrm{Pr}(f_{\lambda_j})$ is the prior probability that f_{λ_j} is the true model. Specification of $\mathrm{Pr}(f_{\lambda_j})$, the prior distribution over competing models, is challenging. And, the integrals implicit in $\mathrm{Pr}(T \mid f_{\lambda_j})$ can in general be hard to compute.

In this paper, we also estimate the posterior probability of each model f_{λ_j} in $\mathcal{M}_{com}$ using the BIC criterion, and refer readers to [20] for a detailed tutorial. In practice, one can compute the BIC for each model along with the LARS algorithm, which can compute the entire Lasso solution path with the cost of a single least squares fit.

4.4   Computational Complexity

The main computational burden of the Baysian model combination for Lasso centers on building the Lasso regularization path. The approximate computational complexity of the Lasso regularization path algorithm is $O(p^3 + np^2)$ [12], where n is the size of the training data and p is the number of predictors. The approximate computational complexity of the Occam's Window is $O(p^3)$. Thus, the total computational complexity of the Baysian model combination for Lasso on regularization path is $O(p^3 + np^2)$. The algorithm has the advantage of a low computational complexity.

5 Numerical Results

In the following experiments, we consider five benchmark data sets[1]. Table 1 contains a summary of these data sets.

Table 1. Summary of the five benchmark data sets.

Dataset	n	p	Dataset	n	p
cpusmall	8292	12	pyrim	74	27
slice_localization	53500	385	triazines	186	60
housing	566	13			

Since the usual goal of regression analysis is to minimize the predicted squared-error loss, the prediction error is defined as

$$\text{PredE} = \frac{1}{m}\sum_{i=1}^{m}(y_i - \bar{f}(\bar{\boldsymbol{\beta}}, X_i))^2,$$

where m is the size of test data set.

The R codes for Bayesian model combination of Lasso are implemented by our own R codes based on LARS [12].

5.1 Combination Consistency

Firstly, we describe a small simulation with data set *slice_localization* and its result to investigate how well the asymptotical result on the model combination of Lasso works for small to moderate samples sizes. We considered $n \in \{8000, 20000, 40000, 60000\}$.

Since the usual goal of regression analysis is to minimize the predicted squared-error loss, we use the following integrated testing mean-squared error to measure how well the Lasso estimates approximate the true output

$$\text{iTMS}_{\tilde{f}} := \frac{1}{R}\sum_{r=1}^{R}\left[\frac{1}{m}\sum_{i=1}^{m}(\tilde{f}(\tilde{\boldsymbol{\beta}}, X_i^{(r)}) - y_i)^2\right],$$

where m is the size of test samples, R is the number of replications, y is the true value of the output, $\tilde{f} = \bar{f}$ denotes the final model with Bayesian model

---

[1] Datasets Available: http://archive.ics.uci.edu/ml/datasets.html.

combination, and $\tilde{f} = \hat{f}$ for the Lasso fitted model with $\lambda = \frac{1}{500}n^{-1/3}$. Here, we set $m = 2000$, $R = 50$.

The simulation results are summarized in Table 2. Here we see that Bayesian model combination estimates perform quite well, because the integrated testing mean-squared error of Bayesian model combination has relatively smaller values and the values decrease with increasing train sample sizes. The simulation indicates that the combination consistency results derived in Sect. ?? can be useful even for small to moderate sample sizes.

Table 2. Simulation results for the integrated testing mean-squared error for Bayesian model combination of Lasso and Lasso with prespecified regularization parameter on data set *slice_localization* with different size of training set.

n	iTMS$_{\tilde{f}}$	iTMS$_{\hat{f}}$
8000	7.5681e-10(2.0311e-11)	7.9301e-10(2.1210e-11)
20000	7.4221e-10(1.9381e-11)	7.8001e-10(2.1193e-11)
40000	7.2023e-10(1.5023e-11)	7.5303e-10(2.0239e-11)
60000	7.1810e-10(1.4198e-11)	7.5002e-10(2.0112e-11)

5.2   Combination Performance

We investigate the performance of the selective Bayesian model combination of Lasso by comparing the prediction performance with the classical BIC-based model selection and Bagging regression The R codes for Bagging regression are download from online web[2]. For data set *cpusmall* we randomly sample 1000 examples from the 8292 examples, and for data set *slice_localization* we randomly sample 8500 examples from the 53500 examples. For each data set, we randomly split the data into training and test sets, with the training set comprising 80 % of the data. We repeated this process 50 times and computed the average prediction errors and their corresponding standard deviations. The results are summarized in Table 3, where PredE_{BIC} denotes the experimental results of model selection using BIC criterion, PredE_{com} denotes the experimental results of Bayesian model combination with model set $\mathcal{M}_{com}$, and PredE_{Bag} denotes the prediction error of Bagging regression.

From the tables we find that the model combination methods perform better on these benchmark data sets, and Bayesian model combination of Lasso on regularization path performs better than the Bagging regression on data sets *housing*, *pyrim* and *triazines*.

---

[2] http://CRAN.R-project.org/package=ipred.

Table 3. Comparison of the prediction error on benchmark data for model selection and model combination.

Dataset	PredE$_{BIC}$	PredE$_{Bag}$	PredE$_{com}$
cpusmall	34.5482	27.068	27.6415
	(4.2134)	(3.9667)	(3.5132)
housing	21.5513	19.36	17.6217
	(4.1402)	(4.0911)	(3.0041)
pyrim	0.0073	0.0110	0.0049
	(0.0021)	(0.0004)	(0.0016)
triazines	0.0180	0.0169	0.0159
	(0.0051)	(0.0048)	(0.0043)
slice_localization	7.9331e-10	7.5263e-10	7.5365e-10
	(2.1293e-11)	(2.0301e-11)	(1.9531e-11)

6   Conclusion

In this paper, we have investigate the consistent model combination of Lasso, which takes advantage of the regularization path. We have extended some of the theoretical results of the Lasso to the consistency of Lasso regularization path and model combination of Lasso, which provides the theory to assure the consistent model combination. Regularization path records all possible models, ranging from the least regularized model to the most regularized model, such that it facilitates the selection of base models. We obtain the initial model set according to the regularization path, applying the Occam's Window method to reduce the number of candidate models, then implemented Bayesian model combination of Lasso. The experiments show that, the proposed Bayesian combination model algorithm is theoretically sound and practically effective.

Acknowledgement. Project supported by the Natural Science Foundation of Heilongjiang Province (No. F2015020, E2016008), the Beijing Postdoctoral Research Foundation (No. 2015ZZ-120), Chaoyang District of Beijing Postdoctoral Foundation (No. 2014ZZ-14), the Natural Science Foundation of China (No. 51104030, 61170019), the Northeast Petroleum University Cultivation Foundation (No. XN2014102).

References

1. Draper, D.: Assessment and propagation of model uncertainty. J. Roy. Stat. Soc. B **57**, 45–97 (1995)
2. Nilsen, T., Aven, T.: Models and model uncertainty in the context of risk analysis. Reliab. Eng. Syst. Saf. **79**(3), 309–317 (2003)
3. Raftery, A.E., Madigan, D., Hoeting, J.A.: Bayesian model averaging for linear regression models. J. Am. Stat. Assoc. **92**(437), 179–191 (1997)
4. Yang, Y.: Adaptive regression by mixing. J. Am. Stat. Assoc. **96**(454), 574–588 (2001)

5. Yuan, Z., Yang, Y.: Combining linear regression models: when and how. J. Am. Stat. Assoc. **100**(472), 1202–1214 (2005)

6. Kittler, J.: Combining classifiers: a theoretical framework. Pattern Anal. Appl. **1**, 18–27 (1998)

7. Wang, M., Liao, S.: Model combination for support vector regression via regularization path. In: Anthony, P., Ishizuka, M., Lukose, D. (eds.) PRICAI 2012. LNCS (LNAI), vol. 7458, pp. 649–660. Springer, Heidelberg (2012). doi:10.1007/978-3-642-32695-0_57

8. Wang, M., Song, K., Lv, H., Liao, S.: Consistent model combination for svr via regularization path. J. Comput. Inf. Syst. **10**(22), 9609–9617 (2014)

9. Lugosi, G., Vayatis, N.: On the Bayes-risk consistency of regularized boosting methods. Ann. Stat. **32**(1), 30–55 (2004)

10. Steinwart, I.: Consistency of support vector machines and other regularized kernel classifiers. IEEE Trans. Inf. Theory **51**(1), 128–142 (2005)

11. Tibshirant, R.: Regression shrinkage, selection via the lasso. J. Roy. Stat. Soc. B (Methodol.) **58**(1), 267–288 (1996)

12. Efron, B., Hastie, T., Johnstone, I., Tibshirani, R.: Least angle regression. Ann. Stat. **32**, 407–499 (2004)

13. Zou, H., Hastie, T., Tibshirani, R.: On the "degrees of freedom" of the lasso. Ann. Stat. **35**(5), 2173–2192 (2007)

14. Fraley, C., Hesterberg, T.: Least angle regression and lasso for large datasets. Stat. Anal. Data Min. **1**(4), 251–259 (2009)

15. Hoeting, J.A., Madigan, D., Raftery, A.E., Volinsky, C.T.: Bayesian model averaging: a tutorial. Stat. Sci. **14**, 382–417 (1999)

16. Polikar, R.: Ensemble based systems in decision making. IEEE Circuits Syst. Mag. **6**(3), 21–45 (2006)

17. Wolpert, D.H.: Stacked generalization. Neural Netw. **5**(2), 241–259 (1992)

18. Breiman, L.: Bagging predictors. Mach. Learn. **24**(2), 123–140 (1996)

19. Schapire, R.E., Freund, Y., Bartlett, P., Lee, W.S.: Boosting the margin: a new explanation for the effectiveness of voting methods. Ann. Stat. **26**(5), 1651–1686 (1998)

20. Wang, M., Liao, S.: Model combination of lasso on regularization path. J. Comput. Inf. Syst. **10**(2), 755–762 (2014)

21. Wang, M., Liao, S.: Thress-step bayesian combination of SVM on regularization path. J. Comput. Res. Dev. **50**(9), 1855–1864 (2013)

22. Chatterjee, S.: Assumptionless consistency of the lasso, pp. 1–10 (2014). arXiv:1303.5817v5 [math.ST]

23. Madigan, D., Raftery, A.E.: Model selection and accounting for model uncertainty in graphical models using Occam's window. J. Am. Stat. Assoc. **89**(428), 1535–1546 (1994)

Using Feature Correlation Measurement to Improve the Kernel Minimum Squared Error Algorithm

Zizhu Fan[1(✉)] and Zuoyong Li[2]

[1] School of Basic Science, East China Jiaotong University, Nanchang, China
zzfan3@163.com
[2] Fujian Provincial Key Laboratory of Information Processing
and Intelligent Control, Fuzhou, China

Abstract. The kernel minimum squared error (KMSE) is less computationally efficient when applied to large datasets. In this paper, we propose IKMSE, an algorithm which improves the computational efficiency of KMSE by using just a part of the training set, key nodes, as a certain linear combination of key nodes in the feature space can be used to approximate the discriminant vector. Our algorithm includes three steps. The first step is to measure the correlation between the column vectors in the kernel matrix, known as the feature correlation, using the cosine distance between them. The second step is to determine the key nodes using the following requirement: two arbitrary column vectors of the kernel matrix that correspond to the key nodes should have a small cosine distance value. In the third step, we use the key nodes to construct the KMSE model and classify the testing samples. There are usually many fewer key nodes than training samples and this is the basis of producing the efficiency of feature extraction in our method. Experimental results show that our improved method has low computational complexity as well as high classification accuracy.

Keywords: Kernel minimum squared error (KMSE) · Feature extraction · Feature correlation measurement · Cosine distance

1 Introduction

Kernel minimum squared error (KMSE) [1–3] is one of the kernel-based methods [4, 5] that have been used with success in image feature extraction and other areas. The kernel-based methods operate by transforming the input space into the feature space through a nonlinear mapping, and then performing the feature extraction in the feature space. They are usually nonlinear feature extraction methods that use the "kernel trick" [6], first used in the support vector machine (SVM), and this makes them computationally more efficient than the conventional nonlinear methods. When the number of samples tends to be infinite, the KMSE model would approximate the Bayesian discriminant function in the feature space with minimum squared error [1]. Other kernel methods include kernel principal component analysis (KPCA) [7, 8], kernel discriminant analysis (KDA) [9, 10], and the kernel-based online learning algorithms [11, 12].

© Springer Nature Singapore Pte Ltd. 2016
T. Tan et al. (Eds.): CCPR 2016, Part I, CCIS 662, pp. 563–573, 2016.
DOI: 10.1007/978-981-10-3002-4_46

To extract features from a testing sample, KMSE must compute the kernel functions between all the training samples and the testing sample. This algorithm is not computationally efficient on large-scale datasets. As we know from the theory of reproducing kernels [13], the KMSE discriminant vector in the feature space is the linear combination of all samples in the space. This means that the features extracted from one sample are a linear combination of the kernel functions of its source sample and all the other training samples. The efficiency of extracting sample features is thus inverse to the size of the training set, so that the larger the scale of the training set, the lower the computational efficiency of the KMSE feature extraction.

A number of reformulated algorithms for the kernel have recently been proposed to improve the computational efficiency of KMSE or its variants. Zheng et al. grouped the training samples to improve computational efficiency but did not provide any analysis to explain how the optimal learning results were obtained [14]. Liang proposed a fast method to solve kernel Fisher discriminant analysis which first used the QR method to maximize the between-class scatter and then solved the generalized eigenvalue problems [15]. However, their algorithm requires decomposing several matrices, which is time-consuming. Xiong et al. also proposed an efficient kernel Fisher discriminant using QR decomposition [16], but their algorithm suffers from the same problem as the method in [15]. Wang developed an efficient algorithm for generalized discriminant analysis using incomplete Cholesky decomposition [17]. These algorithms choose some of the training samples as a basis of the space generated by the total training set in the feature space. However, the methods proposed in [17, 18] become impractical when they deal with the high-dimensional data, e.g. the high-dimensional image data. Xu et al. proposed a fast KMSE using "significant nodes" [19] selected by a complex criterion yet the training procedure in this approach is still time-consuming.

In this research, we propose an improved KMSE (IKMSE) that improves the computational efficiency of the conventional KMSE algorithm. The approach first measures the correlation between the column vectors in the kernel matrix, referred to as the feature correlation measurement (FCM), to determine key nodes from the whole training set, and then uses them to classify a testing sample by computing the kernel functions between the key nodes and the testing sample. Since the key nodes are usually much fewer than all the training samples, the feature extraction efficiency of the IKMSE method is much more computationally efficient than that of the KMSE method. Moreover, our approach is very flexible. It has two implementation schemes and provides the convenience for the applications. Theoretical analysis and experimental results show that the feature correlation measurement (FCM) can effectively determine the key nodes.

The organization of this paper is as follows: In Sect. 2, we describe the conventional KMSE algorithm. Section 3 describes our improved KMSE (IKMSE) algorithm, in particular how we use the feature correlation measurement (FCM) to determine the key nodes and how the algorithm operates, and finally provides an analysis of its computational complexity. Section 4 describes our experiments and results. Section 5 offers our Conclusion.

2   Kernel Minimum Squared Error (KMSE) Algorithm

The KMSE algorithm is the kernel version of the classic minimum squared error algorithm. In this work, we consider two-class classification problems. Let c_1 and c_2 denote the two classes, respectively. Suppose there are n training samples $x_i \in R^d (i = 1, 2, \ldots, n)$; in class c_1 we place the first n_1 samples $x_1, x_2, \cdots, x_{n_1}$; in class c_2 we place the residual n_2 samples $x_{n_1+1}, x_{n_1+2}, \cdots, x_n$; and the class label of each sample in the class c_1 is "+1" while the class label of each sample in the class $c_2 (n_1 + n_2 = n)$ is "-1". We use a nonlinear mapping $\phi(\bullet)$ to transform the original samples into the high-dimensional (possibly infinite dimensional) feature space, giving the samples $\phi(x_1), \phi(x_2), \ldots, \phi(x_{n_1})$ and $\phi(x_{n_1+1}), \phi(x_{n_1+2}), \ldots, \phi(x_n)$, and use the following formula to learn the relation between the samples in the feature space and their class labels:

$$XW = B \tag{1}$$

where,

$$X = \begin{pmatrix} 1 & \phi(x_1)^T \\ 1 & \phi(x_2)^T \\ \vdots & \vdots \\ 1 & \phi(x_n)^T \end{pmatrix}, \quad W = \begin{pmatrix} w_0 \\ w \end{pmatrix}, \quad B = \begin{pmatrix} 1 \\ \vdots \\ -1 \end{pmatrix}$$

We call w the discriminant vector and w_0 the threshold. The numbers of 1 s and -1 s in the column vector B are respectively equal to n_1 and n_2. According to the theory of reproducing kernels, W can be regarded as the linear combination between the threshold w_0 and the samples in the feature space. So it can be written as follows [19]:

$$W = \begin{pmatrix} w_0 \\ \sum_{i=1}^{n} \alpha_i \phi(x_i) \end{pmatrix}. \tag{2}$$

Employing the kernel function $k(x, x_i) = (\phi(x) \bullet \phi(x_i))$, we get

$$KA = B \tag{3}$$

where,

$$K = \begin{pmatrix} 1 & k(x_1, x_1) & \cdots & k(x_1, x_n) \\ 1 & k(x_2, x_1) & \cdots & k(x_2, x_n) \\ \vdots & \vdots & \vdots & \vdots \\ 1 & k(x_n, x_1) & \cdots & k(x_n, x_n) \end{pmatrix}, \quad A = \begin{pmatrix} w_0 \\ \alpha_1 \\ \vdots \\ \alpha_n \end{pmatrix}, \quad B = \begin{pmatrix} 1 \\ \vdots \\ -1 \end{pmatrix}$$

where A is called the discriminant direction. The least-squares solution of Eq. (3) is as follows:

$$A = (K^T K)^{-1} K^T B. \tag{4}$$

Because $K^T K$ is an ill-conditioned matrix, we must introduce a regulation μ in Eq. (4) which is then rewritten as

$$A = (K^T K + \mu I)^{-1} K^T B \tag{5}$$

where I is the identity matrix. We compute A according to Eq. (5) and then perform the classification. Suppose there is a new testing pattern $\phi(x)$ in the feature space, then its projection on W in Eq. (2) is

$$l_p(x) = w_0 + \sum_{i=1}^{n} \alpha_i (\phi^T(x_i)\phi(x)) = w_0 + \sum_{i=1}^{n} \alpha_i k(x, x_i) \tag{6}$$

If $l_p(x) > 0$, then x is classified as the first class and its class label is "+1"; otherwise, its class label is "−1".

3 Improved Kernel Minimum Squared Error (IKMSE) Algorithm

In this section, we describe how we measure the feature correlation using cosine distance and determine the key nodes in our IKMSE algorithm, then how the algorithm operates.

Conventional KMSE extracts the features from a sample by computing the kernel functions between all the training samples and the testing sample. This is not efficient when the training set is large-scale. We seek to improve on this by using just a part of the training set, which we refer to as key nodes. Note that discriminant vector is a linear combination of all the training samples in the feature space. In this vector, different samples play different roles in feature extraction. From lots of experiments, we observe that many samples make little contribution when performing classification. If we find these samples and discard them from the discriminant vector. The classification performance is almost not affected by discarding them. By contrast, the remaining training samples make more contribution in the classification, and they are used as the key nodes. In this sense, the linear combination of these nodes can approximate the discriminant vector. Since the number of the key nodes is usually many fewer than the number of the total training samples, the computational complexity of feature extraction using the key nodes is much lower than that using the whole training set. In the determining process, we can freely specify the number of the key nodes. We refer to this as Scheme 1. The determining process is as follows:

Step 1: Initialization

Compute the kernel matrix; denote the set of the key nodes by S which is initialized as empty; let the set X represent all the training samples $x_i(i = 1, 2, \ldots, n)$ and $r(1 \leqslant r \leqslant n)$ be the specified number of key nodes;

Step 2:   Determine the first key node

The column vector of the i-th sample $x_i (i = 1, 2, \ldots, n)$ in the kernel matrix is $k_i = (k(x_1, x_i), k(x_2, x_i), \ldots, k(x_n, x_i))^T$. We refer to this column vector as the kernel vector. The matrix K_I of the kernel vector k_i is given by

$$K_1 = \begin{pmatrix} 1 & k(x_1, x_i) \\ 1 & k(x_2, x_i) \\ \vdots & \vdots \\ 1 & k(x_n, x_i) \end{pmatrix}.$$

A is calculated using Eq. (5) and denoted by A_i. R_i is obtained by $R_i = (\mu\|A_i\|_2^2 + \|KA_i - B\|_2^2)^{1/2}$. After all the A_i and $R_i (i = 1, 2, \ldots, n)$ have been computed, the sample that corresponds to the minimal R_i is identified as the first key node, denoted by x_1'. We put x_1' into the subset S and $S = \{x_1'\}$.
Step 3:   Determine the j-th key node $(2 \leqslant j \leqslant r)$

After $j - 1$ key nodes, $x_1', x_2', \ldots, x_{j-1}'$, have been determined, $S = \{x_1', x_2', \ldots, x_{j-1}'\}$ and the corresponding matrix K_{j-1} is

$$K_{j-1} = \begin{bmatrix} 1 & k(x_1, x_1') & \cdots & k(x_1, x_{j-1}') \\ 1 & k(x_2, x_1') & \cdots & k(x_2, x_{j-1}') \\ \vdots & \vdots & \vdots & \vdots \\ 1 & k(x_n, x_1') & \cdots & k(x_n, x_{j-1}') \end{bmatrix}$$

For $\forall x \in X - S$, the tentative kernel vector k_x is: $k_x = (k(x_1, x), k(x_2, x), \ldots, k(x_n, x))^T$. We then define the cosine distance between the kernel vector k_x and the matrix K_{j-1}:

$$\cos dis(k_x, K_{j-1}) = \frac{1}{j-1} \sum_{s=1}^{j-1} \cos(k_x, k_s) \tag{7}$$

where k_s is the $(s + 1)$-th column vector in the matrix K_{j-1}: $k_s = (k(x_1, x_s'), k(x_2, x_s'), \ldots, k(x_n, x_s'))^T$. For all $x \in X - S$, we can determine the minimum of $\cos dis(k_x, K_{j-1})$ using Eq. (7). The sample that corresponds to the minimal $\cos dis(k_x, K_{j-1})$ is denoted by x_j' as the j-th key node. x_j' is added into the subset S. The kernel vector k_j of the sample x_j' will be put into the matrix K_{j-1} as the last column vector, and K_{j-1} then becomes K_j.

We repeat Step 3 until $j = r$. And we have obtained all the key nodes and matrix K_r. Finally, we use Eq. (5) to compute A corresponding of K_r and can then perform the classification.

We have another method for determining the key nodes in which we do not specify the number of the key nodes, but do require a threshold ε as the terminal condition of the iteration and a criterion or objective function to control the determining process.

Applying Eq. (5), we determine the key nodes using the following criterion: $J = \mu\|A\|_2^2 + \|KA - B\|_2^2$. The second scheme (Scheme 2) is as follows:

Step 1:   Initialization

Initialize $S = \{\}$; $X = x_i (i = 1, 2, \ldots, n)$; compute the kernel matrix; and choose the μ and the threshold ε.

Step 2:   Determine the first key node

K_1, S, A_i and R_i are computed using the same method described in Step 2 in the first scheme   (Scheme 1).   That   is, A_i is   calculated   using   Eq. (5),   and $R_i = (\mu\|A_i\|_2^2 + \|KA_i - B\|_2^2)^{1/2}$. The minimal R_i is denoted by $R^{(1)}$, and its corresponding solution A_i is denoted by $A^{(1)}$.

Step 3:   Determine the j-th key node

As in Step 3 of Scheme 1, we compute K_j ($j \in [2, n]$). Then apply K_j to compute its corresponding A_j and R_j as in Step 2. The minimal R_j in this step is denoted by $R^{(j)}$.

We repeat Step 3 until $\left|R^{(j)} - R^{(j-1)}\right| < \varepsilon$. The A_j of the last key node is taken as the near-optimal solution of discriminant direction A. Notice that this scheme can automatically determine the number of the key nodes.

4   Experiments

In this section, we describe two experiments to show the classification accuracy and computational efficiency of our IKMSE algorithm. The first experiment is conducted using Scheme 1 which allows us to flexibly specify the number of the key nodes. In the second experiment, we use Scheme 2 which needs the threshold ε. This scheme can automatically determine the optimal number of the key nodes. All the experiments are run on the platform with 2.5 GHz CPU and 2.0 GB RAM by Matlab 7.0 software.

Our experiments use the five two-class benchmark datasets adopted in [20]. Each dataset is randomly partitioned into 100 parts. Each part is made up of four subsets: training, training label, testing, and testing label. For each dataset, the training dataset is the first training subset, whereas testing is implemented on 100 test subsets. We employ the Gaussian kernel function $k(x, y) = \exp(-\|x - y\|^2 / 2\sigma^2)$. To select a suitable kernel parameter σ for each dataset, we employ five methods to compute the kernel parameter in advance: (1) the norm of covariance matrix C computed by using the training data; (2) σ is set to be norm$(C)^2$; (3) σ is set to be the largest singular value of X; (4) σ is set to be $\frac{1}{2n} (\sum_{i,j} \|x_i - x_j\|_2^2)^{1/2}$; and (5) σ is set to be $\frac{1}{nd} \sum_{i=1}^{n} \sum_{j=1}^{d} (x_{ij} - \overline{x_i})^2$ where x_{ij} denotes the jth element of the sample x_i and $\overline{x_i}$ is the mean of the ith sample vector x_i. Thus, we obtain five candidate parameters for each dataset. We run our algorithm using five candidate parameters (for each dataset, the first training subset is used for training, and the first testing subset is used for testing), and select the candidate parameter that yields the highest classification accuracy as the best kernel parameter for the dataset. Two other parameters in our algorithm, μ in Eq. (5) and ε in Scheme 2, are both set to 0.001.

4.1 Experiment 1

In this experiment, we use Scheme 1 to determine the key nodes. This scheme is very flexible. We can freely choose the number of the key nodes, and view the relationship between the classification effectiveness and efficiency and the number of the key nodes. Figure 1 shows the key nodes using different numbers in the *banana* data set. The blue nodes represent the training samples and the red nodes are the key nodes chosen by Scheme 1. As expected, the key nodes distribute separately, in accordance with the true characteristic of the feature correlation, because, according to the definition of the feature correlation, if the feature correlation between the vectors is small, then the Euclidean distance between them is large.

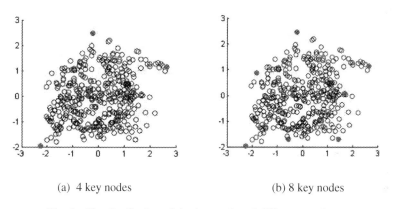

    (a) 4 key nodes   (b) 8 key nodes

Fig. 1. The distribution of the key nodes of different numbers

Table 1 shows the classification error rates and deviations corresponding to different numbers of the key nodes. Table 2 illustrates the training time. The ratio of the key nodes to all the training samples is usually no more than 10 %. So, in general, the numbers of the key nodes is manually specified below 10 percent of the training samples. In the experiment, we focus on the classification error rate and training time. From these two tables, we can conclude that the more the number of the key nodes, the more time the training process needs; and we can see that the classification error rate tends to decrease when the number of the key nodes increases. We take the data set *banana* as an example. When the number of the key nodes is 4, the classification error rate and training time are 25.5 % and 3.219 s, respectively. While the number of the key nodes is 32, the classification error rate and training time are 12.8 % and 7.265 s, respectively.

Table 1. The classification error rates and deviations of different numbers of the key nodes

Banana ($\sigma = 0.36$)	Titanic ($\sigma = 2.68$)	German ($\sigma = 48$)	B. cancer ($\sigma = 8.67$)	F. solar ($\sigma = 0.76$)
25.5 ± 2.3(4)	22.5 ± 3.2(2)	30.1 ± 2.0(2)	25.2 ± 4.6(3)	35.4 ± 1.7(3)
19.8 ± 1.8(8)	22.5 ± 3.2(4)	26.8 ± 1.9(4)	23.9 ± 4.2(6)	35.4 ± 1.7(6)
15.3 ± 1.5(12)	22.5 ± 3.2(6)	27.3 ± 2.0(6)	21.6 ± 3.9(9)	35.4 ± 1.7 (9)
14.9 ± 1.5(16)	22.5 ± 3.2(8)	27.7 ± 2.0(8)	21.6 ± 3.9(12)	34.4 ± 1.8(12)
14.3 ± 1.5(20)	22.5 ± 3.2(10)	27.2 ± 1.8(10)	21.6 ± 3.9(15)	34.9 ± 1.7(15)
12.5 ± 1.3(24)	21.6 ± 3.2(12)	26.9 ± 1.7(12)	21.6 ± 3.9(18)	34.9 ± 1.7(18)
13.0 ± 1.4(28)	22.5 ± 3.2(14)	23.6 ± 2.2(14)	21.6 ± 3.9(21)	34.9 ± 1.7(21)
12.8 ± 1.4(32)	22.5 ± 3.2(16)	23.7 ± 2.0(16)	21.5 ± 3.7(24)	34.8 ± 1.7(24)

The values in parentheses in each column represent different numbers of key nodes.

Table 2. Training times when using the key nodes of different numbers

Banana ($\sigma = 0.36$)	Titanic ($\sigma = 2.68$)	German ($\sigma = 48$)	B. cancer ($\sigma = 8.67$)	F. solar ($\sigma = 0.76$)
3.219 s (4)	0.468 s (2)	9.813 s (2)	0.828 s (3)	8.844 s (3)
3.359 s (8)	0.485 s (4)	9.953 s (4)	0.875 s (6)	9.016 s (6)
3.719 s (12)	0.516 s (6)	10.109 s (6)	0.953 s (9)	9.672 s (9)
4.188 s (16)	0.532 s (8)	10.390 s (8)	1.046 s (12)	9.968 s (12)
4.766 s (20)	0.563 s (10)	10.734 s (10)	1.187 s (15)	10.781 s (15)
5.468 s (24)	0.625 s (12)	11.171 s (12)	1.328 s (18)	11.184 s (18)
6.313 s (28)	0.672 s (14)	11.625 s (14)	1.516 s (21)	12.563 s (21)
7.265 s (32)	0.734 s (16)	12.157 s (16)	1.750 s (24)	13.719 s (24)

The values in parentheses of each column represent different numbers of key nodes.

4.2   Experiment 2

We used Scheme 2 to compute the classification error rates and deviations on the seven datasets shown in Table 3. Table 3 also provides the results when using other methods, specifically KMSE, naïve kernel-based nonlinear method (NKNM), Fast kernel-based nonlinear method and (FKNM) [19]. The number in parentheses in the last column in Table 3 is the parameter of the Gaussian kernel function for each data set. Each parameter is the optimal parameter selected by the aforementioned five methods for computing kernel parameters. The parameters of the KMSE are the same as those of the IKMSE.

For each data set, the bold italics in Table 3 highlight the best classification result of the six methods. Although KMSE has the highest average classification accuracy because it uses all the training samples in the training and testing process, it is noted that IKMSE gets the best results of all methods on the *banana* dataset.

Table 4 shows the number of the key nodes obtained using FKNM and IKMSE. The numbers in brackets are the ratio of the number of key nodes to that of the total training samples. According to this table, the feature extraction efficiency of IKMSE is nearly equal to that of FKNM, since there has no distinct difference in the numbers of key nodes of these two methods.

Table 3. The classification error rate and deviation

Dataset	KMSE	NKNM	FKNM	IKMSE
Banana	11.6 ± 1.4	13.7 ± 0.1	13.5 ± 0.1	*11.3 ± 1.3 (0.36)*
Titanic	*22.5 ± 3.2*	25.5 ± 0.3	22.7 ± 0.3	22.5 ± 3.2 (2.68)
German	*17.4 ± 2.0*	21.3 ± 2.1	22.6 ± 2.1	22.7 ± 2.1 (48)
b. cancer	*15.1 ± 3.8*	22.8 ± 4.4	19.5 ± 3.9	21.6 ± 3.9 (8.67)
f. solar	*31.6 ± 1.9*	32.2 ± 1.6	32.1 ± 1.8	34.9 ± 1.7 (0.76)

The results for models AB and SVM are taken from [2].

Table 4. Quantity of key nodes

	Banana	Titanic	German	b. cancer	f. solar
Quantity	400	150	700	200	666
Quantity using FKNM	14(3.5 %)	3(2 %)	8(1.1 %)	22(11 %)	6(0.9 %)
Quantity using IKMSE	45(11 %)	6(4 %)	28(4 %)	18(9 %)	19(2.9 %)

Each value in parentheses illustrates the ratio of the quantity of key nodes to the quantity of all training samples.

Table 5 presents the classification times of each data set using the IKMSE and KMSE algorithms. Each datum in parentheses represents the ratio of the classification time of the IKMSE to that of the KMSE for every data set. IKMSE classifies much faster than KMSE and this advantage would be even greater on larger datasets. A further advantage of this scheme is that it can automatically calculate the quantity of key nodes.

Table 5. Classification time of **KMSE and IKMSE**

	Banana	Titanic	German	b. cancer	f. solar
KMSE	164.8 s	26.0 s	217.8 s	16.2 s	271.2 s
IKMSE	23.6 s (14.3 %)	4.0 s (15.4 %)	11.4 s (5.2 %)	1.8 s (11.1 %)	9.3 s (3.4 %)

5 Conclusion

This work proposes IKMSE a novel efficient kernel minimum squared error algorithm that improves the computational efficiency of conventional KMSE. The novel algorithm is based on the feature correlation measurement (FCM) which is measured by using the cosine distance. This paper describes how to apply kernel vectors in the feature space and provides an analysis of the FCM. IKMSE determines the key nodes according to a simpler method than FKNM [19]. Our experiments show that IKMSE is both computationally efficient and accurate. Our algorithm is simpler, more flexible, and more practical than the similar algorithms proposed in [17, 18, 21] and should be applicable in other areas such as clustering and RBF. In future work we will seek to employ it in other fields.

Acknowledgements. This article is partly supported by Natural Science Foundation of China (NSFC) under grants Nos. 61472138, 61263032 and 61362031, and Jiangxi Provincial Natural Science Foundation of China under Grant 20161BAB202066, as well as Science and Technology Foundation of Jiangxi Transportation Department of China (2015D0066).

References

1. Xu, J., Zhang, X., Li, Y.: Kernel MSE algorithm: a unified framework for KFD, LS-SVM and KRR. In: International Joint Conference on Neural Networks 2001, Washington DC, USA, pp. 1486–1491 (2001)
2. Billings, S.A., Lee, K.L.: Nonlinear Fisher discriminant analysis using a minimum squared error cost function and the orthogonal least squares algorithm. Neural Netw. **15**(2), 263–270 (2002)
3. Zhu, Q.: Reformative nonlinear feature extraction using kernel MSE. Neurocomputing **73** (16), 3334–3337 (2010)
4. Muller, K.R., Mika, S., Ratsch, G., Tsuda, K., et al.: An introduction to kernel-based learning algorithms. IEEE Trans. Neural Netw. **12**(2), 181–201 (2001)
5. Jenssen, R.: Kernel entropy component analysis. IEEE Trans. Pattern Anal. Mach. Intell. **32** (5), 847–860 (2010)
6. Wang, Y., Guan, L., Venetsanopoulos, A.N.: Kernel cross-modal factor analysis for information fusion with application to bimodal emotion recognition. IEEE Trans. Multimed. **14**(3), 597–607 (2012)
7. Schölkopf, B., Smola, A., Muller, K.R.: Nonlinear component analysis as a kernel eigenvalue problem. Neural Comput. **10**(5), 1299–1319 (1998)
8. Xu, Y., Lin, C., Zhao, W.: Producing computationally efficient KPCA-based feature extraction for classification problems. Electron. Lett. **46**(6), 452–453 (2010)
9. Xu, Y., Yang, J., Lu, J., Yu, D.: An efficient renovation on kernel Fisher discriminant analysis and face recognition experiments. Pattern Recogn. **37**(10), 2091–2094 (2004)
10. Wang, J.: Kernel supervised discriminant projection and its application for face recognition. Int. J. Pattern Recogn. Artif. Intell. **27**(02), 1356003 (2013)
11. Honeine, P.: Online kernel principal component analysis: a reduced-order model. IEEE Trans. Pattern Anal. Mach. Intell. **34**(9), 1814–1826 (2012)
12. Diethe, T., Girolami, M.: Online learning with (multiple) kernels: a review. Neural Comput. **25**(3), 567–625 (2013)
13. Mika, S., Ratsch, G., Weston, J., Scholkopf, B., et al.: Fisher discriminant analysis with kernels. In: Proceedings of the 1999 IEEE Signal Processing Society Workshop on Neural Networks for Signal Processing IX, pp. 41–48 (1999)
14. Zheng, W., Zou, C., Zhao, L.: An improved algorithm for kernel principal component analysis. Neural Process. Lett. **22**(1), 49–56 (2005)
15. Liang, Z., Shi, P.: An efficient and effective method to solve kernel Fisher discriminant analysis. Neurocomputing **61**, 485–493 (2004)
16. Xiong, T., Ye, J., Li, Q., Janardan, R., et al.: Efficient kernel discriminant analysis via QR decomposition. In: Advances in Neural Information Processing Systems, pp. 1529–1536 (2004)
17. Wang, H., Hu, Z., Zhao, Y.E.: An efficient algorithm for generalized discriminant analysis using incomplete Cholesky decomposition. Pattern Recogn. Lett. **28**(2), 254–259 (2007)
18. Ishii, T., Ashihara, M., Abe, S.: Kernel discriminant analysis based feature selection. Neurocomputing **71**(13), 2544–2552 (2008)

19. Xu, Y., Zhang, D., Jin, Z., Li, M., et al.: A fast kernel-based nonlinear discriminant analysis for multi-class problems. Pattern Recogn. **39**(6), 1026–1033 (2006)
20. Kim, J.S., Scott, C.D.: L2 kernel classification. IEEE Trans. Pattern Anal. Mach. Intell. **32**(10), 1822–1831 (2010)
21. Keerthi, S.S., Chapelle, O., DeCoste, D.: Building support vector machines with reduced classifier complexity. J. Mach. Learn. Res. **7**, 1493–1515 (2006)

Robust Supervised Hashing

Tongtong Yuan and Weihong Deng[✉]

Beijing University of Posts and Telecommunications, Beijing, China
{yuantt,whdeng}@bupt.edu.cn

Abstract. Hashing methods on large scale image retrieval have been extensively in attention. These methods can be roughly categorized as supervised and unsupervised. Unsupervised hashing methods mainly search for a projection matrix of the original data to preserve the Euclidean distance similarity, while supervised hashing methods aim to preserve the label similarity. However, most hashing methods propose a complicated objective function and search for optimized or relaxed solutions. Some methods will consume much time to train a good binary code. This paper is not focusing on formulating a complex solution like the previous state-of-art methods. Contrarily, we firstly propose a simple objective function on supervised hashing as far as we have learned. And we devise a novel solution which uses a maximum and equal Hamming distance code to construct the label information. This method keeps a comparable accuracy with the state-of-the-art supervised hashing methods.

Keywords: Image retrieval · Supervised hashing · Hadamard code · Hamming distance · Robustness

1 Introduction

In recent years, hashing methods based on hamming distance have been popular in computer vision, such as content based image retrieval, i.e. CBIR. This method dubbed as hashing is mainly due to the binary code and Hamming distance measurement. Hashing methods have two main steps: (i) quantization - the feature space is partitioned into a number of non-overlapping cells with a unique index (code) for each cell; and (ii) distance computation based on the indices [7]. Hamming distance measurement has got significant gains in storage and speed since it has been proposed in [9].

Hashing methods can be summarized into unsupervised hashing and supervised hashing according to the use of label information in the training procedure. Unsupervised hashing methods deal with the original data without any supervised information. The most famous unsupervised hashing is Locality Sensitive Hashing (LSH) [3]. This method constructs several random hash functions to project the data and achieves a good performance. A simple but effective method is Iterative Quantization (ITQ) [5]. This method learns an optimized rotation matrix to project the real data. ITQ method outperforms other methods by this rotation transformation. Spherical Hashing (SPH) [4] solves the space partition problem with a novel hypersphere and redefined Hamming distance. Density sensitive hashing (DSH) [6] can be regarded as an extension of LSH by exploring the geometric structure of the data to avoid the

© Springer Nature Singapore Pte Ltd. 2016
T. Tan et al. (Eds.): CCPR 2016, Part I, CCIS 662, pp. 574–585, 2016.
DOI: 10.1007/978-981-10-3002-4_47

purely random projections selection. All of these unsupervised methods try to preserve Euclidean similarity, but it turns out to be much weaker in classification than the supervised methods.

Supervised methods obtain a good binary code by taking advantage of the label information or the affinity matrix. One representative work is Semi-Supervised Hashing (SSH) [1]. SSH minimizes the empirical error on the labeled data by maximizing entropy of the generated hash bits over the unlabeled data with pair-wised affinity label matrix. Another method is CCA-ITQ [5]. This method combines the CCA with ITQ and results in a better precision than the uncompressed Euclidean measurement. Supervised Discrete Hashing (SDH) [8] jointly learns a binary embedding and a linear classifier, then proposes a discrete cyclic coordinate (DCC) descent algorithm to generate the hash codes bit by bit. CCA-ITQ and SDH perform better than other previous supervised methods.

Some hashing methods usually generate nonlinear hash functions in a kernel space, including Binary Reconstructive Embedding (BRE) [2], Kernel-Based Supervised Hashing (KSH) [10], Inductive Manifold Hashing (IMH) [15], etc. Most hashing methods propose a complex objective function and look for an optimized solution. These methods can get a pretty good performance but usually consume a lot of time.

Here, we propose a simple and effective method which is comparable with ITQ method [5] and SDH method [8] in supervised hashing. These methods use the point-wise label matrix as supervised information input. This simple method usually performs worse than the complex one, but our improvement on this simple method is effective in hashing image retrieval. One of our main contributions is proposing a simple method based on label information reconstruction. Another is introducing a maximum and equal hamming distance code to improve the simple method. And our hashing method has lower storage cost and less time cost.

In this paper, we present the related work of hashing methods in Sect. 2. Then we formulate an objective function with a linear-regression form in Sect. 3. This method is easy to be understood but performs not well as ITQ or SDH. So we introduce an equal Hamming distance code as supervised information input to replace the old one in Sect. 3. Next, we compare our method with typical supervised and unsupervised hashing methods. Finally, the results show an obvious improvement after changing the label code. This method is proved to be simple but effective by our experiments.

2   Background and Related Work

We have pointed that supervised hashing methods can get a pretty good performance by using label information directly [5, 8] or generating similarity matrix [1]. Here, we focus on the direct label information to deal with the image hashing because similarity matrix usually costs more time and occupies more storage [19]. We enumerate the CCA-ITQ [5] and SDH [8] in this section. Hashing methods need to formulate an objective function to partition the space [8] or minimize the quantization error [5] and a good objective function is the foundation of a good binary code. Here we will compare these supervised hashing objective functions.

For a given set of data points $X = \{x_1, x_2, \ldots, \ldots, x_n\}$ with $x_i \in R^d$, the goal of hashing is to learn a binary code $B = \{b_1, b_2, \ldots, \ldots, b_n\}$ with $b_i = \{0, 1\}^c$, where the ith column b_i is the c-bits binary codes for x_i. The label data is directly denoted as 1-of-k code, i.e. $Y = \{y_1, y_2, \ldots, \ldots, y_n\}$ with $y_i = \{0, 1\}^t$, where t is the total number of labels, and y_{ik} is 1 if the ith image belongs to the class k.

2.1   Iterative Quantization (CCA-ITQ)

The first step of CCA-ITQ is getting a supervised project martix W by (1).

$$C(w_k, u_k) = \frac{w_k^T X^T Y u_k}{\sqrt{w_k^T X^T X w_k u_k^T Y^T Y u_k}} \tag{1}$$

$$s.t. w_k^T X^T X w_k = 1, u_k^T Y^T Y u_k = 1$$

The goal of CCA (Canonical Correlation Analysis) is to find projection directions w_k and u_k for feature and label vectors to maximize the correlation between the projected data Xw_k and Yu_k. Then learn an optimized rotation martix to rotate the data by iterative quantization method, i.e. minimize the quantization error as (2)

$$Q(B, R) = \|B - XWR\|_F^2 \tag{2}$$

R is orthogonal rotation matrix. Solving (2) by fixing R and updating B and let $B = sgn(XWR)$, when fixing B and updating R, ITQ uses the classic Orthogonal Procrustes problem to compute the SVD of the $c \times c$ matrix $B^T V$ as $S\Omega S_1^T$ then let $R = S_1 S^T$.

2.2   Supervised Discrete Hashing (SDH)

$$\min_{B,W,F} \sum_{i=1}^{n} L(y_i, W^T b_i) + \lambda \|W\|^2 + \upsilon \sum_{i=1}^{n} \|b_i - F(x_i)\|^2 \tag{3}$$

$$s.t. b_i \in \{-1, 1\}^L$$

$$F(x) = P^T \phi(x) \tag{4}$$

This objective function is proposed in SDH [8], where y denotes label information, W is a projection matrix to connect the obtained code with the label data. $F(x)$ is defined in (4). To minimize (3), SDH uses a three-step method to separately fix the different variables and loops until converge or reach maximum iterations. In B-step, SDH iteratively learns binary code bit by bit with the DCC method. It avoids a relaxed solution by DCC method. However, the computation complexity of the SDH to deal with the labeled data is huge.

3   Equal Hamming Distance Code

3.1   Our Objective Function

$$\min \ \mathcal{Q}(W) = Y - XW_{\mathrm{F}}^2 \tag{5}$$

$$B = sgn(XW) \tag{6}$$

Differently, we try to deal with the same problem in a simple and direct way. We choose (5) as our objective function. It has only single variable W while CCA-ITQ has W, R and SDH has B, W, F(x). The form of this function is simple and the meaning is easy to be understood. Function (5) attempts to find a suitable matrix W to project X and minimizes the error between original data with label data. This procedure can be seen as minimizing the regression error to classify the images. To get a minimum error in (5), we can easily get W by the following equation.

$$W = (X^T X + \lambda I)^{-1} X^T Y \tag{7}$$

in which λ is the regularization parameter to prevent a trivial solution. And we empirically set λ to 1. After getting the W, we use (6) to generate the binary code like the majority of hashing methods.

This simple way (linear regression) to deal with supervised hashing problem works not well as the complicated methods. We can see the comparison on CIFAR10 dataset in Fig. 1. This simple method is much worse than CCA-ITQ and SDH and performs closely to the unsupervised methods (SPH, DSH). This poor performance may be the reason why it has not been used in hashing method.

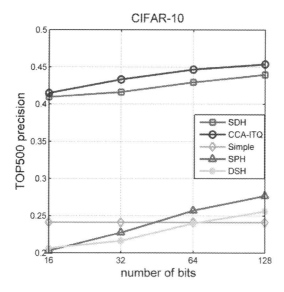

Fig. 1. Top 500 precision on CIFAR10

3.2   Introduce the Question

A general way to encode the label data: the label data is denoted as $Y = \{y_1, y_2, \ldots, \ldots, y_n\}$ with $y_i = \{0, 1\}^t$, where t is the total number of labels, and y_{ik} is 1 if the ith image belongs to the class k. For example, there is a CIFAR10 dataset which has ten different classes. An image label y_i which belongs to the first class can be denoted as $\{1,0,0,0,0,0,0,0,0,0\}$, while the second class can be denoted as $\{0,1,0,0,0,0,0,0,0,0\}$. The hamming distance of this 1-of-k code is zero when two images belong to the same class and equals 2 otherwise.

$$d_h = \begin{cases} 0, when\ y_i = y_j \\ 2, when\ y_i \neq y_j \end{cases} \tag{8}$$

The label code of CIFAR10 dataset is $Y = \{0, 1\}^{n \times 10}$, and the code length is 10 bits while the maximum hamming distance between different classes is 2. If we assume our simple method in Sect. 3.1 can get an ideal zero-error image code, the hamming distance between different images should be same but small. This method cannot achieve ideal retrieval precision because it can easily be influenced by the quantization error so it cannot make an effective classification. We need a maximum Hamming distance between different labels under a given-length code for training a robust classifier. To solve this problem, we formulate a robust label code by enlarging the distance between different classes. The main measure is replacing the common label code with an equal and maximum Hamming distance code which is predefined.

Equal Hamming distance code is a special linear block code which has equal and maximum hamming distance under a given code length. All of these features can be useful to enhance the robustness of label information. The equal Hamming distance preserves the similarity in the Euclidean space. Maximum hamming distances can enlarge the distance of different classes.

Here, we define the code with the following features.

1. There should be maximum hamming distance between different labels to enlarge the inter-class distance
2. There must be an equal hamming distance between different labels to avoid the unbalance of different labels.
3. This code should be easy to generate.

There is a Hadamard Code which satisfies all the features.

3.3   Hadamard Code

Background. Inspired by these demands, we try to find a target code to improve the performance of simple method. Surprisingly, there exists a suitable code which is named as Hadamard Code. This code has been used in [11] to construct a new image class representation in the deep learning network. This special code is a popular error-correcting code which has been used as target coding for its capability of error-correcting (Langford and Beygelzimer in [18]). Because of its unique mathematical properties, the Hadamard code is widely used in coding theory, mathematics, and theoretical computer science. The Hadamard code is also an example of a linear code over a binary alphabet that maps messages of length k to codes of length 2^k. It is unique in that each non-zero codeword has a Hamming weight of exactly 2^{k-1}, which implies that the distance of the code is also 2^{k-1} [16].

When the code length is 2^k, the equal and maximum hamming distance between different codes is 2^{k-1}, and it will generate 2^k kinds of codes. A linear block code denoted as (n, k, d) can be shown as $(2^k, k, 2^{k-1})$ in this Hadamard code.

Constructions. There is a method called Sylvester's method can generate the Hadamard matrix proposed by the Hedayat and Wallis in [17]. This can construct a new Hadamard matrix from the old one by the Kronecker product. A matrix $H \in \{+1, -1\}^{m \times m}$ is called the Hadamard matrix if $HH^T = mI$ where I is an identity matrix.

For example, given the constraints that columns and rows must be orthogonal, we make a Hadamard matrix: $H^2 = \{+1, +1; +1, -1\}$, then $H^4 = H^2 \otimes H^2$, $H^8 = H^4 \otimes H^4$. $\otimes$ stands for Kronecker product. Hadamard matrix is easy to construct by Matlab tools.

$$H^4 = \begin{bmatrix} +1 & +1 \\ +1 & -1 \end{bmatrix} \otimes \begin{bmatrix} +1 & +1 \\ +1 & -1 \end{bmatrix} = \begin{bmatrix} +1 & +1 & +1 & +1 \\ +1 & -1 & +1 & -1 \\ +1 & +1 & -1 & -1 \\ +1 & -1 & -1 & +1 \end{bmatrix} \tag{9}$$

In this case, the Hamming distance between different ones can achieve a maximum value in the given code length like in (10). That is, when we use H^{16}, the margin of different labels is 8, which is large enough to make a good classification and our following experiment proves this truth.

$$d_h = \begin{cases} 0, & when\ y_i = y_j \\ 2^{k-1}, & when\ y_i \neq y_j \end{cases} \tag{10}$$

Adjusted Version. We notice that the first column and row are composed of +1. In order to get a balanced code (the number of +1 and -1 is roughly equal), we must abandon the first row and column though its influence is small to the experimental result. The rest codes (*) are used in our codebook just like (11). So our code length will be $2^k - 1$.

$$\begin{bmatrix} +1 & \cdots & +1 \\ \vdots & * & * \\ +1 & * & * \end{bmatrix} \tag{11}$$

This changed code also has all the three features we have defined above. It has the equal and maximum hamming distance. Therefore, main purpose of equal and maximum Hamming distance is enlarging the distance between classes and getting a better robustness than the defined label matrix in previous papers [5, 8]. The changed simple method which only uses a short bit length can perform as well as state-of-art methods.

4   Experiment

We compare our method with some state-of-art methods (supervised and unsupervised) on three different image dataset: CIFAR10 [12], MNIST[1], SUN15[2]. These methods includes CCA-ITQ [5], SDH [8], SSH [1], DSH [6], SPH [4]. We use the similar experiment criterions in supervised hashing methods [5, 8]. For SDH, we set λ to 1 and randomly sampled 1000 anchor points. And in our EHH, because the purpose of our method is to reconstruct the label information, we set the binary code of training data with Hadamard code.

4.1   CIFAR10

CIFAR10 consists of 60,000 images that have been manually labeled into 10 classes (airplane, automobile, bird, cat, deer, dog, frog, horse, ship and truck). The tiny images are 32*32 pixels. We extract 512 dimensions GIST descriptors [13] to form a feature dataset. We divide the 60000 images into training set and testing set. We randomly choose 1000 images as testing set and the rest as training data. We compare our proposed method EHH (Equal Hadamard Hashing) with the original simple method (Simple) in Fig. 2 (Left). We show the final hashing code length as 16, 32, 64 and 128. Though our EHH code length is $2^k - 1$, we also show the same bit value as ITQ and SDH for convenience. We choose top 500 retrieved images to calculate average precision which is used in [5]. TOP500 images on the retrieved data refer to the 500 samples of minimum Hamming distance with the query. EHH performs an absolute advantage than other methods. The precision of modified methods has been improved by 27 %.

We need to point that the bit length of EHH can be shortened to 15 because that the Hamming distance is long enough to classify the different classes. And the results show that the precision of EHH changes a little along with the increase of bit length in Fig. 2 (Left). And other experiments have the similar result. So we fix the code length of our EEH method to show our advantage in code length. We use 15 bits to code the data and this method achieves better result than the long bits (32, 64, 128 bits) in CCA-ITQ and SDH. Therefore, if we don't emphasize, the default code length of EHH is 15 in the following experiments.

We also compare the mean average precision of these methods which is used in SDH [8]. Figure 2 (Right) shows MAP of many hashing methods. MAP is mean of average precision, which is also named as Hamming ranking. It measures the search

---

[1] http://yann.lecun.com/exdb/mnist/. Accessed 20 May 2016.
[2] http://vision.princeton.edu/projects/2010/SUN/. Accessed 20 May 2016.

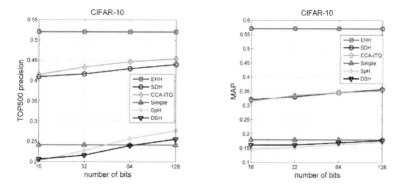

Fig. 2. Comparisons of many hashing methods on Cifar10

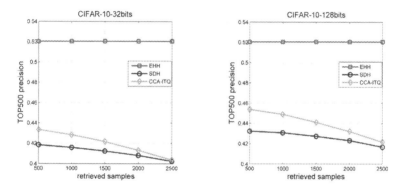

Fig. 3. Precision varies with retrieved samples. The number of retrieved samples is 500, 1000, 1500, 2000 and 2500. We report the precision when using 32bit and 128bit.

quality through ranking the retrieved samples according to their Hamming distances to a specific query [14]. MAP stands for the whole performance of the hashing methods. Our EHH has the highest precision and outperforms other state-of-art methods by a large margin (more than 20 %) on CIFAR10. Notice that we use only 15 bits. Unsupervised methods show a bad performance, so we abandon these methods in the following comparisons. We also report the precision along with retrieved samples in Fig. 3. EHH has no obvious changes with the increasing number, while SDH and CCA-ITQ have a slight decrease.

4.2 MNIST

MNIST is composed of 70,000 hand-written digit images that also have been grouped into 10 classes. We use the original 784-D feature in our experiment. 1,000 images of the whole dataset act as testing data. We only compare the supervised hashing methods (SDH, CCA-ITQ, SSH, Simple, EHH). SSH [1] use a similarity matrix rather than direct label matrix. We randomly choose 1,000 images from training images to

Table 1. Training time on MNIST data set

Method time/s bit	EHH	CCA-ITQ	SDH
16	3.43	5.12	42.17
32	3.47	6.73	54.26
64	4.04	12.20	116.72
128	4.72	24.67	295.41

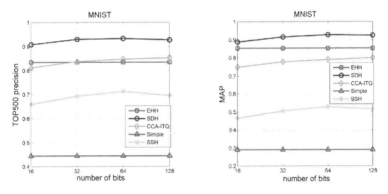

Fig. 4. Top 500 precision and MAP on MNIST dataset

generate the relative similarity matrix S. $S_{ij} = +1$ when the ith image and j^{th} images have the same label, $S_{ij} = -1$ otherwise. Other methods use a direct label matrix as supervised information. MNIST is an easier dataset compared with CIFAR10 and SUN15. All the precision of listed hashing methods is high and SDH performs better than the CCA-ITQ and our EHH, but EHH has a short bit length. EHH performs better than the SSH which uses a pair-wise label information. To better illustrate the efficiency of our EHH, we compare the training time among EHH, SDH and CCA-ITQ in Table 1. Here, our code length of EHH is not limited into 15 bits but the same as SDH and CCA-ITQ in order to a fair comparison. Table 1 shows that SDH costs most time and EHH is the fastest one (Fig. 4).

4.3 SUN15

SUN15 is a sub-sample of SUN397 which has 108,754 images of 397 well-sampled categories. SUN15 is composed of 4,485 images of 15 well-sampled categories (bedroom, kitchen, store, mountain, etc.). We also extract 512-D Gist feature. We use 1000 samples as testing data and the rest 3,485 images as training set. Although this data set has fewer images, it has more classes. The use of SUN 15 dataset is persuasiveness. Because it has 15 classes, we just choose H^{16} to generate the suitable code.

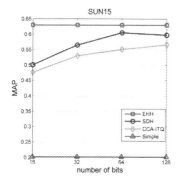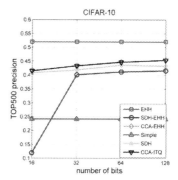

Fig. 5. Results of MAP on SUN15 dataset (Left). Comparisons of SDH and CCA-ITQ coupled with EHH (SDH-EHH, CCA-EHH) on CIFAR10 dataset (Right)

We compare the mean average precision with good supervised methods. Figure 5 (Left) shows our EHH method performs best. This proves our method can fit more classes and this Hadamard code is flexible.

4.4   Experiment Summary

Our method named as EHH (Equal Hadamard hashing) performs a best result in the extensive experiments. The simple method has the minimum precision and the retrieval precision is lower than the unsupervised hashing. The introduction of Hadamard code brings a huge leap to the simple method. This changed supervised method (EHH) performs much better than the typical unsupervised methods (SPH and DSH). From the results of CIFAR10, we can see that EHH shows the best performance. On MNIST dataset, we can see the precision of SDH is obviously higher than CCA-ITQ and EHH. But CCA-ITQ and EHH have a near precision in the comparison of TOP 500 precision. The MAP of EEH is higher than CCA-ITQ. So our EHH is also comparable with the two state-of-art supervised methods on MNIST. And our EHH performs best on SUN15 dataset. Therefore, Equal Hadamard hashing is useful to enlarge the distance of different classes. Using this method to code the label will highly improve the precision of the simple one.

We also try to couple the CCA-ITQ and SDH with the Equal Hadamard code. Our method is based on the reconstruction of label matrix while other methods are different from this purpose. So this code only improves this simple regression problem but have no effect on these complex methods. We report this result in Fig. 5 (Right).

In summary, our EHH method satisfies the hashing demands in both speed and precision. We evaluate these methods under different standards. The final result shows our robust hashing EHH is superior to state-of-art methods. It is a good hashing method which has a fast training speed and a high precision to fit different evaluation standards. We present a new direction of image hashing using label information which is simple but effective.

5   Conclusion

This paper starts with an ordinary objective function (linear regression). Then introduce an improvement algorithm to replace the empirical label matrix with an Equal Hamming distance code, i.e. Hadamard code. The introduction of Equal Hadamard code is different to all of the supervised methods and it observably improves the simple regression method. Making a good code to the label information is a simple but effective method to obtain a high retrieval precision. Our robust method EHH can be applied to image retrieval system because of its fast speed, short code length and high retrieval precision. Hashing based on label reconstruction is promising in the future. Our next work will continue to search for a more effective way to change the way of hashing.

Acknowledgments. This work was partially sponsored by supported by the NSFC (National Natural Science Foundation of China) under Grant No. 61375031, No. 61573068, No. 61471048, and No.61273217, the Fundamental Research Funds for the Central Universities under Grant No. 2014ZD03-01, This work was also supported by Beijing Nova Program, CCF-Tencent Open Research Fund, and the Program for New Century Excellent Talents in University.

References

1. Wang, J., Kumar, S., Chang, S.-F.: Semi-supervised hashing for large scale search. IEEE TPAMI **34**(12), 2393–2406 (2012)
2. Kulis, B., Darrell, T.: Learning to hash with binary reconstructive embeddings. In: NIPS, vol. 22 (2009)
3. Andoni, A., Indyk, P.: Near-optimal hashing algorithms for approximate nearest neighbor in high dimensions. ACM (2008)
4. Heo, J.-P., Lee, Y., He, J., Chang, S.-F., Eui Yoon, S.: Spherical Hashing. In: Proceedings of IEEE Conference on Computer Vision and Pattern Recognition, pp. 2957–2964 (2012)
5. Gong, Y., Lazebnik, S., Gordo, A., Perronnin, F.: Iterative quantization: A procrustean approach to learning binary codes for large-scale image retrieval. IEEE TPAMI **35**(12), 2916–2929 (2013)
6. Jin, Z., Li, C., Lin, Y., Cai, D.: Density sensitive hashing. IEEE Trans. Cybern. **PP**(99), 1 (2013)
7. He, K., Wen, F., Sun, J.: K-means hashing: an affinity preserving quantization method for learning binary compact codes. In: CVPR (2013)
8. Shen, F., Shen, C., Liuand, W., Shen, H.: Supervised discrete hashing. In: CVPR (2015)
9. Torralba, A., Fergus, R., Weiss, Y.: Small codes and large databases for recognition. In: Proceedings of CVPR (2008)
10. Liu, W., Wang, J., Ji, R., Jiang, Y.-G., Chang, S.-F.: Supervised hashing with kernels. In: Proceedings of CVPR (2012)
11. Yang, S., Luo, P., Loy, C.C., Shum, K.W., Tang, X.: Deep representation learning with target coding. In: AAAI (2015)
12. Krizhevsky. Learning multiple layers of features from tiny images. Technical report, University of Toronto (2009)
13. Oliva, A., Torralba, A.: Modeling the shape of the scene: a holistic representation of the spatial envelope. IJCV **42**, 145–175 (2001)

14. Lin, G., Shen, C., Suter, D., van den Hengel, A.: A general two-step approach to learning-based hashing. In: Proceedings of International Conference on Computer Vision (ICCV) (2013)
15. Shen, F., Shen, C., Shi, Q., van den Hengel, A., Tang, Z., Shen, H.T.: Hashing on nonlinear manifolds. IEEE TIP **24**(6), 1839–1851 (2015)
16. Hadamard code. https://en.wikipedia.org/wiki/Hadamard_code. Accessed 20 May 2016
17. Hedayat, A., Wallis, W.D.: Hadamard matrices and their applications. Ann. Stat. **6**, 1184–1238 (1978)
18. Langford, J., Beygelzimer, A.: Sensitive error correcting output codes. In: COLT (2005)
19. Li, W.J., Zhou, Z.H.: Learning to hash for big data: current status and future trends. Chin. Sci. Bull. **60**, 485–490 (2015). doi:10.1360/N972014-00841. (in Chinese)

Joint Learning of Distance Metric and Kernel Classifier via Multiple Kernel Learning

Weiqi Zhang, Zifei Yan$^{(\boxtimes)}$, Hongzhi Zhang, and Wangmeng Zuo

Harbin Institute of Technology, Harbin 150001, China
`cszfyan@gmail.com`

Abstract. Both multiple kernel learning (MKL) and support vector metric learning (SVML) were developed to adaptively learn kernel function from training data, and have been proved to be effective in many challenging applications. Actually, many MKL formulations are based on either the max-margin or radius-margin based principles, which in spirit is consistent with the optimization principle of the between/within class distances adopted in SVML. This motivates us to investigate their connection and develop a novel model for joint learning distance metric and kernel classifier. In this paper, we provide a new parameterization scheme for incorporating the squared Mahalanobis distance into the Gaussian RBF kernel, and formulate kernel learning into a GMKL framework. Moreover, radius information is also incorporated as the supplement for considering the within-class distance in the feature space. We demonstrate the effectiveness of the proposed algorithm on several benchmark datasets of varying sizes and difficulties. Experimental results show that the proposed algorithm achieves competitive classification accuracies with both state-of-the-art metric learning models and representative kernel learning models.

Keywords: Metric learning · Multiple kernel learning · Support vector machine · Mahalanobis distance · Gaussian Radial Basis Function kernel

1 Introduction

Kernel methods, such as Support Vector Machines (SVM), have achieved great success in many learning tasks, and have been widely adopted in various challenging real-world applications ranged from computer vision [1–3] to medical imaging analysis [4]. For these tasks, the performance strongly depends on the choice of the kernels in the learning model. Hand-tuning of kernel functions and parameters, however, is exhausting and cannot guarantee the generalization ability. Kernel learning aims to learn proper kernels from the training data for specific task, and thus can provide a principled solution to this issue.

Recently, a number of kernel learning approaches has been proposed. Chapelle et al. [5] presented several radius-margin principles to tune parameters in kernel functions. Rather than simply tuning parameters for a specific kernel, multiple kernel learning (MKL) suggested to construct a composite kernel by combining a number of basis kernels, where the combination coefficients

© Springer Nature Singapore Pte Ltd. 2016
T. Tan et al. (Eds.): CCPR 2016, Part I, CCIS 662, pp. 586–600, 2016.
DOI: 10.1007/978-981-10-3002-4_48

can be learned from the training data. Many MKL formulations have been proposed in the literatures. Lanckriet et al. [6] formulated MKL as a quadratically constrained quadratic programming problem, using a l_1-norm constraint to promote sparse combinations. Considering that the model in [6] is non-smooth, Bach et al. [7] proposed a smoothed version and proposed a SMO-like algorithm for solving the problem. Sonnenburg et al. [8] reformulated the problem as a semi-infinite linear program and addressed the problem by iteratively solving a classical SVM problem. Rakotomamonjy et al. [9] further developed a SimpleMKL algorithm, and demonstrated that the training time could be further reduced by nearly an order of magnitude on some standard UCI datasets when the number of kernels was large. To further improve the representative ability of combined multiple kernels, Varma et al. [10] developed the Generalized Multiple Kernel Learning (GMKL) formulation, which allowed fairly general kernel parameterization, including both linear and non-linear kernel combinations, together with general regularizations on the kernel parameters. Wu et al. [11] proposed a pre-selecting base kernel method called MRMRKA, using data set to select base kernels from kernel candidates and feed into the process of MKL.

Most existing MKL approaches employ the objective function used in SVM, aiming at finding kernels maximizing the largest margin of the SVM classifier combined with an acceptable empirical loss, which can be explained as maximizing the between-class distance of the input data from a metric learning view. In order to take also the within-class distance of input data into consideration, the radius of the minimum enclosing ball (MEB) of data in the feature space endowed with the learned kernel was integrated into MKL in recent works [12,13].

In the test stage, the MKL approach usually should compute the kernel inner products between the test sample and the support vectors for each basis kernel with non-zero coefficient. When the number of basis kernels is high, the computational cost will be heavy. Fortunately, when the learned kernel can be represented as a kernel function on the linear transformed space, the efficiency in the test stage can be greatly improved in the test stage. To this end, Xu et al. [14] incorporated the distance metric learning idea into a single kernel learning framework and proposed an efficient kernel classifier, SVML, which seamlessly combined the learning of a Mahalanobis metric with the training of the RBF-SVM parameters.

In this paper, we borrow the advantages of both MKL and SVML, and suggest a novel method for joint learning distance metric and kernel classifier. We analyze this problem from the MKL perspective, and formulate it into a GMKL framework. Benefitted from this formulation, we can adopt GMKL for joint learning distance metric and kernel classifier. While in the test stage we can utilize the composition of linear transformation and SVM for classification. The complexity of our model is independent with the number of basis kernels, and thus can avoid the computational burden issue of conventional MKL approach. Moreover, radius information is also incorporated as the supplement of considering the within-class distance of input data. Our extensive experiments on UCI dataset classification clearly demonstrate the effectiveness of the proposed methods.

The contribution of this paper is two-fold. (i) We formulate the Mahalanobis distance into the Gaussian RBF kernel, which behaves as the basis kernel inborn with the thinking of metric learning. With this formulation, the involved kernel matrix can be obtained by computing the pairwise distance over given pairs of instances in the original feature space, instead of transforming to complicated high dimension feature spaces. (ii) The learning of the distance metric kernel can be effortlessly incorporated into the Generalized MKL framework, and the flexibility of the selection of kernel size can tremendously reduce the computation overhead of whole MKL framework.

The remainder of this paper is organized as follows. We review the related work in Sect. 2. In Sect. 3 we propose a novel MKL framework from the viewpoint of metric learning. In Sect. 4, we present experiments on several benchmark datasets, and finally conclude in Sect. 5.

2 Related Work

Denote by $\mathcal{X} = \{(\mathbf{x}_i, y_i)_{i=1}^n\}$ a training set of n samples, where $\mathbf{x}_i \in \mathbb{R}^d$ denotes the ith training sample, $y_i \in \{+1, -1\}$ denotes the class label of $\mathbf{x}_i$. Let $\varphi_p : \mathbb{R}^d \to \mathcal{H}_p$ be the pth feature mapping, inducing the corresponding basis kernel $K_p(\cdot, \cdot)$ in the Hilbert space $\mathcal{H}_p$, where $p = 1, ..., m$. In the MKL framework based on SVM, each sample $\mathbf{x}$ is mapped onto m feature spaces by $\varphi(\mathbf{x}; \boldsymbol{\gamma}) = [\sqrt{\gamma_1}\varphi_1(\mathbf{x}), ..., \sqrt{\gamma_m}\varphi_m(\mathbf{x})]^T$, where γ_p is the weight of the pth basis kernel. Therefore, the employed kernel can be expressed as a linear or nonlinear combination of p basis kernels, expressed as $K(\cdot, \cdot; \boldsymbol{\gamma}) = \sum_{p=1}^m \gamma_p K_p(\cdot, \cdot)$ or $K(\cdot, \cdot; \boldsymbol{\gamma}) = \prod_{p=1}^m (K_p(\cdot, \cdot))^{\gamma_p}$, the formulation of which depends on the needs of different applications. To seek the optimal combination weight γ_p for each basis kernel, most MKL approaches [8,9,15] are suggested to solve the following objective function:

$$
\begin{aligned}
\min_{\mathbf{w}, b, \xi} \quad & \frac{1}{2}\|\mathbf{w}\|_2^2 + C\sum_i \xi_i, \\
s.t. \quad & y_i(\mathbf{w}^T \varphi(\mathbf{x}_i; \boldsymbol{\gamma}) + b) \geq 1 - \xi_i, \forall i, \\
& \xi_i \geq 0, i = 1, 2, \cdots, n, \|\boldsymbol{\gamma}\|_1 = 1, \boldsymbol{\gamma} \succ 0.
\end{aligned}
\tag{1}
$$

where $\mathbf{w}$ is the normal of the separating hyperplane, b is the bias term, ξ_i is the slack variable aiming at maximizing the margin with respect to $\boldsymbol{\gamma}$.

Recently, Gai et al. [13] proved that, the aforementioned large margin principle will lead to scaling and initialization problems. A large margin can be arbitrarily achieved by scaling $\boldsymbol{\gamma}$ to $\tau\boldsymbol{\gamma}(\tau > 1)$, even the kernel may result in poor performance. Although a norm constraint can be imposed on the kernel parameters, the final solution can still be greatly affected by the initial scaling of basis kernels. To remove or mitigate these issues, recent work on MKL has considered incorporating the radius of the minimum enclosing ball (MEB) of data in the feature space endowed with the learned kernel into the traditional formulation, with the manner of using the margin-and-radius ratio. In [13], a tri-level optimization problem was proposed:

$$\min_{\boldsymbol{\gamma}} \mathbf{g}(\boldsymbol{\gamma}) \tag{2}$$

where

$$\mathbf{g}(\boldsymbol{\gamma}) = \{\max_{\alpha_i} \sum_i \alpha_i - \frac{1}{2R^2(\boldsymbol{\gamma})} \sum_{i,j} \alpha_i \alpha_j y_i y_j k_{i,j}, s.t. \sum_i \alpha_i y_i = 0, 0 \le \alpha_i \le C\} \tag{3}$$

$$R^2 = \{\max_{\beta_i} \sum_i \beta_i k_{i,j}(\boldsymbol{\gamma}) - \sum \beta_i k_{i,j}(\boldsymbol{\gamma})\beta_j, s.t. \sum_i \beta_i = 1, \boldsymbol{\beta} \ge 0\} \tag{4}$$

In this tri-level optimization structure, $R^2(\boldsymbol{\gamma})$ is computed by solving the quadratic programming (QP) in (4) with a given $\boldsymbol{\gamma}$, and then taken into (3) to solve another QP to calculate $\mathbf{g}(\boldsymbol{\gamma})$. Thus an extra QP was solved at each iteration comparing with traditional MKL algorithms, which considerably increased the computation cost of SVM-based MKL. According to literature [16], the trace of the total scatter matrix $\text{tr}(\mathbf{S}_T)$ in feature space can be shown as an approximation of $R^2(\boldsymbol{\gamma})$, which can be explicitly expressed in the kernel-induced feature space as

$$\text{tr}(\mathbf{S}_T) = \text{tr}(\mathbf{K}(\boldsymbol{\gamma})) - \frac{1}{n}\mathbf{1}^T\mathbf{K}(\boldsymbol{\gamma})\mathbf{1} = \sum_{p=1}^m \gamma_p a_p \tag{5}$$

where $a_p \triangleq \text{tr}(\mathbf{K}_p - (1/n)\mathbf{1}^T\mathbf{K}_p\mathbf{1})$, $\mathbf{K}_p$ is the kernel matrix computed with $K_p(\cdot,\cdot)$ on training set $\mathcal{X}$. To shorten the kernel learning time and avoid the extra QP for computing the radius, Liu et al. [17] substituted $R^2(\boldsymbol{\gamma})$ with $\text{tr}(\mathbf{S}_T)$ to incorporate the radius information and reformulated the SVM-based MKL problem as:

$$\min_{\boldsymbol{\gamma}} \mathbf{g}(\boldsymbol{\gamma})$$

where

$$\mathbf{g}(\boldsymbol{\gamma}) = \{\min_{\mathbf{w},b} \frac{1}{2}\text{tr}(\mathbf{S}_T)\|\mathbf{w}\|^2, \ s.t. \ y_i(\mathbf{w}^T\varphi(\mathbf{x}_i;\boldsymbol{\gamma})+b) \ge 1, \forall i\} \tag{6}$$

In literature [18], they also proved that with this substitution, not only the extra level of quadratic optimization needed to compute the radius value can be avoided, the refined MKL framework can also behave more robust to outliers.

Instead of using the ratio, Do et al. [19] proved that the sum of the radius and the inverse of the margin can achieve the same optimal solution under proper parameter choices. Their R-SVM$_\mu^+$ is formulated as:

$$\min_{\mathbf{w},b,\boldsymbol{\mu},R_{\boldsymbol{\mu}},x_0} \quad \frac{1}{2}\|\mathbf{w}\|^2 + \lambda R_{\boldsymbol{\mu}}^2 + C\sum_{i=1}^{l}\xi_i,$$

$$s.t. \quad y_i(\langle \mathbf{w}, D_{\sqrt{\boldsymbol{\mu}}}\mathbf{x}_i\rangle + b) \geq 1 - \xi_i, \forall i, i = 1, 2, \cdots, n, \tag{7}$$

$$\sum_{k=1}^{d}\mu_k = 1, \boldsymbol{\mu}, \boldsymbol{\xi} \geq 0,$$

$$\|D_{\sqrt{\boldsymbol{\mu}}}\mathbf{x}_i - D_{\sqrt{\boldsymbol{\mu}}}\mathbf{x}_0\| \leq R_{\boldsymbol{\mu}}^2, \forall i.$$

where $\sqrt{\boldsymbol{\mu}}$ is a vector scaling the transformed feature space; $D_{\sqrt{\boldsymbol{\mu}}}$ is a diagonal linear transformation matrix, whose diagonal elements are given by $\sqrt{\boldsymbol{\mu}}$, thus $D_{\sqrt{\boldsymbol{\mu}}}\mathbf{x}$ gives the image of an instance $\mathbf{x}$ in the transformed feature space; $R_{\boldsymbol{\mu}}$ denotes the radius of the scaled feature space.

Their optimization problem solved SVM from the view point of metric learning, and approximated the radius of the smallest sphere enclosing the data by the maximum pairwise distance over all pairs of instances in the transformed feature space, which may result in very inefficient to solve large-scale problems.

3   Proposed JLDM-MKL Algorithm

In this section, by extending Gaussian RBF kernel, we formulate the joint distance metric and kernel classifier learning problem into the Generalized MKL (GMKL) framework. We then present the learning algorithm for solving the proposed model, and further incorporate the radius information to improve our model.

3.1   Extension of Gaussian RBF Kernel

Gaussian Radial Basis Function (RBF) kernel is a kernel function which has been widely adopted in various kernel methods. Given two samples $\mathbf{x}$ and $\mathbf{y}$, the Gaussian RBF kernel is defined as

$$K(\mathbf{x},\mathbf{y}) = \exp\left(-\frac{\|\mathbf{x}-\mathbf{y}\|_2^2}{\sigma^2}\right) = \exp(-d_\sigma^2(\mathbf{x},\mathbf{y})) \tag{8}$$

where $d_\sigma(\mathbf{x},\mathbf{y}) = \|\mathbf{x}-\mathbf{y}\|_2^2/\sigma^2$ can be explained as a scaled version of squared Euclidean distance. Here we further extend Gaussian RBF kernel by replacing the scaled Euclidean distance $d_\sigma(\mathbf{x},\mathbf{y})$ with the Mahalanobis distance,

$$d_{\mathbf{M}}^2(\mathbf{x},\mathbf{y}) = (\mathbf{x}-\mathbf{y})^T\mathbf{M}(\mathbf{x}-\mathbf{y}) \tag{9}$$

where the matrix $\mathbf{M} \in \mathcal{R}^{d\times d}$ is semi-positive definite. The resulting extended kernel function is then defined follows:

$$K_{\mathbf{M}}(\mathbf{x},\mathbf{y}) = \exp\left(-d_{\mathbf{M}}^2(\mathbf{x},\mathbf{y})\right) = \exp(-(\mathbf{x}-\mathbf{y})^T\mathbf{M}(\mathbf{x}-\mathbf{y})) \tag{10}$$

It is easy to see that $K_{\mathbf{M}}(\mathbf{x}, \mathbf{y})$ is a kernel function and satisfies the Mercer condition [22].

Rather than directly learning $\mathbf{M}$, Xu et al. [14] parameterized $\mathbf{M} = \mathbf{L}^T\mathbf{L}$, and suggested a gradient descent algorithm to learn $\mathbf{M}$ from the training data. Inspired by the metric learning with multiple kernel embedding proposed by Lu et al. [20] and Doublet-SVM metric learning methods proposed by Wang et al. [21], we parameterize $\mathbf{M}$ as,

$$\mathbf{M} = \sum_{ij}^{N_K} \beta_{ij}(\mathbf{x}_i - \mathbf{x}_j)(\mathbf{x}_i - \mathbf{x}_j)^T = \sum_{ij}^{N_K} \beta_{ij}\mathbf{X}_{ij}, \beta_{ij} \geqslant 0 \qquad (11)$$

where $\mathbf{X}_{ij} = (\mathbf{x}_i - \mathbf{x}_j)(\mathbf{x}_i - \mathbf{x}_j)^T$, β_{ij} is the weight describing the contribution of sample pair $(\mathbf{x}_i, \mathbf{x}_j)$ to make up the metric matrix $\mathbf{M}$. With Eq. (10), the kernel can be reformulated as:

$$\begin{aligned} K_{\mathbf{M}}(\mathbf{x}, \mathbf{y}; \boldsymbol{\beta}) &= \prod_{ij}^{N_K} \exp(-(\mathbf{x} - \mathbf{y})^T \beta_{ij}\mathbf{X}_{ij}(\mathbf{x} - \mathbf{y})) \\ &= \prod_{ij}^{N_K} (\exp(-(\mathbf{x} - \mathbf{y})^T \mathbf{X}_{ij}(\mathbf{x} - \mathbf{y})))^{\beta_{ij}} \qquad (12) \\ &= \prod_{ij}^{N_K} (K_{ij}(\mathbf{x}, \mathbf{y}))^{\beta_{ij}} \end{aligned}$$

By defining the basis kernel $K_{ij}(\mathbf{x}, \mathbf{y}) = \exp(-(\mathbf{x}-\mathbf{y})^T\mathbf{X}_{ij}(\mathbf{x}-\mathbf{y}))$, the kernel function in Eq. (12) can also be explained for the multiple kernel perspective [22]. If we take all the sample pairs into account, the number of basis kernels $N_K = N(N-1)/2$, which will be too huge for large scale dataset. To address this, one can adopt the following strategies to reduce N_K: (i) We can refer to the metric learning methods [21] by only using the nearest similar pairs and the nearest dissimilar pairs to construct the set of basis kernels. (ii) After selecting pairs based on (i), some clustering methods (e.g., K-means) can be adopted to further reduce the number of pairs for constructing basis kernels.

3.2   JLDM-MKL Formulation

Denote by $\varphi_{\mathbf{M}}(\mathbf{x}; \boldsymbol{\beta})$ the feature mapping associated with the kernel function $K_{\mathbf{M}}(\mathbf{x}, y; \boldsymbol{\beta})$. Our objective is to learn a function of the form $f(\mathbf{x}) = \mathbf{w}^T\varphi_{\mathbf{M}}(\mathbf{x}_i; \boldsymbol{\beta}) + b$. Give $\varphi_{\mathbf{M}}(\mathbf{x}; \boldsymbol{\beta})$, one can adopted the SVM solver to learn the global optimal values of $(\mathbf{w}, b)$ from the training data $\{(\mathbf{x}_i, y_i)_{i=1}^n\}$. If we want to jointly learning $\boldsymbol{\beta}$ and $(\mathbf{w}, b)$, we should consider the MKL framework. Therefore, we adopt the generalized MKL formulation in [10],

$$\min_{\mathbf{w}, b, \boldsymbol{\beta}} \quad \frac{1}{2}\mathbf{w}^t\mathbf{w} + \sum_i l(y_i, f(\mathbf{x}_i)) + r(\boldsymbol{\beta}) \qquad (13)$$

$$s.t. \quad \boldsymbol{\beta} \geq 0.$$

where both the regularizer $r(\beta)$ and the kernel should be differentiable with respect to β and l could be some loss functions, e.g., hinge loss and logistic loss. In order to learn the classifier and the Mahalanobis distance metric jointly, we formulate the problem as:

$$
\begin{aligned}
\{\mathbf{w}, b, \mathbf{M}, \boldsymbol{\beta}\} = \quad &\arg\min_{\mathbf{w}, b, \mathbf{M}} = \frac{1}{2}\|\mathbf{w}\|_2^2 + C\sum_i \xi_i + \lambda r(\boldsymbol{\beta}) \\
&s.t. \quad y_i(\langle \mathbf{w}, \varphi_{\mathbf{M}}(\mathbf{x}_i; \boldsymbol{\beta})\rangle + b) \geq 1 - \xi_i, \\
&\quad \boldsymbol{\beta} \geq 0, \xi_i \geq 0, \forall i.
\end{aligned}
\tag{14}
$$

Reformulating above primal as a nested two step optimization, the kernel is learned by optimizing over β in the outer loop, while the kernel is fixed and the SVM parameters are learnt in the inner loop. This can be achieved by rewriting the primal as follows:

$$
\min_{\boldsymbol{\beta}} J(\boldsymbol{\beta}), s.t. \quad \beta_{ij} \geq 0
$$

with

$$
\begin{aligned}
J(\boldsymbol{\beta}) = \max \mathbf{1}^T\boldsymbol{\alpha} - \frac{1}{2}\boldsymbol{\alpha}^T\mathbf{Y}\mathbf{K_M}(\boldsymbol{\beta})\mathbf{Y}\boldsymbol{\alpha} + \lambda r(\boldsymbol{\beta}) \\
s.t. \quad \mathbf{1}^T\mathbf{Y}\boldsymbol{\alpha} = 0, 0 \leq \boldsymbol{\alpha} \leq C
\end{aligned}
\tag{15}
$$

where $\mathbf{Y}$ is a diagonal matrix with the labels on the diagonal, λ is a tradeoff to balance the regularization part. As to $r(\boldsymbol{\beta})$, its derivative should exist and be continues. For example, the non-negative l_1-norm regularization $r(\boldsymbol{\beta}) = \mathbf{1}^T\boldsymbol{\beta}$ could be used for learning sparse solutions, or the l_2-norm regularization $r(\boldsymbol{\beta}) = (\boldsymbol{\beta} - \boldsymbol{\mu})^T\boldsymbol{\Sigma}^{-1}(\boldsymbol{\beta} - \boldsymbol{\mu})$ if prior knowledge is available. If $\nabla_{\boldsymbol{\beta}}\mathbf{K_M}$ and $\nabla_{\boldsymbol{\beta}}r$ are smoothly varying function of $\boldsymbol{\beta}$, we can utilize gradient descent in the outer loop, and $J(\boldsymbol{\beta})$ has derivatives given by

$$
\frac{\partial J}{\partial \beta_k} = \lambda\frac{\partial r}{\partial \beta_k} - \frac{1}{2}\boldsymbol{\alpha}^{*T}\frac{\partial \mathbf{Y}\mathbf{K_M}\mathbf{Y}}{\partial \beta_k}\boldsymbol{\alpha}^*
\tag{16}
$$

In order to take a gradient step, in the inner loop, all we need to do is to obtain $\boldsymbol{\alpha}^*$, which can be solved by any off-the-shelf SVM optimization package. To solve the non-convex formulation results from regularizing $\boldsymbol{\beta}$, the projection gradient can be adopted to update it, and the proposed JLDM-MKL algorithm is summarized in Algorithm 1.

3.3   Incorporating Radius Information into JLDM-MKL

Considering the fact that the generalization error of SVM is actually a function of the ratio of radius and margin [23], especially for joint learning of kernels for feature transformation and classifier, the radius information should be valuable and cannot be ignored. Due to the close relationship with the radius of MEB, instead of incorporating the radius directly, the trace of the total scattering

---

Algorithm 1. JLDM-MKL Algorithm

1: Input: Training samples $\mathcal{X} = \{(\mathbf{x}_i, y_i)_{i=1}^n\}$
2: Output: $\boldsymbol{\beta}$
3: $n \leftarrow 0$,
4: Randomly initialize $\boldsymbol{\beta}^0$ from $[0,1]$.
5: **repeat**
6:     Compute distance metric matrix $\mathbf{M}$ and kernel matrix $\mathbf{K_M}$ from $\boldsymbol{\beta}^n$:
7:     $\mathbf{K_M} \leftarrow \mathbf{K_M}(\boldsymbol{\beta}^n)$
8:     Call LibSVM to obtain $\boldsymbol{\alpha}^*$,
9:     Compute gradient $\frac{\partial J}{\partial \boldsymbol{\beta}^n}$ and update on $\boldsymbol{\beta}$:
10:     $\beta_k^{n+1} = \beta_k^n - step * \left[\lambda \frac{\partial r}{\partial \beta_k} - \frac{1}{2}\boldsymbol{\alpha}^{*T}\frac{\partial \mathbf{Y K_M Y}}{\partial \beta_k}\boldsymbol{\alpha}^*\right]$
11:     Project β_k^{n+1} onto feasible set if any constrains are violated:
12:     $\beta_k^{n+1} = \max(\beta_k^{n+1}, 0)$
13:     $n \leftarrow n + 1$
14: **until** $\boldsymbol{\beta}$ and $\boldsymbol{\alpha}^*$ converge

---

matrix of training data is integrated into the proposed JLDM-MKL framework, with the manner of summation rather than ratio as proposed in [19]. Therefore, the proposed model can be reformulated as

$$\{\mathbf{w}, b, \mathbf{M}, \boldsymbol{\beta}\} = \quad \arg\min_{\mathbf{w}, b, \mathbf{M}} = \frac{1}{2}\|\mathbf{w}\|_2^2 + \rho\,\mathrm{tr}(\mathbf{S}_T^{\boldsymbol{\beta}}) + C\sum_i \xi_i + \lambda r(\boldsymbol{\beta})$$

$$s.t. \quad y_i(\langle \mathbf{w}, \varphi_\mathbf{M}(\mathbf{x}_i; \boldsymbol{\beta})\rangle + b) \geq 1 - \xi_i, \tag{17}$$

$$\boldsymbol{\beta} \geq 0, \xi_i \geq 0, \forall i$$

Thus the corresponding dual problem changes to

$$\min_{\boldsymbol{\beta}} J(\boldsymbol{\beta}), \quad s.t. \quad \beta_{ij} \geq 0$$

where

$$J(\boldsymbol{\beta}) = \max \mathbf{1}^T\boldsymbol{\alpha} - \frac{1}{2}\boldsymbol{\alpha}^T\mathbf{Y K_M}(\boldsymbol{\beta})\mathbf{Y}\boldsymbol{\alpha} + \rho\,\mathrm{tr}(\mathbf{S}_T^{\boldsymbol{\beta}}) + \lambda r(\boldsymbol{\beta})$$

$$s.t. \; \mathbf{1}^T\mathbf{Y}\boldsymbol{\alpha} = 0, 0 \leq \boldsymbol{\alpha} \leq C \tag{18}$$

where $\mathrm{tr}(\mathbf{S}_T^{\boldsymbol{\beta}})$ denotes the trace of the total scatter matrix in feature space mapped via $\mathbf{K_M}(\boldsymbol{\beta})$, which can be explicitly expressed in the kernel-induced feature space as

$$\mathrm{tr}(\mathbf{S}_T^{\boldsymbol{\beta}}) = \mathrm{tr}(\mathbf{K_M}(\boldsymbol{\beta})) - \frac{1}{n}\mathbf{1}^T\mathbf{K_M}(\boldsymbol{\beta})\mathbf{1} = \sum_{p=1}^m \beta_p a_p \tag{19}$$

where $a_p \triangleq \mathrm{tr}(\mathbf{K_M}(\boldsymbol{\beta})) - (1/n)\mathbf{1}^T\mathbf{K_M}(\boldsymbol{\beta})\mathbf{1}$, and $\mathbf{K_M}(\boldsymbol{\beta})$ is the kernel matrix formulated in Eq. (12) which can be obtained by computing the pairwise distance over all pairs or given pairs of instances in the original feature space on training set $\mathcal{X}$. After incorporating the radius term, the derivatives of $J(\boldsymbol{\beta})$ in Algorithm 1 becomes

$$\frac{\partial J}{\partial \beta_k} = \lambda \frac{\partial r}{\partial \beta_k} - \frac{1}{2} \boldsymbol{\alpha}^{*T} \frac{\partial \mathbf{Y} \mathbf{K_M} \mathbf{Y}}{\partial \beta_k} \boldsymbol{\alpha}^* + \rho \frac{\partial \mathrm{tr}(\mathbf{S}_T^{\boldsymbol{\beta}})}{\partial \boldsymbol{\beta}_k} \tag{20}$$

4   Experimental Results

In the experiments, we evaluate the proposed JLDM-MKL using the UCI [24] datasets. We compare the proposed method with three representative and state-of-the-art metric learning models and two multiple kernel learning models, i.e., LMNN[1] [25], ITML[2] [26], SVML[3], GMKL[4], and SimpleMKL[5], also the fundamental SVM method, in term of classification accuracy.

4.1   Experiments on UCI Dataset

We use twelve datasets from the UCI Machine Learning repository of varying size, dimension and task description. The datasets are: Australian Credit Approval (ACA), Contraceptive Method Choice (CMC), Mammographic Mass (MM), Musk, Ionosphere (Iono), Heart, Sonar, Pima, Vote, Wpbc, and Zoo. All the experiments use the following setting: for each dataset, 80% of the data is used for training and the rest is used for test. All data sets have been normalized to have zero mean and unit variance on each feature. We use libSVM[6] to solve the SVM dual problem of our method and the parameter C is chosen from $\{0.1, 1, 10, 10^2\}$. The regularization parameter λ and ρ is chosen from $\{1, 10^2, 10^3, 10^4\}$, and the Gaussian RBF kernel parameter σ^2 is chosen from $\{0.1, 1, 10, 10^2\}$. As JLDM-MKL is not particularly sensitive to the exact choice of λ, which is the regularization parameter in Eqs. (14) and (17), during the experiment we find that for better results it is always chosen set to 10^4 and ρ is often set to 10^2, so we set them for default.

For better comparison, the n-fold results of SVM are also included. The ITML and LMNN results are chosen by adjusting the parameters to the best result. For simplicity, we restrict our evaluation to the binary case and convert multi-class problems to binary ones by grouping labels into two sets. Results for our method and others are all obtained by using 5-fold cross validation and averaging over 20 runs. For small datasets, we report the average accuracy after 30 to 40 runs.

All the experiments are executed in a PC with eight Intel Cores Xeon E3-1230 V2 CPU (3.3 GHz) and 32 GB RAM. The results are listed in Table 1,

---

[1] http://www.cs.cornell.edu/~kilian/code/code.html.
[2] http://www.cs.utexas.edu/~pjain/itml.
[3] http://www.cse.wustl.edu/~xuzx/research/code/code.html.
[4] http://research.microsoft.com/en-us/um/people/manik/code/GMKL/download.html.
[5] http://asi.insa-rouen.fr/enseignants/~arakoto/code/mklindex.html.
[6] https://www.csie.ntu.edu.tw/~cjlin/libsvm/.

where the highest accuracy and the proposed methods whose differences from the highest accuracy are not statistically significant are shown in bold for each dataset, respectively. We do not report the accuracy of GMKL on the Musk dataset, and SimpleMKL on the Musk and Msplice datasets, because these two methods require too large memory space on these datasets and cannot be run in our PC. Besides, the released SVML code always collapsed when run the wpbc dataset, we do not report this either.

To compare the classification performance of these models, we list the average ranks of these models in the last row of Table 1. One each dataset, we rank the methods based on their classification accuracy rates, i.e., we assign rank 1 to the best method and rank 2 to the second best method, and so on. The average rank is defined as the mean rank of one method over the 12 datasets, which can provide a fair comparison of the algorithms [27]. Methods, whose classification accuracy are unavailable, are out of consideration for this ranking.

From Table 1, we can observe that JLDM-MKL with l_1 regularization and radius information achieves the best average rank and JLDM-MKL with l_1 regularization and without radius information achieves the fourth best average rank. Training time of different methods on Table 1 is also shown in Fig. 1. As observed, our algorithm is faster than GMKL and SimpleMKL. And it performs better on some datasets such as Sonar, Heart, and ACA compared with LMNN and ITML. It is efficient to use several kernels in our method instead of a large number of base kernels of different types and with different parameters. But the projected gradient descent optimizer is still a limitation. The results validate that, by introducing the distance metric into kernel construction, the proposed JLDM-MKL framework can lead to very competitive classification accuracy with both state-of-the-art metric learning methods and multiple kernel learning methods under acceptable training time.

Table 1. Classification accuracy on UCI dataset.

Dataset (examples, features)	LMNN	ITML	SVM 5-fold	SVML	GMKL	Simplee MKL	JLDM		
							With l_1	With l_2	VO radius
ACA (690, 14)	86.11	85.29	86.52	**86.77**	67.77	85.36	**85.90**	84.92	85.65
CMC (962, 9)	73.32	72.33	74.22	**74.75**	70.55	73.9	**74.08**	66.94	73.76
MM (830, 5)	81.90	81.80	81.81	83.23	77.82	82.65	**83.86**	80.60	83.22
Musk (6598, 166)	97.03	96.80	99.61	97.07	N/A	N/A	**99.44**	90.62	**99.68**
Iono (351, 34)	88.14	85.20	**95.15**	94.01	93.41	92.59	94.67	90.62	94.09
Msplice (3175, 240)	96.41	86.87	**97.61**	95.46	77.89	N/A	**97.32**	60.38	96.57
Heart (303, 13)	80.01	76.57	81.16	82.56	77.96	**84.21**	**83.85**	76.54	83.79
Sonar (208, 60)	80.52	72.11	**87.92**	81.74	85.96	83.65	87.59	71.46	87.57
Pima (768, 8)	75.20	67.97	74.61	76.44	69.52	76.39	**77.23**	65.80	77.09
Vote (435, 16)	94.60	91.72	94.25	94.71	94.13	**95.40**	94.68	91.95	94.04
wpbc (198, 33)	74.34	73.74	**78.85**	N/A	78.37	76.19	**76.33**	76.26	76.26
Liver (345, 6)	63.20	62.61	68.12	71.30	65.19	65.44	**72.51**	63.33	72.28
Average rank	5.67	7.58	3.00	3.18	6.27	4.20	**2.25**	7.67	3.42

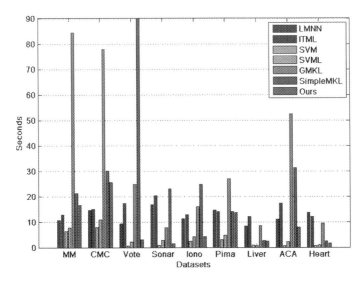

Fig. 1. Training time of different methods. The timing includes 5-fold cross validation, metric learning, SVM training and k-means for our method.

4.2 Construction of Distance Metric Matrix M

In process of choosing sample pairs to construct the distance metric matrix **M**, instead of using all pairs of training samples, we consider following three strategies to construct the doublet sample set.

(1) Random Selection: Given a training set of n samples, randomly select sample pairs from all $n(n-1)$ possible sample pairs, and use their differences to get $X_{i,j}$ in Eq. (11).

(2) K-means I Selection: For all training samples, we first use K-means to find cluster centers, the number of which equal to kernel size, and then for each cluster center, we construct $m_1 + m_2$ doublets $(\mathbf{x}_{c_i}, \mathbf{x}^s_{c_i,1}),...,(\mathbf{x}_{c_i}, \mathbf{x}^s_{c_i,m_1}),(\mathbf{x}_{c_i}, \mathbf{x}^d_{c_i,1}),\ ...,(\mathbf{x}_{c_i}, \mathbf{x}^d_{c_i,m_2})$, where $\mathbf{x}^s_{c_i,k}$ denotes the kth similar nearest neighbor of cluster center $\mathbf{x}_{c_i}$, and $\mathbf{x}^d_{c_i,k}$ denotes the kth dissimilar nearest neighbor of $\mathbf{x}_{c_i}$. After that we compute the difference of these doublets to get $\mathbf{X}_{i,j}$.

(3) K-means II Selection: For each training sample $\mathbf{x}_i$, we construct $m_1 + m_2$ doublets $(\mathbf{x}_i, \mathbf{x}^s_{i,1}),...,(\mathbf{x}_i, \mathbf{x}^s_{i,m_1})\ ,(\mathbf{x}_i, \mathbf{x}^d_{i,1}),\ ...,(\mathbf{x}_i, \mathbf{x}^d_{i,m_2})$, where $\mathbf{x}^s_{i,k}$ denotes the kth similar nearest neighbor of $\mathbf{x}_i$, and $\mathbf{x}^d_{i,k}$ denotes the kth dissimilar nearest neighbor of $\mathbf{x}_i$. After that we compute the difference of these doublets then use K-means to find cluster centers, whose number equals to kernel size. And then we use the cluster centers to get $\mathbf{X}_{i,j}$.

Table 2 lists the classification accuracy of constructing different distance metric matrix **M** by using random selection, K-means I and K-means II strategies to compose doublet sample sets. The K-means II selection outperforms others on

Table 2. Classification results on UCI dataset using random selection strategy, K-means I strategy and K-means II strategy.

Dataset	Heart	Sonar	Pima	Vote	Wpbc	Mammo
Kernelsize	$k = 4$					
Random choose	83.59	87.59	77.39	94.31	76.26	83.56
K-means I	**83.86**	**87.62**	77.18	94.50	**76.28**	83.58
K-means II	83.85	87.61	**77.43**	**94.53**	**76.28**	**83.97**
Kernelsize	$k = 8$					
Random choose	83.46	87.54	77.32	94.28	**76.34**	83.50
K-means I	**83.89**	**87.60**	77.12	**94.55**	76.29	83.52
K-means II	83.76	87.51	**77.43**	94.53	76.30	**83.96**
Kernelsize	$k = 12$					
Random choose	83.80	87.40	**77.42**	94.23	**76.28**	83.49
K-means I	**83.86**	**87.63**	77.18	**94.58**	76.21	83.50
K-means II	83.62	87.52	77.22	94.48	76.26	**83.75**
Kernelsize	$k = 16$					
Random choose	**83.86**	87.50	**77.40**	94.28	76.24	83.43
K-means I	**83.86**	**87.63**	77.19	**94.51**	76.26	83.50
K-means II	83.55	87.51	77.09	94.50	**76.29**	**83.83**

four out of six UCI datasets, and the K-means I selection outperforms others on the left two UCI datasets. To sum up, the doublet selection involved K-means method can always lead to better classification performance.

4.3 Influence of Kernel Size

From aforementioned experiments, one possible bottleneck of the proposed JLDM-MKL seems to be the selection of kernel size, which decides both the performance and the computation overhead of the whole algorithm. To validate the flexibility of this selection, we conduct following experiments. Different numbers of kernel size, ranging from 4 to 40, are adopted to initialize the JLDM-MKL. We choose several UCI datasets, whose feature dimensions are of different magnitude, and test the influence on the classification accuracy when varying the size of kernel. The curve of classification accuracy versus kernel size for the proposed JLDM-MKL are shown in Fig. 2.

One can see that, there are no big obvious fluctuations when choosing different size of kernels. When select a bigger kernel size, the classification accuracy may improve may also decline, but all of them are within a small variance of the average accuracy. This reveal that, one can select a very small kernel size to achieve the result comparable to that of the best choice for any dataset, which dramatically reduce the computation load of the proposed algorithm.

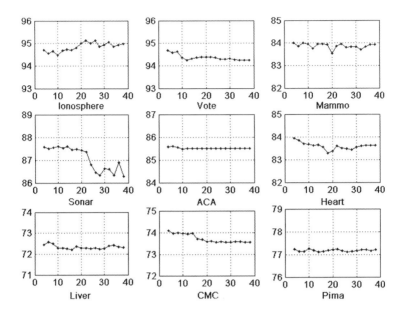

Fig. 2. Classification accuracy using different number of kernels.

5   Conclusion

In this paper, we propose an effective MKL framework for joint learning of distance metric and kernel classifier from the viewpoint of metric learning, referred to as JLDM-MKL. For the formulation of JLDM-MKL, the Gaussian RBF kernel reformulated with the squared Mahalanobis distance is introduced as the basis kernel, radius information is also incorporated as the supplement of considering the within-class distance of input data. By embedding to a GMKL framework, the proposed algorithm is easy to implement. Owing to the fact that the complexity of JLDM-MKL is independent with the number of basis kernels, initializing it with a small kernel size can tremendously reduce the computational burden suffered as conventional MKL approach.

As experimentally demonstrated, our algorithm has given overall the best classification performance among the competing algorithms. One interesting observation is that, the optimization result with or without considering the radius information always achieve comparable performance, which may indicate that the regularizer $r(\beta)$ also imposes constraint on the radius of training samples in the feature space. In the future, we will investigate other forms of kernel learning models from the perspective of metric learning, and extend our model to challenging real applications.

Acknowledgment. This work is partly support by the National Science Foundation of China (NSFC) project under the contract No. 61102037, 61271093, and 61471146.

References

1. Lampert, C.H.: Kernel methods in computer vision. Found. Trends Comput. Graph. Vis. **4**(3), 193–285 (2009)
2. Vedaldi, A., Gulshan, V., Varma, M., Zisserman, A.: Multiple kernels for object detection. In: IEEE International Conference on Computer Vision (ICCV), pp. 606–613 (2009)
3. Chen, L., Chen, C., Lu, M.: A multiple-kernel fuzzy c-means algorithm for image segmentation. IEEE Trans. Syst. Man Cybern. B Cybern. **41**(5), 1263–1274 (2011)
4. Ye, J., Chen, K., Wu, T., Li, J., Zhao, Z., Patel, R., Bae, M., Janardan, R., Liu, H., Alexander, G., Reiman, E.: Heterogeneous data fusion for alzheimer's disease study. In: ACM SIGKDD International Conference on Knowledge Discovery and Data Mining, pp. 1025–1033 (2008)
5. Chapelle, O., Vapnik, V., Bousquet, O., Mukherjee, S.: Choosing multiple parameters for support vector machines. Mach. Learn. **46**(1), 131–159 (2002)
6. Lanckriet, G., Cristianini, N., Bartlett, P., Chaoui, L.E., Jordan, M.: Learning the kernel matrix with semidefinite programming. J. Mach. Learn. Res. **5**(1), 27–72 (2004)
7. Bach, F., Lanckriet, G., Jordan, M.: Multiple kernel learning, conic duality, and the SMO algorithm. In: International Conference on Machine Learning, pp. 6–13 (2004)
8. Sonnenburg, S., Rätsch, G., Schäfer, C., Schölkopf, B.: Large scale multiple kernel learning. J. Mach. Learn. Res. **7**, 1531–1565 (2006)
9. Rakotomamonjy, A., Bach, F., Canu, S., Grandvalet, Y.: SimpleMKL. J. Mach. Learn. Res. **9**, 2491–2521 (2008)
10. Varma, M., Babu, B.R.: More generality in efficient multiple kernel learning. In: International Conference on Machine Learning (ICML), pp. 1065–1072 (2009)
11. Wu, P., Duan, F., Guo, P.: A pre-selecting base kernel method in multiple kernel learning. Neurocomputing **165**, 46–53 (2015)
12. Do, H., Kalousis, A., Woznica, A., Hilario, M.: Margin and radius based multiple kernel learning. In: Buntine, W., Grobelnik, M., Mladenić, D., Shawe-Taylor, J. (eds.) ECML PKDD 2009, Part I. LNCS, vol. 5781, pp. 330–343. Springer, Heidelberg (2009)
13. Gai, K., Chen, G., Zhang, C.: Learning kernels with radiuses of minimum enclosing balls. In: Neural Information Processing Systems (NIPS), pp. 649–657 (2010)
14. Xu, Z., Weinberger, K., Chapelle, O.: Distance metric learning for kernel machines. arXiv preprint arXiv:1208.3422 (2012)
15. Xu, Z., Jin, R., King, I., Lyu, M.: An extended level method for efficient multiple kernel learning. In: Neural Information Processing Systems (NIPS), pp. 1825–1832 (2009)
16. Wang, L.: Feature selection with kernel class separability. IEEE Trans. Pattern Anal. Mach. Intell. **30**(9), 1534–1546 (2008)
17. Liu, X., Wang, L., Yin, J., Zhu, E., Zhang, J.: An efficient approach to integrating radius information into multiple kernel learning. IEEE Trans. Cybern. **43**(2), 557–569 (2013)
18. Liu, X., Wang, L., Yin, J., Liu, L.: Incorporation of radius-info can be simple with SimpleMKL. Neurocomputing **89**, 30–38 (2012)
19. Do, H., Kalousis, A.: Convex formulation of radius-margin based Support Vector Machines. In: International Conference on Machine Learning (ICML), pp. 169–177 (2013)

20. Lu, X., Wang, Y., Zhou, X., Ling, Z.: A method for metric learning with multiple-kernel embedding. Neural Process. Lett. **43**, 905–921 (2016)
21. Wang, F., Zuo, W., Zhang, L., Meng, D., Zhang, D.: A kernel classification framework for metric learning. IEEE Trans. Neural Netw. Learn. Syst. **26**(9), 1950–1962 (2014)
22. Shawe-Taylor, J., Cristianini, N.: Kernel Methods for Pattern Analysis. Cambridge University Press, Cambridge (2004)
23. Vapnik, V., Chapelle, O.: Bounds on error expectation for support vector machines. Neural Comput. **12**(9), 2013–2036 (2000)
24. Frank, A., Asuncion, A.: UCI machine learning repository (2010). http://archive.ics.uci.edu/ml
25. Weinberger, K.Q., Blitzer, J., Saul, L.K.: Distance metric learning for large margin nearest neighbor classification. In: Neural Information Processing Systems (NIPS), pp. 1473–1480 (2005)
26. Davis, J.V., Kulis, B., Jain, P., Sra, S., Dhillon, I.S.: Information-theoretic metric learning. In: International Conference on Machine Learning (ICML), pp. 189–198 (2007)
27. Demšar, J.: Statistical comparisons of classifiers over multiple data sets. J. Mach. Learn. Res. **7**, 1–30 (2006)

Semi-supervised Sparse Subspace Clustering on Symmetric Positive Definite Manifolds

Ming Yin[1(✉)], Xiaozhao Fang[2], and Shengli Xie[1]

[1] School of Automation, Guangdong University of Technology, Guangzhou 510006, People's Republic of China
{yiming,shlxie}@gdut.edu.cn
[2] School of Computer Science and Technology, Guangdong University of Technology, Guangzhou 510006, China
xzhfang168@126.com

Abstract. The covariance descriptor which is a symmetric positive definite (SPD) matrix, has recently attracted considerable attentions in computer vision. However, it is not trivial issue to handle its non-linearity in semi-supervised learning. To this end, in this paper, a semi-supervised sparse subspace clustering on SPD manifolds is proposed, via considering the intrinsic geometric structure within the manifold-valued data. Experimental results on two databases show that our method can provide better clustering solutions than the state-of-the-art approaches thanks to incorporating Riemannian geometry structure.

Keywords: Subspace clustering · Sparse subspace clustering · Graph · Semi-supervised · SPD matrix

1 Introduction

In the past two decades, a large quantity of approaches to subspace clustering have been proposed [5,9,21,25,30], to conduct the tasks from the machine learning and data mining communities. However, clustering is an intrinsically ill-posed problem as its unsupervised nature. The given data can be separated in many ways according to user's intent and goal due to some partitions are equally valid [11]. To this end, semi-supervised clustering is developed to tune the clustering algorithm towards finding the data partition sought by the user [11,20,32,33]. Usually, these methods can effectively utilize the relatively limited *side information*, such as the pairwise similarity between the data points, label information, etc.

Although the fact that the majority of subspace clustering methods have been successfully applied to various applications, the similarity among data points is often measured in the raw space of data. In particular, this similarity is recently computed by the sparse [4] or low-rank representations [14] of data points, by exploiting the so-called *self-expressive* property of the data [5], in terms of Euclidean distance or alike. In general, the representation for each

© Springer Nature Singapore Pte Ltd. 2016
T. Tan et al. (Eds.): CCPR 2016, Part I, CCIS 662, pp. 601–611, 2016.
DOI: 10.1007/978-981-10-3002-4_49

data point is a vector of its linear regression coefficients on the rest of the data subject to sparsity or low-rank constraint. Even for some higher-order signals like images (2D, 3D or higher), one has to handle primarily by vectorizing them and applying any of the available vector techniques. Unfortunately, such type vector features cannot efficiently characterize the high-dimensional data in computer vision, machine learning and medical image analysis [29]. Moreover, the linear combination assumption may not be always true for many high-dimensional data in real world where data will be better modeled by non-linear manifolds [8,12,16]. That is, in this case, the *self-expressive* based algorithms, such as Sparse Subspace Clustering (SSC) [5] and Low Rank Subspace Clustering (LRSC) [26], are no longer applicable. For example, the human facial images can be regarded as samples from a non-linear submanifold [13,18].

To this end, a few solutions have been recently proposed to address sparse coding problems on Riemannian manifolds, such as [3,10,22]. While for subspace clustering, a nonlinear LRR model is proposed to extend the traditional LRR from Euclidean space to Stiefel manifold [29], SPD manifold [6] and abstract Grassmann manifold [27] respectively. The common strategy employed in these work is to use the logarithm mapping "projecting" data onto the tangent space at each data point where a "normal" Euclidean linear reconstruction of a given sample is well defined [24] so that the self-expressive principle works. This idea was first explored by Ho *et al.* [10] and they proposed a nonlinear generalization of sparse coding to handle the non-linearity of Riemannian manifolds, by flattening the manifold using a fixed tangent space.

While for manifold clustering, to address the non-linearity issue, the traditional SSC method has been extended to manifold clustering via Gaussian kernels [18]. Furthermore, Yin *et al.* [31] proposed a kernel sparse subspace clustering for SPD manifold, via embedding the SPD matrices into a Reproducing Kernel Hilbert Space (RKHS). As such, the intrinsic geometric structure among SPD data is sufficiently exploited to perform clustering. However, the prior methods mostly focused on clustering in unsupervised way such that sometimes poor results are achieved. In fact, there always exist some limited, or a handful of, label information by which we apply to improving the clustering performance. Motivated by this intuition, we propose a novel Semi-supervised Sparse Subspace Clustering on Symmetric Positive Definite Manifolds in this paper. Our intention is roughly interpreted from two aspects. One is to perform clustering on the manifold-valued data via considering the intrinsic geometric structure. The other is to concretely improve the clustering performance by both labeled and unlabeled data.

The rest of the paper is organized as follows. In Sect. 2, we give a notation and brief review on related work. Section 3 is dedicated to introducing the novel kernel sparse subspace clustering on Riemannian manifold. The experimental results are given in Sect. 4 and Sect. 5 concludes the paper with a brief summary.

2   Related Work

2.1   Notation and Terminology

Before introducing the proposed method, in this section, we briefly review the recent development of subspace clustering methods [5] and the analysis of Riemannian geometry of SPD Manifold [19]. Throughout the paper, capital letters denote matrices (e.g., X) and bold lower-case letters denote column vectors (e.g., $\mathbf{x}$). x_i is the i-th element of vector $\mathbf{x}$. Similarly, X_{ij} denotes the (i,j)-th entry of matrix X. $\|\mathbf{x}\|_1 = \sum_i |x_i|$ and $\|\mathbf{x}\|_2 = \sqrt{\mathbf{x}^T\mathbf{x}}$ are the ℓ_1 and ℓ_2 norms respectively, where ^{T} is the transpose operation. $\|\cdot\|_F$ is the matrix Frobenius norm defined as $\|X\|_F = \sqrt{\sum_{ij} |X_{ij}|^2}$. The manifold of $d \times d$ SPD matrices is denoted by $\mathcal{S}_d^+$. The tangent space at a point X on $\mathcal{S}_d^+$ is defined by $T_X\mathcal{S}_d^+$, which is a vector space including the tangent vectors of all possible curves passing through X.

2.2   Spare Representation on SPD Matrices

Since SPD matrices belong to a Lie group which is a Riemannian manifold [1], it cripples many methods that rely on linear reconstruction. Generally, there are two methods to deal with the non-linearity of Riemannian manifolds. One is to locally flatten the manifold to tangent spaces [24]. The underlying idea is to exploit the geometry of the manifold directly. The other is to map the data into a feature space usually a Hilbert space [12]. Precisely, it is to project the data into RKHS through kernel mapping [7]. Both of these methods are seeking a transformation so that the linearity emerges.

A typical example of the former method is the one in [10]. Let X be a SPD matrix and hence a point on $\mathcal{S}_d^+$. $\mathbb{D} = \{D_1, D_2, ..., D_N\}, D_i \in \mathcal{S}_d^+$ is a dictionary. An optimization problem for sparse coding of X on a manifold $\mathcal{M}$ is formulated as follows

$$\min_{\mathbf{c}} \ \lambda\|\mathbf{c}\|_1 + \left\| \sum_{i=1}^N c_i \mathbf{log}_X(D_i) \right\|_X^2, \quad \text{s.t.} \ \sum_{i=1}^N c_i = 1, \tag{1}$$

where $\mathbf{log}_X(\cdot)$ denotes Log map from SPD manifold to a tangent space at X, $\mathbf{c} = [c_1, c_2, ..., c_N]$ is the sparse vector and $\|\cdot\|_X$ is the norm associated with $T_X\mathcal{S}_d^+$. Because $\mathbf{log}_X(X) = \mathbf{0}$, the second term in Eq. (1) is essentially the error of linearly reconstructing $\mathbf{log}_X(X)$ by others on the tangent space of X, As this tangent space is a vector space, this reconstruction is well defined. As a result, the traditional sparse representation model can be performed on Riemannian manifold.

However, it turns out that quantifying the reconstruction error is not at all straightforward. Although ℓ_2-norm is commonly used in the Euclidean space, using Riemannian metrics would be better in $\mathcal{S}_d^+$ since they can accurately measure the intrinsic distance between SPD matrices. In fact, a natural way to measure closeness of data on a Riemannian manifold is geodesics, i.e. curves analogous to straight lines in $\mathbb{R}^n$. For any two data points on a manifold, geodesic distance is the length of the shortest curve on the manifold connecting them. For this reason, the affine invariant Riemannian metric (AIRM) is probably the most popular Riemannian metric defined as follows [19]. Given $X \in \mathcal{S}_d^+$, the AIRM of two tangent vectors $\mathbf{v}, \mathbf{w} \in T_X \mathcal{S}_d^+$ is defined as

$$\langle \mathbf{v}, \mathbf{w} \rangle = \langle X^{-1/2} \mathbf{v} X^{-1/2}, X^{-1/2} \mathbf{w} X^{-1/2} \rangle$$
$$= \mathrm{tr}(X^{-1} \mathbf{v} X^{-1} \mathbf{w}).$$

The geodesic distance between points $X, Y \in \mathcal{S}_d^+$ induced from AIRM is then

$$\delta_g(X, Y) = \|\mathbf{log}(X^{-1/2} Y X^{-1/2})\|_F. \tag{2}$$

where $\mathbf{log}(\cdot)$ is the principal matrix logarithm operator.

2.3 Semi-supervised Clustering via Gaussian Fields and Harmonic Functions (GFHF)

Generally, semi-supervised learning task aims to assign labels to unlabeled data, according to the labeled data and the overall data distribution. Among a large number of graph based semi-supervised learning method, the GFHF [33], may be the more popular, if not the most popular, where the predicted label matrix $F \in \mathbb{R}^{n \times c}$ is propagated on the graph w.r.t. the label fitness and manifold smoothness. Thus, the graph Laplacian matrix has to be built beforehand. Given $Q = [\mathbf{q}^1; \mathbf{q}^2; ...; \mathbf{q}^n] \in \mathbb{R}^{n \times c}$ is the initial label indicator binary matrix, defined as follows: the k-th entry of $\mathbf{q}^i$ is 1 and its other entries are all 0 if $\mathbf{x}_i$ is a labeled data in class k. When $\mathbf{x}_i$ is unlabeled, then $\mathbf{q}^i = 0$. Mathematically, the graph based semi-supervised learning model is formulated as following,

$$\min_F \ \mathrm{tr}(F^T L F) + \mathrm{tr}((F - Q)^T U (F - Q)). \tag{3}$$

where L is the graph Laplacian matrix and usually it will be normalized to input. U aims to control the impact of the initial label $\mathbf{q}^i$ of $\mathbf{x}_i$, which is a diagonal matrix with the i-th diagonal element U_{ii}.

3   Semi-supervised SSC on SPD Matrices

In this section, we propose a novel semi-supervised SSC algorithm aiming to handle data on Riemannian manifold by incorporating the intrinsic geometry of the manifold. Let a data set $\mathcal{X} = [X_1, X_2, ..., X_N]$ on SPD manifold, we seek its sparse representation via exploiting the *self-expressive* property of the data. Thus, the corresponding objective function can be given as follows,

$$\min_C \ \lambda \|\mathrm{C}\|_1 + \sum_{i=1}^{N} \left\| \phi(X_i) - \sum_{j=1}^{N} c_{ij} \phi(X_j) \right\|_F^2, \ \text{s.t. } \mathrm{diag}(\mathrm{C}) = 0. \tag{4}$$

where $\phi(\cdot)$ denotes a feature mapping function that projects SPD matrices into RKHS such that $\langle\phi(X),\phi(Y)\rangle = \kappa(X,Y)$ where $\kappa(X,Y)$ is a positive definite (*p.d.*) kernel. diag(C) denotes the vector of the diagonal elements of C.

3.1 Log-Euclidean Kernels for SPD Matrices

In problem (4), how to choose a good kernel is not a trivial issue. Although some commonly kernels, such as polynomial kernel and Gaussian kernel, are widely used [18], the intrinsic geometric structure within SPD data is not carefully considered. As such, linear reconstruction of SPD matrices, by this way, may not be as natural as in Euclidean space such that the errors may be incurred. The recent work in [8] shows that the Stein divergence is akin to AIRM. Furthermore, a *p.d.* kernel can be derived from Stein divergence under some conditions [23]. Concretely, a Stein metric [23], also known as Jensen-Bregman LogDet divergence (JBLD) [2], derived from Bregman matrix divergence is given by,

$$\delta_s(X,Y) = \log\left|\frac{X+Y}{2}\right| - \frac{1}{2}\log|XY|,$$

where $|\cdot|$ denotes determinant. Accordingly a kernel function based on Stein divergence for SPD $\mathcal{S}_d^+$ can be defined as $\kappa_s(X,Y) = \exp\{-\beta\delta_s(X,Y)\}$, though it is guaranteed to be positive definite only when $\beta \in \{\frac{1}{2},1,...,\frac{d-1}{2}\}$ or $\beta > \frac{d-1}{2}$ [23].

However, Stein divergence is actually not applicable to our problem since it is only an approximation to Riemannian metric, and cannot be a *p.d.* kernel without more restricted conditions [13,31]. To this end, a family of Log-Euclidean kernels were proposed in [13], tailored to model data geometry more accurately. These Log-Euclidean kernels were proven to well characterize the true geodesic distance between SPD matrices, especially the Log-Euclidean Gaussian kernel

$$\kappa_g(X,Y) = \exp\{-\gamma\|\log(X) - \log(Y)\|_F^2\},$$

which is a *p.d.* kernel for any $\gamma > 0$. Owing to its superiority, in this paper, we select Log-Euclidean Gaussian kernel to transform the SPD matrices into RKHS.

3.2 Optimization

Through expanding the ℓ_2-norm term and some algebra manipulations, we will consider a set of the equivalent problems that have a same solution to problem (4). Let $\mathcal{X}_{-i} = [X_1, X_2, ..., X_{i-1}, X_{i+1}, ..., X_N]$, we have,

$$\min_{\mathbf{c}_i} \|\mathbf{c}_i\|_1 + \frac{\lambda}{2}\|\bar{\mathbf{x}} - \bar{D}\mathbf{c}_i\|_2^2, \quad \text{s.t. } c_{ii} = 0,$$

where $\bar{\mathbf{x}} = \Sigma^{-1/2}V^T\kappa(X_i, \mathcal{X}_{-i})$ and $\bar{D} = \Sigma^{-1/2}V^T$, given the SVD of $\kappa(\mathcal{X}_{-i}, \mathcal{X}_{-i})$ is $V\Sigma V^T$. For clarity and completeness, the detailed derivation can be found in the appendix.

As such, we are able to adopt Homotopy optimizer [17] to calculate the sparse representation for each sample. After obtaining the coefficients of all samples, the sparse coefficient matrix is formed by $C = [\mathbf{c}_1, \mathbf{c}_2, ..., \mathbf{c}_N]$. The computational complexity of using Homotopy optimizer is $\mathcal{O}(t(N^2m^2 + mN^3))$, where N is the number of samples, m is the dimension of $\bar{\mathbf{x}}$ and t denotes the number of iterations of Homotopy optimizer.

3.3   Semi-supervised Subspace Clustering

As discussed earlier, C is actually a new representation learned from data. Once C is obtained by solving problem (4), the next step is to find the final subspace clusters in semi-supervised learning way. Here we construct a weighted graph L defined by a weight $W = \frac{1}{2}(|C| + |C^T|)$. That is, $L = H - W$, H denotes a degree matrix calculated by $H_{ii} = \sum_j W_{ij}$. Finally, we compute the soft label matrix F^* by substituting L into problem (3), i.e., $F^* = (L + U)^{-1}UY$. In our work, we set $U_{ii} = 1$.

4   Experimental Results

To verify the effectiveness of our proposed method, in this section, we conducted several experimental test on texture images and human faces. Two well-known database, Brodatz [13] and FERET database[1], are used in our test. Some sample images of test databases are shown in Fig. 1.

For texture clustering, a subset of the Brodatz database, i.e., 16-texture ('16c') mosaic, was chosen for clustering performance evaluation. There are 16 objects in this subset in which each class contains only one image. Before clustering, we downsampled each image to 256×256 and then split into 64 regions of size 32×32. To obtain their region covariance matrices (RCM), a feature vector $f(x, y)$ for any pixel $I(x, y)$ was extracted, e.g., $f(x, y) =$

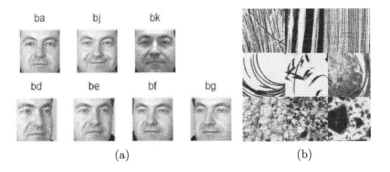

(a)   (b)

Fig. 1. Samples on the FERET (a) and Brodatz (b) database.

---

[1] http://www.nist.gov/itl/iad/ig/colorferet.cfm.

$(I(x,y), |\frac{\partial I}{\partial x}|, |\frac{\partial I}{\partial y}|, |\frac{\partial^2 I}{\partial x^2}|, |\frac{\partial^2 I}{\partial y^2}|)$. Then, each region can be characterized by a 5×5 covariance descriptor. Totally, 1024 RCM were collected, where all test data lie on Riemannian manifold.

As for face clustering, we used the "b" subset of FERET database which consists of 1400 images with the size of 80×80 from 200 subjects (7 images each subject). The images of each individuals were taken under different expression and illumination conditions, marked by 'ba', 'bd', 'be', 'bf', 'bg', 'bj', and 'bk'. To represent a facial image, similar to [28], we created a 43×43 region covariance matrix, i.e., a specific SPD matrix, which is summarised over intensity value, spatial coordinates, 40 Gabor filters at 8 orientations and 5 scales.

In this work, two criteria, i.e. Normalized Mutual Information (NMI) and subspace clustering accuracy, are exploited to quantify the clustering performance more precisely. Meanwhile, to extensively assess how the proposed algorithm improves the performance of data clustering, the following four state-of-the-art graph construction methods are compared against:

- SSC-graph [5].
- LRR-graph [15].
- LRSC-graph (Low-Rank Subspace Clustering graph) [26], which seeks a low-rank representation by decomposing the corrupted data matrix as the sum of a clean and self-expressive dictionary.
- KSSCE-graph (Kernel SSC on Euclidean space graph) [18], which embeds data onto to a nonlinear manifold by using the kernel trick and then apply SSC based on Euclidean metric.

We randomly label some images from each class, in which the percentage of labeled samples ranges from 5 % to 60 %. There is, at least, one labeled point for each class and no bias in the labeling process. The final performance scores were computed by averaging the scores from 20 trials. The detailed clustering results are summarized in Tables 1 and 2. The best results are highlighted in boldface. The penalty parameters λ is selected as $\frac{1}{\mu}$ ($\mu \triangleq \min_i \max_{j \neq i} |\mathbf{k}_i^T \mathbf{k}_j|$) as suggested by [5]. How to choose the suitable parameters is not a trivial issue. Thus, for

Table 1. Clustering results on Brodatz database in terms of accuracy and NMI (%).

Label	Accuracy					NMI				
Ratio (%)	SSC	LRSC	LRR	KSSCE	Ours	SSC	LRSC	LRR	KSSCE	Ours
60	89.84	62.30	78.03	90.04	**96.09**	85.43	56.81	71.80	85.54	**94.01**
50	84.47	52.25	70.90	86.13	**92.38**	78.58	46.56	66.04	79.96	**90.24**
40	79.30	43.16	68.16	82.71	**90.04**	71.68	37.24	62.32	75.36	**87.09**
30	74.41	33.98	68.16	79.00	**88.18**	66.54	27.90	61.50	70.97	**85.81**
20	68.16	24.61	55.96	70.80	**82.32**	59.77	18.44	51.9	63.32	**77.99**
10	53.22	15.23	44.24	59.57	**62.89**	42.87	9.02	39.51	50.19	**55.74**
5	47.56	10.74	44.24	51.76	**57.71**	39.73	4.50	43.71	43.04	**53.16**

Table 2. Clustering results on FERET database in terms of accuracy and NMI (%).

Label	Accuracy					NMI				
Ratio (%)	SSC	LRSC	LRR	KSSCE	Ours	SSC	LRSC	LRR	KSSCE	Ours
60 %	88.57	86.64	86.50	88.93	**91.93**	93.71	92.71	92.56	93.78	**95.27**
50 %	82.50	79.14	79.14	82.14	**85.57**	91.01	87.94	87.93	90.94	**92.15**
40 %	74.29	70.00	70.00	75.00	**77.50**	87.01	81.10	81.11	87.32	**89.14**
30 %	63.93	58.14	58.14	63.14	**67.57**	81.76	73.08	73.15	81.16	**85.46**
20 %	52.50	44.29	44.14	51.43	**58.79**	74.45	62.56	62.32	75.08	**80.87**
10 %	28.71	25.79	25.93	27.36	**32.50**	58.23	47.24	47.29	60.31	**69.17**
5 %	18.36	13.71	13.79	18.00	**20.07**	49.95	32.88	32.84	52.54	**58.67**

Brodatz, we empirically set $\gamma = 0.5$ while for FERET $\gamma = 0.01$. From the results, we can observe that the proposed method consistently achieves the best performance on Brodatz and FERET, compared to other graph-based methods, even with low labeling percentages. Compared to KSSCE-graph, our approach achieves better score in terms of accuracy and NMI. This is owing to the non-linearity in Euclidean space without considering Riemannian metric. In addition, the clustering method based on KSSCE-graph achieves the second best results as respecting a non-linear manifold via a kernel trick.

5   Conclusion

In this paper, we proposed a novel semi-supervised sparse subspace clustering on SPD manifolds, where the intrinsic geometric structure within the manifold-valued data is considered. By the new model, the subspace clustering algorithm is performed in semi-supervised way to concretely improve the segmentation performance. Experimental results show that our method can provide better clustering solutions than the state-of-the-art approaches owing to incorporating Riemannian geometry structure.

Acknowledgement. The Project was supported in part by the Guangdong Natural Science Foundation under Grant (No. 2014A030313511), in part by the Scientific Research Foundation for the Returned Overseas Chinese Scholars, State Education Ministry, China.

Appendix

Given a least-squares problem as following,

$$\min_{\mathbf{c}} \left\| \phi(X) - \sum_{i=1}^{N} c_i \phi(Y_i) \right\|_2^2. \tag{5}$$

where data $\mathcal{Y} = [Y_1, Y_2, ..., Y_N]$ and X are on SPD manifold $\mathcal{S}_d^+$. This problem can be rewritten as a least-squares problem on Euclidean space. That is,

$$\min_{\mathbf{c}} \|\bar{\mathbf{x}} - \bar{D}\mathbf{c}\|_2^2. \tag{6}$$

where $\bar{\mathbf{x}} = \Sigma^{-1/2} V^T \kappa(X, \mathcal{Y})$ and $\bar{D} = \Sigma^{-1/2} V^T$, given the SVD of $\kappa(\mathcal{Y}, \mathcal{Y})$ is $V \Sigma V^T, V V^T = \mathbf{I}$.

Proof 1. By expanding the ℓ_2-norm term in problem (5), we have the following formulation,

$$\begin{aligned}
\min_{\mathbf{c}} & \left\| \phi(X) - \sum_{i=1}^{N} c_i \phi(Y_i) \right\|_2^2 \\
= \min_{\mathbf{c}} & \ \mathbf{c}^T \kappa(\mathcal{Y}, \mathcal{Y}) \mathbf{c} - 2\mathbf{c}^T \kappa(X, \mathcal{Y}) + f(X) \\
= \min_{\mathbf{c}} & \ \mathbf{c}^T V \Sigma V^T \mathbf{c} - 2\mathbf{c}^T V \Sigma^{-1/2} \Sigma^{1/2} V^T \kappa(X, \mathcal{Y}) \\
& + \kappa(X, \mathcal{Y})^T V \Sigma^{-1/2} \Sigma^{-1/2} V^T \kappa(X, \mathcal{Y}) \\
= \min_{\mathbf{c}} & \ \|\Sigma^{-1/2} V^T \kappa(X, \mathcal{Y}) - \Sigma^{-1/2} U^T \mathbf{c}\|_2^2.
\end{aligned} \tag{7}$$

Let $\bar{\mathbf{x}} = \Sigma^{-1/2} V^T \kappa(X, \mathcal{Y})$ and $\bar{D} = \Sigma^{-1/2} V^T$, then the formulation (6) is recognized.

References

1. Arsigny, V., Fillard, P., Pennec, X., Ayache, N.: Geometric means in a novel vector space structure on symmetric positive-definite matrices. SIAM J. Matrix Anal. Appl. **29**(1), 328–347 (2007)
2. Cherian, A., Sra, S., Banerjee, A., Papanikolopoulos, N.: Jensen-bregman logdet divergence with application to efficient similarity search for covariance matrices. IEEE Trans. Pattern Anal. Mach. Intell. **35**(9), 2161–2174 (2013)
3. Cherian, A., Sra, S.: Riemannian sparse coding for positive definite matrices. In: Fleet, D., Pajdla, T., Schiele, B., Tuytelaars, T. (eds.) ECCV 2014, Part III. LNCS, vol. 8691, pp. 299–314. Springer, Heidelberg (2014). doi:10.1007/978-3-319-10578-9_20
4. Donoho, D.L., Elad, M., Temlyakov, V.: Stable recovery of sparse overcomplete representations in the presence of noise. IEEE Trans. Inf. Theory **52**(1), 6–18 (2006)
5. Elhamifar, E., Vidal, R.: Sparse subspace clustering: algorithm, theory, and applications. IEEE Trans. Pattern Anal. Mach. Intell. **35**(11), 2765–2781 (2013)
6. Fu, Y., Gao, J., Hong, X., Tien, D.: Low rank representation on Riemannian manifold of symmetric positive deffinite matrices. In: Proceedings of SDM (2015). doi:10.1137/1.9781611974010.36
7. Harandi, M., Sanderson, C., Hartley, R., Lovell, B.: Sparse coding, dictionary learning for symmetric positive definite matrices: a kernel approach. In: Fitzgibbon, A., Lazebnik, S., Perona, P., Sato, Y., Schmid, C. (eds.) ECCV 2012. LNCS, vol. 7573, pp. 216–229. Springer, Heidelberg (2012). doi:10.1007/978-3-642-33709-3_16

8. Harandi, M.T., Hartley, R., Lovell, B.C., Sanderson, C.: Sparse coding on symmetric positive definite manifolds using bregman divergences. IEEE Trans. Neural Netw. Learn. Syst. (2015). doi:10.1109/TNNLS.2014.2387383

9. He, R., Wang, L., Sun, Z., Zhang, Y., Li, B.: Information theoretic subspace clustering. IEEE Trans. Neural Netw. Learn. Syst. **PP**(99), 1–13 (2015). doi:10.1109/TNNLS.2015.2500600

10. Ho, J., Xie, Y., Vemuri, B.C.: On a nonlinear generalization of sparse coding and dictionary learning. In: Proceedings of ICML, vol. 28, pp. 1480–1488 (2013)

11. Jain, A., Jin, R., Chitta, R.: Semi-supervised clustering. In: Hennig, C., Meila, M., Murtagh, F., Rocci, R. (eds.) Handbook of Cluster Analysis, pp. 1–35. Chapman & Hall, CRC Press (2015). http://www.crcpress.com

12. Jayasumana, S., Hartley, R., Salzmann, M., Li, H., Harandi, M.T.: Kernel methods on the Riemannian manifold of symmetric positive definite matrices. In: Proceedings of CVPR, pp. 73–80, June 2013

13. Li, P., Wang, Q., Zuo, W., Zhang, L.: Log-Euclidean kernels for sparse representation and dictionary learning. In: Proceedings of ICCV, pp. 1601–1608, December 2013

14. Liu, G., Lin, Z., Yan, S., Ju, S., Ma, Y.: Robust recovery of subspace structures by low-rank representation. IEEE Trans. Pattern Anal. Mach. Intell. **35**(1), 171–184 (2013)

15. Liu, G., Lin, Z., Yu, Y.: Robust subspace segmentation by low-rank representation. In: Proceedings of ICML, pp. 663–670 (2010)

16. Nguyen, H., Yang, W., Shen, F., Sun, C.: Kernel low-rank representation for face recognition. Neurocomputing **155**, 32–42 (2015)

17. Osborne, M.R., Presnell, B., Turlach, B.A.: A new approach to variable selection in least squares problems. IMA J. Numer. Anal. **20**(3), 389 (2000)

18. Patel, V.M., Vidal, R.: Kernel sparse subspace clustering. In: Proceedings of ICIP, pp. 2849–2853, October 2014

19. Pennec, X., Fillard, P., Ayache, N.: A Riemannian framework for tensor computing. Int. J. Comput. Vis. **66**, 41–66 (2006)

20. Shang, F., Liu, Y., Wang, F.: Learning spectral embedding for semi-supervised clustering. In: IEEE 11th International Conference on Data Mining, pp. 597–606 (2011)

21. Shi, J., Malik, J.: Normalized cuts and image segmentation. IEEE Trans. Pattern Anal. Mach. Intell. **22**, 888–905 (1997)

22. Sivalingam, R., Boley, D., Morellas, V., Papanikolopoulos, N.: Tensor sparse coding for positive definite matrices. IEEE Trans. Pattern Anal. Mach. Intell. **36**(3), 592–605 (2014)

23. Sra, S.: A new metric on the manifold of kernel matrices with application to matrix geometric means. In: Pereira, F., Burges, C.J.C., Bottou, L., Weinberger, K.Q. (eds.) Proceedings of NIPS, pp. 144–152 (2012)

24. Tuzel, O., Porikli, F., Meer, P.: Pedestrian detection via classification on Riemannian manifolds. IEEE Trans. Pattern Anal. Mach. Intell. **30**(10), 1713–1727 (2008)

25. Vidal, R.: Subspace clustering. IEEE Signal Process. Mag. **28**(2), 52–68 (2011)

26. Vidal, R., Favaro, P.: Low rank subspace clustering (LRSC). Pattern Recogn. Lett. **43**(1), 47–61 (2014)

27. Wang, B.Y., Hu, Y.L., Gao, J., Sun, Y.F., Yin, B.C.: Low rank representation on Grassmann manifolds: an extrinsic perspective. arXiv preprint arXiv:1504.01807

28. Yang, M., Zhang, L., Shiu, S.C.K., Zhang, D.: Gabor feature based robust representation and classification for face recognition with Gabor occlusion dictionary. Pattern Recogn. **46**(7), 1865–1878 (2013)
29. Yin, M., Gao, J., Guo, Y.: Nonlinear low-rank representation on Stiefel manifolds. Electron. Lett. **51**(10), 749–751 (2015)
30. Yin, M., Gao, J., Lin, Z., Shi, Q., Guo, Y.: Dual graph regularized latent low-rank representation for subspace clustering. IEEE Trans. Image Process. **24**(12), 4918–4933 (2015)
31. Yin, M., Guo, Y., Gao, J., He, Z., Xie, S.: Kernel sparse subspace clustering on symmetric positive definite manifolds. In: Proceedings of CVPR, pp. 5157–5164 (2016)
32. Zhu, X.: Semi-supervised learning literature survey. Technical report 1530, Computer Sciences, University of Wisconsin-Madison (2005)
33. Zhu, X., Ghahramani, Z., Lafferty, J.: Semi-supervised learning using gaussian fields and harmonic functions. In: Proceedings of ICML, pp. 912–919 (2003)

Combination of Multiple Classifier Using Feature Space Partitioning

Xia Yingju[⊠], Hou Cuiqin, and Sun Jun

Information Technology Laboratory,
Fujitsu Research & Development Center Co., Ltd., Beijing, China
{yjxia, houcuiqin, sunjun}@cn.fujitsu.com

Abstract. Combination of Multiple Classifier has been consider as the approach of improving the classification performance. The popular diversification approach named local specialization is based on the simultaneous partitioning of the feature space and an assignment of a compound classifier to each of the sub-space. This paper presents a novel feature space partitioning algorithm for the combination of multiple classifier. The proposed method uses pairwise measure to get the diversity between classifiers and selects the complementary classifiers to get the pseudo labels. Based on the pseudo labels, it splits the feature space into constituents and selects the best classifier committee from the pool of available classifiers. The partitioning and selection are taken place simultaneously as part of a compound optimization process aimed at maximizing system performance. Evolutionary methods are used to find the optimal solution. The experimental results show the effectiveness and efficiency of the proposed method.

Keywords: Multiple classifier · Feature space partitioning

1 Introduction

Multiple classifier systems (MCS) focus on the combination of classifiers from heterogeneous or homogeneous modeling background to give the final decision [1, 2]. MCS is primarily used to improve the classification performance of a model, or reduce the likelihood of an unfortunate selection of a poor one. Dietterich [1] summarized the benefits of MCS: (a) allowing to filter out hypothesis that, though accurate, might be incorrect due to a small training set. (b) Combining classifiers trained starting from different initial conditions could overcome the local optima problem. (c) The true function may be impossible to be modeled by any single hypothesis, but combinations of hypotheses may expand the space of representable functions.

Generally, individual classifier has different performance in different regions of the feature space, or different classifiers are preferred in different regions of the feature space, especially for complex problems. Hence, the local specialization technique is adopted in building multiple classifier systems. In local specialization, it selects the best single classifier or a subset of classifiers from a pool of classifiers trained over each partition of or the entire feature space [3]. Then the selected classifiers make decisions for the new data coming from the corresponding regions. Several methods have been

© Springer Nature Singapore Pte Ltd. 2016
T. Tan et al. (Eds.): CCPR 2016, Part I, CCIS 662, pp. 612–624, 2016.
DOI: 10.1007/978-981-10-3002-4_50

proposed. Kuncheva's [4] clustering and selection algorithm partitions the feature space by a clustering algorithm, and selects the best individual classifier for each cluster according to its local accuracy. Adaptive Splitting and Selection algorithm (AdaSS) in [5] put partitioning the feature space and assigning classifiers to each partition into one integrated process. Additionally, the majority voting or more sophisticated rules are proposed as combination method of area classifiers [6].

One of the main problem of these methods is the large feature space makes the optimization very difficult which has been shown in the AdaSS method [5] and its modification (MAD) [6]. The other main problem is that the stationary assumption does not always hold in the real applications [7, 8]. For many learning tasks where data is collected over an extended period of time, its underlying distribution is likely to change. The drift in the underling distribution may result in a change in the learning problem.

This paper proposes a novel feature space partitioning algorithm using pseudo feedback. It tries to catch the property of the test set and make the feature space partitioning more efficient.

The rest of this paper is organized as follows. Section 2 presents the multiple classifier systems method and the pseudo feedback feature space partitioning algorithm. Section 3 shows the experimental results. Section 4 gives several conclusions and future works.

2   Method

The classification task is to classify an object to one of the predefined categories based on the observation of its features. All data concerning the object and its attributes will be presented as d dimensional vector of features marked in the following way [5]:

$$X = \{X^{(1)},\ X^{(2)}, \cdots,\ X^{(d)}\} \in R^d$$

Assume that we have n classifiers $\psi^{(1)},\ \psi^{(2)}, \cdots,\ \psi^{(d)}$. For a given object x, each individual classifier decides whether it belongs to the class $i \in M = \{1, \cdots, M\}$ based on the values of discriminants. Let $F^{(l)}(i, x)$ denotes a function that is assigned to class i for a give value of x, and that is used by the *l-th* classifier $\psi^{(l)}$. The combined classifier ψ uses the following decision rule [6]:

$$\psi(x) = i \ if \ \hat{F}(i, x) = \max_{k \in M} \hat{F}(k, x) \tag{1}$$

Where:

$$\hat{F}(i, x) = \frac{\sum_{l=1}^{n} w^{(l)} F^{(l)}(i, x)}{\sum_{l=1}^{n} w^{(l)}(i)} \tag{2}$$

In the local specialization technique [5, 6], the feature space χ is divided into a set of H constituents:

$$\chi = \bigcup_{h=1}^{H} \hat{\chi}_h \tag{3}$$

$$\hat{\chi}_h \cap \hat{\chi}_l = \phi, \quad \forall k, l \in \{1, \cdots, H\}, k \neq l$$

Where $\hat{\chi}_h$ denotes the h-th constituent. ψ_h is a combined classifier assigned to the h-th constituent:

$$\psi(x) = i \Leftrightarrow \psi_h(x) = i \text{ and } x \in \hat{\chi}_h \tag{4}$$

2.1 Feature Space Partitioning

As mentioned above, there are two main steps in the local specialization technique: feature space partitioning and classifier committee selection. The partitioning can be consider as searching the splitting borders on the feature space. Figure 1 shows a classification task with two features. The x-axis shows the value of feature 1 while y-axis shows the feature 2. The red line in Fig. 1 shows one possible partitioning border which splits the feature space into two sub feature space. The method of [6] proposed to search the whole feature space by adding a random disturbance on each border at every iteration step. This make the searching very difficult when there are large feature number and complex feature space.

As the main step, the quality of feature space partition has a strong influence on the system capabilities. The better feature partitioning, the better classifier selection and the better system performance. In this study, we propose a pseudo feedback feature space partitioning method which splits the feature space into some bins and searches the optimal feature border on these bins. Take Fig. 1 for example, the feature 1(x-axis) is split into 10 bins ($[0, f_{1,1}), [f_{1,1}, f_{1,2}), \ldots, [f_{1,9}, f_{1,10}]$) and feature 2 ($y$-axis) is split into 9 bins ($[0, f_{2,1}), [f_{2,1}, f_{2,2}), \ldots, [f_{2,8}, f_{2,9}]$). The whole feature space is such split into 90 blocks. The local specialization approach will only need to search on these 90 blocks.

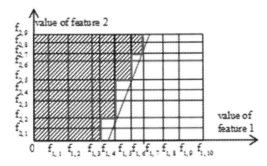

Fig. 1. Feature space partitioning for MCS (Color figure online)

Comparing with searching in the whole space, this will make effective searching anchors in the case of perfect feature space partitioning. It will also lead to the different borders which shown in Fig. 1 (the red line is the conventional border and the blue lines are the borders of proposed feature partitioning method).

Generally, the feature space partitioning methods include unsupervised and supervised ways. Unsupervised methods do not consider the class label whereas supervised ones do. Comprehensive listings of these techniques can be found in the works of [9]. The main drawback of all these previous work is their difficulty to accurately handle the gap between the training set and test set. Once the test set changes or concept drift happened, the previous trained model cannot catch the property of the new test set.

Figure 2 shows the pseudo feedback feature space partitioning method comparing with the typical unsupervised method (Equal-width) and supervised method (Bayesian) [10]. The binary classification problem is taken as example. In Fig. 2, the circle indicates objects (instances) for one class while the triangle for the other class. The x-axis shows the feature values for one feature dimension. The task of feature space partitioning is to put these feature values into several bins. The feature number will be reduced since the number of bins is generally less than the feature number.

The typical unsupervised method such as the equal-width method, does not make use of object labels. As shown in the Fig. 2, the features are put into several equal sized bins. The supervised methods try to utilize the distribution of the classes in the training set to supervise the feature partitioning procedure. In Fig. 2, we can see that the equal-width method has the risk that merges values that are strongly associated with different classes into the same bin. The EW method partitions the feature space into 7 equal width bins, each bin contains 3 feature values (for discrete values) or 3 feature intervals (for continuous value). The objects in the first bin are mainly circle class. The second bin is mixed with objects belong to circle and triangle class (objects on the first two values are circle while the third one are mainly triangle). The Bayesian method avoids this problem by estimating the condition probability in the training set. In Bayesian method, the first 5 feature value are put in one bin. But when the distribution

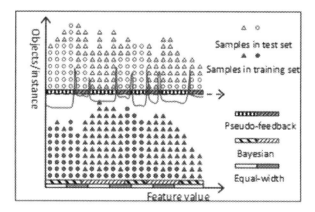

Fig. 2. The pseudo feedback feature space partitioning (Color figure online)

of training set is different with the test set (or concept drift happened), the Bayesian method will fail. Here, we use the hollow circle and triangle to indicate the objects in the test set. According to the test set, the first bin in Bayesian method should be split into two parts. The first one has four feature values and mainly contains circle objects. The second one cannot be merged into the first one since it contain more objects belong to triangle class.

Notice that we cannot get the real labels of the test set. We can only get the pseudo labels using other classifiers. These pseudo labels may not consist with the real labels and can only be trusted with some confidence. This is the main risk to use these pseudo labels.

In the proposed pseudo feedback feature partitioning method, the pseudo labels are gotten from other classifiers' prediction on test set. The distribution distance between the training set and test set (the red curve in Fig. 2) is calculated using pseudo labels on test set and real labels on training set. The distribution distance is adopted to modify the feature partitioning procedure. More concretely, for each adjacent interval, the pseudo feedback method calculates the distribution distance and decide whether merge these intervals or not.

Since the pseudo labels are the crucial to the feature space partitioning, the issue that how to select the candidate classifiers for getting the pseudo labels arise. The intuition is that the mutually complementary classifiers which are characterized by high diversity and accuracy should be selected to get the pseudo labels for each other. Actually, the diversity has been recognized as a very important characteristic in classifier combination. Empirical results have illustrated that there exists positive correlation between accuracy of the ensemble and diversity among the base classifiers [11–13]. Further, most of the existing ensemble learning algorithms [14, 15] can be interpreted as building diverse base classifiers implicitly. However, the problem of measuring classifier diversity and so using it effectively for building better classifier ensembles is still an open topic. Most researchers discuss the concept of diversity in terms of correct/incorrect outputs [13, 16, 17]. The [refa6] divides the diversity measures into pairwise diversity measures and non-pairwise diversity measures. For pairwise diversity measure, the Q statistics, the correlation coefficient, the disagreement measure and the double-fault measure are most commonly used. The previous experimental studies have shown that most diversity measures perform similarly [13, 17]. In this study, the disagreement measure [18, 19] is adopted. Empirical results have illustrated that there exists positive correlation between accuracy of the ensemble and diversity among the base classifiers [12, 20].

The disagreement measure of classifier i and k is defined:

$$Dis_{i,k} = \frac{N^{01} + N^{10}}{N^{00} + N^{01} + N^{10} + N^{11}}$$

Where N^{00}, N^{01}, N^{10} and N^{11} are derived from Table 1.

Suppose we have gotten the L classifiers which have high diversity with the target classifier for feature space partitioning. The straightforward way is to use the classifier with highest diversity to make the pseudo labels. However, this method does not consider the accuracy of the classifiers been selected. How about the result if classifier

Table 1. The relationship between a pair of classifiers

	D_k correct(1)	D_k wrong(0)
D_i correct(1)	N^{11}	N^{10}
D_i wrong(0)	N^{01}	N^{00}

with the highest diversity does not performance well? Actually, beside the diversity, the accuracy of the classifier and the classification confidence are also key factors for the pseudo labels getting. The accuracy of classifier can be explicitly expressed by the weight of classifier. The classification confidence, which was theoretically proved to be a key factor on the generalization performance [21], has been utilized in certain ensemble learning algorithms [22–25].

In this study, we extract the pseudo labels by combining the ensemble margin [26] and classification confidence [23].

Let:

h_j ($j = 1,2, ..., L$): the selected classifiers with high diversity.
$X = \{(x_i, y_i), i = 1,2, ..., n\}$: the data set
y_i: the class label of the sample x_i
$\bar{y}_{ij}$: the classification decision of xi estimated by the classifier hj
c_{ij}: the classification confidence of xi estimated by the classifier h_j
define the margin as:

$$m(X_i) = \sum_{j=1}^{L} W_j \gamma_{ij} C_{ij}$$

$$s.t.\ w_j \geq 0,\ \sum_{j=1}^{L} W_j = 1 \tag{5}$$

where the w_j is the weight of the classifier h_j and

$$\gamma_{ij} = \begin{cases} 1 & if\ y_i = \bar{y}_{ij} \\ -1 & if\ y_i \neq \bar{y}_{ij} \end{cases} \tag{6}$$

We can get the optimal $W = [w_1, ..., w_L]$ by minizing the objective function below:

$$w = \arg \min_{w} \|U - TW\|_2^2 + \lambda \|W\|_2 \tag{7}$$

Where

$$U = [1, ..., 1]_{n+1}^T, \quad T = [\gamma_{ij} C_{ij}]_{n*L}$$

$$\|U - TW\|_2^2 = \sum_{i=1}^{n} (1 - m(X_i))^2 \tag{8}$$

The proposed method searches the whole feature space by a fixed moving step. For each adjacent interval, it calculates the divergence between the training set and test set. The adjacent intervals which have small change in the distribution will be merged. By elaborately selected moving step and the distribution distance threshold, the feature space will finally partitioned into several sub-space.

In this study, the KL divergence is adopted to measure the distribution difference between training set and test set.

$$D(P_{tr}||P_{ts}) = \sum_{y}\sum_{i} P_{tr}(y|f_i) \log \frac{P_{tr}(y|f_i)}{P_{ts}(y|f_i)} \tag{9}$$

Here the $D(P_{tr} \| P_{ts})$ is the divergence between the training and test set in the given interval, the f_i denotes the feature i, the $P_{tr}(y|f_i)$ and $P_{ts}(y|f_i)$ are the probability of the output label under the condition f_i in the training set and test set respectively.

The algorithm is shown below:

INITIALIZE:
```
  Select the classifiers for getting the pseudo labels
using diversity
  Get the pseudo labels using the selected classifiers
  Set the step for interval merge: T
  Set feature range [MIN, MAX]
  Set cursor lowb=MIN
  Set threshold θb for Bayesian-measure Bp(previous),
Bc(current)
  Set threshold θd for KL Divergence Dp(previous),
Dc(current)
```
BEGIN
```
  upb=lowb+T
  While upb < MAX
    Calculate the Bp, Dp
    upb = lowb+T
    Calculate the Bc, Dc
    if(|Bc-Bp|> θb or |Dc-Dp| > θd)
      Segment the interval as [lowb, upb-T)
      lowb = upb-T
    else
      upb += T
    end if
  end while
```
END

2.2 Optimization

The next main step of the local specialization technique is classifier committee selection which will assure the lowest possible mistake level. The optimization criterion is defined as:

$$Q(\hat{\psi}) = \frac{1}{n} \sum_{n=1}^{N} (\delta(\hat{\psi}_{member(c,X_n)}(X_n), j_n)) \tag{10}$$

Where the member (C,x) denote the function that returns the cluster index to which a given x belongs.

In order to solve the optimization task, many optimization algorithms widely used in theory and practice can be adopted. In this study, the evolutionary algorithm [27] is adopted as the optimization algorithm. The key points of the evolutionary algorithm are shown below.

- Representation (chromosome)
  The chromosome consists of two components. The first one embodies a set of centroids C (the centroids of the H constituents in the feature space). The other includes definitions of the combined classifiers for each of the space partitions.
  CW = [C,W]

Where C is centroid vector for the H feature partitions.

$$C = \{C_1, C_2, \cdots, C_H\}$$

And the W is the weights assigned to each individual classifier on each feature partitions.

$$W = \{W_1, W_2, \cdots, W_H\}$$

$$W_i = \begin{bmatrix} w_i^1(1) & w_i^2(1) & \cdots & w_i^n(1) \\ w_i^1(2) & w_i^2(2) & \cdots & w_i^n(2) \\ \vdots & \vdots & \ddots & \vdots \\ w_i^1(M) & w_i^2(M) & \cdots & w_i^n(M) \end{bmatrix}$$

- Fitness function
  Each chromosome corresponds to a given realization of the combination of multiply classifiers, the quality or its fitness function can be obtained by (10).
- Mutation
  Since the chromosome consists of two components, each component can be altered with certain probability that is changing along with the optimization progress. The mutation involves adding a vector of numbers randomly generated according to the normal density distribution. This is the conventional way for mutation operator. By employing the proposed pseudo feedback feature space partitioning method, the feature space has been segment into several bins. The mutation can be taken in the following two ways:

(1) Adding the vector of the random number, then find the nearest bin for the centroids.
(2) Directly shift on the bins by random steps.

- Crossover

  The crossover operator generate one offspring member on the basis of two parents. All the chromosomes are paired up and with a probability cross over according to the conventional two-point rule. Notice that the centroid part (C) needs a mapping to the bins after the crossover.

With the above defined key operations, the evolutionary algorithm can find the optimal solution.

3   Evaluation

The performance of the proposed method is evaluated on 20 UCI datasets (UCI Machine Learning Repository: http://archive.ics.uci.edu/ml).

The datasets are shown in Table 2. Here, the '#O' denotes the number of objects/instances, the '#F' denotes the feature number and '#C' gives the number of classes. In the 'Name' field, the 'Internet' is the abbreviation for 'Internet Advertisements', 'Letter' for 'Letter Recognition', 'Magic' for 'Magic Gamma Telescope', 'Mammographic' for 'Mammographic Mass', 'Molecular' for 'Molecular Biology', 'Ozone' for 'Ozone Level Detection', 'Page' for 'Page Blocks Classification', 'Pima' for 'Pima Indians Diabetes'.

These datasets cover some high-dimensional sets, some large sets, some small sets and some typical/balanced sets. More detailed information can be found on the UCI website.

The classifier pool includes Random Forest, Decision Tree, Gradient boosting, Maximum Entropy and Naïve Bayes. Every classifier uses the pseudo labels gotten from others classifiers to make the feature partitioning.

We compared our method (PF) with the CS method [4] and the modified AdaSS method (MAD) [6]. We also compared our method with the conventional feature partitioning method: equal-width (EW) and Bayesian (Bayes) method [10]. The experimental results are shown in Table 3.

From the experimental results, we see that the pseudo feedback feature partitioning method outperforms the CS and MAD method in all the datasets. This indicates that the

Table 2. The datasets

Name	#O	#F	#C	Name	#O	#F	#C
Abalone	4177	8	28	Breast cancer	286	9	2
Audiolog	226	69	23	Car evaluation	1728	6	4
Census	199523	40	2	Mammographic	961	6	2
Ecoli	336	8	8	Nursery	12960	8	5
Internet	3297	1558	2	Ozone	2536	73	2
Iris	150	4	3	Page	5473	10	5
Letter	20000	16	26	Pima	768	8	2
Magic	19020	11	2	Spectf heart	267	44	2
Molecular	3190	61	3	Statlog	946	18	4
Musk	476	168	2	Yeast	1484	8	10

Table 3. Experimental results

Methods datasets	CS	MAD	Feature partitioning method		
			EW	Bayes	PF
Abalone	88.01	89.29	88.86	90.2	**90.38**
Audiolog	53.35	60.01	60.13	60.4	**60.99**
Breast cancer	92.5	92.86	92.6	92.85	**93.01**
Car evaluation	85.16	86.06	85.24	88.02	**89.7**
Census	85.1	86.28	85.34	90.19	**92.33**
Ecoli	74.98	79.25	79.5	80.35	**80.59**
Internet	63.2	65.53	65.09	65.64	**66.35**
Iris	94.00	95.0	**95.5**	94.25	94.25
Letter	88.75	90.33	89.59	90.21	**90.64**
Magic	88.5	90.45	89.72	90.52	**91.09**
Mammographic	68.01	69.04	68.6	69.27	**70.21**
Molecular	71.64	72.53	72.19	72.94	**73.73**
Musk	84.46	85.55	85.42	86.31	**88.04**
Nursery	85.5	86.03	85.59	86.79	**88.25**
Ozone	70.5	74.66	74.02	75.26	**75.75**
Page	80.65	85.65	84.72	86.16	**87.73**
Pima	69.2	70.31	69.67	70.52	**70.91**
Spectf heart	76.46	81.85	81.57	81.72	**82.19**
Statlog	89.3	90.29	90.05	90.35	**92.21**
Yeast	61.8	62.73	61.99	63.84	**64.35**

proposed feature partitioning method can give better feature split border. Comparing our pseudo-feedback method with the conventional feature partitioning method (EW and Bayes), we see that our method get the best performance in almost all dataset expect the 'Iris'. By analysis of the size of dataset, we found that it will impact the performance. Take the 'Iris' as example, there are only 150 objects in this dataset which lead to a small feature space (only 22 unique values for the first feature). It is very difficult to put them into several bins.

To further investigate the performance on different data size. A set of experiments on 'Census' dataset is conducted. The sub-datasets range from 50 to 190,000 are extracted from the whole dataset. The experiments is intend to compare the performance of EW, Bayesian and PF methods. The experimental results are shown as the relative difference with the baseline method. The relative difference is calculated as:

$$relative\ difference = \frac{Accuracy_{ref} - Accuracy_{baseline}}{Accuracy_{baseline}}$$

Here, the $Accuracy_{baseline}$ is the accuracy of EW on each dataset. The $Accuracy_{ref}$ is the accuracy of Bayesian and PF method.

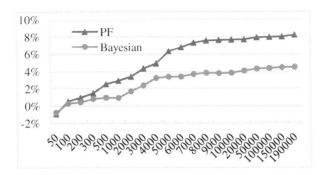

Fig. 3. The effect of data size

Figure 3 shows the experimental results. The *x*-axis shows the size of each sub-datasets. The *y*-axis shows the relative difference. When the data size is small, both PF and Bayesian method cannot get good performance. For example, when the data size is less than 100, the PF and Bayesian methods are worse than EW. It is because that the Bayesian method needs to make statistic on the training set. The PF method need more data to calculate the distribution difference between training set and test set. From the Fig. 3, we can see that, even there are about 1,000 samples, the PF method cannot get great enhancement in comparison with the Bayesian method. The PF method is worse than Bayesian method when the data size is small than 200. With the bigger dataset, the PF method performance better, about 8 % enhancement can be achieved.

We also investigated the method to get the pseudo labels. As shown in Sect. 2, we extract the pseudo labels using the classification confidence instead of the classification decision of the candidate classifiers. We conducted experiments on different data set to compare the pesudo label getting method with classification decision (CD) and classification confidence (CC). The Fig. 4 shows the relative difference of the classification confidence method versus the baseline classification decision method. The classification confidence method outperforms the classification decision method in most cases. The larger the dataset, the better the classification confidence method.

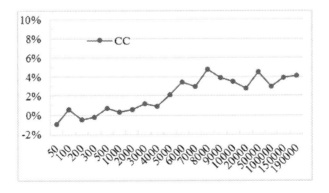

Fig. 4. Pseudo labels getting method with classification confidence

4   Conclusions and Future Works

This paper deals with the combination of multiple classifier and presents a novel feature space partitioning algorithm using pseudo feedback approach. The proposed method gets the pseudo labels by using the pseudo feedback from other classifiers which selected by diversity measure of disagreement. It uses the pseudo labels to split the feature space into constituents and select the best classifier committee from the pool of available classifiers. The proposed method gives better anchors of feature space searching. The partitioning and selection are taken place simultaneously as part of a compound optimization process aimed at maximizing system performance. The experiments conducted on different type of datasets compared the proposed method with the conventional combination method and some feature space partitioning methods. The experimental results shown the effectiveness and efficiency of the proposed method.

The future work includes improving the performance on small datasets. The measurement on distribution difference between training set and test set also need to further explored since it's the key point to make the feature space partitioning.

References

1. Dietterich, T.G.: Ensemble methods in machine learning. In: Kittler, J., Roli, F. (eds.) MCS 2000. LNCS, vol. 1857, pp. 1–15. Springer, Heidelberg (2000)
2. Wozniak, M., Grana, M., Corchado, E.: A survey of multiple classifier systems as hybrid systems. Inf. Fusion **16**, 3–17 (2014)
3. Baruque, B., Porras, S.: Hybrid classification ensemble using topology-preserving clustering. New Gener. Comput. **29**(3), 329–344 (2011)
4. Kuncheva, L.: Clustering-and-selection model for classifier combination. In: The Fourth International Conference on Knowledge-Based Intelligent Engineering Systems and Allied Technologies, vol. 1, pp. 185–188 (2000)
5. Jackowski, K., Wozniak, M.: Algorithm of designing compound recognition system on the basis of combining classifiers with simultaneous splitting feature space into competence areas. Pattern Anal. Appl. **12**(4), 415–425 (2009)
6. Wozniak, M., Krawczyk, B.: Combined classifier based on feature space partitioning. Int. J. Appl. Math. Comput. Sci. **22**(4), 855–866 (2012)
7. Bai, Q.X., Lam, H., Sclaroff, S.: A Bayesian framework for online classifier ensemble. In: The 31st International Conference on Machine Learning, pp. 1584–1592, Beijing, China (2014)
8. Gama, J., Zliobaite, I., Bifet, A., Pechenizkiy, M., Bouchachia, A.: A survey on concept drift adaptation. ACM Comput. Surv. **46**(4), 44 (2014)
9. Garcia, S., Luengo, J., Sez, J., Lpez, V., Herrera, F.: A survey of discretization techniques: taxonomy and empirical analysis in supervised learning. IEEE Trans. Knowl. Data Eng. **25**, 734–750 (2013)
10. Ferreira, A., Figueiredo, M.: An supervised approach to feature discretization and selection. Pattern Recogn. **45**, 3048–3060 (2012)
11. Dietterich, T.: An experimental comparison of three methods for constructing ensembles of decision trees: bagging, boosting and randomization. Mach. Learn. **40**(1), 1–22 (2000)

12. Kuncheva, L., Whitaker, C.: Measures of diversity in classifier ensembles and their relationship with the ensemble accuracy. Mach. Learn. **51**, 181–207 (2003)
13. Tang, E.K., Suganthan, P.N., Yao, X.: An analysis of diversity measures. Mach. Learn. **65**, 247–271 (2006)
14. Breiman, L.: Bagging predictors. Mach. Learn. **24**(2), 123–140 (1996)
15. Liu, Y., Yao, X., Higuchi, T.: Evolutionary ensembles with negative correlation learning. IEEE Trans. Evol. Comput. **4**, 380–387 (2000)
16. Brown, G., Wyatt, J., Harris, R., Yao, X.: Diversity creation methods: a survey and categorization. J. Inf. Fusion **6**(1), 5–20 (2005)
17. Kuncheva, L., Whitaker, C.: Measures of diversity in classifier ensembles and their relationship with the ensemble accuracy. Mach. Learn. **51**, 181–207 (2003)
18. Ho, T.: The random space method for constructing decision forests. IEEE Trans. Pattern Anal. Mach. Intell. **20**(8), 832–844 (1998)
19. Skalak, D.: The sources of increased accuracy for two proposed boosting algorithms. In: Proceedings of American Association for Artificial Intelligence, AAAI-96, Integrating Multiple Learned Models Workshop
20. Kuncheva, L.: Combining Pattern Classifiers: Methods and Algorithms, 2edn., pp. 248–280. Wiley Hoboken (2014)
21. Shawe-Taylor, J., Cristianini, N.: Robust bounds on generalization from the margin distribution. In: The 4th European Conference on Computational Learning Theory (1999)
22. Freund, Y., Schapire, R.E.: A decision-theoretic generalization of on-line learning and an application to boosting. J. Comput. Syst. Sci. **55**(1), 119–139 (1997)
23. Li, L., Hu, Q., Wu, X., Yu, D.: Exploration of classification confidence in ensemble learning. Pattern Recogn. **47**(9), 3120–3131 (2014)
24. Quinlan, J.R.: Bagging, boosting, and C4. 5. In: AAAI/IAAI, vol. 1, pp. 725–730 (1996)
25. Schapire, R.E., Singer, Y.: Improved boosting algorithms using confidence-rated predictions. Mach. Learn. **37**(3), 297–336 (1999)
26. Schapire, R.E., Freund, Y., Bartlett, P., Lee, W.S.: Boosting the margin: a new explanation for the effectiveness of voting methods. Ann. Stat. **26**, 1651–1686 (1998)
27. Gabrys, B., Ruta, D.: Genetic algorithm in classifier fusion. Appl. Soft Comput. J. **6**(4), 337–347 (2006)

On Tightening the M-Best MAP Bounds

Qiang Cheng[1,2(✉)], Li Chen[1,2], Yuanjian Xing[1,2], and Yuhao Yang[1,2]

[1] Nanjing Research Institute of Electronics Technology, Nanjing 210039, China
421263669@qq.com
[2] Key Laboratory of IntelliSense Technology, CETC, Nanjing 210039, China

Abstract. We consider the problem of finding the M assignments with the maximum probabilities (or equivalently, the M-best MAP assignments) on a probabilistic graphical model. The covering graph approximation method provides an upper bound on each of the true M-best MAP costs. However, the tightness of these bounds is closely related to how to split the parameters of the duplicate nodes. We propose a monotonic algorithm to tighten the M-best MAP bounds by finding the optimal splitting of these parameters. Experimental results on synthetic and real problems show that our algorithm provides much tighter bounds than those provided by uniformly splitting the parameters.

Keywords: M-best MAP inference · Probabilistic graphical models · Covering graph

1 Introduction

Probabilistic graphical models, such as Markov random fields (MRFs) and Bayesian networks, have been widely studied and used as powerful tools in the fields of artificial intelligence, computer vision, bioinformatics, signal processing, and many others. The problems in these applications are often formed as the task of finding the most probable assignment, also known as the maximum a posteriori (MAP) assignment. In many of these applications, one can benefit from finding the M most probable assignments, or equivalently, the M-best MAP assignments [1, 2]. For example, in computer vision, the diverse M most probable assignments can provide better results than the most probable assignment [1]. In protein design problems, the ensemble of the M most stable amino acid sequences often provides more accurate predictions [3].

The M-best MAP assignments can be solved sequentially or simultaneously. The sequential algorithms [2, 4–6] find the M-best MAP assignments one by one, by alternating between partitioning the solution set and solving a MAP problem on a subset. In addition, the M-best MAP assignments can be solved simultaneously by using the dynamic programming method [7]. However, the computational complexity of this method is exponential in the treewidth of the graphs, and is feasible only for graphical models with small treewidth. The covering graph [8] is an approximation to the original graph, where we keep all the edges of the original graph, and introduce duplicate copies of some nodes to reduce the treewidth. The M-best MAP costs of a

© Springer Nature Singapore Pte Ltd. 2016
T. Tan et al. (Eds.): CCPR 2016, Part I, CCIS 662, pp. 625–637, 2016.
DOI: 10.1007/978-981-10-3002-4_51

covering graph yield the M-best upper bounds over the M-best MAP costs of the original graph. The M-best MAP assignments decoded from the covering graph are greatly influenced by the tightness of the bounds. However, the tightness of these bounds is closely related to how to split the parameters of the duplicate nodes, and is much too loose if the parameters are split arbitrarily or uniformly.

In this paper, we propose a monotonic algorithm to find the optimal splitting of the parameters of the duplicate variables, which, as far as we know, is the first algorithm on tightening the M-best MAP bounds. In our method, we first compute the mth-max-marginals of the duplicate variables, and then use the coordinate descend method to optimize the splitting of the parameters. With tighter bounds, we can decode much better M-best MAP assignments from the covering graph. Moreover, the tightened bounds can significantly reduce the number of branches when using the branch and bound search algorithm to solve the M-best MAP assignments.

2 The M-Best Max-Product

We introduce some necessary concepts and preliminaries in this section. We denote by $G = (\mathcal{V}, \mathcal{E})$ an undirected graph, with nodes set $\mathcal{V}$ and edges set $\mathcal{E}$. Each node $i \in \mathcal{V}$ is associated with a random variable X_i, which takes values in space $\mathcal{X}_i$. Define a unary potential function $\theta_i(\cdot) : \mathcal{X}_i \rightarrow \mathbb{R}$ for each node i and a pairwise potential function $\theta_{ij}(\cdot, \cdot) : \mathcal{X}_i \times \mathcal{X}_j \rightarrow \mathbb{R}$ for each edge ij. The parameter vector of a pairwise MRF can be denoted by $\boldsymbol{\theta} = \left\{ \{\boldsymbol{\theta}_i\}_{i \in \mathcal{V}}, \{\boldsymbol{\theta}_{ij}\}_{(i,j) \in \mathcal{E}} \right\}$, where $\boldsymbol{\theta}_i = \{\theta_i(x_i), \forall x_i \in \mathcal{X}_i\}$ and $\boldsymbol{\theta}_{ij} = \left\{ \theta_{ij}(x_i, x_j), \forall x_i \in \mathcal{X}_i, x_j \in \mathcal{X}_j \right\}$. A pairwise Markov random field (MRF) is defined by the graph G and the potential functions. The task of the MAP problem is to maximize the MRF energy $E(\boldsymbol{\theta}, \mathbf{x})$ w.r.t $\mathbf{x} \in \mathcal{X}$,

$$\max_{\mathbf{x} \in \mathcal{X}} \left\{ E(\boldsymbol{\theta}, \mathbf{x}) \overset{def}{=} \sum_{i \in \mathcal{V}} \boldsymbol{\theta}_i(x_i) + \sum_{(i,j) \in \mathcal{E}} \boldsymbol{\theta}_{ij}(x_i, x_j) \right\} \tag{1}$$

The M-best MAP problem is to find M assignments $\left\{ \mathbf{x}^1, \mathbf{x}^2, \ldots \mathbf{x}^M \right\}$, which satisfies

$$\mathbf{x}^m = \underset{\mathbf{x} \in \mathcal{X} \setminus \left\{ \mathbf{x}^1, \mathbf{x}^2, \ldots \mathbf{x}^{m-1} \right\}}{\arg \max} E(\boldsymbol{\theta}, \mathbf{x}), \forall m \leq M \tag{2}$$

The M energies w.r.t $\left\{ \mathbf{x}^1, \mathbf{x}^2, \ldots \mathbf{x}^M \right\}$ are denoted by $\left\{ E^1(\boldsymbol{\theta}), E^2(\boldsymbol{\theta}), \ldots, E^M(\boldsymbol{\theta}) \right\}$, where $E^m(\boldsymbol{\theta}) = E(\boldsymbol{\theta}, \mathbf{x}^m)$.

For tree graphs, the M-best MAP problem can be solved using a message passing algorithm, which is similar to the max-product algorithm for solving the MAP problem:

$$\pi_{i \rightarrow j}\left(x_j\right) = M\underset{x_i}{max}\left(\boldsymbol{\theta}_{ij}(x_i, x_j) \oplus \boldsymbol{\theta}_i(x_i) \left(\underset{l \in N(i) \setminus j}{\oplus} \pi_{l \rightarrow i}(x_i) \right) \right), \tag{3}$$

where $\pi_{i \to j}\left(x_j = k\right)$ is a set of the M-best MAP costs passed from x_i to x_j conditional on $x_j = k$. The operation $\oplus$ is a set sum operation, defined by

$$A \oplus B = \{a + b | \forall a \in A, b \in B\}$$

for the two sets A and B. Note the number of elements in $A \oplus B$ is $|A| \cdot |B|$. The operation "Mmax" finds the M maximal elements of a set, that is, given a set $A = \{a_1, \dots, a_L\}$ where the elements follow a descend order, i.e., $a_1 \geq a_2 \geq \dots \geq a_L$,

$$Mmax(A) = \{a_1, a_2, \dots, a_{\min\{M,L\}}\}.$$

We define the M-best max-marginal on $x_i = k$ as

$$q_i^M\left(x_i = k\right) = Mmax\left(\boldsymbol{\theta}_i\left(x_i = k\right) \underset{j \in N(i)}{\oplus} \pi_{j \to i}\left(x_i = k\right)\right)$$

that is, for $x_i = k$, $q_i^M\left(x_i = k\right)$ denotes the set of the M-best MAP costs conditional on $x_i = k$. The M-best max-marginal $q_i^M\left(x_i\right)$ is a vector of such sets, i.e., $q_i^M\left(x_i\right) = \left[q_i^M\left(x_i = 1\right), \dots, q_i^M\left(x_i = |\mathcal{X}_i|\right)\right]^T$. The "Mmax" operation over x_i, i.e., $Mmax_{x_i}$, is defined as finding the M maximal elements in the union of the sets in $q_i^M\left(x_i\right)$. The M-best MAP costs can be obtained by taking Mmax on $q_i^M\left(x_i\right)$ over x_i, i.e., $Mmax_{x_i}\left(q_i^M\left(x_i\right)\right)$.

We call this message passing algorithm for the M-best MAP problem the M-best max-product algorithm. This M-best max-product algorithm is essentially a dynamic programming method [7]. Equation (3) reduces to the max-product algorithm when $M = 1$. We give an example to illustrate the M-best max-product algorithm.

Example 1: The graphical model is shown in Fig. 1, and the 2-best MAP costs are to be computed. The message from x_1 to x_2 is computed as

$$\pi_{1 \to 2}\left(x_2\right) = Mmax_{x_1}\left(\boldsymbol{\theta}_{12}\left(x_1, x_2\right) \oplus \boldsymbol{\theta}_1\left(x_1\right)\right) = \begin{bmatrix} \{5, 2\} \\ \{4, 3\} \end{bmatrix}.$$

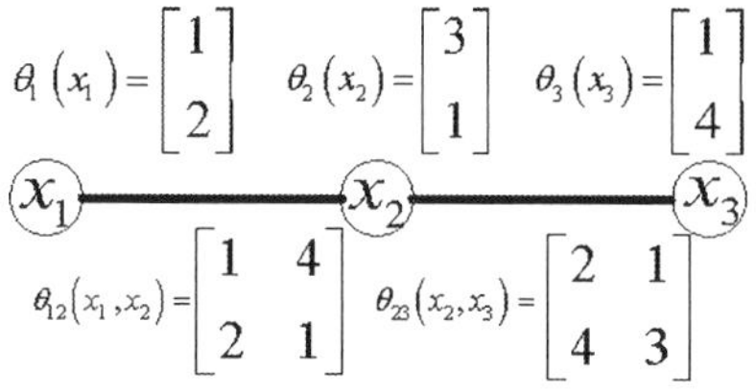

$$\theta_1(x_1) = \begin{bmatrix} 1 \\ 2 \end{bmatrix} \quad \theta_2(x_2) = \begin{bmatrix} 3 \\ 1 \end{bmatrix} \quad \theta_3(x_3) = \begin{bmatrix} 1 \\ 4 \end{bmatrix}$$

$$(x_1) \quad\rule{2cm}{0.4pt}\quad (x_2) \quad\rule{2cm}{0.4pt}\quad (x_3)$$

$$\theta_{12}(x_1,x_2) = \begin{bmatrix} 1 & 4 \\ 2 & 1 \end{bmatrix} \quad \theta_{23}(x_2,x_3) = \begin{bmatrix} 2 & 1 \\ 4 & 3 \end{bmatrix}$$

Fig. 1. A simple chain MRF.

The message from x_2 to x_3 is computed as

$$\pi_{2\rightarrow 3}(x_3) = Mmax_{x_2}\left(\boldsymbol{\theta}_{23}(x_2, x_3) \oplus \boldsymbol{\theta}_2(x_2) \oplus \pi_{1\rightarrow 2}(x_2)\right) = \begin{bmatrix} \{10, 9\} \\ \{9, 8\} \end{bmatrix}.$$

Thus, the M-best max-marginal on x_3 is

$$q_3^M(x_3) = \begin{bmatrix} Mmax\left(\boldsymbol{\theta}_3(x_3 = 1) \oplus \pi_{2\rightarrow 3}(x_3 = 1)\right) \\ Mmax\left(\boldsymbol{\theta}_3(x_3 = 2) \oplus \pi_{2\rightarrow 3}(x_3 = 2)\right) \end{bmatrix} = \begin{bmatrix} \{11, 10\} \\ \{13, 12\} \end{bmatrix}$$

That is, $q_3^M(x_3 = 1) = \{11, 10\}$, and $q_3^M(x_3 = 2) = \{13, 12\}$. The M-best costs are $Mmax_{x_3} q_3^M(x_3) = Mmax\{13, 12, 11, 10\} = \{13, 12\}$.   □

3 The Mth-Max-Marginals

In addition to the M-best max-marginals, we are also interested in the mth-max-marginals, which are the max-marginals on the reduced solution set (generated by cutting the $(m-1)$-best MAP assignments from the solution set).

We first give the definition of the mth-max-marginal on variable:

$$b_i^m(x_i) = \max_{\mathbf{x}\backslash x_i, \mathbf{x}\in\mathcal{X}\backslash\{\mathbf{x}^1,\ldots,\mathbf{x}^{m-1}\}} E(\boldsymbol{\theta}, \mathbf{x}) \tag{4}$$

$\{\mathbf{x}^1, \ldots, \mathbf{x}^{m-1}\}$ are the $(m-1)$-best MAP assignments. $b_i^m(x_i)$ reduces to the max-marginal of the MAP problem when $m = 1$.

3.1 Relations of the M-Best Max-Marginals and the Mth-Max-Marginals

The M-best max-marginals and the mth-max-marginals are closely related. They are equivalent in the sense of computing the M-best MAP costs and assignments.

First, the mth-max-marginals $b_i^m(x_i)$ can be easily obtained from the M-best max-marginals $q_i^M(x_i)$. Given the M-best max-marginals $q_i^M(x_i)$. Recall that $q_i^M(x_i = k)$ is a set of the M-best MAP costs, and $q_i^M(x_i)$ is a vector of such sets. We first construct an auxiliary max-marginals $\bar{q}_i^M(x_i)$ by removing the $(m-1)$-best MAP costs from $q_i^M(x_i)$. Then the mth-max-marginals $b_i^m(x_i)$ can be computed as:

$$b_i^m(x_i = k) = \max \bar{q}_i^M(x_i = k)$$

Second, given all the mth-max-marginals $\{b_i^1(x_i), \ldots, b_i^M(x_i)\}$, the M-best max-marginals can be recovered. We construct $\tilde{q}_i^M(x_i)$ in such a way:

$$\tilde{q}_i^M (x_i = k) = \{b_i^1 (x_i = k), \ldots, b_i^M (x_i = k)\}$$

The constructed $\tilde{q}_i^M (x_i)$ is not exactly the same as $q_i^M (x_i)$. Their difference lies in the $\geq (M + 1)$-best costs and assignments, but they are the same on the M-best costs and assignments. Thus we say they are equivalent from the sense of computing the M-best MAP costs and assignments. In the following, we give an example to compute $b_i^m (x_i)$ from $q_i^M (x_i)$ and recover $q_i^M (x_i)$ from $b_i^m (x_i)$.

Example 2: In Example 1, $b_3^1 (x_3)$ can be computed as $b_3^1 (x_3) = \begin{bmatrix} 11 \\ 13 \end{bmatrix}$. When computing $b_3^2 (x_3)$, we first construct $\bar{q}_3^M (x_3) = \begin{bmatrix} \{11, 10\} \\ \{12\} \end{bmatrix}$ by remove the element 13 from $q_3^M (x_3)$. Then $b_3^2 (x_3 = 1)$ is the maximal element in $\bar{q}_3^2 (x_3 = 1) = \{11, 10\}$, and $b_3^2 (x_3 = 2)$ is the maximal element in $\bar{q}_3^2 (x_3 = 2) = \{12\}$, i.e., $b_3^2 (x_3) = \begin{bmatrix} 11 \\ 12 \end{bmatrix}$.

The $\tilde{q}_3^M (x_3)$ is recovered from $b_3^m (x_3)$ as

$$\tilde{q}_3^M (x_3) = \begin{bmatrix} \{b_3^1 (x_3 = 1), b_3^2 (x_3 = 1)\} \\ \{b_3^1 (x_3 = 2), b_3^2 (x_3 = 2)\} \end{bmatrix} = \begin{bmatrix} \{11, 11\} \\ \{13, 12\} \end{bmatrix}. \qquad \Box$$

The maximum of $b_i^m (x_i)$ is equal to the mth maximum of $q_i^M (x_i)$, and the M-best costs can be recovered from both the M-best max-marginals $q_i^M (x_i)$ and the mth-max-marginals $b_i^m (x_i)$.

3.2   The Mth-Best Reparameterization

The max-marginals of the MAP problem form a reparameterization of the graphical model, which guarantees

$$\sum_{i \in V} \theta_i(x_i) + \sum_{(i,j) \in \mathcal{E}} \theta_{ij}(x_i, x_j) = \sum_{i \in V} b_i^1(x_i) + \sum_{(i,j) \in \mathcal{E}} \left(b_{ij}^1(x_i, x_j) - b_i^1(x_i) - b_j^1(x_j) \right), \forall \mathbf{x} \in \mathcal{X}$$

However, this property only holds for the 1st-max-marginals, and does not hold for $m > 1$. An interesting property is that, when the mth-max-marginals are used to "reparameterize" a tree-structured model, the MAP assignment and cost of the new model are equal to the mth-best MAP assignment and cost of the original model. We define the mth-best reparameterization as:

Definition 1. Model A is an mth-best reparameterization of model B, if the MAP assignment and cost of model A are equal to the mth-best MAP assignment and cost of model B.

Theorem 1. The mth-max-marginals yield an mth-best reparameterization of a tree-structured graphical model.

Proof: The mth-max-marginals satisfy the consistent constraint between the edge marginals and the node marginals, that is, $b_i^m(x_i) = \sum_{x_j} b_{ij}^m(x_i, x_j)$. Thus, the model constructed by the mth-max-marginals has $\max_{x_i} b_i^m(x_i)$ as the MAP cost, which equals the mth-best MAP cost of the original model. Furthermore, the MAP assignment decoded from the mth-max-marginals is the same as the mth-best MAP assignment. □

The mth-best reparameterization property only holds for tree-structured graphs. For loopy graphs, we use the covering graph approximation to obtain the upper bounds of the M-best MAP costs, and then use the mth-max-marginals to tighten these upper bounds.

4   The Covering Graph Approximation

The computational complexity of the M-best max-product algorithm is exponential in the treewidth of the graphs, and it is intractable for the graphs with large treewidth. The covering graph with small treewidth provides a feasible approximation to this problem.

The covering graph is such a graph that includes each edge in the original graph exactly once, but includes duplicate copies of some nodes. Figure 2 gives an example of a 3×3 grid and its two covering graphs with different treewidths. The covering graph in Fig. 2(b) has treewidth of 1, while that in Fig. 2(c) has treewidth of 2. Given a particular treewidth, the covering graph can be constructed using the mini-bucket elimination method [9].

With a slight abuse of notation, we will use X_i^c to refer to the copies of variable X_i. Let $\mathcal{C}$ be the set of all duplicate variables, and let $\mathcal{C}_i$ be the indices set of the duplicate variables of X_i, i.e., $\mathcal{C}_i = \{c | X_i^c \in \mathcal{C}\}$. $\mathcal{C}_i = \{1\}$ if X_i has no duplicate copy.

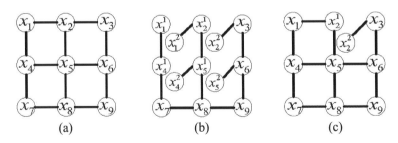

Fig. 2. The original graph (a), and its two covering graphs (b) and (c).

The covering graph idea has been explored to provide an upper bound on the MAP cost [8]. Similarly, the covering graph also provides the M-best upper bounds on the M-best MAP costs. The M-best upper bounds is defined as [7].

Definition 2. Let $A = \{a_1, \ldots a_M\}$ and $B = \{b_1, \ldots b_M\}$ be two sets with $a_i \geq a_{i+1}, \forall i \leq M - 1$ and $b_i \geq b_{i+1}, \forall i \leq M - 1$. A is the **M-best upper bounds** on B iff $\forall 1 \leq i \leq M, a_i \geq b_i$.

We represent the energy function of a covering graph as

$$E\left(\boldsymbol{\theta}_{CG}, \mathbf{x}_{CG}\right) = \sum_{i \in \mathcal{V}, c \in \mathcal{C}_i} \boldsymbol{\theta}_i^c\left(x_i^c\right) + \sum_{(i,j) \in \mathcal{E}} \boldsymbol{\theta}_{ij}\left(x_i^{c_{ij}}, x_j^{c_{ij}}\right)$$

where $x_i^{c_{ij}}$ denotes the copy of node i which is adjacent to a copy of node j by the edge (i,j) in the covering graph. $\boldsymbol{\theta}_{CG}$ denotes the set of the parameters:

$$\boldsymbol{\theta}_{CG} = \left\{ \boldsymbol{\theta}_i^c\left(x_i^c\right), \boldsymbol{\theta}_{ij}\left(x_i^{c_{ij}}, x_j^{c_{ij}}\right) \middle| \begin{array}{l} \sum_{c \in \mathcal{C}_i} \boldsymbol{\theta}_i^c\left(x_i^c\right) = \boldsymbol{\theta}_i\left(x_i\right), \forall i \in \mathcal{V}, \\ \boldsymbol{\theta}_{ij}\left(x_i^{c_{ij}}, x_j^{c_{ij}}\right) = \boldsymbol{\theta}_{ij}\left(x_i, x_j\right) \forall (i,j) \in \mathcal{E} \end{array} \right\}$$

$\mathbf{x}_{CG} = \left\{x_i^c | \forall i \in \mathcal{V}, c \in \mathcal{C}_i\right\}$. The M-best energies of the covering graph are denoted by $\left\{E^1\left(\boldsymbol{\theta}_{CG}\right), E^2\left(\boldsymbol{\theta}_{CG}\right), \ldots, E^M\left(\boldsymbol{\theta}_{CG}\right)\right\}$.

The covering graph approximation is equivalent to the mini-bucket elimination method [7], and they both provides the M-best upper bounds on the true M-best MAP costs. However, the tightness of the bounds is greatly related on how to construct the covering graph and how to split the parameters of the duplicate variables. In the next section, we will introduce algorithms to tighten the M-best upper bounds by optimizing the parameters of the duplicate variables.

5   Tightening the M-Best Upper Bounds

We introduce a monotonic algorithm to find the optimal splitting to the parameters of the duplicate variables.

The objective function for tightening the M-best upper bounds is

$$\min_{\boldsymbol{\theta}_{CG} \in \boldsymbol{\theta}_{CG}} \max_{\mathbf{x}_{CG} \in \mathcal{X}_{CG}^m} E(\boldsymbol{\theta}_{CG}, \mathbf{x}_{CG}), \forall m \leq M \tag{5}$$

where $\mathcal{X}_{CG}^m = \mathcal{X}_{CG} \backslash \left\{\mathbf{x}_{CG}^1, \mathbf{x}_{CG}^2, \ldots \mathbf{x}_{CG}^{m-1}\right\}$, and $\mathbf{x}_{CG}^m$ is the mth-best MAP assignment of the covering graph.

With the mth-max-marginals on the duplicate variables, we give an algorithm to optimize Eq. (5), as shown in Algorithm 1. In essence, Algorithm 1 is a coordinate descent algorithm. We will show that the updating scheme given in Algorithm 1 monotonically decreases the M-best upper bounds.

Algorithm 1 Tightening the M-best Upper Bounds
_____

Input: A pairwise MRF $G = (\mathcal{V}, \mathcal{E})$, a covering graph with $\mathcal{C}$, $\mathcal{C}_i$ and initial parameter $\boldsymbol{\theta}_i^c$ (split arbitrarily or uniformly).

From m=1 to m=M
  Iterate the following steps until convergence
    1: Run the M-best max-product algorithm to obtain the mth-max-marginals
       $b_i^m \left(x_i \right)$ for each variable in $\mathcal{C}$;
    2: Update the parameter $\boldsymbol{\theta}_i^c$ in $\mathcal{C}$ as

$$\overline{\boldsymbol{\theta}}_i^c \left(x_i^c \right) = \boldsymbol{\theta}_i^c \left(x_i^c \right) - \frac{1}{|\mathcal{C}|} \left(b_i^m \left(x_i^c \right) - \frac{1}{|\mathcal{C}_i|} \sum_{k \in \mathcal{C}_i} b_i^m \left(x_i^k \right) \right) \qquad (6)$$

Output: The upper bounds on the M-best MAP costs and the assignments.
_____

Theorem 2. The updating scheme in Eq. (6) strictly decreases the upper bounds: $E^m \left(\overline{\boldsymbol{\theta}}_{CG} \right) \leq E^m (\boldsymbol{\theta}_{CG}), \forall 1 \leq m \leq M$ ($\overline{\boldsymbol{\theta}}_{CG}$ denotes the parameter after updating, while $\boldsymbol{\theta}_{CG}$ denotes the parameter before updating).

Proof: This theorem can be proven by using Theorem 1 in Sect. 3 and Theorem 4.1 in [8]. For m = 1, the above theorem reduces to Theorem 4.1 in [8]. For m > 1, the mth-max-marginals form an mth-best reparameterization of the original model. The updating scheme Eq. (6) decreases the energy of the reparameterized model, which also decreases the mth-best energy of the original model.   □

The block coordinate descent updating may get stuck at local optimums. The subgradient method can be used to overcome this problem, with the updating scheme similar to Eq. (6). Moreover, in practice, we can increase the step size of Eq. (6) (i.e., $\frac{1}{|\mathcal{C}|}$) to obtain faster convergence.

6   M-Best Decoding

In this section, we discuss how to decode the M-best MAP assignments from the covering graph with optimized parameters. We introduce two methods on decoding the M-best MAP assignments: one approximates the M-best MAP assignments by using the M-best max-product algorithm on the spanning graph, and another obtains the true M-best MAP assignments by using the branch and bound search algorithm.

6.1   Decoding by the M-Best Max-Product Algorithm

Using the M-best max-product algorithm, the true M-best MAP assignments can be decoded for junction trees. Similarly, we can decode the M-best MAP assignments for covering graphs using the M-best max-product algorithm. However, in the covering graph case, we may get different assignments for the duplicate variables. Assigning the duplicate variables arbitrarily may result in identical assignment in the M assignments. So the problem is how to obtain M different assignments from the covering graphs. Our method is to decode M different assignments from a spanning graph which is constructed from the covering graph.

Given a covering graph, a spanning graph can be generated in such a way: for each variable that has duplicates, we keep one duplicate and take max operation to eliminate all the other duplicates. The M-best max-product algorithm always returns M different assignments on a spanning graph. The spanning graphs of Fig. 2(b) and (c) are shown in Fig. 3.

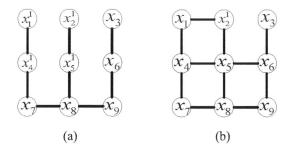

(a)   (b)

Fig. 3. The spanning graphs of Fig. 2(b) and (c).

6.2   Decoding by the Branch and Bound Algorithm

Given the upper bounds and the assignments obtained from the covering graph, the branch and bound method searches for the exact M-best assignments by iteratively applying the branching and bounding strategies.

In the branch and bound algorithm, we restrict the search space to $\mathcal{X}_{V_d}$, where $\mathbf{X}_{V_d}$ is the set of variables that have duplicates. In the branching step, we select a variable X_i in $\mathbf{X}_{V_d}$, and split the states of X_i into two parts, which generates two search spaces.

7   Experiments

In this section, we conduct experiments on synthetic and real problems to verify the effectiveness of our algorithms. For synthetic problems, we test the algorithms on the 10×10 grid, with each variable having 2 states. The parameters $\boldsymbol{\theta}_s(x_s)$ are drawn independently from $\mathcal{N}(0, \sigma^2)$, and the parameters $\boldsymbol{\theta}_{st}(x_s, x_t)$ are set as follows:

$$\boldsymbol{\theta}_{st}(x_s = i, x_t = j) = \begin{cases} 0 \ \ if \ i = j, \\ \lambda_{ij} \ \ if \ i \neq j, \end{cases}$$

where λ_{ij} is generated as $\mathcal{N}(0, 1)$. The parameter σ is set to be 0.1 and 0.5 to obtain different graphical models. The results are obtained after averaging 10 trials. For real problems, we test the algorithms on the benchmark problems of the probabilistic inference challenge 2011.

7.1   Synthetic Data

We compare the 5-best upper bounds and the decoded 5-best assignments of different algorithms on the 10×10 grid model. The results are shown in Fig. 4, where "MBE" denotes the mini-bucket elimination method [7], and "CG" denotes the covering graph method using Algorithm 1. The only difference of these two methods lies in the splitting of the parameters of the duplicate nodes, and the MBE method uniformly splits the parameters. We plot the errors of the upper bounds (denoted by "Upper Errors") of different algorithms and different treewidths. The error of the kth upper

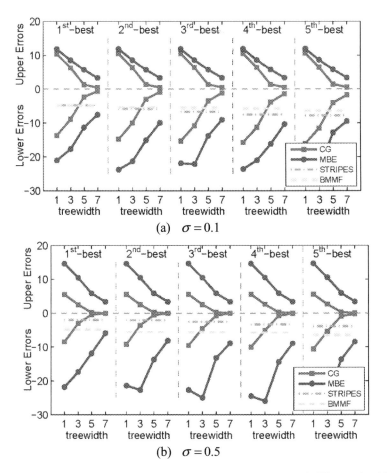

Fig. 4. Comparison of the M-best upper bounds and lower bounds of different algorithms.

bound is computed as $err = E^k(\theta_{upper}) - E^k(\theta)$. The results show that Algorithm 1 can significantly tighten the M-best upper bounds.

When comparing the decoded M-best assignments, we also test the results of the "STRIPES" algorithm [5] and the "BMMF" algorithm [2]. The errors of the decoded M-best assignments are shown in Fig. 4, which are denoted by "Lower Errors". The error of the kth assignment $\mathbf{x}^k$ is computed as $err = E(\theta, \mathbf{x}^k) - E^k(\theta)$. "STRIPES" and "BMMF" are sequential algorithms, thus they have only "Lower Errors" and these errors have no relationship with treewidth. The results show that the decoded M-best assignments of the covering graph method are much better than those of the mini-bucket elimination method on models with different parameters σ. The results also show that the covering graph method performs better than the "STRIPES" algorithm and the "BMMF" algorithm for $\sigma = 0.1$ and $\sigma = 0.5$ when using the treewidth larger than 3.

The branch and bound search method is tested on the grid model to find the true M-best MAP assignments. The treewidth of the covering graphs is set to be 7. The number of branches used for finding the optimal assignments is shown in Table 1. The comparison shows that using the optimized covering graph, the branch and bound search algorithm takes much less branches to find the true M-best MAP assignments. Figure 5 shows the curves of the estimated M costs (averaged) w.r.t. the number of branches on model with $\sigma = 0.5$. The result shows that "BB+CG" takes less than 100 branches to find the true M-best MAP assignments, while "BB+MBE" takes almost 1000 branches.

Table 1. The number of branches (averaged) of the branch and bound (BB) algorithm

	BB+CG	BB+MBE
$\sigma = 0.1$	634.8	2992.8
$\sigma = 0.5$	136.4	2394.8

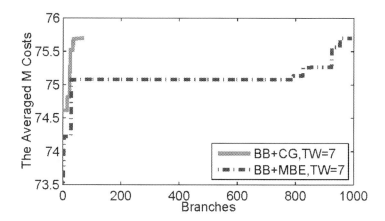

Fig. 5. The estimated M costs w.r.t. the number of branches for synthetic data with $\sigma = 0.5$.

7.2   Real Data

We also compare the M-best bounds on the benchmark problems of the probabilistic inference challenge 2011[1]. The data sets "Promedas", "Pedigree.tgz" and "Segmentation" are tested. For these data, we find the 3-best MAP assignments and costs. The results in Fig. 6 show that Algorithm 1 can obtain much tighter upper bounds. The "STRIPES" algorithm and the "BMMF" algorithm are not tested for some models, since "STRIPES" cannot solve large models, and the codes of "BMMF" only support the pairwise models.

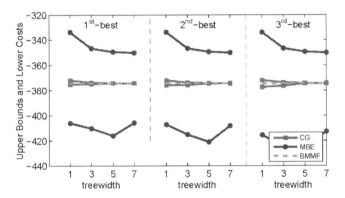

Fig. 6. Comparison of the M-best upper bounds and lower bounds of different algorithms on model "8_25_s.binary".

We also compare the branch and bound search algorithm, with the treewdith of the covering graph being 8. The number of branches is shown in Table 2. The results show the branches can be significantly reduced when using the optimized upper bounds, and thus the computational complexity can be reduced. The curves of the estimated M costs (averaged) w.r.t. the branches of the BB algorithm on the "or_chain_182.fg" model are shown in Fig. 7. The curves show that the optimized covering graph not only provides tighter bounds, but also provides better initialization. "BB+CG" takes only 1039 branches to find the true M-best MAP assignments, while "BB+MBE" takes 2105 branches.

Table 2. The number of branches of the branch and bound (BB) algorithm

	BB+CG	BB+MBE
or_chain_85.fg	51	87
or_chain_182.fg	1039	2105
18_11_s.binary	83	1043
18_1_s.binary	121	9057

---

[1] http://www.cs.huji.ac.il/project/PASCAL/.

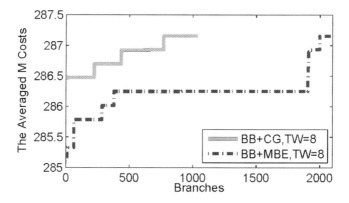

Fig. 7. The estimated M costs w.r.t. the number of branches for "or_chain_182.fg".

8 Conclusions

We propose a monotonic algorithm to tighten the M-best upper bounds for the M-best MAP inference on a probabilistic graphical model. Our algorithm can significantly tighten the M-best upper bounds, and decode better M-best assignments. Experimental results on synthetic and real problems verify the effectiveness of our algorithm.

References

1. Batra, D., Yadollahpour, P., Shakhnarovich, G.: Diverse M-best solutions in MRFs. In: ECCV (2012)
2. Yanover, C., Weiss, Y.: Finding the M most probable configurations using loopy belief propagation. In: NIPS (2004)
3. Fromer, M., Yanover, C.: Accurate prediction for atomic-level protein design and its application in diversifying the near-optimal sequence space. Proteins Struct. Funct. Bioinform. **75**, 682–705 (2008)
4. Nilsson, D.: An efficient algorithm for finding the M most probable configurations in probabilistic expert systems. Stat. Comput. **8**, 159–173 (1998)
5. Fromer, M., Globerson, A.: An LP View of the M-best MAP problem. In: NIPS (2009)
6. Batra, D.: An efficient message passing algorithm for the M-best MAP problem. In: UAI (2012)
7. Flerova, N., Rollon, E., Dechter, R.: Bucket and mini-bucket schemes for M best solutions over graphical models. In: Croitoru, M., Rudolph, S., Wilson, N., Howse, J., Corby, O. (eds.) GKR 2011. LNCS, vol. 7205, pp. 91–118. Springer, Heidelberg (2012)
8. Yarkony, J., Fowlkes, C., Ihler, A.: Covering trees and lower-bounds on quadratic assignment. In: CVPR (2010)
9. Dechter, R., Rish, I.: Mini-buckets: a general scheme for bounded inference. J. ACM **50**, 107–153 (2003)

A Data Cleaning Method and Its Application for Earthen Site Data Monitored by WSN

Yun Xiao[1(✉)], Xuanhong Wang[2], Xin Wang[1,3], Pengfei Xu[1],
Xiaojiang Chen[1], Dingyi Fang[1], and Baoying Liu[4]

[1] School of Information Science and Technology,
Northwest University, Xi'an, China
{yxiao,pfxu,xjchen,dyf}@nwu.edu.cn, wxh@xupt.edu.cn
[2] Department of Communication,
Xi'an University of Posts and Telecommunications, Xi'an, China
xcwang@ucalgary.ca
[3] Department of Geomatics Engineering,
University of Calgary, Calgary, Canada
[4] Cultural Heritage Department, University of Salento, Lecce, Italy
ficre_2008@hotmail.com

Abstract. This paper focuses on a data cleaning method to denoise and detect outliers of earthen site monitoring data with wireless sensor network (WSN). A data cleaning method, named DC_ESVS is proposed, which is based on the temporal and spatial characteristics of monitoring data with WSN. Using the cubic exponential smoothing algorithm and voting strategy, it can denoise and detect outliers of earthen site monitoring data based on the decision rule. We conduct various experiments on the dataset of the monitoring data of Xi'an Tang Hanguangmen city wall site with WSN to show detection accuracy of the presented method. Experimental results on anther dataset of the monitoring data of the Ming Great Wall in Shaanxi also show good performance of the proposed method.

Keywords: Data cleaning · Earthen site · Exponential smoothing · Voting strategy

1 Introduction

Earthen sites refer to the left traces of production, livelihood and other actives in human history that use the earth as the main construction material. The wireless sensor network (WSN) is a technology with the advantage of the low price, easy deployable which is suitable for the remote, real-time and long-term monitoring, once deployed without manual intervention. It has been widely used in environmental monitoring of earthen sites. For example, Abrardo and Rodriguez-Sanchez used the technology of WSN to collect the environmental information of heritage [1, 2]. As a result of the sensor network itself inherent problems [3], monitoring data is inevitable with the noise and outliers. Noise refers to real data modification, as showed in Fig. 1. Outlier refers to the data deviate from the larger data, as showed in Fig. 2.

© Springer Nature Singapore Pte Ltd. 2016
T. Tan et al. (Eds.): CCPR 2016, Part I, CCIS 662, pp. 638–649, 2016.
DOI: 10.1007/978-981-10-3002-4_52

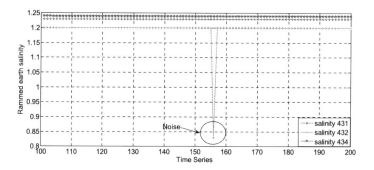

Fig. 1. Noise in rammed earth salinity data of earthen site

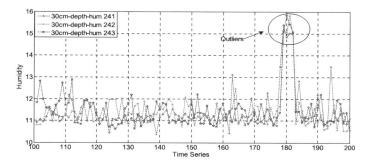

Fig. 2. Outliers in 30 cm depth humidity data of earthen site

According to the existing data warehouse development experience, the data cleaning process accounts for roughly 80 % time and costs in the whole process of data warehouse development [4]. Despite a lot of data cleaning method, but there is still no common cleaning method [5]. Because they disregard the fact that a lot of errors are systematic, inherent to the process that produces the data, and thus will keep occurring unless the problem is corrected at its source [3]. Different system determines the different characteristics of the data, and the characteristics of the data also reflect the nature of the system. So the data cleaning method must be based on the data characteristics, namely the system nature.

In this paper, we first analyze the temporal and spatial characteristics of monitoring data with WSN and the characteristics of different earthen site monitoring indicators. On the basis of analysis of the characteristics, a data cleaning method, called DC_ESVS, is proposed using the cubic exponential smoothing algorithm and voting strategy. Based on the decision rule, the DC_ESVS method can denoise and detect outliers of earthen site monitoring data with WSN.

The rest of the paper is: an overview of related work on data cleaning method is presented in Sect. 2, followed by introduction to analysis of the characteristics in Sect. 3. We then describe the detail of our proposed method DC_ESVS in Sect. 4. The experimental results are presented in Sect. 5. Finally, we conclude the paper in Sect. 6.

2 Related Work

Applications rely heavily on data, which makes data quality a detrimental for their function. Data management research has long recognized the importance of data quality, and has developed an extensive arsenal of data cleaning approaches based on integrity constraints [6–10], statistics [11], or machine learning [12]. Unfortunately, despite their applicability and generality, they are best-effort approaches that cannot ensure the accuracy of the repaired data. Due to their very nature, these methods do not have enough evidence to precisely identify and update errors.

To increase the accuracy of the above methods, a natural approach is to use external information in tabular master data [13] and domain experts [13–16]. However, these resources may be scarce and are usually expensive to employ.

The monitoring data of earthen sites with WSN are the typical time-series data. According to monitoring indicators of earthen sites, the time-series monitoring data show two characteristics. The first is a smoothness of time-series data, such as rammed earth salinity. The second is gradual change regularity of time-series data, such as rammed earth temperature.

Inspired by the literature thoughts [17], if we can select an appropriate prediction method, then the data cleaning problem is converted into a decision problem. If the prediction value is far from measured value, then measured value will be the noise or outlier according to a specially designed decision rules.

The method of time series prediction uses the historical data of target, which are arranged in a sequence with time elapse, to analyze the changing trend over time. Meanwhile, corresponding mathematical model for extrapolation of a quantitative prediction methods are established [18]. It is much convenient in the prediction of objects with much impacts and complex relationships [19]. As to the prediction for the monitoring data of earthen sites, it can be used to determine the variation and change trend with time series prediction method. The exponential smoothing method is one of the methods in time series prediction. It is evolved from the moving average method. By introducing a weighting factor, that is the smoothing coefficient α (in the range [0,1]), with a certain amount of time series prediction model, it can achieve forecasting future values of time sequence. It's often used in time series forecasting of short cycle trend forecasting.

In this paper, we use the cubic exponential smoothing method to accurately forecast data of different features, and then design a voting strategy to further process based on the temporal and spatial characteristics of the monitoring data. At last, through a specially designed decision rules, the test data can be determined quickly and efficiently.

3 The Characteristics of Monitoring Data

The proposed data cleaning method DC_ESVS is based on the data characteristics of the analysis. Part of the Ming Great Wall in northern Shaanxi, called Shaanxi Ming Great Wall, is a typical large earthen site. We deployed about 300 nodes in Shaanxi Ming Great Wall to monitor the environmental and ontology information. The characteristics of monitoring data of Ming Great Wall in Shaanxi are as following:

3.1   The Temporal Characteristics of Monitoring Data

The temporal characteristic of monitoring data is that the monitoring data with the same sensor nodes in the adjacent time present a strong consistency and stability. Figure 3 shows the spatial characteristics of the monitoring data of different monitoring indicators by calculating the deviation of two related data. As shown in Fig. 3(a), we can see that 87 percent of deviation of two humidity data at adjacent time is focused on a point value between 0 and 1. As shown in Fig. 3(b), 98 percent of deviation of two salinity data at adjacent time is focused on a point value between 0 and 0.1. This is because the monitoring indicators change slowly and continuously.

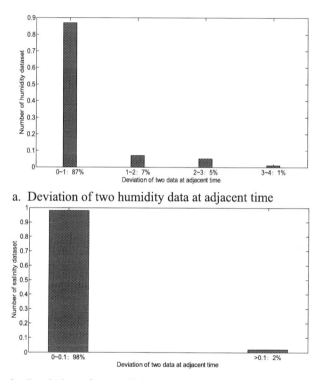

a. Deviation of two humidity data at adjacent time

b. Deviation of two salinity data at adjacent time

Fig. 3. The temporal characteristic of monitoring data with same sensor node

3.2   The Spatial Characteristics of Monitoring Data

The spatial characteristic of monitoring data is that the monitoring data present a strong consistency and stability in the same local space and same time. As shown in Fig. 4, we can see that 69 percent of data deviation of two nodes at local space in same time is focused on a point value between 0 and 1, and only 1 percent of data deviation is larger than 3. The reason is that the same local space has same or similar environmental and ontology conditions.

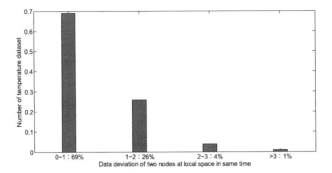

Fig. 4. The spatial characteristic of monitoring data at the local space in same time

4   DC_ESVS Method

4.1   The Cubic Exponential Smoothing Method

Using the cubic exponential smoothing method, our proposed DC_ESVS method can accurately forecast data of different features. The main idea of the cubic exponential smoothing method is described as following:

Supposed $x_1, x_2, \ldots, x_i, \ldots, x_n$ is a time series of n real value, $S_1^{(1)}, S_2^{(1)}, \ldots, S_n^{(1)}$ is the forecasted value in the time. The linear exponential smoothing model is:

$$S_t^{(1)} = \alpha x_t + (1 - \alpha)S_{t-1}^{(1)} \tag{1}$$

Among them, α is the smooth coefficient, t is the time stamp. By the formula (1), the forecast value of the time stamp t $S_t^{(1)}(t = 1, 2, \ldots n)$ is equals to the forecast value of the previous time stamp $t-1$ $S_{t-1}^{(1)}$ with an error correction value in addition. In general, the forecast accuracy of the linear exponential smoothing method is affected by the smooth coefficient a. So it applies to the time series prediction with horizontal trend. Especially, there is a certain trend in the time series, the prediction results tend to lag obviously, and then cause errors. Therefore, a higher order exponential smoothing method is needed.

Considering all linear exponential smoothing value $S_t^{(1)}(t = 1, 2, \ldots n)$ as a new time series, the exponential smoothing method is used once again, secondary exponential smoothing is get with the value of the original time series. It can be used to a linear trend prediction. The model can be expressed as:

$$S_t^{(2)} = \alpha S_t^{(1)} + (1 - \alpha)S_{t-1}^{(2)} \tag{2}$$

Similarly, if the $S_t^{(2)}$, (t = 1, 2,...n) is considered as a new time series, and the values of cubic exponential smoothing can be calculated. The model is expressed as:

$$S_t^{(3)} = \alpha S_t^{(2)} + (1 - \alpha)S_{t-1}^{(3)} \tag{3}$$

Suppose $\hat{x}$ to be the predicted value,

$$\hat{x}_{i+m} = a_i + b_i m + c_i m^2 \tag{4}$$

In Eq. (4), m is the prediction step, and m can be 1, 2, 3, etc. Prediction parameters a_i, b_i and c_i can be calculated using Eq. (5):

$$\begin{cases} a_i = 3S_i^{(1)} - 3S_i^{(2)} + S_i^{(3)} \\ b_i = \frac{\alpha}{2(1-\alpha)^2}[(6-5\alpha)S_i^{(1)} - (10-8\alpha)S_i^{(2)} + (4-3\alpha)S_i^{(3)}] \\ c_i = \frac{\alpha^2}{2(1-\alpha)^2}(S_i^{(1)} - 2S_i^{(2)} + S_i^{(3)}) \end{cases} \tag{5}$$

Using the cubic exponential smoothing method, the smoothing initial value $S_1^{(1)}$, $S_1^{(2)}$, $S_1^{(3)}$ and smoothing coefficient α are needed to be set. When the data volume is small, the initial value's influence on the predicted value is greater, so the smoothing initial value $S_1^{(1)}$, $S_1^{(2)}$, $S_1^{(3)}$ can be set with Eq. (6).

$$S_1^{(1)} = S_1^{(2)} = S_1^{(3)} = (x_1 + x_2 + x_3)/3 \tag{6}$$

When the data volume is large, the initial value's influence on the predicted value is smaller, so the smoothing initial value $S_1^{(1)}$, $S_1^{(2)}$, $S_1^{(3)}$ can be set with Eq. (7).

$$S_1^{(1)} = S_1^{(2)} = S_1^{(3)} = x_1 \tag{7}$$

As for smoothing coefficient α, it reflects the historical data in the proportion of exponential smoothing values at different periods. The prediction accuracy of exponential smoothing forecasting model largely depends on the value of smoothing coefficient α. In this paper, the smoothing coefficient α is calculated according to the historical time series of weight with Eq. (8).

$$\alpha = 1 - e^{\left[\frac{\ln(1-w)}{h}\right]} \tag{8}$$

where w denotes the weight of the historical data, and h is the number of the historical time series.

4.2 Voting Strategy

In order to detect outlier and noise, the temporal and spatial characteristics of monitoring data are needed to use. The adjacent node set is selected as the reference nodes to use the temporal and spatial characteristics according to the WSN topology. Directly using the adjacent node as reference node to determine whether the test data is noise or outlier could be problematic since a single reference node may have data errors, which will result in the outlier found false positives and reduce the accuracy of the test results.

In order to solve the above problem, a voting strategy based on probability is proposed to modify the results in this paper. Before discussing the specific steps of voting strategy, we first give the definition of the outlier probability.

Definition 1. Outlier probability $p_{i\_outlier}$ is defined as the ratio of the number of the references nodes and the number of nodes in the reference node set which determine the test data as an outlier.

$$p_{i\_outlier} = \frac{K_i}{L_i} \tag{9}$$

where L_i denotes the number of the reference nodes of node i, and K_i denotes the number of nodes in the reference node set which determine the test data as an outlier.

First each node in the reference node set determines the test data of node i as a noise or an outlier according to the noise or outlier threshold T_{Noi_Out}. If the test data is determined as an outlier, so K_i is added by one. If $p_{i\_outlier}$ is greater than the outlier probability threshold T_{Op}, then the under test data is labeled as an outlier, else it is labeled as a noise.

4.3   DC_ESVS Method

The DC_ESVS method is described with the following steps.

Step 1. Choose the appropriate reference node set, and form the input data set of the test node and reference node.
Step 2. Choose the appropriate weight w and the number of the historical time series h, calculate the smoothing coefficient α with Eq. (8).
Step 3. Input the normal threshold T_{Nor_Ab}, the noise or outlier threshold T_{Noi_Out}, and the outlier probability threshold T_{Op}.
Step 4. Calculate the predicted value $\hat{x}_i$ according t o Eq. (4).
Step 5. Use the following decision rule to judge the test data x_i as a normal data, noise or outlier.

If abs($\hat{x}_i$-x_i) < T_{Nor_Ab} then x_i is labeled as a normal data, go to step 4
Else if $p_{i\_outlier}$ < T_{Op} then x_i is labeled as a noise, update $x_i = \hat{x}_i$, go to step 4
Else x_i is labeled as an outlier, go to step 4

The thresholds in the DC_ESVS method can be set according to the largest proportion in the deviation of real data. Using the cubic exponential smoothing method, the predicted value can be calculated accurately. With the decision rule, the test data can be accurately judged to be a normal data, noise, or outlier.

4.4   Evaluation Index

In order to evaluate the performance of the DC_ESVS method, we define the detection accuracy.

Definition 2. Detection accuracy A_{noi_out}

$$A_{noi\_out} = \text{Num}_{det\_noi\_out}/\text{Num}_{real\_noi\_out}, \tag{10}$$

where $\text{Num}_{det\_noi\_out}$ denotes the number of the detected noises and outliers with DC_ESVS, and $\text{Num}_{real\_noi\_out}$ denotes the number of the real noises and outliers in test dataset.

5 Experimental Results

To evaluate the proposed method, we applied it on two datasets. The first dataset, called the Tang dataset, is from the monitoring data of Xi'an Tang Hanguangmen city wall site with WSN. The Xi'an Tang Hanguangmen city wall site is an indoor earthen site. We have monitored environmental data in Xi'an Tang Hanguangmen city wall site since 2009. Part of the node deployment situation is shown in the Fig. 5(a). We chose 200 humidity data to form the Tang dataset. As an indoor earthen site, Xi'an Tang Hanguangmen city wall is in an artificial environment, and workers in the Xi'an Tang Hanguangmen city wall museum use air conditioners, dryers and humidifiers to adjust the environment of the earthen site, thus the data we got appears smooth. We labeled the noises and outliers according to the daily record of Xi'an Tang Hanguangmen city wall site. These labeled data are used to test the accuracy of the proposed method.

a. Part of Tang Hanguangmen city wall   b. Part of Ming Great Wall

Fig. 5. Part of the node deployment situation for two datasets

The second dataset, called the Ming dataset, is from the monitoring data of the Ming Great Wall in Shaanxi with WSN. It includes rammed earth surface salinity, rammed earth vertical salinity, rammed earth temperature and humidity data in different depths, i.e. 5 cm, 15 cm and 30 cm, from November 1, 2015 to November 30, 2015. Part of the node deployment situation is shown in Fig. 5(b).

From the actual deployment diagram, we can easily get the space adjacent reference node set of a certain node. The DC_ESVS method is implemented in Matlab 2011b, and all experiments are run on a Windows 10 platform with CPU speed of 2.5 GHz*2, and 8 GB RAM.

For Tang dataset, we set the normal or abnormal threshold T_{Nor_Ab} to 0.3, the noise or outlier threshold T_{Noi_Out} to 0.1 according to the largest proportion in the deviation of two air humidity data at adjacent time. And the outlier probability threshold T_{Op} is set to 0.5. The experimental results are showed in Figs. 6 and 7. Figure 6 shows the relationship between smoothing coefficient α and detection precision A_{noi_out} running on air humidity of dataset A with a fixed w and a varying h. When weight w is set to 0.9, smoothing coefficient α is calculated with h from 3 to 20. From this figure, we can obtain that the best smoothing coefficient α is from 0.15 to 0.32, and the best detection accuracy A_{noi_out} is 100 %. Figure 7 shows the relationship between smoothing coefficient α and detection precision A_{noi_out} running on air humidity of dataset A with a fixed h and a varying w. When h, the historical air humidity data is set to 0.9, smoothing coefficient α is calculated with w from 0.1 to 0.98. From this figure, we can obtain that the best smoothing coefficient α is from 0.16 to 0.30, and the best detection accuracy A_{noi_out} is 100 %. So in order to get the best detection accuracy A_{noi_out}, the best smoothing coefficient α is from 0.16 to 0.30.

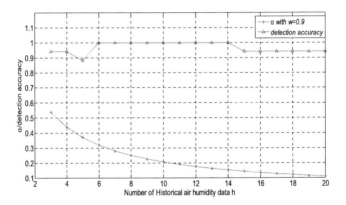

Fig. 6. The relationship between smoothing coefficient α and detection accuracy A_{noi_out} with a fixed w and a varying h

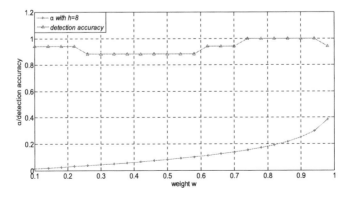

Fig. 7. The relationship between smoothing coefficient α and detection accuracy A_{noi_out} with a fixed h and a varying w

For Ming dataset, we set the normal or abnormal threshold T_{nor_ab}, the noise or outlier threshold T_{noi_out} according to the largest proportion in the deviation of two related data at adjacent time. Table 1 shows the detail parameters of the DC_ESVS. And the typical experimental results are shown in Figs. 8, 9 and 10. From Fig. 8, we can obtain that the three noises are detected by our proposed method on the rammed earth vertical salinity data of node 131. Figure 9 shows some detected noises and outliers on the rammed earth humidity data of 30 cm depth of node 241. Figure 10 shows the noises detection result on the rammed earth temperature data of 30 cm depth of node 468.

Table 1. All parameters of the DC_ESVS

Monitoring indicators	Parameters			
	T_{Nor_Ab}	T_{Noi_Out}	T_{Op}	α ($h = 8$, $w = 0.9$)
Rammed earth surface salinity	0.3	0.1	0.5	0.25
Rammed earth vertical salinity	0.3	0.1	0.5	0.25
Rammed earth temperature in depth of 5 cm	6	2	0.5	0.25
Rammed earth temperature in depth of 15 cm	5	2	0.5	0.25
Rammed earth temperature in depth of 30 cm	4	2	0.5	0.25
Rammed earth humidity in depth of 5 cm	3	1	0.5	0.25
Rammed earth humidity in depth of 15 cm	3	1	0.5	0.25
Rammed earth humidity in depth of 30 cm	2	1	0.5	0.25

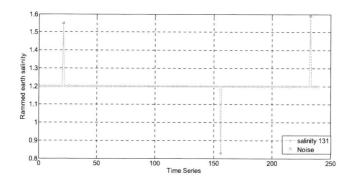

Fig. 8. Noises detection result on the rammed earth vertical salinity data of node 131

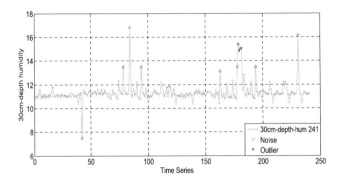

Fig. 9. Noises and outliers detection result on the rammed earth humidity data of 30 cm depth of node 241

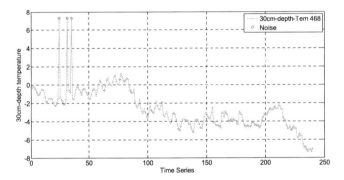

Fig. 10. Noises detection result on the rammed earth temperature data of 30 cm depth of node 468

6   Conclusion

In this paper, we proposed a data cleaning method, named DC_ESVS, for detecting noise and outlier in dataset. The proposed method is based on the temporal and spatial characteristics of monitoring data with WSN. Using the cubic exponential smoothing algorithm and voting strategy, and based on the decision rule, it can denoise and detect outliers of earthen site monitoring data. Moreover, we conducted a set of experiments to show accuracy of the presented method. For future, we plan to do more experiments for examining other aspects of the method. In addition, parameter optimization will be considered in order to improve the performance of the method.

Acknowledgements. This work was partially supported by International cooperation and exchange program of Shaanxi Province (No. 2016KW-034), Industrial Science and technology project of Shaanxi Province (No. 2015GY013), the Education Department of Shaanxi Province Natural Science Foundation, China (Grant No. 12JK0937, 15JK1742).

References

1. Abrardo, A., Balucanti, L., Belleschi, M., et al.: Health monitoring of architectural heritage: the case study of San Gimignano. In: 2010 IEEE Workshop on Environmental Energy and Structural Monitoring Systems (EESMS)(2010)
2. Rodriguez-Sanchez, M.C., Borromeo, S., Hernández-Tamames, J.A.: Wireless sensor networks for conservation and monitoring cultural assets. IEEE Sens. J. **11**, 1382–1389 (2011)
3. Wang, X., Dong, X.L., Meliou, A.: Data X-ray: a diagnostic tool for data errors. In: Proceedings of the 2015 ACM SIGMOD International Conference on Management of Data (2015)
4. Sean, K., Jeffrey, H., Catherine, P., et al.: Research directions in data wrangling: visualizations and transformations for usable and credible data. Inf. Vis. **10**(4), 271–288 (2011)
5. Wang, M.J., Pan, Q.M., Liu, Z., Chen, W.: Survey of visualization data cleaning. J. Image Graph. **20**(4), 0468–0482 (2015)
6. Bohannon, P., Fan, W., Flaster, M., Rastogi, R.: A cost-based model and effective heuristic for repairing constraints by value modification. In: SIGMOD (2005)
7. Chiang, F., Miller, R.J.: A unified model for data and constraint repair. In ICDE (2011)
8. Chu, X., Ilyas, I.F., Papotti, P.: Holistic data cleaning: putting violations into context. In: ICDE (2013)
9. Geerts, F., Mecca, G., Papotti, P., Santoro, D.: The LLUNATIC data-cleaning framework. In: VLDB (2013)
10. Song, S., Cheng, H., Yu, J.X., Chen, L.: Repairing vertex labels under neighborhood constraints. In: VLDB (2014)
11. May_eld, C., Neville, J., Prabhakar, S.: ERACER: a database approach for statistical inference and data cleaning. In: SIGMOD (2010)
12. Yakout, M., Berti-Equille, L., Elmagarmid, A.K.: Don't be SCAREd: use SCalable Automatic REpairing with maximal likelihood and bounded changes. In: SIGMOD (2013)
13. Fan, W., Li, J., Ma, S., Tang, N., Yu, W.: Towards certain fixes with editing rules and master data. In: VLDB (2012)
14. Raman, V., Hellerstein, J.M.: Potter's wheel: an interactive data cleaning system. In: VLDB (2001)
15. Volkovs, M., Chiang, F., Szlichta, J., Miller, R.J.: Continuous data cleaning. In: ICDE (2014)
16. Yakout, M., Elmagarmid, A.K., Neville, J., Ouzzani, M., Ilyas, I.F.: Guided data repair. In: VLDB (2011)
17. Zhang, W., Yang, Y., Wang, Q.: Handling missing data in software effort prediction with naive Bayes and EM algorithm. In: Proceedings of the 7th International Conference on Predictive Models in Software Engineering (2011)
18. Liming, Wang, Lian, Wang: Application of time series analysis. Fudan University Press, Shanghai (2009)
19. Lu, J., Xu, F.: Research on prediction model of landslide based on the exponential smoothing and regression analysis. J. Wuhan Univ. Technol. **33**(10), 88–91 (2011)

Analysis of Convergence Properties for Gath-Geva Clustering Using Jacobian Matrix

Chaomurilige Wang[1], Jian Yu[1(✉)], and Jie Zhu[1,2]

[1] Beijing Key Lab of Traffic Data Analysis and Mining,
School of Computer and Information Technology,
Beijing Jiaotong University, Beijing 100044, China
jianyu@bjtu.edu.cn
[2] Department of Information Management,
The Central Institute for Correctional Police, Baoding 071000, China

Abstract. The GG algorithm is one of the most frequently used clustering algorithms. Just like FCM and other clustering algorithms, the coincident clustering result (all membership functions are equal to 1/c where c is the cluster number and all cluster centers are the mass center of data set) is a fixed point of GG clustering algorithm. In this paper we reveal the relation between the stable fixed points of the GG algorithm and the data sets using Jacobian matrix analysis. Unlike the Hessian matrix analysis, The Jacobian matrix analysis can be applied to the clustering algorithms which only have update equation. We demonstrate that the coincident clustering result is not a stable fixed point of GG clustering algorithm and discuss the relationship between convergence rate of clustering algorithm and fuzziness index m. Some experimental results verify the effectiveness of our theoretical results.

Keywords: GG clustering algorithm · Jacobian matrix · Convergence properties analysis

Clustering is the unsupervised classification of patterns (observations, data items, or feature vectors) into groups (clusters) [1]. Data clustering has been widely applied in many fields, including machine learning, data mining, pattern recognition, image analysis and bioinformatics. The main idea of the clustering algorithm is to organize data into clusters based on similarity. Intuitively, data within a cluster are more similar to each other than they are to data belonging to a different cluster [1]. The K-means clustering algorithm, one of the most popular and simple clustering algorithms, was first proposed in 1955. Since Zadeh [11] introduced fuzzy set theory, the hard clustering algorithm was extended to the FCM (Fuzzy C-means) clustering algorithm in [3]. The first extension of the FCM algorithm is the well-known GK clustering algorithm. Gaustafson and Kessel replaced the Euclidean distance with the Mahalanobis distance, thus the GG clustering algorithm can detect clusters of different shapes [4, 5].

The GG clustering algorithm is one of the most famous extension of FCM clustering algorithm [6]. Different from other fuzzy clustering algorithms, the GG algorithm has no objective function. The GG clustering algorithm cluster data by iteratively updating the cluster center and the partition matrix. The dissimilarity measurement in

© Springer Nature Singapore Pte Ltd. 2016
T. Tan et al. (Eds.): CCPR 2016, Part I, CCIS 662, pp. 650–662, 2016.
DOI: 10.1007/978-981-10-3002-4_53

the GG algorithm is similar to the EM algorithm, and the update equation of the cluster center is same as the GK clustering algorithm. Several work have been proposed to improve the performance of GG clustering algorithm and apply it to more application, such as [12, 13]. However there is no work focus on the parameter selection and convergence property analysis of the GG clustering algorithm. It is obvious that the value of fuzziness index m may influence the behavior of the GG clustering algorithm as well as the GK clustering algorithm and other fuzzy clustering algorithms. We have addressed the parameter selection of the GK clustering algorithm by using Jacobian matrix analysis in [7]. Thus we discuss the parameter selection problem of the GG clustering algorithm by using Jacobian matrix analysis in this paper.

We also propose a method for convergence rate estimation of fuzzy clustering algorithms. Yu et $al.$ [10] presented a novel test to judge the local optimality of the FCM fixed points with Hessian matrix and made analysis of the fuzziness index m in the FCM [7]. The Hessian matrix analysis also can be used to estimate the convergence rate of clustering algorithm. However, the Hessian matrix is too complicated and complex in practical applications. In this paper, we focus on the Jacobian matrix analysis. By apply the Jacobian matrix analysis method to the GG clustering algorithm, we demonstrate that the Jacobian matrix analysis can be used to analyze any kind of iteration-based clustering algorithm whether there exist an objective function.

The rest of the paper is organized as follows: In Sect. 1, we give a brief review of the GG clustering algorithm. In Sect. 2, we make the Jacobian matrix analysis and give the theoretical parameter selection for the GG algorithm. In Sect. 3, we validate our theoretical results by presenting several experimental results. Finally, the conclusions are stated in Sect. 4.

1   The Gath-Geva Clustering Algorithm

In this section, we give a brief description of the GG (Gath-Geva) fuzzy clustering algorithm. Cluster analysis is based on partitioning a collection of data points into a number of subgroups according to the partition matrix [2]. The traditional clustering algorithms provide a hard clustering result that each data point is assigned to the most similar cluster center. However, in some case the data point is belong to several clusters. Since Zadeh [11] introduced fuzzy set theory, the hard clustering algorithm is extended to the fuzzy clustering algorithm, such as FCM (Fuzzy C-means) clustering algorithm [3]. The GG clustering algorithm is one of the most famous extension of FCM clustering algorithm. Different from other fuzzy clustering algorithms, the GG algorithm has no objective function.

Let $X = \{x_1, x_2, \ldots, x_n\} \in R^S$ be a data set where R^S is an s-dimensional Euclidean space. The clustering algorithm aimed to cluster n data point to c clusters with its cluster centers $V = \{v_i\}, i = 1, 2, \ldots, c$. Let $U = [\mu_{ik}]_{c \times n} \in M_{fc}$ is the fuzzy c-partition of data set where $M_{fc} = \left\{ U = [\mu_{ik}]_{c \times n} \middle| \forall i, \forall k, \mu_{ik} \geq 0, n > \sum_{k=1}^{n} \mu_{ik} > 0, \sum_{i=1}^{c} \mu_{ik} = 1 \right\}$. And $A = \{\alpha_i\}, i = 1, 2, \ldots, c$ is the priori probability matrix, $F = \{F_i\}, i = 1, 2, \ldots, c$ is the cluster covariance matrix.

The GG clustering algorithm is iterated through following update equations:

$$v_i = \frac{\sum_{k=1}^{n} (\mu_{ik})^m x_k}{\sum_{k=1}^{n} (\mu_{ik})^m} \tag{1}$$

$$\mu_{ik} = \frac{d_{ik}}{\sum_{i=1}^{c} d_{ik}} \tag{2}$$

$$\alpha_i = \frac{1}{n} \sum_{k=1}^{n} \mu_{ik} \tag{3}$$

$$F_i = \frac{\sum_{k=1}^{n} \mu_{ik}(x_k - v_i)(x_k - v_i)^T}{\sum_{k=1}^{n} \mu_{ik}} \tag{4}$$

where $d_{ik} = \alpha_i (\det F_i)^{-\frac{1}{2}} \exp\left[-\frac{1}{2}(x_k - v_i)^T F_i^{-1}(x_k - v_i)\right]$. Thus we have the GG clustering algorithm:

GG Algorithm.

Step 1: Fix $2 \leq c \leq n$ and fix any $\varepsilon > 0$
         Give initials $U^{(0)}$ and let $t = 1$.
Step 2: Compute the cluster centers $V^{(t)}$ with $U^{(t-1)}$ using Eq. (1).
Step 3: Compute matrix $A^{(t)}$ and cluster covariance matrix $F^{(t)}$ using Eqs. (3) and (4).
Step 4: Update $U^{(t)}$ with $V^{(t)}$ using Eq. (2).
Step 5: Compare $U^{(t)}$ to $U^{(t-1)}$ in a convenient matrix norm $\|\cdot\|$.
         IF $\left\| U^{(t)} - U^{(t-1)} \right\| < \varepsilon$, STOP
         ELSE $t = t + 1$ and return to step 2.

The GG clustering algorithm generates the cluster result by an iterative updating of cluster center and partition matrix. Thus we interpreted the GG clustering algorithm as a map: $U^{(t)} = \theta(U^{(t-1)}) = H(G(U^{(t-1)}))$, where $G : U = [\mu_{ik}]_{c \times n} \in M_{fcn} \mapsto V = (v_1, \cdots v_c)^T \in R^{cs}$ and $H : V = (v_1, \cdots, v_c)^T \in R^{cs} \mapsto U = [\mu_{ik}]_{c \times n} \in M_{fcn}$. It is easy to know that, if $\forall i, v_i = \sum_{k=1}^{n} x_k/n = \bar{x}$, then $U = [c^{-1}]_{c \times n}$ and $\alpha_i = c^{-1}, \forall i$. In the next iteration, the coincident cluster center $(\bar{x}, U = [c^{-1}]_{c \times n})$ may be outputted by GG algorithm again. Thus the coincident clustering result $(\bar{x}, U = [c^{-1}]_{c \times n})$ is the fixed point of the GG clustering algorithm. In order to get a meaningful clustering result, the mass center of data $\bar{x} = \sum_{k=1}^{n} x_k/n$ should be avoided to be outputted by algorithm, the coincident clustering result $(\bar{x}, U = [c^{-1}]_{c \times n})$ should not be a stable fixed point of GG clustering algorithm.

We have discussed about the parameter selection of the GK clustering algorithm in [7]. The GG clustering result may influenced by the fuzziness index m as well as GK clustering algorithm. In next section, we analysis the parameter selection of GG clustering algorithm using Jacobian matrix analysis.

2 Jacobian Matrix Analysis on Parameter Selection for the Gath-Geva Clustering Algorithm

If a data set is clustered into c clusters, different clusters should have different cluster centers than each other. If the mass center of data $\bar{x} = \sum_{k=1}^{n} x_k / n$ is outputted as the cluster center for each cluster, $\forall i, v_i = \bar{x}$, the partition matrix may become $U = [c^{-1}]_{c \times n}$. In this case, it is hard to cluster data set according to the partition matrix because all the data points are belong to one cluster. Thus $(\bar{x}, U = [c^{-1}]_{c \times n})$ should not be a stable fixed point of GG clustering algorithm.

First, we rewrite the GG clustering algorithm as a mapping $U^{(t)} = \theta(U^{(t-1)})$:

$$v_i^{(t)} = \frac{\sum_{k=1}^{n} \left(\mu_{ik}^{(t-1)} \right)^m x_k}{\sum_{k=1}^{n} \left(\mu_{ik}^{(t-1)} \right)^m} \tag{5}$$

$$\alpha_i^{(t)} = \frac{1}{n} \sum_{k=1}^{n} \mu_{ik}^{(t-1)} \tag{6}$$

$$F_i^{(t)} = \frac{\sum_{k=1}^{n} \mu_{ik}^{(t-1)} \left(x_k - v_i^{(t)} \right) \left(x_k - v_i^{(t)} \right)^T}{\sum_{k=1}^{n} \mu_{ik}^{(t-1)}} \tag{7}$$

$$\mu_{ik}^{(t)} = \frac{d_{ik}^{(t)}}{\sum_{i=1}^{c} d_{ik}^{(t)}} \tag{8}$$

where $d_{ik}^{(t)} = \alpha_i^{(t)} \left(\det F_i^{(t)} \right)^{-\frac{1}{2}} \exp \left[-\frac{1}{2} \left(x_k - v_i^{(t)} \right)^T \left(F_i^{(t)} \right)^{-1} \left(x_k - v_i^{(t)} \right) \right]$. The GG clustering result my heavily value influenced by value of fuzziness index m. We address the parameter selection problem using Jacobian matrix analysis. From Olver's Corollary [8] we know that if spectral radius of the Jacobian matrix $\frac{\partial \theta(U)}{\partial U}$ at point $U = [c^{-1}]_{c \times n}$ is less than 1, then $(\bar{x}, U = [c^{-1}]_{c \times n})$ is the stable fixed point of GG clustering algorithm.

Olver's Corollary (Olver [8], p. 143). If the Jacobian matrix $g'(\mu^*) = \frac{dg(\mu)}{d\mu} \big|_{\mu = \mu^*}$ is a convergent matrix, meaning that its spectral radius (i.e. the maximum of absolute eigenvalues of the Jacobian matrix) $\lambda(g'(\mu^*))$ satisfies $\lambda(g'(\mu^*)) < 1$, then μ^* is an asymptotically stable fixed point.

The element $\frac{\partial \theta_{ik}}{\partial \mu_{jr}}, i = 1, \cdots, c, k = 1, \cdots, n, j = 1, \cdots, c-1, r = 1, \cdots, n$ of Jacobian matrix is obtained by taking the derivations of θ_{ik} with respect to μ_{jr}.

Lemma 1. For $i = 1, \cdots, c, \; k = 1, \cdots, n$ and $j = 1, \cdots, c-1, \; r = 1, \cdots, n$, each element $\frac{\partial \theta_{ik}}{\partial \mu_{jr}}$ of Jacobian matrix is

$$\frac{\partial \theta_{ik}}{\partial \mu_{jr}} = \frac{\delta_{ij}\mu_{ik}}{n\alpha_i} - \frac{\delta_{ij}\mu_{ik}}{2\sum_{k=1}^{n}\mu_{ik}}\left((x_r - v_i)^T F_i^{-1}(x_r - v_i) - s\right)$$

$$+ \frac{\delta_{ij}\mu_{ik}m\mu_{ir}^{m-1}}{\sum_{k=1}^{n}\mu_{ik}^m}(x_r - v_i)^T F_j^{-1}(x_k - v_i)$$

$$- \frac{\delta_{ij}\mu_{ik}}{2}\left[(x_k - v_i)^T\left(-F_i^{-1}\frac{(x_r - v_i)(x_r - v_i)^T}{\sum_{k=1}^{n}\mu_{ik}}F_i^{-1} + \frac{F_i^{-1}}{\sum_{k=1}^{n}\mu_{ik}}\right)(x_k - v_i)\right]$$

$$- \frac{\mu_{ik}\mu_{jk}}{n\alpha_j} + \frac{\mu_{ik}\mu_{jk}}{2\sum_{k=1}^{n}\mu_{jk}}\left((x_r - v_j)^T F_j^{-1}(x_r - v_j) - s\right)$$

$$- \frac{\mu_{ik}\mu_{jk}m\mu_{jr}^{m-1}}{\sum_{k=1}^{n}\mu_{jk}^m}(x_r - v_j)^T F_j^{-1}(x_k - v_j)$$

$$+ \frac{\mu_{ik}\mu_{jk}}{2}\left[(x_k - v_j)^T\left(-F_j^{-1}\frac{(x_r - v_j)(x_r - v_j)^T}{\sum_{k=1}^{n}\mu_{jk}}F_j^{-1} + \frac{F_j^{-1}}{\sum_{k=1}^{n}\mu_{jk}}\right)(x_k - v_j)\right]$$

$$+ \frac{\mu_{ik}\mu_{ck}}{n\alpha_c} - \frac{\mu_{ik}\mu_{ck}}{2\sum_{k=1}^{n}\mu_{ck}}\left((x_r - v_c)^T F_c^{-1}(x_r - v_c) - s\right)$$

$$+ \frac{\mu_{ik}\mu_{ck}m\mu_{cr}^{m-1}}{\sum_{k=1}^{n}\mu_{ck}^m}(x_r - v_c)^T F_c^{-1}(x_k - v_c)$$

$$- \frac{\mu_{ik}\mu_{ik}}{2}\left[(x_k - v_c)^T\left(-F_c^{-1}\frac{(x_r - v_c)(x_r - v_c)^T}{\sum_{k=1}^{n}\mu_{ck}}F_c^{-1} + \frac{F_c^{-1}}{\sum_{k=1}^{n}\mu_{ck}}\right)(x_k - v_c)\right]$$

$$(9)$$

where $\delta_{ij} = \begin{cases} 1, & if\ i = j \\ 0, & if\ i \neq j \end{cases}$ is the kronecker delta function.

Proof. The partition matrix of GG clustering algorithm satisfy $\sum_{i=1}^{c}\mu_{ik} = 1$, thus $\frac{\partial \theta_{ik}}{\partial \mu_{jr}}$ can be obtained by taking the derivations of θ_{ik} with respect to μ_{jr} as follows:

$$\frac{\partial \theta_{ik}}{\partial \mu_{jr}} = \frac{\left(\sum_{i=1}^{c}d_{ik}\right)\frac{\partial d_{ik}}{\partial \mu_{jr}}}{\left(\sum_{i=1}^{c}d_{ik}\right)^2} - \frac{d_{ik}\left(\frac{\partial \sum_{i=1}^{c}d_{ik}}{\partial \mu_{jr}}\right)}{\left(\sum_{i=1}^{c}d_{ik}\right)^2} = \frac{\delta_{ij}\frac{\partial d_{ik}}{\partial \mu_{jr}}}{\left(\sum_{i=1}^{c}d_{ik}\right)} - \frac{d_{ik}\frac{\partial \sum_{i=1}^{c-1}d_{ik}}{\partial \mu_{jr}}}{\left(\sum_{i=1}^{c}d_{ik}\right)^2} + \frac{d_{ik}\frac{\partial d_{ck}}{\partial \mu_{cr}}}{\left(\sum_{i=1}^{c}d_{ik}\right)^2}$$

$$(10)$$

We can get following result by simple computation:

$$\frac{\partial d_{jk}}{\partial \mu_{jr}} = \frac{\partial \alpha_j (\det F_j)^{-\frac{1}{2}} \exp\left[-\frac{1}{2}(x_k - v_j)^T F_j^{-1}(x_k - v_j)\right]}{\partial \mu_{jr}}$$

$$= (\det F_j)^{-\frac{1}{2}} \exp\left[-\frac{1}{2}(x_k - v_j)^T F_j^{-1}(x_k - v_j)\right] \frac{\partial \alpha_j}{\partial \mu_{jr}}$$

$$+ \alpha_j \exp\left[-\frac{1}{2}(x_k - v_j)^T F_j^{-1}(x_k - v_j)\right] \frac{\partial (\det F_j)^{-\frac{1}{2}}}{\partial \mu_{jr}}$$

$$+ \alpha_j (\det F_j)^{-\frac{1}{2}} \frac{\partial \exp\left[-\frac{1}{2}(x_k - v_j)^T F_j^{-1}(x_k - v_j)\right]}{\partial \mu_{jr}}$$

$$= \frac{1}{n\alpha_j} d_{jk} - \frac{1}{2}(\det F_j)^{-1} d_{jk} \frac{\partial \det F_j}{\partial \mu_{jr}}$$

$$- \frac{1}{2} d_{jk} \left[\frac{\partial (x_k - v_j)^T}{\partial \mu_{jr}} F_j^{-1}(x_k - v_j) + (x_k - v_j)^T \frac{\partial F_j^{-1}}{\partial \mu_{jr}}(x_k - v_j) \right.$$

$$\left. + (x_k - v_j)^T F_j^{-1} \frac{\partial (x_k - v_j)}{\partial \mu_{jr}} \right]$$

Similarity,

$$\frac{\partial F_j}{\partial \mu_{jr}} = \frac{(x_r - v_j)(x_r - v_j)^T - F_j}{\sum_{k=1}^n \mu_{jk}}$$

$$\frac{\partial F_j^{-1}}{\partial \mu_{jr}} = -F_j^{-1} \frac{\partial F_j}{\partial \mu_{jr}} F_j^{-1} = -F_j^{-1} \frac{(x_r - v_j)(x_r - v_j)^T}{\sum_{k=1}^n \mu_{jk}} F_j^{-1} + \frac{F_j^{-1}}{\sum_{k=1}^n \mu_{jk}}$$

$$\frac{\partial \det F_j}{\partial \mu_{jr}} = (\det F_j)\, tr(F_j^{-1} \frac{\partial F_j}{\partial \mu_{jr}}) = (\det F_j)\, tr(\frac{F_j^{-1}(x_r - v_j)(x_r - v_j)^T}{\sum_{k=1}^n \mu_{jk}} - \frac{I_s}{\sum_{k=1}^n \mu_{jk}})$$

$$= (\det F_j)\, (\frac{(x_r - v_j)^T F_j^{-1}(x_r - v_j)}{\sum_{k=1}^n \mu_{jk}} - \frac{s}{\sum_{k=1}^n \mu_{jk}})$$

Thus we have

$$\frac{\partial d_{jk}}{\partial \mu_{jr}} = \frac{1}{n\alpha_j} d_{jk} - \frac{d_{jk}}{2\sum_{k=1}^n \mu_{jk}} \left((x_r - v_j)^T F_j^{-1}(x_r - v_j) - s\right) + \frac{m\mu_{jr}^{m-1} d_{jk}}{\sum_{k=1}^n \mu_{jk}^m}(x_r - v_j)^T F_j^{-1}(x_k - v_j)$$

$$- \frac{1}{2} d_{jk} \left[(x_k - v_j)^T \left(-F_j^{-1} \frac{(x_r - v_j)(x_r - v_j)^T}{\sum_{k=1}^n \mu_{jk}} F_j^{-1} + \frac{F_j^{-1}}{\sum_{k=1}^n \mu_{jk}}\right)(x_k - v_j) \right]$$

$$(11)$$

Then we substituting Eq. (11) into (10). Finally, get Eq. (9).

The proof is completed.   □

Next, we define a notation and consider the Jacobian matrix $\frac{\partial\theta(U)}{\partial U}$ of the GG clustering algorithm at the special point $U = [c^{-1}]_{c\times n}$.

Notation. For any $p \times q$ matrix $M = (m_1, \cdots, m_q)$, $vec(M)$ is used to denote the vector obtained by stacking the column vectors of the matrix M. That is, $vec(M) = (m_1^T, \cdots, m_q^T)^T$.

Theorem 1. Jacobian matrix $\frac{\partial\theta(U)}{\partial U}$ of the GG clustering algorithm at the special point $U = [c^{-1}]_{c\times n}$

$$\left.\frac{\partial\theta_{ik}}{\partial\mu_{jr}}\right|_{\forall i,k,\mu_{ik}=c^{-1}} = \frac{\delta_{ij}}{2n}(H)_{kr} = \frac{\delta_{ij}}{2n}(A^T A)_{kr} \tag{12}$$

where

$$(H)_{kr} = \left[s - (x_r - \bar{x})^T\sigma_x^{-1}(x_r - \bar{x}) - (x_k - \bar{x})^T\sigma_x^{-1}(x_k - \bar{x}) + (x_k - \bar{x})^T\right.$$
$$\left.\sigma_x^{-1}(x_r - \bar{x})(x_r - \bar{x})^T\sigma_x^{-1}(x_k - \bar{x}) + 2m(x_r - \bar{x})^T\sigma_x^{-1}(x_k - \bar{x}) + 2\right]$$

$$A = \begin{bmatrix} \sqrt{2} & \cdots & \sqrt{2} \\ \sqrt{2m}\sigma_x^{-1/2}(x_1 - \bar{x}) & \cdots & \sqrt{2m}\sigma_x^{-1/2}(x_n - \bar{x}) \\ vec\left(\sigma_x^{-1/2}(x_1 - \bar{x})(x_1 - \bar{x})^T\sigma_x^{-1/2} - I_s\right) & \cdots & vec\left(\sigma_x^{-1/2}(x_n - \bar{x})(x_n - \bar{x})^T\sigma_x^{-1/2} - I_s\right) \end{bmatrix}$$

and $\bar{x} = \sum_{k=1}^n x_k/n$, $\sigma_x = n^{-1}\sum_{k=1}^n (x_k - \bar{x})(x_k - \bar{x})^T$, δ_{ij} is the kronecker delta function.

Proof. If $U = [c^{-1}]_{c\times n}$, then we can get $\alpha_i = \mu_{ik} = \frac{1}{c}$, $v_i = \frac{\sum_{k=1}^n x_k}{n} = \bar{x}$ and $F_i = \frac{\sum_{k=1}^n (x_k - \bar{x})(x_k - \bar{x})^T}{n} = \sigma_x$ by simple computation. Thus

$$\frac{\partial\theta_{ik}}{\partial\mu_{jr}} = \frac{\delta_{ij}}{n} - \frac{\delta_{ij}}{2n}\left((x_r - \bar{x})^T\sigma_x^{-1}(x_r - \bar{x}) - s\right) + \frac{\delta_{ij}m}{n}(x_r - \bar{x})^T\sigma_x^{-1}(x_k - \bar{x})$$

$$- \frac{\delta_{ij}}{2n}\left[(x_k - \bar{x})^T\left(-\sigma_x^{-1}(x_r - \bar{x})(x_r - \bar{x})^T\sigma_x^{-1} + \sigma_x^{-1}\right)(x_k - \bar{x})\right]$$

$$= \frac{\delta_{ij}}{2n}\left[s - (x_r - \bar{x})^T\sigma_x^{-1}(x_r - \bar{x}) - (x_k - \bar{x})^T\sigma_x^{-1}(x_k - \bar{x}) + (x_k - \bar{x})^T\right.$$
$$\left.\sigma_x^{-1}(x_r - \bar{x})(x_r - \bar{x})^T\sigma_x^{-1}(x_k - \bar{x}) + 2m(x_r - \bar{x})^T\sigma_x^{-1}(x_k - \bar{x}) + 2\right]$$

Set

$$(H)_{kr} = \left[s - (x_r - \bar{x})^T\sigma_x^{-1}(x_r - \bar{x}) - (x_k - \bar{x})^T\sigma_x^{-1}(x_k - \bar{x}) + (x_k - \bar{x})^T\right.$$
$$\left.\sigma_x^{-1}(x_r - \bar{x})(x_r - \bar{x})^T\sigma_x^{-1}(x_k - \bar{x}) + 2m(x_r - \bar{x})^T\sigma_x^{-1}(x_k - \bar{x}) + 2\right]$$

We have that

$$\frac{\partial\theta_{ik}}{\partial\mu_{jr}}\bigg|_{\forall i,k,\mu_{ik}=c^{-1}} = \frac{\delta_{ij}}{n} - \frac{\delta_{ij}}{2n}\left((x_r-\bar{x})^T\sigma_x^{-1}(x_r-\bar{x})-s\right) + \frac{\delta_{ij}m}{n}(x_r-\bar{x})^T\sigma_x^{-1}(x_k-\bar{x})$$

$$-\frac{\delta_{ij}}{2n}\left[(x_k-\bar{x})^T\left(-\sigma_x^{-1}(x_r-\bar{x})(x_r-\bar{x})^T\sigma_x^{-1}+\sigma_x^{-1}\right)(x_k-\bar{x})\right]$$

$$=\frac{\delta_{ij}}{2n}\left[-\left((x_r-\bar{x})^T\sigma_x^{-1}(x_r-\bar{x})-s\right)\right.$$

$$-\left((x_k-\bar{x})^T\left(-\sigma_x^{-1}(x_r-\bar{x})(x_r-\bar{x})^T\sigma_x^{-1}+\sigma_x^{-1}\right)(x_k-\bar{x})\right)$$

$$\left.+2m(x_r-\bar{x})^T\sigma_x^{-1}(x_k-\bar{x})+2\right]$$

$$=\frac{\delta_{ij}}{2n}\left[tr\left(\left(\sigma_x^{-\frac{1}{2}}(x_r-\bar{x})(x_r-\bar{x})^T\sigma_x^{-\frac{1}{2}}-I_s\right)^T\left(\sigma_x^{-\frac{1}{2}}(x_k-\bar{x})(x_k-\bar{x})^T\sigma_x^{-\frac{1}{2}}-I_s\right)\right)\right.$$

$$\left.+2m(x_r-\bar{x})^T\sigma_x^{-1}(x_k-\bar{x})+2\right]$$

$$=\frac{\delta_{ij}}{2n}\left[vec\left(\sigma_x^{-1/2}(x_k-\bar{x})(x_k-\bar{x})^T\sigma_x^{-1/2}-I_s\right)^T\right.$$

$$\left.\times vec\left(\sigma_x^{-1/2}(x_r-\bar{x})(x_r-\bar{x})^T\sigma_x^{-1/2}-I_s\right)+2m(x_r-\bar{x})^T\sigma_x^{-1}(x_k-\bar{x})+2\right]$$

$$=\frac{\delta_{ij}}{2n}\left[s-(x_r-\bar{x})^T\sigma_x^{-1}(x_r-\bar{x})-(x_k-\bar{x})^T\sigma_x^{-1}(x_k-\bar{x})\right.$$

$$\left.+(x_k-\bar{x})^T\sigma_x^{-1}(x_r-\bar{x})(x_r-\bar{x})^T\sigma_x^{-1}(x_k-\bar{x})+2m(x_r-\bar{x})^T\sigma_x^{-1}(x_k-\bar{x})+2\right]$$

$$=\frac{\delta_{ij}}{2n}(H)_{kr}$$

It is easy to know that

$$(A^TA)_{kr}=\left[vec\left(\sigma_x^{-1/2}(x_k-\bar{x})(x_k-\bar{x})^T\sigma_x^{-1/2}-I_s\right)^T\times vec\left(\sigma_x^{-1/2}(x_r-\bar{x})(x_r-\bar{x})^T\sigma_x^{-1/2}-I_s\right)\right.$$

$$\left.+2m(x_r-\bar{x})^T\sigma_x^{-1}(x_k-\bar{x})+2\right]$$

That is

$$\frac{\partial\theta_{ik}}{\partial\mu_{jr}}\bigg|_{\forall i,k,\mu_{ik}=c^{-1}} = \frac{\delta_{ij}}{2n}(H)_{kr}=\frac{\delta_{ij}}{2n}(A^TA)_{kr}.$$

The proof is completed.   □

It is widely known that the eigenvalues of matrix A^TA are equal to the eigenvalues of matrix AA^T, so we can compute the spectral radius of matrix AA^T instead of A^TA. Following result can be obtained by simple calculation:

$$\frac{1}{2n}(A\times A^T)=\frac{1}{2n}\begin{pmatrix}L_{11}&L_{12}&L_{13}\\L_{21}&L_{22}&L_{23}\\L_{31}&L_{32}&L_{33}\end{pmatrix}=\begin{pmatrix}1&0&0\\0&m\times I_s&L_{23}/2n\\0&L_{32}/2n&L_{33}/2n\end{pmatrix} \quad (13)$$

where $L_{11}=2n,$ $L_{12}=2\sqrt{m}\sum_{k=1}^n(x_k-\bar{x})^T\sigma_x^{-1/2}=0,$ $L_{13}=\sqrt{2}\sum_{k=1}^n$

$vec\left[\sigma_x^{-1/2}(x_k-\bar{x})(x_k-\bar{x})^T\sigma_x^{-1/2}-I_s\right]^T=0,$ $L_{12}=L_{21}^T,$ $L_{13}=L_{31}^T,$ $L_{23}=L_{32}^T,$

$L_{22} = 2m \sum_{k=1}^{n} \sigma_X^{-1/2}(x_k - \bar{x})(x_k - \bar{x})^T \sigma_X^{-1/2} = 2mn \times I_s, \qquad L_{23} = \sqrt{2m} \sum_{k=1}^{n} \sigma_x^{-1/2}$
$(x_k - \bar{x}) \times vec[\sigma_x^{-1/2}(x_k - \bar{x})(x_k - \bar{x})^T \sigma_x^{-1/2} - I_s]^T, \qquad L_{33} = \sum_{k=1}^{n} vec[\sigma_x^{-1/2}(x_k - \bar{x})$
$(x_k - \bar{x})^T \sigma_x^{-1/2} - I_s] \times vec[\sigma_x^{-1/2}(x_k - \bar{x})(x_k - \bar{x})^T \sigma_x^{-1/2} - I_s]^T.$

We have mentioned that if spectral radius of the Jacobian matrix $\frac{\partial \theta(U)}{\partial U}$ at point $U = [c^{-1}]_{c \times n}$ is less than 1, then $(\bar{x}, U = [c^{-1}]_{c \times n})$ is the stable fixed point of GG clustering algorithm. However, the spectral radius of matrix $\frac{1}{2n}(A \times A^T)$ is larger than 1, so $(\bar{x}, U = [c^{-1}]_{c \times n})$ is not a stable fixed point of GG clustering algorithm. In this case, the GG clustering algorithm will not output meaningless clustering results in any situation.

Theorem 2. Let λ^* denote the spectral radius of Jacobian matrix $\frac{\partial \theta}{\partial U}|_{\forall i,k,\mu_{ik}=c^{-1}}$, then we have that $\lambda^* \geq 1$.

Proof. The spectral radius of Jacobian matrix $\frac{\partial \theta}{\partial U}|_{\forall i,k,\mu_{ik}=c^{-1}}$ is equal to the spectral radius of matrix $\frac{1}{2n}(A \times A^T)$ computed by Eq. (13). We have that if B is a symmetric matrix, then $\lambda_{\max}\left(\frac{1}{2n}(A \times A^T)\right) = \max_{x \neq 0} \frac{x^T\left(\frac{1}{2n}(A \times A^T)\right)x}{x^T x}$, where $x = e^{(i)}$ is a vector in which all elements are zero except that the ith element is one. Also we have that $\max_{x \neq 0} \frac{x^T\left(\frac{1}{2n}(A \times A^T)\right)x}{x^T x} \geq \frac{[e^{(i)}]^T\left(\frac{1}{2n}(A \times A^T)\right)e^{(i)}}{[e^{(i)}]^T e^{(i)}} = \left(\frac{1}{2n}(A \times A^T)\right)_{ii}$. In other words, we have $\lambda^* \geq 1$.

The proof is completed. □

The Jacobian matrix also can be used to estimate the convergence rate of GG clustering algorithm. Suppose the GG clustering algorithm is convergent to a point U^*, that is $U^* = \theta(U^*)$. Then the convergence rate of clustering algorithm can be computed as follows:

$$r = \lim_{t \to \infty} \|U^{(t+1)} - U^*\| / \|U^{(t)} - U^*\| \qquad (14)$$

It can be easily proved that if the clustering algorithm is convergent, then $r < 1$ and a large value of r mean low convergence rate. We can obtain a Taylor expansion

$$U^{(t+1)} - U^* \approx \left(U^{(t)} - U^*\right)\frac{\partial \theta(U)}{\partial U}|_{U=U^*} \qquad (15)$$

under assumption that the mapping is differentiable in a neighborhood of U^*. That is, the convergence rate can be estimated by Jacobian matrix. We will verifies the correctness of theoretical deduction in next section by experiment results.

3 Experimental Results

In this section, we use the experimental results to demonstrates several facts as follows: (1) the coincident clustering result $(\bar{x}, U = [c^{-1}]_{c \times n})$ is not the stable fixed point of the GG clustering algorithm; (2) the convergence rate of the GG clustering algorithm can be estimated by computing the spectral radius of Jacobian matrix.

3.1   Experimental Data Set

In this section, we use some artificial data sets as well as real data sets to demonstrate the correctness of our theoretical results. The artificial data sets are generated from Mixture Gaussian Models, and the mixing proportions, means values and variances are listed in Table 1. The data sets generated by Mixture Gaussian Models are shown in Fig. 1. The real data sets are come from UCI database (University of California Irvine) and we list these data sets in Table 2.

Table 1. Mixing proportions, means values and variances of Gaussian Mixture Models

Dataset	Mixing proportions	Means values	Variances
Data1	$\alpha_1 = 0.5$	$m_1 = (1,1)$	$\Sigma_1 = \begin{bmatrix} 1 & 0 \\ 0 & 1 \end{bmatrix}$
	$\alpha_2 = 0.5$	$m_2 = (1,2)$	$\Sigma_2 = \begin{bmatrix} 0.5 & 0 \\ 0 & 0.1 \end{bmatrix}$
Data2	$\alpha_1 = 0.2$	$m_1 = (1,1,2)$	$\Sigma_1 = \begin{bmatrix} 1 & 0 & 0 \\ 0 & 1 & 0 \\ 0 & 0 & 1 \end{bmatrix}$
	$\alpha_2 = 0.3$	$m_2 = (5,3,0.5)$	$\Sigma_2 = \begin{bmatrix} 0.5 & 0 & 0 \\ 0 & 0.1 & 0 \\ 0 & 0 & 1 \end{bmatrix}$
	$\alpha_3 = 0.5$	$m_2 = (2,6,5)$	$\Sigma_3 = \begin{bmatrix} 0.5 & 0 & 0 \\ 0 & 0.1 & 0 \\ 0 & 0 & 2 \end{bmatrix}$

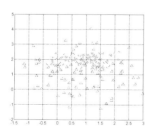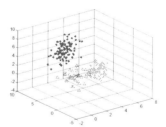

Fig. 1. Data sets used in the experiments

Table 2. UCI Data sets used in the experiments

Datasets	Data	Feature no.	Cluster no.
Iris	150	4	3
Wine	178	13	3
Winequality-red	1599	11	3
Sonar	208	60	2
Ionosphere	351	34	2

It should be mentioned that, in order to analyze the relationship between the convergence rate and the fuzziness index m, we use the K-means clustering algorithm to initialize the GG clustering algorithm. Thus we can discuss about the influence of fuzziness index value under the same initialization.

3.2   Jacobian Matrix Analysis of the GG Clustering Algorithm

We have theoretically prove that the coincident clustering result $\left(\bar{x}, U = [c^{-1}]_{c \times n}\right)$ is not a stable fixed point of the GG clustering algorithm. Next, we use several experiments to verify the correctness of theoretical results.

We compute the spectral radius of Jacobian matrix of the GG clustering algorithm by using Eqs. (12) and (13), the result is shown in Table 3.

We have prove the spectral radius of Jacoboian matrix at the special point $\left(\bar{x}, U = [c^{-1}]_{c \times n}\right)$ is not less than 1, $\lambda^* \geq 1$. That is, $\left(\bar{x}, U = [c^{-1}]_{c \times n}\right)$ is not a stable fixed point of the GG clustering algorithm. The spectral radius of Jacoboian matrix at the special point $\left(\bar{x}, U = [c^{-1}]_{c \times n}\right)$ is actually independent of the initialization and depends only on the data points and the fuzziness index m. It can be seen from the Table 3 is that, the spectral radius of Jacoboian matrix at the special point $\left(\bar{x}, U = [c^{-1}]_{c \times n}\right)$ is larger than 1 for any fuzziness index value. That is, no matter how to initialize the GG clustering algorihtm, it can avoid to output the coincident clustering result $\left(\bar{x}, U = [c^{-1}]_{c \times n}\right)$.

Next we use the non-fuzzy index (NFI)

$$NFI(c, \mu, m) = (c/(n \times (c-1))) \sum_{i=1}^{c} \sum_{k=1}^{n} \mu_{ik}^2 - 1/(c-1) \tag{16}$$

proposed by Roubens [14] to evaluate the performance of the GG algorithm and demonstrate that the coincident clustering result is actually not a stable fixed point of the GG algorihtm. We initialize the GG clustering algorithm with K-means clustering algorithm and compute the non-fuzzy index of the GG clustering result. These NFI values for different values of m with different data sets are shown in Table 4.

We mention that $\{NFI(c, \mu, m) = 0$ if $\mu = [c^{-1}]_{c \times n}$ if $v = \bar{x}\}$, and $\{NFI(c, \mu, m) = 1$ if hard clustering results$\}$ [7]. It can be seen from Table 4 that the GG clustering algorithm can avoid to output a bad clustering result $\mu = [c^{-1}]_{c \times n}$ for any fuzziness index value.

Table 3. Spectral radius of Jacobian matrix at special point $\left(\bar{x}, U = [c^{-1}]_{c \times n}\right)$

	m = 1.5	m = 1.6	m = 1.7	m = 1.8	m = 1.9	m = 2	m = 2.1	m = 2.2	m = 2.3	m = 2.4	m = 2.5
Data1	2.3524	2.4311	2.5116	2.5938	2.6775	2.7626	2.8489	2.9364	3.0248	3.1141	3.2042
Data2	2.8761	2.9695	3.0632	3.1574	3.2519	3.3468	3.442	3.5374	3.6332	3.7291	3.8253
Iris	2.4617	2.5429	2.6263	2.7115	2.7982	2.8863	2.9754	3.0656	3.1566	3.2483	3.3407
Wine	10.7905	10.8311	10.8723	10.9142	10.9566	10.9996	11.0433	11.0875	11.1324	11.1779	11.224
Winequality-red	25.5103	25.5893	25.6684	25.7476	25.827	25.9065	25.9862	26.0659	26.1458	26.2259	26.306
Sonar	46.1391	46.2062	46.2735	46.3409	46.4085	46.4763	46.5442	46.6122	46.6804	46.7484	46.8172
Ionosphere	52.0116	52.0662	52.121	52.176	52.2311	52.2864	52.3419	52.3975	52.4533	52.5092	52.5654

Table 4. NFI values corresponding to different m for different data sets

	m = 1.5	m = 1.6	m = 1.7	m = 1.8	m = 1.9	m = 2	m = 2.1	m = 2.2	m = 2.3	m = 2.4	m = 2.5
Data1	0.9013	0.9031	0.9045	0.9056	0.9065	0.9073	0.908	0.9086	0.9091	0.9096	0.91
Data2	0.9979	0.9979	0.9979	0.9979	0.9979	0.9979	0.9979	0.9979	0.9979	0.9979	0.9979
Iris	0.9761	0.9762	0.9762	0.9763	0.9763	0.9764	0.9764	0.9765	0.9765	0.9565	0.9765
Wine	0.9917	0.9917	0.9917	0.9917	0.9904	0.9903	0.9903	0.9902	0.9902	0.9901	0.9901
Winequality - red	0.9144	0.9147	0.915	0.9152	0.9154	0.9155	0.9157	0.9158	0.9159	0.916	0.9161
Sonar	0.9995	0.9995	0.9995	0.9995	0.9995	0.9995	0.9995	0.9995	0.9995	0.9995	0.9995
Ionosphere	0.9999	0.9999	0.9999	0.9999	0.9999	0.9999	0.9999	0.9999	0.9998	0.9998	0.9998

In our experiment, we initialize the GG clustering algorithm with the K-means clustering algorithm to reduce the influence of the initialization. That is, we get not bad clustering results before using the GG clustering algorithm, finally we get relatively hard clustering results. In fact, the GG clustering algorithm can avoid the coincident clustering result with any initialization.

3.3   Convergence Properties of the GG Clustering Algorithm

In this section, we discuss about the relationship between fuzziness index value and convergence rate of the GG clustering algorithm. We also initialize the GG clustering algorithm with the K-means clustering algorithm. We compute the spectral radius of Jacobian matrix at convergence point (V^*, U^*), then estimate the convergence rate of clustering algorithm. It can be seen from Table 5 that a larger fuzziness index value lead to a lower convergence rate. This result is coincident with most experimental results for the GG clustering algorithm.

Table 5. Convergence rate of GG clustering algorithm corresponding to different m

	m = 1.5	m = 1.6	m = 1.7	m = 1.8	m = 1.9	m = 2	m = 2.1	m = 2.2	m = 2.3	m = 2.4	m = 2.5
Data1	0.8271	0.8255	0.8251	0.8253	0.8263	0.8274	0.8287	0.8301	0.8316	0.8331	0.8347
Data2	0.2154	0.2153	0.2155	0.2159	0.2165	0.2173	0.2183	0.2193	0.2204	0.2215	0.2228
Iris	0.5366	0.5352	0.5345	0.5343	0.5346	0.5352	0.5362	0.5374	0.5388	0.5404	0.5421
Wine	0.5389	0.5401	0.5417	0.5437	0.639	0.6503	0.6614	0.6723	0.6831	0.6937	0704
Winequality-red	0.803	0.8057	0.808	0.8101	0.8119	0.8134	0.8148	0.816	0.8171	0.818	0.8187
Sonar	0.4152	0.4164	0.4177	0.4192	0.4208	0.4224	0.4242	0.4259	0.4277	0.4295	0.4313
Ionosphere	0.0588	0.0589	0.0591	0.0592	0.0593	0.0594	0.0595	0.0597	0.0598	0.0599	0.1137

4   Conclusion

Since Gath and Geva proposed the GG clustering algorithm in 1989, it has become one of the most popular clustering algorithm. In this paper, we address the parameter selection and convergence analysis of GG clustering algorithm using Jacobian matrix analysis. Our main contribution in this paper is that we illustrate following facts: 1) the coincident clustering result $\left(\bar{x}, U = [c^{-1}]_{c \times n}\right)$ is not the stable fixed point of the GG clustering algorithm; 2) the convergence rate of the GG clustering algorithm can be estimated by compute the spectral radius of Jacobian matrix. We find that for any parameter value, coincident clustering result $\left(\bar{x}, U = [c^{-1}]_{c \times n}\right)$ will not outputted by the GG clustering algorithm. And the fuzziness index value may heavily influence the convergence properties that a larger fuzziness index value lead to a lower convergence rate.

In general, we propose a theoretical method to analysis the convergence properties and parameter selection for partition-based clustering algorithm. The Jacobian matrix analysis method can be applied to any kind of partition-based clustering algorithm including these clustering algorithms without objective functions. In the future, we may analyze more partition-based clustering algorithm by using Jacobian matrix analysis.

Acknowledgments. This work was supported by the Fundamental Research Funds for the Central Universities (Grant No. K16JB00070).

References

1. Jain, A.K., Murty, M.N., Flynn, P.J.: Data clustering: a review. ACM Comput. Surv. (CSUR) **31**(3), 264–323 (1999)
2. Everitt, B.S.: Cluster Analysis. Halstead Press, London (1974)
3. Dunn, J.C.: A fuzzy relative of the ISODATA process and its use in detecting compact well separated clusters. J. Cybern. **3**(3), 32–57 (1974)
4. Gustafson, D.E., Kessel, W.C.: Fuzzy clustering with a fuzzy covariance matrix. In: Proceedings of IEEE CDC, San Diego, CA, pp. 761–766 (1979)
5. Krishnapuram, R., Kim, J.: A note on the Gus-tafson-Kessel and adaptive fuzzy clustering algorithms. IEEE Trans. Fuzzy Syst. **7**, 453–461 (1999)
6. Gath, I., Geva, A.B.: Unsupervised optimal fuzzy clustering. IEEE Trans. Pattern Anal. Mach. Intell. **11**(7), 773–780 (1989)
7. Wang, C., Yu, J., Yang, M.-S.: Analysis of parameter selection for Gustafson-Kessel fuzzy clustering using Jacobian matrix. IEEE Trans. Fuzzy Syst. **23**(6), 2329–2342 (2015)
8. Olver, P.J.: Lecture Notes on Numerical Analysis (2008). http://www.math.umn.edu/ ~ olver/num.html
9. Bezdek, J.C.: Pattern Recognition with Fuzzy Objective Function Algorithms. Kluwer Academic Publishers, New York (1981)
10. Yu, J., Cheng, Q., Huang, H.: Analysis of the weighting exponent in the FCM. IEEE Trans. Syst. Man Cybern. Part B Cybern. **34**, 634–639 (2004)
11. Zadeh, L.A.: Fuzzy sets. Inf. Control **8**, 338–356 (1965)
12. Abonyi, J., Babuska, R., Szeifert, F.: Modified Gath-Geva fuzzy clustering for identification of Takagi-Sugeno fuzzy models. IEEE Trans. Syst. Man Cybern. Part B (Cybern.) **32**(5), 612–621 (2002)
13. Wang, N., Liu, X., Yin, J.: Improved Gath-Geva clustering for fuzzy segmentation of hydrometeorological time series. Stochast. Environ. Res. Risk Assess. **26**(1), 139–155 (2012)
14. Roubens, M.: Pattern classification problems and fuzzy sets. Fuzzy Sets Syst. **1**, 239–253 (1978)

Subspace Clustering by Capped l_1 Norm

Quanmao Lu[1,2], Xuelong Li[1], Yongsheng Dong[1(✉)], and Dacheng Tao[3]

[1] Center for OPTical IMagery Analysis and Learning (OPTIMAL),
State Key Laboratory of Transient Optics and Photonics,
Xi'an Institute of Optics and Precision Mechanics,
Chinese Academy of Sciences, Xi'an 710119, Shaanxi, People's Republic of China
dongyongsheng98@163.com
[2] University of the Chinese Academy of Sciences,
19A Yuquanlu, Beijing 100049, People's Republic of China
[3] Centre for Quantum Computation and Intelligent Systems,
Faculty of Engineering and Information Technology, University of Technology Sydney,
81 Broadway Street, Ultimo, NSW 2007, Australia

Abstract. Subspace clustering, as an important clustering problem, has drawn much attention in recent years. State-of-the-art methods generally try to design an efficient model to regularize the coefficient matrix while ignore the influence of the noise model on subspace clustering. However, the real data are always contaminated by the noise and the corresponding subspace structures are likely to be corrupted. In order to solve this problem, we propose a novel subspace clustering algorithm by employing capped l_1 norm to deal with the noise. Consequently, the noise term with large error can be penalized by the proposed method. So it is more robust to the noise. Furthermore, the grouping effect of our method is theoretically proved, which means highly correlated points can be grouped together. Finally, the experimental results on two real databases show that our method outperforms state-of-the-art methods.

Keywords: Subspace clustering · Capped l_1 norm · Grouping effect

1 Introduction

Nowadays, available data are increasing in an exponential manner. They are usually produced from multiple modalities with high-dimensional. In practice, however, high-dimensional data are often drawn from multiple low-dimensional subspaces and the dimension of each subspace is unknown. For instance, face images of one person under various illumination conditions lie in a linear subspace. Therefore, a technique is needed to simultaneously find the multiple low-dimensional subspaces and cluster the data into different subspaces. This problem, known as subspace clustering [5,10,22,25,28], has drawn much attention in recent years and has been utilized for many real applications in image processing (e.g. image representation and compression [8]) and computer vision (e.g. motion segmentation [3] and face clustering [7]).

© Springer Nature Singapore Pte Ltd. 2016
T. Tan et al. (Eds.): CCPR 2016, Part I, CCIS 662, pp. 663–674, 2016.
DOI: 10.1007/978-981-10-3002-4_54

During the past two decades, many subspace clustering approaches have been proposed [4,6,20,24,27], including iterative methods [30], algebraic methods [2], statistical methods [23] and spectral clustering based methods [12,14,15]. Among them, spectral clustering based methods show their superiority in dealing with the real data. Spectral clustering based methods consist of two main steps: find an affinity matrix and employ an affinity matrix based clustering method to cluster the data into different groups. Obviously, constructing a proper affinity matrix is the key step to spectral clustering based methods. Reviewing the existing works, spectral clustering based methods always use the self expression model to learn an affinity matrix. The self expression model regards the original data as the dictionary and can be formulated as

$$\min_{Z,E} \varphi(E) + \delta(Z), \ s.t. \ X = XZ + E, \tag{1}$$

where X is the original data matrix, Z denotes the coefficient matrix and E is the noise matrix. The functions of $\varphi(E)$ and $\delta(Z)$ are designed for restricting E and Z respectively. Ideally, the coefficient matrix Z should be block-diagonal, which means that the affinities between-cluster are all zeros.

The state-of-the-art methods usually choose specific norm to regularize the coefficient matrix, such as Sparse Subspace Clustering (SSC) [3], Low-Rank Representation (LRR) [17] and Correlation Adaptive Subspace Segmentation (CASS) [19]. SSC uses the ℓ_1 norm to find the sparsest representation for each point. LRR employs the nuclear norm to find a low-rank coefficient matrix. CASS adopts trace Lasso to balance SSC and LRR adaptively. These methods pay much attention on constructing the coefficient matrix and overlook the influence of the noise matrix on subspace clustering. However, the real data are always contaminated by the noise [21] and the statistical distribution of the noise is very complicated. As pointed out in [13,16], if we can not describe the noise in an efficient way, the coefficient matrix will be unable to capture the relationship between points, which can reduce the performance of subspace clustering.

To solve this problem, we propose a robust subspace clustering algorithm by using capped ℓ_1 norm based loss function to penalize the noise term in this paper. The definition of capped ℓ_1 norm [11,31] is

$$l_\varepsilon(r_i) = min(|r_i|, \varepsilon), \tag{2}$$

where r_i is the residual of the i-th data with its estimated value and ε is a parameter to control the upper bound of the function. Compared with the traditional ℓ_1 norm, the value of capped ℓ_1 norm will be a constant when $|r_i|$ is larger than ε, which is illustrated in Fig. 1. It means that the noise term with large error can be penalized by capped ℓ_1 norm. Therefore, capped ℓ_1 norm is more robust to noise than ℓ_1 norm. Considering that our method is to verify the influence of noise model on subspace clustering, we simply employ the Frobenius norm to regularize the coefficient matrix Z. Besides, using the Frobenius norm on the coefficient matrix makes the problem easy to be solved. Furthermore, the grouping effect of our method is theoretically proved in this paper, which can

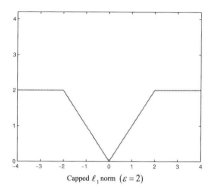

Capped ℓ_1 norm ($\varepsilon = 2$)

Fig. 1. An example of capped ℓ_1 norm

ensure that highly correlated samples can be grouped together. It is very important for subspace clustering to keep the structure information between points. Finally, we demonstrate the effectiveness of our method on two real databases. From the experimental results, we can see that our method outperforms the state-of-the-art methods. To summarize, our work has two contributions:

1. We propose a novel subspace clustering algorithm by adopting capped ℓ_1 norm on the noise term. It is more robust to the noise.
2. We prove the grouping effect of our proposed method which can keep highly correlated points together.

The rest of the paper is arranged as follows. Section 2 introduces the related works about spectral clustering based methods. In Sect. 3, we describe our subspace clustering method and prove the grouping effect of the proposed method. Then comparison experiments are conducted in Sect. 4. Finally, the paper is briefly concluded in Sect. 5.

2   Related Work

Considering that our proposed method is a kind of spectral clustering based method, we mainly review the most recent and related works.

Sparse Subspace Clustering (SSC) [3] was proposed to find a sparse representation for each point, and the corresponding problem is

$$\min_{Z,E} \|E\|_F^2 + \lambda\|Z\|_0, \ s.t. \ X = XZ + E, \ diag(Z) = 0, \tag{3}$$

where X is the data matrix, Z is the coefficient matrix and E denotes the noise matrix. $\lambda > 0$ is a parameter. $diag(Z) = 0$ is used to avoid the solution Z being an identity matrix. As we all known, this problem is NP hard and can be approximated using the following ℓ_1-minimization problem

$$\min_{Z,E} \|E\|_F^2 + \lambda\|Z\|_1, \ s.t. \ X = XZ + E, \ diag(Z) = 0, \tag{4}$$

It has been shown that the solution of SSC may be too sparse to capture the structure information of the points in the same subspace.

Low-Rank Representation (LRR) [17] aims to seek the lowest rank representation of all data jointly. The objective function of LRR is given below

$$\min_{Z,E} \|E\|_{2,1} + \lambda \|Z\|_*, \ \ s.t. \ X = XZ + E, \tag{5}$$

where $\|Z\|_*$ is the nuclear norm of Z, i.e. the sum of singular values of Z. Lu et al. [18] proposed Least Squares Regression (LSR) which adopts the Frobenius norm to handle the coefficient matrix and the noise matrix simultaneously by solving the following problem

$$\min_{Z} \|X - XZ\|_F^2 + \lambda \|Z\|_F^2, \ \ diag(Z) = 0. \tag{6}$$

It can be shown that when the data are drawn from independence subspaces and noise free, LRR and LSR can obtain a block-diagonal coefficient matrix. However, the real data are always contaminated by noise or outliers and the corresponding solutions may be very dense and far from block-diagonal.

In order to avoid finding a too sparse or dense coefficient matrix, Correlation Adaptive Subspace Segmentation (CASS) [19] was proposed by using trace Lasso to regularize the matrix Z, which can choose ℓ_1 norm or nuclear norm adaptively. The definition of CASS is

$$\min_{Z,E} \|E\|_F^2 + \lambda \sum_{i=1}^{n} \|Xdiag(z_i)\|_*, \ \ s.t. \ X = XZ + E, \tag{7}$$

where n is the number of points and $\|Xdiag(z_i)\|_*$ is trace lasso. Note that trace Lasso involves the data matrix X, so it can be adaptive to the correlation of the data.

3   Subspace Clustering by Capped ℓ_1 Norm

In this section, we introduce the whole framework of our subspace clustering method and prove its grouping effect.

3.1   Problem Formulation

As illustrated in Fig. 1, the value of capped ℓ_1 norm does not grow linearly with the increasing of the error and will be a constant when $|r_i|$ is larger than ε. Therefore, it can reduce the influence of large noise on subspace clustering. Due to its robustness to the noise, we use capped ℓ_1 norm based loss function to penalize the noise term in our model and the corresponding problem of our robust subspace clustering method can be formulated as

$$\min_{Z} \sum_{i=1}^{n} \min(\|x_i - Xz_i\|_2, \varepsilon) + \lambda \|Z\|_F^2, \tag{8}$$

Algorithm 1. Re-weighted Algorithm

Input: data matrix X, parameters λ and ε.
Initialize: $s_i = 1$ for $i = 1, ..., n$.
Output: the final coefficient matrix Z^*.
Repeat

1. Update z_i for $i = 1, ..., n$ as

$$z_i = \begin{cases} \left(X^T X + \frac{\lambda}{s_i} I\right)^{-1} X^T x_i, & if \ s_i > 0 \\ 0, & otherwise \end{cases}$$

2. Update s_i for $i = 1, ..., n$ as

$$s_i = \begin{cases} \frac{1}{2\|x_i - X z_i\|_2}, & if \ \|x_i - X z_i\|_2 \leq \varepsilon \\ 0, & otherwise \end{cases}$$

Until Converge

where n is the number of points, x_i is the i-th sample point of the data matrix X and z_i denotes the corresponding representation vector. $\lambda > 0$ is a parameter to balance the effect of two terms. Note that we simply use the Frobenius norm to regularize the coefficient matrix for verifying the influence of the noise model on subspace clustering. From the formulation (8), we can see that if the value of ε tends to infinite, the first term of our objective function becomes $\ell_{2,1}$ norm.

Note that our objective function consists of a concave function and a convex function, which can be solved using the re-weighted algorithm proposed in [11]. Therefore, we can construct an auxiliary variable s_i and update Z and s_i alternately by using the following rules:

$$Z = \arg \min_Z \sum_{i=1}^{n} s_i \|x_i - X z_i\|_2^2 + \lambda \|Z\|_F^2 , \tag{9}$$

$$s_i = \begin{cases} \frac{1}{2\|x_i - X z_i\|_2}, & if \ \|x_i - X z_i\|_2 \leq \varepsilon \\ 0, & otherwise \end{cases} . \tag{10}$$

From the formulation (9), we can see that it is very similar to LSR. The main difference between our method and LSR is that there is a weight factor to penalize the noise term, where a larger weight will be given to a lower reconstruction error. Therefore, our method is more robust to the noise and can be regarded as a weighted subspace clustering algorithm.

To solve problem (9), we can decompose it into n independent subproblem as follows

$$z_i = \arg \min_z s_i \|x_i - X z\|_2^2 + \lambda \|z\|_2^2 . \tag{11}$$

when $s_i \neq 0$, the above problem can be rewritten as

$$z_i = \arg \min_z \|x_i - X z\|_2^2 + \frac{\lambda}{s_i} \|z\|_2^2 . \tag{12}$$

Obviously, the problem (12) is a convex function. After setting the derivative of the problem (12) with respect to z to zero, we have

$$z_i = \left(X^T X + \frac{\lambda}{s_i} I\right)^{-1} X^T x_i, \tag{13}$$

where I is an identity matrix.

Algorithm 1 gives the iteration procedures of solving the problem (8).

3.2 Convergence Analysis

Theorem 1. *The re-weighted algorithm proposed in Algorithm 1 guarantees that the objective function value of (8) is monotone decreasing in iterations until it converges.*

Proof. Let

$$L_x(s, X, z) = \begin{cases} s\|x - Xz\|_2^2 + \frac{1}{4s}, & if\ \|x - Xz\|_2 < \varepsilon \\ s\|x - Xz\|_2^2 - s\varepsilon^2 + \varepsilon, & otherwise \end{cases}, \tag{14}$$

where $s \geq 0$. Then we have

$$\inf_{s \geq 0} L_x(s, X, z) = \begin{cases} \|x - Xz\|_2, & if\ \|x - Xz\|_2 < \varepsilon \\ \varepsilon, & otherwise \end{cases}. \tag{15}$$

So the problem (8) is equivalent to

$$\min_Z \sum_{i=1}^n \inf_{s_i \geq 0} L_x(s_i, X, z_i) + \lambda \|z_i\|_2^2. \tag{16}$$

Assume that

$$J(Z, s) = \sum_{i=1}^n L_x(s_i, X, z_i) + \lambda \|z_i\|_2^2, \tag{17}$$

then the formulation (16) can be re-written as

$$\min_{Z, s \geq 0} J(Z, s) = \min_{Z, s \geq 0} \sum_{i=1}^n L_x(s_i, X, z_i) + \lambda \|z_i\|_2^2. \tag{18}$$

So Algorithm 1 can be regarded as a two stage optimization method to update Z and s alternately. Z is updated by solving the problem

$$\min_Z J(Z, s), \tag{19}$$

and s is updated by solving

$$\min_s J(Z, s). \tag{20}$$

Both stages can decrease the value of $J(Z, s)$, hence the objective function value of the original objective function (8) is monotone decreasing until it converges.

Algorithm 2. Capped ℓ_1 Norm for Subspace Clustering

Input: data matrix X, number of subspaces k.
Output: k subspaces.

1. Obtain the coefficient matrix Z^* by solving the problem (8) using the Alg. 1.
2. Construct the affinity matrix by using $(|Z^*| + |(Z^*)^T|)/2$.
3. Group the data into k subspaces by Normalized Cuts.

3.3   The Grouping Effect

Theorem 2. *Given a sample point $x \in R^d$, the normalized data matrix X, a parameter λ and auxiliary variable s, let z^* be the optimal solution of the problem*

$$\min_{z} \|x - Xz\|_2^2 + \frac{\lambda}{s} \|z\|_2^2 . \tag{21}$$

Then we have

$$\frac{\left|z_i^* - z_j^*\right|}{\|x\|_2} \leq \frac{s}{\lambda}\sqrt{2(1-r)}, \tag{22}$$

where $r = \cos(x_i, x_j)$, x_i and x_j are the i-th and j-th columns of the data matrix X. z_i^* and z_j^* are the i-th and j-th elements of z^*.

The proof of Theorem 2 is similar to LSR which is given in Appendix.

From Theorem 2, we can see that when x_i and x_j are very correlated, the corresponding value r is close to 1. Then the value of $\left|z_i^* - z_j^*\right|$ is close to 0, which means that highly correlated data points have the similar representation coefficient. So x_i and x_j can be grouped together.

3.4   Capped ℓ_1 Norm for Subspace Clustering

Similar to other spectral clustering based methods, we will employ spectral clustering to group data into different clusters after obtaining the coefficient matrix Z^*. Firstly, we construct the affinity matrix A by using $(|Z^*| + |(Z^*)^T|)/2$. Then we choose a kind of spectral clustering method, i.e. Normalized Cuts, to segment the data into k groups. The whole procedures of our proposed method are summarized in Algorithm 2.

4   Experimental Results

In order to verify the effectiveness of our proposed method, we conduct the experiments on two real databases: Hopkins 155 motion segmentation database [26] and Extended Yale B database [9,32]. We compare our method with five representative methods, i.e. SSC [3], LRR [17], LSR [18], CASS [19] and MoG Regression [13] which are all the state-of-the-art methods. SSC, LRR, LSR and CASS have been introduced in Sect. 1. MoG regression employs mixture of Gaussian model to tackle the noise term and chooses the Frobenius norm to regularize the coefficient matrix. Clustering accuracy [1,29] is employed to evaluate the performance of subspace clustering. In all experiments, we first use PCA to reduce the dimension of the data and keep nearly 98 % energy. Then different methods are employed for comparison. For each method, we manually tune its parameter for achieving the best performance and report the corresponding clustering results. From the experimental results, we can see that our method outperforms state-of-the-art methods.

4.1   Hopkins 155 Motion Segmentation Database

The first experiment is conducted on the Hopkins 155 motion segmentation database which contains 155 video sequences. Among them, 120 of the videos have two motions and the remaining 35 videos have three motions. For each video, feature trajectories have been extracted for motion segmentation and each motion belongs to one subspace. A single video can be regarded as a subspace clustering task, so there are totally 155 tasks to be implemented. Figure 2 shows some frames in the videos.

Fig. 2. Some frames of the Hopkins 155 motion segmentation database.

Table 1 reports the experimental results on the Hopkins 155 database. From the Table 1, we can see that our proposed method can achieve the best performance for all the situations. In particular, for the case of 3 motions, our average value is 4.1 % better than the second best result, and the corresponding Median is still 100 %. The total average of our method is more than 99 % which means our method can almost segment the whole videos precisely. Although LSR is similar to our method, it gives a worse performance than our method for almost all the cases expect the median of 2 motions. So adding the weight for the noise term is very effective for subspace clustering.

4.2   Extended Yale B Database

The Extended Yale B database has 2414 frontal face images with 38 different subjects. Each subject has about 64 face images under different angle and illumination situations. The size of each face image is 32×32. Similar to previous

Table 1. Experimental results on the Hopkins 155 database. MoG represents the MoG regression method. The best results are in bold font.

Method		SSC	LRR	LSR	CASS	MoG	Ours
2 motions Acc. (%)	Average	86.42	96.54	97.12	97.60	94.07	**99.45**
	Median	91.75	99.65	**100.00**	**100.00**	97.74	**100.00**
3 motions Acc. (%)	Average	76.03	88.78	91.27	94.35	89.74	**98.45**
	Median	76.30	92.62	92.66	96.49	90.10	**100.00**
Total Acc. (%)	Average	84.10	94.74	95.85	96.83	93.17	**99.17**
	Median	89.36	98.91	99.47	**100.00**	97.40	**100.00**

works, we construct 3 subspace clustering tasks by choosing the first 5, 8 and 10 subjects. For each task, we first use PCA to reduce the dimension of the corresponding data. Figure 3 shows same examples of the database. From Fig. 3, we can see that this database has large noise, which can be hardly distinguished even with human eyes for some face images. So we use this database for verifying the effect of the noise model on subspace clustering furthermore.

Fig. 3. Some faces images in the Extended Yale B database. Each column belongs to the same subject.

Table 2 gives subspace clustering results on Extended Yale B database. From the results, we can see that our method outperforms state-of-the-art methods for these three subspace clustering tasks. Especially for the 10 subjects, our method is higher than the second best result by 4.37%. Comparing with MoG, which

Table 2. The clustering accuracies of different algorithms on the Extended Yale B database. The best results are in bold font.

Method	SSC	LRR	LSR	CASS	MoG	Ours
5 subjects	80.00	80.63	91.56	85.00	84.06	**92.50**
8 subjects	70.90	61.52	79.29	72.07	83.59	**84.96**
10 subjects	64.06	60.62	67.34	78.91	68.28	**83.28**

uses mixture of Gaussian model to deal with the noise, our method can achieve a better performance. Therefore, our proposed method is more robust to the noise and is more effective on subspace clustering.

5   Conclusion

This paper presents a robust subspace clustering method by using capped l_1-norm to deal with the noise term. Capped l_1-norm can penalize the noise term with large error, so it is more robust to noise. For the coefficient matrix, we simply choose the Frobenius norm for verifying the influence of the noise model on subspace clustering. Furthermore, we can prove the grouping effect of our method, which means highly correlated points can be grouped together. Finally, we conduct the experiments on two real database. The corresponding results show the effectiveness of our proposed method.

Acknowledgments. This work was supported in part by the National Natural Science Foundation of China under Grant 61301230, in part by the International Science and Technology Cooperation Project of Henan Province under Grant 162102410021, in part by the Key Research Program of the Chinese Academy of Sciences under Grant KGZD-EW-T03, in part by the State Key Laboratory of Virtual Reality Technology and Systems under Grant BUAA-VR-16KF-04, and in part by the Key Laboratory of Optoelectronic Devices and Systems of Ministry of Education and Guangdong Province under Grant GD201605.

Appendix

Proof of Theorem 2

Proof. Let

$$L(z) = \|x - Xz\|_2^2 + \frac{\lambda}{s} \|z\|_2^2 . \tag{23}$$

Since $z^* = \arg\min_z L(z)$. We have

$$\left. \frac{\partial L(z)}{\partial z} \right|_{z=z^*} = 0 . \tag{24}$$

It gives

$$- 2x_i^T (x - Xz^*) + 2\frac{\lambda}{s} z_i^* = 0, \tag{25}$$

$$-2x_j^T (x - Xz^*) + 2\frac{\lambda}{s} z_j^* = 0. \tag{26}$$

Equations (25) and (26) give

$$z_i^* - z_j^* = \frac{s}{\lambda}(x_i^T - x_j^T)(x - Xz^*). \tag{27}$$

Note that each column of the data X is normalized, we get $\left\|x_i^T - x_j^T\right\|_2 = 2\sqrt{1-r}$, where $r = x_i^T x_j$. Since z^* is the optimal solution of the problem (21), we have

$$\|x - Xz^*\|_2^2 + \frac{\lambda}{s}\|z^*\|_2^2 = L(z^*) \leq L(0) = \|x\|_2^2. \tag{28}$$

Thus $\|x - Xz^*\|_2^2 \leq \|x\|_2^2$. Finally, we get

$$\frac{\left|z_i^* - z_j^*\right|}{\|x\|_2} \leq \frac{s}{\lambda}\sqrt{2(1-r)}. \tag{29}$$

References

1. Cao, X., Zhang, C., Fu, H., Liu, S., Zhang, H.: Diversity-induced multi-view subspace clustering. In: IEEE Conference on Computer Vision and Pattern Recognition, pp. 586–594 (2015)
2. Costeira, J.P., Kanade, T.: A multibody factorization method for independently moving objects. Int. J. Comput. Vis. **29**(3), 159–179 (1998)
3. Elhamifar, E., Vidal, R.: Sparse subspace clustering. In: IEEE Conference on Computer Vision and Pattern Recognition, pp. 2790–2797 (2009)
4. Elhamifar, E., Vidal, R.: Sparse subspace clustering: algorithm, theory, and applications. IEEE Trans. Pattern Anal. Mach. Intell. **35**(11), 2765–2781 (2013)
5. Gao, H., Nie, F., Li, X., Huang, H.: Multi-view subspace clustering. In: IEEE International Conference on Computer Vision, pp. 4238–4246 (2015)
6. Ghaemi, R., Sulaiman, M.N., Ibrahim, H., Mustapha, N., et al.: A survey: clustering ensembles techniques. World Acad. Sci. Eng. Technol. **50**, 636–645 (2009)
7. Ho, J., Yang, M.H., Lim, J., Lee, K.C., Kriegman, D.: Clustering appearances of objects under varying illumination conditions. In: IEEE Conference on Computer Vision and Pattern Recognition, vol. 1, p. I-11 (2003)
8. Hong, W., Wright, J., Huang, K., Ma, Y.: Multiscale hybrid linear models for lossy image representation. IEEE Trans. Image Process. **15**(12), 3655–3671 (2006)
9. Hu, H., Lin, Z., Feng, J., Zhou, J.: Smooth representation clustering. In: IEEE Conference on Computer Vision and Pattern Recognition, pp. 3834–3841 (2014)
10. Jiang, W., Liu, J., Qi, H., Dai, Q.: Robust subspace segmentation via nonconvex low rank representation. Information Sciences (2016)
11. Jiang, W., Nic, F., Huang, H.: Robust dictionary learning with capped l1 norm. In: International Joint Conferences on Artificial Intelligence, pp. 3590–3596 (2015)
12. Lee, M., Lee, J., Lee, H., Kwak, N.: Membership representation for detecting block-diagonal structure in low-rank or sparse subspace clustering. In: IEEE Conference on Computer Vision and Pattern Recognition, pp. 1648–1656 (2015)
13. Li, B., Zhang, Y., Lin, Z., Lu, H.: Subspace clustering by mixture of gaussian regression. In: IEEE Conference on Computer Vision and Pattern Recognition, pp. 2094–2102 (2015)
14. Li, C.G., Vidal, R.: Structured sparse subspace clustering: a unified optimization framework. In: IEEE Conference on Computer Vision and Pattern Recognition, pp. 277–286 (2015)
15. Li, Q., Sun, Z., Lin, Z., He, R., Tan, T.: Transformation invariant subspace clustering. Pattern Recognition (2016)

16. Liu, G., Lin, Z., Yan, S., Sun, J., Yu, Y., Ma, Y.: Robust recovery of subspace structures by low-rank representation. IEEE Trans. Pattern Anal. Mach. Intell. **35**(1), 171–184 (2013)

17. Liu, G., Lin, Z., Yu, Y.: Robust subspace segmentation by low-rank representation. In: International Conference on Machine Learning, pp. 663–670 (2010)

18. Lu, C.-Y., Min, H., Zhao, Z.-Q., Zhu, L., Huang, D.-S., Yan, S.: Robust and efficient subspace segmentation via least squares regression. In: Fitzgibbon, A., Lazebnik, S., Perona, P., Sato, Y., Schmid, C. (eds.) ECCV 2012. LNCS, vol. 7578, pp. 347–360. Springer, Heidelberg (2012). doi:10.1007/978-3-642-33786-4_26

19. Lu, C., Feng, J., Lin, Z., Yan, S.: Correlation adaptive subspace segmentation by trace lasso. In: IEEE International Conference on Computer Vision, pp. 1345–1352 (2013)

20. Lu, L., Vidal, R.: Combined central and subspace clustering for computer vision applications. In: International Conference on Machine Learning, pp. 593–600 (2006)

21. Lu, Y., Lai, Z., Xu, Y., You, J., Li, X., Yuan, C.: Projective robust nonnegative factorization. Inf. Sci. **364**, 16–32 (2016)

22. Pang, Y., Ye, L., Li, X., Pan, J.: Moving object detection in video using saliency map and subspace learning. arXiv preprint arXiv:1509.09089 (2015)

23. Rao, S.R., Tron, R., Vidal, R., Ma, Y.: Motion segmentation via robust subspace separation in the presence of outlying, incomplete, or corrupted trajectories. In: IEEE Conference on Computer Vision and Pattern Recognition, pp. 1–8 (2008)

24. Soltanolkotabi, M., Elhamifar, E., Candes, E.J., et al.: Robust subspace clustering. Annal. Stat. **42**(2), 669–699 (2014)

25. Tang, K., Dunson, D.B., Su, Z., Liu, R., Zhang, J., Dong, J.: Subspace segmentation by dense block and sparse representation. Neural Netw. **75**, 66–76 (2016)

26. Tron, R., Vidal, R.: A benchmark for the comparison of 3-d motion segmentation algorithms. In: IEEE Conference on Computer Vision and Pattern Recognition, pp. 1–8 (2007)

27. Vidal, R., Favaro, P.: Low rank subspace clustering. Pattern Recogn. Lett. **43**, 47–61 (2014)

28. Xu, Y., Wu, J., Li, X., Zhang, D.: Discriminative transfer subspace learning via low-rank and sparse representation. IEEE Trans. Image Process. **25**(2), 850–863 (2016)

29. Yang, S., Yi, Z., He, X., Li, X.: A class of manifold regularized multiplicative update algorithms for image clustering. IEEE Trans. Image Process. **24**(12), 5302–5314 (2015)

30. Zhang, T., Szlam, A., Lerman, G.: Median k-flats for hybrid linear modeling with many outliers. In: 12th IEEE International Conference on Computer Vision Workshops (ICCV Workshops), pp. 234–241 (2009)

31. Zhang, T.: Multi-stage convex relaxation for feature selection. Bernoulli **19**(5B), 2277–2293 (2013)

32. Zhang, Z., Xu, Y., Yang, J., Li, X., Zhang, D.: A survey of sparse representation: algorithms and applications. IEEE Access **3**, 490–530 (2015)

Clique-Based Locally Consistent Latent Space Clustering for Community Detection

Zhuanlian Ding[1], Dengdi Sun[1,2], Xingyi Zhang[1,2], and Bin Luo[1,2(✉)]

[1] School of Computer Science and Technology, Anhui University,
Hefei 230601, People's Republic of China
{dingzhuanlian,sundengdi,luobinahu}@163.com, xyzhanghust@gmail.com
[2] Key Lab of Industrial Image Processing & Analysis of Anhui Province,
Hefei 230039, People's Republic of China

Abstract. Community structure is one of the most important properties of complex networks and a keypoint to understanding and exploring real-world networks. One popular technique for community detection is matrix-based algorithms. However, existing matrix-based community detection models, such as nonnegative matrix factorization, spectral clustering and their variants, fit the data in a Euclidean space and have ignored the local consistency information which is crucial when discovering communities. In this paper, we propose a novel framework of latent space clustering to cope with community detection, by incorporating the clique-based locally consistency into the original objective functions to penalize the latent space dissimilarity of the nodes within the clique. We evaluate the proposed methods on both synthetic and real-world networks and experimental results show that our approaches significantly improve the accuracy of community detection and outperform state-of-the-art methods, especially on networks with unclear structures.

Keywords: Community detection · Local consistency · Graph regularization · Nonnegative Matrix Factorization (NMF) · Spectral Clustering (SC)

1 Introduction

Community structure is a natural characteristic in many real networks, such as social networks, biological networks and technological networks [1–3]. Although no general and widely-accepted definition of community structure has been agreed upon, it is commonly believed that a community is a group of nodes with more internal than external connections [4]. For example, authors from the same institution in collaboration networks, proteins with the same functionality in biochemical networks and the collections of pages on a single topic on the Web. It's worth noting that community structure in complex networks is often critical in understanding the natural structure of networks and revealing the network functions [5–7]. Thus, how to detect and extract these community

© Springer Nature Singapore Pte Ltd. 2016
T. Tan et al. (Eds.): CCPR 2016, Part I, CCIS 662, pp. 675–689, 2016.
DOI: 10.1007/978-981-10-3002-4_55

structures becomes an significant and challenging problem in the study of network systems.

The identification of community structure has recently received enormous amounts of attention in various scientific fields and many methods have been proposed and applied successfully to some specific complex networks. These methods are from different perspectives, such as the hierarchical clustering [4], label propagation [8] and optimization based algorithms [9]. Besides these methods, matrix-based community detection algorithms [3,10–12] have gained great success at uncovering the community structure, such as nonnegative matrix factorization (NMF) [11,12], spectral clustering (SC) [3,10] and their variants. Although NMF and SC have different formulations and meanings, these algorithms can be generally interpreted as the process of clustering in the latent space [7]. Specifically, the adjacency matrix which encodes the topology information is used as input and then the new property representation in the latent space is obtained by minimizing an objective function. Finally, clustering (e.g. k-means clustering) is based on the new property representation. Even though these traditional matrix-based community detection algorithms exhibit good performance on some community detection applications, they do have drawbacks. These approaches always succeed when the community structure is clear, but they fail significantly when one gradually increases the number of external edges between communities. Generally, the traditional NMF methods always approximate the original adjacency matrix in value using either the conventional least squares error or the generalized KL divergence as objective function, resulting in the indiscriminate penalty problem. The same problem arises for traditional SC methods. For example, assume that there are two linked pairs (i,j) and (i,k), where vertex i and j belong to the same community while vertex i and k do not. Since vertex i and k both have positive weights in some community c, $h_i h_k^{\mathrm{T}}$ (H is the indicator matrices, i.e. node-community membership matrix) is positive. However, existing NMF and SC approaches will not penalize $h_i h_k^{\mathrm{T}}$ for being positive since $A_{ik} = 1$. Thus, there is no difference between vertex j and k with respect to i, which is against the intuition that for vertex i, vertex j in the same community is more preferable than vertex k outside its community. In fact, it is reasonable that $h_i h_j^{\mathrm{T}}$ is higher than $h_i h_k^{\mathrm{T}}$, and indiscriminately penalizing the two pairs are problematic.

In this paper, our goal is to overcome the drawback of the traditional the matrix-based approaches. It is easy to find that the nodes belonging to the same community should have similar representations in latent space and the nodes belonging to the different communities should have dissimilar representations. If we have known that vertices i and j are likely to belong to the same community, we expect the new representations h_i and h_j in the indicator matrices H are similar. Thus, we propose a clique-based locally consistent regularization term into the original objective functions to penalize the latent space dissimilarity of the nodes within the clique. As our algorithms exploit clique-based locally consistency of the data and incorporate it as an additional graph regularization term, the quality of obtained communities is improved. Extensive experiments illustrate the competitive performance in terms of NMI compared to state-of-the-art algorithms.

The rest of this paper is organized as follows. In Sect. 2, we formally define the community detection problem and give a brief review of existing latent space clustering, including NMF and SC methods for community detection. Section 3 describes community detection using clique-based locally consistent latent space clustering in details. Extensive experiments on synthetic and real-world networks are presented in Sect. 4. Finally, Sect. 5 gives the conclusion of this paper.

2   Related Work

A network can be modeled as a graph $G(V, E)$, in which V is a set of N vertices and E is the set of M edges. In this paper, we focus on undirected unweighted networks. The adjacency matrix of G can be denoted as a nonnegative symmetric binary matrix A. The entry $A_{ij} = 1$ if and only if there is an edge between vertices i and j and $A_{ij} = 0$ otherwise. Besides, we assume there are K communities in the network, and K is known as a priori. In this section, we first briefly introduce two representative matrix based methods: NMF and SC, and then give a unified interpretation to these kinds of algorithms.

2.1   Nonnegative Matrix Factorization for Community Detection

NMF is an approximate factorization of the matrix V into a pair of matrices W and H. Note that none of the matrices is permitted to have negative entries is the unique feature of the NMF algorithm. Here, we transform community detection problem to the NMF problem $A \approx WH^T$. The factorization is carried out with a particular rank K so that the latent variables $W = [w_{ik}] \in R_+^{N \times K}$ and $H = [h_{ik}] \in R_+^{N \times K}$ whose elements w_{ik} and h_{ik} represent the probability that node i generates an in-edge and an out-edge that belong to the community k, respectively. Since matrix A is symmetric, W and H^T can be considered equivalent in a scale view. In this paper, we employ H to determine the nodes membership. Moreover, the factorization could be viewed as a representation of the data in a new space of lower dimensionality K. From the viewpoint of clustering, we can regard the factorization process as projecting the N dimension feature in the adjacent matrix into a K dimension latent space.

There are two commonly used loss functions that quantify the quality of the approximation. The first one is the square of the Euclidean distance between two matrices [11]:

$$\mathcal{L}_{LSE}(A, H) = \|A - WH^T\|_F^2. \tag{1}$$

The second one is the divergence between two matrices:

$$\mathcal{L}_{KL}(A, H) = KL(A\|WH^T). \tag{2}$$

Since they are convex in either W or H, Lee & Seung [13] presented iterative updating algorithms minimizing the objective function $\mathcal{L}_{LSE}$ as follows:

$$w_{ik} \leftarrow w_{ik} \frac{(AH)_{ik}}{(WH^TH)_{ik}}, \quad h_{jk} \leftarrow h_{jk} \frac{(A^TH)_{jk}}{(HW^TW)_{jk}}. \tag{3}$$

Similarly, the algorithms minimizing the objective function $\mathcal{L}_{KL}$ are:

$$w_{ik} \leftarrow w_{ik} \frac{\sum_j (a_{ij} h_{jk} / \sum_k w_{ik} h_{jk})}{\sum_j h_{jk}},$$

$$h_{jk} \leftarrow h_{jk} \frac{\sum_i (a_{ij} w_{ik} / \sum_k w_{ik} h_{jk})}{\sum_i w_{ik}}. \tag{4}$$

2.2 Spectral Clustering for Community Detection

As discussed in Ref. [3], spectral analysis achieves great success in uncovering the community structure based on the adjacency matrix A ($\mathcal{L}_{ADJ} = -\mathbf{Tr}(H^{\mathrm{T}} AH)$) and standard Laplacian matrix $L_S = D - A$ ($\mathcal{L}_{LAP} = \mathbf{Tr}(H^{\mathrm{T}}(D-A)H)$), where D is a diagonal matrix with the ith diagonal element. Actually, the community structure can be further identified using the eigenvectors of these matrices. Existing studies indicate that the spectrum of these matrices sheds light on the community structure of network.

Generally speaking, only several eigenvectors are utilized to project each node into a low-dimensional node vectors, and then the community structure is identified through clustering the node vectors using k-means clustering method. Specifically, these selected eigenvectors are stacked as columns of a matrix and the transpose of the ith row of this matrix is taken as the projected node vector corresponding to the node i. The community structure is then detected through clustering the projected node vectors.

2.3 Latent Space Clustering

Although why NMF and SC make sense and how they work in community detection are very different, these types of algorithms can be general interpreted from the viewpoint of latent space clustering. The input of community detection algorithms is the adjacent matrix A, which shows the topology information of the network. Many community detection algorithms, such as NMF and SC methods, first obtain a new matrix from adjacent matrix by minimizing an objective function. The obtained new matrix can be regarded as a representation in the latent space. Then the nodes are divided by clustering rows in the new matrix using k-means or other algorithms. Thus we can summarize these methods as the process of clustering in the latent space, called latent space clustering, as shown in Fig. 1. In general, we take the adjacency matrix as the input. Next, we obtain the new low-dimensional representation in the latent space. Finally, the community structure is discovered by k-means clustering based on the new property representation.

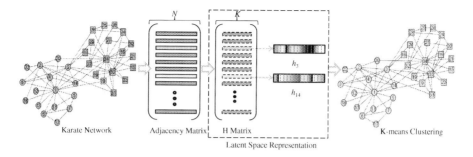

Fig. 1. The process of clustering in the latent space.

3   The Proposed Algorithms

In this section, we first depict the main idea of our methods using a simple example; then we specify the proposed clique-based locally consistent regularization; Finally, the overview of community detection using clique-based locally consistent latent space clustering is presented and the updating rules minimizing the corresponding objective function in our algorithms are provided.

3.1   The Main Idea

In this section, we propose a framework of latent space graph regularized community detection, which integrates the local consistency and global consistency to improve the performance of community detection. The main idea is displayed in Fig. 2. Its known that the traditional matrix-based methods always result in the indiscriminate penalty problem. For instance, directly connecting the nodes belonging to the same community cannot guarantee that they are classified into the same community, such as the vertex 3 and 14 in Fig. 2, and directly connecting the nodes belonging to different communities are always misclassified into the same community, such as the vertex 3 and 10 in Fig. 2. Inspired by a nature idea that the nodes in the same clique are more likely to belong to the same community, so we incorporate the clique-based locally consistency into the original objective functions. The pairwise closeness is obtained based on the statistics of all pairwise vertexes belonging to the same clique. It is easy to find that the nodes belonging to the same community should have similar representations in latent space. If we have known that the vertices i and j belong to the same community with great possibility, then h_i and h_j in the indicator matrices H should be similar. Therefore by minimizing the difference between h_i and h_j, we can assign them into the same community.

Now, we give an illustrative example of our approaches on the Zachary's karate club network, which has been widely used to evaluate the community detection methods. The network and its real social fission are depicted in above Fig. 1. The network has two reference communities, which is represented by two different shapes, circle and square.

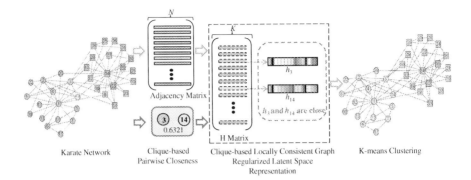

Fig. 2. The process of clique-based locally consistent latent space clustering for community detection.

Figure 3(a) shows the projected low-dimensional node vectors in the original latent space by SC-LAP method. Figure 3(b) presents the projected node vectors in the new latent space by OurSC-LAP, which incorporates clique-based locally consistency information. The community structure is then detected by the k-means clustering method. Apparently, the result of OurSC-LAP shown in Fig. 2, differentiated with different colors, matches its ground-truth at all. However, it's also clearly that SC-LAP misclassified the vertex 3 (Fig. 3(a)), which has correct label in OurSC-LAP (Fig. 3(b)). The results demonstrate the effectiveness of the proposed latent space graph regularization encoding clique-based locally consistency. Different from the traditional latent space clustering, which sees only the global topology information and ignores the local consistency, we balance the

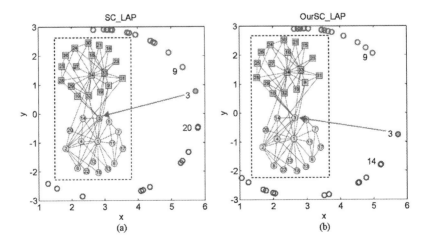

Fig. 3. Projected low-dimensional node vectors in the original latent space and the new latent space.

contributions of global consistency and local consistency in a seamless way. In real world, global topology information alone is often inadequate to accurately find community structure.

3.2   Clique-Based Locally Consistent Regularization

To measure the similarity between the two low dimensional vectors which denote the new representations of vertexes i and j, we can use either square distance

$$\mathcal{D}_{LSE}(h_i, h_j) = \|h_i - h_j\|_F^2, \tag{5}$$

or KL-divergence

$$\mathcal{D}_{KL}(h_i\|h_j) = \sum_{k=1}^{K} (h_{ik} \log \frac{h_{ik}}{h_{jk}} - h_{ik} + h_{jk}). \tag{6}$$

We use a weight matrix $C = [c_{ij}] \in R_+^{N \times N}$ to formulate the clique-based local consistency information, where c_{ij} represents pairwise closeness that vertexes i and j belong to the same community. We define $c_{ij} = 1 - e^{-s(i,j)^2}$, where $s(i, j)$ denotes the co-occurrence score, i.e., the number of two vertexes detected in the same cliques, motivated by the observation that cliques are one of the characteristic structures contained within communities. Definition of the weight matrix C on the graph is inspired by heat kernel weighting [14]. As clique detection in a graph is generally computationally expensive, we employ the centred cliques proposed in Ref. [2]. For each vertex x in the network, a clique is built using a greedy polynomial algorithm. As long as a clique is produced, vertices adjacent to x are added in decreasing order of their relative degree. The resulting centred clique is not necessarily the maximal clique.

The constraints from this information can be formulated as

$$\begin{aligned} \mathcal{R}_{LSE}(C, H) &= \frac{1}{2} \sum_{i=1}^{N} \sum_{j=1}^{N} c_{ij} \mathcal{D}_{LSE} \\ &= \sum_{i=1}^{N} h_i^{\mathrm{T}} h_i d_{ii} - \sum_{i,j=1}^{N} h_i^{\mathrm{T}} h_j c_{ij} \\ &= \mathbf{Tr}(H^{\mathrm{T}} D H) - \mathbf{Tr}(H^{\mathrm{T}} C H) \\ &= \mathbf{Tr}(H^{\mathrm{T}} L H), \end{aligned} \tag{7}$$

where $\mathbf{Tr}(\cdot)$ denotes the trace of a matrix and D is a diagonal matrix whose entries are column sums of C, $d_{ii} = \sum_k c_{ik}$. L is called graph Laplacian, $L = D - C$. Similarly, the KL-divergence based constraints can be written as

$$\mathcal{R}_{KL}(C, H) = \frac{1}{2} \sum_{i=1}^{N} \sum_{j=1}^{N} (\mathcal{D}_{KL}(h_i\|h_j) + \mathcal{D}_{KL}(h_j\|h_i)). \tag{8}$$

By minimizing $\mathcal{R}_{LSE}$ or $\mathcal{R}_{KL}$, we expect the new representations of two nodes h_i and h_j are close if we have some information indicating they might belong to the same community, i.e., the corresponding element c_{ij} is high.

3.3 Community Detection Using Clique-Based Locally Consistent Latent Space Clustering

By incorporating clique-based locally consistency information into the original objective function of matrix factorization, we propose the novel framework of latent space graph regularized community detection. Our algorithms minimize the objective function as follows:

$$\mathcal{O}_{\alpha\beta}(H|A,C) = \mathcal{L}_{\alpha}(A,H) + \lambda\mathcal{R}_{\beta}(C,H), \tag{9}$$

where $\alpha \in \{LSE, KL, ADJ, LAP\}$ and $\beta \in \{LSE, KL\}$ as described above and regularization parameter λ controls the smoothness of the new representation.

A. The Proposed Algorithm for Nonnegative Matrix Factorization. According to above formula, we obtain the following objective functions of our NMF methods

$$\mathcal{O}_{LSE}(H|A,C) = \|A - WH^{\mathrm{T}}\|_F^2 + \lambda\mathbf{Tr}(H^{\mathrm{T}}LH). \tag{10}$$

$$\mathcal{O}_{KL}(H|A,C) = KL(A\|WH^{\mathrm{T}} + \frac{1}{2}\sum_{i=1}^{N}\sum_{j=1}^{N}c_{ij}(\mathcal{D}_{KL}(h_i\|h_j) + \mathcal{D}_{KL}(h_j\|h_i))$$

$$= \sum_{i=1}^{N}\sum_{j=1}^{N}(a_{ij}\log\frac{a_{ij}}{\sum_{k=1}^{K}w_{ik}h_{jk}} - a_{ij} + \sum_{k=1}^{K}w_{ik}h_{jk}) \tag{11}$$

$$+ \frac{\lambda}{2}\sum_{i=1}^{N}\sum_{j=1}^{N}\sum_{k=1}^{K}c_{ij}(h_{ik}\log\frac{h_{ik}}{h_{jk}} + h_{jk}\log\frac{h_{jk}}{h_{ik}}).$$

In this section, we develop corresponding iterative algorithms as in Ref. [14], which are proved that the updating rules can find the local minima of these objective functions. We obtain the following updating rules minimize $\mathcal{O}_{LSE}(H|A,C)$

$$w_{ik} \leftarrow w_{ik}\frac{(AH)_{ik}}{(WH^{\mathrm{T}}H)_{ik}}, \quad h_{jk} \leftarrow h_{jk}\frac{(A^{\mathrm{T}}H + \lambda CH)_{jk}}{(HW^{\mathrm{T}}W + \lambda DH)_{jk}}. \tag{12}$$

Similarly, we obtain the following updating rule to minimize $\mathcal{O}_{KL}(H|A,C)$

$$w_{ik} \leftarrow w_{ik}\frac{\sum_j(a_{ij}h_{jk}/\sum_k w_{ik}h_{jk})}{\sum_j h_{jk}},$$

$$h_{jk} \leftarrow (\sum_i w_{ik}I + \lambda L)^{-1}\begin{bmatrix} h_{1k}\sum_i(a_{i1}w_{ik}/\sum_k w_{ik}h_{1k}) \\ \cdots \\ h_{Nk}\sum_i(a_{iN}w_{ik}/\sum_k w_{ik}h_{Nk}) \end{bmatrix}, \tag{13}$$

where I is an $N \times N$ identity matrix. When λ equals to zero, the updating rules reduce to the updating rules of standard NMF methods.

B. The Proposed Algorithm for Spectral Clustering. According to above formula, we obtain the following objective functions of our SC methods.

$$
\begin{aligned}
\mathcal{O}_{ADJ}(A, C) &= -\mathbf{Tr}(H^{\mathrm{T}}AH) + \lambda\mathbf{Tr}(H^{\mathrm{T}}LH) \\
&= \mathbf{Tr}(H^{\mathrm{T}}(-A + \lambda L)H).
\end{aligned}
\tag{14}
$$

$$
\begin{aligned}
\mathcal{O}_{LAP}(A, C) &= \mathbf{Tr}(H^{\mathrm{T}}(D - A)H) + \lambda\mathbf{Tr}(H^{\mathrm{T}}LH) \\
&= \mathbf{Tr}(H^{\mathrm{T}}(D - A + \lambda L)H).
\end{aligned}
\tag{15}
$$

Based on Rayleigh quotient, these problems above can be minimized by finding the eigenvectors corresponding to the smallest K eigenvalues of matrix $-A + \lambda L$ and $D - A + \lambda L$, respectively.

3.4 Complexity Analysis

In our algorithms, firstly we need $O(k^2 N)$ computational complexity to compute weight matrix C, where k is the average degree of nodes in the network. Then we analyze the computational complexity of our NMF and our SC as follows.

A. Complexity Analysis for the Proposed Nonnegative Matrix Factorization. In this section, we analyze the computational complexity of our NMF based on Frobenius norm distance metrics. In our algorithms, each iteration in the updating process needs $O(N^2 K)$ floating point operations by considering the number of communities $K \ll N$. Therefore, the proposed framework does not increase the complexity of the original NMF methods.

B. Complexity Analysis for the Proposed Spectral Clustering. Our algorithm is the same as that for common SC, i.e., eigenvalue decomposition. In theory, it has the same complexity of matrix multiplication whose upper bound is $O(N^3)$.

4   Experiments

In this section, both synthetic and real-world networks are applied to test the quality of obtained communities. The synthetic networks allow us to test the viability of different methods for known community detection under controlled conditions, while the real-world networks allow us to observe their capabilities under practical conditions. To evaluate the quality of obtained overlapping communities, we employ the widely used normalized mutual information (NMI) [7] as the quality metric.

Further, we compared the performance of our methods with representative approaches: the original NMF [11] and SC methods [10], Jin's NMF [5], BNMTF [15] and MEAs-SN [9]. For each algorithm, the final results are obtained after having optimized the algorithm parameters to yield the best possible results as measured by NMI. Note that all the experiments here are conducted on a PC with a 3.0GHz Pentium(R) Dual-Core CPU and the Windows 7 SP1 32 bit operating system. Our programming environment is MATLAB 2010.

4.1 Synthetic Networks

We empirically use the well-known GN benchmark networks and LFR benchmark networks, to test the performance of community detection methods.

In the following experiments, GN benchmark networks are generated based on Newman model $RN(C, s, d, P_{in})$ [4], where C is the number of communities, s is the number of nodes in each community, d is the degree of nodes and Pin is the density of the connections intra the community. In our study, the widely used benchmark network $RN(4, 32, 16, P_{in})$ is employed. We generate 11 sets of networks with parameter P_{in} ranging from 0.4 to 0.9 with interval 0.05. When P_{in} is small, there are few edges connecting the nodes intra the community and the community structure is very vague. The community structure becomes clear with P_{in} growing larger.

Account for the heterogeneity in the distributions of vertex degrees and of community sizes, LFR benchmark is proposed by Lancichinetti et al. [16]. In our experiments, we generate 8 sets of networks with the mixing parameter μ ranging from 0 to 0.8 with interval 0.1. With μ growing larger, the community structure becomes vaguer. Each network contains 1000 vertices, with an average degree of 20 and a max degree of 50. The community size ranges from 10 to 50, vertex degrees and community sizes are controlled by power-law distribution with exponents $\tau_1 = 2$ and $\tau_2 = 1$ respectively.

Here, we conduct two parts of experiments. The first part is designed to evaluate the influence of clique-based locally consistent regularization, and the second part focuses on the performance of the proposed algorithm via comparative experiments against other algorithms. For each parameter set generated via GN benchmark or LFR benchmark, we generated 20 instantiations.

Figures 4 and 5 shows the results of the first part of experiments both on GN benchmark networks and LFR benchmark networks, respectively, to evaluate the

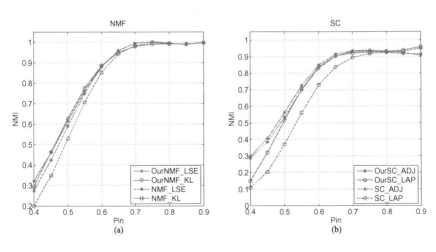

Fig. 4. The influence of the proposed clique-based locally consistent regularization on GN benchmark networks.

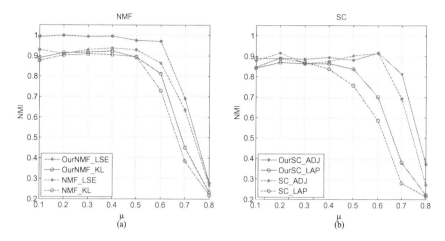

Fig. 5. The influence of the proposed clique-based locally consistent regularization on LFR benchmark networks.

influence of the newly designed clique-based locally consistent regularization. In general, increasing μ (decreasing P_{in}) typically results in poorer performance for all methods, due to the vague community structure. From Figs. 4 and 5 we can get the following two observations. One is that OurNMF and OurSC outperform the original NMF and SC approaches respectively, and the other is such superiority becomes more significant when the community structure tend to vague (low P_{in} or high μ). The main reason is attributed to the incorporating of clique-based locally consistent regularization into the original objective functions, making penalizing the two linked pairs (i, j) and (i, k) discriminately by exploiting the local geometrical structure and local invariance of the network. The other observation is that generally OurNMF-LSE achieves the best performance in all NMF-based methods and OurSC-ADJ is the best of all SC methods.

Figure 6(a) and (b) present the results of the second comparative experiments against other algorithms on GN benchmark networks and LFR benchmark networks respectively. Here, we select OurNMF-LSE and OurSC-ADJ as OurNMF and OurSC which exhibit excellent performance as described above. From Fig. 6 we conclude that OurNMF and Jin's NMF perform much better than their counterparts, and in most cases OurNMF is quite competitive to Jin's NMF algorithm with NMI being even slightly better. These two algorithms can find partitions close to the true ones for GN networks with $P_{in} \geq 0.7$ and for LFR networks with $\mu \leq 0.6$. This further implies the efficiency of our local consistency information encoding strategy. However, the performance of OurNMF degrades when $\mu > 0.6$, maybe due to the fact that limited local consistency information is employed and even wrong local consistency information is also unavoidable in these networks. In this case, OurSC exhibits satisfying results.

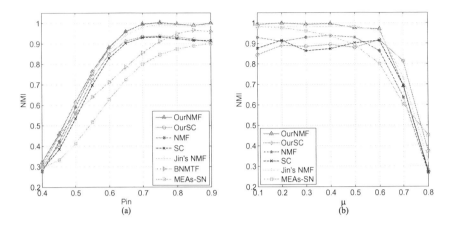

Fig. 6. Results of all algorithms on GN benchmark networks and LFR benchmark networks. Note that BNMTF model needs to update the element in the matrix one by one, resulting not suitable for large networks as high computational complexity, so here BNMTF model is not suitable for LFR networks.

4.2   Parameter Setting

To illustrate the effect of the tradeoff parameter and discuss how to determine it, we evaluate the role of balancing parameter λ in formula (10). Here, we take NMF-KL and SC-LAP on GN benchmark networks as an example. Figure 7 shows the performance of our algorithms as the λ varies from 0 to 3 with interval 0.5 with different parameter P_{in}. From the curves in Fig. 7 we obtain the following three findings and conclusions. The first finding is that compared with the original NMF and SC approaches ($\lambda = 0$), the impacts of λ are positive in

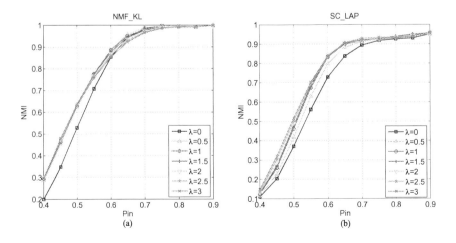

Fig. 7. Performance of our methods with different parameter λ on GN networks.

most cases, especially when the network structure is unclear. This also means the proposed clique-based locally consistent regularization plays a more important role in detecting on complicated networks. The second conclusion is that if the network structure is quite vague, increasing λ can slightly and consistently improve the performance, and if the network structure is just a little vague, the performance may degrade slightly when we use a large λ. The last observation is that for networks with clear structure, the performance of our methods is generally not very sensitive to parameter λ. In summary, if we have some knowledge that the structure of network is very vague, we can increase λ appropriately, and vice versa. In our experiments, the final results are obtained after having optimized the algorithm parameter λ to yield the best possible results measured by NMI.

4.3   Real-World Networks

In this subsection, we test clustering quality on five real networks, including Karate network, Dolphin network, Football network, Polbooks network, and Word network listed in Table 1. All the datasets are available on Newmans homepage (http://www-personal.umich.edu/~mejn/netdata/). Table 2 illustrates the NMI of all the methods. From the results we can conclude that our methods outperform the other algorithms in terms of NMI. This confirms that for real-world networks with complicated organizational structures, our methods exhibited even better relative performance to all the other methods. This further implies the efficiency of the proposed clique-base locally consistent regularization in our methods. The observation is in agreement with the fact that our algorithms can achieve better performance on networks with vague community

Table 1. Real-world networks used in the experiments

Network	Node	Edges	Number of communities	Description
Karate	34	78	2	Zachary's karate club
Dolphins	62	159	2	Dolphins social network
Football	115	613	12	American college football
Polbooks	105	441	3	Krebs books on US politics
Word	112	425	2	Word network

Table 2. Comparison of NMI by different algorithms on real-world networks

Dataset	NMF-LSE		NMF-KL		SC-ADJ		SC-LAP		Jin's NMF	BNMTF	MEAs-SN
	Original	Ours	Original	Ours	Original	Ours	Original	Ours			
Karate	1.000	1.000	1.000	1.000	1.000	1.000	0.837	1.000	1.000	0.553	0.782
Dolphins	0.814	0.858	0774	0.794	0.142	0.504	0.889	**0.889**	0.814	0.590	0.437
Football	0.913	**0.925**	0.880	0.914	0.872	0.865	0.870	0.862	0.924	0.867	0.924
Polbooks	0.519	0.540	0.530	0.567	0.364	0.559	0.574	**0.581**	0.540	0.505	0.449
Word	0.341	**0.344**	0.326	0.330	0.001	0.018	0.002	0.002	0.001	0.321	0.183

structure as shown in Fig. 6. Besides, on clear structure networks where original methods can achieve good performance, e.g., Karate network and Football network, our methods also obtain competitive results.

5   Conclusion

In this paper, we propose a novel framework of community detection based on latent space similarity, which combines the global network topology with clique-based local consistency. Extensive experiments on synthetic and real-world networks illustrate the robustness and effectiveness of our framework on encoding clique based local consistency information. In the future, we may conduct research in the following two directions. Firstly, it would be interesting to integrate network topology and semantic information on nodes, e.g., node attributes into the objection functions for community detection. Secondly, we will investigate how to design parallel algorithms to make our framework suitable for large-scale networks.

Acknowledgments. This work is supported by the National High Technology Research and Development Program (863 Program) of China (2014AA015104), the National Natural Science Foundation of China (61402002, 61272152 and 61472002), the Natural Science Foundation of Anhui Province (1408085QF120), the Natural Science Foundation of Anhui Higher Education Institutions (KJ2016A040), Open Project of IAT Collaborative Innovation Center of Anhui University (ADXXBZ201511) and Public Sentiment and Regional Image Research Center of Anhui University (Y01002364).

References

1. Newman, M.E.J.: Communities, modules and large-scale structure in networks. Nat. Phys. **8**, 25–31 (2012)
2. Becker, E., Robisson, B., Chapple, C.E., Gunoche, A., Brun, C.: Multifunctional proteins revealed by overlapping clustering in protein interaction network. Bioinformatics **28**, 84–90 (2012)
3. Shen, H.W., Cheng, X.Q.: Spectral methods for the detection of network community structure: a comparative analysis. J. Stat. Mech. Theory Exp. **10**, P10020 (2010)
4. Girvan, M., Newman, M.E.J.: Community structure in social and biological networks. Proc. Natl. Acad. Sci. **99**, 7821–7826 (2002)
5. Jin, D., Gabrys, B., Dang, J.: Combined node and link partitions method for finding overlapping communities in complex networks. Sci. Rep. **5**, 8600 (2015)
6. Ding, Z.L., Zhang, X.Y., Sun, D.D., Luo, B.: Overlapping community detection based on network decomposition. Sci. Rep. **6**, 24115 (2016)
7. Yang, L., Cao, X., Jin, D., Wang, X., Meng, D.: A unified semi-supervised community detection framework using latent space graph regularization. IEEE Trans. Cybern. **45**(11), 2585–2598 (2015)
8. Frey, B.J., Dueck, D.: Clustering by passing messages between data points. Science **315**(5814), 972–976 (2007)

9. Liu, C., Liu, J., Jiang, Z.: A multiobjective evolutionary algorithm based on similarity for community detection from signed social networks. IEEE Trans. Cybern. **44**(12), 2274–2287 (2014)
10. Newman, M.E.J.: Finding community structure in networks using the eigenvectors of matrices. Phys. Rev. E **74**(3), 036104 (2006)
11. Zhang, S., Wang, R.S., Zhang, X.S.: Uncovering fuzzy community structure in complex networks. Phys. Rev. E **76**(4), 046103 (2007)
12. Wang, F., Li, T., Wang, X., Zhu, S., Ding, H.Q.C.: Community discovery using nonnegative matrix factorization. Data Min. Knowl. Disc. **22**(3), 493–521 (2011)
13. Lee, D.D., Seung, H.S.: Algorithms for non-negative matrix factorization. In: 13th Advances in Neural Information Processing Systems, pp. 556–562. Denver (2000)
14. Cai, D., He, X., Han, J., Huang, T.S.: Graph regularized nonnegative matrix factorization for data representation. IEEE Trans. Pattern Anal. Mach. Intell. **33**(8), 1548–1560 (2011)
15. Zhang, Y., Yeung, D.: Overlapping community detection via bounded nonnegative matrix tri-factorization. In: 18th ACM SIGKDD International Conference on Knowledge Discovery and Data Mining, pp. 606–614. ACM, Beijing (2012)
16. Lancichinetti, A., Fortunato, S.: Benchmarks for testing community detection algorithms on directed and weighted graphs with overlapping communities. Phys. Rev. E **80**, 016118 (2009)

An Improved Self-adaptive Regularization Method for Mixed Multiplicative and Additive Noise Reduction

Ziling Wu[1,2], Hongxia Gao[1,2], Ge Ma[1,2(✉)], and Lixuan Wu[1,2]

[1] School of Automation Science and Engineering,
South China University of Technology,
Guangzhou 510641, Guangdong, People's Republic of China
m.ge.tina@gmail.com
[2] Engineering Research Centre for Manufacturing Equipment of Ministry
of Education, South China University of Technology, Guangzhou 510641,
Guangdong, People's Republic of China

Abstract. The noise in micro focus X-ray images is complicated with low signal-to-noise-ratio (SNR) and can be described as mixed multiplicative and additive noise. Nevertheless, the present self-adaptive regularization methods for smoothing such mixed noise remain scarce. Thus, this paper proposes an improved self-adaptive regularization method to reduce the mixed multiplicative and additive noise in micro focus X-ray images. A novel scheme to adaptively select the regularization operator and regularization parameter based on local variance is presented, in which a p-Laplace function is used as the regularization operator with self-adaptive p and the regularization parameter is designed according to a barrier function. Experiment results demonstrate that the proposed method can achieve a better balance between noise-reducing and edge-preserving, which effectively improve the denoising quality.

Keywords: Mixed multiplicative and additive noise · Micro focus X-ray images · Self-adaptive regularization

1 Introduction

Integrated circuit (IC) detection is common in industrial and there are lots of fine structures in the nano chips. And the micro focus X-ray detector is used for such kind of chips detection, whose focal spot is smaller than 5 μm. However, limited to the low-photon counting characteristic, the noise in micro focus X-ray images is complicated with low signal-to-noise ratio (SNR). Among the present researches with respect to the low-photon imaging mechanism, some established the noise model in X-ray images as Gaussian. For example, [1] used the Gaussian noise model and employed an EM (Expectation-Maximization) scheme to restore the X-ray images. Differently, [2, 3] adopted that Poisson noise model for low-photon counting systems and proposed transformation techniques for the Poisson noise. In recent year, more work considered the mixed Gaussian-Poisson noise model for such imaging mechanism [4–7]. Among

© Springer Nature Singapore Pte Ltd. 2016
T. Tan et al. (Eds.): CCPR 2016, Part I, CCIS 662, pp. 690–702, 2016.
DOI: 10.1007/978-981-10-3002-4_56

them, [4] introduced a wavelet-based method (PURE-LET) to reduce the mixed Gaussian-Poisson noise, which was considered state-of-the-art.

In 2014, Gao et al. [8] developed a novel idea for the noise in micro focus X-ray images. By describing the intrinsic thermal and electronic fluctuations of the acquisition devices as additive component and the signal-dependent noise as multiplicative model, they established the noise model for micro focus X-ray images as mixed multiplicative and additive noise. Such mixed noise model is more universal. Besides, the objective function was set with total variation (TV) and wavelet-based sparse constraints. Then it was solved via explicit difference and gradient projection. Experimental results showed that it was able to achieve good results in denoising micro focus X-ray images.

However, the above method is limited to the non-adaptive TV regularization. Image denoising is actually an ill-posed problem. And regularization method is an effective approach to solve such problem by imposing a priori information constraint on the restored image. Generally, regularized image restoration methods can be classified as non-adaptive methods and self-adaptive methods [9]. Since the non-adaptive regularization methods employ invariant and sole regularization operator and regularization parameter for the whole images, it can't reach a good balance between noise-reducing and edge-preserving. To overcome the difficulty, self-adaptive regularization methods select different regularization operator or regularization parameter according to different images features. Thus, they are now considered state-of-the-art and receive much attention.

Specifically, self-adaptive regularization methods can be classified as adaptive regularization operator methods and adaptive regularization parameter methods. For example, a natural approach to improve the classical TV model is to use a p-Laplace function as the regularization operator [10, 11], of which p is self-adaptive based on different image features [12–15]. And most of the related work focused on designing adaptive regularization parameter in the principle that a small parameter should be used for edges and a large one for smooth regions [16–20]. All these researches have shown the superiority of the self-adaptive regularization methods. Nevertheless, the self-adaptive regularization methods for mixed multiplicative and additive noise reduction remain scarce.

In this paper, we propose an adaptive regularization method to reduce the mixed multiplicative and additive noises in micro focus X-ray images. We improve the TV via a p-Laplace variation function. And the regularization parameter is designed based on a barrier function, which can preserve more fine details. Experimental results indicate that our proposed method can not only suppress the mixed noise in micro focus X-ray images, but also preserve more fine structures, which effectively improve the denoising quality.

The remainder of the paper is organized as follows: Section 2 reviews the background of the mixed multiplicative and additive noise model in [8] and the regularization methods. And our proposed method is presented in Sect. 3. Section 4 displays the experiment results and analysis. Conclusions are given in Sect. 5.

2 Background

2.1 Mixed Multiplicative and Additive Noise Model

Generally, additive noise is frequently used to describe the thermal noise and its degraded model is

$$b = u + r, \tag{1}$$

where $b \in \mathbb{R}^n$ is the observed image, $u \in \mathbb{R}^n$ is the unknown true image, $r \in \mathbb{R}^n$ represents additive noise.

Multiplicative noise is signal-dependent and is described as

$$b = Mu + u, \tag{2}$$

where M is an multiplicative operator.

Hence, the mixed multiplicative and additive noise model is given as (3)

$$b = Mu + u + r. \tag{3}$$

And (3) can be transformed to be:

$$b = v + r, \tag{4}$$

$$v = Mu + u, \tag{5}$$

where $v \in \mathbb{R}^n$. And b in formula (4) is regarded as the observation of v corrupted by additive noise. In formula (5), v represents the observed image degraded by multiplicative noise.

2.2 Regularization Method

The general expression of regularization method is

$$min\ T(u) = F(b|u) + \lambda R(u), \tag{6}$$

where $F : \mathbb{R}^n \rightarrow \mathbb{R}$ is the data fidelity term, $\lambda > 0$ is the regularization parameter which provides a tradeoff between fidelity to the measurements and noise sensitivity, and $R : \mathbb{R}^n \rightarrow \mathbb{R}$ represents the regularization operator.

The fidelity term is usually derived from the noise models. Based on (4), the fidelity term $F(b|v)$ takes the following form:

$$F(b|v) = \|v - b\|_2^2. \tag{7}$$

And for degraded model of multiplicative noise in formula (5), the RLO (Rudin-Lions-Osher) model [21] is considered that

$$F(v|u) = \int \left(\frac{v}{u} - 1\right)^2 dxdy. \tag{8}$$

The regularization operator $R(u)$ is designed based on the priori assumptions on u. At first, Tikhonov regularization [22] was widely used, which added a quadratic penalty to the objective function, resulting in isotropic diffusion and over-smooth. To overcome the disadvantage, TV regularization was proposed by Rudin et al. [23] in 1992. The TV operator is described as

$$R(u) = \|\nabla u\|_1, \tag{9}$$

where ∇ is the gradient operator. TV is now widely used in many image processing problems and works well in preserving edges.

However, TV will suffer from staircase effects. Therefore, Blomgren et al. [10] proposed the l_p norm of gradient ($p \in [1, 2]$) as formula (10) to improve the performance of TV (we refer it as p-Laplace).

$$R(u) = \|\nabla u\|_p, \tag{10}$$

Generally, if $p \to 1$, the p-Laplace function tends to preserve edges and if $p \to 2$, it can strongly smooth noise.

Nowadays, it is regarded one of the state-of-the-art methods to penalize the l_1 norm of wavelet-domain representation coefficients as the sparsity constraint. Therefore, $R(u)$ is expressed as

$$R(u) = \|Wu\|_1. \tag{11}$$

where W is the discrete wavelet basis

3   The Proposed Self-adaptive Regularization Method

In this section, we will introduce an improved self-adaptive scheme under the basis of [8] to reduce the mixed multiplicative and additive noise in micro focus X-ray images, in which both regularization operator and regularization parameter are decided based on local variance. It can ensure a better effectiveness both in noise-reducing and detail-preserving.

Firstly, to restore v from the observed b in formula (4), the objective minimization function takes the following form:

$$min \, T_1(v) = min\|v - b\|_2^2 + \lambda(x, y)\|\nabla v\|_{p(x,y)}. \tag{12}$$

Then, the RLO model combined with wavelet-based sparsity constrains of u is used to reduce multiplicative noise in (5), leading to the following minimization problem:

$$\min T_2(\boldsymbol{u}) = \min \int \left(\frac{\boldsymbol{v}}{\boldsymbol{u}} - 1\right)^2 dxdy \tag{13}$$

$$s.t. \ \|\boldsymbol{u} - \boldsymbol{W}\boldsymbol{\zeta}\|_2^2 + \lambda_2 \|\boldsymbol{\zeta}\|_1 \le \varepsilon.$$

where $\lambda_2 > 0$ is scaling parameter, $\boldsymbol{\zeta}$ is the wavelets coefficients on $\boldsymbol{W}$, $\varepsilon \in \mathbb{R}^+$ and $\varepsilon \to 0$.

3.1 Self-adaptive Regularization

Usually, adaptive regularization methods select different regularization operator and regularization parameter according to different types of image features. Consequently, we first raise a 4-step scheme to segment the image:

Step 1: Smooth the observed image $\boldsymbol{b}$ with a Gaussian filter, we denote the filtered image as $\tilde{\boldsymbol{b}}$.

Step 2: Compute the local variance of $\tilde{\boldsymbol{b}}$

$$\sigma^2(x, y) = \frac{1}{(2K+1)(2L+1)} \sum_{k=-K}^{k=K} \sum_{l=-L}^{l=L} \left[\tilde{\boldsymbol{b}}(x+k, y+l) - \bar{\tilde{\boldsymbol{b}}}(x, y)\right]^2, \tag{14}$$

$$\bar{\tilde{\boldsymbol{b}}}(x, y) = \frac{1}{(2K+1)(2L+1)} \sum_{k=-K}^{k=K} \sum_{l=-L}^{l=L} \left[\tilde{\boldsymbol{b}}(x+k, y+l)\right], \tag{15}$$

where K and L are the window size, usually $K = L = 2$.

Step 3: Compute the related binary image $\tilde{\boldsymbol{b}}_B$ of the local variance image via Otsu [24]. The background pixels set is denoted as

$$\emptyset := \left\{(x, y)|\tilde{\boldsymbol{b}}_B(x, y) = 0\right\}. \tag{16}$$

Step 4: Define

$$\Omega := \left\{\sigma_\emptyset^2(x, y)|(x, y) \in \emptyset\right\}, \tag{17}$$

where $\sigma_\emptyset^2(x, y)$ is the corresponding local variance of the pixels in $\emptyset$. And calculate the mean value Th of Ω.

Then, the image can be segmented as:

$$\begin{cases} (x, y) \in detail \ regions, & if \ \sigma^2(x, y) > Th \\ (x, y) \in smooth \ regions, & if \ \sigma^2(x, y) \le Th \end{cases}. \tag{18}$$

With the segmentation results, $\boldsymbol{p}(x, y)$ in formula (12) is defined as

$$\begin{cases} p(x, y) = 1, & if \ \sigma^2(x, y) > Th \\ p(x, y) = 2, & if \ \sigma^2(x, y) \le Th \end{cases}. \tag{19}$$

The underlying philosophy of (19) is that a TV norm ($p = 1$) is applied in the detail regions to preserve details and a H^1 norm ($p = 2$) is used in smooth regions to reduce noise.

As for adaptive regularization parameter, the present rule mainly chose small parameters in detail regions and larger ones in smooth regions, such as inverse proportion to gradient. But on this occasion, weak edges with small gradient and relatively large parameter may be over smoothened. Therefore, we propose a novel scheme for the adaptive parameter based on a barrier function to preserve more weak edges:

$$\lambda(x, y) = \begin{cases} \lambda_d(x, y) = \left\{ \ln\left[\frac{\sigma^2(x,y)}{Th}\right] + \frac{Th}{\sigma^2(x,y)} \right\}, & \text{if } \sigma^2(x, y) > Th \\ \lambda_s(x, y) = 1.5 * \max\{\lambda_d\}, & \text{if } \sigma^2(x, y) \leq Th \end{cases} \tag{20}$$

Generally, a largest parameter is used in smooth regions to strongly reduce the noise. And in detail regions, weak edges with small local variance correspond to small regularization parameters to give mild smoothness. Furthermore, for large local variance values, namely strong edges, the regularization parameters will be slightly larger for moderately smoothing.

3.2   The Proposed Method

According to the analysis carried out above, our proposed method for mixed multiplicative and additive noise reduction is introduced in this section.

First, the minimization problem (12) can be solved via many existing methods, such as gradient descent algorithm [25].

Second, based on the perspective of alternative direction method, the minimization problem (13) is divided into subproblems (21), (22).

$$\boldsymbol{u}_k = \arg\min_{\boldsymbol{u}} \int \left(\frac{\boldsymbol{v}}{\boldsymbol{u}} - 1\right)^2 dxdy, \tag{21}$$

$$\boldsymbol{\zeta}_k = \arg\min_{\boldsymbol{\zeta}} \|\boldsymbol{u}_k - \boldsymbol{W}\boldsymbol{\zeta}\|_2^2 + \lambda_2\|\boldsymbol{\zeta}\|_1, \tag{22}$$

$$\boldsymbol{u}_{k+1} = \boldsymbol{W}^{-1}\boldsymbol{\zeta}_k. \tag{23}$$

(21) can also be solved by gradient descent algorithm. And (22) is a wavelet-based $l_2 - l_1$ problem, it can be easily solved by gradient projection algorithm [26].

Finally, the general procedure of the proposed improved self-adaptive regularization method for mixed multiplicative and additive noise reduction is as follows (Table 1):

Table 1. The proposed improved self-adaptive regularization method for mixed multiplicative and additive noise reduction

---

Algorithm 1.

Initialization: define $\boldsymbol{p}$ and $\boldsymbol{\lambda}$, choose $\boldsymbol{u}^0$, $\lambda_2 > 0$, $\tau > 0$, $k = 0$.

Repeat:

Step 1: $\boldsymbol{v} = \underset{\boldsymbol{v}}{argmin} \ \|\boldsymbol{v} - \boldsymbol{b}\|_2^2 + \lambda(x,y)\|\nabla \boldsymbol{v}\|_{\boldsymbol{p}(x,y)}$;

Step 2: $\boldsymbol{u}_k = \underset{\boldsymbol{u}}{argmin} \ \int \left(\frac{\boldsymbol{v}}{\boldsymbol{u}} - 1\right)^2 dxdy$;

Step 3: $\boldsymbol{\zeta}_k = \underset{\boldsymbol{\zeta}}{argmin} \ \|\boldsymbol{u}_k - \boldsymbol{W}\boldsymbol{\zeta}\|_2^2 + \lambda_2\|\boldsymbol{\zeta}\|_1$;

Step 4: $\boldsymbol{u}_{k+1} = \boldsymbol{W}^{-1}\boldsymbol{\zeta}_k$;

Step 5: $\boldsymbol{b} = \boldsymbol{u}_{k+1}$;

Step 6: update $k = k + 1$;

End while

$$\frac{\|\boldsymbol{u}_{k+1} - \boldsymbol{u}_k\|_2^2}{\|\boldsymbol{u}_k\|_2^2} \le \tau .$$

---

4   Experiments

In this section, we present experiments to demonstrate the superiority of the proposed method in deducing mixed multiplicative and additive noise in micro focus X-ray images. The test images in Fig. 1 are micro focus X-ray images of nano chips with high noise. They are generated by the X-ray detector whose focal spot is smaller than 5 μm. PURE-LET [4] for mixed Poisson-Gaussian noise (PG+PURE-LET), GAT [27] +BLS-GSM [28] for mixed Poisson-Gaussian noise (PG+GAT+BLS-GSM), the denoising method in [8] (MA+TV-L1) are employed to compare with the proposed method. Objective evaluation indexes including Mean to Standard Deviation Ratio (MSR) [29] and Laplacian Sum (LS) [30] are used to evaluate the denoising quality objectively. Generally, the higher MSR and LS are, the better performance is.

$$LS(\boldsymbol{u}) =$$
$$\frac{1}{(M-2)(N-2)} \times \sum_{i=1}^{M-1} \sum_{j=1}^{N-1} |8 \times \boldsymbol{u}(i,j) - \boldsymbol{u}(i,j-1) - \boldsymbol{u}(i-1,j) - \boldsymbol{u}(i,j+1) - \quad (24)$$
$$\boldsymbol{u}(i-1,j-1) - \boldsymbol{u}(i-1,j+1) - \boldsymbol{u}(i+1,j-1) - \boldsymbol{u}(i+1,j+1)|,$$

where M, $N \in \mathbb{R}^+$, are the image sizes.

The denosing results are demonstrated in Fig. 2. To compare more clearly, the local variance results can be found in Fig. 3. Also, the objective evaluation indexes are given in Table 2.

From the results in Fig. 2, it is obvious that PG+PURE-LET and PG+GAT+ BLS-GSM will suffer from over-smooth. Moreover, the artefacts are severe in the denoising results of PG+PURE-LET and PG+GAT+BLS-GSM; see Fig. 2(a), (b) and

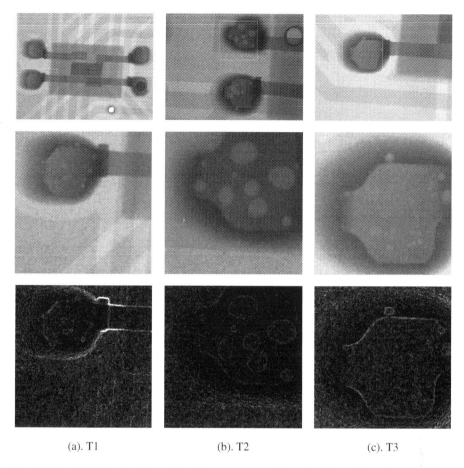

(a). T1   (b). T2   (c). T3

Fig. 1. Three typical micro focus X-ray images generated by micro focus X-ray detector whose focal spot is smaller than 5 μm for detecting nano chip. From left to right: we name them as T1, T2 and T3 respectively. From top to bottom: the first row: original micro focus X-ray images; the second row: enlarged parts of the original images; the third row: local variance of the enlarged parts.

(c). MA+TV-L1 gives relatively better results that most noise is suppressed, but the weak edges like the bubble defects on the micro focus X-ray images are blurred. The proposed method achieves best results. Noise in the smooth regions is strongly reduced and most strong and weak edges are preserved.

This is clearer in Fig. 3. PG+PURE-LET and PG+GAT+BLS-GSM lose lots of details and MA+TV-L1 can restore strong edges. Compared with MA+TV-L1, the proposed method improves a lot. It not only reduces the noise in the smooth regions, but also preserves most weak edges, even if the small and dense clusters of bubbles or the weak contours; see Fig. 3(a), (b) and (c). Consequently, the proposed method outperforms the other three methods in visual effects. And the objective evaluation indexes in

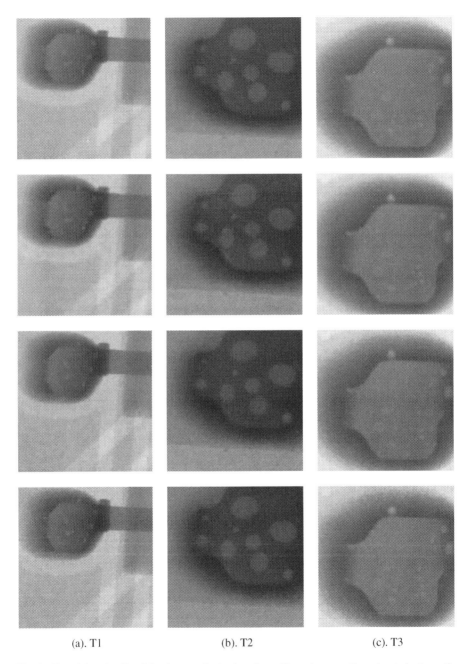

(a). T1   (b). T2   (c). T3

Fig. 2. Denoising details of the three typical micro focus X-ray images. From top to bottom: the first row: PG+PURE-LET; the second row: PG+GAT+BLS-GSM; the third row: MA+TV-L1; the fourth row: The proposed method. From left to right: (a) T1; (b) T2; (c) T3.

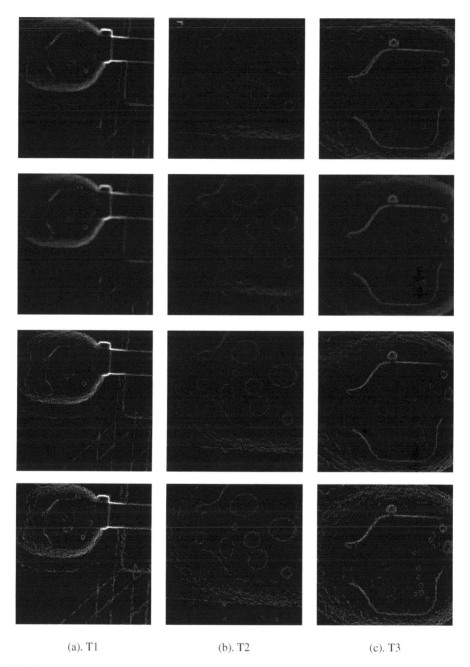

(a). T1   (b). T2   (c). T3

Fig. 3. Local variance of Fig. 2. From top to bottom: the first row: PG+PURE-LET; the second row: PG+GAT+BLS-GSM; the third row: MA+TV-L1; the fourth row: The proposed method. From left to right: (a) T1; (b) T2; (c) T3.

Table 2. The comparison of MSR and LS.

Image		PG+PURE-LET	PG+GAT+BLS-GSM	MA+TV-L1	The proposed
T1	MSR	54.77	27.92	54.42	**55.36**
	LS	2.55	2.55	2.07	**4.93**
T2	MSR	68.06	67.96	64.41	**70.78**
	LS	2.58	2.41	2.38	**4.07**
T3	MSR	47.29	29.33	44.19	**47.90**
	LS	2.73	2.44	2.20	**4.31**

Table 3. The comparison of MSR and LS of another 5 micro focus X-ray images.

Image		PG+PURE-LET	PG+GAT+BLS-GSM	MA+TV-L1	The proposed
E1	MSR	26.27	14.48	26.35	**26.38**
	LS	2.30	2.21	2.64	**3.75**
E2	MSR	33.21	30.39	34.69	**35.69**
	LS	2.56	2.75	2.82	**4.26**
E3	MSR	19.62	9.60	19.47	**20.60**
	LS	2.12	2.00	2.60	**3.38**
E4	MSR	13.90	8.54	13.92	**13.93**
	LS	2.83	2.28	2.74	**2.99**
E5	MSR	56.16	20.61	56.05	**57.33**
	LS	7.08	4.83	6.90	**7.64**

Table 2 also show that the proposed method can reach a better balance between noise-reducing and edge-preserving.

Besides, to have a better comparison, we test another 5 micro focus X-ray images, and their objective evaluation indexes are displayed in Table 3. It can be seen that the proposed method attains better results for both MSR and LS, which reaches the similar conclusion that the proposed method can better reduce the mixed noise in micro focus X-ray images.

To sum up, our proposed method can improve the denoising quality and achieve better results in both visual effects and objective evaluation.

5   Conclusions

In this paper, we focus on the mixed multiplicative and additive noise in micro focus X-ray images. And an improved denoising method is developed to reduce the mixed noise, in which both regularization operator and regularization parameter are self-adaptive according to the local variance. Specifically, a p-Laplace function with adaptive p is used as regularization operator. And we raise a new scheme for the self-adaptive regularization parameter based on a barrier function, which tends to preserve more weak edges. Experimental results show that the proposed method outperforms the contrast methods in both visual effects and objective evaluation indexes. And the proposed method can achieve a better balance between noise-reducing and detail-preserving, which effectively improve the denoising quality.

Acknowledgements. This work was supported by the National Natural Science Foundation of China under Grant 61403146 and the Fundamental Research Funds for the Central Universities (x2zd-D2155120).

References

1. Zhang, Y.K., Zhang, J.Y., Lu, H.B.: Noise reduction of low-dose CT sinograms based on EM algorithm. Tien Tzu Hsueh Pao/acta Electronica Sinica. **40**(1), 27–34 (2012)
2. Makitalo, M., Foi, A.: On the inversion of the anscombe transformation in low-count poisson image denoising. In: IEEE International Workshop on Local and Non-Local Approximation in Image Processing (LNLA), pp. 26–32 (2011)
3. Harmany, Z.T., Marcia, R.F., Willett, R.M.: SPIRAL out of convexity: sparsity-regularized algorithms for photon-limited imaging. In: SPIE - The International Society for Optical Engineering, pp. 75330R–75330R-12 (2010)
4. Luisier, F., Blu, T., Unser, M.: Image denoising in mixed poisson-gaussian noise. IEEE Trans. Image Process. **20**(3), 696–708 (2011)
5. Chakrabarti, A., Zickler, T.: Image restoration with signal-dependent camera noise. arXiv preprint arXiv:1204.2994 (2012)
6. Gil-Rodrigo, E., Portilla, J., Miraut, D., et al.: Efficient joint poisson-gauss restoration using multi-frame l_2-relaxed-l_0 analysis-based sparsity. In: 18th IEEE International Conference on Image Processing(ICIP), pp. 1385–1388 (2011)
7. Makitalo, M., Foi, A.: Poisson-gaussian denoising using the exact unbiased inverse of the generalized anscombe transformation. In: 37th IEEE International Conference on Acoustics, Speech, & Signal Processing (ICASSP), pp. 1081–1084 (2012)
8. Gao, H.X., Wu, L.X., Xu, H., et al.: Denoising method of micro-focus X-ray images corrupted with mixed multiplicative and additive noises. Opt. Precis. Eng. **22**(11), 3100–3113 (2014)
9. Wu, X., Wang, R., Wang, C.: Regularized image restoration based on adaptively selecting parameter and operator. In: 17th IEEE International Conference on Pattern Recognition (ICPR), vol. 3, pp. 662–665 (2004)
10. Blomgren, P., Chan, T.F., Mulet, P, et al.: Total variation image restoration: numerical methods and extensions. In: IEEE International Conference on Image Processing (ICIP), vol. 3, pp. 384–387 (1997)
11. Bing, S.: Topics in variational PDE image segmentation, inpainting and denoising. Doctoral dissertation. University of California Los Angeles, USA (2003)
12. Zhang, H.Y., Peng, Q.C.: Adaptive image denoising model based on total variation. Opt. Eng. **33**(3), 50–53 (2006)
13. Xu, Y.B., Song, X.Z., Dong, F., et al.: An adaptive total variation regularization method for electrical resistance tomography. In: IEEE International Conference on Imaging Systems and Techniques (IST), pp. 127–131 (2013)
14. Wang, Y., Yang, J., Yin, W., et al.: A new alternating minimization algorithm for total variation image reconstruction. SIAM J. Imaging Sci. **1**(3), 248–272 (2015)

15. Jidesh, P.: A convex regularization model for image restoration. Comput. Electr. Eng. **40**(8), 66–78 (2014)
16. Fan, Z., Gao, R.X.: An adaptive total variation regularization method for electrical capacitance tomography. In: 29th IEEE International Instrumentation and Measurement Technology Conference (I2MTC), pp. 2230–2235 (2012)
17. Yuan, Q., Zhang, L., Shen, H.: Regional spatially adaptive total variation super-resolution with spatial information filtering and clustering. IEEE Trans. Image Process. **22**(6), 2327–2342 (2013)
18. Lagendijk, R.L., Biemond, J., Boekee, D.E.: Regularized iterative image restoration with ring reduction. IEEE Trans. Acoust. Speech Signal Process. **36**(12), 1874–1887 (1988)
19. Katsaggelos, A.K., Biemond, J., Schafer, R.W., et al.: A regularized iterative image restoration algorithm. IEEE Trans. Signal Process. **39**(4), 914–929 (1991)
20. Reeves, S.J.: Optimal space-varying regularization in iterative image restoration. IEEE Trans. Image Process. **3**(3), 319–324 (1994)
21. Rudin, L., Lions, P.L., Osher, S.: Multiplicative denoising and deblurring: theory and algorithms. In: Osher, S., Paragios, N. (eds.) Geometric Level Set Methods in Imaging, Vision, and Graphics 2003, pp. 103–119. Springer, New York (2003)
22. Tikhonov, A.N., Arsenin, V.Y.: Solution of Ill-posed problems. Math. Comput. **32**(144), 491 (1978)
23. Rudin, L.I., Osher, S., Fatemi, E.: Nonlinear total variation based noise removal algorithms. Physica D: Nonlinear Phenom. **60**(1–4), 259–268 (1992)
24. Otsu, N.: A threshold selection method from gray-level histograms. IEEE Trans. Syst. Man Cybern. **9**(1), 62–66 (1979)
25. Boyd, S., Vandenberghe, L.: Convex Optimization. Cambridge University Press, Cambridge (2004)
26. Figueiredo, M.A.T., Nowak, R.D., Wright, S.J.: Gradient projection for sparse reconstruction: application to compressed sensing and other inverse problems. IEEE J. Sel. Top. Signal Process. **1**(4), 586–597 (2007)
27. Starck, J.L., Murtagh, F., Bijaoui, A.: Image processing and data analysis. The multiscale approach. J. Am. Stat. Assoc. **94**(448), 297 (1998)
28. Portilla, J., Strela, V., Wainwright, M.J., et al.: Image denoising using scale mixtures of gaussians in the wavelet domain. IEEE Trans. Image Process. Publ. IEEE Signal Process. Soc. **12**(11), 1338–1351 (2003)
29. Geronimo, J.S., Hardin, D.P., Massopust, P.R.: Fractal functions and wavelet expansions based on several scaling functions. J. Approx. Theor. **78**(3), 373–401 (1994)
30. Tang, X.: Research of regularization method in image restoration. Doctoral dissertation. Huazhong University of Science and Technology, China (2006)

The GEPSVM Classifier Based on L1-Norm Distance Metric

A. He Yan, B. Qiaolin Ye, C. Ying'an Liu$^{(\boxtimes)}$, and D. Tian'an Zhang

School of Information Science and Technology,
Nanjing Forestry University, Nanjing, China
{yanhecom, lyastat}@163.com

Abstract. The proximal support vector machine via generalized eigenvalues (GEPSVM) is an excellent classifier for binary classification problem. However, the distance of GEPSVM from the point to the plane is measured by L2-norm, which emphasizes the role of outliers by the square operation. To optimize this, we propose a robust and effective GEPSVM classifier based on L1-norm distance metric, referred to as L1-GEPSVM. The optimization goal is to minimize the intra-class distance dispersion and maximize the inter-class distance dispersion simultaneously. It is known that the application of L1-norm distance is often considered as a simple and powerful way to reduce the impact of outliers, which improves the generalization ability and flexibility of the model. In addition, we design an effective iterative algorithm to solve the L1-norm optimal problems, which is easy to actualize and its convergence to a logical local optimum is theoretically ensured. Thus, the classification performance of L1-GEPSVM is more robust. Finally, the feasibility and effectiveness of L1-GEPSVM are proved by extensive experimental results on both UCI datasets and artificial datasets.

Keywords: GEPSVM · L1-GEPSVM · L1-norm · L2-norm · Outliers

1 Introduction

Support Vector Machine [1–3] (SVM) plays a very important role in data classification and regression. Its main idea is to find an optimal separating plane by maximizing the margin between two parallel support planes. Thus, as an effective classification tool, SVM has been successfully used in many practical problems.

Nevertheless, there are two main troubles in the original SVM: XOR problem and the complex Quadratic Programming Problems (QPP) [4]. In order to solve these two problems above, Mangasarian and Wild proposed a fast classifier for binary classification problem, termed as proximal SVM via generalized eigenvalues (GEPSVM) [5], which is an extension of proximal SVM (PSVM) [6]. GEPSVM relaxes the requirement of SVM that the planes should be parallel, and tries to find two nonparallel planes by solving a pair of generalized eigenvalue problems instead of a complex QPP, which deals with the XOR problem smoothly and has better generalization ability than SVM. The geometric interpretation of GEPSVM is that each plane is closer to one of the two classes and away from the other class as far as possible [7]. A new data point will be

© Springer Nature Singapore Pte Ltd. 2016
T. Tan et al. (Eds.): CCPR 2016, Part I, CCIS 662, pp. 703–719, 2016.
DOI: 10.1007/978-981-10-3002-4_57

assigned to which class (Class 1 or Class -1) based on its proximity to the two planes. Currently, the research about GEPSVM is still in the ascendant, and many improved methods of GEPSVM have been developed based on the idea of GEPSVM. Ye [8] proposed a new algorithm via singular value decomposition (IGEPSVM) to deal with the singular problem in GEPSVM by solving a simple eigenvalue problem rather than generalized eigenvalues. Besides, Ye [8] also proposed an algorithm of IDGEPSVM to overcome bad performance when influenced by noise data and longer training time problem in GEPSVM. Guarracino et al. [9] reformulated the optimization problems of GEPSVM by using the regularization technique to solve the generalized eigenvalue problem, and proposed a regularized general eigenvalue classifier (ReGEC). Later, Guarracino et al. [10] proposed an algorithm of MultiReGEC, which is a new technique to extend ReGEC to multiclass classification problems, and is based on statistical and geometrical considerations. Shao et al. [11] proposed an improved version of GEPSVM (IGEPSVM) by using the standard eigenvalue decomposition instead of the generalized eigenvalue decomposition, which can solve the singular problem. Marghny et al. [12] reformulated the optimization problems of GEPSVM by using Differential Search Algorithm to find near optimal values of the GEPSVM parameters and its kernel parameters, and proposed an improved version of GEPSVM, called DSA-GEPSVM for short, which overcomes the influence of error or noise in real world. To facilitate the robustness and generalization of nonparallel proximal support vector machine (L1-NPSVM), Li et al. [13] reformulated the optimization problems of PSVM by using L1-norm distance instead of L2-norm distance, and put forward a gradient ascending (GA) iterative algorithm to solve the objective function, which is simple to carry out but may not ensure the optimality of the solution due to both the need of introduction of a non-convex surrogate function and the difficult selection of step-size [14].

However, it should be noted that GEPSVM is sensitive to outliers, because the distance of GEPSVM from the point to the plane is measured by L2-norm, which may exaggerate the effect of outliers by the square operation [15]. To address this, we propose a robust GEPSVM based on L1-norm distance metric for binary classification, termed as L1-GEPSVM. The application of L1-norm distance is often considered as a simple and powerful way to reduce the impact of noises [16, 17]. Thus, it is necessary to establish a robust GEPSVM model using L1-norm distance. L1-GEPSVM is to seek two nonparallel optimal planes by solving a pair of QPPs instead of generalized eigenvalue problems, whose goal is to make each plane closest to the samples of its own class and at the same time furthest from other classes. In summary, our L1-GEPSVM owns the following several compelling properties:

(1) L1-GEPSVM converts the generalized eigenvalue problem into a convex programming problem. We implement a simple and impactful iterative algorithm to solve the L1-norm optimal problems, and its convergence to a reasonable local optimum is theoretically ensured.
(2) The application of L1-norm distance makes L1-GEPSVM more robust to outliers than L2-norm distance, and it can effectively decrease the impact of the outliers even if the ratio of outliers is large.

(3) Extensive experimental results on both UCI datasets and artificial datasets confirm that, compared with GEPSVM, IGEPSVM and TWSVM, L1-GEPSVM increases the generalization ability and flexibility of the model.

(4) Last but not the least, it is worth pointing out that the method which we propose can be easily extended to solve other improved methods of GEPSVM. It is our future work to research these.

The rest of the article is organized as follows: Sect. 2 briefly introduces the GEPSVM. Section 3 proposes L1-GEPSVM with its feasibility and theoretical analysis. All the experiments are shown in Sect. 4 and conclusions are given in Sect. 5.

In this paper, all vectors are column vectors unless transformed to row vectors by a prime superscript T. The vectors $\mathbf{e}_1$ and $\mathbf{e}_2$ of appropriate dimension are represented by an identity column vector. Besides, we denote $\mathbf{I}$ as an identity matrix of appropriate dimension.

2   Related Works

In the n dimensional real space R^n, the set of training sample is indicated by $T = \left\{ \left(\mathbf{x}_j^{(i)}, y_i \right) | i = 1, 2, j = 1, 2, \ldots, m_i \right\}$, where $\mathbf{x}_j^{(i)} \in R^n$ and $y_j \in \{-1, 1\}$, $\mathbf{x}_j^{(i)}$ denotes the i-th class and j-th sample. We suppose that matrix $\mathbf{A} = \left[\mathbf{A}_1^{(1)}, \mathbf{A}_2^{(1)}, \ldots, \mathbf{A}_{m_1}^{(1)} \right]^T$ with size of $m_1 \times n$ represents the data points of Class 1, while matrix $\mathbf{B} = \left[\mathbf{B}_1^{(2)}, \mathbf{B}_2^{(2)}, \ldots, \mathbf{B}_{m_2}^{(2)} \right]^T$ with size of $m_2 \times n$ represents those of Class -1, and matrices $\mathbf{A}$ and $\mathbf{B}$ represent all training data points, where $m_1 + m_2 = m$. In the following, we review a well-known nonparallel proximal classifiers: GEPSVM [6].

The GEPSVM classifier aims to seek two nonparallel proximal optimal planes:

$$x\mathbf{w}_1^T + \mathbf{b}_1 = 0, x\mathbf{w}_2^T + \mathbf{b}_2 = 0 \tag{1}$$

Where $\mathbf{w}_1, \mathbf{w}_2 \in R^n, \mathbf{b}_1, \mathbf{b}_2 \in R$. The aim is to minimize the Euclidean distance of the planes from the data points of Class 1 and Class -1 respectively. This produces the following two objective problems of GEPSVM:

$$\min_{(\mathbf{w}_1, \mathbf{b}_1) \neq 0} \frac{\|\mathbf{A}\mathbf{w}_1 + \mathbf{e}_1\mathbf{b}_1\|_2^2 \Big/ \|(\mathbf{w}_1\mathbf{b}_1)^T\|_2^2}{\|\mathbf{B}\mathbf{w}_1 + \mathbf{e}_2\mathbf{b}_1\|_2^2 \Big/ \|(\mathbf{w}_1\mathbf{b}_1)^T\|_2^2} \tag{2}$$

$$\min_{(\mathbf{w}_2, \mathbf{b}_2) \neq 0} \frac{\|\mathbf{B}\mathbf{w}_2 + \mathbf{e}_2\mathbf{b}_2\|_2^2 \Big/ \|(\mathbf{w}_2\mathbf{b}_2)^T\|_2^2}{\|\mathbf{A}\mathbf{w}_2 + \mathbf{e}_1\mathbf{b}_2\|_2^2 \Big/ \|(\mathbf{w}_2\mathbf{b}_2)^T\|_2^2} \tag{3}$$

Where $|| \cdot ||_2$ denotes the L2-norm. This is implicitly assumed that $(\mathbf{w}_1 \mathbf{b}_1) \neq 0 \Rightarrow \mathbf{Bw}_1 + \mathbf{e}_2 \mathbf{b}_1 \neq 0$ and $(\mathbf{w}_2 \mathbf{b}_2) \neq 0 \Rightarrow \mathbf{Aw}_2 + \mathbf{e}_1 \mathbf{b}_2 \neq 0$. The original problems (2) and (3) can be optimized in the following form:

$$\min_{\mathbf{w}_1, \mathbf{b}_1} \frac{||\mathbf{Aw}_1 + \mathbf{e}_1 \mathbf{b}_1||_2^2}{||\mathbf{Bw}_1 + \mathbf{e}_2 \mathbf{b}_1||_2^2} \tag{4}$$

$$\min_{\mathbf{w}_2, \mathbf{b}_2} \frac{||\mathbf{Bw}_2 + \mathbf{e}_2 \mathbf{b}_2||_2^2}{||\mathbf{Aw}_2 + \mathbf{e}_1 \mathbf{b}_2||_2^2} \tag{5}$$

The positive semi-definite matrix may be involved in the computation when solving the generalized eigenvalue equations, which may cause singularity problem. Therefore, formula (4) and (5) can be regularized by introducing Tikhonov regularization terms, shown as followed:

$$\min_{\mathbf{w}_1, \mathbf{b}_1} \frac{||\mathbf{Aw}_1 + \mathbf{e}_1 \mathbf{b}_1||_2^2 + \delta ||(\mathbf{w}_1 \mathbf{b}_1)^T||_2^2}{||\mathbf{Bw}_1 + \mathbf{e}_2 \mathbf{b}_1||_2^2}, \tag{6}$$

$$\min_{\mathbf{w}_2, \mathbf{b}_2} \frac{||\mathbf{Bw}_2 + \mathbf{e}_2 \mathbf{b}_2||_2^2 + \delta ||(\mathbf{w}_2 \mathbf{b}_2)^T||_2^2}{||\mathbf{Aw}_2 + \mathbf{e}_1 \mathbf{b}_2||_2^2}, \tag{7}$$

Where $\delta || (\mathbf{w}_1 \mathbf{b}_1)^T ||_2^2$ and $\delta || (\mathbf{w}_2 \mathbf{b}_2)^T ||_2^2$ are regularization terms, δ is a regularization factor, the regularization term can improve the stability and classification accuracy of GEPSVM. However, it no longer has the original geometric meaning. Formula (6) and (7) are equivalent to:

$$\min \frac{\mathbf{z}_1^T \mathbf{E} \mathbf{z}_1}{\mathbf{z}_1^T \mathbf{F} \mathbf{z}_1}, \tag{8}$$

$$\min \frac{\mathbf{z}_2^T \mathbf{L} \mathbf{z}_2}{\mathbf{z}_2^T \mathbf{M} \mathbf{z}_2}, \tag{9}$$

Where $\mathbf{H} = [\mathbf{A} \ \mathbf{e}_1]$, $\mathbf{G} = [\mathbf{B} \ \mathbf{e}_2]$, $\mathbf{E} = \mathbf{H}^T \mathbf{H} + \delta \mathbf{I}$, $\mathbf{F} = \mathbf{G}^T \mathbf{G}$, $\mathbf{L} = \mathbf{G}^T \mathbf{G} + \delta \mathbf{I}$, $\mathbf{M} = \mathbf{H}^T \mathbf{H}$, $\mathbf{z}_1 = (\mathbf{w}_1 \mathbf{b}_1)^T$, $\mathbf{z}_2 = (\mathbf{w}_2 \mathbf{b}_2)^T$.

Both $\mathbf{H}$ and $\mathbf{G}$ are symmetric, while formula (8) and (9) are Rayleigh quotient problems. It is easy to obtain the solutions of them by solving the generalized eigenvalue problem.

$$\mathbf{Gz}_1 = \lambda_1 \mathbf{Hz}_1, \mathbf{z}_1 \neq 0 \tag{10}$$

$$\mathbf{Hz}_2 = \lambda_2 \mathbf{Gz}_2, \mathbf{z}_2 \neq 0 \tag{11}$$

The minimum of (8) is achieved at an eigenvector corresponding to the smallest eigenvalue λ_1 of (10). Thus, if $\mathbf{z}_1$ denotes the eigenvector corresponding to λ_1, then

$\mathbf{z}_1 = (\mathbf{w}_1 \mathbf{b}_1)^T$, the first d components are weight w_1 of the first planes, and the last component is the deviation b_1. $\mathbf{z}_1$ determines the plane $x\mathbf{w}_1^T + \mathbf{b}_1 = 0$, which is close to data points of Class 1. And $\mathbf{z}_2 = (\mathbf{w}_2 \mathbf{b}_2)^T$ determines the plane $x\mathbf{w}_2^T + \mathbf{b}_2 = 0$, which is close to those of Class -1.

3   GEPSVM Based on L1-Norm Distance

It can be seen that the distance of GEPSVM is measured by L2-norm. To obtain the minimum value of the objective function, GEPSVM emphasizes the role of outliers remote from the sample by the square operation, which is easy to exaggerate their impact and reduce the classification accuracy. To alleviate this, we propose a GEPSVM classifier based on L1-norm distance metric, termed as L1-GEPSVM. L1-GEPSVM solves a pair of convex programming problems instead of generalized eigenvalue problems. In addition, L1-GEPSVM inherits the advantages of GEPSVM of solving the XOR problem. The two objective problems of L1-GEPSVM are shown as followed:

$$\min_{\mathbf{w}_1,\mathbf{b}_1} \frac{||\mathbf{A}\mathbf{w}_1 + \mathbf{e}_1\mathbf{b}_1||_1 + \delta||(\mathbf{w}_1\mathbf{b}_1)^T||_2^2}{||\mathbf{B}\mathbf{w}_1 + \mathbf{e}_2\mathbf{b}_1||_1}, \tag{12}$$

$$\min_{\mathbf{w}_2,\mathbf{b}_2} \frac{||\mathbf{B}\mathbf{w}_2 + \mathbf{e}_2\mathbf{b}_2||_1 + \delta||(\mathbf{w}_2\mathbf{b}_2)^T||_2^2}{||\mathbf{A}\mathbf{w}_2 + \mathbf{e}_1\mathbf{b}_2||_1}, \tag{13}$$

Where $|| \cdot ||_1$ denotes the L1-norm, $\delta || (\mathbf{w}_1\mathbf{b}_1)^T ||_2^2$ and $\delta || (\mathbf{w}_2\mathbf{b}_2)^T ||_2^2$ are regularization terms, and δ is a regularization factor. The aim of L1-GEPSVM is to make points of the same class as compact as possible while as far as possible from the other class [6], which guarantees the objective function to be minimized.

The original problems can be optimized in the following form:

$$\min_{z_1} \frac{||\mathbf{H}\mathbf{z}_1||_1 + \delta\mathbf{z}_1^T\mathbf{z}_1}{||\mathbf{G}\mathbf{z}_1||_1} \tag{14}$$

$$\min_{z_2} \frac{||\mathbf{G}\mathbf{z}_2||_1 + \delta\mathbf{z}_2^T\mathbf{z}_2}{||\mathbf{H}\mathbf{z}_2||_1} \tag{15}$$

Where $\mathbf{H} = [\mathbf{A} \ \mathbf{e}_1]$, $\mathbf{G} = [\mathbf{B} \ \mathbf{e}_2]$, $\mathbf{z}_1 = (\mathbf{w}_1\mathbf{b}_1)^T$, $\mathbf{z}_2 = (\mathbf{w}_2\mathbf{b}_2)^T$. We can get two nonparallel optimal planes by solving formula (14) and (15):

$$x\mathbf{w}_1^T + \mathbf{b}_1 = 0, x\mathbf{w}_2^T + \mathbf{b}_2 = 0 \tag{16}$$

Next, we solve formula (14), whose objective function is invariant to the order of magnitude of w_1. Then, we can scale the z_1 so that the denominator of formula (14) is equal to 1, that is $||\mathbf{G}\mathbf{z}_1||_1 = 1$. So formula (14) can be rewritten as:

$$\min_{z_1} \|\mathbf{H}\mathbf{z}_1\|_1 + \delta\mathbf{z}_1^T\mathbf{z}_1$$

$$\text{s.t. } \|\mathbf{G}\mathbf{z}_1\|_1 = 1 \tag{17}$$

Where $\mathbf{H} = (\mathbf{h}_1, \mathbf{h}_2, \ldots, \mathbf{h}_{m_1})^T \in \mathbf{R}^{m_1 \times (n+1)}$, $\mathbf{G} = (\mathbf{g}_1, \mathbf{g}_2, \ldots, \mathbf{g}_{m_2})^T \in \mathbf{R}^{m_2 \times (n+1)}$, $\mathbf{h}_i, \mathbf{g}_i \in \mathbf{R}^n (i = 1, 2, \ldots, n)$ denotes the i-th column of matrix $\mathbf{H}$ and $\mathbf{G}$ separately. Thus, formula (17) is equivalent to:

$$\min_{z_1} \sum_{i=1}^{n} \left|\mathbf{h}_i^T\mathbf{z}_1\right| + \delta\mathbf{z}_1^T\mathbf{z}_1$$

$$\text{s.t. } \sum_{i=1}^{n} \left|\mathbf{g}_i^T\mathbf{z}_1\right| = 1 \tag{18}$$

Where $|.|$ is the absolute value operation. According to the relevant knowledge of Mathematics, we can know that,

$$\sum_{i=1}^{n} \left|\mathbf{h}_i^T\mathbf{z}_1\right| = \sum_{i=1}^{n} \left|\mathbf{z}_1^T\mathbf{h}_i\right| = \mathbf{z}_1^T \left(\sum_{i=1}^{n} \frac{\mathbf{h}_i\mathbf{h}_i^T}{|\mathbf{z}_1^T\mathbf{h}_i|}\right)\mathbf{z}_1 \tag{19}$$

$$\sum_{i=1}^{n} \left|\mathbf{g}_i^T\mathbf{z}_1\right| = \sum_{i=1}^{n} \left|\mathbf{z}_1^T\mathbf{g}_i\right| = \sum_{i=1}^{n} sign(\mathbf{z}_1^T\mathbf{g}_i)(\mathbf{z}_1^T\mathbf{g}_i) \tag{20}$$

Where $sign(\cdot)$ is a symbolic function: when the value of the bracket is greater than 0, the value is 1, otherwise -1. In this way, formula (17) can be described by the following equivalence model:

$$\min_{z_1} \mathbf{z}_1^T \left(\sum_{i=1}^{n} \frac{\mathbf{h}_i\mathbf{h}_i^T}{|\mathbf{z}_1^T\mathbf{h}_i|}\right)\mathbf{z}_1 + \delta\mathbf{z}_1^T\mathbf{z}_1$$

$$\text{s.t. } \sum_{i=1}^{n} sign(\mathbf{z}_1^T\mathbf{g}_i)(\mathbf{z}_1^T\mathbf{g}_i) = 1 \tag{21}$$

Formula (21) contains absolute value operation. To solve this, we propose an iterative convex optimization strategy. Specifically, the basic idea of this method is to iteratively update the weight vector $\mathbf{z}_1$ until it converges to a local optimal solution. Assuming that $\mathbf{z}_1^{(p)}$ is the optimal solution for the iteration of p. Then, the optimal solution of $\mathbf{z}_1^{(p+1)}$ for the iteration of $p+1$ is defined as the solution to the following problems:

$$\min_{z_1} \mathbf{z}_1^T \left(\sum_{i=1}^n \frac{\mathbf{h}_i \mathbf{h}_i^T}{\left|\mathbf{z}_1^{(p)^T} \mathbf{h}_i\right|} \right) \mathbf{z}_1 + \delta \mathbf{z}_1^T \mathbf{z}_1 \tag{22}$$

$$\text{s.t.} \quad \sum_{i=1}^n sign(\mathbf{z}_1^{(p)^T} \mathbf{g}_i)(\mathbf{z}_1^T \mathbf{g}_i) = 1$$

Where $\left|\mathbf{z}_1^{(p)^T} \mathbf{h}_i\right| \neq 0$, it is easy to prove that $sign(\mathbf{z}_1^{(p)^T} \mathbf{g}_i)(\mathbf{z}_1^T \mathbf{g}_i)$ is a first order Taylor expansion of $\left|\mathbf{g}_i^T \mathbf{z}_1\right|$ at the point $\mathbf{z}_1^{(p)}$. L1-norm often requires some elements in $\mathbf{z}_1^{(p)}$ to approach zero, instead of being exactly zero, so the situation of $\left|\mathbf{z}_1^{(p)^T} \mathbf{t}_i\right| = 0$ does not appear in most cases. If $\left|\mathbf{z}_1^{(p)^T} \mathbf{t}_i\right| = 0$, our strategy is to set $\mathbf{z}_1^{(p)} + \Delta \rightarrow \mathbf{z}_1^{(p)}$, where Δ is a very small number. So formula (22) are rewritten as:

$$\min_{z_1} \mathbf{z}_1^T \left(\mathbf{\Gamma}^{(p)} + \delta \mathbf{I} \right) \mathbf{z}_1 \tag{23}$$

$$\text{s.t.} \quad \mathbf{s}^{(p)} \mathbf{G} \mathbf{z}_1 = 1$$

$$\mathbf{\Gamma}^{(p)} = \mathbf{H}^T diag \left(1 / \left(\left|\mathbf{z}_1^{(p)^T} \mathbf{h}_1\right| \right), 1 / \left(\left|\mathbf{z}_1^{(p)^T} \mathbf{h}_2\right| \right), \ldots, 1 / \left(\left|\mathbf{z}_1^{(p)^T} \mathbf{h}_{m_1}\right| \right) \right) \mathbf{H}, \ \mathbf{s}^{(p)}$$
$$= sign(\mathbf{z}_1^{(p)^T} \mathbf{G}^T)$$

The algorithm we design is described in Algorithm 1. In each iteration, $\mathbf{z}_1^{(P+1)}$ is calculated with current $\mathbf{z}_1^p$, and then $\mathbf{z}_1^p$ is updated based on the current result $\mathbf{z}_1^{(P+1)}$. The iteration procedure is repeated until the algorithm converges.

Algorithm 1: a simple iterative algorithm to solve the problem in formula (12).

Data: $\mathbf{H} \in R^{m_1 \times (n+1)}, \mathbf{G} \in R^{m_2 \times (n+1)}$.

Result: $\mathbf{z}^{(p+1)} \in R^{(n+1) \times 1}$.

Set $t = 0$, Initialize $\mathbf{z}_1^p \in R^{(n+1) \times 1}$, set $\mathbf{z}_1^p$ equal to $\mathbf{z}$ which is a standard solution of GEPSVM.

Repeat:

Calculate $\mathbf{z}_1^{(p+1)} = \dfrac{\left(\mathbf{\Gamma}^{(p)} + \delta \mathbf{I} \right)^{-1} \mathbf{G}^T \mathbf{s}^{(p)^T}}{\mathbf{s}^{(p)} \mathbf{G} \left(\mathbf{\Gamma}^{(p)} + \delta \mathbf{I} \right)^{-1} \mathbf{G}^T \mathbf{s}^{(p)^T}}$, $t = t + 1$.

If $\left\| z_1^p - z_1^{(p+1)} \right\|_{l_1} \leq 0.001$ or $t = 50$, stop. Otherwise, go on.

Until Converges

Algorithm 1 makes the original problem in formula (12) monotonically decrease in each iteration. The proof procedure is shown as following:

Definition. Define any two vectors $\mathbf{a}$ and $\mathbf{p}$, assume the following inequality is established:

$$Q(\mathbf{a}) = \frac{\mathbf{a}^T \mathbf{S}_3 \mathbf{a}}{\mathbf{a}^T (\mathbf{S}_1 + \beta \mathbf{I})\mathbf{a} + \lambda \|\mathbf{a}\|_1}$$

$$\geq \frac{\mathbf{p}^T \mathbf{S}_3 \mathbf{p}}{\mathbf{p}^T (\mathbf{S}_1 + \beta \mathbf{I})\mathbf{p} + \lambda \|\mathbf{p}\|_1} = Q(\mathbf{p}) \tag{24}$$

Then, we suppose that $\mathbf{a}$ is a better solution than $\mathbf{p}$, so we can make the conclusion that $\mathbf{z}_1^{(p+1)}$ is a better solution than $\mathbf{z}_1^{(p)}$ in the iterative algorithm. This conclusion is given by Theorem 1. To prove it, we first introduce the following Lemma 1.

Lemma 1. For any vector $c = [c_1, c_2, \ldots, c_n]^T \in \mathbb{R}^n$, the following equality is established:

$$\|c\|_1 = \min_{v \in \mathbb{R}^n_+} \frac{1}{2} \sum_{i=1}^{n} \frac{c_i^2}{v_i} + \frac{1}{2} \|v\|_1 \tag{25}$$

The minimum is uniquely arrived at $v_i = |c_i|$ for $i = 1, 2, \ldots, n$, where $v = [v_1, v_2, \ldots, v_n]^T$.

Theorem 1. Assume that $\mathbf{z}_1^{(p)}$ is a vector, which makes the equation $sign\left(z_1^{(p)^T} G^T\right) G z_1 = 1$. By formula (23) we can obtain the solution $\mathbf{z}_1^{(p+1)}$, which is better than $\mathbf{z}_1^{(p)}$. The proof procedure of Theorem 1 is as follows.

From the definition of $\mathbf{z}_1^{(p+1)}$ in formula (23), we have that

$$\sum_{j=1}^{m_2} s_j^{(p)} g_j z_1^{(p+1)} = 1 \tag{26}$$

Let

$$J(z_1) = \frac{1}{2} z_1^T \left(\mathbf{\Gamma}^{(p)} + \delta \mathbf{I}\right) z_1 + \frac{1}{2} \left\|z_1^{(p)^T} H\right\|_1 \tag{27}$$

Then, from the physical meaning of $\mathbf{z}_1^{(p+1)}$, we have that

$$J\left(z_1^{(p+1)}\right) = \frac{1}{2} \sum_{i=1}^{m_1} \frac{\left[h_i z_1^{(p+1)^T}\right]^2}{\left|h_i z_1^{(p)^T}\right|} + \frac{1}{2} \left\|z_1^{(p)^T} H\right\|_1 \leq J\left(z_1^{(p)}\right) = \sum_{i=1}^{m_1} \left|h_i z_1^{(p)^T}\right| \tag{28}$$

In addition, from Lemma 1, we have that

$$J\left(z_1^{(p+1)}\right) = \frac{1}{2}\sum_{i=1}^{m_1}\frac{\left[h_i z_1^{(p+1)^T}\right]^2}{\left|h_i z_1^{(p)^T}\right|} + \frac{1}{2}\left\|z_1^{(p)^T}H\right\|_1$$

$$\geq \frac{1}{2}\sum_{i=1}^{m_1}\frac{\left[h_i z_1^{(p+1)^T}\right]^2}{\left|h_i z_1^{(p+1)^T}\right|} + \frac{1}{2}\left\|z_1^{(p+1)^T}H\right\|_1 = \sum_{i=1}^{m_1}\left|h_i z_1^{(p+1)^T}\right| \tag{29}$$

Combining Eqs. (28) and (29), we can obtain

$$\sum_{i=1}^{m_1}\left|h_i z_1^{(p)^T}\right| \geq \sum_{i=1}^{m_1}\left|h_i z_1^{(p+1)^T}\right| \tag{30}$$

Combining Eqs. (26) and (30), we can get

$$Q\left(z_1^{(p+1)}\right) = \frac{\sum_{j=1}^{m_2}\left|g_j z_1^{(p+1)}\right|}{\sum_{i=1}^{m_1}\left|h_i z_1^{(p+1)^T}\right|} \geq \frac{\sum_{j=1}^{m_2}\mathbf{s}_j^{(p)}g_j z_1^{(p+1)}}{\sum_{i=1}^{m_1}\left|h_i z_1^{(p+1)^T}\right|} = \frac{1}{\sum_{i=1}^{m_1}\left|h_i z_1^{(p+1)^T}\right|} \geq \frac{1}{\sum_{i=1}^{m_1}\left|h_i z_1^{(p)^T}\right|} \tag{31}$$

From the equality $\sum_{j=1}^{m_2}\left|g_j \mathbf{z}_1^{(p)}\right| = 1$, we can obtain

$$Q\left(z_1^{(p)}\right) = \frac{\sum_{j=1}^{m_2}\left|g_j z_1^{(p)}\right|}{\sum_{i=1}^{m_1}\left|h_i z_1^{(p)^T}\right|} = \frac{1}{\sum_{i=1}^{m_1}\left|h_i z_1^{(p)^T}\right|} \tag{32}$$

Combining Eqs. (31) and (32), we have that

$$Q\left(z_1^{(p+1)}\right) \geq Q\left(z_1^{(p)}\right) \tag{33}$$

So, $\mathbf{z}_1^{(p+1)}$ is better than $\mathbf{z}_1^{(p)}$.

Formula (23) is a convex optimization problem with equality constraints, and it has a close-form solution. Now, we can set up the Lagrange function of formula (23) to solve this objective problem, as shown in the following:

$$L(\mathbf{z}_1, \kappa) = \mathbf{z}_1^T\left(\mathbf{\Gamma}^{(p)} + \delta\mathbf{I}\right)\mathbf{z}_1 - \kappa\left(\mathbf{s}^{(p)}\mathbf{Gz}_1 - 1\right) \tag{34}$$

Where κ is a Lagrange multiplier. Taking the derivative of $L(z_1, k)$ w.r.t $\mathbf{z}_1$, and setting the derivative to zero, we can easily get the following equation:

$$L(\mathbf{z}_1, \kappa) = \left(\mathbf{\Gamma}^{(p)} + \delta\mathbf{I}\right)\mathbf{z}_1 - \kappa\mathbf{G}^T\mathbf{s}^{(p)^T} = 0 \tag{35}$$

The solution of $\mathbf{z}_1^{(p+1)}$ can be obtained by Eq. (35).

$$\mathbf{z}_1^{(p+1)} = \kappa\left(\mathbf{\Gamma}^{(p)} + \delta\mathbf{I}\right)^{-1}\mathbf{G}^T\mathbf{s}^{(p)^T} \tag{36}$$

Bring Eq. (36) into $\mathbf{s}^{(p)}\mathbf{G}\mathbf{z}_1 = 1$, we can get an expression about κ, shown as following:

$$\kappa\mathbf{s}^{(p)}\mathbf{G}\left(\mathbf{\Gamma}^{(p)} + \delta\mathbf{I}\right)^{-1}\mathbf{G}^T\mathbf{s}^{(p)^T} = 1 \Rightarrow \kappa = 1 \Big/ \left(\mathbf{s}^{(p)}\mathbf{G}\left(\mathbf{\Gamma}^{(p)} + \delta\mathbf{I}\right)^{-1}\mathbf{G}^T\mathbf{s}^{(p)^T}\right) \tag{37}$$

Combining Eqs. (36) and (37), we can obtain

$$\mathbf{z}_1^{(p+1)} = \frac{\left(\mathbf{\Gamma}^{(p)} + \delta\mathbf{I}\right)^{-1}\mathbf{G}^T\mathbf{s}^{(p)^T}}{\mathbf{s}^{(p)}\mathbf{G}\left(\mathbf{\Gamma}^{(p)} + \delta\mathbf{I}\right)^{-1}\mathbf{G}^T\mathbf{s}^{(p)^T}} \tag{38}$$

Increase p until $\mathbf{z}_1^{(p+1)}$ converges to a fixed value. Since the problem in formula (23) is a convex problem, then $\mathbf{z}_1^{(p+1)}$ is a sound optimal solution that we seek. Further, weight vectors $\mathbf{w}_1$ and deviations $\mathbf{b}_1$ can be obtained, that is, $\mathbf{z}_1^{(p+1)} = (\mathbf{w}_1\mathbf{b}_1)^T$. Using the same optimization strategy, $\mathbf{z}_2^{(p+1)} = (\mathbf{w}_2\mathbf{b}_2)^T$.

From the above we can know that GEPSVM needs to solve the generalized eigenvalue problem, however, the matrices $\mathbf{F}$ and $\mathbf{M}$ of formula (8) and (9) can only guarantee positive semi-definite, so we may get an inaccurate or unstable solution. As we have seen, Eq. (38) contains inverse operation, which does not suffer from the singular problem, that is because $\mathbf{\Gamma}^{(p)} + \delta\mathbf{I}$ is positive. According to its decision function $f(x) = \arg\min\limits_{1,2}\left(\left|\mathbf{w}_{1,2}^T x + \mathbf{b}_{1,2}\right| \Big/ \|\mathbf{w}_{1,2}\|\right)$, the distance of a new data point to the two nonparallel optimal planes determines which class (Class 1 or Class -1) it belongs to. Next, the effectiveness and classification accuracy of L1-GEPSVM are demonstrated by the experimental results on the artificial dataset and UCI dataset [18, 19].

4   Experimental Results

To verify the classification performance of five algorithms (GEPSVM, IGEPSVM, TWSVM, L1-NPSVM and L1-GEPSVM), compared on the artificial dataset and UCI dataset, which reflect the performance of the algorithm [20, 21]. Experimental environment: Windows 10 operating system, PC with an Intel(R) Core(TM) i5-5200u, quad core processor (2.2 GHz), 4 GB of RAM, and five kinds of classification algorithms are implemented in MATLAB 7.1. The experimental data contains only two classifications, experimental parameters are obtained by cross validation method [22, 23] (10-fold), and the classification accuracy is the average value of test results for 10 times.

To illustrate that L1-GEPSVM has good robustness to outliers, extensive experiments have been done. Moreover, we reduced the difference between the characteristics of different samples, normalized all the sample data in the interval $[-1, 1]$. The classification accuracy and training time are given in the following table, the black bold marked is the best classification accuracy, and * represents that L1-GEPSVM is the best. As is known, parameters have a certain influence on the classification accuracy, to obtain the best generalization performance, all the parameters are obtained from the following. The parameters c_1 and c_2 are in the range of $\{2^i | i = -12, -11, -10, \ldots, 12\}$., while parameter δ is in the range of $\{10^i | i = -10, -9, \ldots, 10\}$.

4.1 Experiments on Artificial Datasets

To prove the effectiveness of L1-GEPSVM to deal with outliers, we consider a two-dimensional XOR dataset (called Crossplans1 (230×2)), on Crossplans1 we introduce two extra outliers. We believe that the more outliers we introduce, the more obvious influence on the classification performance. Here, we do the experiments to verify our ideas. The data distribution of Crossplans1 with outliers as shown in Fig. 1, and the classification results of GEPSVM and L1_GEPSVM on this polluted XOR dataset are given in Fig. 2 respectively. Figure 2 illustrates the ability of L1-GEPSVM and GEPSVM to deal with outliers, which indicates L1-GEPSVM is more robust.

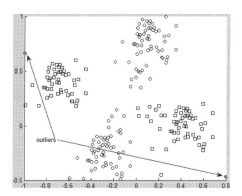

Fig. 1. Data distribution of Crossplans1 with outliers value in the iterative process

The classification accuracy of GEPSVM and L1-GEPSVM are 65.00 % and 84.54 % respectively, which reveals that the classification ability of L1-GEPSVM is better after introducing outliers, and effectively explains that classifiers based on L2-norm distance are sensitive to outliers. However, the L1-norm distance can powerful suppress the influence of outliers, the experiment proves our idea. The reason is that the distance of GEPSVM is measured by L2-norm, if the sample data has outliers,

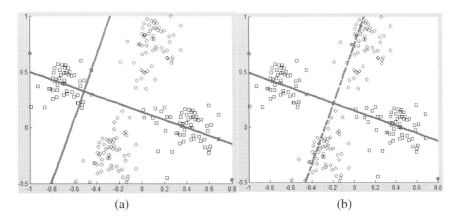

(a)   (b)

Fig. 2. Two classification planes on Crossplans1 (a) by GEPSVM (b) by L1-GEPSVM. Annotation: red line is the optimal plane of ○ sample, while blue line is the optimal plane of □ sample. (Color figure online)

GEPSVM will result in the biased results. However, the distance of L1-GEPSVM is measured by L1-norm, which is more robust to outliers than L2-norm distance [13, 16, 17, 24, 25].

4.2   Experiments on UCI Datasets

We design an algorithm which monotonically decreases in each iteration, and iteratively update the weight vector z_1 until it converges to a fixed value, which is a reasonable optimal solution that we seek. The convergence process is shown in Fig. 3, horizontal axis represents the number of iterations, and vertical axis represents the weight value.

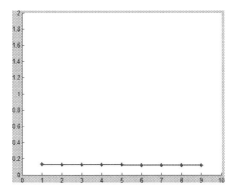

Fig. 3.   The curve graph of the weight

We choose twelve commonly used data from the UCI dataset to test the effectiveness and robustness of L1-GEPSVM. Table 1 is the comparison about classification accuracy of five algorithms, while Table 2 shows the comparison about the five algorithms on the twelve commonly used data where 5 % Gaussian noise was introduced respectively. Table 3 is the comparison about the classification accuracy of the five algorithms on the data where 10 % Gaussian noise was introduced respectively.

From Table 1 we find that the classification accuracy of L1-GEPSVM is much higher than other four algorithms (only in Monks2, Wpbc, Cancer, Ionodata, Clevedata and Housingdata datasets, the classification accuracy of L1-GEPSVM is slightly lower). However, the training time is higher than them, because the iterative algorithm need to iteratively calculate the optimum solution of $\mathbf{z}_1^{(p+1)}$, that is why L1-GEPSVM demands more computation time. The results are the same as we expected that the classification performance of L1-GEPSVM is the best.

Table 1. Test results of GEPSVM, IGEPSVM, TWSVM, L1-NPSVM and L1-GEPSVM

Dataset	GEPSVM	IGEPSVM	TWSVM	L1-NPSVM	L1-GEPSVM
$(N \times n)$	Accuracy (%) Time (s)	Accuracy (%) Time (s)	Accuracy (%) Time (s)	Accuracy (%) Time (s)	Accuracy (%) Time (s)
Monks1 (561 × 6)	78.42 ± 3.2764 0.3062	74.15 ± 6.2449 0.2526	71.65 ± 5.4678 0.1641	79.33 ± 2.9492 0.4701	**82.00 ± 3.3390**[*] 1.4682
Monks2 (601 × 6)	**68.72 ± 2.1208** 0.2178	64.56 ± 4.8576 0.2291	62.73 ± 4.0855 0.0299	68.39 ± 2.9216 0.3773	68.22 ± 3.8875 1.1470
Monks3 (554 × 6)	80.14 ± 3.7623 0.2464	82.85 ± 3.9813 0.2602	80.86 ± 3.0458 0.1285	84.28 ± 5.4189 0.3789	**86.64 ± 3.6113**[*] 3.4061
Wpbc (194 × 33)	72.74 ± 12.5525 3.6359	**75.89 ± 9.9627** 0.3214	75.29 ± 6.7128 0.0070	75.74 ± 7.0030 0.4727	73.74 ± 11.6601 10.9200
Cancer (683 × 9)	**95.76 ± 2.6617** 0.3211	95.17 ± 2.7104 0.3394	92.54 ± 0.9527 0.0787	86.23 ± 10.2011 0.6949	95.17 ± 3.6013 3.9370
Ticdata (958 × 9)	65.87 ± 3.1758 1.3943	66.69 ± 4.7198 1.4278	67.54 ± 2.9682 0.0245	66.81 ± 2.8124 1.5159	**68.90 ± 3.4579**[*] 11.1298
Pidd (768 × 8)	73.83 ± 5.7464 0.6419	69.67 ± 3.5404 0.6303	75.39 ± 2.3068 0.0393	74.87 ± 5.0051 0.7953	**75.65 ± 5.8309**[*] 2.5100
ClaveVectors (963 × 19)	73.41 ± 3.0727 1.6742	70.72 ± 3.9230 1.6084	73.52 ± 2.1024 0.4164	73.31 ± 3.3311 1.8152	**74.35 ± 3.4548**[*] 17.1704
Sonar (208 × 60)	74.12 ± 8.1330 1.0288	73.12 ± 8.7744 0.8470	68.70 ± 5.5519 0.0114	74.14 ± 9.9677 1.3207	**76.55 ± 11.660**[*] 1.0920
Pimadata (768 × 8)	73.83 ± 5.7464 1.0951	69.67 ± 3.5404 0.6294	74.39 ± 2.3068 0.0383	73.70 ± 4.0178 0.8215	**74.74 ± 4.9625**[*] 2.8318
Ionodata (351 × 34)	78.07 ± 5.5333 1.6083	78.33 ± 8.7197 0.5270	**85.75 ± 5.6590** 0.0132	80.06 ± 6.3632 1.4292	84.89 ± 5.1429 5.0568
Clevedata (297 × 13)	85.89 ± 4.1119 0.1794	**86.56 ± 4.1556** 0.1976	83.85 ± 3.7094 0.0210	84.53 ± 5.1730 0.4316	84.16 ± 5.2295 2.3229

We know that noise is one of the standards to measure the robustness of the algorithm. The classification accuracy changes smoothly with the increase of the noise, which indicates the algorithm has good robustness and anti-noise ability. Next, we prove this through experiments. To begin with, we introduce 5 % Gaussian noise on the twelve experimental datasets respectively, the results of five algorithms are as follows:

From the data in Table 2 we see clearly that the classification accuracy of L1-GEPSVM is much higher than other four algorithms after introducing 5 % Gaussian noise. Comparing Table 1 with Table 2, we find that after the introduction of Gaussian noise, the classification accuracy of L1-GEPSVM and L1-NPSVM have fewer changes. That is because the distance of them is measured by L1-norm, which reduces the effect of noise on the classification performance. The results effectively testify our viewpoint. To further prove our idea, we introduce 10 % Gaussian noise on the datasets respectively, the results are as follows:

Table 2. Test results of five algorithms when 5 % Gaussian noise is introduced

Dataset ($N \times n$)	GEPSVM Accuracy (%) Time (s)	IGEPSVM Accuracy (%) Time (s)	TWSVM Accuracy (%) Time (s)	L1-NPSVM Accuracy (%) Time (s)	L1-GEPSVM Accuracy (%) Time (s)
Monks1 (561 × 6)	77.89 ± 3.3395 0.2935	73.61 ± 6.3089 0.2359	70.76 ± 4.2628 0.1613	79.51 ± 3.0580 0.4825	**82.00 ± 3.3390*** 1.4763
Monks2 (601 × 6)	68.57 ± 4.1842 0.2074	66.23 ± 4.4156 0.2227	62.06 ± 3.4920 0.0341	68.74 ± 7.4466 0.3487	**69.56 ± 6.5125*** 1.4038
Monks3 (554 × 6)	79.43 ± 4.5188 0.2954	82.67 ± 4.3019 0.2309	83.93 ± 2.7805 0.1380	83.92 ± 5.5612 0.3507	**87.34 ± 4.4941*** 3.4682
Wpbc (194 × 33)	74.74 ± 7.4924 0.3351	73.82 ± 10.0994 0.2787	74.78 ± 6.0456 0.0059	75.92 ± 10.6852 0.4579	**79.97 ± 8.9260*** 1.9077
Cancer (683 × 9)	**95.32 ± 3.1385** 0.3773	**95.32 ± 2.6935** 0.3144	92.39 ± 0.7271 0.0724	84.49 ± 9.7413 0.7077	**95.32 ± 3.2698*** 4.0261
Ticdata (958 × 9)	64.41 ± 3.4015 1.3513	66.37 ± 4.6023 1.3663	65.35 ± 3.7157 0.1491	67.12 ± 1.9471 1.4799	**68.69 ± 3.5225*** 11.2433
Pidd (768 × 8)	**75.00 ± 4.7898** 1.0760	67.18 ± 6.5636 0.5992	74.74 ± 1.9113 0.0660	74.48 ± 4.2571 0.7487	73.83 ± 5.4931 4.4335
ClaveVectors (963 × 19)	73.31 ± 3.3731 1.5313	70.82 ± 3.7983 1.4882	72.69 ± 1.5601 0.2715	73.63 ± 3.1242 2.0841	**74.66 ± 3.1691*** 19.3030
Sonar (208 × 60)	73.17 ± 7.5338 0.9698	73.57 ± 8.0115 0.7940	71.59 ± 4.4405 0.0113	72.17 ± 8.1054 1.3278	**75.12 ± 8.8225*** 2.8268
Pimadata (768 × 8)	74.48 ± 4.1763 0.5961	66.81 ± 4.5995 0.5971	**76.55 ± 3.1741** 0.0335	75.00 ± 4.0728 0.7921	72.53 ± 3.9032 4.9072
Ionodata (351 × 34)	77.78 ± 3.9955 0.6941	76.93 ± 7.0280 0.5067	85.18 ± 4.5878 0.0140	80.90 ± 8.8797 1.2372	**85.46 ± 5.2025*** 4.3907
Clevedata (297 × 13)	85.90 ± 5.4790 0.2166	**86.56 ± 4.1556** 0.1888	84.19 ± 3.6981 0.0205	85.21 ± 5.3670 0.3922	84.83 ± 5.4995 2.0966

Table 3. Introduce 10 % Gaussian noise, test results of five algorithms

Dataset ($N \times n$)	GEPSVM Accuracy (%) Time (s)	IGEPSVM Accuracy (%) Time (s)	TWSVM Accuracy (%) Time (s)	L1-NPSVM Accuracy (%) Time (s)	L1-GEPSVM Accuracy (%) Time (s)
Monks1 (561 × 6)	78.07 ± 2.9143 0.2798	75.40 ± 6.6230 0.2341	70.22 ± 4.5290 0.0780	78.95 ± 6.6978 0.4537	**82.17 ± 2.7858*** 1.4950
Monks2 (601 × 6)	67.73 ± 3.9638 0.2018	64.55 ± 4.7449 0.2082	62.39 ± 3.4785 0.0182	**68.24 ± 6.9758** 0.3423	68.05 ± 3.0355 1.4580
Monks3 (554 × 6)	77.61 ± 4.5604 0.2276	81.42 ± 3.9631 0.2325	80.14 ± 2.3286 0.0463	84.31 ± 4.2190 0.3540	**87.18 ± 3.4687*** 3.5403
Wpbc (194 × 33)	74.16 ± 8.5101 0.3238	77.00 ± 9.8557 0.2851	76.33 ± 5.8256 0.0069	74.87 ± 12.2888 0.4520	**79.00 ± 11.7634*** 1.9145
Cancer (683 × 9)	**95.46 ± 2.9649** 0.3056	95.31 ± 2.4456 0.3171	92.10 ± 0.6979 0.0723	85.94 ± 8.8914 0.6776	**95.46 ± 3.2510*** 4.0085
Ticdata (958 × 9)	65.14 ± 2.9734 1.3438	66.90 ± 4.5647 1.3501	64.83 ± 2.2365 0.1390	66.60 ± 2.8931 1.4517	**68.79 ± 3.6503*** 10.8815
Pidd (768 × 8)	74.48 ± 4.5257 0.5849	66.28 ± 4.1949 0.6094	73.57 ± 1.5978 0.0682	**75.39 ± 3.7020** 0.7366	74.09 ± 6.0323 5.1581
ClaveVectors (963 × 19)	73.51 ± 2.9302 1.5131	70.61 ± 3.7631 1.4792	73.73 ± 1.8125 0.3496	73.73 ± 2.7286 1.9368	**74.76 ± 3.2035*** 18.6983
Sonar (208 × 60)	74.12 ± 7.8492 0.9384	75.57 ± 9.5096 0.7939	68.75 ± 4.8806 0.0115	74.64 ± 10.5609 1.2851	**76.52 ± 7.0632*** 2.8723
Pimadata (768 × 8)	75.13 ± 4.4184 0.5866	66.54 ± 3.8219 0.5993	74.86 ± 4.6422 0.0303	**75.39 ± 3.7020** 0.7418	73.18 ± 5.1963 5.2519
Ionodata (351 × 34)	79.21 ± 6.3443 0.6749	77.51 ± 5.5551 0.5038	86.32 ± 4.1192 0.0129	81.17 ± 8.2317 1.1045	**86.60 ± 6.7773*** 4.3480
Clevedata (297 × 13)	85.54 ± 4.2138 0.1698	**85.90 ± 4.5969** 0.1853	75.76 ± 7.7899 0.0068	85.53 ± 5.1674 0.3948	84.51 ± 5.2542 2.0584

From the data we see that the classification accuracy of L1-GEPSVM is also higher than other four algorithms after introducing 10 % Gaussian noise. By comparing the data above, we find that the classification accuracy of GEPSVM, IGEPSVM and TWSVM decrease sharply with the increase of noise. However, L1-GEPSVM and L1-NPSVM have a little change. The results indicate that other three algorithms are more susceptible to noise than L1-GEPSVM and L1-NPSVM, and the classification performance of L1-GEPSVM is more robust, especially when Gaussian noise is introduced.

5   Conclusions

A robust L1-norm distance based on GEPSVM for binary classification is proposed in this paper, named L1-GEPSVM for short. The application of L1-norm distance makes L1-GEPSVM improve the generalization ability and flexibility of the model. Also, we design a simple and powerful iterative algorithm to solve the L1-norm optimal problems of L1-GEPSVM, which is easy to implement and its convergence to a local optimum is theoretically ensured. Thus we can obtain the nonparallel optimal planes. In general, the classification performance of L1-GEPSVM is more robust, especially when Gaussian noise is introduced. Finally, the effectiveness and robustness of L1-GEPSVM is proved through extensive experiments.

At present, L1-GEPSVM is only effective for the binary classification problem, and our future work is to extend the multiclass L1-GEPSVM and its application.

Acknowledgement. This work was supported in part by the National Foundation for Distinguished Young Scientists under Grant 31125008, the Scientific Research Foundation for Advanced Talents and Returned Overseas Scholars of Nanjing Forestry University under Grant 163070679, and the National Science Foundation of China under Grants 61101197, 61272220, and 61401214.

References

1. Vapnik, V.N.: Statistical learning theory. Encycl. Sci. Learn. **41**, 3185 (2010)
2. Burges, C.J.C.: A tutorial on support vector machines for pattern recognition. Data Min. Knowl. Discov. **2**, 121–167 (1998)
3. Deng, N., Tian, Y., Zhang, C.: Support Vector Machines. Optimization Based Theory, Algorithms, and Extensions. CRC Press, Boca Raton (2012)
4. Lin, Y.H., Chen, C.H.: Template matching using the parametric template vector with translation, rotation and scale invariance. Pattern Recogn. **41**, 2413–2421 (2008)
5. Mangasarian, O.L., Wild, E.W.: Multisurface proximal support vector machine classification via generalized eigenvalues. IEEE Trans. Pattern Anal. Mach. Intell. **28**, 69–74 (2006)
6. Fung, G.M., Mangasarian, O.L.: Proximal support vector machine classifiers. Mach. Learn. **59**, 77–97 (2005)
7. Jayadeva, Khemchandani, R., Chandra, S.: Twin support vector machines for pattern classification. IEEE Trans. Pattern Anal. Mach. Intell. **29**, 905–910 (2007)
8. Ye, Q., Ye, N.: International Joint Conference on Computational Sciences and Optimization, CSO 2009, Sanya, Hainan, China, 24–26 April, pp. 705–709 (2009)
9. Guarracino, M.R., Cifarelli, C., Seref, O., Pardalos, P.M.: A classification method based on generalized eigenvalue problems. Optim. Methods Softw. **22**, 73–81 (2007)
10. Guarracino, M.R., Irpino, A., Verde, R.: International Conference on Complex, Intelligent and Software Intensive Systems, pp. 1183–1185
11. Shao, Y.H., Deng, N.Y., Chen, W.J., Wang, Z.: Improved generalized eigenvalue proximal support vector machine. IEEE Signal Process. Lett. **20**, 213–216 (2013)
12. Marghny, M.H., Elaziz, R.M.A., Taloba, A.I.: Differential search algorithm-based parametric optimization of fuzzy generalized eigenvalue proximal support vector machine. Int. J. Comput. Appl. **108**, 38–46 (2015)
13. Li, C.N., Shao, Y.H., Deng, N.Y.: Robust L1-norm non-parallel proximal support vector machine. Optimization **65**, 1–15 (2015)
14. Kwak, N.: Principal component analysis by Lp-norm maximization. IEEE Trans. Cybern. **44**, 594–609 (2014)
15. Kwak, N.: Principal component analysis based on L1-norm maximization. IEEE Trans. Pattern Anal. Mach. Intell. **30**, 1672–1680 (2008)

16. Wang, H., Lu, X., Hu, Z., Zheng, W.: Fisher discriminant analysis with L1-norm. IEEE Trans. Cybern. **44**, 828–842 (2014)
17. Li, C.N., Shao, Y.H., Deng, N.Y.: Robust L1-norm two-dimensional linear discriminant analysis. Neural Netw. Off. J. Int. Neural Netw. Soc. **65C**, 92–104 (2015)
18. Chen, X., Yang, J., Ye, Q., Liang, J.: Recursive projection twin support vector machine via within-class variance minimization. Pattern Recogn. **44**, 2643–2655 (2011)
19. Bache, K., Lichman, M.: UCI Machine Learning Repository (2013)
20. Yang, X., Chen, S., Chen, B., Pan, Z.: Proximal support vector machine using local information. Neurocomputing **73**, 357–365 (2009)
21. Xue, H., Chen, S.: Glocalization pursuit support vector machine. Neural Comput. Appl. **20**, 1043–1053 (2011)
22. Ye, Q., Zhao, C., Gao, S., Zheng, H.: Weighted twin support vector machines with local information and its application. Neural Netw. Off. J. Int. Neural Netw. Soc. **35**, 31–39 (2012)
23. Ding, S., Hua, X., Yu, J.: An overview on nonparallel hyperplane support vector machine algorithms. Neural Comput. Appl. **25**, 975–982 (2013)
24. Lin, G., Tang, N., Wang, H.: Locally principal component analysis based on L1-norm maximisation. IET Image Process. **9**, 91–96 (2015)
25. Zhong, F., Zhang, J.: Linear discriminant analysis based on L1-norm maximization. IEEE Trans. Image Process. Publ. IEEE Signal Process. Soc. **22**, 3018–3027 (2013)

Kernel Learning with Hilbert-Schmidt Independence Criterion

Tinghua Wang$^{(\boxtimes)}$, Wei Li, and Xianwen He

School of Mathematics and Computer Science,
Gannan Normal University, Ganzhou 341000, P.R. China
`wthgnnu@163.com`

Abstract. Measures of statistical independence between random variables have been successfully applied in many learning tasks, such as independent component analysis, feature selection and clustering. The success is based on the fact that many existing learning tasks can be cast into problems of dependence maximization (or minimization). Motivated by this, we introduce a unifying view of kernel learning with the Hilbert-Schmidt independence criterion (HSIC) which is a kernel method for measuring the statistical dependence between random variables. The key idea is that good kernels should maximize the statistical dependence, measured by the HSIC, between the kernels and the class labels. As a special case of kernel learning, we also propose an effective Gaussian kernel optimization method for classification by maximizing the HSIC, where the spherical kernel is considered. The proposed approach is demonstrated with several popular UCI machine learning benchmark examples.

Keywords: Kernel learning · Hilbert-Schmidt independence criterion (HSIC) · Statistical dependence · Gaussian kernel optimization · Kernel method

1 Introduction

Measuring dependence of random variables is one of the main concerns of statistical inference. A typical example is the inference of a graphical model, which expresses the relations among variables in terms of independence and conditional independence [1]. Kernel methods are broadly established as a useful way of constructing nonlinear algorithms from linear ones, by embedding data examples into a higher dimensional reproducing kernel Hilbert space (RKHS) [2]. A generalization of this idea is to embed probability distributions into RKHSs, giving us a linear method to infer properties of the distributions, such as independence and homogeneity [3]. In the last decade, various kernel statistical dependence measures, which differ in the way they summarize the covariance operator spectrum and the normalization they use, have been proposed for statistical analysis. Among these measures, the most well known one is the Hilbert-Schmidt independence criterion (HSIC) [4], which is defined as the Hilbert-Schmidt norm of the cross covariance operator between RKHSs. With several key advantages over other classical metrics on distributions, namely easy computability, fast convergence and low bias of finite sample estimates, HSIC has been successfully applied in statistical test of independence [3]. More interestingly, HSIC is

© Springer Nature Singapore Pte Ltd. 2016
T. Tan et al. (Eds.): CCPR 2016, Part I, CCIS 662, pp. 720–730, 2016.
DOI: 10.1007/978-981-10-3002-4_58

a very useful tool for many machine learning problems. For instance, clustering can be viewed as a problem where one strives to maximize the dependence between the observations and a discrete set of labels [5]. If labels are given, feature selection can be achieved by finding a subset of features in the observations which maximize the dependence between features and labels [6].

Although HSIC has been widely applied in machine learning, there is as yet no application in kernel learning and optimization (i.e. the way of choosing an appropriate kernel and setting its parameters), which is one of the central interests in kernel methods [7]. The core objective of kernel learning is to guarantee good generalization performance of the learning machine (predictor). Kernel learning is usually implemented by minimizing the generalization error, which can be estimated either via testing on some unused data (hold-out testing or cross validation) or with theoretical bounds [8]. The notion bounding the generalization error provides approaches not only for selecting single optimal kernel but also for combining multiple base kernels [9]. There are mainly two limitations for these methods. Firstly, they are dependent of the predictor. For instance, the radius-margin bound [8] is only applicable to support vector machines (SVM) [2]. Second, they require the whole learning process for evaluation many times [10]. Many universal kernel evaluation measures have been proposed to address these limitations, such as kernel alignment [7, 11, 12]. From a geometric point of view, kernel learning with these measures actually searches for an optimal RKHS, in which data points associated with the same class come close while those belonging to different classes go apart.

In this paper, we investigate the kernel learning problem from a different point of view, i.e., the statistical viewpoint: given the input data (such as features) and outputs (such as labels), we aim to find a kernel such that the statistical dependence between the input data and outputs is maximized. Specifically, we first introduce a general kernel learning framework with the HSIC based on the observation that rich constraints on the outputs can be defined in RKHSs. We will see that this framework is directly applicable to classification, clustering and other learning models. As a special case of kernel learning, we also propose an effective Gaussian kernel optimization method in this framework. The relationship between the proposed approach and the centered kernel alignment method [12], which is an improved kernel evaluation technique based on the popular kernel target alignment method [11], is also discussed.

2   Hilbert-Schmidt Independence Criterion (HSIC)

Let X and Y be two domains from which we draw a set of samples $D = \{(x_i, y_i)\}_{i=1}^{n}$ jointly from some probability distribution P_{xy}. The HSIC [4] measures the dependence (or independence) between x and y by computing the norm of the cross-covariance operator over the domain $X \times Y$ in RKHS. Formally, let F and G be the RKHSs on X and Y with feature maps $\phi : X \to F$ and $\varphi : Y \to G$, respectively. The associated reproducing kernels are defined as $k(x, x') = \langle \phi(x), \phi(x') \rangle$ for any $x, x' \in X$ and $l(y, y') = \langle \varphi(y), \varphi(y') \rangle$ for any $y, y' \in Y$, respectively. The cross-covariance operator between feature maps ϕ and φ is defined as a linear operator $C_{xy} : G \to F$, such that:

$$C_{xy} = E_{xy}\{(\phi(x) - E_x[\phi(x)]) \otimes (\phi(y) - E_y[\phi(y)]\} \tag{1}$$

where $\otimes$ is the tensor product, and the expectations E_{xy}, E_x, and E_y are taken according to some probability distribution P_{xy} and the marginal probability distributions P_x and P_y, respectively. The HSIC is then defined as the square of the Hilbert-Schmidt norm of C_{xy}:

$$
\begin{aligned}
HSIC(F, G, P_{xy}) = \|C_{xy}\|_{HS}^2 &= E_{xx'yy'}[k(x, x')l(y, y')] \\
&\quad - 2E_{xy}\{E_{x'}[k(x, x')]E_{y'}[l(y, y')]\} \\
&\quad + E_{xx'}[k(x, x')]E_{yy'}[l(y, y')]
\end{aligned} \tag{2}
$$

where $E_{xx'yy'}$ is the expectation over both $(x, y) \sim P_{xy}$ and an additional pair of variables $(x', y') \sim P_{xy}$ drawn independently according to the same law. It is easy to see that if both feature maps are linear (i.e., $\phi(x) = x$ and $\varphi(y) = y$), HSIC is equivalent to the square of the Frobenius norm of the cross-covariance matrix. Given D, an empirical estimator of HSIC is given by

$$
\begin{aligned}
HSIC(F, G, D) &= \frac{1}{n^2} \text{tr}(\mathbf{KL}) - \frac{2}{n^3} \mathbf{e}^\mathsf{T} \mathbf{KLe} + \frac{1}{n^4} \mathbf{e}^\mathsf{T} \mathbf{Kee}^\mathsf{T} \mathbf{Le} \\
&= \frac{1}{n^2} \left[\text{tr}(\mathbf{KL}) - \frac{1}{n} \text{tr}(\mathbf{KLee}^\mathsf{T}) - \frac{1}{n} \text{tr}(\mathbf{LKee}^\mathsf{T}) + \frac{1}{n^2} \text{tr}(\mathbf{Lee}^\mathsf{T} \mathbf{Kee}^\mathsf{T}) \right] \\
&= \frac{1}{n^2} \left\{ \text{tr} \left[\mathbf{KL} \left(\mathbf{I} - \frac{1}{n} \mathbf{ee}^\mathsf{T} \right) \right] - \frac{1}{n} \text{tr} \left[\mathbf{Kee}^\mathsf{T} \mathbf{L} \left(\mathbf{I} - \frac{1}{n} \mathbf{ee}^\mathsf{T} \right) \right] \right\} \\
&= \frac{1}{n^2} \text{tr} \left[\mathbf{K} \left(\mathbf{I} - \frac{1}{n} \mathbf{ee}^\mathsf{T} \right) \mathbf{L} \left(\mathbf{I} - \frac{1}{n} \mathbf{ee}^\mathsf{T} \right) \right] \\
&= \frac{1}{n^2} \text{tr}(\mathbf{KHLH})
\end{aligned} \tag{3}
$$

where tr is the trace operator, $\mathbf{e} = (1, \cdots, 1)^\mathsf{T} \in \mathbf{R}^n$, and $\mathbf{K}, \mathbf{L} \in \mathbf{R}^{n \times n}$ (R denotes the set of real numbers) are respectively the kernel matrices defined as $\mathbf{K}_{i,j} = k(x_i, x_j)$ and $\mathbf{L}_{i,j} = l(y_i, y_j)$. Moreover, $\mathbf{H} = \mathbf{I} - \mathbf{ee}^\mathsf{T}/n \in \mathbf{R}^{n \times n}$ is a centering matrix, where $\mathbf{I} \in \mathbf{R}^{n \times n}$ is the identity matrix.

The attractiveness of HSIC stems from the fact that the empirical estimator can be expressed completely in terms of kernels. For a particular class of kernels, i.e. the so-called universal or characteristic kernels [13] such as Gaussian and Laplace kernels, HSIC is equal to zero if and only if two random variables are statistically independent. Note that non-universal and non-characteristic kernels can also be used for HSIC, although they may not guarantee that all dependence is detected [6]. In general, the larger HSIC is, the larger the dependence between two random variables.

3   Kernel Learning with HSIC

3.1   A General Kernel Learning Framework

As a typical kernel method, HSIC used by previous work requires us to choose kernels manually because no objective model selection approaches is available. In practice, using Gaussian kernel with width parameter set to the median distance between samples is a popular heuristic [4], although such a heuristic does not always work well [14]. Different from previous work, we here use the HSIC as an evaluation criterion to assess the quality of a kernel for kernel learning and optimization. For machine learning, X and Y can be seen as the input space and output space, respectively. Correspondingly, $\mathbf{K}$ and $\mathbf{L}$ are the kernel matrices for input data and output labels, respectively. Intuitively, good kernels should maximize the statistical dependence between the kernel for input data and the kernel for output labels. Hence kernel learning can be cast into the dependence maximization framework as follows:

$$\mathbf{K}^* = \max_{\mathbf{K}} \ \mathrm{tr}\mathbf{KHLH} \tag{4}$$

$$\text{s.t. constraints on } \mathbf{K} \text{ and } \mathbf{L}$$

That is to say, kernel learning and optimization can be viewed as maximizing the empirical HISC subject to constraints on $\mathbf{K}$ and $\mathbf{L}$, for particular Hilbert spaces on the inputs and labels.

There are several advantages of this framework. First, the empirical HSIC is stable with respect to different splits of the data since it is sharply concentrated around its expected value (the empirical HSIC asymptotically converges to the true HSIC with $O(1/\sqrt{n})$ [4]). This means that the same kernels should be consistently selected to achieve high dependence when the data are repeatedly drawn from the same distribution. Second, the empirical HSIC is easy to compute (it can be computed in $O(n^2)$ time [4, 6]) since only the kernel matrices $\mathbf{K}$ and $\mathbf{L}$ are needed and no density estimation is involved. Furthermore, evaluating the empirical HSIC is dependent of the specific learning machines. These mean that we can use only the training data to select the good kernels in an efficient way prior to any computationally intensive training of the kernel machines. Finally and most importantly, rich choices of kernels can be directly applicable to the inputs and labels. This freedom of choosing kernels allows us to generate a family of kernel learning models via simply defining appropriate kernels (in some sense, these kernels incorporate prior knowledge of the leaning tasks at hand) on the inputs and outputs, respectively. Some examples are given as follows:

- **Kernels on inputs**. Kernels on the input data can be either the popular used kernels, such as polynomial kernel and Gaussian kernel, or kernels defined on non-vectorial data, such as string kernel, tree kernel and graph kernel. Moreover, instead of using a single kernel, we can use kernel combinations. A popular technique addressing this issue is the multiple kernel learning (MKL), which aims to learn an optimal combination of a set of predefined base kernels [9]. If we consider the convex combination of the base kernels, the problem of kernel learning can be transformed to the problem of determining the combination coefficients:

$$\boldsymbol{\mu}^* = \max_{\boldsymbol{\mu}} \ \mathrm{tr}\mathbf{KHLH}$$

$$\mathrm{s.t.}\mathbf{K} = \sum_{i=1}^{m} \mu_i\mathbf{K}_i, \ \mu_i \geq 0, \ \sum_{i=1}^{m} \mu_i = 1, \tag{5}$$

$$\text{and constraints on } \mathbf{L}$$

where $\mathbf{K}_i$ ($i = 1, \cdots, m$) are the base kernels and $\boldsymbol{\mu} = (\mu_1, \cdots, \mu_m)^\mathrm{T}$ is the weight coefficient vector of the given base kernels.

- **Kernels on outputs**. Generally speaking, kernels on the output labels can be as general as those defined on the input data. However, in kernel learning, prior knowledge of the learning models should be more considered to define such kernels. We can define different kernels according to different leaning models, such as classification and clustering. For classification, the kernel can be defined as

$$l(y_i, y_j) = \begin{cases} -1 & y_i = y_j \\ +1 & y_i \neq y_j \end{cases} \tag{6}$$

This definition reveals the ideal pairwise similarities between samples, i.e. the similarities from the same class are set to +1 while those from different classes are −1. For clustering, since the labels are not provided, the definition of kernels is somewhat more complex. Song et al. [6] proposed a method to define such kernels. They first chose a symmetric positive semidefinite matrix A of size $b \times b$ (b denotes the number of clusters and $b \ll n$) defining the similarities between samples in Y, and then used a partition matrix $\boldsymbol{\Pi}$ of size $n \times b$ to parameterize the kernel matrix of the outputs as

$$\mathbf{L} = \boldsymbol{\Pi}\mathbf{A}\boldsymbol{\Pi}^\mathrm{T} \tag{7}$$

Each row of $\boldsymbol{\Pi}$ contains all zeros but a single entry of 1. Actually, the partition matrix $\boldsymbol{\Pi}$ constrains us to assign each sample to a particular cluster by putting 1 in an appropriate column, which means each sample should be assigned to one and only one cluster.

To summarize, our formula (4) is very general: we can obtain a family of kernel learning algorithms by combining a kernel for the input space and another for the output space. For instance, we can have a model of multiple kernel classification [12] by combining the formulas (5) and (6), and obtain an algorithm of multiple kernel clustering [15] by combining the formulas (5) and (7).

3.2   Gaussian Kernel Optimization for Classification

In this section, we will illustrate a special case of kernel learning in the proposed framework, i.e. the kernel on the inputs is the Gaussian kernel and that on the outputs is defined as (6). Our objective is to learn the parameters of the Gaussian kernel for classification. We here only consider the spherical Gaussian kernel $k(x_i, x_j) = \exp(-\|x_i - x_j\|^2/2\sigma^2)$, where σ $(\sigma > 0)$ is the kernel width parameter.

Let $\bar{\mathbf{L}} = \mathbf{HLH}$, The optimal σ^* can be obtained by

$$
\begin{aligned}
\sigma^* &= \max_{\sigma}\ \mathrm{tr}\mathbf{KHLH} = \max_{\sigma}\ \mathrm{tr}\mathbf{K}\bar{\mathbf{L}} \\
&= \max_{\sigma} \sum_{i=1}^{n}\sum_{j=1}^{n} \mathbf{K}_{i,j}\bar{\mathbf{L}}_{i,j} = \max_{\sigma} \sum_{i=1}^{n}\sum_{j=1}^{n} k(x_i, x_j)\bar{\mathbf{L}}_{i,j}
\end{aligned}
\tag{8}
$$

The first and second derivatives of $\mathrm{tr}\mathbf{KHLH}$ with respect to σ can be respectively formulated as (9) and (10):

$$
\frac{\partial \mathrm{tr}\mathbf{KHLH}}{\partial \sigma} = \sum_{i=1}^{n}\sum_{j=1}^{n} \frac{\partial k(x_i, x_j)}{\partial \sigma}\bar{\mathbf{L}}_{i,j} = \sum_{i=1}^{n}\sum_{j=1}^{n} \left[\frac{\|x_i - x_j\|^2}{\sigma^3}\exp\left(-\frac{\|x_i - x_j\|^2}{2\sigma^2}\right)\right]\bar{\mathbf{L}}_{i,j}
\tag{9}
$$

$$
\begin{aligned}
\frac{\partial^2 \mathrm{tr}\mathbf{KHLH}}{\partial \sigma^2} &= \sum_{i=1}^{n}\sum_{j=1}^{n} \frac{\partial^2 k(x_i, x_j)}{\partial \sigma^2}\bar{\mathbf{L}}_{i,j} \\
&= \sum_{i=1}^{n}\sum_{j=1}^{n} \left[\left(\frac{\|x_i - x_j\|^4}{\sigma^6} - \frac{3\|x_i - x_j\|^2}{\sigma^4}\right)\exp\left(-\frac{\|x_i - x_j\|^2}{2\sigma^2}\right)\right]\bar{\mathbf{L}}_{i,j}
\end{aligned}
\tag{10}
$$

According to (8)–(10), the optimal kernel parameter σ^* can be found by using gradient-based optimization techniques.

3.3   Relation to Centered Kernel Alignment Method

The notion of kernel alignment, which measures the degree of agreement between a kernel and a learning task, is widely used for kernel learning and optimization due to its simplicity, efficiency and theoretical guarantee [7, 11]. Mathematically, with two kernel matrices $\mathbf{K}$ and $\mathbf{L}$, kernel alignment is given by

$$
KA(\mathbf{K}, \mathbf{L}) = \frac{<\mathbf{K}, \mathbf{L}>_F}{\sqrt{<\mathbf{K}, \mathbf{K}>_F <\mathbf{L}, \mathbf{L}>_F}}
\tag{11}
$$

where $<\bullet, \bullet>_F$ denotes the Frobenius inner product between two matrices. A limitation of kernel alignment is that it doesn't consider the unbalanced class distribution which may cause the sensitivity of the measure to drop drastically. Cortes et al. [12] proposed to center kernels (or kernel matrices) before computing the alignment measure to cancel the effect of unbalanced class distribution. Let $\bar{\mathbf{K}} = \mathbf{HKH}$, built upon the kernel alignment, centered kernel alignment (CKA) is defined as

$$CKA(\mathbf{K}, \mathbf{L}) = \frac{<\bar{\mathbf{K}}, \bar{\mathbf{L}}>_F}{\sqrt{<\bar{\mathbf{K}}, \bar{\mathbf{K}}>_F <\bar{\mathbf{L}}, \bar{\mathbf{L}}>_F}} \tag{12}$$

Although this improved definition of alignment may appear to be a technicality, it is actually a critical difference. Without that centering, the definition of alignment does not correlate well with the performance of learning machines [12].

Since $<\bar{\mathbf{K}}, \bar{\mathbf{L}}>_F = <\bar{\mathbf{K}}, \mathbf{L}>_F = <\mathbf{K}, \bar{\mathbf{L}}>_F = \text{tr}\mathbf{K}\bar{\mathbf{L}} = \text{tr}\mathbf{KHLH}$, comparing (3) and (12), the CKA is simply a normalized version of HSIC. However, for computational convenience the normalization is often omitted in practice [16]. Despite this similarity between HSIC and CKA, CKA has mainly been used for kernel learning, areas of application rather dissimilar to the applications of HSIC mentioned in Sect. 1.

4   Experiments

This section evaluates the effectiveness and efficiency of the proposed Gaussian kernel optimization method for classification. Since this method is a special case of the proposed general kernel learning framework, such evaluation can also demonstrate the benefits of the proposed general kernel learning framework to a certain degree.

4.1   Experimental Setup

Data sets. We selected 10 popular data sets, i.e., *Zoo, Sonar, Ecoli, Ionosphere, Dermatology, Australian Credit Approval* (*Australia* for short), *Vehicle Silhouettes* (*Vehicle* for short), *Yeast, Image Segmentation* (*Image* for short) and *Waveform*, from the UCI repository [17]. Among them, *Sonar, Ionosphere* and *Australian* are binary-class data sets and the others are multiclass data sets. Table 1 provides the statistics of these data sets. It presents, for each data set, the number of samples, the number of features, the number of classes, and the minimum, maximum and average

Table 1. Statistics of the selected ten data sets from UCI

Data set	#samples	#features	#classes	#min/max/average
Zoo	101	16	7	4/41/14.4
Sonar	208	60	2	97/111
Ecoli	336	7	8	2/143/42
Ionosphere	351	34	2	126/225
Dermatology	366	34	6	20/112/61
Australian	690	14	2	307/383
Vehicle	846	18	4	199/218/211.5
Yeast	1484	8	10	5/463/148.4
Image	2310	19	7	330/330/330
Waveform	5000	21	3	1647/1696/1666.7

numbers of samples per class. All the benchmark examples are small data sets ranging in sample number from 101 to 5000, and in dimension from 7 to 60. Some of them are characterized by unbalanced class distribution, such as *Zoo*, *Ecoli*, *Dermatology* and *Yeast*. These data sets form a good test bed for evaluating different algorithms.

For each data set, we partitioned it into a training set and a test set by stratified sampling (by which the object generation follows the class prior probabilities): 50 % of the data set serves as training set and the left 50 % as test set. For the training set of each data set, feature values along each dimension were linearly scaled to [0, 1]. The test set was also scaled accordingly.

Optimization Issues. There are many choices of optimization techniques to maximize the HSIC. In this experiment, the Broyden–Fletcher–Goldfarb–Shanno (BFGS) quasi-Newton algorithm was used because it generally takes a smaller number of iterations before convergence [8]. To avoid the constraint of $\sigma > 0$, the criterion was optimized according to $\ln(\sigma)$, where $\ln(\cdot)$ denotes the natural logarithm. Once the optimum is reached, the value of the parameter can be obtained by taking the exponential operation, which is always positive. Thus, the maximization of HSIC becomes an unconstrained optimization problem. The initial value of σ was set as $1/num\_features$, where *num_features* is the number of the features. Suppose BFGS starts an optimization iteration at σ, then successfully completes a line search and reaches the next point $\bar{\sigma}$. Optimization is terminated when the inequality $f(\bar{\sigma}) - f(\sigma) \leq 10^{-5} f(\sigma)$ is met, where $f(\sigma)$ and $f(\bar{\sigma})$ are the corresponding values of the objective function, respectively. In other words, the optimization will be terminated if the difference of the values of the objective function in two consecutive iterations is less than a predetermined tolerance. Finally, the BFGS quasi-Newton algorithm was implemented by using the function *fminunc*() in Matlab.

Comparison Approaches. For kernel methods, potentially any kernel can work with any kernel-based algorithm. However, the SVM is employed as the classifier in this experiment, due to its great popularity in kernel methods and delivering very impressive performance in real applications [2]. The LIBSVM software [18] was used to train and test the SVM classifier. The following four kernel learning approaches are considered:

- The cross-validation method [8] denoted as 'CV', which is regarded as a benchmark here since it is probably the most simplest and prominent approach to general kernel learning and optimization. There are two parameters, i.e. the kernel width parameter σ and the regularization parameter C, to be optimized. We performed the 5-fold cross-validation to find the best σ and C on the training set.
- The kernel alignment (KA) [11] and BFGS quasi-Newton method. We first optimized the kernel parameter using the BFGS quasi-Newton method with KA on the training set. After the kernel optimization, the regularization parameter C was selected using the 5-fold cross-validation.
- The centered kernel alignment (CKA) [12] and BFGS quasi-Newton method. We first optimized the kernel parameter using the BFGS quasi-Newton method with

CKA on the training set. After the kernel optimization, the regularization parameter C was selected using the 5-fold cross-validation.
- The proposed HSIC and BFGS quasi-Newton method. We first optimized the kernel parameter using the BFGS quasi-Newton method with HSIC on the training set. After the kernel optimization, the regularization parameter C was selected using the 5-fold cross-validation.

All experiments are conducted on a PC with 2.6 GHz CPU and 4 GB RAM.

4.2   Results and Discussion

The average classification error rates (test error rates) and running time over 10 trials are summarized in Tables 2 and 3, respectively. The bold font denotes the best result across the approaches compared. For more reliable comparison, we performed two-tailed t-test [19] with a significant level of 0.05 to determine whether there is a significant difference between the proposed method and other approaches. Based on the t-test, the win-tie-loss (W-T-L) summarizations are also attached at the bottoms of tables. A win or a loss means that the proposed method is better or worse than other method on a data set. A tie means both methods have the same performance.

Table 2. Classification error rates of the compared four approaches

Data set	Classification error (%)			
	CV	KA	CKA	HSIC
Zoo	**7.7295**	8.2106	7.8924	7.9218
Sonar	**10.0623**	11.3634	11.6357	11.2246
Ecoli	13.4467	14.6871	13.6736	**13.2213**
Ionosphere	**8.1843**	10.0398	10.1321	9.2145
Dermatology	**2.7716**	3.8246	2.9863	3.0526
Australian	25.2587	24.9827	**24.1253**	24.5760
Vehicle	**16.7913**	18.8105	17.9753	17.8924
Yeast	**39.9421**	43.1327	41.7740	41.9306
Image	**3.4268**	4.3531	4.5249	4.4375
Waveform	11.8733	12.0217	**11.6402**	11.7089
W-T-L	0-7-3	3-7-0	0-10-0	–

From Table 2, it is seen that the proposed method 'HSIC' gives rise to the classification performance comparable to that obtained by using the 5-fold cross-validation method 'CV'. Although the 'CV' method always gives the lowest accuracy error rate compared with other approaches, the difference between the 'CV' method and the proposed method 'HSIC' is not significant in most cases. Compared with the 'KA' method, it is found that 'HSIC' gives a better or comparable performance on all data sets. There is no significant performance difference between the 'CKA' method and the

Table 3. Running time of the compared four approaches

Data set	Running time (second)			
	CV	KA	CKA	HSIC
Zoo	26.5720	0.5943	0.6186	**0.3295**
Sonar	72.3589	0.8391	0.8653	**0.5922**
Ecoli	63.1945	0.6327	0.5969	**0.3873**
Ionosphere	90.1433	2.3015	2.8438	**1.2564**
Dermatology	274.7760	7.9376	10.1605	**6.3941**
Australian	362.3681	16.7325	17.9854	**15.3386**
Vehicle	551.0466	30.3647	36.0775	**18.6812**
Yeast	853.6207	50.9156	48.1163	**23.1945**
Image	2362.3390	123.8351	226.5194	**89.3783**
Waveform	5498.4162	372.1274	432.4635	**241.3056**
W-T-L	10-0-0	9-1-0	10-0-0	–

'HSIC' method on all data sets. In terms of the running time, from Table 3, it is found that the proposed method 'HSIC' gives the shortest time on all the data sets. Take the *Zoo* data set for example: 'HSIC' achieves the running time of 0.3295 s, which is significant shorter than 26.5720, 0.5943 and 0.6186 s obtained by 'CV', 'KA' and 'CKA' methods, respectively. The reason why 'HSIC' method is more computationally efficient than 'CKA' method may be that 'HSIC' simplifies the 'CKA' criterion by ridding the latter of its denominator, making the evaluation considerably easier. In a nutshell, although 'HSIC' yields almost the same classification performance as those achieved by other baseline approaches, it takes significantly less computational time. This suggests that our proposed method is both effective and efficient for Gaussian kernel optimization.

5 Conclusion

We have explored the kernel learning problem from the statistical dependence maximization viewpoint. We first proposed a general kernel learning framework with the HSIC which is probably the most popular kernel statistical dependence measure for statistical analysis. This framework has been demonstrated that it can be directly applied in classification, clustering and other learning models. In this framework, we also presented a Gaussian kernel optimization method for classification. Extensive experimental study on multiple benchmark data sets verifies the effectiveness and efficiency of the proposed method. Future investigation will focus on the validation of the use of the proposed general kernel learning framework for multiple kernel learning (MKL), domain transfer learning [20] and so on. As a final remark, it should be emphasized that, in general, by treating various learning models under a unifying framework and elucidating their relations, we also expect our work to benefit practitioners in their specific applications.

Acknowledgements. This work is supported in part by the National Natural Science Foundation of China (No. 61562003) and the Natural Science Foundation of Jiangxi Province of China (Nos. 20151BAB207029 and 20161BAB202070).

References

1. Fukumizu, K., Gretton, A., Sun, X., Schölkopf, B.: Kernel measures of conditional dependence. In: Advances in Neural Information Processing Systems 20, pp. 489–496 (2007)
2. Shawe-Taylor, J., Cristianini, N.: Kernel methods for pattern analysis. Cambridge University Press, New York (2004)
3. Chwialkowski, K., Gretton, A.: A kernel independence test of random process. In: Proceedings of the 31th International Conference on Machine Learning, Beijing, China, pp. 1422–1430 (2014)
4. Gretton, A., Bousquet, O., Smola, A., Schölkopf, B.: Measuring statistical dependence with Hilbert-Schmidt norms. In: Proceedings of the 16th International Conference on Algorithmic Learning Theory, Singapore, pp. 63–77 (2005)
5. Song, L., Smola, A., Gretton, A., Borgwardt, K.: A dependence maximization view of clustering. In: Proceedings of the 24th International Conference on Machine Learning, Corvallis, USA, pp. 823–830 (2007)
6. Song, L., Smola, A., Gretton, A., Bedo, J., Borgwardt, K.: Feature selection via dependence maximization. J. Mach. Learn. Res. **13**, 1393–1434 (2012)
7. Wang, T., Zhao, D., Tian, S.: An overview of kernel alignment and its applications. Artif. Intell. Rev. **43**(2), 179–192 (2015)
8. Keerthi, S.S.: Efficient tuning of SVM hyperparameters using radius/margin bound and iterative algorithms. IEEE Trans. Neural Netw. **13**(5), 1225–1229 (2002)
9. Gönen, M., Alpaydın, E.: Multiple kernel learning algorithms. J. Mach. Learn. Res. **12**, 2211–2226 (2011)
10. Wang, T., Tian, S., Huang, H., Deng, D.: Learning by local kernel polarization. Neurocomputing **72**(13–15), 3077–3084 (2009)
11. Cristianini, N., Shawe-Taylor, J., Elisseeff, A., Kandola, J.: On kernel-target alignment. In: Advances in Neural Information Processing Systems 14, pp 367–373 (2001)
12. Cortes, C., Mohri, M., Rostamizadeh, A.: Algorithms for learning kernels based on centered alignment. J. Mach. Learn. Res. **13**, 795–828 (2012)
13. Steinwart, I.: On the influence of the kernels on the consistency of support vector machines. J. Mach. Learn. Res. **2**, 67–93 (2001)
14. Sugiyama, M.: On kernel parameter selection in Hilbert-Schmidt independence criterion. IEICE Trans. Inf. Syst. **E95-D**(10), 2564–2567 (2012)
15. Lu, Y., Wang, L., Lu, J., Yang, J., Shen, C.: Multiple kernel clustering based on centered kernel alignment. Pattern Recogn. **47**(11), 3656–3664 (2014)
16. Neumann, J., Schnörr, C., Steidl, G.: Combined SVM-based feature selection and classification. Mach. Learn. **61**(1–3), 129–150 (2005)
17. Lichman, M.: UCI machine learning repository. Irvine, CA: University of California, School of Information and Computer Science (2013). http://archive.ics.uci.edu/ml/
18. Chang, C.-C., Lin, C.-J.: LIBSVM: a library for support vector machines. ACM Trans. Intell. Syst. Technol. **2**(3), Article No. 27 (2011). http://www.csie.ntu.edu.tw/~cjlin/libsvm
19. Demšar, J.: Statistical comparisons of classifiers over multiple data sets. J. Mach. Learn. Res. **7**, 1–30 (2006)
20. Lu, J., Behbood, V., Hao, P., Zuo, H., Xue, S., Zhang, G.: Transfer learning using computational intelligence: a survey. Knowl.-Based Syst. **80**, 14–23 (2015)

Key Course Selection in Academic Warning with Sparse Regression

Min Yin, Xijiong Xie, and Shiliang Sun$^{(\boxtimes)}$

Department of Computer Science and Technology, East China Normal University,
500 Dongchuan Road, Shanghai 200241, People's Republic of China
xjxie11@gmail.com, slsun@cs.ecnu.edu.cn

Abstract. Many colleges and universities are paying more attention to academic warning which warns large numbers of students who have unsatisfactory academic performance. Academic warning becomes a new part in the teaching management constitution but lacks of unified and scientific standards under the establishment of this stipulation at present. This paper solves the current setting of academic warning through well-known methods lasso and ℓ_1-norm support vector regression with ϵ-insensitive loss function which can select key courses based on the failed credits in one semester. The experiments are made on our collected academic warning datasets which are incomplete data. We impute them with one nearest neighbor method. The experimental results show that sparse regression is effective for colleges and universities to remind the students of key courses.

Keywords: Academic warning · Lasso · ℓ_1-norm support vector regression · Sparse regression

1 Introduction

With the continuous improvement of the whole education system, a large number of colleges and universities adopt academic warning systems in student academic management. Academic warning has been one of computational education science and problems [1]. Academic warning can monitor and supervise the students' study and promote them to learn consciously. As we know, different courses have correspondingly different credits and the score of each course decides whether the student can obtain the relevant credit. Whether a student is warned can be based on the sum of the credits of all failed courses. In general, each college or university warns the student by setting a credit line reasonably. If the setting can be obtained through machine learning methods [2–4], it will improve the performance of academic warning. Simultaneously, key course selection is also important. Key course selection is easily accomplished by traditional statistical methods. However, some hidden information in data may be ignored. For example, the failed student numbers of some courses are few but their scores are low. Machine learning methods can take advantage of this information.

The first author and the second author contributed equally to this work.

© Springer Nature Singapore Pte Ltd. 2016
T. Tan et al. (Eds.): CCPR 2016, Part I, CCIS 662, pp. 731–741, 2016.
DOI: 10.1007/978-981-10-3002-4_59

A vastly popular and successful approach in statistical modeling is to use regularization penalties in model fitting. The use of ℓ_1 regularization for statistical inference has become very popular over the last two decades. Tibshirani [5] proposed the least absolute shrinkage and selection operator (lasso) technique which uses an ℓ_1-penalized likelihood for linear regression with independent Gaussian noise, which involves minimizing the usual sum of squared error loss with ℓ_1 regularization. The lasso has become the standard tool for sparse regression for which its ℓ_1 penalty leads to sparse solutions. That is, there are few nonzero estimates. Sparse models are more interpretable and often preferred in the natural and social sciences. Much of the early effort has been dedicated to solving the optimization problem efficiently [6–8]. Support vector machine is the most popular algorithm for classification and regression. There is an important model called ℓ_1-norm support vector regression (SVR) [9–11], which is used to identify the critical features for regression. It can deal with the case where a lot of noisy and redundant features are present. ℓ_1-norm SVR can be regarded as a linear programming (LP) problem. In fact, the ℓ_1-norm SVR with Gaussian loss function is equivalent to a lasso problem. In this paper, we select key courses through well-known methods lasso and ℓ_1-norm support vector regression with ϵ-insensitive loss function on our collected academic warning datasets. The experimental results show that sparse regression can provide a new universal method for colleges and universities to remind the students of key courses.

The structure of the paper is organized as follows. In Sect. 2, we introduce two models ℓ_1-norm SVR and lasso. Then data collection is also depicted in this section. After reporting experimental results in Sect. 3, we give conclusions in Sect. 4.

2   Model and Data Collection

In this section, we introduce two models ℓ_1-norm SVR and lasso. Then we introduce our collected academic warning datasets.

2.1 ℓ_1-norm SVR

We give a brief outline of ℓ_1-norm SVR. Suppose $f(\bar{x})\colon \mathbb{R}^d \to \mathbb{R}$ that transforms the input vector $\bar{x} \in \mathbb{R}^d$ to a real number $f(\bar{x})$. SVR aims to estimate $f(\bar{x})$ by observing n training examples. Here we consider the linear case, $f(\bar{x}) = \bar{w}^\top \bar{x} + b$, where $\bar{w} \in \mathbb{R}^d$, $b \in \mathbb{R}$. Define that $x = \begin{pmatrix} \bar{x} \\ 1 \end{pmatrix}$, $w = \begin{pmatrix} \bar{w} \\ b \end{pmatrix}$. Then above formulation can be written in the homogeneous form $f(x) = w^\top x$. The optimization objective of ℓ_1-norm SVR is

$$
\begin{aligned}
&\min \ \|w\|_1 \\
&\text{s.t.} \ \ y_i = w^\top x_i, \ i = 1, 2, \ldots, n.
\end{aligned}
\tag{1}
$$

In practical applications, the output y_i may be corrupted by some noise. A number of loss functions $L(x, y, f(x))$ and a penalty term $e^\top q$ can be introduced into (1) to deal with the noise. The formulation (1) can be updated to

$$\min \ \|w\|_1 + Ce^\top q$$
$$\text{s.t.} \ \ L(x_i, y_i, f(x_i)) \le q_i, \ i = 1, 2, \ldots, n, \ q_i \ge 0, \tag{2}$$

where C is non-negative constant controlling the tradeoff between the norm regularization and penalty. Different loss functions are suitable for different problems. The ϵ-insensitive loss function

$$L(x, y, f(x)) = \max\{|y - f(x)| - \epsilon, 0\} \tag{3}$$

is one of the most commonly used loss functions, where ϵ is a parameter which needs to be set in advance. The optimization objective is specified as

$$\min \ \|w\|_1 + Ce^\top q_i$$
$$\text{s.t.} \ \ |y_i - w^\top x_i| \le q_i, \ i = 1, 2, \ldots, n, \ q_i \ge 0. \tag{4}$$

This formulation can be solved as an LP problem. The Gaussian loss function

$$L(x, y, f(x)) = \frac{1}{2}(y - f(x))^2 \tag{5}$$

is appropriate to deal with the Gaussian-type noise. Then the optimization objective is specified as

$$\min \ \|w\|_1 + Ce^\top q_i$$
$$\text{s.t.} \ \ (y_i - w^\top x_i)^2 \le q_i, \ i = 1, 2, \ldots, n, \ q_i \ge 0. \tag{6}$$

The optimization objective cannot be regard as an LP problem but there exist many efficient methods for solving it. One noteworthy fact is that with the Gaussian noise, the notion of support vector is then meaningless.

2.2 Lasso

Given an input $X \in \mathbb{R}^{n \times d}$, each row of X represents an example, and an output $y \in R^n$, the lasso is least square regression with ℓ_1-norm regularization, which attempts to solve the following optimization problem

$$\min \ \|w\|_1 + C\|y - Xw\|_2^2 \tag{7}$$

where w is sparse, which means that most elements of w are zero. In fact, the ℓ_1-norm SVR with the Gaussian loss function is equivalent to a lasso problem.

2.3 Data Collection and Imputation

Academic warning datasets are collected by ourselves from a certain university. They contain 28 datasets which represent the scores of students of class 1 and class 2 of grade 2010 and grade 2011 in the seven semesters (one row represents one student and one column represents one course). The number of students are 47, 51, 23 and 52 in four classes. The label represents the failed credits of one student in one semester. However, there are a number of miss values in these datasets. We use one nearest neighbor (1NN) to impute these missing values. The 1NN method replaces the missing value in the data matrix with the corresponding value from the nearest row. That is to say, it can identify the most similar score to the current one with a missing value, and use the score as a guess for the missing one. Dataset 201011 represents the scores of students of class 1 of grade 2010 in the first semester. Dataset 201012 represents the scores of students of class 1 of grade 2010 in the second semester. The names of other datasets are analogous to the above ones.

3   Experiments and Results

We use lasso and ℓ_1-norm SVR with the ϵ-insensitive loss function to select features which have the most information. The nonzero values of w represent key courses and the absolute value of element in w represents the importance of course. If the absolute value of element in w is bigger, the corresponding course is more important. In the lasso method, we use the lasso function in matlab and select number of non-zero coefficients from small to large in the range of integers until w emerges. In the ℓ_1-norm SVR, we use the code in reference [11] and select optimal parameter by grid search strategy until the sparseness of w is smallest. The experimental results are list in Figs. 1, 2, 3, 4, 5, 6, 7, 8, 9, 10, 11, 12, 13, 14, 15, 16, 17, 18, 19, 20, 21, 22, 23, 24, 25, 26, 27 and 28 (the horizontal coordinate represents courses and the vertical coordinate represents vaules of elements in w). The results of the lasso are represented by the white bar and the results of ℓ_1-norm SVR are represented by the green bar. For simplicity, we analyze the results of the first four datasets in detail.

The dataset 201011 contains C programming, sports, military theory, psychology, ideological and moral cultivation and legal basis, linear algebra, English, introduction to computer science, experiments of introduction to computer and advanced mathematic. In this semester, we use lasso to select key courses. The results are advanced mathematic, English and C programming in descending order. Then we use ℓ_1-norm support vector regression to select key courses. The results are advanced mathematic, introduction to computer science, C programming, ideological and moral cultivation and legal basis, military theory and experiments of introduction to computer. The true importance descending order of courses is C programming, experiments of introduction to computer, advanced mathematic, psychology, linear algebra according to the failed student number of courses. We can find that the lasso method finds out key courses advanced mathematic and C programming while ℓ_1-norm support vector regression finds out key courses advanced mathematic, C programming and experiments of introduction to computer.

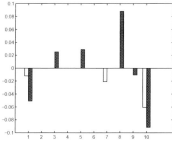

Fig. 1. Results of dataset 201011

Fig. 2. Results of dataset 201012

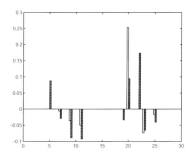

Fig. 3. Results of dataset 201013

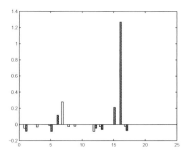

Fig. 4. Results of dataset 201014

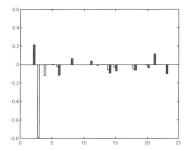

Fig. 5. Results of dataset 201015

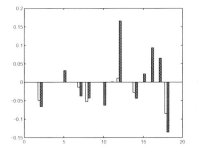

Fig. 6. Results of dataset 201016

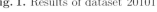

    The dataset 201012 contains outline of modern Chinese history, sports, public elective course, college English, teaching media and technology, education, module one, life sciences, computer programming practice, object-oriented programming (based on C++), object-oriented programming (based on java) and advanced mathematic. We use lasso to select key courses. The results are life sciences, advanced mathematic, college English, object-oriented programming (based on java), computer programming practice and teaching media and technology in descending order. Then we use ℓ_1-norm support vector regression to select key courses. The results are outline of modern Chinese history, advanced

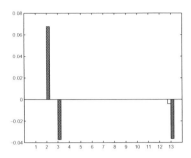

Fig. 7. Results of dataset 201017

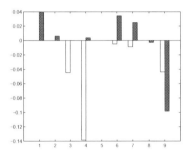

Fig. 8. Results of dataset 201021

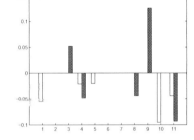

Fig. 9. Results of dataset 201022

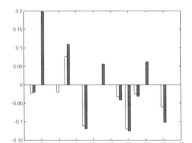

Fig. 10. Results of dataset 201023

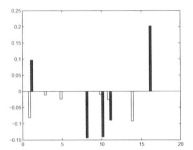

Fig. 11. Results of dataset 201024

mathematic, college English, public elective course, module one, computer programming practice in descending order. The true importance descending order of courses is advanced mathematic, computer programming practice, object-oriented programming (based on java), object-oriented programming (based on C++) and college English according to the failed student number of courses. We can find that the lasso method finds out key courses advanced mathematic, object-oriented programming (based on java), college English while ℓ_1-norm support vector regression finds out key courses advanced mathematic, computer programming practice, college English.

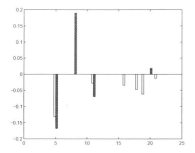

Fig. 12. Results of dataset 201025

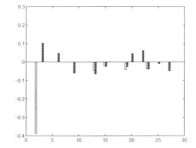

Fig. 13. Results of dataset 201026

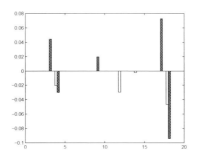

Fig. 14. Results of dataset 201027

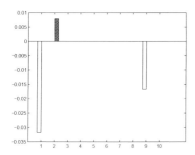

Fig. 15. Results of dataset 201111

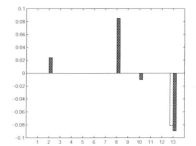

Fig. 16. Results of dataset 201112

The dataset 201013 contains C programming, sports, information technology curriculum and teaching theory, confucianism and modern society, public elective course, the outline of history, classical Chinese culture, college English level 4, college Chinese, psychology, teachers spoken language, digital logic and experiment, data structure, module one, an introduction to Mao Zedong thought and the theory system of socialism with Chinese characteristics, material science, discrete mathematics, linear algebra, basic principle of Marxism and advanced mathematic. We use lasso to select key courses. The results are classical Chinese culture, digital logic and experiment, C programming, information technology

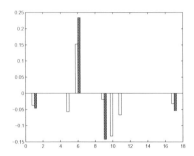

Fig. 17. Results of dataset 201113

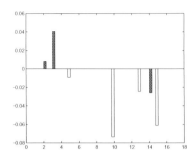

Fig. 18. Results of dataset 201114

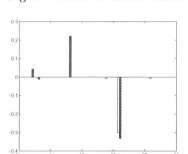

Fig. 19. Results of dataset 201115

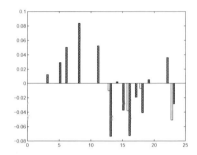

Fig. 20. Results of dataset 201116

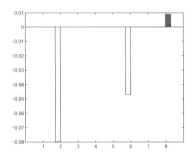

Fig. 21. Results of dataset 201117

curriculum and teaching theory, public elective course, data structure, college English level 4 and discrete mathematics in descending order. Then we use ℓ_1-norm support vector regression to select key course. The results are material science, an introduction to Mao Zedong thought and the theory system of socialism with Chinese characteristics, the outline of history, public elective course, C programming, discrete mathematics, data structure, digital logic and experiment in descending order. The true importance descending order of courses is C programming, discrete mathematics, public elective course, digital logic and experiment, data structure, information technology curriculum and teaching theory, college English level 4, module one and material science according to

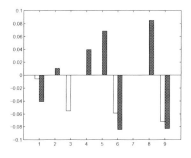

Fig. 22. Results of dataset 201121

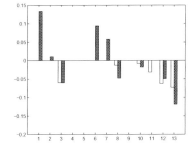

Fig. 23. Results of dataset 201122

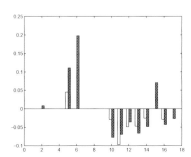

Fig. 24. Results of dataset 201123

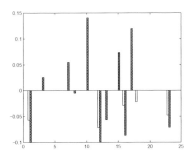

Fig. 25. Results of dataset 201124

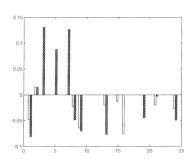

Fig. 26. Results of dataset 201125

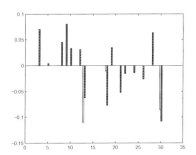

Fig. 27. Results of dataset 201126

the failed student number of courses. We can find that the lasso method finds out key courses C programming, discrete mathematics, public elective course, digital logic and experiment, data structure, information technology curriculum and teaching theory, college English level 4 while ℓ_1-norm support vector regression finds out key courses C programming, discrete mathematics, public elective course, digital logic and experiment, data structure, material science.

The dataset 201014 contains web application technology, windows application design, sports, introduction to information system security, public elective course, college English, college English (advanced), operating system,

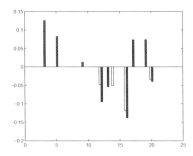

Fig. 28. Results of dataset 201127

educational probation, probability theory and mathematical statistics, module one, material science, algorithm analysis and design, computer composition and structure, computer aided education, object-oriented programming (based on C++), object-oriented programming (based on java), and advanced mathematic. We use lasso to select key courses. The results are advanced mathematic, operating system, windows application design, computer composition and structure, college English (advanced) and material science in descending order. We use ℓ_1-norm support vector regression to select key courses. The results are material science, advanced mathematic, object-oriented programming (based on C++), windows application design, object-oriented programming (based on java), probability theory and mathematical statistics, computer composition and structure, operating system, college English (advanced), public elective course, computer aided education and module one in descending order. The true importance descending order of courses is windows application design, computer composition and structure, operating system, probability theory and mathematical statistics, advanced mathematic, public elective course, college English (advanced), algorithm analysis and design, sports and object-oriented programming (based on java) according to the failed student number of courses. We can find that the lasso method finds out key courses windows application design, computer composition and structure, operating system, advanced mathematic, college English (advanced) while ℓ_1-norm support vector regression finds out key courses windows application design, computer composition and structure, operating system, public elective course, college English (advanced), object-oriented programming (based on java).

Above all, we can conclude that the two methods can obtain comparatively front key courses. In addition, the two methods can find out other courses for which many students obtained low scores but they cannot be found by simple statistics. This is a special advantage of machine learning methods that can explore the hidden information of data. From all figures, ℓ_1-norm SVR with ϵ-insensitive loss function can select more key courses compared with the lasso method in most cases.

4   Conclusion

In this paper, we use lasso and ℓ_1-norm SVR to select key courses on our collected academic warning datasets. The experimental results show that the two methods can obtain comparatively accurate key courses.

Acknowledgments. The corresponding author Shiliang Sun would like to thank support by the National Natural Science Foundation of China under Project 61370175, and Shanghai Knowledge Service Platform Project (No. ZF1213).

References

1. Sun, S.: Computational education science and ten research directions. Communications of the Chinese Association for Artificial Intelligence **5**, 15–16 (2015)
2. Dai, J., Li, M., Li, W., Xia, T., Zhang, Z.: Application of Monte Carlo simulation in college and university academic warning. Advanced Materials Research **955–959**, 1817–1824 (2014)
3. Dai, J., Li, M., Li, W., Xia, T., Zhang, Z.: Setting of academic warning based on multivariate copula functions. Applied Mechanics & Materials **571–572**, 156–163 (2014)
4. Taylor, J., Lawrence, J.: Making students AWARE: an online strategy for students given academic warning. Studies in Learning Evaluation Innovation & Development **4**, 39–52 (2007)
5. Tibshirani, R.: Regression shrinkage and selection via the lasso: a retrospective. Journal of the Royal Statistical Society **73**, 273–282 (2011)
6. Meinshausen, N., Bhlmann, P.: High-dimensional graphs and variable selection with the Lasso. Annals of Statistics **34**, 1436–1462 (2006)
7. Vidaurre, D., Bielza, C., Larrañaga, P.: A survey of $L1$ regression. International Statistical Review **81**, 361–387 (2013)
8. Sun, S., Huang, R., Gao, Y.: Network-scale traffic modeling and forecasting with graphical lasso and neural networks. Journal of Transportation Engineering **138**, 1358–1367 (2012)
9. Shawe-Taylor, J., Sun, S.: A review of optimization methodologies in support vector machines. Neurocomputing **74**, 3609–3618 (2011)
10. Shawe-Taylor, J., Sun, S.: Kernel methods and support vector machines. Book Chapter for E-Reference Signal Processing, Elsevier, (2013). doi:10.1016/B978-0-12-396502-8.00026-7
11. Zhang, Q., Hu, X., Zhang, B.: Comparison of ℓ_1-norm SVR and sparse coding algorithms for linear regression. IEEE Transactions on Neural Networks and Learning Systems **26**, 1828–1833 (2015)

The Necessary and Sufficient Conditions for the Existence of the Optimal Solution of Trace Ratio Problems

Guoqiang Zhong$^{(\boxtimes)}$ and Xiao Ling

Department of Computer Science and Technology, Ocean University of China,
238 Songling Road, Qingdao 266100, China
gqzhong@ouc.edu.cn, mumubuguai@vip.qq.com

Abstract. Many dimensionality reduction problems can be formulated as a trace ratio form, i.e. $\mathrm{argmax}_{\mathbf{W}} Tr(\mathbf{W}^T \mathbf{S}_p \mathbf{W})/Tr(\mathbf{W}^T \mathbf{S}_t \mathbf{W})$, where $\mathbf{S}_p$ and $\mathbf{S}_t$ represent the (dis)similarity between data, $\mathbf{W}$ is the projection matrix, and $Tr(\cdot)$ is the trace of a matrix. Some representative algorithms of this category include principal component analysis (PCA), linear discriminant analysis (LDA) and marginal Fisher analysis (MFA). Previous research focuses on how to solve the trace ratio problems with either (generalized) eigenvalue decomposition or iterative algorithms. In this paper, we analyze an algorithm that transforms the trace ratio problems into a series of trace difference problems, i.e. $\mathrm{argmax}_{\mathbf{W}} Tr[(\mathbf{W}^T(\mathbf{S}_p - \lambda \mathbf{S}_t)\mathbf{W}]$, and propose the necessary and sufficient conditions for the existence of the optimal solution of trace ratio problems. The correctness of this theoretical result is proved. To evaluate the applied algorithm, we tested it on three face recognition applications. Experimental results demonstrate its convergence and effectiveness.

Keywords: Trace ratio problems · Dimensionality reduction · Convergence · Necessary and sufficient conditions

1 Introduction

In current era of big data, data in many applications, such as image, audio and video processing, are generally described in a high dimensional form. In most cases, there are serious redundances in these high dimensional data that have the potential to be represented in a low dimensional space. In addition, the noise in the high dimensional data may increase the amount of computation in the practical pattern recognition, data mining, computer vision and machine learning tasks. Hence, how to find a compact representation of data is an important problem in many related research areas.

To solve the above problem, many algorithms have been proposed in the literature [10]. In order to find the essential features of data, unsupervised dimensionality reduction approaches are widely used. Two popular unsupervised dimensionality reduction algorithms are PCA (short for principal component analysis)

© Springer Nature Singapore Pte Ltd. 2016
T. Tan et al. (Eds.): CCPR 2016, Part I, CCIS 662, pp. 742–751, 2016.
DOI: 10.1007/978-981-10-3002-4_60

and MDS (short for multidimensional scaling), respectively. However, for classification applications, unsupervised dimensionality reduction approaches cannot perform very well.

Two representative supervised dimensionality reduction algorithms are LDA (short for linear discriminant analysis) and MFA (short for marginal Fisher analysis) [8]. LDA minimizes the within-class scattering, and at the same time, maximizes the between-class scattering. It's learning criterion is defined as

$$J_{Fisher}(\varphi) = \max \frac{\varphi^T \mathbf{S}_b \varphi}{\varphi^T \mathbf{S}_w \varphi}, \tag{1}$$

where $\mathbf{S}_w$ is the within-class scattering matrix and $\mathbf{S}_b$ is the between-class scattering matrix. To approximately solve Problem (1), $J_{Fisher}(\varphi)$ is generally converted into

$$\mathbf{W}_{opt} = \operatorname{argmax} \frac{|\mathbf{W}^T \mathbf{S}_b \mathbf{W}|}{|\mathbf{W}^T \mathbf{S}_w \mathbf{W}|}. \tag{2}$$

Problem (2) can be solved using the generalized eigenvalue decomposition method.

MFA is a graph embedding method [8], which describes the structure of data using an intrinsic graph $\mathbf{S}$ and a penalty graph $\mathbf{S}^p$. Its learning criterion is

$$\operatorname*{argmin}_{\mathbf{W}} \frac{\sum_{i \neq j} \|\mathbf{W}^T \mathbf{x}_i - \mathbf{W}^T \mathbf{x}_j\|^2 \mathbf{S}_{ij}}{\sum_{i \neq j} \|\mathbf{W}^T \mathbf{x}_i - \mathbf{W}^T \mathbf{x}_j\|^2 \mathbf{S}_{ij}^p}, \tag{3}$$

where $\mathbf{W}$ is the projection matrix. Problem (3) can be transformed into the form of Trace Ratio, i.e.

$$\operatorname*{argmax}_{\mathbf{W}} \left\{ \frac{Tr(\mathbf{W}^T \mathbf{X} \mathbf{L}^p \mathbf{X}^T \mathbf{W})}{Tr(\mathbf{W}^T \mathbf{X} \mathbf{L} \mathbf{X}^T \mathbf{W})} = \frac{Tr(\mathbf{W}^T \mathbf{S}_p \mathbf{W})}{Tr(\mathbf{W}^T \mathbf{S}_t \mathbf{W})} \right\}, \tag{4}$$

where $\mathbf{S}_p = \mathbf{X} \mathbf{L}^p \mathbf{X}^T$ and $\mathbf{S}_t = \mathbf{X} \mathbf{L} \mathbf{X}^T$. Due to the generality of the graph embedding framework, many algorithms, such as PCA, LDA or MFA, and their corresponding kernel methods can be formulated as trace ratio problems [1,2,4,7,9], even tensor dimensionality reduction ones [8,9].

Moreover, many metric learning and semi-supervised learning models can be also formulated as trace ratio problems, such as that in [6] and [3]. Hence, how to optimally solve trace ratio problems is an important issue to many learning problems. However, previous work only focuses on solving trace ratio problems with either (generalized) eigenvalue decomposition or iterative algorithms. In this paper, we aim at proposing the necessary and sufficient conditions for the existence of the optimal solution of trace ratio problems, and prove their correctness.

The rest of this paper is organized as follows. Section 2 introduces some related work on dimensionality reduction and solutions of trace ratio problems. Section 3 presents an iterative algorithm to solve trace ratio problems. In Sect. 4, we give the necessary and sufficient conditions for the existence of the optimal solution of trace ration problems, and prove their correctness. The experimental results are reported in Sect. 5, and finally, we summarize this paper in Sect. 6.

2   Related Work

In the literature, many dimensionality reduction algorithms have been proposed, such as PCA, LDA and MFA. The essence of PCA is to represent the original data with a series of uncorrelated low dimensional features. By PCA, we can obtain a projection matrix $\mathbf{W}$ by applying eigenvalue decomposition on the covariance matrix of the original data. And the high dimensional vector $\mathbf{x}$ can be mapped into a low dimensional subspace, i.e. $\mathbf{y} = \mathbf{W}^T\mathbf{x}$. LDA aims at learning a projection matrix with good separability. In order to do this, two matrices $\mathbf{S}_b$ and $\mathbf{S}_w$ are defined, where $\mathbf{S}_b$ is the between-class scattering matrix and $\mathbf{S}_w$ is the within-class scattering matrix:

$$\mathbf{S}_b = \sum_{i=1}^{C} n_i(\mathbf{u}_i - \mathbf{u})(\mathbf{u}_i - \mathbf{u})^T \tag{5}$$

$$\mathbf{S}_w = \sum_{i=1}^{C} \sum_{\mathbf{x}_j \in c_i} (\mathbf{u}_i - \mathbf{x}_j)(\mathbf{u}_i - \mathbf{x}_j)^T \tag{6}$$

where $\mathbf{u}_i$ is the sample mean of class c_i, $\mathbf{u}$ is the overall sample mean and C is the number of classes. The learning criterion of LDA is

$$J_{Fisher}(\varphi) = \max \frac{\varphi^T \mathbf{S}_b \varphi}{\varphi^T \mathbf{S}_w \varphi} \tag{7}$$

where φ is an arbitrary d-dimensional column vector. Hence, we have

$$J_{Fisher}(\varphi) = \max \frac{\sum_{i=1}^{C} n_i \varphi^T (\mathbf{u}_i - \mathbf{u})(\mathbf{u}_i - \mathbf{u})^T \varphi}{\sum_{i=1}^{C} \sum_{\mathbf{x}_j \in class c_i} \varphi^T (\mathbf{u}_i - \mathbf{x}_j)(\mathbf{u}_i - \mathbf{x}_j)^T \varphi}. \tag{8}$$

Problem (8) can be approximately solved with

$$W_{opt} = \operatorname*{argmax}_{\mathbf{W}} \frac{|\mathbf{W}^T \mathbf{S}_b \mathbf{W}|}{|\mathbf{W}^T \mathbf{S}_w \mathbf{W}|}. \tag{9}$$

Yan et al. [8] claimed that many traditional dimensionality reduction methods, either linear or nonlinear, can be transformed into a unified structure, called graph embedding. The graph preserving criterion can be defined as follows

$$\mathbf{y}^* = \operatorname*{argmin}_{\mathbf{w}^T \mathbf{C} \mathbf{w} = \mathbf{I}} \sum_{i \neq j} \|\mathbf{w}_i - \mathbf{w}_j\|^2 \mathbf{S}_{ij} = \operatorname*{argmin}_{\mathbf{w}^T \mathbf{C} \mathbf{w} = \mathbf{I}} \mathbf{w}^T \mathbf{S} \mathbf{w}, \tag{10}$$

where $\mathbf{C}$ and $\mathbf{S}$ are two symmetric and positive semi-definite matrices. Yan et al. also proposed the MFA algorithm, which is a special linear graph embedding method. Under the graph embedding framework, many dimensionality reduction algorithms, including PCA, LDA and MFA, can be re-formulated as the following trace ratio form:

$$\mathbf{W} = \operatorname*{argmax}_{\mathbf{w}^T \mathbf{w} = \mathbf{I}} \frac{Tr(\mathbf{W}^T \mathbf{S}_p \mathbf{W})}{Tr(\mathbf{W}^T \mathbf{S}_t \mathbf{W})}, \tag{11}$$

where $\mathbf{S}_p$ and $\mathbf{S}_t$ are two symmetric and positive semi-definite matrices, $Tr(\cdot)$ is the trace of a matrix, and $Tr(\mathbf{X}) = \sum \mathbf{X}_{ii}$.

To get a closed-form solution, Problem (11) are generally transformed into a Ratio Trace form, $\mathrm{argmax}_{\mathbf{W}} Tr[(\mathbf{W}^T \mathbf{S}_t \mathbf{W})^{-1} (\mathbf{W}^T \mathbf{S}_p \mathbf{W})]$. Considering the solution of Ratio Trace form is only an approximation to that of Problem (11), Wang et al. [5] put forward an iterative algorithm, in which the projection matrix $\mathbf{W}$ and the value of the trace ratio λ are obtained by the iterative steps $\mathrm{argmax}_{\mathbf{W}} Tr[\mathbf{W}^T (\mathbf{S}_p - \lambda \mathbf{S}_t) \mathbf{W}]$. Then, Nie et al. [4] proposed a new method to find the global optimal solution by directly optimizing the projection matrix. In [3], the authors proposed a fast algorithm for trace ratio problems. Besides, based on this fast algorithm, a new semi-supervised method has been proposed, by which the unlabeled data can be mapped into the learned subspace. Jia et al. [2] put forward a novel Newton-Raphson method for trace ratio problems, which can be proved to be convergent. Recently, Huang et al. [5] introduced a semi-supervised method to solve dimensionality reduction problems based on the Trace Ratio formulation. Afterwards, Zhong and Cheriet [9] proposed a novel framework for tensor representation learning, where the problem can also be transformed into a trace ratio problem. In addition, Zhong, Shi and Cheriet [10] have presented a novel method called relational Fisher analysis, which is based on the trace ratio formulation and sufficiently exploits the relational information of data.

However, the above approaches only formulate the learning problem in the trace ratio form, and solve it with generalized eigenvalue decomposition or iterative procedures. In this paper, we present the necessary and sufficient conditions for the existence of the optimal solution of trace ratio problems, and prove their correctness.

3   An Iterative Algorithm for Trace Ratio Problems

In order to analyze the necessary and sufficient conditions for the existence of the optimal solution of trace ratio problems, we use the following iterative procedures to solve the trace ratio problems [7,9].

Algorithm 1: the Iterative Procedures for Trace Ratio Problems

(1) Initialize the matrix $\mathbf{W}^0$ with an arbitrary orthogonal matrix

(2) For $n = 1, 2, \ldots, N_{max}, Do$

a. Calculate the value of the objective function, λ^n, based on the projection matrix $\mathbf{W}^{n-1}$:

$$\lambda^n = \frac{Tr(\mathbf{W}^{n-1}{}^T \mathbf{S}_p \mathbf{W}^{n-1})}{Tr(\mathbf{W}^{n-1}{}^T \mathbf{S}_t \mathbf{W}^{n-1})} \tag{12}$$

b. Define the trace difference problem

$$\mathbf{W}^n = \underset{\mathbf{W}^T \mathbf{W} = \mathbf{I}_d}{\mathrm{argmax}} \; Tr[\mathbf{W}^T (\mathbf{S}_p - \lambda^n \mathbf{S}_t) \mathbf{W}], \tag{13}$$

where d is the dimensionality of the target space.

c. Obtain the target matrix $\mathbf{W}^n$ by the method of eigenvalue decomposition,

$$(\mathbf{S}_p - \lambda\mathbf{S}_t)\gamma_k = \tau_k\gamma_k, \tag{14}$$

where τ_k is the k-th maximal eigenvalues and γ_k is the corresponding eigenvectors, and $\mathbf{W}^n = [\gamma_1, \gamma_2, \ldots, \gamma_d]$.

    d. If $\|\mathbf{W}^n - \mathbf{W}^{n-1}\| < \varepsilon (\varepsilon = 10^{-4})$, end loop.

(3) Output $\mathbf{W} = \mathbf{W}^n$.

4 Sufficient and Necessary Conditions for the Existence of the Optimal Solution of Trace Ratio Problems

In this section, we prove the sufficient and necessary conditions for the existence of the optimal solution of trace ratio problems.

4.1 The Main Theoretical Result

The theoretical result about the sufficient and necessary conditions for the existence of the optimal solution of trace ratio problems can be described as follows.

Theorem 1. *The sufficient and necessary conditions for the existence of the optimal solution of trace ratio problems are that there is a sequence* $\{\lambda_1^*, \lambda_2^*, \ldots, \lambda_n^*\}$ *which converges to* λ^* *as* $n \to +\infty$, *where* λ^* *is the optimal value of the trace ratio in Problem (11).*

4.2 Proof

For the necessary condition for the existence of the optimal solution of trace ratio problems, we have

Lemma 1. *If* λ^* *is the optimal value of the trace ratio in Problem (11), there exists a sequence* $\{\lambda_1^*, \lambda_2^*, \ldots, \lambda_n^*\}$ *which converges to* λ^* *as* $n \to +\infty$.

Proof. We use a constructive method to prove this lemma.

    The Lagrangian function of Problem (11) is

$$L(W) = \frac{Tr(\mathbf{W}^T\mathbf{S}_p\mathbf{W})}{Tr(\mathbf{W}^T\mathbf{S}_t\mathbf{W})} - Tr[\gamma(\mathbf{W}^T\mathbf{W} - \mathbf{I})], \tag{15}$$

where γ is a symmetric matrix. Suppose $\gamma = \mathbf{U}\Gamma\mathbf{U}^T$, where Γ is the eigenvalue matrix of γ, and $\mathbf{U}$ is the corresponding eigenvector matrix.

    We calculate the gradient of Eq. (15) with respect to $\mathbf{W}$ and let it be equal to 0,

$$\frac{Tr(\mathbf{W}^T\mathbf{S}_t\mathbf{W})\mathbf{S}_p\mathbf{W} - Tr(\mathbf{W}^T\mathbf{S}_p\mathbf{W})\mathbf{S}_t\mathbf{W}}{Tr(\mathbf{W}^T\mathbf{S}_t\mathbf{W})^2} - \mathbf{W}\gamma = 0, \tag{16}$$

$$\Rightarrow (\mathbf{S}_p - \frac{Tr(\tilde{\mathbf{W}}^T\mathbf{S}_p\tilde{\mathbf{W}})}{Tr(\tilde{\mathbf{W}}^T\mathbf{S}_t\tilde{\mathbf{W}})}\mathbf{S}_t)\tilde{\mathbf{W}} = \tilde{\mathbf{W}}\tilde{\Gamma} \tag{17}$$

Where $\tilde{\mathbf{W}} = \mathbf{W}\mathbf{U}$, $\tilde{\Gamma} = Tr(\mathbf{W}^T\mathbf{S}_t\mathbf{W})\Gamma$.

Denote $\lambda = \frac{Tr(\tilde{\mathbf{W}}^T\mathbf{S}_p\tilde{\mathbf{W}})}{Tr(\tilde{\mathbf{W}}^T\mathbf{S}_t\tilde{\mathbf{W}})}$, which results in $(\mathbf{S}_p - \lambda\mathbf{S}_t)\tilde{\mathbf{W}} = \tilde{\mathbf{W}}\tilde{\Gamma}$. Furthermore, as we know, the trace ratio problem (11) has the optimal solution λ^*, i.e. $\lambda^* = \frac{Tr(\mathbf{W}^T\mathbf{S}_p\mathbf{W})}{Tr(\mathbf{W}^T\mathbf{S}_t\mathbf{W})}$.

Based on the iterative algorithm introduced in Sect. 3, we construct such a sequence $\{\lambda_1^*, \lambda_2^*, \cdots, \lambda_n^*\}$, and each λ_i^* satisfies

$$(1) \quad \frac{\partial L(\mathbf{W}, \lambda_0^*)}{\partial \mathbf{W}} = 0 \Rightarrow \lambda_1^* \tag{18}$$

$$(2) \quad \frac{\partial L(\mathbf{W}, \lambda_1^*)}{\partial \mathbf{W}} = 0 \Rightarrow \lambda_2^* \tag{19}$$

$$\vdots$$

$$(n) \quad \frac{\partial L(\mathbf{W}, \lambda_{n-1}^*)}{\partial \mathbf{W}} = 0 \Rightarrow \lambda_n^* \tag{20}$$

until $\|\mathbf{W}^n - \mathbf{W}^{n-1}\| < \varepsilon$, where λ_0^* is computed based on the initialized $\mathbf{W}^0$.

In the above iterative procedures, we have $(\mathbf{S}_p - \lambda_{k-1}^*\mathbf{S}_t)\mathbf{W}_{k-1} = \mathbf{W}_{k-1}\Gamma$ and $\lambda_k^* = \frac{tr(\mathbf{W}_{k-1}^T\mathbf{S}_p\mathbf{W}_{k-1})}{tr(\mathbf{W}_{k-1}^T\mathbf{S}_t\mathbf{W}_{k-1})}$, that is, the sequence $\{\lambda_1^*, \lambda_2^*, \cdots, \lambda_n^*\}$ is generated by Algorithm 1 as shown in Sect. 3. Due to the convergence of Algorithm 1, which will be proved in the sufficiency part, we know $\{\lambda_1^*, \lambda_2^*, \cdots, \lambda_n^*\}$ converges to the optimal value of the trace ratio, λ^*, of Problem (11).   □

Next, we prove the sufficient condition for the existence of the optimal solution of trace ratio problems.

Define the objective function of Problem (11) as $f(\mathbf{W}) = \frac{Tr(\mathbf{W}^T\mathbf{S}_p\mathbf{W})}{Tr(\mathbf{W}^T\mathbf{S}_t\mathbf{W})}$, we only need to prove that λ_i^* learned from Eqn. (12)-(14) monotonically increases and converges to the optimal value, λ^*, of the trace ratio problem.

Lemma 2. $\{\lambda_1^*, \lambda_2^*, \cdots, \lambda_n^*\}$ *monotonically increases, i.e.* $f(\mathbf{W}^i) \geqslant f(\mathbf{W}^{i-1})$ *and* $\lambda_{i+1}^* \geqslant \lambda_i^*$, *and Algorithm 1 converges to the optimal solution of Problem (11).*

Proof. Define $g_i(\mathbf{W}) = Tr[\mathbf{W}^T(\mathbf{S}_p - \lambda_i^*\mathbf{S}_t)\mathbf{W}]$, then $g_i(\mathbf{W}^{i-1}) = 0$.

Because the matrix $\mathbf{W}^i$ obtained by Algorithm 1 is orthogonal, that is $\mathbf{W}^T\mathbf{W} = \mathbf{I}_d$, we have

$$\sup_{\mathbf{W}^T\mathbf{W}=\mathbf{I}_d} g_i(\mathbf{W}) = \sum_{k=1}^{d} \tau_k, \tag{21}$$

where τ_k is the k-th largest eigenvalue of $(\mathbf{S}_p - \lambda_i^*\mathbf{S}_t)$. So, $g_i(\mathbf{W}^i) \geqslant g_i(\mathbf{W}^{i-1}) = 0$.

That is $Tr[\mathbf{W}^{iT}(\mathbf{S}_p - \lambda_i^*\mathbf{S}_t)\mathbf{W}^i] \geqslant 0$. Since matrix $\mathbf{S}_t$ is positive definite, we have

$$\frac{Tr(\mathbf{W}^{iT}(\mathbf{S}_p - \lambda_i^*\mathbf{S}_t)\mathbf{W}^i)}{Tr(\mathbf{W}^{iT}\mathbf{S}_t\mathbf{W}^i)} \geqslant 0, \tag{22}$$

$$\frac{Tr(\mathbf{W}^{iT}\mathbf{S}_p\mathbf{W}^i) - \lambda_i^* Tr(\mathbf{W}^{iT}\mathbf{S}_t\mathbf{W}^i)}{Tr(\mathbf{W}^{iT}\mathbf{S}_t\mathbf{W}^i)} \geqslant 0, \tag{23}$$

$$\frac{Tr(\mathbf{W}^{iT}\mathbf{S}_p\mathbf{W}^i)}{Tr(\mathbf{W}^{iT}\mathbf{S}_t\mathbf{W}^i)} \geqslant \lambda_i^*. \tag{24}$$

So, $f(\mathbf{W}^i) \geqslant f(\mathbf{W}^{i-1})$, that is $\lambda_{i+1}^* \geqslant \lambda_i^*$.

That is, λ_i^* is monotonically increasing. Since the objective function of Problem (11) has upper bound, $\{\lambda_1^*, \lambda_2^*, \cdots, \lambda_n^*\}$ will converge to its optimal solution.

□

5   Experiments

In this section, we show the experimental results on three face recognition tasks to demonstrate the convergence and effectiveness of the adopted trace ratio algorithm.

5.1   Data Sets

In the experiment, we use three data sets that are commonly used in face recognition research.

CMU-PIE CMU-PIE has more than 41000 images from 68 volunteers under the conditions of different poses, illuminations and expressions. In the experiment, the eye position of the human face is fixed, and we intercept the face images into the size of 32×32 and give them corresponding label tags. For this dataset, the first 15 face images from all 68 subjects are used.

Yale Yale is provided by the Vision and Control Center, Yale University. It contains 165 pictures of 15 volunteers, by different illuminations, expressions and posture conditions. In the experiment, the eye position of the human face is fixed, and we intercept the face images into the size of 32×32 and give them corresponding label tags. For this dataset, all the face images are used.

YaleB The Yale Face Database B (YaleB) contains 2432 images of 38 human subjects under 64 illumination conditions. In the experiment, the eye position of the human face is fixed, and we intercept the face images into the size of 32×32 and give them corresponding label tags. For this dataset, the first 15 face images from all 38 subjects are used.

5.2   Classification Algorithm – KNN

In the experiment, we use k-Nearest Neighbors (KNN) classifier for classification. KNN is a relatively simple algorithm. Given a training data set which has been successfully classified, for a test sample, we select the nearest $k(k = 7)$ training instances, and based on a majority vote method, choose the class which has most of the instances. We break ties by selecting the smallest number among the classes. Particularly, we use the Euclidean distance for the distance metric in our experiment.

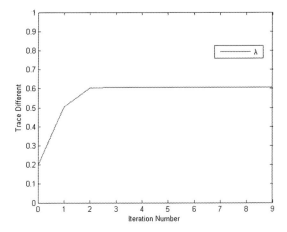

Fig. 1. Value of λ_i^* obtained in the iteration process.

5.3 Convergence of the Sequence $\{\lambda_1^*, \lambda_2^*, \ldots, \lambda_n^*\}$

In this section, we show that $\{\lambda_1^*, \lambda_2^*, \ldots, \lambda_n^*\}$ converges to the optimal solution, λ^*, efficiently. The used data set is CMU-PIE. During the iterative processes, the decision step $\|\mathbf{W}^n - \mathbf{W}^{n-1}\| < \varepsilon$ $(\varepsilon = 10^{-4})$ is omitted. The iteration is carried out for 10 times, and the value obtained at each time is recorded in Fig. 1.

From Fig. 1, we can see that the value of λ_i^* is gradually increased, and finally converges to a maximum value. This also proves the convergence of the iterative method for optimizing the Trace Ratio problems.

5.4 Recognition Rate

In this section, we compare 3 methods mentioned in Sect. 2, i.e. PCA, LDA and MFA, with the convergent algorithm for solving the Trace Ratio problems (Trace Ratio), where PCA, LDA and MFA is solved with (generalized) eigenvalue decomposition. For PCA, the low-dimensional representations with dimensionalities between 1 and $C - 1$ are evaluated, where C is the number of classes, and the best result is reported. For LDA, MFA and Trace Ratio, we first project the data to a subspace of dimensionality C using PCA, where C is the number of classes, and then evaluate the compared methods for all possible dimensionalities of the low-dimensional representations. As mentioned before, we use KNN with $k = 7$ for the classification. Furthermore, for MFA, we set the number of nearest neighbors for the intra-class compactness k_1 to 5 and that for the inter-class separability k_2 to 20. For the Trace Ratio algorithm, we use a same formulation as LDA except for the conjugate orthogonal constraints. In addition, "GxPy" means that each person's x pictures are used for training, and the rest y pictures for testing. Tables 1 and 2 show the recognition rates obtained on the CMU-PIE and Yale faces, where the dimensionality that results in the best result is shown in the brackets. We can see that the adopted Trace Ratio method greatly improves previous approaches.

Table 1. Recognition rate on the CMU-PIE data set.

	G4P11	G5P10	G6P9
PCA	0.1698 (65)	0.1985 (46)	0.2369 (51)
LDA	0.4265 (13)	0.4618 (15)	0.5948 (11)
MFA	0.3904 (10)	0.4353 (13)	0.5229 (11)
Trace Ratio	**0.4278 (15)**	**0.4838 (16)**	**0.5980 (17)**

Table 2. Recognition rate on the Yale data set.

	G4P7	G5P6	G6P5
PCA	0.4286 (11)	0.4667 (7)	0.5067 (9)
LDA	0.6190 (9)	0.6778 (8)	0.7067 (8)
MFA	0.6095 (8)	0.6444 (6)	0.6933 (8)
Trace Ratio	**0.6762 (6)**	**0.7000 (7)**	**0.7333 (8)**

To evaluate the effect of different dimensionalities for the low-dimensional representations, we compare the adopted Trace Ratio method with LDA and MFA on the YaleB data set. In the experiments, we first reduce the dimensionality of data to $(C =)38$ using PCA, and then apply LDA, MFA and the convergent Trace Ratio algorithm to further project the PCA features to subspaces with different dimensionalities. From Table 3, we can see that the Trace Ratio method outperforms other compared approaches consistently with regard to the recognition rate. This proves the effectiveness of the adopted Trace Ratio method.

Table 3. Recognition rate on the YaleB data set.

Dimension	LDA	MFA	Trace Ratio
10	0.9375	0.9211	**0.9605**
15	0.9507	0.9474	**0.9671**
20	0.9243	0.9178	**0.9474**
optimum	0.9671 (19)	0.9605 (17)	**0.9737 (17)**

6   Conclusion

In this paper, we put forward the necessary and sufficient conditions for the existence of the optimal solution of trace ratio problems. The correctness of the theoretical result is proved. Experiments on three face recognition tasks demonstrate the convergence and effectiveness of the adopted algorithm.

Acknowledgments. This work was supported by the National Natural Science Foundation of China (NSFC) under Grant No. 61403353, the Open Project Program of the National Laboratory of Pattern Recognition (NLPR) and the Fundamental Research Funds for the Central Universities of China.

References

1. Huang, Y., Xu, D., Nie, F.: Semi-supervised dimension reduction using trace ratio criterion. IEEE Trans. Neural Netw. Learn. Syst. **23**(3), 519–526 (2012)
2. Jia, Y., Nie, F., Zhang, C.: Trace ratio problem revisited. IEEE Trans. Neural Netw. **20**(4), 729–735 (2009)
3. Nie, F., Xiang, S., Jia, Y., Zhang, C.: Semi-supervised orthogonal discriminant analysis via label propagation. Pattern Recogn. **42**(11), 2615–2627 (2009)
4. Nie, F., Xiang, S., Jia, Y., Zhang, C., Yan, S.: Trace ratio criterion for feature selection. In: AAAI, pp. 671–676 (2008)
5. Wang, H., Yan, S., Xu, D., Tang, X., Huang, T.: Trace ratio vs. ratio trace for dimensionality reduction. In: CVPR (2007)
6. Xiang, S., Nie, F., Zhang, C.: Learning a Mahalanobis distance metric for data clustering and classification. Pattern Recogn. **41**(12), 3600–3612 (2008)
7. Yan, S., Tang, X.: Trace quotient problems revisited. In: Leonardis, A., Bischof, H., Pinz, A. (eds.) ECCV 2006. LNCS, vol. 3952, pp. 232–244. Springer, Heidelberg (2006)
8. Yan, S., Xu, D., Zhang, B., Zhang, H.J., Yang, Q., Lin, S.: Graph embedding and extensions: A general framework for dimensionality reduction. IEEE Trans. Pattern Anal. Mach. Intell. **29**(1), 40–51 (2007)
9. Zhong, G., Cheriet, M.: Tensor representation learning based image patch analysis for text identification and recognition. Pattern Recogn. **48**(4), 1211–1224 (2015)
10. Zhong, G., Shi, Y., Cheriet, M.: Relational Fisher analysis: A general framework for dimensionality reduction. In: IJCNN (2016)

Robust Multi-label Feature Selection
with Missing Labels

Qian Xu, Pengfei Zhu$^{(\boxtimes)}$, Qinghua Hu, and Changqing Zhang

School of Computer Science and Technology, Tianjin University, Tianjin, China
{xuqian912,zhupengfei,huqinghua,zhangchangqing}@tju.edu.cn

Abstract. With the fast development of social networks, high-dimensionality is becoming an intractable problem in many machine learning and computer vision tasks. This phenomenon also exists in the field of multi-label classification. So far many supervised or semi-supervised multi-label feature selection methods have been proposed to reduce the feature dimension of training samples. However, almost all existing feature selection works focus on multi-label learning with complete labels. In fact, labels are very expensive to obtain and the training instances usually have an incomplete/partial set of labels (some labels are randomly missed). Very few researchers pay attention to the problem of multi-label feature selection with missing labels. In this paper, we propose a robust model to solve the above problem. We recover the missing labels by a linear regression model and select the most discriminative feature subsets simultaneously. The effective $l_{2,p}$-norm $(0 < p \leq 1)$ regularization is imposed on the feature selection matrix. The iterative reweighted least squares (IRLS) algorithm is used to solve the optimization problem. To verify the effectiveness of the proposed method, we conduct experiments on five benchmark datasets. Experimental results show that our method has superior performance over the state-of-the-art algorithms.

Keywords: Feature selection · Multi-label learning · Missing labels

1 Introduction

Multi-label learning refers to that one instance can be assigned to several labels simultaneously. In many real-world applications such as semantic annotation of images, text categorization and gene function classification, we would be faced with many tasks involved with multi-label learning. In text categorization, a text document from the hospital may be assigned to more than one labels, e.g., *diabetes* and *gravida*. For image and video annotations, one video may be annotated with several tags, including *grass* and *sheep*. Unfortunately, in real applications, the labels are usually incomplete because it is hard to label a sample accurately, especially for automatic image annotation. Due to the boom of mass media, high-dimensional data bring about many difficulties because of great storage burden

© Springer Nature Singapore Pte Ltd. 2016
T. Tan et al. (Eds.): CCPR 2016, Part I, CCIS 662, pp. 752–765, 2016.
DOI: 10.1007/978-981-10-3002-4_61

and high time complexity in multi-label learning tasks. Moreover, the parameters of learning machines may increase exponentially with feature dimensions. Too many parameters can lead to over-fitting and decrease the generalization ability of learning machines [19,22].

To alleviate the curse of dimensionality, feature selection is used to choose the most discriminative features, and remove the redundant and noisy ones at the same time [7]. According to the availability of sample labels, feature selection methods can be categorized into supervised [16], semi-supervised [3] and unsupervised [1,11,15] cases. For supervised feature selection, the training data are completely labeled and we can make full use of these labels. But for unsupervised feature selection, the task becomes very challenging for the lack of labels. Generally, there are three types of feature selection algorithms, i.e., filter, wrapper and embedding ones. Filter methods evaluate the discriminative capacity of features by defining various measurements (e.g., variance [7], consistency [4], Laplacian score [8]). Wrapper methods mainly rely on the performance of the classification or clustering algorithms to rank or select a subset of features. However, the dependence on the classification or clustering algorithms can lead to bad generalization ability. For embedding methods, the features are selected during the process of model construction [18]. Sparse regularizations are usually imposed on the feature selection matrices such as $l_{2,1}$-norm, which has been proved can effectively select features.

Multi-label feature selection algorithms are developed to reduce the dimensionality of multi-label data. Two key factors for multi-label feature selection are the metrics to evaluate the importance of candidate features and the search strategy to solve the optimization function. Zhang et al. proposed a multi-label feature selection method which adopts a two-stage filter-wrapper feature selection strategy [24]. Principal component analysis is used in the first stage and genetic algorithms are used to choose the most discriminative features. Cai et al. proposed a graph structured sparsity model for multi-label learning, which can also apply to multi-label feature selection tasks [2]. Ma et al. proposed a subspace-sparsity collaborated feature selection model for multi-label learning [13]. Lee et al. proposed to utilize mutual information between selected features and the label set [9]. Lin et al. proposed to select features by maximizing the dependency and minimizing the redundancy simultaneously [12]. The feature dependency evaluates the contribution of candidate features and the feature redundancy represents the information overlap. Zhang et al. proposed to transform the original data into a low-dimensional feature space by maximizing the dependence between the original feature description and the associated class labels [25].

The existing multi-label feature selection methods require that instances are completely labeled. Since the cost to get amounts of labeled data is extremely expensive, we can hardly get fully labeled data. Thus we have some missing labels missed for certain instances [17,20,21]. For example, for the task of image annotation, when the candidate targets are large or there is a large ambiguity between classes, it is hard to give all the right labels for the human annotators.

As for a label which has not been assigned to an instance, we have no idea whether this label belongs to the sample or not. This kind of problem is obvious different from the supervised or semi-supervised multi-label feature selection. This kind of multi-label problem is called multi-label learning with missing labels (MLML) [20].

Under the MLML circumstance, since the label information is incomplete, it is hard for us to exploit the relations between labels and features. Moreover, missing labels would arise class imbalance for there are multiple labels for each instance. The key factor for MLML is how to handle and address the information of missing labels. Some methods treat the missing labels as negative labels directly. It is simple but crude and usually brings label bias into the objective function. Some approaches treat the recovery of missing labels as a matrix completion (MC) problem based on the low rank assumption [6]. Zhou et al. proposed WELL that considers the inherent class imbalance by assuming that the classification boundary for each label goes across low density regions [17]. Xu et al. defined a supplementary label matrix which captures rich information regarding label dependence to address the issue of incomplete label matrices [21]. Wu et al. defined three label states, including positive labels as $+1$, negative labels as -1, and missing labels as 0 to avoid the label bias [20]. Wu et al. proposed an inductive model based on the framework of regularized logistic regression [20]. It takes label consistency and label smoothness into account simultaneously and the problem of MLML is formulated as a mixed graph to encode a network of label dependencies.

While earlier researches pay much attention to multi-label feature selection and multi-label learning, no efforts are made to select features for multi-label tasks when the labels are incomplete. In this paper, we proposed an algorithm to solve the problem of multi-label feature selection with missing labels (MLMLFS). We consider the problem of multi-label learning, feature selection and missing labels simultaneously. The missing labels are recovered by a linear regression model. We iteratively update the label matrix and the parameters of the linear regression model. The effective $l_{2,p}$-norm $(0 < p \leq 1)$ regularization is imposed on the feature selection matrix to select the most discriminative feature subsets. The iterative reweighted least squares (IRLS) algorithm is used to solve the optimization problem. Experiment results on benchmark multi-label datasets show that MLMLFS outperforms the state-of-the-art multi-label feature selection algorithms, especially when the percentage of the missing labels is high.

The organization of this paper is listed as follows. In Sect. 2, we give the basic notations and introduce the objective of multi-label feature selection. In Sect. 3, we propose our robust multi-label feature selection model to deal with missing labels. Section 4 conducts experiments and Sect. 5 concludes.

2   Problem Statement

Here, we denote $\mathbf{X} = \{x_1, \cdots, x_n\}$ as the training sample matrix, where $x_i \in \mathbb{R}^d$ represents the i-th instance and n is the number of training samples.

$\mathbf{Y} = \{y_1, \cdots, y_n\}$ denotes the label matrix where $y_i \in \mathbb{R}^c$ is the label vector of i-th sample and c is the number of classes. $\mathbf{Y}_{i,j}$ represents the association between i-th sample and j-th label. $\mathbf{Y}_{i,j} = +1, -1$ represent that the i-th sample is assigned to label j or has no relationship with label j. Particularly, $\mathbf{Y}_{i,j} = 0$ means we could not conclude whether this sample belongs to label j or not and it can be seen as a missing label. We define a predicted label matrix $\mathbf{F} \in \mathbb{R}^{n \times c}$ and set $\mathbf{F}_l = \mathbf{Y}_l$ for all the assigned labels. For missing labels, we initially set the value of labels to zeros.

As previously stated, when labels are randomly missed, the key issues are to capture the relations between labels and features, and alleviate the class imbalance problem. To tackle these problems, we expect to recover the missing labels and conduct feature selection using the constructed completely labeled matrix. The general model can be written as

$$\min_{f, \mathbf{F}_l = \mathbf{Y}_l} \sum_{i=1}^{n} loss(f(x_i), y_i) + \mu \Omega(f) \tag{1}$$

where $loss(\cdot)$ is a loss function and $\Omega(f)$ is the regularization term with μ is a parameter.

Missing labels can be recovered by different means, e.g., matrix completion, linear or logistic regression [3], label propagation [20], deep learning [10], etc. In this paper, we aim to solve the curse of dimensionality problem in multi-label tasks with missing labels. Hence, we consider the linear or logistic regression models.

3   The Proposed Model

3.1   Motivation

In this paper, three aspects including multi-label learning, feature selection and the recovery of missing labels are simultaneously considered. We expect to select the most discriminative feature subsets when recovering the missing labels. Our model can be written as:

$$\min_{\mathbf{W}, \mathbf{b}, \mathbf{Y}} \left\| \mathbf{X}^T \mathbf{W} + \mathbf{1}\mathbf{b}^T - \mathbf{Y} \right\|_{2,1} + \lambda \left\| \mathbf{W} \right\|_{2,p}^p \tag{2}$$

where $\mathbf{X}$ is the data matrix and $\mathbf{Y}$ is the label matrix with randomly missing labels. $\mathbf{1} \in \mathbb{R}^{n \times 1}$ is a column vector with all elements being 1. Here, we use $l_{2,1}$-norm loss to remove outliers and impose a $l_{2,p}$-norm constraint on the feature selection matrix $\mathbf{W}$. $l_{2,p}$-norm is the generalization of $l_{2,1}$-norm first introduced by [5].

$\|W\|_{2,p}^p$ is defined as

$$\|W\|_{2,p}^p = \sum_{i=1}^{d} \left(\sum_{j=1}^{c} w_{ij}^2 \right)^{p/2} = \sum_{i=1}^{d} \|w_i\|^p \tag{3}$$

where w_i is the i-th row of $\mathbf{W}$, d is the feature dimension and c is the class number. When the values of λ increases to some certain value, most rows of $\mathbf{W}$ become zeros. The energy of each row reflects the discriminative ability of the corresponding feature. Hence, we can perform feature selection by ranking features with $\|w_i\|_2$. According to [23,26], the degree of sparsity on $\mathbf{W}$ increases when the value of p decreases. Despite that $\|\mathbf{W}\|_{2,p}^p$ $(p < 1)$ is non-convex, the selected features under some good local optimal solutions perform better than those selected at $p = 1$. That is, the best performance of selected features does not always lie in $p = 1$ cases. In this paper, we use the above model to implement multi-label feature selection with missing labels and analysis the influence of different p values $(0 < p \leq 1)$ on feature selection performance under diverse missing ratios.

3.2   Optimization and Algorithms

Since the objective function is non-smooth and difficult to solve, we propose to solve it as follows

$$tr((\mathbf{X}^T\mathbf{W} + \mathbf{1}\mathbf{b}^T - \mathbf{Y})^T\mathbf{G}_0(\mathbf{X}^T\mathbf{W} + \mathbf{1}\mathbf{b}^T - \mathbf{Y})) + \lambda\|\mathbf{W}\|_{2,p}^p \qquad (4)$$

where $\mathbf{G}_0$ is a diagonal matrix which is defined as $\mathbf{G}_0 = \frac{1}{2\|x_i^T\mathbf{W}+b_i^T-y_i\|}$. Firstly, we set the derivative of Eq. (4) $w.r.t.\,\mathbf{b}$ to 0, and we can get

$$\mathbf{b} = \frac{1}{m}\mathbf{Y}^T\mathbf{G}_0\mathbf{1} - \frac{1}{m}\mathbf{W}^T\mathbf{X}\mathbf{G}_0^T\mathbf{1} \qquad (5)$$

where $m = \mathbf{1}^T\mathbf{G}_0\mathbf{1}$. Substituting the expression of $\mathbf{b}$ into Eq. (4), the original problem can be transformed into

$$\min_{\mathbf{W},\mathbf{b},\mathbf{Y}} tr(((\mathbf{I} - \frac{1}{m}\mathbf{1}\mathbf{1}^T\mathbf{G}_0)\mathbf{X}^T\mathbf{W} - (\mathbf{I} - \frac{1}{m}\mathbf{1}\mathbf{1}^T\mathbf{G}_0)\mathbf{Y})^T\mathbf{G}_0$$
$$((\mathbf{I} - \frac{1}{m}\mathbf{1}\mathbf{1}^T\mathbf{G}_0)\mathbf{X}^T\mathbf{W} - (\mathbf{I} - \frac{1}{m}\mathbf{1}\mathbf{1}^T\mathbf{G}_0)\mathbf{Y})) + \lambda\|\mathbf{W}\|_{2,p}^p \qquad (6)$$

where $\mathbf{I}$ is an identity matrix. Here, we denote $\mathbf{H} = \mathbf{I} - \frac{1}{m}\mathbf{1}\mathbf{1}^T\mathbf{G}_0$ as a centering matrix which satisfies the property of $\mathbf{H}^T = \mathbf{H}$. Then we can rewrite Eq. (6) as

$$\min_{\mathbf{W},\mathbf{b},\mathbf{Y}} tr((\mathbf{H}\mathbf{X}^T\mathbf{W} - \mathbf{H}\mathbf{Y})^T\mathbf{G}_0(\mathbf{H}\mathbf{X}^T\mathbf{W} - \mathbf{H}\mathbf{Y})) + \lambda\|\mathbf{W}\|_{2,p}^p \qquad (7)$$

As for $\|\mathbf{W}\|_{2,p}^p$, when $p = 1$, it is the standard $l_{2,1}$-norm and the problem is convex but non-smooth. When $0 < p \leq 1$, the problem is non-convex, we can solve the problem by using iterative reweighted least squares (IRLS). Given the current $\mathbf{W}^{t+1}$, we can define the diagonal weighting matrices $\mathbf{G}^t$ as:

$$g_j^t = \frac{p}{2}\|w_j^t\|_2^{p-2} \qquad (8)$$

where g_j^t is the j-th diagonal element of $\mathbf{G}_1^t$ and w_j^t is the j-th row of $\mathbf{W}^t$. Then we can update the value of $\mathbf{W}^{t+1}$ by solving the following weighted least squares problem:

$$\mathbf{W}^{t+1} = \arg \min_{\mathbf{W}} \ Q(\mathbf{W} \mid \mathbf{W}^t)$$
$$= \arg \min_{\mathbf{W}} \ tr((\mathbf{H}\mathbf{X}^T\mathbf{W} - \mathbf{H}\mathbf{Y})^T \mathbf{G}_0(\mathbf{H}\mathbf{X}^T\mathbf{W} - \mathbf{H}\mathbf{Y})) + \lambda tr(\mathbf{W}^T\mathbf{G}_1^t\mathbf{W})$$
(9)

Let $\dfrac{\partial Q(\mathbf{W}\mid\mathbf{W}^t)}{\partial \mathbf{W}} = 0$, we can get

$$\mathbf{X}\mathbf{H}^T\mathbf{G}_0(\mathbf{H}\mathbf{X}^T\mathbf{W} - \mathbf{H}\mathbf{Y}) + \lambda \mathbf{G}_1^t\mathbf{W} = 0 \tag{10}$$

Then the closed-form solution of $\mathbf{W}^{t+1}$ can be written as:

$$\mathbf{W}^{t+1} = (\mathbf{X}\mathbf{H}^T\mathbf{G}_0\mathbf{H}\mathbf{X}^T + \lambda \mathbf{G}_1^t)^{-1}\mathbf{X}\mathbf{H}^T\mathbf{G}_0\mathbf{H}\mathbf{Y} \tag{11}$$

Here, in Eq. (11), we need to compute the inverse of $(\mathbf{X}\mathbf{H}^T\mathbf{G}_0\mathbf{H}\mathbf{X}^T + \lambda \mathbf{G}_1^t)$. While in some applications such as gene expression when the feature dimension is much larger than the numbers of samples, the computation complexity would be very high. Hence, we deal with the inverse according to different situations.

When the feature dimension is smaller than the number of samples, we can compute $\mathbf{W}$ using Eq. (11). For the case that the feature dimension is larger than the number of samples, we can compute $\mathbf{W}$ with the help of Woodbury Matrix Identity. The general form of Woodbury Matrix Identity is:

$$(\mathbf{A} + \mathbf{B}\mathbf{C}\mathbf{D})^{-1} = \mathbf{A}^{-1} - \mathbf{A}^{-1}\mathbf{B}(\mathbf{C}^{-1} + \mathbf{D}\mathbf{A}^{-1}\mathbf{B})^{-1}\mathbf{D}\mathbf{A}^{-1} \tag{12}$$

As for the equation to compute $\mathbf{W}$, let $\mathbf{K} = \mathbf{X}\mathbf{H}^T$, then we have

$$\mathbf{W}^{t+1} = (\mathbf{K}\mathbf{G}_0\mathbf{K}^T + \lambda\mathbf{G}_1^t)^{-1}\mathbf{K}\mathbf{G}_0\mathbf{H}\mathbf{Y} \tag{13}$$

According to Woodbury Matrix Identity, we can further have

$$\mathbf{W}^{t+1} = (\lambda\mathbf{G}_1^t + \mathbf{K}\mathbf{G}_0\mathbf{K}^T)^{-1}\mathbf{K}\mathbf{G}_0\mathbf{H}\mathbf{Y}$$
$$= (\mathbf{G}_1^t)^{-1}\mathbf{K}(\lambda\mathbf{I} + \mathbf{G}_0\mathbf{K}^T(\mathbf{G}_1^t)^{-1}\mathbf{K})^{-1}\mathbf{G}_0\mathbf{H}\mathbf{Y} \tag{14}$$

After $\mathbf{W}$ is updated, the diagonal matrix $\mathbf{G}$ is updated by Eq. (8). To get a stable solution, a sufficiently small tolerance value is introduced by defining

$$g_j^t = \frac{p}{2\max(\|w_j^t\|_2^{2-p}, \varepsilon)} \tag{15}$$

By iteratively updating $\mathbf{W}_t$, $\mathbf{G}_0$ and $\mathbf{G}_t$, the objective value of Eq. (2) monotonically decreases and guarantees to converge to a fixed point.

After we get $\mathbf{W}$ and $\mathbf{b}$, we can compute the prediction label matrix $\tilde{\mathbf{F}} = \mathbf{X}^T\mathbf{W} + \mathbf{1}\mathbf{b}^T$. After each iteration, we adjust the label values of missing positions as:

$$\mathbf{F}_{ij} = \begin{cases} 0, & if\ \tilde{\mathbf{F}}_{ij} \leq 0 \\ \tilde{\mathbf{F}}_{ij}, & if\ 0 \leq \tilde{\mathbf{F}}_{ij} \leq 1 \\ 1, & if\ \tilde{\mathbf{F}}_{ij} \geq 1 \end{cases} \tag{16}$$

Here, we restrict the value of F_{ij} in the range of 0 to 1 to avoid trivial solution. By constantly iterating, the algorithm can converge to a stable point. Then we have the final $\mathbf{F}$ and we can use simple operation to have discrete label values for the missing labels. After the recovery, we get completely labeled data. This is extremely informative to the further feature selection. The algorithm of MLMLFS is summarized in Algorithm 1.

3.3   Time Complexity

The main complexity of the proposed model is the computation of $\mathbf{W}$ in each iteration. Note that in multi-label classification, the feature dimension and the number of samples are usually larger than the number of classes. When $d < n$, the time complexity of MLMLFS is $O(Td^2n)$. Otherwise, the time complexity is $O(Tn^2d)$, where T is the iteration number.

3.4   Convergence Analysis

For the model in Eq. (2), when $p = 1$, the objective function is convex. The convergence of the $l_{2,1}$-norm optimization problem has been well studied and proved [14]. Hence, when $p = 1$, the model in Eq. (2) is sure to converge. When $0 < p < 1$, the proposed model can converge to a stationary point, which has been

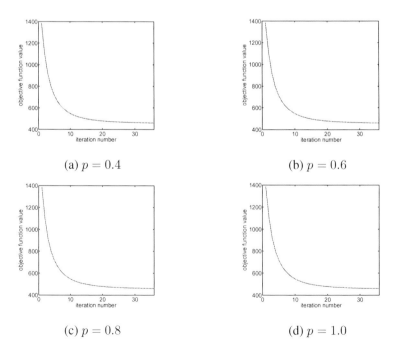

(a) $p = 0.4$ (b) $p = 0.6$

(c) $p = 0.8$ (d) $p = 1.0$

Fig. 1. The convergence curve of MLMLFS with different p values.

Algorithm 1. Robust Multi-label Feature Selection with Missing Labels (MLMLFS)

Input:
    Training data $\mathbf{X} \in \mathbb{R}^{n \times d}$
    Training label matrix with missing labels $\mathbf{Y} \in \mathbb{R}^{n \times c}$
    Parameter λ, p, missing percent a
1: Set $t = 0$ and initialize $\mathbf{W}_0 \in \mathbb{R}^{d \times c}$;
2: **repeat**
3:     Compute the diagonal matrices $\mathbf{G}_0$ and $\mathbf{G}_1$;
4:     Compute $\mathbf{W}_{t+1}$;
5:     **if** $d < n$ **then**
6:         $\mathbf{W}^{t+1} = (\mathbf{X}\mathbf{H}^T\mathbf{G}_0\mathbf{H}\mathbf{X}^T + \lambda\mathbf{G}_1^t)^{-1}\mathbf{X}\mathbf{H}^T\mathbf{G}_0\mathbf{H}\mathbf{Y}$;
7:     **else**
8:         let $\mathbf{K} = \mathbf{X}\mathbf{H}^T$;
9:         $\mathbf{W}^{t+1} = (\mathbf{G}_1^t)^{-1}\mathbf{K}(\lambda\mathbf{I} + \mathbf{G}_0\mathbf{K}^T(\mathbf{G}_1^t)^{-1}\mathbf{K})^{-1}\mathbf{G}_0\mathbf{H}\mathbf{Y}$;
10:     **end if**
11:     Compute $\mathbf{b}_{t+1}$ according to $\mathbf{b}_{t+1} = \frac{1}{m}\mathbf{Y}^T\mathbf{G}_0\mathbf{1} - \frac{1}{m}\mathbf{W}^T\mathbf{X}\mathbf{G}_0^T\mathbf{1}$;
12:     Compute $\tilde{\mathbf{F}}_{t+1}$ according to $\tilde{\mathbf{F}}_{t+1} = \mathbf{X}^T\mathbf{W} + \mathbf{1}\mathbf{b}^T$;
13:     Adjust $\mathbf{F}$ according to Eq. (16);
14: **until** Convergence criterion satisfied.
Output:
    Feature selection matrix $\mathbf{W} \in \mathbb{R}^{d \times c}$
    Global optimized predicted label matrix $\mathbf{F} \in \mathbb{R}^{n \times c}$

---

proved [26]. We fix the value of λ and get the convergence curves of MLMLFS with different p values on dataset Artificial in Fig. 1. We can see that MLMLFS converges rapidly with different p values.

4 Experimental Analysis

4.1 Datasets and Evaluation Metrics

In this paper, we use five benchmark datasets including Artificial, Birds, Emotion, Health and Scene to validate the effectiveness of our model. These datasets are from different application areas and we give the detailed information about the five datasets in Table 1.

The evaluation measures used in multi-label feature selection is different from that used in single-label cases. Five evaluation metrics included Average Precision, Hamming Loss, Ranking Loss, Coverage and One Error are the most common used metrics to evaluate the performance of multi-label feature selection. In this paper, we just report Average Precision (AP) results for concise.

Average Precision computes the average fraction of labels ranked above a particular label $\gamma \in y_i$.

$$AP = \frac{1}{n}\sum_{i=1}^{n}\frac{1}{|y_i|}\sum_{\gamma \in y_i}\frac{|\gamma' \in y_i : r_i(\gamma') \leq r_i(\gamma)|}{r_i(\gamma)} \tag{17}$$

Table 1. Descriptions of the benchmark datasets

Data	Samples	Features	Classes	Training	Testing	Domains
Artificial	5000	462	26	2000	3000	Text
Birds	645	260	19	322	323	Audio
Emotion	593	72	6	391	202	Music
Health	5000	612	32	2000	3000	Text
Scene	2407	294	6	1211	1196	Image

where $r_i(l)$ stands for the rank of label $l \in \mathbf{L}$ predicted by the algorithm for a given instance x_i. Greater value indicates better performance.

4.2   Comparison Methods

To verify the effectiveness of MLMLFS, we compare its performance with five baseline algorithms given by MLNB, MDDMspc, MDDMproj, PMU and SFUS.

- MDDMspc [25]: MDDM is a multi-label dimensionality reduction method. Attempting to project the original data into a lower-dimensional feature space maximizing the dependence between the original feature description and the associated class labels.It has two kinds of projection strategies, MDDMspc is the first kind of projection strategy.
- MDDMproj [25]: MDDMproj is another kind of projection strategy given by MDDM.
- PMU [9]: PMU naturally derives from mutual information between selected features and the label set. It considers label interactions in evaluating the dependency of given features without resorting to problem transformation.
- SFUS [13]: It selects the most relevant features by using a sparsity-based model and uncover the shared subspace of original features.

4.3   Parameter Setting

We tune all the parameters in the range of $\{10^{-6}, 10^{-4}, 10^{-2}, 10^{0}, 10^{2}, 10^{4}, 10^{6}\}$ for each algorithm and the best results are reported. The four comparison algorithms are for supervised multi-label feature selection which requires the whole labels are available. But for MLMLFS, we need to use training label matrices with certain percentage of missing labels. For we have primitively validated that MLMLFS performs relatively better when the missing percentage is high. Here, we set the percentage of missing labels as 50 %, 80 % and 95 % to compare the performance of MLMLFS and other algorithms. Especially for MLMLFS, we have another parameter p which influences the ability of feature selection. To analyze the impacts of different p values, we conduct experiments on different p values including 0.4, 0.6, 0.8, 1 and record the results under each circumstance. Then, the Average Precision is used to evaluate the performance.

4.4   Feature Selection Results

We present the results measured by Average Precision. The performance under 50%, 80% and 95% percent missing labels are listed in Tables 2, 3, 4, respectively. From the results of experiments, we can observe that

- Under one certain percentage of missing labels, MLMLFS outperforms the other four comparison algorithms. This can be explained by the following reasons. Firstly, we use a linear regression model to recover missing labels while other methods do no operations before feature selection. This exactly indicates that the recovery of missing labels is of much benefit to the process of feature selection. Secondly, we impose a $l_{2,p}$-norm $(0 < p \le 1)$ constraint on the feature selection matrix.
- For different p values, the best result does not always lies to $p = 1$. In fact, the degree of sparsity increases when the value of p decreases. It is the sparsity that influence the ability of selecting useful features while removing irrelevant and noisy ones. Hence, proper p values should be chosen to boost the performance.
- With the increase of missing percentage, the performance of all methods decreases to some extent. However, our method still gets better performance than the state-of-the-arts. And especially when the missing percentage is high, our method performs much better than others. This shows the robustness of our method.

Figure 2 shows that the values of Average Precision (AP) vary with different number of selected features. We can observe that when the number of selected features is small, the AP is quite small. When we increase the feature number to a certain extent, the AP reaches a peak value. And if we continue to increase the feature number, the AP decreases gradually or changes subtly. In addition, MLMLFS consistently outperforms other methods on all the datasets under different percentage of missing labels.

Table 2. Average Precision of different methods under 50% percent missing labels on benchmark datasets

Data	MDDMproj	MDDMspc	PMU	SFUS	MLMLFS $(p = 0.4)$	MLMLFS $(p = 0.6)$	MLMLFS $(p = 0.8)$	MLMLFS $(p = 1)$
Artificial	46.27	44.89	47.73	49.24	53.84	53.89	**53.90**	53.74
Birds	62.38	62.51	63.19	65.35	64.68	66.48	**67.86**	66.97
Emotion	77.86	77.99	76.89	77.26	79.02	79.14	79.14	**79.16**
Health	63.52	63.56	65.85	67.39	72.35	72.25	72.38	**72.50**
Scene	79.78	78.36	79.47	80.46	**83.58**	82.85	82.59	82.53

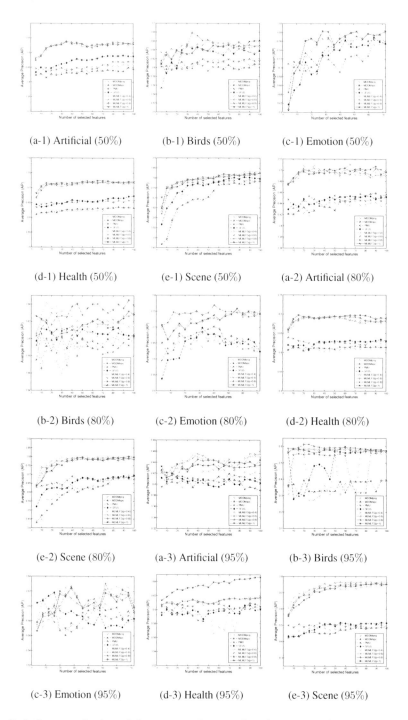

(a-1) Artificial (50%)   (b-1) Birds (50%)   (c-1) Emotion (50%)

(d-1) Health (50%)   (e-1) Scene (50%)   (a-2) Artificial (80%)

(b-2) Birds (80%)   (c-2) Emotion (80%)   (d-2) Health (80%)

(e-2) Scene (80%)   (a-3) Artificial (95%)   (b-3) Birds (95%)

(c-3) Emotion (95%)   (d-3) Health (95%)   (e-3) Scene (95%)

Fig. 2. Influence of selected feature number on five datasets under various missing percentages.

Table 3. Average Precision of different methods under 80 % percent missing labels on benchmark datasets

Data	MDDMproj	MDDMspc	PMU	SFUS	MLMLFS $(p = 0.4)$	MLMLFS $(p = 0.6)$	MLMLFS $(p = 0.8)$	MLMLFS $(p = 1)$
Artificial	44.68	46.15	45.42	46.00	**51.07**	51.02	50.60	50.68
Birds	60.63	61.05	60.91	61.55	64.16	63.08	**64.50**	64.12
Emotion	69.85	71.03	69.85	70.86	76.94	76.73	**77.37**	76.54
Health	62.81	61.69	64.14	63.42	69.53	69.69	69.56	**69.92**
Scene	69.86	69.48	69.13	70.31	80.12	80.28	79.30	**80.56**

Table 4. Average Precision of different methods under 95 % percent missing labels on benchmark datasets

Data	MDDMproj	MDDMspc	PMU	SFUS	MLMLFS $(p = 0.4)$	MLMLFS $(p = 0.6)$	MLMLFS $(p = 0.8)$	MLMLFS $(p = 1)$
Artificial	42.64	42.45	42.53	43.10	45.46	**45.60**	45.11	44.41
Birds	57.29	52.20	57.05	57.69	59.40	58.93	**61.56**	60.00
Emotion	61.77	61.77	60.89	62.03	62.96	63.14	63.11	**63.41**
Health	61.10	61.10	61.15	60.71	61.80	61.73	61.82	**64.78**
Scene	49.34	50.68	51.52	51.68	75.23	**77.10**	76.34	75.89

5   Conclusions and Future Work

In this paper, we proposed a novel feature selection model for multi-label feature selection with missing labels (MLMLFS). Although many efforts are made to develop multi-label feature selection algorithms, few researchers focus on feature selection when labels are incomplete. A linear regression model is utilized to recover the missing labels. The label matrix and parameters of the linear regression model are iteratively updated until the proposed model converges. Experiments on benchmark multi-label datasets show that MLMLFS outperforms the state-of-the-art multi-label feature selection algorithms, especially when the percentage of missing labels is very high. In multi-label classification tasks, label dependency is very important and widely used in multi-label learning. In the future work, we will extent MLMLFS by adding the label dependency.

References

1. Cai, D., Zhang, C., He, X.: Unsupervised feature selection for multi-cluster data. In: ACM SIGKDD International Conference on Knowledge Discovery and Data Mining, Washington, DC, USA, July, pp. 333–342 (2010)
2. Cai, X., Nie, F., Cai, W., Huang, H.: New graph structured sparsity model for multi-label image annotations. In: Proceedings of the IEEE International Conference on Computer Vision, pp. 801–808 (2013)

3. Chang, X., Nie, F., Yang, Y., Huang, H.: A convex formulation for semi-supervised multi-label feature selection. In: Twenty-Eighth AAAI Conference on Artificial Intelligence, the Twenty-Sixth Innovative Applications of Artificial Intelligence Conference, the Symposium on Educational Advances in Artificial Intelligence, pp. 1171–1177 (2014)
4. Dash, M., Liu, H.: Consistency-based search in feature selection. Artif. Intell. **151**(1–2), 155–176 (2003)
5. Ding, C., Zhou, D., He, X., Zha, H.: R1-pca: rotational invariant L1-norm principal component analysis for robust subspace factorization. In: Proceedings of the 23rd International Conference on Machine Learning, pp. 281–288. ACM (2006)
6. Goldberg, A.B., Zhu, X., Recht, B., Xu, J., Nowak, R.D.: Transduction with matrix completion: three birds with one stone. In: Conference on Neural Information Processing Systems 2010, Proceedings of a Meeting Held 6–9 December 2010, Vancouver, British Columbia, Canada, vol. 23, pp. 757–765 (2010)
7. Guyon, I., Elisseeff, A.: An introduction to variable and feature selection. Appl. Phys. Lett. **3**(6), 1157–1182 (2002)
8. He, X., Cai, D., Niyogi, P.: Laplacian score for feature selection. Adv. Neural Inf. Process. Syst. **18**, 507–514 (2005)
9. Lee, J., Kim, D.W.: Feature selection for multi-label classification using multivariate mutual information. Pattern Recogn. Lett. **34**(3), 349–357 (2013)
10. Li, X., Zhao, F., Guo, Y.: Conditional restricted boltzmann machines for multi-label learning with incomplete labels. In: AISTATS (2015)
11. Li, Z., Yang, Y., Liu, J., Zhou, X., Lu, H.: Unsupervised feature selection using nonnegative spectral analysis. In: Proceedings of the National Conference on Artificial Intelligence, pp. 1026–1032 (2012)
12. Lin, Y., Hu, Q., Liu, J., Duan, J.: Multi-label feature selection based on max-dependency and min-redundancy. Neurocomputing **168**, 92–103 (2015)
13. Ma, Z., Nie, F., Yang, Y., Uijlings, J.R.R., Sebe, N.: Web image annotation via subspace-sparsity collaborated feature selection. IEEE Trans. Multimedia **14**(4), 1021–1030 (2012)
14. Nie, F., Huang, H., Cai, X., Ding, C.H.: Efficient and robust feature selection via joint l2, 1-norms minimization. In: Advances in Neural Information Processing Systems, pp. 1813–1821 (2010)
15. Qian, M., Zhai, C.: Robust unsupervised feature selection. In: International Joint Conference on Artificial Intelligence, pp. 1621–1627 (2013)
16. Song, L., Smola, A., Gretton, A., Borgwardt, K.M., Bedo, J.: Supervised feature selection via dependence estimation. In: ICML, pp. 823–830 (2007)
17. Sun, Y.Y., Zhang, Y., Zhou, Z.H.: Multi-label learning with weak label. In: Twenty-Fourth AAAI Conference on Artificial Intelligence, AAAI 2010, Atlanta, Georgia, USA, July (2010)
18. Wang, S., Tang, J., Liu, H.: Embedded unsupervised feature selection (2015)
19. Wolf, L., Shashua, A.: Feature selection for unsupervised and supervised inference: the emergence of sparsity in a weight-based approach. J. Mach. Learn. Res. **6**(3), 1855–1887 (2005)
20. Wu, B., Lyu, S., Hu, B.G., Ji, Q.: Multi-label learning with missing labels for image annotation and facial action unit recognition. Pattern Recogn. **48**(7), 2279–2289 (2015)
21. Xu, L., Wang, Z., Shen, Z., Wang, Y., Chen, E.: Learning low-rank label correlations for multi-label classification with missing labels, vol. 36, pp. 1067–1072 (2014)

22. Zhai, Y., Ong, Y.S., Tsang, I.W.: The emerging "big dimensionality". IEEE Comput. Intell. Mag. **9**(3), 14–26 (2014)
23. Zhang, M., Ding, C., Zhang, Y., Nie, F.: Feature selection at the discrete limit. In: AAAI Conference on Artificial Intelligence (2014)
24. Zhang, M.L., Pena, J.M., Robles, V.: Feature selection for multi-label naive bayes classification. Inf. Sci. **179**(19), 3218–3229 (2009)
25. Zhang, Y., Zhou, Z.H.: Multilabel dimensionality reduction via dependence maximization. ACM Trans. Knowl. Discov. Data **4**(3), 1503–1505 (2010)
26. Zhu, P., Hu, Q., Zhang, C., Zuo, W.: Coupled dictionary learning for unsupervised feature selection. In: AAAI (2016)

Fractional Orthogonal Fourier-Mellin Moments for Pattern Recognition

Huaqing Zhang, Zongmin Li[(⊠)], and Yujie Liu

China University of Petroleum (Huadong), 66# Changjiang West Road,
Qingdao Economic and Technological Development Zone,
Beijing 266580, Shandong, China
zhang.huaqing@163.com,
{lizongmin,liuyujie}@upc.edu.cn

Abstract. In this paper, we generalize the orthogonal Fourier-Mellin moments (OFMMs) to the fractional orthogonal Fourier-Mellin moments (FOFMMs), which are based on the fractional radial polynomials. We propose a new method to construct FOFMMs by using a continuous parameter t ($t > 0$). The fractional radial polynomials of FOFMMs have the same number of zeros as OFMMs with the same degree. But the zeros of FOFMMs polynomial are more uniformly distributed than which of OFMMs and the first zero is closer to the origin. A recursive method is also given to reduce computation time and improve numerical stability. Experimental results show that the proposed FOFMMs have better performance.

Keywords: Fractional orthogonal Fourier-Mellin moments · Orthogonal Fourier-Mellin moments · Moment invariant · Pattern recognition

1 Introduction

Moment methods effectively describe an image in terms of form and shape and possess excellent invariant to image translation, rotation and scaling. They have been used in a multitude of applications in image analysis, such as object representation and recognition [1–5], image registration [6, 7], robot navigation [8], image retrieval [9, 10], medical imaging [11, 12] and watermarking [13–15].

Hu [16] introduced moment invariants in 1961, based on methods of algebraic invariants. However, higher order moments are vulnerable, and the recovery of the image from these moments is considered to be extremely difficult [17]. Teague [18] has proposed the concept of orthogonal moments to recover the image from moments based on the theory of orthogonal polynomials, and has introduced Zernike moments. The most remarkable quality of orthogonal moments, is their ability of full description of an object, with least redundancy [19, 20]. Consequently, the reconstruction of an object by a finite number of moments is possible. Furthermore, Teh and Chin [17] proved that the orthogonal moments perform better than non-orthogonal moments in image representation and are more robust to noise. Other families of orthogonal moments were proposed through the years, such as Lengendre [21], Chebyshev [11], Pseudo-Zernike [11] and orthogonal Fourier-Mellin moments [23–28].

© Springer Nature Singapore Pte Ltd. 2016
T. Tan et al. (Eds.): CCPR 2016, Part I, CCIS 662, pp. 766–778, 2016.
DOI: 10.1007/978-981-10-3002-4_62

The most frequently employed orthogonal families are the Zernike, Pseudo-Zernike and Fourier-Mellin moment. Ref. [17, 29–31] have shown that Zernike moments have outstanding performance as image descriptors. But, when they are used for scale-invariant pattern recognition, Zernike moments have difficulty in describing images of small size. Sheng and Shen [27] introduced the orthogonal Fourier-Mellin moments (OFMMs) as the generalized Zernike moments or the orthogonal complex moments, and demonstrated OFMMs have better performance than ZMs in regard to image reconstruction errors and signal-to-noise ratios, especially in describing images of small size. OFMMs have been widely used in image processing. Kan and Srinath [32] applied the OFMMs to character recognition. Ref. [33] proposed a method based on OFMM to edge location. Ref. [27, 34] proposed OFMMs for pattern recognition and the latter demonstrated a general approach to construct a complete set of OFMM invariants. In recent years, some research progress on OFMMs, especially on the computation of OFFMs, has been achieved [20, 24, 31, 35–42]. However, higher order moments are more sensitive to noise, which prevents the applications in image representation and pattern recognition. On the other hand, it is shown that only the moments of higher orders carry the fine detail of an image [17]. How to construct superior moments and moment invariants is still a valuable research issue.

In this paper, we generalize the OFMMs to fractional Fourier-Mellin moments (FOFMMs) by letting it depending on a continuous parameter $t(t > 0)$. The traditional OFMMs is proved to be a special case of FOFMMs. We first propose a new method to construct FOFMMs, then discuss the properties and performance of FOFMMs. Experimental results show that FOFMMs perform better than OFMMs, especially when t takes a value with $0 < t < 1$.

The paper is organized as follows. In Sect. 2, the definitions of OFMMs are given. In Sect. 3, definitions of FOFMMs are presented. Section 4 discusses the properties and performance of FOFMMs. Section 5 proposes a fast recursive algorithm for the computation of the FOFMMs. Section 6 provides the experimental results and the comparative analysis in terms of the image reconstruction capability, recognition accuracy and robustness to noise. Finally, some concluding remarks are given in Sect. 7.

2   Traditional Orthogonal Fourier-Mellin Moments(OFMMs)

Firstly, we introduce the definition of OFMMs described in Ref. [27], which is based on a set of radial polynomials.

Let $f(r, \theta)$ be a two-dimensional (2D) image function defined in the polar coordinate system (r, θ) with $|r| \leq 1$. The 2D OFMM with order p and repetition q, O_{pq}, is defined as

$$O_{pq} = \frac{p+1}{\pi} \int_0^{2\pi} \int_0^1 U_{pq}(r, \theta) f(r, \theta) r dr d\theta \qquad (1)$$

where p is a non-negative integer and represents the order of Mellin radial transform, q is the circular harmonic order and $q = 0, \pm 1, \pm 2, \ldots$. $U_{pq}(r, \theta)$ is a set of complex polynomials defined as

$$U_{pq}(r,\theta) = Q_p(r)e^{-iq\theta} \tag{2}$$

where $Q_p(r)$ represents a set of orthogonal radial polynomials given by

$$Q_p(r) = \sum_{k=0}^{p}(-1)^{p+k}\frac{(p+k+1)!}{(p-k)!k!(k+1)!}r^k \tag{3}$$

It has been verified in [27] that the set $Q_p(r)$ is orthogonal over the range $0 \le r \le 1$.

$$\int_0^1 Q_p(r)Q_s(r)rdr = \frac{1}{2(p+1)}\delta_{ps}, \tag{4}$$

where δ_{ps} is Kronecker delta defined as $\delta_{ps} = \begin{cases} 0, p \ne s \\ 1, p = s \end{cases}$. And it can be proved that the polynomials of (2) possess the orthogonality

$$\int_0^{2\pi}\int_0^1 U_{pq}(r,\theta)U_{st}^*(r,\theta)rdrd\theta = \frac{\pi}{p+1}\delta_{qt}\delta_{ps}. \tag{5}$$

The discrete version of OFMMs can be expressed in rectangle coordinates (x,y) as

$$O_{pq} = \frac{p+1}{\pi}\sum_x\sum_y f(x,y)U_{pq}(r,\theta)\Delta x\Delta y. \tag{6}$$

where $x^2 + y^2 \le 1$. Here we employ the method of mapping transformation proposed in Ref. [34].

Suppose that the image is rotated through angle φ, and $\hat{O}_{pq}$ is the OFMMs after rotation, then

$$\hat{O}_{pq} = O_{pq}e^{-iq\varphi}. \tag{7}$$

Therefore $|O_{pq}|$, the moduli of the OFMMs, can be taken as rotation-invariant features for recognition.

3 Fractional Orthogonal Fourier-Mellin Moments(FOFMMs)

The idea of this paper comes from the fractional calculus. Based on fractional order integral, FOFMMs are a set of moments based on the fractional orthogonal Fourier-Mellin polynomials. In this section, definition of a set of fractional orthogonal polynomials is provided, which differs from the definition in Ref. [42]. FOFMMs and FOFMMs invariants are also introduced in this section.

3.1   The Fractional Radial Polynomials of the FOFMMs

Let $t > 0$, the orthogonal radial polynomials set $Q_p(r)$ is generalized to fractional form $Q_p^t(r)$, which is defined as

$$Q_p^t(r) = \begin{cases} 0, & r = 0 \\ \sqrt{t}r^{t-1}Q_p(r^t), & 0 < r \leq 1 \end{cases} \tag{8}$$

namely

$$Q_p^t(r) = \begin{cases} 0, & r = 0 \\ \sqrt{t}\sum_{k=0}^{p}(-1)^{p+k}\frac{(p+k+1)!}{(p-k)!k!(k+1)!}r^{t(k+1)-1}, & 0 < r \leq 1 \end{cases} \tag{9}$$

It is true $Q_p^t(r) = Q_p(r)$ if $t = 1$. The fractional polynomials set is defined as

$$U_{pq}^t(r, \theta) = Q_p^t(r)e^{-iq\theta} \tag{10}$$

From Eq. (4) we obtain Eq. (11)

$$\int_0^1 Q_p^t(r)Q_s^t(r)rdr = \frac{1}{2(p+1)}\delta_{ps} \tag{11}$$

Equation (11) is satisfied if $r = 0$. Consequently, the fractional radial polynomials $Q_p^t(r)$ are orthogonal over the interval $0 \leq r \leq 1$.

Based on Eqs. (10) and (11), we have the following derivation

$$\begin{aligned} &= \int_0^{2\pi} \int_0^1 U_{pq}^t(r, \theta)U_{sv}^{t*}(r, \theta)rdrd\theta \\ &= \int_0^{2\pi} e^{-iq\theta}e^{iv\theta}d\theta \int_0^1 Q_p^t(r)Q_s^t(r)rdr \\ &= \frac{\pi}{p+1}\delta_{qv}\delta_{ps} \end{aligned}$$

We obtain

$$\int_0^{2\pi} \int_0^1 U_{pq}^t(r, \theta)U_{sv}^{t*}(r, \theta)rdrd\theta = \frac{\pi}{p+1}\delta_{qv}\delta_{ps} \tag{12}$$

The Eq. (12) is satisfied if $r = 0$. The fractional polynomials $U_{pq}^t(r, \theta)$ are orthogonal over the interval $0 \leq r \leq 1$.

3.2   The Definition of FOFMMs

We define FOFMMs using the orthogonal polynomials set $U_{pq}^t(r, \theta)$ in Polar system as

$$F_{pq}^t = \frac{p+1}{\pi} \int_0^{2\pi} \int_0^1 U_{pq}^t(r, \theta)f(r, \theta)rdrd\theta \tag{13}$$

where t is a non-negative real number, p is a non-negative integer,$q = 0, \pm 1, \pm 2, \ldots,$ are the moment orders. It is easy to see that F^t_{pq} is O_{pq} when $t = 1$.

Like OFMMs, the FOFMMs are orthogonal complex moments, therefore FOFMMs may be computed in the rectangle coordinate system. The discrete version of FOFMMs can be expressed in rectangle coordinates (x, y) as

$$F^t_{pq} = \frac{p+1}{\pi} \sum_x \sum_y U^t_{pq}(r, \theta) f(x, y) \Delta x \Delta y. \tag{14}$$

where $x^2 + y^2 \leq 1$ and employ the as method of mapping transformation as in OFMM. FOFMMs are defined in polar coordinates (r, θ) with $|r| \leq 1$, so the computation of FOFMMs needs a transformation of the image coordinates to a domain inside a unit circle. Here we use the mapping transformation proposed Ref. [34].

3.3   FOFMMs Invariants

If an image $f(r, \theta)$ is rotated by an angle φ, and $\hat{F}^t_{pq}$ is the FOFMMs after rotation, then

$$\begin{aligned}
\hat{F}^t_{pq} &= \frac{p+1}{\pi} \int_0^{2\pi} \int_0^1 U^t_{pq}(r, \theta) f(r, \theta + \emptyset) r \, dr \, d\theta \\
&= \frac{p+1}{\pi} \int_0^{2\pi} \int_0^1 U^t_{pq}(r, \theta) f(r, \theta) r \, dr \, d\theta \, e^{-iq\emptyset} \\
&= F^t_{pq} e^{-iq\emptyset}
\end{aligned}$$

Therefore the moduli of FOFMMs $\left| F^t_{pq} \right|$, can be taken as rotation-invariant features for recognition.

4   Properties of FOFMM

The number of zeros of the radial polynomials corresponds to the ability of the polynomials to describe high-spatial-frequency components of an image. The distribution and positions of zeros correspond to the sampling positions in the image and the capability of the polynomials to catch the images information respectively. For OFMMs, the equation $Q_p(r) = 0$ has p real and distinct roots in the range of $0 \leq r \leq 1$ [27]. Correspondingly, in FOFMMs, it's easy to see that the equation $Q^t_p(r) = 0$ also has p real and distinct roots in the range of $0 \leq r \leq 1$.

Figure 1(a) shows the plots of radial polynomials $Q_p(r)$ of OFMMs with $0 \leq r \leq 1$, p = 0,1,2,···,10, namely the plots of $Q^t_p(r)$ of FOFMM when t = 1; Fig. 1(b) shows the plots of radial polynomials $Q^t_p(r)$ of FOFMMs with $t = 0.7$, $0 \leq r \leq 1$, $p = 0, 1, 2, \cdots, 10$; Fig. 1(c) shows the plots of radial polynomials $Q_p(r)$ of OFMMs with $0 \leq r \leq 0.1$, p = 0, 1, 2, ···, 10; Fig. 1(d) shows the plots of radial polynomials $Q^t_p(r)$ of FOFMMs with $t = 0.7, 0 \leq r \leq 0.1$, p = 0,1,2,···, 10. As illustrated in Fig. 1 (a) and (b), the zeros of the $Q^t_p(r)$ are more uniformly distributed than the $Q_p(r)$ over the interval $0 \leq r \leq 1$. This property shows that center and margin of an image have the same contribution to the calculation of FOFMMs.

Figure 1(a) and (b) also show that the first zero of $Q_p^t(r)$ is nearer the origin than which of $Q_p(r)$. The first zero of $Q_{10}(r)$ is at r = 0.030029. The first zero of $Q_{10}^{0.6}(r)$ is at r = 0.00290114, as shown in Table 1. Sheng and Shen [27] proposed that this difference is important, since in scale-invariant pattern recognition the object sizes are unknown a priori and the moments of objects of different size should be computed with the same basis functions. The first zero's location affects the description capability of moments. Hence, FOFMMs are better than OFMMs in describing small images, as we show in Sect. 6.

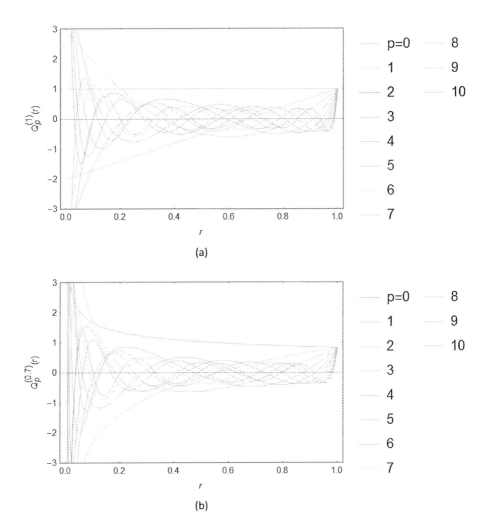

Fig. 1. (a) Radial polynomials $Q_p(r)$ of OFMMs with $0 \leq r \leq 1$, $p = 0, 1, 2, \cdots, 9$; (b) Radial polynomials $Q_p^t(r)$ of FOFMMs with $t = 0.7$, $0 \leq r \leq 1$, $p = 0, 1, 2, \cdots, 9$.

Table 1. Positions of the First Zeros of $Q_p(r)$ and $Q_p^t(r)$

	$Q_p(r)$	$Q_p^{0.5}(r)$	$Q_p^{0.6}(r)$	$Q_p^{0.7}(r)$	$Q_p^{0.8}(r)$
p	r	r	r	r	r
1	0.666667	0.444444	0.508762	0.560326	0.602401
2	0.355051	0.126061	0.178026	0.227802	0.274071
3	0.212341	0.0450885	0.0755767	0.1093	0.144142
4	0.13976	0.0195328	0.0376389	0.0601337	0.0854532
5	0.0985351	0.00970916	0.0210209	0.0364983	0.0552063
6	0.0730543	0.00533693	0.0127666	0.0238033	0.0379802
7	0.0562626	0.00316548	0.008261	0.0163908	0.0274015
8	0.044634	0.0822684	0.0056162	0.0117747	0.0205155
9	0.0362578	0.0139426	0.00397194	0.00874988	0.0158217
10	0.030029	0.00966073	0.00290114	0.00668435	0.0125005

5 Computation of FOFMMs

Although FOFMMs have superior properties as discussed above, the direct computation of $Q_p^t(r)$ is a very time-consuming task. The same problem also exists in the case of OFMMs. A factorial-free algorithm for the computation, that reduce or even eliminate the factorial calculations, have been introduced by Papakostas [36]. In this paper, a recursive algorithm for the computation of the FOFMMs, as in the case of OFMMs in Ref. [36], is proposed.

Equation (14) shows the approximate computation of FOFMMs. The orthogonal radial polynomials $Q_p^t(r)$ are computed recursively using the following equations:

$$Q_p^t(r) = (T_1 r^t + T_2)Q_{p-1}^t(r) + T_3 Q_{p-2}^t(r), \ p = 2, 3, 4, \ldots, M \qquad (14a)$$

with

$$Q_0^t(r) = \sqrt{t} \qquad (14b)$$

$$Q_1^t(r) = -2\sqrt{t}r^t + 3\sqrt{t}r^{t+1} \qquad (14c)$$

where

$$T_1 = \frac{4p+2}{p+1} \qquad (15a)$$

$$T_2 = -2p + \frac{p(p-1)}{2p-1}T_1 \qquad (15b)$$

$$T_3 = (2p-1)(p-1) - \frac{(p-1)(p-2)}{2}T_1 + 2(p-1)T_2 \qquad (15c)$$

where M is the maximum order of FOFMMs. Eq. (14a) is free from factorial terms and powers of p. Therefore, it is numerically more stable and efficient compared with direct computation methods. Using Eq. (15a), (15b) and (15c), the recursive algorithm eliminates the calculation of the fractional terms in $Q_p^t(r)$. A fast computation of trigonometric functions is also employed which is introduced in Ref. [36]. Based on them, we proposed a similar algorithm in this paper, which is a fast, accurate and memory efficient method.

6 Experimental Results

In this section, we compare the performance of the proposed FOFMMs with that of the OFMMs in image reconstruction and object classification.

6.1 Image Reconstruction

Based on the principle of the orthogonal theory, the reconstructed image $\hat{f}(r, \theta)$ is approximately expressed by a truncated series

$$\hat{f}(r, \theta) \approx \sum_{p=0}^{p_{max}} \sum_{q=-q_{max}}^{q_{max}} F_{pq}^t Q_p^t(r) e^{iq\theta}$$

where p_{max}, q_{max} are the highest orders of the FOFMMs used in image reconstruction. We use the mean square reconstruction error, which has been defined in [47], to measure the performance of reconstruction. The MSRE is defined as

$$\varepsilon = \frac{\sum_{x=0}^{N-1} \sum_{y=0}^{M-1} \left| f(x, y) - \hat{f}(x, y) \right|^2}{\sum_{x=0}^{N-1} \sum_{y=0}^{M-1} f^2(x, y)} \tag{17}$$

where $f(x, y)$ is the original image, $\hat{f}(x, y)$ is the reconstructed image and ε is the mean square constructing error. Figure 2 shows a set of characters of size 16×16, which is used as test images. FOFMMs are obtained from the images shown in Fig. 2, and the maximum order of the moments used to reconstruct images is 12. The results are given in Fig. 3. Comparative construction results of the same letter "B" by OFMMs and FOFMMs are demonstrated in Fig. 4. It can be observed that FOFMMs have better performance.

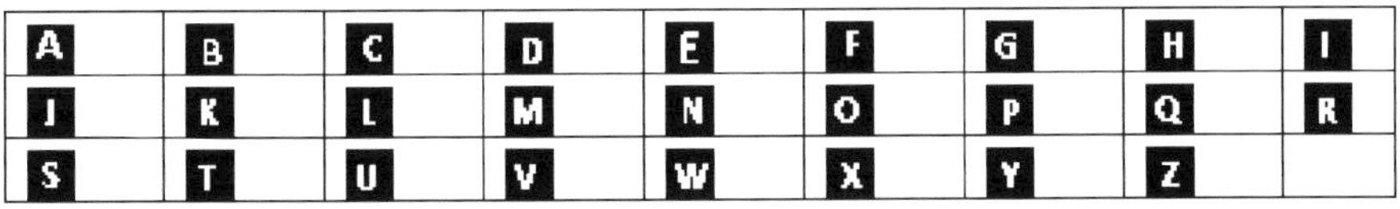

A	B	C	D	E	F	G	H	I
J	K	L	M	N	O	P	Q	R
S	T	U	V	W	X	Y	Z	

Fig. 2. Set of uppercase English letters used as test images, each of size 16×16.

	A	**B**	**C**	**D**	**E**	**F**	**G**	**H**	**I**
MSRE	0.0322	0.0000	0.0000	0.0833	0.0000	0.0000	0.2058	0.0000	0.0000
	J	**K**	**L**	**M**	**N**	**O**	**P**	**Q**	**R**
MSRE	0.0000	0.0625	0.0000	0.2857	0.0285	0.2941	0.0000	0.3333	0.1764
	S	**T**	**U**	**V**	**W**	**X**	**Y**	**Z**	
MSRE	0.1600	0.0000	0.0000	0.0000	0.1964	0.0000	0.0000	0.0000	

Fig. 3. Reconstructed Images by FOFMMs up to 12th order of uppercased English letters, each of size 16×16.

B
(a)

Order of Moments	3	4	5	6	7
Reconstructed image					
MSRE of OFMM	1	1	0.892857	0.785714	0.785714
Reconstructed image with t=0.7					
MSRE of FOFMM	0.928571	0.7500000	0.7500000	0.642857	0.607143

Order of Moments	8	9	10	11	12
Reconstructed image					
MSRE of OFMM	0.642857	0.571429	0.428571	0.5000000	0.178571
Reconstructed image with t=0.7					
MSRE of FOFMM	0.5000000	0.392857	0.214286	0.25	0.0000000

(b)

Fig. 4. (a) Binary character image of size 16×16. (b) Comparative reconstruction results of the letter "B" of size 16×16.

6.2 Character Recognition

The magnitude of FOFMMs is invariant to image rotation, so we take it as rotation invariant feature. Different size images are tested. Figure 5 shows two binary images of size 16×16 and 64×64 respectively, which are used as test images. Tables 2 and 3 show the invariant and discriminative abilities of the second and third order of FOFMMs.

We also build a database consists of 16×16 binary images of all 26 English characters. Per character consists of ten images, which are transformed by scaling and rotating the original set with random scale and random rotation angle. The method is applied for recognizing characters in the database and the correct probability of recognition arrives at 89.1 %. To test the robustness to noises, a salt-and-pepper noise

Fig. 5. (a) Binary character image of size 16×16. (b) Binary number image of size 64×64.

Table 2. Proposed rotational invariants of character "B" of size 16×16 with $t = 0.7$

	$\varphi_{20}^{0.3}$	$\varphi_{21}^{0.3}$	$\varphi_{22}^{0.3}$	$\varphi_{30}^{0.3}$	$\varphi_{31}^{0.3}$	$\varphi_{32}^{0.3}$	$\varphi_{33}^{0.3}$
0°	0.340389	0.0248282	0.0058609	0.189838	0.073089	0.0620696	0.0075823
31°	0.355257	0.0057526	0.0231535	0.275571	0.0595053	0.054652	0.0205546
79°	0.340614	0.0186866	0.0260691	0.246493	0.0687777	0.0605162	0.0183208
87°	0.351211	0.0184053	0.029083	0.223959	0.0782477	0.064685	0.0153476
176°	0.306554	0.024581	0.0185391	0.184066	0.0517776	0.0846756	0.0270668
328°	0.35033	0.0084456	0.0413356	0.281069	0.0096005	0.0652299	0.0171987

Table 3. Proposed rotational invariants of number "5" of size 64×64 with $t = 0.7$

	$\varphi_{20}^{0.7}$	$\varphi_{21}^{0.7}$	$\varphi_{22}^{0.7}$	$\varphi_{30}^{0.7}$	$\varphi_{31}^{0.7}$	$\varphi_{32}^{0.7}$	$\varphi_{33}^{0.7}$
0°	0.090912	0.0242071	0.021758	0.0151125	0.0330562	0.0358022	0.0050103
64°	0.093253	0.0200913	0.019120	0.0259979	0.0239509	0.0448683	0.0109889
144°	0.100475	0.0064395	0.026155	0.0178789	0.0136821	0.0475496	0.0078181
201°	0.098669	0.0050788	0.025642	0.0190703	0.0070835	0.0480788	0.0056094
277°	0.094091	0.0088955	0.027966	0.0188434	0.0174077	0.0471151	0.016351
317°	0.101177	0.0072479	0.028280	0.0206633	0.0054734	0.0506043	0.0039457

has been added. Table 4 demonstrates the comparative recognition results with different noise densities. It can be seen that FOFMMs method is more robust than OFMMs method for noisy images.

The proposed algorithm has been implemented using C++ and tested on a PC with an Intel(R) Core(TM) i5 480 M 2.67 GHz processor and 2 GB memory. The average recognition time is about 0.05 s.

Table 4. Classification accuracy of the character images with different noise densities

	Noise-free(%)	4 %	8 %	12 %
FOFMMs	100 %	100 %	100 %	65.3 %
OFMMs	100 %	100 %	92.3 %	61.5 %

7   Conclusion

In this paper, we propose a new set of orthogonal moments based on the fractional radial polynomials, named FOFMMs, which are generalization of OFMMs. A new method to construct FOFMMs by using a continuous parameter $t(t > 0)$ is introduced. The fractional radial polynomials of FOFMMs have the same number of zeros as OFMMs of the same degree. But the zeros of the FOFMMs polynomials are more uniformly distributed than which of OFMMs. Moreover, the first zero is closer to the origin. That is an important property. Experimental results show that the proposed FOFMMs have better descriptive capability and more robust to noise, especially for images of small size.

Acknowledgments. This work is partly supported by National Natural Science Foundation of China (Grant no. 61379106), the Shandong Provincial Natural Science Foundation (Grant nos. ZR2013FM036, ZR2015FM011), the Open Project Program of the State Key Lab of CAD&CG (Grant no. A1315), Zhejiang University, the Fundamental Research Funds for the Central Universities (Grant nos. 14CX02032A, 14CX02031A).

References

1. Flusser, J., Suk, T.: Classification of degraded signals by the method of invariants. Signal Process. **60**, 243–249 (1997). http://www.sciencedirect.com/science/journal/01678655
2. Bigiin, J., Hans du Buf, J.M.: N-folded symmetries by complex moments in Gabor space and their application to unsupervised texture segmentation. IEEE Trans. Pattern Anal. Mach. Intell. **16**, 80–87 (1994)
3. Reeves, P., Prokop, R.J.: Three-dimensional shape analysis using moments and Fourier descriptors. IEEE Trans. Pattern Anal. Mach. Intell. **10**, 937–943 (1988)
4. Gope, C., Kehtarnavaz, N., Hillman, G., Wursig, B.: An affine invariant curve matching method for photo-identification of marine mammals. Pattern Recogn. **38**, 125–132 (2005)
5. Mokhtarian, F., Abbasi, S.: Robust automatic selection of optimal views in multi-view free-form object recognition. Pattern Recogn. **38**, 1021–1031 (2005)
6. Goshtasby, A.: Template matching in rotated images. IEEE Trans. Pattern Anal. Mach. Intell. **7**, 338–344 (1985)
7. Zitova, B., Flusser, J.: Image registration methods: a survey. Image Vis. Comput. **21**, 977–1000 (2003)
8. Lee, J., Ko, H.: Gradient-based local affine invariant feature extraction for mobile robot localization in indoor environments. Pattern Recogn. Lett. **29**, 1934–1940 (2008)
9. Wei, C.H., Li, Y.: Trademark image retrieval using synthetic features for describing global shape and interior structure. Pattern Recogn. **42**, 386–394 (2009)
10. Fu, X., Li, Y.: Content-based image retrieval using Gabor-Zernike features. In: 18th International Conference on Pattern Recognition (ICPR 2006), vol. 2, pp. 417–420 (2006)
11. Flusser, J., Suk, T., Zitova, B.: Moments and Moment Invariants in Pattern Recognition, pp. 165–209. Wiley, New York (2009)
12. Dai, X.B., Shu, H.Z., et al.: Reconstruction of tomographic images from limited range projections using discrete radon transform and Tchebichef moments. Pattern Recogn. **43**, 1152–1164 (2010)
13. Xin, Y.Q., Liao, S.: Circularly orthogonal moments for geometrically robust image watermarking. Pattern Recogn. **40**, 3740–3752 (2007)
14. Zhang, H., Shu, H.Z.: Affine Legendre moment invariants for image watermarking robust to geometric distortion. IEEE Trans. Image Process. **20**, 2189–2199 (2011)
15. Shao, Z.H., Shu, H.Z.: Quaternion Bessel-Fourier moments and their invariant descriptors for object reconstruction and recognition. Pattern Recogn. **47**, 603–611 (2014)
16. Hu, M.K.: Visual pattern recognition by moment invariants. IRE Trans. Inf. Theory **8**, 179–187 (1962)
17. Teh, C.H., Chin, R.T.: On image analysis by the method of moments. IEEE Trans. Pattern Anal. Mach. Intell. **10**, 496–513 (1988)
18. Teague, M.R.: Image analysis via the general theory of moments. J. Opt. Soc. Am. A: Opt. Image Sci. **70**, 920–930 (1980)

19. Mostafa, Y.A., Psaltis, D.: Recognitive aspects of moment invariants. IEEE Trans. Pattern Anal. Mach. Intell. **6**, 689–706 (1984)
20. Papakostas, G.A., et al.: Fast numerically stable computation of orthogonal Fourier-Mellin moments. IET Comput. Vis. **1**, 11–16 (2007)
21. Shu, H.Z., et al.: A new fast method for computing Legendre moments. Pattern Recogn. **33**, 341–348 (2000)
22. Pawlak, M.: Image analysis by moments: reconstruction and computational aspects. Oficyna Wydawnicza Politechniki Wroclawskiej (2006)
23. Terrillon, J.C., et al.: Invariant neural-network based face detection with orthogonal Fourier–Mellin moments. In: The 15th IEEE International Conference on Pattern Recognition, vol. 2, pp. 993–1000 (2000)
24. Wang, X., et al.: Scaling and rotation invariant analysis approach to object recognition based on Radon and Fourier-Mellin transforms. Pattern Recogn. **40**(12), 3503–3508 (2007)
25. Andrew, T.B.J., David, N.C.L.: Integrated wavelet and Fourier-Mellin invariant feature in fingerprint verification system. In: ACM SIGMM Workshop on Biometrics Methods and Applications, pp. 82–88 (2003)
26. Sheng, Y., Arsenault, H.H.: Experiments on pattern recognition using invariant Fourier-Mellin descriptors. J. Opt. Soc. Am. A: Opt. Image Sci. **3**(6), 771–776 (1986)
27. Sheng, Y., Shen, L.: Orthogonal Fourier-Mellin moments for invariant pattern recognition. J. Opt. Soc. Am. A: Opt. Image Sci. **11**, 1748–1757 (1994)
28. Papakostas, G.A., et al.: Fast computation of orthogonal Fourier–Mellin moments using modified direct method. In: Proceedings of 6th EURASIP Conference Focused on Speech and Image Processing, Multimedia Communications and the 14th International Workshop of Systems, Signals and Image Processing, pp. 153–156 (2007)
29. Bailey, R.R., Srinath, M.: Orthogonal moment features for use with parametric and nonparametric classifiers. IEEE Trans. Pattern Anal. Mach. Intell. **18**, 389–399 (1996)
30. Khotanzad, A., Hong, Y.H.: Invariant image recognition by Zernike moments. IEEE Trans. Pattern Anal. Mach. Intell. **12**, 489–497 (1990)
31. Papakostas, G.A., Boutalis, Y.S., Karras, D.A., Mertzios, B.G.: Efficient computation of Zernike and Pseudo-Zernike moments for pattern classification applications. Pattern Recogn. Image Anal. **20**, 56–64 (2010)
32. Kan, C., Srinath, M.D.: Invariant character recognition with Zernike and orthogonal Fourier-Mellin moments. Pattern Recogn. **35**, 143–154 (2002)
33. Bin, T.J., et al.: Subpixel edge location based on orthogonal Fourier-Mellin moments. Image Vis. Comput. **26**, 563–569 (2008)
34. Zhang, H., Shu, H.Z., et al.: Construction of a complete set of orthogonal Fourier-Mellin moment invariants for pattern recognition applications. Image Vis. Comput. **28**(1), 38–44 (2010)
35. Hosny, K.M., et al.: Fast computation of orthogonal Fourier-Mellin moments in polar coordinates. J. Real-Time Image Proc. **6**(2), 73–80 (2011)
36. Walia, E., Singh, C., Goyal, A.: On the fast computation of orthogonal Fourier-Mellin moments with improved numerical stability. J. Real-Time Image Proc. **7**(4), 247–256 (2012)
37. Singh, C., Upneja, R.: Accurate computation of orthogonal Fourier-Mellin moments. J. Math. Imaging Vis. **44**, 411–431 (2012)
38. Wang, X.Y., Liao, S.: Image reconstruction from orthogonal Fourier-Mellin moments. Image Anal. Recogn. **7950**, 687–694 (2013)
39. Götze, N., Drüe, S., Hartmann, G.: Invariant object recognition with discriminant features based on local Fast-Fourier Mellin transform. Int. Conf. Pattern Recogn. **1**, 948–951 (2000)

40. Derrode, S., Ghorbel, F.: Shape analysis and symmetry detection in gray-level objects using the analytical Fourier-Mellin representation. Sig. Process. **84**, 25–39 (2004)
41. Guo, L.Q., Zhu, M.: Quaternion Fourier-Mellin moments for color images. Pattern Recogn. **44**, 187–195 (2011)
42. Mendlovic, D., Ozaktas, H.M.: Fractional Fourier transforms and their optical implementation: I. J. Optical Soc. Am. A: Optics, Image Sci. **10**(9), 1875–1881 (1993)
43. Hu, H.T., et al.: Orthogonal moments based on exponent functions: exponent-Fourier moments. Pattern Recogn. **47**(8), 2596–2606 (2014)

Author Index